Introduction to Quadratic Forms over Fields

GRADUATE STUDIES IN MATHEMATICS 67

Introduction to Quadratic Forms over Fields

T.Y. Lam

AMERICAN MATHEMATICAL SOCIETY
Providence, Rhode Island

2000 *Mathematics Subject Classification.* Primary 11Exx; Secondary 15A63.

For additional information and updates on this book, visit
www.ams.org/bookpages/gsm-67

Library of Congress Cataloging-in-Publication Data

Lam, T. Y. (Tsit-Yuen), 1942–
Introduction to quadratic forms over fields / T.Y. Lam.
p. cm. – (Graduate studies in mathematics, ISSN 1065-7339 ; v. 67)
Includes bibliographical references and index.
ISBN 0-8218-1095-2 (alk. paper) | Softcover ISBN 1-4704-8017-4
1. Forms, Quadratic I. Title. II. Series

QA243.L252 2004
512.7′4 22 2004062281

∞ The paper used in this book is acid-free and falls within the guidelines
established to ensure permanence and durability.
Visit the AMS home page at `http://www.ams.org/`

11 10 9 8 7 6 5 4 3 30 29 28 27 26 25

In Memory of My Parents

Lam Shiu Fan (1908–1969)
Lam Tsui Shau Hah (1908–1974)

Contents

Preface

The algebraic theory of quadratic forms over fields originated from a classical (1937) paper of Witt [Wi]. However, while quadratic form theory over local fields and global fields flourished steadily through the middle of the 20th century, Witt's theory of quadratic forms over general fields seemed to have gone into dormancy. This situation changed dramatically with the appearance of Pfister's work [Pf_1, Pf_2, Pf_3] in 1965-66. In these seminal papers, Pfister introduced the powerful method of multiplicative forms in quadratic form theory, proved the first significant structural results on the Witt ring, and established a fundamental local-global principle for quadratic forms over a general field. Pfister's contributions in 1965-66 not only revived the general quadratic form theory so ingeniously conceived by Witt in 1937, but also brought it into fruitful contact with the algebraic theory of formally real fields and real-closed fields invented in a different context by Artin and Schreier [AS] in 1927. In the early 1970s, Arason and Pfister succeeded in proving the "Hauptsatz" in quadratic form theory (a Krull intersection theorem for the ideal of even-dimensional forms in the Witt ring). Aside from its intrinsic importance, the Hauptsatz of Arason and Pfister in [AP_1] also turned out to be a harbinger for the powerful use of the method of function fields of quadratic forms in the algebraic theory of Witt rings.

Against the above historical backdrop, I wrote my Benjamin book "The Algebraic Theory of Quadratic Forms" (hereafter referred to as "ATQF") in 1972. Only five years beyond my Ph.D. and looking for something new to do in research, I was quickly caught up in the atmosphere of excitement then prevailing in the new field of the algebraic theory of quadratic forms. My decision to write "ATQF" was perhaps based in part on the ill-advised excuse that "to learn a new subject you write a book about it". Looking back, I certainly don't see anything that could have qualified me to author such a book other than a copious dosage of youthful invincibility. Anyway, my debut as a mathematical author took hardly more than a year. I still remember churning out chapter after chapter of the book in my mother's small apartment in West Vancouver when I had a prolonged visit with her in the summer of 1972; somehow, one tends to write very fast when one is very young!

In "ATQF" (which came out in 1973), I started with two chapters introducing Witt's algebraic theory of quadratic forms and the basic facts about Witt rings of fields (of characteristic not 2). This was followed by a self-contained exposition on quaternion algebras, Brauer-Wall groups, simple graded algebras, and Clifford algebras. A chapter on the rich quadratic form theory over local and global fields served as a reminder of (as well as an introduction to) the classical origins of the subject. The book then progressed into the treatments of quadratic forms under algebraic and transcendental field extensions, with an intermittent coverage of the quadratic form theory over formally real fields and pythagorean fields (highlighting Pfister's local-global principle and key structural results on the Witt ring). A penultimate chapter featured Pfister's theory of multiplicative forms and the Arason-Pfister Hauptsatz, culminating in a final chapter dealing with the quadratic form theoretic invariants of fields, such as the level, the Pythagoras number, and the u-invariant, etc. The book closed with a list of eight open problems.

The wonderful reception given to "ATQF" by the mathematical community came to me as a total surprise. A rather informal introduction to quadratic form theory based on my lecture courses at Berkeley turned out to be a welcome entry text for students learning the theory for the first time, and in the meantime, the research community in the theory of quadratic forms quickly accepted the book as a convenient reference for the basic results in the area. In retrospect, the success of the Benjamin book obviously owed little to its author, but was solely a result of the fortuitous circumstance that it just happened to have been the right book written at the right time.

Starting from the early 1970s, the algebraic theory of quadratic forms experienced a tremendous growth. A dramatic illustration of this growth is given in a special chart prepared by W. Scharlau in the bibliography section of his book [Sc_4] (c. 1985), which showed a spectacular jump of hundreds of pages of published research in the theory of quadratic forms in the period 1970-1980. Throughout this decade (and thereafter), the theory of quadratic forms made contact with a large number of other research topics in algebra, including the theory of central simple algebras and their involutions, linear algebraic groups, algebraic geometry (especially Chow groups), algebraic K-theory, Galois theory and Galois cohomology, the theory of ordered fields and valuated fields, axiomatic geometry, real commutative algebra, and semialgebraic geometry. These interactions with other fields have greatly enriched the scope of the research in the algebraic theory of quadratic forms, and have permanently established this theory as a significant and vibrant branch of modern abstract algebra.

While "ATQF" served its function well in the 1970s, it went out of print by the end of the decade. A second printing with revisions issued by Benjamin in 1980 kept the book in the market for a few more years, but "ATQF" finally succumbed to the fate of orphanhood as Benjamin later became defunct! No author can completely escape the unspeakable feeling that his/her book might have killed its publisher, but the practical effect of the demise of my publisher was clearly that "ATQF" would thereafter survive only in beaten up copies in private collections and university libraries.

This situation would have been tolerable if "ATQF" had outlived its use. However, although at least a few other books on the subject of quadratic forms have been written in the intermittent years, "ATQF" continued to be a textbook of choice for students in quadratic form theory, and a frequent reference for researchers in the field. This trend finally firmed up my resolve to make the book available again to the mathematical community. My primary choice of a publisher was the American Mathematical Society, since the AMS has graciously honored the book with the award of a Leroy P. Steele Prize in Mathematical Exposition in 1982. Plans for reissuing the book went underway in 1998.

From the very start of the republication process, it was clear that "ATQF" could not, and should not, just reappear in its original form. Many new results and new viewpoints have been obtained in quadratic form theory; the 1980 version of the book would have simply looked as outdated as its typography produced then by an IBM Selectric typewriter. But, short of writing a brand new book on the subject, how could one even begin to transform a 20-year past into a relatively satisfying present?

After much vacillation, I decided that the best course of action is to keep the main organization of "ATQF" intact, but rewrite as many chapters in it as possible to accommodate the new results and new viewpoints in the field. Furthermore, two chapters of new material are added to the original text, making the book into one with thirteen chapters. As the result of these expansions, the book has more than doubled in size. To tell it apart from the Benjamin 1973/1980 versions, I have renamed the book "Introduction to Quadratic Forms over Fields". The main focus of the book is still on the algebraic theory of quadratic forms over fields of characteristic not 2;

discussions on the interactions of this theory with other parts of algebra are deliberately kept to a minimum. By limiting the scope of the present book in this way, I hope to have preserved the suitability of this book both as a general reference work and as an introductory text to quadratic form theory. The many exciting ongoing mathematical developments at the interface of quadratic form theory and other branches of algebra (especially algebraic geometry and various cohomology theories) certainly will merit a detailed exposition in the near future, but such an ambitious project is better left to the pen of a more capable author.

Once the boundaries for this book were set, the rewriting of "ATQF" started in earnest in 1999. Needless to say, I fully recognized that being able to rewrite a book after a passage of thirty years is a rare privilege granted to very few authors. After all, how many can boast about publishing two books on the very same subject in two different centuries? And mustn't there have been some fateful "principle of symmetry" at work that I started the first book five years after my Ph.D., and now finish the second one about five years before my retirement? With all of these poignant thoughts on my mind, I returned to write about the subject of my youthful love! But sadly enough, age has taken its toll. The young author who so gallantly churned out chapter after chapter of "ATQF" in a Vancouver apartment has now metamorphosed into a foot-dragging writer who took weeks to draft a single section. With the century mark quietly slipping by, I bore hapless witness to the harsh reality that the resuscitation of my 12-month maiden project in 1972 slowly turned into a six-year arduous writing ordeal.

It was with a tremendous sense of satisfaction and joy that the rewriting of "ATQF" was brought to a conclusion in September of this year. What lies ahead is the finished product, under its new name. The two books spanned much of my professional career, and were in large part the result of my efforts in teaching, research, and mathematical exposition. I am delighted that my maiden work in 1972 has now a second lease in life, and humbly offer this new version of it for the use of the international mathematical community in the years to come.

Needless to say, a book of this nature owes much to the work of others. I hereby thank all the authors whose work I have used in my exposition, implicitly or explicitly. This includes, but is by no means limited to, all authors whose books and papers are cited in the bibliography at the end of this book. Special thanks are due to Richard Elman and Adrian Wadsworth, who were two of my principal collaborators in quadratic form theory. In preparing this book for publication, I have also greatly profited from the input and help from Detlev Hoffmann, David Leep, and Jan Mináč; for their kind and generous assistance, I remain grateful. Incredibly, it was almost forty years ago when I first took an introductory course in quadratic form theory at Columbia University from Professor Hyman Bass. I hope he is pleased to see that his teaching four decades ago has continued to have a substantial impact up to this very day.

It gives me special pleasure to acknowledge the role of the American Mathematical Society in making this book possible. I could not have found a better or more suitable place than the Society's Graduate Studies in Mathematics series for my book to appear in. I thank Dr. Sergei Gelfand for acquiring my book for this series, and for having unflagging faith for six long years that I would one day finish my book. Without his frequent encouragement (and occasional prodding), I probably would have taken even more years. The production of the book at AMS Headquarters was handled in the utmost professional manner by Ralph Sizer and Mary Letourneau. To both of them, I express my heartfelt appreciation. Very sadly, however, the fact that Ralph did not live to see the completion of this book will remain the deepest regret of my authorship.

Last, first, and always, I owe the greatest debt to members of my family. Chee King has lovingly stood by my side for 34 years; I could not have asked for anything more. When "ATQF" was written, we were virtually newlyweds; as this book goes to press, we are the proud parents of four grown children. Juwen, Fumei, Juleen and Tsai Yu are the focus and joy of our lives; whatever I have accomplished (or can accomplish) is in large measure a result of their love, devotion, and unstinting support.

T.Y.L.

Berkeley, California
October, 2004

Notes to the Reader

The main text of this book consists of thirteen chapters, each containing a number of sections. The chapters are referred to in roman numerals, such as I, II, III, etc. A cross-reference such as XI.6 refers to Section 6 in Chapter XI. Within a given chapter, a reference such as 6.21 refers to the result (lemma, theorem, example, or remark) so labelled in Section 6 of that chapter, while, globally, X.5.6 refers to the result 5.6 in Chapter X. The running heads offer the quickest and most convenient way to tell what chapter and what section a particular page belongs to. This should make it very easy to find any result, such as X.5.6.

Each chapter concludes with a set of exercises that are consecutively numbered. "Exercise 10" refers to the exercise so numbered in a given chapter, whereas a reference such as XI.Exercise 10 refers to Exercise 10 at the end of Chapter XI. Many exercises belong rightfully to the folklore of the subject, while a number of others are adapted from original results published in the quadratic form literature. Some (by no means all) of the harder exercises come with hints toward their solutions.

Throughout the text, a good grounding in graduate level algebra is assumed. In particular, facts in field theory and Galois theory will be used rather freely. For the chapter on local fields and global fields (Ch. VI), some familiarity with number theory will be helpful, although it is not absolutely essential. In a couple of places (in discussing the Brauer group), we also assume Wedderburn's classification theorem on finite-dimensional central simple algebras over a field. Results of this nature are usually well covered in a beginning course in graduate algebra.

The title of this book is a slight misnomer, in that we treat here only the theory of quadratic forms over fields *of characteristic not* 2. Ideally, a book on quadratic forms over fields should cover the case of characteristic 2 as well, so that the theory would apply truly to *all* fields. However, there is an all-too-well-known predicament to this, which was perhaps best expressed through the following humorous limerick by an anonymous Irish poet:

A mathematician said "Who
Can quote me a theorem that's true?
For the ones that I know
Are simply not so
When the characteristic is two!"

But, while this poet's lament was solidly grounded and frequently echoed, the real truth is actually somewhere in between. Theorems on quadratic forms over fields of characteristic not 2 usually become problematic (and sometimes meaningless) when they are transferred verbatim to the characteristic 2 case. However, experience has shown that *many* such theorems do have complete, suitably formulated analogues for fields of characteristic 2. What one needs to do is to *find* such analogues, and to *devise new proofs for them*! Thus, *each* theorem would require extra work. For a book of this size with hundreds of results, the total amount of extra work needed to cover the characteristic 2 case would have been truly staggering.

With the preservation of his sanity uppermost on his mind, this senior-aged author has made his clear and unequivocal choice. Unless explicitly stated to the contrary, *all fields over which quadratic forms are considered in this book will be assumed to have characteristic not equal to* 2. Readers interested in learning the theory of quadratic forms in a wider setting (including the case of quadratic forms over rings) may consult some of the existing literature on the subject, such as the books of Baeza [Bae], Knus [Knu], and Milnor-Husemoller [MH].

Partial List of Notations

$\mathbb{Z}$	the ring of integers
$\mathbb{Q}$	the field of rational numbers
$\mathbb{R}$	the field of real numbers
$\mathbb{C}$	the field of complex numbers
$\mathbb{Q}_p$	the field of p-adic numbers
$\mathbb{F}_q$	the finite field of q elements
$\mathbb{Z}_n$	the ring (or the cyclic group) $\mathbb{Z}/n\mathbb{Z}$
$\mathcal{H}$	Hamilton's quaternion algebra
$\mathbb{H}$	the hyperbolic plane
$\dot{F}$	multiplicative group of the field F
$\dot{F}/\dot{F}^2$	square class group of F
$Q(F)$	extended square class group of F
$\sigma(F)$	sums of squares in F
$\dot{\sigma}(F)$	nonzero sums of squares in F
$\overline{F}$	algebraic closure of F
$\widetilde{F}$	quadratic closure of F
F_{py}	pythagorean closure of F
$F_{\mathfrak{p}}$	$\mathfrak{p}$-adic completion of a global field F
$W(F)$	Witt ring of F
$\widehat{W}(F)$	Witt-Grothendieck ring of F
$W_t(F)$	torsion subgroup of $W(F)$
$W(K/F)$	kernel of $W(F) \to W(K)$
$\mathrm{Gal}(K/F)$	Galois group of K/F
$\langle K \rangle$	trace form of an extension K/F
IF	fundamental ideal of $W(F)$
I^nF	n th power of IF

$BW(F)$	Brauer-Wall group of F
$B(F)$	Brauer group of F
${}_2B(F)$	subgroup of elements of exponent ≤ 2 in $B(F)$
$\mathrm{Quat}(F)$	subgroup of $B(F)$ generated by the quaternion algebras
$\left(\frac{a,b}{F}\right)$	quaternion algebra determined by $a, b \in \dot{F}$
K_nF	Milnor's K_n-group of F (also denoted by $K_n^M(F)$)
k_nF	the factor group $K_nF/2K_nF$
X_F	space of orderings on F
$\mathrm{C}(X_F, \mathbb{Z})$	ring of continuous functions from X_F to $\mathbb{Z}$
$s(F)$	level of F
$h(F)$	height of F
$u(F)$	u-invariant of F
$u_n(F)$	nth system u-invariant of F
$P(F)$	Pythagoras number of F
$R(F)$	Kaplansky radical of F
$F(t)$	rational function field over F
$F((t))$	Laurent series field over F
$\varphi \cong \psi$	φ is isometric to ψ
$\varphi \sim \psi$	φ is Witt-similar to ψ
$\varphi \perp \psi$	orthogonal sum of φ and ψ
$\varphi \otimes \psi$	tensor product of φ and ψ
$F[\psi]$	(big) function field of ψ
$F(\psi)$	(small) function field of ψ
$\varphi > \psi$	φ becomes isotropic over $F[\psi]$
$\varphi \gg \psi$	φ becomes hyperbolic over $F[\psi]$
$\langle a_1, \ldots, a_n \rangle$	the diagonal form $a_1x_1^2 + \cdots + a_nx_n^2$
$\langle\langle a_1, \ldots, a_n \rangle\rangle$	the n-fold Pfister form $\langle 1, a_1 \rangle \otimes \cdots \otimes \langle 1, a_n \rangle$
φ'	pure subform of the Pfister form φ
$\dim\, q$	dimension of q
$\dim_{\mathrm{es}}\, q$	essential dimension of q
$G_F(q)$	group of similarity factors of an F-form q
$D_F(q)$	set of nonzero values of an F-form q
$D_F(n)$	nonzero sums of n squares in F
$D_F(\infty)$	nonzero sums of squares in F (same as $\dot{\sigma}(F)$)
q_{an}	the anisotropic part of q
$c \cdot q$	the tensor product $\langle c \rangle \otimes q$
$r\, q$	$q \perp \cdots \perp q$ (r times)
$i(q)$	Witt index of q
$i_1(q)$	first Witt index of q
$C(q)$	Clifford algebra of q
$C_0(q)$	even Clifford algebra of q
$C^{p,q}$	Clifford algebra of $p\langle -1 \rangle \perp q\langle 1 \rangle$

$c(q)$	Witt invariant of q
$s(q)$	Hasse invariant of q
$w(q)$	(second) Stiefel-Whitney invariant of q (in k_2F)
$w_\pm(q)$	"signed" version of $w(q)$
$\Gamma(q)$	Clifford invariant of q
$d(q)$	determinant of q (also denoted by $\det(q)$)
$d_\pm(q)$	signed determinant of q
$\mathrm{sgn}_\alpha(q)$	signature of q with respect to an ordering α
$\mathrm{sgn}(q)$	total signature of q
$\partial_i(q)$	i th residue form of q $(i = 1, 2)$
$\mathrm{O}(q)$	orthogonal group of q
$\mathrm{SO}(q)$	special orthogonal group of q
$\mathrm{len}_F(x)$	the (sums-of-squares) length of $x \in F$
$\mathrm{N}(x)$	norm of the quaternion x
$\mathrm{T}(x)$	trace of the quaternion x
$\mathrm{N}_{K/F}(x)$	field norm of $x \in K$
$\mathrm{T}_{K/F}(x)$	field trace of $x \in K$ (also denoted by $\mathrm{tr}_{K/F}(x)$)
$s_*(U)$	transfer of U with respect to the functional s
$\mathrm{Spec}(A)$	prime spectrum of the commutative ring A
$\mathrm{U}(A)$	group of units of the ring A
$\mathbb{M}_n(A)$	$n \times n$ matrix ring over the ring A
$\widehat{\mathbb{M}}_n(A)$	$\mathbb{M}_n(A)$ with the checker-board grading
$\mathrm{GL}_n(A)$	$n \times n$ general linear group over the ring A
$(\ ,\ )_p$	p-adic Hilbert symbol
$\left(\frac{p}{q}\right)$	Legendre symbol of p and q

Chapter I

Foundations

1. Quadratic Forms and Quadratic Spaces

Throughout this book, a field always means a field of characteristic different from 2, unless it is stated otherwise.

An (n-ary) *quadratic form* over a field F is a polynomial f in n variables over F that is homogeneous of degree 2. It has the general form

$$f(X_1, \dots, X_n) = \sum_{i,j=1}^{n} a_{ij} X_i X_j \in F[X_1, \dots, X_n] = F[X].$$

To render the coefficients symmetric, it is customary to rewrite f as

$$f(X) = \sum_{i,j} \tfrac{1}{2}(a_{ij} + a_{ji}) X_i X_j = \sum_{i,j} a'_{ij} X_i X_j,$$

where $a'_{ij} = \frac{1}{2}(a_{ij} + a_{ji})$. In this way, f determines uniquely a *symmetric* matrix (a'_{ij}), which we shall denote by M_f. In terms of matrix notations, we have

$$f(X) = (X_1, \dots, X_n) \cdot M_f \cdot \begin{pmatrix} X_1 \\ \vdots \\ X_n \end{pmatrix} = X^t \cdot M_f \cdot X \quad (t = \text{ transpose})$$

where X is viewed as a column vector.

Let f and g be n-ary quadratic forms. We say that f is *equivalent* to g ($f \cong g$) if there exists an invertible matrix $C \in \mathrm{GL}_n(F)$ such that $f(X) = g(C \cdot X)$. This means that there exists a nonsingular, homogeneous linear substitution of the variables $X_1, \dots, X_n$ that takes the form g to the form f. Since

$$g(C\cdot X) = (C\cdot X)^t\cdot M_g\cdot (C\cdot X) = X^t\cdot (C^t\cdot M_g\cdot C)\cdot X,$$

the equivalence condition $f(X) = g(C \cdot X)$ stipulated above amounts to a matrix equation[(1)]

$$M_f = C^t\cdot M_g\cdot C.$$

Thus, *equivalence* of forms amounts to *congruence* of the associated symmetric matrices. For example, if we perform the homogeneous linear substitution, $X_1 \mapsto X_1 + X_2$, $X_2 \mapsto X_1 - X_2$, the binary form $g(X_1, X_2) = X_1X_2$ goes to

$$\begin{aligned} g(C \cdot X) &= g(X_1 + X_2, X_1 - X_2) = (X_1 + X_2)(X_1 - X_2) \\ &= X_1^2 - X_2^2. \end{aligned} \tag{1.1}$$

Thus g is equivalent to $f(X_1, X_2) = X_1^2 - X_2^2$. The corresponding matrix equation reads:

$$M_f = \begin{pmatrix} 1 & 0 \\ 0 & -1 \end{pmatrix} = \begin{pmatrix} 1 & 1 \\ 1 & -1 \end{pmatrix} \begin{pmatrix} 0 & 1/2 \\ 1/2 & 0 \end{pmatrix} \begin{pmatrix} 1 & 1 \\ 1 & -1 \end{pmatrix} = C^t \cdot M_g \cdot C.$$

As expected, the equivalence of forms is indeed an equivalence relation.

Let F^n denote the space of n-tuples, given the usual F-vector space structure. Let $e_1, \dots, e_n$ be the standard basis for F^n given by the unit vectors. Any quadratic form, f, gives rise to a map $Q_f : F^n \to F$ defined by sending a column tuple

$$x = \begin{pmatrix} x_1 \\ \vdots \\ x_n \end{pmatrix} = \sum x_i e_i$$

to $Q_f(x) := x^t \cdot M_f \cdot x \in F$. We shall refer to Q_f as the *quadratic map* defined by f. In terms of quadratic maps, the notion of equivalence of forms, $f \cong g$, amounts to the existence of a linear automorphism C of F^n such that $Q_f(x) = Q_g(C \cdot x)$, for every column tuple x.

Note that the quadratic map Q_f determines uniquely the quadratic form f (not only the equivalence class of f). In fact, suppose $Q_f = Q_g$ as maps from F^n to F. For any i, we have

$$(M_f)_{ii} = Q_f(e_i) = Q_g(e_i) = (M_g)_{ii}.$$

For $i \neq j$, we have

$$Q_f(e_i + e_j) = (M_f)_{ii} + (M_f)_{jj} + 2(M_f)_{ij},$$

---

(1) To fully justify this equation, we should note, of course, that $C^t\, M_g\, C$ remains a symmetric matrix.

and a similar equation for $Q_g(e_i + e_j)$. It follows immediately (since F has characteristic not 2) that $(M_f)_{ij} = (M_g)_{ij}$. Thus $M_f = M_g$, and $f = g \in F[X]$.

Let us examine the "quadratic map" Q_f more closely. It has the following remarkable properties:

(1) Q_f is "quadratic" in the sense that

$$Q_f(ax) = a^2 Q_f(x) \quad (\text{for all } x \in F^n, \text{ and } a \in F).$$

This follows since $Q_f(ax) = (ax)^t\, M_f\, (ax) = a^2\, (x^t M_f\, x)$.

(2) If we "polarize" Q_f by setting

$$B_f(x, y) = [Q_f(x + y) - Q_f(x) - Q_f(y)]/2,$$

then $B_f\colon F^n \times F^n \to F$ is a *symmetric bilinear* pairing. Here, symmetry is clear, and bilinearity follows easily from the observation that

$$\begin{aligned} B_f(x,y) &= [(x+y)^t M_f (x+y) - x^t\, M_f\, x - y^t\, M_f\, y]/2 \\ &= [x^t M_f\, y + y^t M_f\, x]/2 = x^t M_f\, y. \end{aligned}$$

Note that, in (2), the quadratic map Q_f may be recaptured from the symmetric bilinear pairing B_f by "depolarization"; that is,

$$Q_f(x) = B_f(x, x), \qquad \text{for any } x \in F^n.$$

This remark motivates a slightly more "geometric" (coordinate-free) approach to the notion of a quadratic form. Let V be any finite-dimensional F-vector space, and $B : V \times V \to F$ a symmetric bilinear pairing on V. We shall call the pair (V, B) a "quadratic space," and associate with it a "quadratic map," $q = q_B : V \to F$, given by $q(x) = B(x, x)$ $(x \in V)$. As in (1) and (2) above, we have $q(ax) = B(ax, ax) = a^2 B(x, x) = a^2 q(x)$, and

$$\begin{aligned} q(x+y) - q(x) - q(y) &= B(x+y, x+y) - B(x,x) - B(y,y) \\ &= B(x,y) + B(y,x) \\ &= 2B(x,y). \end{aligned}$$

Since q and B determine each other, it is legitimate to write (V, q) to represent the quadratic space (V, B). If we coordinatize V, that is, choose a basis $e_1, \ldots, e_n$ for it, then the quadratic space (V, B) gives rise to a quadratic form

$$f(X_1, \ldots, X_n) = \sum_{i,j} B(e_i, e_j) X_i X_j, \quad \text{with } M_f = (B(e_i, e_j)).$$

Note that if we identify V with F^n via the given coordinatization, then $q = q_B$ corresponds precisely to the quadratic map q_f associated with the form f. If we coordinatize V differently, by choosing another basis, $e_1', \ldots, e_n'$, the quadratic form f' resulting from the new coordinatization will be equivalent to f. In fact, if $e_i' = \sum_k c_{ki}e_k$, then

$$\begin{aligned}(M_{f'})_{ij} &= B\left(\sum_k c_{ki}e_k\,,\ \sum_\ell c_{\ell j}e_\ell\right)\\ &= \sum_{k,\ell} c_{ki}\cdot B(e_k, e_\ell)\cdot c_{\ell j}\\ &= (C^t\cdot M_f\cdot C)_{ij},\end{aligned}$$

where $C = (c_{k\ell})$. Thus the quadratic space (V, B) determines uniquely an *equivalence class* of quadratic forms, which we shall denote by (f_B).

If $(V, B), (V', B')$ are quadratic spaces, we say that they are *isometric* ($\cong$) if there exists a linear isomorphism $\tau : V \to V'$ such that

$$B'(\tau(x), \tau(y)) = B(x, y) \quad \text{for all } x, y \in V;$$

that is, τ is "inner product preserving". It is clear that

$$(V, B) \cong (V', B') \iff (f_B) = (f_{B'}).$$

Thus, there is a one-one correspondence between the equivalence classes of n-ary quadratic forms and the isometry classes of n-dimensional quadratic spaces. In proving theorems, *we shall often prefer to argue geometrically with quadratic spaces*, and then pass back freely to quadratic forms, viewing the above one-one correspondence as an identification.

Let (V, B) be a quadratic space, and let M be a symmetric matrix associated to one of the forms in the equivalence class (f_B). The following elementary fact from linear algebra is well-known.

Proposition 1.2. *The following statements are equivalent:*

(1) *M is a nonsingular matrix.*

(2) *$x \mapsto B(\ , x)$ defines an isomorphism $V \to V^*$, where V^* denotes the vector space dual of V.*

(3) *For $x \in V$, $B(x, y) = 0$ for all $y \in V$ implies that $x = 0$.*

If one (and hence all) of these statements holds, we shall say that (V, B) is a *regular* (or nonsingular) quadratic space, or that q_B is a *nonsingular* quadratic form. [Note: The zero quadratic space satisfies (2) and (3) above (forget about (1)!), hence should be called regular, also.]

Let (V, B) be a quadratic space, and S be a subspace of V. Then $(S,\ B|(S \times S))$ is a quadratic space in its own right. As usual, the *orthogonal complement* of S is defined by

$$S^\perp = \{x \in V \mid B(x, S) = 0\}.$$

The orthogonal complement of V itself is called the "radical" of (V, B), denoted by $V^\perp = \operatorname{rad} V$. Observe that (V, B) is regular iff $\operatorname{rad} V = 0$. However, if (V, B) is regular, the subspace S of V need not be regular. (For instance, $B|S \times S$ may be zero.)

Proposition 1.3. *Let (V, B) be a regular quadratic space, and S be a subspace of V. Then:*

(1) (Dimension Formula) $\dim S + \dim S^\perp = \dim V$.

(2) $(S^\perp)^\perp = S$.

Proof. Let $\varphi : V \to V^*$ be the linear isomorphism defined in (2) of the preceding proposition. Then $S^\perp$ is precisely the subspace of V annihilated by the functionals in $\varphi(S)$. By the usual duality theory in linear algebra, we have

$$\begin{aligned}\dim S^\perp &= \dim V^* - \dim \varphi(S)\\ &= \dim V - \dim S,\end{aligned}$$

since φ is an isomorphism. This establishes (1). Applying (1) twice, we get

$$\dim (S^\perp)^\perp = \dim V - (\dim V - \dim S) = \dim S.$$

Since $(S^\perp)^\perp \supseteq S$, (2) follows immediately. □

Note that neither conclusion would hold in 1.3 if the quadratic space (V, B) was not assumed to be *regular*: the case of the zero form ($B \equiv 0$) provides, for instance, an extreme counterexample.

2. Diagonalization of Quadratic Forms

Throughout this book, we shall write $\dot{F}$ to denote the multiplicative group, $F \setminus \{0\}$, of the field F.

Definition 2.1. Let f be a quadratic form over F, and $d \in \dot{F}$. We shall say that f *represents* d if there exist $x_1, \ldots, x_n$ in F such that $f(x_1, \ldots, x_n) = d$. Note that $(x_1, \ldots, x_n)$ is automatically a nonzero vector. We shall write $D_F(f) = D(f)$ to denote the set of values in $\dot{F}$ represented by f. This set clearly depends only on the equivalence class of f.

If (V, B) is any quadratic space corresponding to the equivalence class of f, then

$$D(f) = \{d \in \dot{F} \mid \text{ there exists } v \in V \text{ such that } q_B(v) = d\}.$$

We shall also denote this by $D(V)$ if the pairing B is clear from the context.

If $a, d \in \dot{F}$, then, clearly, $d \in D(f)$ iff $a^2 d \in D(f)$. Thus, $D(f)$ consists of a union of cosets of $\dot{F}$ modulo $\dot{F}^2$. We shall call $\dot{F}/\dot{F}^2$ the *group of square classes* of F. By abuse of notation, we may also regard $D(f)$ as a subset of $\dot{F}/\dot{F}^2$. In $\dot{F}$, the subset $D(f)$ is always closed under inverses, since $d \in D(f)$ implies that $d^{-1} = (d^{-1})^2 d \in D(f)$.

In general, $D(f)$ is *not* a subgroup of $\dot{F}$. For one thing, $D(f)$ may not contain 1. Even if $D(f)$ contains 1, it may fail to be closed under multiplication. For instance, consider the form $f = X^2 + Y^2 + Z^2$ over the field of rational numbers $\mathbb{Q}$. Here, $D(f)$ consists of (nonzero) rational numbers which are sums of three squares of rational numbers. We have $1, 2, 2^{-1}, 14 \in D(f)$, but $2^{-1} \cdot 14 = 7$ is known to be *not* in $D(f)$ in elementary number theory.

If the value set $D(f)$ for a form f happens to be closed under multiplication, then (ignoring the case of the 0-dimensional form) $1 \in D(f)$, and $D(f)$ is a subgroup of $\dot{F}$. In this case, we say f is a *group form* over F. For instance, by the classical 2-square, 4-square and 8-square identities, the quadratic forms $\sum_{i=1}^n X_i^2$ for $n = 1, 2, 4, 8$ *are group forms over any field* F. This result will be substantially generalized in a future chapter. As an interesting case in point, we can cite Lagrange's Theorem which says that any positive integer is a sum of four squares of integers. This beautiful theorem implies readily that $D_{\mathbb{Q}}(W^2 + X^2 + Y^2 + Z^2)$ is the group of positive rational numbers $\dot{\mathbb{Q}}^+$.

Next we want to introduce *orthogonal sums*. If (V_1, B_1), (V_2, B_2) are quadratic spaces, we may form (V, B), where $V = V_1 \oplus V_2$, and B is the pairing $V \times V \to F$ given by

$$B((x_1, x_2), (y_1, y_2)) = B_1(x_1, y_1) + B_2(x_2, y_2).$$

B is clearly symmetric and bilinear, so (V, B) is a new quadratic space. We have $B(V_1, V_2) = 0$, and $B \mid (V_i \times V_i) = B_i$ $(i = 1, 2)$. We shall henceforth denote (V, B) by $V_1 \perp V_2$; this is called the *orthogonal sum* of (V_1, B_1) and (V_2, B_2). Note that

$$\begin{aligned} q_B(x_1, x_2) &= B((x_1, x_2), (x_1, x_2)) \\ &= B_1(x_1, x_1) + B_2(x_2, x_2) \\ &= q_{B_1}(x_1) + q_{B_2}(x_2). \end{aligned} \tag{2.2}$$

This dictates, incidentally, how orthogonal sums should be defined in the category of quadratic forms. For example, if q_1 is the binary form $X^2 + 2Y^2$, and q_2 is the ternary form $5XY - Z^2$, then their orthogonal sum is to be taken as

$$q_1 \perp q_2 = U^2 + 2V^2 + 5XY - Z^2,$$

in the five variables U, V, X, Y, Z.

It is easy to see that the orthogonal sum of two quadratic spaces is regular iff each of them is regular. This fact (especially the "if" part) will be used freely without mention throughout.

For $d \in F$, we shall write $\langle d \rangle$ to denote the isometry class of the 1-dimensional space corresponding to the quadratic form dX^2. Clearly, $\langle d \rangle$ is regular iff $d \in \dot{F}$.

Representation Criterion 2.3. *Let (V, B) be a quadratic space, and $d \in \dot{F}$. Then $d \in D(V)$ iff there exists another quadratic space (V', B') together with an isometry $V \cong \langle d \rangle \perp V'$.*

Proof. If we have $V \cong \langle d \rangle \perp V'$, then $d \in D(\langle d \rangle \perp V') = D(V)$. Conversely, suppose $d \in D(V)$, so there exists $v \in V$ with $q(v) = d$ (where $q = q_B$). We first make a reduction to the case where V is regular. Take any subspace W such that $V = (\operatorname{rad} V) \oplus W = (\operatorname{rad} V) \perp W$. By (2.2), we have $D(V) = D(W)$, and W is clearly regular. We may thus assume that V itself is regular. The quadratic subspace $F \cdot v$ is isometric to $\langle d \rangle$, and $(F \cdot v) \cap (F \cdot v)^\perp = 0$. Since

$$\dim (F \cdot v) + \dim (F \cdot v)^\perp = \dim V$$

by Proposition 1.3, we conclude that $V \cong \langle d \rangle \perp (F \cdot v)^\perp$. □

The first consequence of the above criterion is the existence of "orthogonal bases" in any quadratic space.

Corollary 2.4. *If (V, B) is any quadratic space over F, then there exist scalars $d_1, \ldots, d_n \in F$ such that $V \cong \langle d_1 \rangle \perp \cdots \perp \langle d_n \rangle$. (In other words, any n-ary quadratic form is equivalent to some diagonal form, $d_1 X_1^2 + \cdots + d_n X_n^2$.)*

Proof. If $D(V)$ is empty, then $B \equiv 0$ and V is isometric to an orthogonal sum of $\langle 0 \rangle$'s. If there exists some $d \in D(V)$, then $V \cong \langle d \rangle \perp V'$ for some (V', B'), and the proof proceeds by induction on $\dim V$. □

Notation. In the rest of the book, we shall abbreviate the diagonal form $\langle d_1 \rangle \perp \cdots \perp \langle d_n \rangle$ by $\langle d_1, \ldots, d_n \rangle$. The special n-ary diagonal form $\langle d, \ldots, d \rangle$ will be abbreviated as $n \langle d \rangle$. For instance, $3 \langle a \rangle \perp 2 \langle b \rangle$ means the quinary form $\langle a, a, a, b, b \rangle$.

Corollary 2.5. *If (V, B) is a quadratic space (not necessarily regular) and S is a regular subspace, then:*

(1) $V = S \perp S^\perp$.

(2) *If T is a subspace of V such that $V = S \perp T$, then $T = S^\perp$.*

Proof. (1) $\Rightarrow$ (2). If $V = S \perp T$, then, clearly, $T \subseteq S^{\perp}$. But

$$\dim T = \dim V - \dim S = \dim S^{\perp}$$

by (1) (not by Proposition 1.3!), so we must have $T = S^{\perp}$.

(1) Since $S \cap S^{\perp} = \operatorname{rad} S = 0$, it suffices to show that V is spanned by S and $S^{\perp}$. By 2.4, S has an orthogonal basis $x_1, \ldots, x_p$, and the regularity of S implies that $B(x_i, x_i) \neq 0$ for all i. Given any $z \in V$, consider the vector

$$y = z - \sum_{i=1}^{p} \frac{B(z, x_i)}{B(x_i, x_i)}\, x_i.$$

We have

$$\begin{aligned} B(y, x_j) &= B(z, x_j) - \sum_{i=1}^{p} \frac{B(z, x_i)}{B(x_i, x_i)}\, B(x_i, x_j) \\ &= B(z, x_j) - \frac{B(z, x_j)}{B(x_j, x_j)}\, B(x_j, x_j) = 0. \end{aligned}$$

Thus, $y \in S^{\perp}$, and

$$z = y + \sum_{i=1}^{p} \frac{B(z, x_i)}{B(x_i, x_i)}\, x_i \in S \perp S^{\perp}.$$

□

Corollary 2.6. *Let (V, B) be a regular quadratic space. Then a subspace S is regular iff there exists $T \subseteq V$ such that $V = S \perp T$.*

Proof. For the "only if" part, take $T = S^{\perp}$, and apply (1) of the above corollary. Conversely, if $V = S \perp T$, then $\operatorname{rad} S \subseteq \operatorname{rad} V = 0$, so S is regular (and $T = S^{\perp}$). □

We discuss now the *determinant* of a nonsingular quadratic form f. This is defined to be $d(f) = \det(M_f) \cdot \dot{F}^2$ (an element of $\dot{F}/\dot{F}^2$), where M_f is the symmetric matrix associated with f. Note that if $f \cong g$, then $M_f = C^t M_g C$ for some nonsingular C, and hence

$$d(f) = \det(M_f) \cdot \dot{F}^2 = \det(M_g) \cdot (\det C)^2 \cdot \dot{F}^2 = d(g).$$

This shows that $d(f)$ is an *invariant* of the equivalence class of f. Also, we have clearly

$$d(f_1 \perp f_2) = d(f_1) d(f_2) \in \dot{F}/\dot{F}^2.$$

Let (V, B) be a (regular) quadratic space that corresponds to the equivalence class of f. If $V \cong \langle d_1, \ldots, d_n \rangle$ is a "diagonalization" of V, then $d(f) = d_1 \cdots d_n \cdot \dot{F}^2$. It is sometimes convenient to call this quantity the determinant of V, written $d(V)$.

3. Hyperbolic Plane and Hyperbolic Spaces

In this section, we shall introduce the important notion of a hyperbolic quadratic space. We begin by defining "isotropy" and "anisotropy".

Definition 3.1. Let v be a nonzero vector in a quadratic space (V, B). We say that v is an *isotropic vector* if $B(v, v) = 0$ (or equivalently, if $q(v) = 0$, where $q = q_B$), and say that v is *anisotropic* otherwise. The quadratic space (V, B) is said to be *isotropic* if it contains a (nonzero) isotropic vector, and is said to be *anisotropic* otherwise. (Note that anisotropic spaces are necessarily regular.) Finally, we shall say that (V, B) is *totally isotropic* if all nonzero vectors in V are isotropic (that is, $B \equiv 0$).

Whether or not the zero vector should be taken as anisotropic leads mostly to fruitless debate. It will be essentially harmless to side-step this issue altogether, which is exactly what we are going to do. The important thing to keep in mind is just that when we try to say something interesting about isotropic or anisotropic vectors v, these vectors are understood to be nonzero. Finally, we should note that, according to Definition 3.1, the 0-dimensional quadratic form is technically *anisotropic*.

Theorem 3.2. *Let (V, q) be a 2-dimensional quadratic space. The following four statements are equivalent:*

(1) *V is regular and isotropic.*

(2) *V is regular, with $d(V) = -1 \cdot \dot{F}^2$.*

(3) *V is isometric to $\langle 1, -1 \rangle$.*

(4) *V corresponds to the equivalence class of the binary quadratic form X_1X_2.*

Proof. We have already seen that (3) $\Leftrightarrow$ (4) in 1.1.

(1) $\Rightarrow$ (2): Let x_1, x_2 be an orthogonal basis for V. Regularity of V implies that $q(x_i) = d_i \neq 0$ $(i = 1, 2)$. Let $ax_1 + bx_2$ be an isotropic vector, with (say) $a \neq 0$. Then

$$0 = q(ax_1 + bx_2) = a^2 d_1 + b^2 d_2 \implies d_1 = -(ba^{-1})^2 \cdot d_2$$
$$\implies d(V) = d_1 d_2 \cdot \dot{F}^2 = -1 \cdot \dot{F}^2.$$

(2) $\Rightarrow$ (3): Under the hypothesis (2), we have clearly a diagonalization, $V \cong \langle a, -a \rangle$ for some $a \in \dot{F}$. By the argument in 1.1, we know that $aX_1^2 - aX_2^2$ is equivalent to aX_1X_2. The latter clearly represents all elements in $\dot{F}$. In particular, (V, q) itself represents 1. By the Representation Criterion, we conclude that $V \cong \langle 1, -1 \rangle$.

(3) $\Rightarrow$ (1) is trivial. $\square$

The isometry class of a 2-dimensional quadratic space satisfying the conditions in Theorem 3.2 is called the "hyperbolic plane" (presumably because the graphs of the equations $X_1X_2 = d$ are called hyperbolas). The hyperbolic plane will be denoted by $\mathbb{H}$, and will play a very special role throughout the subsequent developments. An orthogonal sum of hyperbolic planes will be called a *hyperbolic space.* The corresponding quadratic form may be taken either as $(X_1^2 - X_2^2) + \cdots + (X_{2m-1}^2 - X_{2m}^2)$ or as $X_1X_2 + \cdots + X_{2m-1}X_{2m}$.

Definition 3.3. A quadratic form (or quadratic space) is called *universal* if it represents all the nonzero elements of F. (Of course, any such form is a group form over F.)

Theorem 3.4. *Let (V, B) be a regular quadratic space. Then:*

(1) *Every totally isotropic subspace $U \subseteq V$ of positive dimension r is contained in a hyperbolic subspace $T \subseteq V$ of dimension $2r$.*

(2) *V is isotropic iff V contains a hyperbolic plane (necessarily as an orthogonal summand, by Corollary 2.5(1)).*

(3) *V is isotropic $\Rightarrow$ V is universal.*

Proof. We show first that (1) $\Rightarrow$ (2) $\Rightarrow$ (3), and then come back to prove (1). (1) $\Rightarrow$ (2) is clear by putting $r = 1$ in (1). (2) $\Rightarrow$ (3) is also clear because the form X_1X_2 corresponding to $\mathbb{H}$ is obviously universal. We shall now prove (1) by induction on r. Take any basis $x_1, \ldots, x_r$ in U, and let S be the span of $x_2, \ldots, x_r$. Of course, $U^\perp \subsetneq S^\perp$. Since V is regular, we may apply 1.3 to get

$$\dim S^\perp = \dim V - \dim S > \dim V - \dim U = \dim U^\perp.$$

This means that there exists a vector y_1 that is orthogonal to $x_2, \ldots, x_r$, but not orthogonal to x_1. In particular, x_1, y_1 are linearly independent vectors (since x_1 is isotropic). The subspace $H_1 = Fx_1 + Fy_1$ has determinant

$$d(H_1) = \begin{vmatrix} 0 & B(x_1, y_1) \\ B(x_1, y_1) & B(y_1, y_1) \end{vmatrix} \cdot \dot{F}^2 = -1 \cdot \dot{F}^2,$$

so $H_1 \cong \mathbb{H}$ by Theorem 3.2. We have thus a splitting $V = H_1 \perp V'$, where $V' = H_1^\perp$ contains $x_2, \ldots, x_r$ (Corollary 2.5). Since V' is regular (Corollary 2.6), the proof proceeds by induction. □

Remark. (A) Since the hyperbolic plane $\mathbb{H}$ has a diagonalization $\langle 1, -1 \rangle$, the fact that $\mathbb{H}$ is universal amounts to the fact that any (nonzero) element in F is a difference of two squares. This fact can be checked directly by using the following equation:

$$a = \left(\frac{a+1}{2}\right)^2 - \left(\frac{a-1}{2}\right)^2 \quad (\forall a \in F),$$

which will be used many times over in the sequel.[(2)]

(B) If one prefers to give a direct argument for (3), one can proceed as follows. Fix an isotropic vector x_1, and take $y_1 \in V$ such that $B(x_1, y_1) \neq 0$. Then $B(tx_1 + y_1, tx_1 + y_1) = 2tB(x_1, y_1) + B(y_1, y_1)$ clearly assumes all values as t ranges over F.

Corollary 3.5 (First Representation Theorem). *Let q be a regular quadratic form, and $d \in \dot{F}$. Then, $d \in D(q)$ iff $q \perp \langle -d\rangle$ is isotropic.*

Proof. We may assume that $q(X) = a_1X_1^2 + \cdots + a_nX_n^2$. If there exists an equation $d = a_1x_1^2 + \cdots + a_nx_n^2$ ($x_i \in F$), then

$$a_1x_1^2 + \cdots + a_nx_n^2 + (-d)\cdot 1^2 = 0,$$

so the form $q \perp \langle -d\rangle$ is isotropic. Conversely, let $(x_1, \ldots, x_{n+1})$ be an isotropic vector for $q \perp \langle -d\rangle$, so $a_1x_1^2 + \cdots + a_nx_n^2 - dx_{n+1}^2 = 0$. If $x_{n+1} \neq 0$, then

$$d = a_1\left(\frac{x_1}{x_{n+1}}\right)^2 + \cdots + a_n\left(\frac{x_n}{x_{n+1}}\right)^2 \in D(q).$$

If, on the contrary, $x_{n+1} = 0$, then $(x_1, \ldots, x_n) \neq 0$ is an isotropic vector for q itself. By (3) of the theorem, $D(q) = \dot{F}$, so, of course, $d \in D(q)$. □

Corollary 3.6. *Let q_1, q_2 be regular forms of positive dimensions. Then $q := q_1 \perp q_2$ is isotropic iff $D(q_1) \cap -D(q_2) \neq \varnothing$.*

Proof. For the "if" part, take an element $a \in D(q_1) \cap -D(q_2)$. If $q_1(x) = a$ and $q_2(y) = -a$, then $(x, y) \neq 0$ is an isotropic vector for q.

For the "only if" part, we may assume that q_1, q_2 are anisotropic. (If, say q_2 is isotropic, then $D(q_1) \cap -D(q_2) = D(q_1) \neq \varnothing$ by 3.4(3).) Suppose $q(x, y) = 0$, where $(x, y) \neq 0$. Say $x \neq 0$. Then $a := q_1(x) \neq 0$, and we have $a \in D(q_1) \cap -D(q_2)$, as desired. □

Corollary 3.7. *For a positive integer r, the following two statements are equivalent (over a given field F):*

(1) *Any regular quadratic form of dimension r over F is universal.*

(2) *Any quadratic form of dimension $r + 1$ over F is isotropic.*

The easy proof of this is left to the reader. The nice fact expressed in this corollary will form the basis for the investigation of the "u-invariant" of a field in Ch. XI.

[(2)]Of course, it is essential here that $\text{char}(F) \neq 2$. If $\text{char}(F) = 2$, a difference of two squares is simply a square, but we may very well have $F \neq F^2$.

4. Decomposition Theorem and Cancellation Theorem

In this section, we come to some of the most important classical theorems in quadratic form theory, which first appeared in Witt's seminal paper [Wi], ca. 1937. Note that 4.1 and 4.2 below are proved for *arbitrary* quadratic spaces (V, q), without any regularity assumptions on q.

Witt's Decomposition Theorem 4.1. *Any quadratic space (V, q) splits into an orthogonal sum*

$$(V_t, q_t) \perp (V_h, q_h) \perp (V_a, q_a),$$

where V_t is totally isotropic, V_h is hyperbolic (or zero), and V_a is anisotropic ("Witt decomposition"). Furthermore, the isometry types of V_t, V_h, V_a are all uniquely determined.

Proof. For existence, take any subspace V_0 such that

$$V = (\operatorname{rad} V) \oplus V_0 = (\operatorname{rad} V) \perp V_0.$$

Then $V_t = \operatorname{rad} V$ is totally isotropic, and V_0 is regular. If V_0 is isotropic, we may write $V_0 = H_1 \perp V_1$ (by 3.4(2)), where $H_1 \cong \mathbb{H}$. If V_1 is again isotropic, we may further write $V_1 = H_2 \perp V_2$, where $H_2 \cong \mathbb{H}$. After a finite number of steps, we achieve a decomposition

$$V_0 = (H_1 \perp \cdots \perp H_r) \perp V_a \quad (r \geq 0),$$

where $H_1 \perp \cdots \perp H_r = V_h$ is hyperbolic (or zero), and V_a is anisotropic. This proves the existence part. The uniqueness part will be deduced from the Cancellation Theorem, which reads as follows.

Witt's Cancellation Theorem 4.2. *If q, q_1, q_2 are arbitrary quadratic forms, then $q \perp q_1 \cong q \perp q_2 \Longrightarrow q_1 \cong q_2$.*

To establish the uniqueness part in 4.1, suppose V has another "Witt decomposition," $V = V_t' \perp V_h' \perp V_a'$. Since V_t' is totally isotropic and $V_h' \perp V_a'$ is regular, we have

$$\operatorname{rad} V = (\operatorname{rad} V_t') \perp \operatorname{rad}(V_h' \perp V_a') = V_t',$$

so $V_t' = V_t$. By the Cancellation Theorem, we have now $V_h \perp V_a \cong V_h' \perp V_a'$. Write $V_h \cong m \cdot \mathbb{H}$ (orthogonal sum of m copies of $\mathbb{H}$) and $V_h' \cong m' \cdot \mathbb{H}$. By cancelling $\mathbb{H}$ one at a time, we conclude that $m = m'$ since V_a, V_a' are both anisotropic. After all m hyperbolic planes have been cancelled, we arrive at $V_a \cong V_a'$, completing the proof of 4.1.

Definition 4.3. The integer m $(= \frac{1}{2} \dim V_h)$ uniquely determined in the Witt decomposition above is called the *Witt index* of the quadratic space (V, q). The isometry class of V_a is called the *anisotropic part* of (V, q).

Corollary 4.4. *If (V, q) is regular, the Witt index m of V equals the dimension of any maximal totally isotropic subspace of V.*

Proof. Let U be any maximal totally isotropic subspace in V, with $\dim U = r$. By Theorem 3.4, there exists a hyperbolic space $T \supseteq U$ with dimension $2r$. Since T is regular, 2.5 implies that $V = T \perp T^{\perp}$. Here, $T^{\perp}$ must be anisotropic. (If $0 \neq x \in T^{\perp}$ is an isotropic vector, then $U + F \cdot x$ will be a totally isotropic subspace of V properly containing U, a contradiction.) The uniqueness part of 4.1 implies that $T \cong V_h$, so

$$m = \tfrac{1}{2} \dim V_h = \tfrac{1}{2}(2r) = r = \dim U. \qquad \square$$

Our next goal is to establish Witt's Cancellation Theorem 4.2. To present the proof, we need the notion of a "hyperplane reflection." Let (V, B, q) be any quadratic space. We shall write $\mathrm{O}_q(V) = \mathrm{O}(V)$ to denote the group of isometries of (V, q). This so-called *orthogonal group* is the symmetry group which underlies the geometry of our quadratic space. The following important construction associates an element $\tau_y \in \mathrm{O}(V)$ to every *anisotropic* vector $y \in V$. As a self-map from V to itself, τ_y is defined by

$$\tau_y(x) = x - \frac{2B(x, y)}{q(y)}\, y \quad \text{for every } x \in V. \tag{4.5}$$

(1) τ_y is evidently a linear endomorphism.

(2) τ_y is the identity map on $(F \cdot y)^{\perp}$. In fact, in the above formula, if $B(x, y) = 0$, then $\tau_y(x) = x$. Furthermore, if we apply τ_y to y itself, we get

$$\tau_y(y) = y - \frac{2B(y, y)}{q(y)} \cdot y = y - 2y = -y.$$

In particular, τ_y is an involution ($\tau_y^2 = 1$): it leaves the hyperplane $(F \cdot y)^{\perp}$ pointwise fixed, and "reflects" the vector y across $(F \cdot y)^{\perp}$ to $-y$.

(3) $\tau_y \in \mathrm{O}(V)$ by the following straightforward calculation:

$$\begin{aligned} B(\tau_y(x),\, \tau_y(x')) &= B\left(x - \frac{2B(x, y)}{q(y)}\, y,\ x' - \frac{2B(x', y)}{q(y)}\, y\right) \\ &= B(x, x') + \frac{4B(x, y)\, B(x', y)}{q(y)^2}\, B(y, y) \\ &\qquad - \frac{4B(x, y)\, B(x', y)}{q(y)} \\ &= B(x, x') \quad (\text{since } B(y, y) = q(y)). \end{aligned}$$

Alternatively, $\tau_y \in \mathrm{O}(V)$ can also be deduced from the fact that $\tau_y \,|\, F \cdot y$ and $\tau_y \,|\, (F \cdot y)^{\perp}$ are both isometries.

(4) As a linear automorphism, τ_y has determinant -1.

Remark 4.6. The set of hyperplane reflections $\{\tau_y \mid q(y) \neq 0\}$ is closed under conjugation in the orthogonal group $\mathrm{O}(V)$. In fact, if $\sigma \in \mathrm{O}(V)$, then one has $\sigma\tau_y\sigma^{-1} = \tau_{\sigma y}$. The verification is straightforward:

$$\begin{aligned}(\sigma\tau_y\sigma^{-1})(x) &= \sigma[\tau_y(\sigma^{-1}x)]\\ &= \sigma\left[\sigma^{-1}x - \frac{2B(\sigma^{-1}x, y)}{q(y)}\,y\right]\\ &= x - \frac{2B(x,\sigma y)}{q(\sigma y)}\,\sigma y\\ &= \tau_{\sigma y}(x) \quad \text{for every } x \in V.\end{aligned}$$

It follows, in particular, that the subgroup of $\mathrm{O}(V)$ generated by all τ_y (where $q(y) \neq 0$) is normal in $\mathrm{O}(V)$. In Section 7, we shall show that this normal subgroup actually coincides with $\mathrm{O}(V)$.

Proof of 4.2. Suppose it is given that $q \perp q_1 \cong q \perp q_2$.

Step 1. *Cancellation holds if q is totally isotropic and q_1 is regular.* In fact, let M_1, M_2 be the symmetric matrices associated with q_1 and q_2. The hypothesis implies that $\left(\begin{smallmatrix}0 & 0\\ 0 & M_1\end{smallmatrix}\right)$ is congruent to $\left(\begin{smallmatrix}0 & 0\\ 0 & M_2\end{smallmatrix}\right)$, so there exists an invertible matrix $E = \left(\begin{smallmatrix}A & B\\ C & D\end{smallmatrix}\right)$ such that

$$\begin{pmatrix}0 & 0\\ 0 & M_1\end{pmatrix} = E^t\begin{pmatrix}0 & 0\\ 0 & M_2\end{pmatrix}E = \begin{pmatrix}* & *\\ * & D^tM_2D\end{pmatrix}.$$

In particular, $M_1 = D^tM_2D$. Since M_1 is nonsingular, so is D, and hence M_1, M_2 are congruent. This gives $q_1 \cong q_2$.

Step 2. *Cancellation holds if q is totally isotropic.* To see this, diagonalize q_1, q_2 and assume that q_1 has exactly r zero coefficients in the diagonalization, while q_2 has r zeros or more. Rewriting the hypothesis, we have

$$q \perp r\,\langle 0\rangle \perp q_1' \cong q \perp r\,\langle 0\rangle \perp q_2'.$$

Since q_1' is regular, Step 1 implies that $q_1' \cong q_2'$. By tagging on r terms of $\langle 0\rangle$, we conclude that $q_1 \cong q_2$.

Step 3 (*General case*). Let $\langle a_1, \ldots, a_n\rangle$ be a diagonalization of q. Inducting on n, we are reduced to the case $n = 1$. The case $a_1 = 0$ has been handled in Step 2, so we may assume that $q = \langle a_1\rangle$, $a_1 \neq 0$. The hypothesis now reads: $\langle a_1\rangle \perp q_1 \cong \langle a_1\rangle \perp q_2$. The Cancellation Theorem in this case is clearly equivalent to the following result.

Proposition 4.7. *Let (V, q) be a quadratic space, and let x, y be vectors in V such that $q(x) = q(y) \neq 0$. Then there exists an element $\tau \in \mathrm{O}(V)$ such that $\tau(x) = y$.*

Proof. Geometrically, if we consider the reflection with respect to the hyperplane $(F\cdot(x-y))^{\perp}$, then x should be taken to y. But, is $x-y$ necessarily anisotropic? We derive first the "law of parallelogram":

$$\begin{aligned} q(x+y)+q(x-y) &= B(x+y,\,x+y)+B(x-y,\,x-y) \\ &= 2B(x,x)+2B(y,y) \\ &= 4q(x)\neq 0. \end{aligned}$$

This implies that $q(x+y)$, $q(x-y)$ cannot be both zero. Replacing y by $-y$ if necessary, we may assume that $q(x-y)\neq 0$. [If we can find $\tau\in \mathrm{O}(V)$ such that $\tau(x)=-y$, the isometry $-\tau$ will take x to y.] Applying the reflection τ_{x-y} to x, we obtain

$$\tau_{x-y}(x)=x-\frac{2B(x,\,x-y)}{q(x-y)}\cdot(x-y).$$

But

$$\begin{aligned} q(x-y) &= B(x-y,\,x-y) \\ &= B(x,x)+B(y,y)-2B(x,y) \\ &= 2\,[B(x,x)-B(x,y)] \\ &= 2B(x,\,x-y). \end{aligned}$$

Thus, $\tau_{x-y}(x)=x-(x-y)=y$, as we claimed at the beginning of the proof. □

5. Witt's Chain Equivalence Theorem

The theorem in the section title (also originating from Witt's classical paper [Wi]) describes the equivalence of two diagonal quadratic forms in terms of the equivalence of *binary* diagonal forms. First, let us prove the following easy fact for binary forms.

Proposition 5.1. *Let $q=\langle a,b\rangle$, $q'=\langle c,d\rangle$ be regular binary forms. Then $q\cong q'$ iff $d(q)=d(q')$, and q, q' represent a common element $e\in \dot F$.*

Proof. "Only if" is clear. Conversely, assume that $d(q)=d(q')\in \dot F/\dot F^2$ and $e\in D(q)\cap D(q')$. By the Representation Criterion (2.3), we know that $q\cong\langle e,e'\rangle$ for some $e'\in\dot F$. Taking determinants, we have $ab\dot F^2=ee'\dot F^2$, so $q\cong\langle e,\,abe\rangle$. Similarly, $q'\cong\langle e,\,cde\rangle$. But $ab\dot F^2=cd\dot F^2$, so $q\cong q'$. □

We shall now introduce the notion of *simple equivalence* for diagonal forms. Let $q=\langle a_1,\dots,a_n\rangle$ and $q'=\langle b_1,\dots,b_n\rangle$. We shall say that q and q' are *simply-equivalent*, if there exist two indices, i and j, such that

(1) $\langle a_i,\,a_j\rangle\cong\langle b_i,\,b_j\rangle$, and

(2) $a_k=b_k$ whenever k is different from i and j.

(Note: In condition (1) above, if i is equal to j, the expression $\langle a_i,\, a_j\rangle$ is understood to be just $\langle a_i\rangle$.)

More generally, we say that two diagonal forms f and g are *chain-equivalent*, if there exists a sequence of diagonal forms $f_0,\, f_1,\, \dots,\, f_m$ such that $f_0 = f$, $f_m = g$, and each f_i is simply-equivalent to f_{i+1} $(0 \le i \le m-1)$. Chain equivalence is clearly an equivalence relation on all diagonal forms (of a fixed dimension); it will be denoted by the symbol $\approx$. Of course, $f \approx g$ implies that $f \cong g$. It turns out that the converse is also true, and this is the content of the following celebrated result of Witt.

Chain Equivalence Theorem 5.2. *If f and g are arbitrary diagonal forms (of the same dimension), then $f \cong g \Longrightarrow f \approx g$.*

Proof. Say $f = \langle a_1, \dots, a_n\rangle$, $g = \langle b_1, \dots, b_n\rangle$. Note that if σ is a permutation of the indices $\{1, 2, \dots, n\}$, and $f^\sigma = \langle a_{\sigma(1)}, \dots, a_{\sigma(n)}\rangle$, then $f \approx f^\sigma$. This follows from the observation that the full symmetric group on $\{1, \dots, n\}$ is generated by the transpositions. Since $f \cong g$, the two forms, f and g, have the same number of zero terms in their diagonalizations. It is, therefore, sufficient to show that the "regular parts" of f and g are chain-equivalent. We may thus assume that f, g are both regular, that is, a_i, b_j are all nonzero. The argument is by induction on n. There is nothing to prove if $n = 1, 2$, so we consider $n \ge 3$ in the following.

Among all diagonal forms that are chain-equivalent to f, choose an $f' = \langle c_1, \dots, c_n\rangle$ such that some $\langle c_1, \dots, c_p\rangle$ represents b_1, and p is smallest possible. (The existence of f' follows from the Well-Ordering Principle.) *We claim that* $p = 1$. In fact, suppose the contrary. Write $b_1 = c_1 e_1^2 + \cdots + c_p e_p^2$ $(p \ge 2)$. By the minimality of p, no subsum in this summation can be equal to zero. In particular, $d = c_1 e_1^2 + c_2 e_2^2 \ne 0$. By 2.3, $\langle c_1, c_2\rangle \cong \langle d,\, c_1 c_2 d\rangle$. Thus,

$$\begin{aligned} f \approx f' &= \langle c_1,\, c_2,\, c_3,\, \dots,\, c_n\rangle \\ &\approx \langle d,\, c_1 c_2 d,\, c_3,\, \dots,\, c_p,\, \dots,\, c_n\rangle \\ &\approx \langle d,\, c_3,\, \dots,\, c_p,\, \dots,\, c_n,\, c_1 c_2 d\,\rangle, \end{aligned}$$

and $b_1 = d + c_3 e_3^2 + \cdots + c_p e_p^2$ is already represented by $\langle d, c_3, \dots, c_p\rangle$, which has dimension $p-1$, contradicting the choice of p. We have thus shown that $p = 1$. Hence $\langle c_1\rangle \cong \langle b_1\rangle$, and so $f \approx \langle b_1, c_2, \dots, c_n\rangle$. By Witt's Cancellation Theorem,

$$\langle b_1, c_2, \dots, c_n\rangle \cong \langle b_1, b_2, \dots, b_n\rangle \Longrightarrow \langle c_2, \dots, c_n\rangle \cong \langle b_2, \dots, b_n\rangle.$$

By the inductive hypothesis, this implies that $\langle c_2, \dots, c_n\rangle \approx \langle b_2, \dots, b_n\rangle$. So finally, $f \approx \langle b_1, c_2, \dots, c_n\rangle \approx \langle b_1, b_2, \dots, b_n\rangle = g$. $\square$

6. Kronecker Product of Quadratic Spaces

Let (V_1, B_1, q_1), (V_2, B_2, q_2) be two quadratic spaces over F, of dimensions m and n. We define a new vector space $V = V_1 \otimes V_2$ ($\otimes = \otimes_F$), and let $B : V \times V \to F$ be the unique symmetric bilinear pairing satisfying

$$B(v_1 \otimes v_2,\ v_1' \otimes v_2') = B_1(v_1, v_1')\, B_2(v_2, v_2') \qquad (v_i, v_i' \in V_i).$$

The pair (V, B) is a new quadratic space over F with dimension mn, called the *Kronecker product* (or *tensor product*) of the $(V_i,\ B_i)$'s. The associated quadratic map $q = q_B$ satisfies

$$\begin{aligned} q(v_1 \otimes v_2) &= B(v_1 \otimes v_2,\ v_1 \otimes v_2) \\ &= B_1(v_1, v_1) \cdot B_2(v_2, v_2) \\ &= q_1(v_1) \cdot q_2(v_2) \qquad (v_i \in V_i). \end{aligned}$$

We shall denote q by $q_1 \otimes q_2$, or sometimes just $q_1 q_2$. Let us coordinatize V_1 and V_2 by choosing ordered bases, $\{e_1, \ldots, e_m\}$ for V_1 and $\{f_1, \ldots, f_n\}$ for V_2. Let $a_{ij} = B(e_i, e_j)$, $b_{k\ell} = B(f_k, f_\ell)$. Then $M = (a_{ij})$ and $N = (b_{k\ell})$ are the symmetric matrices associated with q_1 and q_2 in the given coordinatizations. Now, consider the ordered basis of $V = V_1 \otimes V_2$ given by

$$\{e_1 \otimes f_1,\ e_1 \otimes f_2,\ \ldots\ ,\ e_1 \otimes f_n;\ \ldots\ ;\ e_m \otimes f_1,\ \ldots\ ,\ e_m \otimes f_n\}.$$

With respect to this coordinatization, the form q gives rise to the symmetric matrix

$$\left(\begin{array}{ccc|ccc|c} a_{11}b_{11} & a_{11}b_{12} & \cdots & a_{12}b_{11} & a_{12}b_{12} & \cdots & \cdots \\ a_{11}b_{21} & a_{11}b_{22} & \cdots & a_{12}b_{21} & a_{12}b_{22} & \cdots & \cdots \\ \vdots & \vdots & \ddots & \vdots & \vdots & \ddots & \ddots \\ \hline a_{21}b_{11} & a_{21}b_{12} & \cdots & & \cdots & & \cdots \\ \vdots & \vdots & \ddots & & \cdots & & \cdots \end{array}\right)$$

$$= \begin{pmatrix} a_{11}N & a_{12}N & \cdots & a_{1m}N \\ a_{21}N & a_{22}N & \cdots & a_{2m}N \\ \vdots & \vdots & \ddots & \vdots \\ a_{m1}N & a_{m2}N & \cdots & a_{mm}N \end{pmatrix},$$

which is precisely the ordinary *Kronecker product* of the two matrices M, N! In particular, $\langle a \rangle \otimes \langle b \rangle \cong \langle ab \rangle$ for all $a, b \in F$.

The Kronecker product operation for quadratic forms satisfies the usual commutative, associative, and distributive laws, as follows.

(1) $q_1 \otimes q_2 \cong q_2 \otimes q_1$.

(2) $(q_1 \otimes q_2) \otimes q_3 \cong q_1 \otimes (q_2 \otimes q_3)$.

(3) $q \otimes (q_1 \perp q_2) \cong (q \otimes q_1) \perp (q \otimes q_2)$.

Using the distributive law, we obtain the following simple rule for forming the product of two *diagonal* forms:

$$\langle a_1, \ldots, a_m \rangle \otimes \langle b_1, \ldots, b_n \rangle \cong \langle a_1 b_1, \ldots, a_i b_j, \ldots, a_m b_n \rangle.$$

The following convenient notation will be used for the rest of the book: If r is a nonnegative integer and f is a form, $r \cdot f$ (or sometimes $r f$) denotes the orthogonal sum of r copies of f.

Corollary 6.1. *If q is any regular quadratic form, then $q \otimes \mathbb{H} \cong (\dim q) \cdot \mathbb{H}$.*

Proof. Inducting on $\dim q$, we are reduced to the case where $q = \langle a \rangle$, $a \neq 0$. But then, $\langle a \rangle \otimes \mathbb{H} = \langle a \rangle \otimes \langle 1, -1 \rangle \cong \langle a, -a \rangle \cong \mathbb{H}$, by 3.2. □

As for the $r \cdot f$ notation (for $r \in \mathbb{N}$), a cautionary note is in order. In the quadratic form literature, for $a \in F$, many authors would write $a \cdot f$ for the Kronecker product $\langle a \rangle \otimes f$. We shall, by and large, follow this practice too, but then an expression like $3 \cdot f$ could become ambiguous. As a rule of thumb, $3 \cdot f$ shall mean $f \perp f \perp f$; the form obtained by "scaling" f by $3 \in F$ can safely be denoted by $\langle 3 \rangle f$ or, most unequivocally, by $\langle 3 \rangle \otimes f$.

7. Generation of the Orthogonal Group by Reflections

The purpose of this section is to present the classical result of H. Cartan and J. Dieudonné on the structure of the orthogonal group of a regular quadratic space. Although this theorem will not be heavily used in the sequel, it is nevertheless an important result in the theory of classical groups. In a nutshell, the theorem says that the orthogonal group is generated by hyperplane reflections. More precisely, we have the following quantitative result.

Cartan-Dieudonné Theorem 7.1. *Let (V, B, q) be a regular quadratic space of dimension n. Then every isometry $\sigma \in O_q(V)$ is a product of at most n hyperplane reflections (hyperplane reflections are defined in (4.5)).*

Before we present the proof of this important theorem, we record a few of its corollaries.

Corollary 7.2. *Suppose σ can be expressed as the product of n hyperplane reflections. Then the first (or, similarly, the last) reflection in the product may be arbitrarily chosen.*

Proof. Say $\sigma = \tau_1\tau_2\cdots\tau_n$, where each τ_i is a hyperplane reflection. Let τ be a given reflection. Consider the isometry $\tau\sigma$. By the theorem, we may write $\tau\sigma = \tau_2'\cdots\tau_r'$ $(r \le n+1)$, where $\tau_2', \ldots, \tau_r'$ are reflections. We have an equation $\det \sigma = (-1)^n = (-1)^r$, so r and n are congruent modulo 2. Thus, the inequality $r \le n+1$ can be strengthened to $r \le n$. Since $\tau^{n-r} = 1$, we may write $\sigma = \tau\tau_2'\cdots\tau_r'\cdot\tau^{n-r}$, which is the desired factorization. □

Corollary 7.3. *If* $\dim V = 2$, *every isometry of determinant* -1 *is a reflection. If* $\dim V \le 3$, *every* $\sigma \in \mathrm{SO}(V)$ *is the product of two reflections. (As usual,* $\mathrm{SO}(V) = \{\sigma \in \mathrm{O}(V) : \det \sigma = 1\}$.*)*

Corollary 7.4 ($\dim V = n$). (1) *If* $\sigma \in \mathrm{O}(V)$ *is the product of* r *reflections* $(r \le n)$, *then the dimension of the subspace of fixed vectors of* σ *is at least* $n - r$.

(2) *If* $\sigma \in \mathrm{O}(V)$ *has no nonzero fixed vectors, then* σ *cannot be written as the product of fewer than* n *reflections.*

Proof. (2) is an obvious consequence of (1). For (1), write $\sigma = \tau_1\tau_2\cdots\tau_r$ $(r \le n)$, and let U_j be the hyperplane pointwise fixed by τ_j. Then $U_1\cap\cdots\cap U_r$ consists of fixed vectors of σ, and $\dim(U_1 \cap \cdots \cap U_r) \ge n - r$. □

We shall now begin the proof of Theorem 7.1. For any given isometry σ, we introduce the following notations:

$\tilde{\sigma} = \sigma - 1$ (a linear endomorphism of V).

$L(\sigma) = \ker(\tilde{\sigma}) = \{v \in V \mid \sigma v = v\}$.

Lemma 7.5. $L(\sigma) = \ker(\tilde{\sigma}) = (\mathrm{Im}(\tilde{\sigma}))^{\perp}$.

Proof. Let $v \in L(\sigma)$ and $w \in V$. Then

$$\begin{aligned} B(v, \tilde{\sigma}w) &= B(v, \sigma w - w) \\ &= B(v, \sigma w) - B(v, w) \\ &= B(\sigma v, \sigma w) - B(v, w) = 0, \end{aligned}$$

so $v \in (\tilde{\sigma}V)^{\perp}$. Conversely, consider $v \in (\tilde{\sigma}V)^{\perp}$. We have, for any $w \in V$,

$$\begin{aligned} B(\sigma v - v, \sigma w) &= B(\sigma v, \sigma w) - B(v, \sigma w) \\ &= B(v, w) - B(v, \sigma w) \\ &= -B(v, \tilde{\sigma}w) = 0. \end{aligned}$$

This shows that $\sigma v - v \in \mathrm{rad}\, V = 0$, that is, $v \in L(\sigma)$. □

Corollary 7.6. (1) $(\ker(\tilde{\sigma}))^{\perp} = \mathrm{Im}(\tilde{\sigma})$.

(2) $\tilde{\sigma}^2 = 0$ *iff* $\mathrm{Im}(\tilde{\sigma})$ *is totally isotropic.*

Proof. (1) follows by taking the orthogonal complement of the equation in Lemma 7.5 (meanwhile using Proposition 1.3(2)). For (2), note that

$$\begin{aligned}\mathrm{Im}(\tilde{\sigma}) \text{ is totally isotropic } &\iff \mathrm{Im}(\tilde{\sigma}) \subseteq (\mathrm{Im}(\tilde{\sigma}))^{\perp} \\ &\iff \mathrm{Im}(\tilde{\sigma}) \subseteq \ker(\tilde{\sigma}) \quad \text{(by Lemma 7.5)} \\ &\iff \tilde{\sigma}^2 = 0. \end{aligned}$$

□

Lemma 7.7. *For any vector $w \in V$, $\tilde{\sigma}w$ is orthogonal to $\tilde{\sigma}w$ iff $\tilde{\sigma}w$ is orthogonal to w.*

Proof. Indeed,

$$\begin{aligned} B(\tilde{\sigma}w,\, \tilde{\sigma}w) = 0 &\iff B((\sigma-1)w,\, (\sigma-1)w) = 0 \\ &\iff B(\sigma w, \sigma w) - B(\sigma w, w) - B(w, \sigma w) + B(w,w) = 0 \\ &\iff 2\,[B(w,w) - B(\sigma w, w)] = 0 \\ &\iff B(\tilde{\sigma}w, w) = 0. \end{aligned}$$

□

Lemma 7.8. *Suppose $\tilde{\sigma}^2 \neq 0$. Then*

(1) *There exists an anisotropic vector $w \neq 0$ such that $z = \tilde{\sigma}(w)$ is anisotropic or possibly zero.*

(2) *If z above is nonzero, and $\sigma_1 = \tau_z \sigma$, then $w \in L(\sigma_1)$.*

This lemma looks somewhat technical, so let us first show how it will be used in proving the theorem.

Proof of Theorem 7.1. Use induction on $n = \dim V$. If $n = 1$, $\mathrm{O}(V) = \{\pm 1\}$ (where -1 represents the unique reflection), and there is nothing to prove. Assume now that the theorem holds for quadratic spaces of dimension $< n$. Suppose there exists $\sigma \in \mathrm{O}(V)$ that is a counterexample to the theorem. *We claim that $\tilde{\sigma}^2 = 0$.* Indeed, if otherwise, consider the vector w in (1) of Lemma 7.8. By the provision of the lemma, we have two possibilities:

(A) $z = \tilde{\sigma}(w) = 0$, that is, $\sigma w = w$. In this case, σ induces an isometry on $(F \cdot w)^{\perp}$, and, by induction, $\sigma |\ (F \cdot w)^{\perp}$ can be written as a product of at most $(n-1)$ reflections in $(F \cdot w)^{\perp}$. Each such reflection extends (uniquely) to a reflection of V that acts as identity on $F \cdot w$. Thus, σ itself is the product of at most $(n-1)$ reflections of V. Contradiction.

(B) $z = \tilde{\sigma}(w)$ is anisotropic (and nonzero). By (2) of Lemma 7.8, $\sigma_1 = \tau_z \cdot \sigma$ fixes the anisotropic vector w. Arguing as in Case (A) above, we know that σ_1 is a product of at most $(n-1)$ reflections of V. By transposition, we obtain σ itself as the product of at most n reflections of V. Contradiction.

The contradictions derived in (A) and (B) simply mean that $\tilde{\sigma}^2 = 0$, as we claimed. Thus $\mathrm{Im}(\tilde{\sigma}) \subseteq \ker(\tilde{\sigma})$. On the other hand, $\ker(\tilde{\sigma})$ $(= L(\sigma))$ must be totally isotropic (otherwise the argument in (A) enables us to express σ as the product of at most $(n-1)$ reflections). Hence

$$\ker(\tilde{\sigma}) \subseteq \ker(\tilde{\sigma})^{\perp} = \operatorname{Im}(\tilde{\sigma}),$$

by 7.6. It follows that the three spaces above are all equal. By the "Dimension Formula" (Proposition 1.3),

$$\begin{aligned} n &= \dim(\ker(\tilde{\sigma})) + \dim(\ker(\tilde{\sigma}))^{\perp} \\ &= \dim L(\sigma) + \dim L(\sigma) \\ &= 2 \cdot \dim L(\sigma), \end{aligned}$$

so n must be even. (The above work, together with 3.4(1), actually implies that V is a hyperbolic space, although we won't need this information.) Now σ acts as the identity on $L(\sigma)$, and induces the identity on $V/L(\sigma)$ $(= V/(\sigma-1)V)$. In particular, $\sigma \in \mathrm{SO}(V)$. Take any reflection τ of V. Then $\tau\sigma \notin \mathrm{SO}(V)$, and, by the arguments presented so far, $\tau\sigma$ cannot be a counterexample to our theorem. This means $\tau\sigma$ is the product of at most n reflections, so σ is the product of at most $(n+1)$ reflections. It cannot be the product of exactly $(n+1)$ reflections, because n is even, and we already know that $\det\sigma = 1$. Consequently, σ is the product of at most n reflections of V, a final contradiction. □

To complete the arguments, we now return to:

Proof of Lemma 7.8. Let us assume that (1) of Lemma 7.8 is false. This means that, for every anisotropic $w \neq 0$, $\tilde{\sigma}(w)$ is always isotropic (in particular, $\neq 0$). In particular, Lemma 7.7 implies that $\tilde{\sigma}(w)$ is orthogonal to w. The binary space $F \cdot w \oplus F(\tilde{\sigma}w)$ is not regular, since $\tilde{\sigma}w$ lies in its radical. But V is regular by assumption, so we have $\dim V \geq 3$. We claim now:

(7.9) *$\tilde{\sigma}(y)$ is orthogonal to y for all $y \in V$.*

For $y = 0$ this is trivial. For y nonzero and anisotropic, this was already observed above. Assume now y is isotropic. Consider some anisotropic $w \neq 0$ that is orthogonal to y (this vector exists, since $\dim V \geq 3$ and we can build y into a hyperbolic plane in V). Write $u = y + \varepsilon w$ $(\varepsilon \in \dot{F})$. We have

$$\begin{aligned} B(u,u) &= B(y+\varepsilon w,\ y+\varepsilon w) \\ &= B(y,y) + \varepsilon^2 B(w,w) \\ &= \varepsilon^2 B(w,w) \neq 0, \end{aligned}$$

so u is anisotropic (and nonzero since y, w are independent). Consequently, $\tilde{\sigma}(u)$ is orthogonal to u, which means that

$$\begin{aligned} 0 &= B(\tilde{\sigma}(y+\varepsilon w),\ y+\varepsilon w) \\ &= B(\tilde{\sigma}y,\ y) + \varepsilon(B(\tilde{\sigma}w,\ y) + B(\tilde{\sigma}y,\ w)) + \varepsilon^2 B(\tilde{\sigma}w,\ w), \end{aligned}$$

where the last term is zero. Since $\varepsilon \in \dot{F}$ is arbitrary (and $|F| > 2$), we conclude that $B(\tilde{\sigma}y,\ y) = 0$, proving (7.9) in all cases. Applying Lemma 7.7 to the statement (7.9), we see that $\mathrm{Im}(\tilde{\sigma})$ is totally isotropic. But now (7.6)(2) gives $\tilde{\sigma}^2 = 0$, in contradiction to the hypothesis of Lemma 7.8. This establishes (1) of Lemma 7.8. It only remains to prove (2). Suppose $z = \widetilde{\sigma}(w) \neq 0$ and $\sigma_1 = \tau_z\sigma$. The claim $w \in L(\sigma_1)$ is clear from geometrical consideration of the "isosceles triangle" ABC:

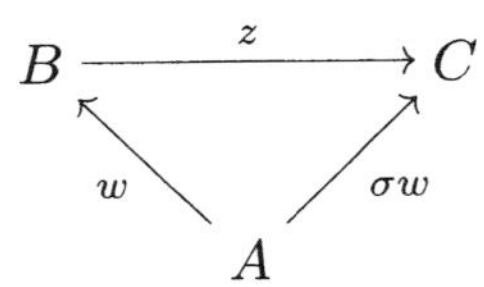

This completes the proof. □

Remark 7.10. In (2) of Lemma 7.8, the conclusion can actually be strengthened to

$$L(\sigma_1) \supseteq L(\sigma) + F\cdot w \supsetneq L(\sigma),$$

although we did not need it in this form.

Exercises for Chapter I

1. Show that the group of self-isometries of the n-dimensional quadratic space $n\langle 1\rangle$ is isomorphic to the group $\mathrm{O}(n)$ of $n \times n$ orthogonal matrices over F.
2. Let $V = \mathbb{M}_n(F)$, viewed as a vector space (of dimension n^2) over F. Show that $B(X,Y) = \mathrm{tr}(XY)$ (for $x, y \in \mathbb{M}_n(F)$) defines a regular quadratic space $(V,\ B)$. Show that $(V,\ B)$ is isometric to $n\langle 1\rangle \perp m\mathbb{H}$ where $m = n(n-1)/2$, and find an orthogonal basis for $(V,\ B)$. Do the same problem for the new form $B'(X,Y) = \mathrm{tr}(X \cdot Y^t)$, and show that $(V,\ B')$ is isometric to $n^2\langle 1\rangle$. (For more background information on trace forms on algebras, see Exercise 29 below.)
3. On $V = \mathbb{M}_n(F)$, define $B_U(X,Y) = \mathrm{tr}(X\cdot UY^tU^{-1})$, where $X,\ Y \in V$, and U is a fixed nonsingular symmetric matrix. Show that B_U defines a nonsingular symmetric bilinear form on V. If U has a diagonalization $\langle a_1, \ldots, a_n\rangle$, show that $(V,\ B_U)$ has a diagonalization $\perp_{i,j} \langle a_ia_j\rangle$ (i.e. isometric to $\langle a_1, \ldots, a_n\rangle \otimes \langle a_1, \ldots, a_n\rangle$).

4. Let $a, b \in \dot{F}$, and let f be a regular quadratic form. Show that $f \perp \langle a \rangle$ represents $-b$ iff $f \perp \langle b \rangle$ represents $-a$.

5. If $a, b \in F$ are such that $a^2 + b^2 = c \neq 0$, show that the 4-dimensional form $\langle 1, 1, -c, -c \rangle$ is hyperbolic.

6. (Extending 3.6.) For (regular) quadratic forms $q_1, \ldots, q_n$, show that the orthogonal sum $q_1 \perp \cdots \perp q_n$ is isotropic iff there exist $a_i \in D(q_i)$ $(1 \leq i \leq n)$ such that $\langle a_1, \ldots, a_n \rangle$ is isotropic.

7. Let f be a regular isotropic diagonal quadratic form over a field of more than five elements. Show that f admits an isotropic vector whose coordinates are *all* nonzero.

8. (This exercise will be used at least a few times in the sequel.)
   (1) Show that, if $\{F_i : i \in I\}$ is a family of subfields of a field K and $F = \bigcap_{i \in I} F_i \subseteq K$, then the natural map $\dot{F}/\dot{F}^2 \to \prod_i \dot{F}_i/\dot{F}_i^2$ is one-to-one.
   (2) Deduce from (1) that, if $|I| < \infty$ and $|\dot{F}_i/\dot{F}_i^2| < \infty$ for all i, then $|\dot{F}/\dot{F}^2| < \infty$.

9. Let A be a UFD, whose group of units is U. If F is the quotient field of A, show that $\dot{F}/\dot{F}^2$ is the direct product of U/U^2 and a $\mathbb{Z}_2$-vector space whose basis consists of the prime elements of A (taken up to associates). If $A = \mathbb{Z}$, and $\{p_1, \ldots, p_n\}$, $\{q_1, \ldots, q_n\}$ are sets of distinct primes, show that $\langle p_1, \ldots, p_n \rangle \cong \langle q_1, \ldots, q_n \rangle$ over $\mathbb{Q}$ iff $p_i = q_i$ for all i (after a permutation).

10. Show that the following conditions are equivalent:

    (1) Every 4-dimensional form over F of determinant -1 is isotropic.
    (2) Every even-dimensional form over F of determinant -1 is isotropic.
    (3) Every 3-dimensional form over F represents its own determinant.
    (4) Every odd-dimensional form over F represents its own determinant.

    (For more information on the four equivalent conditions above, see Ch. X, Exercise 11.)

11. Prove the following "Witt's Extension Theorem." Let V be a regular quadratic space, and U_1, U_2 be two subspaces. If there exists a (bijective) isometry $\sigma : U_1 \to U_2$, show that there exists an isometry σ' of V onto V such that $\sigma' | \, U_1 = \sigma$. (This is essentially an equivalent version of 4.2.)

12. In a hyperbolic space V, a maximal totally isotropic subspace is sometimes called a *Lagrangian*. Show that V is always the sum of two Lagrangians.

13. Show that a regular quadratic space is isotropic iff it has a basis consisting of isotropic vectors.

14. Let U be a (possibly not regular) subspace of dimension $m+r$ in a hyperbolic space $m\mathbb{H}$. Show that $i(U)$ (the Witt index of U) is at least r. (In particular, $\dim U > m \Longrightarrow U$ is isotropic.)
15. Let U be a (possibly not regular) quadratic space of dimension k. Use the last exercise to show that U can be embedded (as a quadratic space) into the hyperbolic space $m\mathbb{H}$ iff $i(U) \geq k-m$.
16. For regular quadratic forms σ and φ, show that

    (1) $i\,(\sigma \otimes \varphi) \geq i\,(\sigma) \cdot \dim \varphi$;
    (2) $i\,(\sigma \perp \varphi) \leq i\,(\sigma) + \dim \varphi$; and
    (3) if σ is isometric to a subform of a regular form τ, then

    $$\dim \tau - i\,(\tau) \geq \dim \sigma - i\,(\sigma).$$

    (This is essentially a slight reformulation of (2).) Deduce that, if $\dim \sigma > \dim \tau - i\,(\tau)$, then σ must be isotropic.
17. Let G be a finite group and $V = FG$ be the group ring of G over F. Let $T : V \to F$ be the linear functional defined by $T\big(\sum_{g\in G} a_g g\big) = a_1$, and let q be the quadratic form on V associated with the (symmetric) bilinear form $(\alpha, \beta) \mapsto T(\alpha\beta)$. Compute the Witt index of q. (**Hint.** The answer is $\big(|G| - r\big)/2$, where $r = \text{Card}\,\{g \in G : \ g^2 = 1\}$.)
18. Let φ be a regular group form. Show that for any regular form σ, $D(\varphi) \cdot D(\varphi \otimes \sigma) = D(\varphi \otimes \sigma)$.
19. (*Inductive Description of Isometry.*) For $n \geq 3$, show that $\langle a_1, \ldots, a_n\rangle \cong \langle b_1, \ldots, b_n\rangle$ iff there exist $a,\, b,\, c_3,\, \ldots\,,\, c_n \in \dot{F}$ such that

    $$\langle a_2, \ldots, a_n\rangle \cong \langle a,\, c_3, \ldots, c_n\rangle\,, \qquad \langle b_2, \ldots, b_n\rangle \cong \langle b,\, c_3, \ldots, c_n\rangle\,,$$

    and $\langle a_1,\, a\rangle \cong \langle b_1,\, b\rangle$.
20. (*Inductive Description of Value Sets.*) For $\varphi = \sigma \perp \tau$, show that

    $$D(\varphi) = \bigcup \{D\,(\langle s,\, t\rangle) : \ s \in D(\sigma),\ t \in D(\tau)\}.$$

    From this, deduce that

    $$D(\langle a\rangle \perp \tau) = \bigcup \{D\,(\langle a,\, t\rangle) : \ t \in D(\tau)\}.$$
21. If $0 \neq a^2+b^2 \neq c^2$ in a field F, show that $\langle a^2+b^2,\, a^2+b^2-c^2\rangle$ always represents 1 over F. (For instance, $1 \in D_{\mathbb{Q}}\langle 17,\, 13\rangle$.)
22. (The Seven-Eleven Problem) What integers from 1 to 20 are represented by $\langle 7,\, 11\rangle$ over $\mathbb{Q}$?
23. Show that $q = \langle 2,\, 3,\, 6\rangle$ does not represent 7 over $\mathbb{Q}$. (**Hint.** Find a chain equivalence from q to the form $\langle 1,\, 1,\, 1\rangle$. The isometry $\langle 2,\, 3,\, 6\rangle \cong \langle 1,\, 1,\, 1\rangle$ also reoccurs in a later calculation over the rationals: see the Example following II.3.3.)

24. For $a,\, b \in \dot{F}$, show that
    (1) $b \in D(\langle 1,\, a\rangle) \Longleftrightarrow b \cdot \langle 1,\, a\rangle \cong \langle 1,\, a\rangle$, and
    (2) $D(\langle 1,\, a\rangle) \cap D(\langle 1,\, b\rangle) \subseteq D(\langle 1,\, -ab\rangle)$.
25. Let $a,\, b \in \dot{F}$. If $\langle 1,\, -a\rangle$ is universal, show that
$$D(\langle 1,\, b\rangle) = D(\langle 1,\, ab\rangle).$$
26. Show that a binary form $\langle 1,\, -a\rangle$ over $\mathbb{Q}$ is universal iff $a \in \dot{\mathbb{Q}}^2$.
27. Give an example of a regular ternary quadratic form $q(x, y, z)$ over a field for which each of the forms $q(0, y, z)$, $q(x, 0, z)$, and $q(x, y, 0)$ has rank 1.
28. Let $q = \sum_{i,j=1}^{n} a_{ij} x_i x_j$ $(a_{ij} = a_{ji})$ be a quadratic form over a field. The rank of q is defined to be the rank of the symmetric matrix (a_{ij}). Show that rank(q) is the largest integer k such that, upon setting a suitable set of $n-k$ of the variables equal to 0, we get a *regular* quadratic form in the remaining k variables.
29. For any finite-dimensional F-algebra A, let $\mathrm{tr}_A : A \to F$ denote the algebra trace on A. Then
$$(x,\, y) \mapsto \mathrm{tr}_A(xy) \quad (x,\, y \in A)$$
defines a symmetric bilinear form on A, denoted by $(A,\, \mathrm{tr}_A)$ (or more precisely, $(A,\, \mathrm{tr}_{A/F})$). (This is called the *trace form* on the F-algebra A.) If B is another finite-dimensional F-algebra, show that:
    (1) $(A \times B,\, \mathrm{tr}_{A\times B}) \cong (A,\, \mathrm{tr}_A) \perp (B,\, \mathrm{tr}_B)$; and
    (2) $(A \otimes B,\, \mathrm{tr}_{A\otimes B}) \cong (A,\, \mathrm{tr}_A) \otimes (B,\, \mathrm{tr}_B)$
30. Let K be a finite field extension of F. If K/F is an inseparable extension, show that the trace form $\mathrm{tr}_{K/F}$ is identically zero. On the other hand, if K/F is a separable extension, show that $\mathrm{tr}_{K/F}$ is a *nonsingular* symmetric bilinear form; in particular, this is always the case if char$(F) = 0$, or if char(F) is prime to $[K : F]$. (**Aside.** From the second part, it follows that $\mathrm{tr}_{A/F}$ is a nonsingular symmetric bilinear form for any commutative étale algebra A over the field F.)
31. Find diagonalizations over $\mathbb{Q}$ for the trace forms on the following number fields:
    (1) $K_1 = \mathbb{Q}(\sqrt{2}, \sqrt{3}\,)$;
    (2) $K_2 = \mathbb{Q}\,(\alpha)$, where $\alpha = \sqrt{2+\sqrt{2}}\,$;
    (3) $K_3 = \mathbb{Q}\,(\zeta)$, where ζ is a primitive 5th root of unity;
    (4) $K_4 = \mathbb{Q}(\sqrt[3]{2}\,)$;
    (5) $K_5 =$ the splitting field of $X^3 - 2$ over $\mathbb{Q}$; and
    (6) $K_6 =$ the splitting field of $X^3 + 3X^2 - X - 1$ over $\mathbb{Q}$.

Chapter II

Introduction to Witt Rings

1. Definition of $\widehat{W}(F)$ and $W(F)$

To define the Witt ring of a field F, we begin by building up a commutative ring from the set $M(F)$ of all isometry classes of (nonsingular)[(1)] quadratic forms over F. The binary operations $\perp$ and $\otimes$ already define the structure of a commutative semiring on $M(F)$. By Witt's Cancellation Theorem (I.4.2), the additive structure ($\perp$) actually makes $M(F)$ into a "cancellation monoid," although no nonzero element in $M(F)$ has an additive inverse. The procedure required to remedy this is the so-called Grothendieck construction.

In general, let M be any commutative cancellation monoid under addition. We define a relation $\sim$ on $M \times M$ by

$$(x, y) \sim (x', y') \iff x + y' = x' + y \in M.$$

The cancellation law in M implies that $\sim$ is an equivalence relation on $M \times M$. We define the *Grothendieck group* of M to be $\text{Groth}(M) = (M \times M)/\sim$ (the set of equivalence classes) with addition induced by

$$(x, y) + (x', y') = (x + x',\, y + y').$$

It is easy to see that this is a well-defined addition on $\text{Groth}(M)$, and that in $\text{Groth}(M)$, the two classes (x, y), (y, x) are additive inverses of each other.

[(1)] For the rest of the book, we shall be dealing primarily with *nonsingular* quadratic forms. Thus, the word "form" may henceforth be taken to mean a nonsingular quadratic form, unless it is stated otherwise.

So, indeed, Groth(M) is a group. The map $i : M \to \text{Groth}(M)$ defined by $i(x) = (x, 0)$ is an injection of M into Groth(M), which may be viewed as an inclusion $M \subseteq \text{Groth}(M)$. Note that $(x, y) = i(x) - i(y) = x - y$, so, in particular, Groth(M) is the additive group generated by M. Any monoid homomorphism f of M into an abelian group G extends uniquely to a group homomorphism $f : \text{Groth}(M) \to G$ by the rule $f(x - y) = f(x) - f(y) \in G$. This is called the "universal property" of Groth(M). Lastly, if M has a (commutative) multiplication which makes it into a semiring, then

$$(x,\, y)\,(x',\, y') = (xx' + yy',\, yx' + xy')$$

induces a (commutative) multiplication on Groth(M) that makes it into a (commutative) ring.

We may now apply the above machinery to the commutative semiring $M = M(F)$.

Definition 1.1. $\widehat{W}(F) = \text{Groth}(M(F))$ is called the *Witt-Grothendieck ring* of quadratic forms over the field F

Every element of $\widehat{W}(F)$ has the formal expression $q_1 - q_2$, where q_1, q_2 are nonsingular quadratic forms, or rather, isometry classes of such forms. Since we have observed that $M(F) \subseteq \widehat{W}(F)$, the two statements $q_1 = q_2 \in \widehat{W}(F)$ and $q_1 \cong q_2$ are synonymous.

Now, consider the dimension map $\dim : M(F) \to \mathbb{Z}$, which is a semiring homomorphism on $M(F)$. This extends uniquely (via the "universal property") to a ring homomorphism $\dim : \widehat{W}(F) \to \mathbb{Z}$, by

$$\dim\,(q_1 - q_2) = \dim\, q_1 - \dim\, q_2.$$

The kernel of this ring homomorphism, denoted by $\widehat{I}F$, is called the *fundamental ideal* of $\widehat{W}(F)$. We have clearly $\widehat{W}(F)/\widehat{I}F \cong \mathbb{Z}$.

Proposition 1.2. *$\widehat{I}F$ is additively generated by the expressions $\langle a \rangle - \langle 1 \rangle$, $0 \neq a \in F$.*

Proof. If $z \in \widehat{I}F$, then $z = q_1 - q_2$, where q_1 and q_2 have the same dimension. Say, $q_1 = \langle a_1, \ldots, a_n \rangle$, $q_2 = \langle b_1, \ldots, b_n \rangle$. Then

$$x = \sum_i (\langle a_i \rangle - \langle b_i \rangle) = \sum_i (\langle a_i \rangle - \langle 1 \rangle) - \sum_i (\langle b_i \rangle - \langle 1 \rangle). \qquad \square$$

We shall now look at another ideal of $\widehat{W}(F)$, namely, $\mathbb{Z} \cdot \mathbb{H}$. This consists of all hyperbolic spaces and their "additive inverses," and they form an ideal by I.6.1.

Definition 1.3. The factor ring $W(F) = \widehat{W}(F)/\mathbb{Z} \cdot \mathbb{H}$ is called the *Witt ring* of F. Using the usual procedure of extension of scalars, it is easy to verify that $\widehat{W}$ and W are both "functors" from fields (of characteristic not 2) to commutative rings.

The commutative ring $W(F)$ was first defined by Witt [Wi] in 1937, though, of course, the name Witt ring and the notation $W(F)$ came only some years later.

Proposition 1.4. (1) *The elements of* $W(F)$ *are in one-one correspondence with the isometry classes of all anisotropic forms.*

(2) *Two (nonsingular) forms* q, q' *represent the same element in* $W(F)$ *iff* $q_a \cong q'_a$ *(see* I.4.1 *for the notation). (In this case,* q *and* q' *are said to be "Witt-similar.")*

(3) *If* $\dim q = \dim q'$, *then* q *and* q' *represent the same element in* $W(F)$ *iff* $q \cong q'$.

Proof. (1) Since the form $\mathbb{H}$ represents the element 0 in $W(F)$, we have $-\langle a\rangle = \langle -a\rangle \in W(F)$ for all $a \in \dot{F}$. In particular, every element of $W(F)$ is represented by a form q. If we write down the Witt decomposition of q, say, $q = q_h \perp q_a$, then q and q_a represent the same element in $W(F)$ (since $q_h = 0 \in W(F)$). Therefore, each element of $W(F)$ is represented by a suitable *anisotropic* form. For the proof of (1), it remains only to show that, if q and q' are anisotropic forms, then $q = q' \in W(F) \Rightarrow q \cong q'$. But $q = q' \in W(F)$ implies that $q = q' + m\mathbb{H} \in \widehat{W}(F)$ for some integer m. Without loss of generality, we may assume that $m \geq 0$. Then we have an isometry $q \cong q' \perp m\mathbb{H}$, which implies that $m = 0$ (since q is anisotropic). Thus, indeed, $q \cong q'$. (2) and (3) follow immediately from (1). □

The image of the ideal $\widehat{I}F$ under the natural projection $\widehat{W}(F) \to W(F)$ will be denoted by IF; this is called the *fundamental ideal* of $W(F)$. Since $\dim \mathbb{H} = 2$, we have clearly $\mathbb{Z} \cdot \mathbb{H} \cap \widehat{I}F = \{0\}$; thus, the natural projection induces an isomorphism $\widehat{I}F \cong IF$.

Proposition 1.5. *A form* q *represents an element in* $IF \subseteq W(F)$ *iff* $\dim q$ *is even.*

Proof. "If" part: We may clearly assume that q is a binary form, say, $q = \langle a, b\rangle$. Then q is the image of $\langle a\rangle - \langle -b\rangle \in \widehat{I}F$ under the natural projection $\widehat{W}(F) \to W(F)$. By definition, this says that $q \in IF \subseteq W(F)$.

"Only if" part: If q represents an element in IF, then there exists an equation $q = q_1 - q_2 + m\mathbb{H} \in \widehat{W}(F)$, where $m \in \mathbb{Z}$ and $\dim q_1 = \dim q_2$. Applying the map "dim," we see that $\dim q = 2m$. □

The ring epimorphism $\dim : \widehat{W}(F) \to \mathbb{Z}$ induces another epimorphism $\widehat{W}(F)/\mathbb{Z}\cdot\mathbb{H} = W(F) \to \mathbb{Z}/2\mathbb{Z}$, which we shall denote by $\dim_0$. By the above proposition, $\ker(\dim_0) = IF$, so we obtain:

Corollary 1.6. $\dim_0$ *defines an isomorphism* $W(F)/IF \cong \mathbb{Z}/2\mathbb{Z}$.

2. Group of Square Classes

At the end of I.2, we defined a monoid homomorphism $d : M(F) \to \dot{F}/\dot{F}^2$. By

$$d(q_1 - q_2) = d(q_1)d(q_2)^{-1} = d(q_1)d(q_2),$$

this extends to a homomorphism d from the additive group $\widehat{W}(F)$ to $\dot{F}/\dot{F}^2$. Since $d(\mathbb{H}) = -1 \cdot \dot{F}^2$, the homomorphism d does not factor through $W(F)$. However, there is a clever way to remedy this.

Let q be a (nonsingular) form of dimension n. We define the "signed determinant" of q by

$$d_\pm(q) = (-1)^{n(n-1)/2} d(q) \in \dot{F}/\dot{F}^2.$$

The obvious advantage of this signed determinant is that $d_\pm(\mathbb{H}) = 1 \cdot \dot{F}^2$. However, the formula $d_\pm(q \perp q') = d_\pm(q) d_\pm(q')$ clearly fails. To restore the homomorphism property, we look at $d_\pm$ *together with* $\dim_0$ (introduced at the end of Section 1), and manufacture a bigger group to receive the combined invariant. This new group is an extension of $\dot{F}/\dot{F}^2$ by $\mathbb{Z}_2$ ($= \mathbb{Z}/2\mathbb{Z} = \{0, 1\}$).[(2)] Namely, we define (*set theoretically*)

$$Q(F) = \mathbb{Z}_2 \times (\dot{F}/\dot{F}^2)$$

and introduce on it the binary operation

$$(e,\, d) \cdot (e',\, d') = (e + e',\, (-1)^{ee'} dd')$$

(not the direct product operation!). This is easily checked to be commutative and associative, with $(0, 1)$ playing the role of the identity element. The inverse of (e, d) is $(e, (-1)^e d)$, since

$$(e,\, d)\,(e,\, (-1)^e d) = (e + e,\, (-1)^{ee}(-1)^e dd) = (0, 1).$$

The inclusion $d \mapsto (0, d)$ identifies $\dot{F}/\dot{F}^2$ as a subgroup of index 2 in $Q(F)$. The pair $(1, 1)$ represents the only nonidentity coset modulo $\dot{F}/\dot{F}^2$, with $(1, 1)^2 = (0, -1)$. From this, it is clear that $Q(F)$ is a split extension of $\dot{F}/\dot{F}^2$ iff -1 is a square in F. Thus, for most fields, $Q(F)$ is a *nonsplit* extension of $\dot{F}/\dot{F}^2$ (by $\mathbb{Z}_2$).

---

(2) What we mean here is that the new group G has a normal subgroup $H \cong \dot{F}/\dot{F}^2$ such that $G/H \cong \mathbb{Z}_2$. In the literature, such G is sometimes also called an extension of $\mathbb{Z}_2$ by $\dot{F}/\dot{F}^2$.

Proposition 2.1. $(\dim_0, d_\pm)$ *defines a monoid epimorphism from* $M(F)$ *to* $Q(F)$. *This extends to a group epimorphism* $\widehat{W}(F) \to Q(F)$. *The latter induces a group isomorphism* $f : W(F)/I^2F \cong Q(F)$.

Proof. The map $M(F) \to Q(F)$ in question sends a form q to

$$(\dim_0 q,\, d_\pm(q)) \in Q(F)$$

(recall that $\dim_0(q)$ is just $\dim(q)$ taken modulo 2). To check that it is a monoid homomorphism, we calculate as follows (where $\dim q = n$, and $\dim q' = n'$):

$$\begin{aligned}
&(\dim_0,\, d_\pm)(q) \cdot (\dim_0,\, d_\pm)(q') \\
&\quad = (n,\, (-1)^{n(n-1)/2} d(q))\,(n',\, (-1)^{n'(n'-1)/2} d(q')) \\
&\quad = (n + n',\, (-1)^{nn'} (-1)^{[n(n-1)+n'(n'-1)]/2} \cdot d(q)d(q')) \\
&\quad = (n + n',\, (-1)^{(n+n')(n+n'-1)/2} \cdot d(q \perp q')) \\
&\quad = (\dim_0,\, d_\pm)(q \perp q') \in Q(F).
\end{aligned}$$

Further, $M(F) \to Q(F)$ is clearly an epimorphism, since

$$(\dim_0,\, d_\pm)(\langle a\rangle) = (1,\, a\cdot\dot{F}^2) \quad \text{and} \quad (\dim_0,\, d_\pm)(\langle 1, -a\rangle) = (0,\, a\cdot\dot{F}^2).$$

By the universal property of $\widehat{W}(F)$, the map $(\dim_0,\, d_\pm)$ extends uniquely to a group epimorphism from $\widehat{W}(F)$ to $Q(F)$. Moreover, since

$$(\dim_0,\, d_\pm)(\mathbb{H}) = (0,\, (-1)\cdot d(\mathbb{H})) = (0, 1)$$

is the identity element of $Q(F)$, we get an induced epimorphism $W(F) \to Q(F)$. We claim that this homomorphism is trivial on I^2F. By 1.2 (or 1.5), IF is additively generated by binary forms $\langle 1, a\rangle$, so I^2F is additively generated by the four-dimensional forms $\langle 1, a\rangle \otimes \langle 1, b\rangle$. But

$$(\dim_0,\, d_\pm)(\langle 1,\, a,\, b,\, ab\rangle) = (0,\, (-1)^0\cdot a\cdot b\cdot ab\dot{F}^2) = (0, 1),$$

so we obtain an epimorphism $f : W(F)/I^2F \to Q(F)$. We shall show that f is an isomorphism, by constructing an inverse $g : Q(F) \to W(F)/I^2F$. We simply set

$$g(0,\, a) = \langle 1,\, -a\rangle \ (\mathrm{mod}\ I^2F), \qquad g(1,\, a) = \langle a\rangle \ (\mathrm{mod}\ I^2F),$$

and carry out the following computation:

$$\begin{aligned} g\left[(0,\, a)(0,\, b)\right] &= g(0,\, ab) = \langle 1,\, -ab\rangle \equiv \langle 1,\, -a,\, 1,\, -b\rangle \\ &\equiv g(0,\, a) + g(0,\, b) \pmod{I^2F}, \\ g\left[(1,\, a)(1,\, b)\right] &= g(0,\, -ab) = \langle 1,\, ab\rangle \equiv \langle a,\, b\rangle \\ &\equiv g(1,\, a) + g(1,\, b) \pmod{I^2F}, \\ g\left[(0,\, a)(1,\, b)\right] &\equiv g(1,\, ab) = \langle ab\rangle \\ &\equiv \langle 1,\, -a,\, b\rangle \equiv g(0,\, a) + g(1,\, b) \pmod{I^2F}. \end{aligned}$$

Hence, g is a group homomorphism. Clearly, $f \circ g$ is the identity map on $Q(F)$, that is, g splits the surjection f. But, by $g(1, a) \equiv \langle a\rangle \pmod{I^2F}$, g is onto. It follows immediately that f and g are inverse isomorphisms of each other. $\square$

Corollary 2.2. (Pfister) *I^2F consists of classes of even-dimensional forms q for which $d(q) = (-1)^{n(n-1)/2}$ (where $n = \dim(q)$).*

Proof. This is just restating that $f : W(F)/I^2F \to Q(F)$ is a monomorphism. $\square$

Corollary 2.3. (Pfister) *The restriction of f induces an isomorphism from IF/I^2F to $\dot{F}/\dot{F}^2$.*

Note that, for an even-dimensional form q of dimension $n = 2r$, the sign $(-1)^{n(n-1)/2}$ simplifies to $(-1)^{2r(2r-1)/2} = (-1)^r$. Thus, the criterion for a form $q \in IF$ to lie in I^2F is $d(q) = 1$ in case $4 \mid \dim q$, and $d(q) = -1$ in case $4 \nmid \dim q$. For instance, the 6-dimensional form

$$q = \langle a,\, b,\, ab,\, -c,\, -d,\, -cd\rangle$$

with $d(q) = -1$ lies in I^2F. In fact, we can rewrite it as

$$q = \langle 1, a\rangle\, \langle 1, b\rangle - \langle 1, c\rangle\, \langle 1, d\rangle \quad \text{in } W(F),$$

where clearly each summand belongs to I^2F.

For later reference, we state one more useful consequence of the results above.

Corollary 2.4. *The following three statements are equivalent:*

(1) *$\widehat{W}(F)$ is a noetherian ring.*

(2) *$W(F)$ is a noetherian ring.*

(3) *$\dot{F}/\dot{F}^2$ is a finite group.*

Proof. (1) $\Rightarrow$ (2) is trivial, since a factor ring of any noetherian ring is noetherian.

(2) $\Rightarrow$ (3). Since $W(F)$ is assumed noetherian, IF is a finitely generated $W(F)$-module, so IF/I^2F is a finitely generated $W(F)/IF$-module. But $W(F)/IF \cong \mathbb{Z}_2$, so IF/I^2F must be finite. It follows from 2.3 that $\dot{F}/\dot{F}^2$ is finite.

(3) $\Rightarrow$ (1). By the diagonalization theorem (I.2.4), $\widehat{W}(F)$ is additively generated by $\langle a \rangle$, $a \in \dot{F}/\dot{F}^2$. Thus, (3) implies that $\widehat{W}(F)$ is a finitely generated abelian group. As a ring, of course, $\widehat{W}(F)$ is then noetherian. $\square$

Note that while the map f in (2.1) is a group homomorphism from $W(F)/I^2F$ to $Q(F)$, the domain $W(F)/I^2F$ of f is actually a (commutative) ring. This suggests that $Q(F)$ should possess a natural structure of a ring, which would make f a ring isomorphism. An easy computation leads to the following description of this ring structure. (Note that we have already used the "dot" notation for the addition in $Q(F)$, so we shall use "$\circ$" below for the multiplication.) For $a, b \in \dot{F}/\dot{F}^2$:

$$
\begin{aligned}
(0,\, a) \circ (0,\, b) &= (0,\, 1), \\
(0,\, a) \circ (1,\, b) &= (0,\, a), \qquad (2.5)\\
(1,\, a) \circ (1,\, b) &= (1,\, ab).
\end{aligned}
$$

This multiplication operation depends only on the group $\dot{F}/\dot{F}^2$. On the other hand, the addition operation depends on $\dot{F}/\dot{F}^2$ as a group "with a distinguished element $-1 \cdot \dot{F}^2$". Thus, if F and K are two fields with an isomorphism $\theta : \dot{F}/\dot{F}^2 \to \dot{K}/\dot{K}^2$ *preserving the square class of* -1, then $Q(F) \cong Q(K)$ as rings. The existence of such an isomorphism θ means precisely that $\dot{F}/\dot{F}^2 \cong \dot{K}/\dot{K}^2$ as groups and that $-1 \in \dot{F}^2$ iff $-1 \in \dot{K}^2$. In this case, we may conclude that $W(F)/I^2F \cong W(K)/I^2K$ (as rings), but not otherwise. For more information on the ring $Q(F)$, see Exercise 11.

On the other hand, it is easy to see that the factor ring $\widehat{W}(F)/\widehat{I}^2F$ is determined entirely by the group $\dot{F}/\dot{F}^2$, since

$$\widehat{W}(F)/\widehat{I}^2F = \mathbb{Z} \oplus (\widehat{I}F/\widehat{I}^2F), \qquad \widehat{I}F/\widehat{I}^2F \cong \dot{F}/\dot{F}^2,$$

and $\widehat{I}F/\widehat{I}^2F$ is an ideal of square zero. Thus, curiously enough, the ring $\widehat{W}(F)/\widehat{I}^2F$ actually carries *a little less* information than the ring $W(F)/I^2F$. As we shall see eventually, the Witt ring $W(F)$ is indeed a nicer and more useful object to work with than the Witt-Grothendieck ring $\widehat{W}(F)$.

3. Some Elementary Computations

In this section, we wish to compute a few Witt rings to illustrate the general theory. We shall begin with fields in which every element is a square: these are called *quadratically closed* fields. The most obvious example of such is, of course, an algebraically closed field. But there are *non* algebraically closed examples too. For instance, if $\mathbb{F}_5$ is the finite field of 5 elements, it is easy to see that its algebraic extension $\bigcup_{n\geq 1} \mathbb{F}_5(\sqrt[2^n]{2})$ is quadratically closed: see Exercise 18.

Proposition 3.1. *F is quadratically closed iff* dim : $\widehat{W}(F) \to \mathbb{Z}$ *is a (ring) isomorphism. In this case,* $W(F) \cong \mathbb{Z}_2$ *(by* $\dim_0$*).*

Proof. If F is quadratically closed, then $\langle a\rangle \cong \langle 1\rangle$, and $q \cong (\dim q)\langle 1\rangle$ for every form q. So, clearly, "dim" is an isomorphism. Conversely, if "dim" is an isomorphism, then $\langle a\rangle \cong \langle 1\rangle$ for every a ($\neq 0$), so every $a \in F$ is a square. □

We shall next deal with $\mathbb{R}$, the field of real numbers. The same calculations will apply to the class of the so-called "euclidean" fields, which we shall define later in §5..

Proposition 3.2. *Let* $F = \mathbb{R}$ *(or any "euclidean" field). Then:*

(1) *There exist exactly two anisotropic forms at each (positive) dimension. For dimension* $n > 0$*, these are* $n\langle 1\rangle$ *and* $n\langle -1\rangle$*.*
(2) $W(F) \cong \mathbb{Z}$.
(3) (Sylvester's Law of Inertia) *Two (nonsingular) forms over F are equivalent iff they have the same dimension and the same signature (the term will be defined in the proof).*
(4) $\widehat{W}(F) \cong \mathbb{Z} \oplus \mathbb{Z}$. *As a ring,* $\widehat{W}(F)$ *is isomorphic to the integral group ring* $\mathbb{Z}[G]$ *of a 2-element group G.*

Proof. We have $\dot{F}/\dot{F}^2 = \{\pm 1\}$. If a form is anisotropic, in its diagonalization we cannot have coefficients of mixed signs. So, (1) is clear. Since elements of $W(F)$ are in one-one correspondence with the anisotropic forms, (2) follows from (1).

For (3), let us first define "signature." We claim that, in a diagonalization of a form q, the number of positive coefficients (hence also the number of negative coefficients) is uniquely determined. In fact, suppose $r\langle 1\rangle \perp (n-r)\langle -1\rangle$ and $s\langle 1\rangle \perp (n-s)\langle -1\rangle$ are two diagonalizations of q ($\dim q = n$), where $s \geq r$. Passing to the *Witt ring* $W(F)$, we have an equation

$$r\langle 1\rangle - (n-r)\langle 1\rangle = s\langle 1\rangle - (n-s)\langle 1\rangle \in W(F),$$

which implies that $2r\langle 1\rangle = 2s\langle 1\rangle \in W(F)$. By (2), it follows immediately that $r = s$. Thus, we may write $n_+ = r$ (number of positive terms), and $n_- = n - r$ (number of negative terms). The *signature* of q is defined to be

$$n_+ - n_- = n_+ - (n - n_+) = 2n_+ - n.$$

Two forms are equivalent iff they have the same n and the same n_+, iff they have the same n and the same signature. This is Sylvester's Law of Inertia (3).

To prove (4), it suffices to show that $\langle 1\rangle$, $\langle -1\rangle$ form a free $\mathbb{Z}$-basis for $\widehat{W}(F)$. They clearly span $\widehat{W}(F)$. To show that they are independent, let $a\langle 1\rangle + b\langle -1\rangle = 0$ in $\widehat{W}(F)$, where $a, b \in \mathbb{Z}$. Passing to $W(F)$, we see that $a = b$. But then, clearly, $a = b = 0$, as desired. □

Remark 3.3. In the above situation:

(A) $\widehat{I}F$ is the free abelian group generated by $\langle 1\rangle - \langle -1\rangle$.

(B) Signature: $M(F) \to \mathbb{Z}$ is a monoid homomorphism, which extends to a group epimorphism $\widehat{W}(F) \to \mathbb{Z}$. The kernel of this is precisely $\mathbb{Z}\cdot\mathbb{H}$. Thus, we may say that the isomorphism $W(F) \to \mathbb{Z}$ in (2) above is induced by the signature map.

For an explicitly given quadratic form over a field F, a diagonalization can be easily computed by a completion of squares process that goes back to Lagrange. In the case where F is real-closed, the signature of the form can be determined accordingly. The following simple example suffices to illustrate this point.

Example. Let $q(x, y, z) = x^2 + 2y^2 + 3z^2 + 4xy + 6xz + 6yz$ over $\mathbb{Q}$. Then

$$\begin{aligned} q &= 3[z^2 + 2z(x+y)] + x^2 + 2y^2 + 4xy \\ &= 3(z+x+y)^2 - 3(x+y)^2 + x^2 + 2y^2 + 4xy \\ &= 3(z+x+y)^2 - (y^2 + 2yx) - 2x^2 \\ &= 3(z+x+y)^2 - (y+x)^2 - x^2. \end{aligned}$$

Thus, $q \cong \langle 3, -1, -1\rangle$ over $\mathbb{Q}$, and q has signature -1 over $\mathbb{R}$. Using the fact that 3 is not a sum of two squares in $\mathbb{Q}$, we see that q is *anisotropic* over $\mathbb{Q}$, though it is isotropic over $\mathbb{R}$. Of course, the completion of squares process leads to many different diagonalizations. For instance, we also have

$$\begin{aligned} q &= [x^2 + 2x(2y+3z)] + 2y^2 + 3z^2 + 6yz \\ &= (x+2y+3z)^2 - (2y+3z)^2 + 2y^2 + 3z^2 + 6yz \\ &= (x+2y+3z)^2 - 2\left[y^2 + 2y\cdot\frac{3z}{2}\right] - 6z^2 \\ &= (x+2y+3z)^2 - 2\left(y + \frac{3z}{2}\right)^2 - 6\left(\frac{z}{2}\right)^2, \end{aligned}$$

which yields a diagonalization $q \cong \langle 1, -2, -6\rangle$ over $\mathbb{Q}$. From this new diagonalization, the anisotropicity of q over $\mathbb{Q}$ is now a little less apparent!

A few more sophisticated problems on diagonalization of quadratic forms over $\mathbb{Q}$ are offered in Exercises 6 and 7 of this chapter.

Next, we shall consider finite fields. Let $F = \mathbb{F}_q$ denote the finite field of q $(= p^m)$ elements $(p \neq 2)$. Since $\dot{F}$ is a cyclic group of even order $(|\dot{F}| = q - 1)$, we can split it into the direct product of an odd order cyclic group and a nontrivial cyclic 2-group. Thus $|\dot{F}/\dot{F}^2| = 2$. We may denote the two square classes by 1 and s. Recall that $-1 \in \dot{F}^2$ iff $q \equiv 1 \pmod 4$, so s may be taken to be -1 iff $q \equiv 3 \pmod 4$.

Proposition 3.4. *Let $F = \mathbb{F}_q$, and $\dot{F}/\dot{F}^2 = \{1, s\}$. Then* (1) *$s$ is a sum of two squares, and* (2) *every (nonsingular) binary form is universal.*

Proof. We first show that (1) $\Rightarrow$ (2). Since 1 and s are the only square classes, there are at most three nonequivalent binary forms:

$$f_1 = \langle 1,\, 1\rangle, \qquad f_2 = \langle s,\, s\rangle, \qquad f_3 = \langle 1,\, s\rangle.$$

Clearly, $D(f_3) = \dot{F}$, and by (1), $D(f_1) = \dot{F}$, $D(f_2) = \dot{F}$. This proves (2). To establish (1), we argue in two cases.

(A) $-1 \in \dot{F}^2$. Then $\langle 1, 1\rangle \cong \langle 1, -1\rangle = \mathbb{H}$, and, in particular, $\langle 1, 1\rangle$ is universal (by I.3.4).

(B) $-1 \notin \dot{F}^2$. Look at the two sets $\dot{F}^2$ and $1 + \dot{F}^2$, which are subsets of F of the same cardinality. These sets are not equal, since $1 \in \dot{F}^2$ but $1 \notin 1 + \dot{F}^2$. Thus, there exists an element of the form $1 + z^2$ that lies outside $\dot{F}^2$. But $1 + z^2 \neq 0$ (lest $-1 \in \dot{F}^2$), so we may take s to be $1 + z^2$, which proves (1). $\square$

It turns out that the above statement (2) alone is sufficient to determine the Witt ring $W(F)$ completely. Therefore, we pass to a slightly more general setting, waiving the assumption that $|F| < \infty$.

Theorem 3.5. *Assume that every binary form over the field F is universal. Then*

(1) *two quadratic forms are isometric iff they have the same dimension and the same determinant;*

(2) $\widehat{I}^2F \cong I^2F = 0$ *and* $\widehat{I}F \cong IF \cong \dot{F}/\dot{F}^2$; *and*

(3) $W(F) \cong Q(F)$ *as rings, and* $\widehat{W}(F) = \mathbb{Z} \oplus \widehat{I}F$ *with trivial multiplication on* $\widehat{I}F$. *(For another description of the ring $Q(F)$, see Exercise 11(2).)*

Proof. Since any binary form $\langle a_1, a_2\rangle$ represents 1 by hypothesis, we have $\langle a_1, a_2\rangle \cong \langle 1, a_1a_2\rangle$. By induction, an arbitrary nonsingular form $q = \langle a_1, \ldots, a_n\rangle$ is equivalent to $\langle 1, \ldots, 1,\, d(q)\rangle$. This proves (1). By 1.2, $\widehat{I}^2F$

is additively generated by

$$(\langle a_1\rangle - \langle 1\rangle)(\langle a_2\rangle - \langle 1\rangle) = \langle a_1a_2\rangle + \langle 1\rangle - \langle a_1\rangle - \langle a_2\rangle = 0,$$

so $\widehat{I}^2F = 0$, proving the first part of (2). It follows that

$$\widehat{I}F \cong IF \cong IF/I^2F \cong \dot{F}/\dot{F}^2,$$

by 2.3. Finally, the isomorphism $W(F) \cong Q(F)$ in (3) follows from 2.1, and the description of $\widehat{W}(F)$ follows from the (split) exact sequence

$$0 \to \widehat{I}F \to \widehat{W}(F) \xrightarrow{\dim} \mathbb{Z} \to 0. \qquad \square$$

Now, back to finite fields.

Corollary 3.6. *Let* $F = \mathbb{F}_q$ $(q = odd)$.

(A) *If* $q \equiv 1 \pmod 4$, *then* $W(F)$ *is ring-isomorphic to the group ring* $\mathbb{Z}_2[\dot{F}/\dot{F}^2]$.

(B) *If* $q \equiv 3 \pmod 4$, *then* $W(F)$ *is ring-isomorphic to* $\mathbb{Z}_4$.

Proof. This follows by simply working out the structure of $Q(F)$, the main observation being that $Q(F)$ is a split extension of $\dot{F}/\dot{F}^2$ in case (A), and a nonsplit extension of $\dot{F}/\dot{F}^2$ in case (B). Actually, the situation is so simple here that we could have ascertained the ring structure of $W(F)$ directly by working with the anisotropic forms. In case (B), the anisotropic forms are $0, \langle 1\rangle, \langle 1, 1\rangle$ and $\langle -1\rangle$ (with $\langle 1, 1, 1\rangle = \langle -1\rangle \in W(F)$), so $W(F)$ is ring-isomorphic to $\mathbb{Z}_4$. In case (A), let s represent the nontrivial square class. Then the anisotropic forms are 0, $\langle 1\rangle$, $\langle s\rangle$, and $\langle 1, s\rangle$. Since

$$2 = \langle 1,\, 1\rangle = 0 \in W(F),$$

$W(F)$ is clearly the group ring $\mathbb{Z}_2[\dot{F}/\dot{F}^2]$, once we identify $\{\langle 1\rangle,\, \langle s\rangle\}$ with $\dot{F}/\dot{F}^2$. $\square$

Note that, according to the above result, we can tell apart the case $q \equiv 1 \pmod 4$ from the case $q \equiv 3 \pmod 4$ by looking at the Witt ring (or even the Witt group) $W(F)$. On the other hand, in *both* cases, the finite fields have the same Witt-Grothendieck ring $\widehat{W}(F)$. This is another explicit instance where $W(F)$ actually reflects the properties of F more accurately than $\widehat{W}(F)$ does..

It turns out that there are also other classes of fields to which Theorem 3.5 applies. For instance, it applies to any field F of transcendence degree 1 over an algebraically closed field k. This follows from the theorem of Tsen (and Lang), for the statement of which the reader may consult the introduction to Ch. XI. In the case when $\operatorname{tr.d}_k F = 1$, the Tsen-Lang Theorem implies that ternary quadratic forms over F are isotropic, which is equivalent to saying that binary quadratic forms over F are universal. To give the reader a feeling for this very useful theorem, we shall prove it below in the special case $F = k(t)$, utilizing the following general fact.

Lemma 3.7. *Over any field F, any binary form $q = \langle 1, a\rangle$ is a* group form; *that is, $D(q)$ is a subgroup of $\dot{F}$.*

Proof. A direct proof results by checking the formula

$$(*) \qquad (x^2+ay^2)(z^2+aw^2) = (xz - ayw)^2 + a(xw+yz)^2,$$

which is a slight generalization of the classical 2-square identity. For a more conceptual proof, we proceed as follows.

Consider the quadratic algebra $K = F[x]/(x^2+a)$, which has an F-basis $\{1,\ \theta\}$ where $\theta^2 = -a$. With respect to this basis, multiplication by $x+y\theta$ on K has the matrix $\begin{pmatrix} x & -ay \\ y & x \end{pmatrix}$. Thus, the "algebra norm" of $x+y\theta \in K$ is given by the determinant of this matrix:

$$\mathrm{N}_{K/F}(x+y\theta) = x^2 + ay^2 \quad (\forall\, x, y \in F).$$

Since the algebra norm is multiplicative, $q = \langle 1,\ a\rangle$ is a group form. Indeed, we have $D(q) = \mathrm{N}_{K/F}(\mathrm{U}(K))$, which is a subgroup of $\dot{F}$. (As the reader will no doubt notice, if we write out the product $(x+y\theta)(z+w\theta)$ and take the norm, we'll get precisely the multiplication formula $(*)$!) □

Note also that, in the above proof, K is a field (namely, the quadratic field extension $F(\sqrt{-a}\,)$) iff $a \notin -\dot{F}^2$, iff q is anisotropic over F. Otherwise, $K \cong F \times F$, q is hyperbolic, and $D(q) = \dot{F}$ by I.3.4(3).

Sneak Preview: Lemma 3.7 will be generalized to the quaternionic norm forms $\langle 1, a\rangle \otimes \langle 1, b\rangle$ in Chapter III, and to the so-called n-fold Pfister forms $\langle 1, a_1\rangle \otimes \cdots \otimes \langle 1, a_n\rangle$ in Chapter X.

Proposition 3.8. *Let $F = k(t)$, where k is any algebraically closed field. Then any binary quadratic form q over F is universal.*

Proof. We may assume that $q = \langle 1, f\rangle$, where $f \in \dot{F}$. It is easy to see that the $\mathbb{F}_2$-vector space $\dot{F}/\dot{F}^2$ has a basis $\{(t-b)\dot{F}^2 : b \in k\}$, so by 3.7, it suffices to show that $t - b \in D(q)$ for any $b \in k$. After a change of variables, we are reduced to showing that $t \in D(q)$, or equivalently, that $\langle 1,\ -t,\ f\rangle$ is isotropic. Another application of the same trick enables us to assume that $f = t - c$, where $c \in k$. For such an f, the isotropy of the form $\langle 1, -t, f\rangle$ follows from the equation $(\sqrt{c}\,)^2 - t + f = 0$. □

Having proved (3.8), we see that Theorem 3.5 can be utilized to calculate $\widehat{W}(F)$ and $W(F)$ for $F = k(t)$. Note that although $\dot{F}/\dot{F}^2$ is quite big (with an $\mathbb{F}_2$-basis $\{(t-b)\dot{F}^2 :\ b \in k\}$), the structures of $\widehat{W}(F)$ and $W(F)$ are easy to understand, in view of Theorem 3.5. Incidentally, 3.5 also applies to any field F of transcendental degree 1 over a "real-closed field" (e.g. $\mathbb{R}$), provided that -1 is a sum of squares in F. For this, the reader may look ahead at XI.1.9 and XI.4.2(2).

4. Presentation of Witt Rings

In this section we are interested in writing down full sets of generators and relations for $\widehat{W}(F)$ in the category of commutative rings, as well as in the category of abelian groups. Once we establish such results, then similar results may be derived for $W(F)$, since $W(F) = \widehat{W}(F)/\mathbb{Z}\cdot\mathbb{H}$.

We first consider $\widehat{W}(F)$ as a commutative ring. The elements $\langle a\rangle$ $(a \in \dot{F})$ generate $\widehat{W}(F)$, and satisfy the following obvious properties:

$(R_0 1)$ $\langle 1\rangle = 1$ (= the identity of the ring);

$(R_0 2)$ $\langle a\rangle \cdot \langle b\rangle = \langle ab\rangle$ $(a,\, b \in \dot{F})$; and

$(R_0 3)$ $\langle a\rangle + \langle b\rangle = \langle a+b\rangle \cdot (1 + \langle ab\rangle)$, where $a,\, b,\, a+b \in \dot{F}$.

(Here $(R_0 3)$ follows from I.5.1.)

Our aim is to prove that these are essentially all the relations among the symbols $\langle a\rangle$, $a \in \dot{F}$. The precise meaning of this statement is the content of the following.

Theorem 4.1. *Let $\mathcal{F}$ be the free commutative ring generated by the symbols $[a]$ $(a \in \dot{F})$. Let $\mathcal{R}$ be the ideal of $\mathcal{F}$ generated by the elements*

$(R1)$ $[1] - 1$,

$(R2)$ $[ab] - [a]\cdot[b]$ $(a, b \in \dot{F})$, and

$(R3)$ $[a] + [b] - [a+b]\cdot(1 + [ab])$ $(a,\, b,\, a+b \in \dot{F})$.

Then, the factor ring $X = \mathcal{F}/\mathcal{R}$ is isomorphic to $\widehat{W}(F)$.

Proof. By the universal property of the free commutative ring $\mathcal{F}$, and by $(R_0 1), (R_0 2), (R_0 3)$, we have a ring surjection $f : X \to \widehat{W}(F)$. We need only show that there exists an inverse. Thus, we try to define first a monoid homomorphism $\varphi : M(F) \to X$. For a given quadratic form q, take any diagonalization of q, say, $\langle a_1, \ldots, a_n\rangle$. We propose to set

$$\varphi(q) = [a_1] + \cdots + [a_n] \in X.$$

We must show, however, that $\varphi(q)$ does not depend on the particular diagonalization of q chosen above. This means that if $\langle b_1, \ldots, b_n\rangle$ is another diagonalization of q, we must show that $\sum [a_i] = \sum [b_i] \in X$. By Witt's Chain Equivalence Theorem (I.5.2), we may suppose that $\langle a_1, \ldots, a_n\rangle$ is actually simply-equivalent to $\langle b_1, \ldots, b_n\rangle$. Without loss of generality, we may assume that $a_i = b_i$ for $i \geq 3$, and $\langle a_1, a_2\rangle \cong \langle b_1, b_2\rangle$. Consequently, it suffices to show that

$$(4.2) \qquad \langle a_1, a_2\rangle \cong \langle b_1, b_2\rangle \implies [a_1] + [a_2] = [b_1] + [b_2] \in X.$$

Before we proceed, we must deduce some consequences of the relations in $(R1), (R2), (R3)$, in order to know more about X. We claim that, for every $a \in \dot{F}$, $[a^2] = 1 \in X$. To see this, we calculate $[a] + [a]$ in two different ways.

(A) Since $a + a = 2a \neq 0$, $(R3)$ implies

$$[a] + [a] = [2a] \cdot (1 + [a^2]) \in X.$$

(B) By $(R2)$ and the distributive law, we have

$$\begin{aligned}[a] + [a] &= [a] \cdot ([1] + [1]) \\ &= [a] \cdot [2] \cdot (1 + [1]) \qquad (\text{by } (R3)) \\ &= [2a] \cdot (1 + [1]) \in X \quad (\text{by } (R2)).\end{aligned}$$

But $(R2)$ implies that each $[b]$ $(b \in \dot{F})$ is a unit in X. Comparison of (A) and (B) then yields the desired information: $[a^2] = 1 \in X$.

Coming back to 4.2, we write $b_1 = a_1x^2 + a_2y^2$, and $a_1a_2 = b_1b_2c^2$ $(c \in \dot{F})$.

Case 1. $x = 0$, or $y = 0$. Suppose, for instance, $x = 0$ ($y = 0$ is similar). Then $b_1 = a_2y^2 \Rightarrow [b_1] = [a_2y^2] = [a_2] \in X$. On the other hand,

$$[a_1] = \left[b_2 \cdot \frac{b_1}{a_2} \cdot c^2\right] = [b_2y^2] = [b_2] \in X.$$

Hence, (4.2) follows.

Case 2. $x \neq 0$, $y \neq 0$. Then, in X, we have

$$\begin{aligned}[a_1] + [a_2] &= [a_1x^2] + [a_2y^2] \\ &= [a_1x^2 + a_2y^2] \cdot (1 + [a_1a_2(xy)^2]) \\ &= [b_1] \cdot (1 + [b_1b_2]) \\ &= [b_1] + [b_2].\end{aligned}$$

Thus, $\varphi : M(F) \to X$ is well-defined, and is clearly a monoid homomorphism. By the universal property of $\widehat{W}(F)$, φ extends to a group homomorphism $\varphi : \widehat{W}(F) \to X$, which is evidently an inverse for $f : X \to \widehat{W}(F)$. The latter is therefore a ring isomorphism. □

We shall now pass to the category of abelian groups. The corresponding result is the following.

Theorem 4.3. *Let $\mathcal{F}'$ be the free abelian group generated by the symbols $\{a\}$ $(a \in \dot{F})$. Let $\mathcal{R}'$ be the subgroup of $\mathcal{F}'$ generated by the elements*

$(R'1)$ $\{ab^2\} - \{a\}$ $(a, b \in \dot{F})$,

$(R'2)$ $\{a\} + \{b\} - \{a+b\} - \{ab(a+b)\}$ $(a, b, a+b \in \dot{F})$.

Then, the factor group $X' = \mathcal{F}'/\mathcal{R}'$ is isomorphic to $\widehat{W}(F)$.

Proof. Same as above! □

It is now easy to derive similar results for $W(F)$. In the category of commutative rings, we need only add the relation $(R4): [1]+[-1]$ to $(R1)$, $(R2)$, $(R3)$; and in the category of abelian groups, we need only add the relation $(R'3): \{1\}+\{-1\}$ to $(R'1)$ and $(R'2)$.

5. Classification of Small Witt Rings

In this section, we shall present a few results on the classification of Witt rings, focusing mainly on fields F with small square class groups $\dot{F}/\dot{F}^2$. Since $\dot{F}/\dot{F}^2$ is an abelian group of exponent ≤ 2, we may view it as an $\mathbb{F}_2$-vector space. In the following, a "basis" for $\dot{F}/\dot{F}^2$ shall always mean an $\mathbb{F}_2$-basis, and such a basis will often be written as $\{x_i\}$ instead of $\{x_i\dot{F}^2\}$.

In computing $W(F)$, it is important to distinguish the following two cases. We say that F is *formally real* if $x_1^2 + \cdots + x_n^2 = 0$ in F implies that all $x_i = 0$, or equivalently, -1 is not a sum of squares in F. Otherwise, F is said to be *nonreal*. The theory of formally real fields will be studied more systematically in Chapter VIII. The computations in this section generally do not require essential material from that chapter; however, some of the constructions of fields (existence proofs) do depend on later techniques.

We begin by looking at fields F with $|\dot{F}/\dot{F}^2| = 2$. If F is formally real, then the two square classes are represented by 1 and -1 respectively. The field F is said to be *euclidean* in this case, and $W(F)$ is clearly isomorphic to the ring $\mathbb{Z}$ (as in 3.2). Examples of euclidean fields include the fields of real numbers, real algebraic numbers, real constructible numbers,[(3)] etc. Now assume F is nonreal. If $-1 \in F^2$, then, as in the case of $\mathbb{F}_q$ with $q \equiv 1 \pmod 4$, we get $W(F) \cong \mathbb{Z}_2[\dot{F}/\dot{F}^2]$. If $-1 \notin F^2$, write $-1 = x_1^2 + \cdots + x_n^2$ with $n \geq 2$ minimal. Then $x_1^2 + x_2^2 \notin F^2$. Since the only square classes are represented by ± 1, we must have $x_1^2 + x_2^2 \in -\dot{F}^2$, and so -1 is a sum of two squares.[(4)] This shows that $\langle 1, 1\rangle \cong \langle -1, -1\rangle$. Thus, $W(F) = \mathbb{Z}\cdot\langle 1\rangle$ is isomorphic to the ring $\mathbb{Z}_4$, as in the case of $\mathbb{F}_q$ with $q \equiv 3 \pmod 4$.

---

(3)For more details on the field of real constructible numbers, see Ch. VIII, Exer. 4.

(4)Note that this argument, carried out for a finite field $\mathbb{F}_q$, actually leads to a new (and very simple) proof for the fact that -1 is a sum of two squares in $\mathbb{F}_q$!

We now proceed to the case where $|\dot{F}/\dot{F}^2| = 4$. We shall postpone the case where F is nonreal (to VI.2.35), and take up in the following proposition the case where F is formally real.

Proposition 5.1. *Let F be a formally real field with $|\dot{F}/\dot{F}^2| = 4$. Then $W(F)$ is isomorphic to either a group ring $\mathbb{Z}[G]$ where $|G| = 2$, or the ring $\mathbb{Z}[t]/(2t,\, t^2)$.*

Proof. First assume that every sum of squares in F is a square. (A field with this property is said to be a *pythagorean field*.) Let $\{-1, x\}$ be a basis of $\dot{F}/\dot{F}^2$. Then $\langle x\rangle$ has infinite additive order in $W(F)/\mathbb{Z}\langle 1\rangle$. In fact, if

$$\langle x, \dots, x\rangle = \pm\langle 1, \dots, 1\rangle \in W(F),$$

then[(5)] x is represented by either $\langle 1, \dots, 1\rangle$ or $\langle -1, \dots, -1\rangle$, and hence $x \in \pm\dot{F}^2$, a contradiction. It follows that $W(F) = \mathbb{Z}\langle 1\rangle \oplus \mathbb{Z}\langle x\rangle$, and this is isomorphic to the integral group ring $\mathbb{Z}[G]$ with G the 2-element group $\{1,\, \langle x\rangle\}$.

Next, assume F is *not* pythagorean. Then there exists an element $a = c^2 + d^2 \notin F^2$, and $\{-1, a\}$ necessarily form a basis of $\dot{F}/\dot{F}^2$. The Witt group $W(F)$ is additively generated by $\langle 1\rangle$, $\langle a\rangle$, and therefore also by $\langle 1\rangle$ and $\alpha := \langle 1,\, -a\rangle$. It is easy to see that $\langle a\rangle \notin \mathbb{Z}\langle 1\rangle$, so we also have $\alpha \notin \mathbb{Z}\langle 1\rangle$. Now $\langle 1, 1\rangle \cong \langle a, a\rangle$ implies that $2\,\alpha = 0 \in W(F)$; hence

$$W(F) = \mathbb{Z}\langle 1\rangle \oplus \mathbb{Z}_2\,\alpha \cong \mathbb{Z} \oplus \mathbb{Z}_2. \tag{5.2}$$

To determine the ring structure on $W(F)$, note that

$$\alpha^2 = \langle 1, -a\rangle\langle 1, -a\rangle \cong \langle 1, 1, -a, -a\rangle \cong 2\mathbb{H},$$

so $\alpha^2 = 0 \in W(F)$. Thus, we have a ring homomorphism

$$\varepsilon :\ \mathbb{Z}[t]/(2t,\, t^2) \longrightarrow W(F)$$

defined by $\varepsilon(t) = \alpha$, and the additive structure obtained for $W(F)$ in (5.2) implies that ε is an isomorphism. (Of course, we could also have identified $W(F)$ in the form $\mathbb{Z}[G]/(2g - 2)$, where $G = \{1, g\}$ is a cyclic group of order 2.) $\square$

Remark 5.3. Both cases in Proposition 5.1 are possible. For instance, the field $F = \mathbb{R}((x))$ consisting of all Laurent series in x with real coefficients is a formally real pythagorean field with $\dot{F}/\dot{F}^2 = \{\pm 1, \pm x\}$ (see VI.1.3). To get a field as in the second case of the proposition, we can use a direct limit construction applied in conjunction with a future result, VII.3.8. What we shall present here is a special case of the Gross-Fischer construction in VII.3.17; we'll only give a sketch here. Start with $\mathbb{Q} \subseteq \mathbb{R}$, and take an

---

(5) Note that both $\langle a, \dots, a\rangle$ and $\langle 1, \dots, 1\rangle$ are anisotropic forms here, since F is formally real.

$\mathbb{F}_2$-basis for $\dot{\mathbb{Q}}/\dot{\mathbb{Q}}^2$ given by -1, 2 and $\{a_i\}$, where each $a_i > 0$. Let $F_1 = \mathbb{Q}(\{\sqrt{a_i}\}) \subseteq \mathbb{R}$. By VII.3.8, -1 and 2 remain independent square classes in F_1, so we can take a basis -1, 2, $\{b_j\}$ for $\dot{F}_1/\dot{F}_1^2$ with $b_j > 0$, and form $F_2 = F_1(\{\sqrt{b_j}\}) \subseteq \mathbb{R}$. Repeating this construction and letting $F = \bigcup_i F_i \subseteq \mathbb{R}$, we see that F is formally real with $\dot{F}/\dot{F}^2 = \{\pm 1, \pm 2\}$. Since $2 = 1^2 + 1^2 \notin \dot{F}^2$, F is not pythagorean.(6)

The technique used for the two cases in (5.1) can be extended to handle the case of *nonpythagorean* formally real fields F with $|\dot{F}/\dot{F}^2| = 8$. We shall try to complete the classification of the Witt rings of such fields below. We start with two crucial examples, (5.4) and (5.7).

Example 5.4. *Let F be a formally real field such that $\dot{F}/\dot{F}^2$ has a basis given by $\{-1, a, c\}$, where $a \in \dot{F}$ is a sum of two squares in F and $c \in \dot{F}$ is a sum of four squares but not a sum of two squares in F. Then $W(F)$ is isomorphic to the ring*

$$A = \mathbb{Z}[t, s]/(2t,\ 4s,\ t^2,\ s^2 - 2s,\ ts - 2s). \tag{5.5}$$

To see this, let $\alpha = \langle 1, -a\rangle$, $\gamma = \langle 1, -c\rangle$ in $W(F)$. Then, as before, $2\alpha = \alpha^2 = 0$. Write $c = b + b'$, where $b, b' \in \dot{F}$ (and hence $bb' \in \dot{F}$) are sums of two squares in F. Thus,

$$\langle 1,\ 1\rangle \cong \langle b,\ b\rangle \cong \langle b',\ b'\rangle \cong \langle bb',\ bb'\rangle,$$

and we have

$$\begin{aligned}\langle 1, 1, 1, 1\rangle &\cong \langle b,\ b',\ b, b'\ \rangle\\ &\cong \langle c,\ cbb'\rangle \perp \langle c,\ cbb'\rangle\\ &\cong \langle c,\ c\rangle \perp \langle c\rangle\, \langle bb',\ bb'\rangle\\ &\cong \langle c,\ c,\ c,\ c\rangle.\end{aligned}$$

This shows that γ has additive order 4 in $W(F)$. (Note that $2\gamma \neq 0$, since c is not a sum of two squares in F.) Next, we compute γ^2 and $\alpha\gamma$ in $W(F)$. The first is easy, since

$$\gamma^2 = \langle 1, -c\rangle\, \langle 1, -c\rangle = \langle 1, -c, 1, -c\rangle = 2\gamma \in W(F).$$

We claim that $\alpha\gamma = 2\gamma \in W(F)$ also. To see this, we use the decomposition $c = b + b'$ introduced earlier. Since

$$\dot{F}^2 \subsetneq D(2\langle 1\rangle) \subsetneq D(4\langle 1\rangle) \subsetneq \dot{F},$$

(6) A less constructive method of getting such a field would be to take a subfield K of $\mathbb{R}$ that is maximal with respect to the property that $2 \notin \dot{K}^2$. Such a field K exists by Zorn's Lemma, and a direct application of VII.3.8 shows that $\dot{K}/\dot{K}^2$ consists of the four square classes $\{\pm 1, \pm 2\}$. This method of construction is a special case of Artin's method of "digging holes in a field." For a more formal presentation of this "digging holes" process, see XII.7.

$D(2\langle 1\rangle)$ consists only of the two square classes $\dot{F}^2$ and $a\dot{F}^2$. If $b, b' \in D(2\langle 1\rangle)$ both belong to $\dot{F}^2$ or to $a\dot{F}^2$, then $c = b+b'$ would belong to $D(2\langle 1\rangle)$, which is not the case. Therefore, we may assume that $b \in \dot{F}^2$ and $b' \in a\dot{F}^2$, and thus $c \in D(\langle 1, a\rangle)$. From this,[(7)] we get $\langle 1, a, -c, -ac\rangle = 0 \in W(F)$, and hence

$$\alpha\gamma \cong \langle 1,\ -c,\ -a,\ ac\rangle \cong \langle 1,\ -c,\ 1,\ -c\rangle \cong 2\gamma,$$

as claimed.

Having now computed α^2, γ^2 and $\alpha\gamma$, we can now determine the structure of $W(F)$. As an additive group, $W(F)$ is generated by $\langle 1\rangle, \langle a\rangle, \langle c\rangle$ and $\langle ac\rangle$. Since $\langle ac\rangle = \langle 1, a, -c\rangle \in W(F)$, we get

$$W(F) = \mathbb{Z}\langle 1\rangle + \mathbb{Z}_2\,\alpha + \mathbb{Z}_4\,\gamma.$$

This is a direct sum since $\mathbb{Z}_2\alpha + \mathbb{Z}_4\gamma$ is a torsion group, and $\alpha \notin I^2F$ implies that $\alpha \neq 2\gamma$ in $W(F)$. This shows that

$$W(F) \cong \mathbb{Z} \oplus \mathbb{Z}_2 \oplus \mathbb{Z}_4 \tag{5.6}$$

as an additive group, and that $W(F)$ is isomorphic to the ring A given in 5.5 by the isomorphism $t \mapsto \alpha,\ s \mapsto \gamma$.

Example 5.7. *Let F be a formally real field such that $\dot{F}/\dot{F}^2$ has a basis $\{-1, a, b\}$, where $a, b \in \dot{F}$ are sums of two squares in F. If $1 \in D\langle a, b\rangle$, then $W(F)$ is isomorphic to the ring*

$$B = \mathbb{Z}\,[t, u]/(2t,\ 2u,\ t^2,\ u^2,\ tu). \tag{5.8}$$

Otherwise, $W(F)$ is isomorphic to the ring

$$C = \mathbb{Z}\,[t, u, v]/(2t,\ 2u,\ 2v,\ t^2,\ u^2,\ v^2,\ tu - t - u - v). \tag{5.9}$$

To see this, let $\alpha = \langle 1, -a\rangle$, $\beta = \langle 1, -b\rangle$ and $\delta = \langle 1, -ab\rangle$ in $W(F)$. As before, we have

$$2\alpha = 2\beta = 2\delta = 0 = \alpha^2 = \beta^2 = \delta^2 \in W(F), \tag{5.10}$$

and an easy calculation shows that

$$\alpha + \beta + \delta = \langle 1,\ -a,\ -b,\ ab\rangle = \alpha\beta \in W(F). \tag{5.11}$$

Case 1. $1 \in D\langle a, b\rangle$. Here, $\langle a, b\rangle \cong \langle 1, ab\rangle$, so $W(F)$ is additively generated by $\langle 1\rangle, \langle a\rangle, \langle b\rangle$, and we have $\alpha\beta = 0$. It follows that

$$W(F) = \mathbb{Z}\,\langle 1\rangle \oplus \mathbb{Z}_2\,\alpha \oplus \mathbb{Z}_2\,\beta,$$

and that $W(F)$ is isomorphic to the ring B in 5.8 with an isomorphism given by $t \mapsto \alpha,\ u \mapsto \beta$. (Alternatively, $W(F)$ is isomorphic to

$$\mathbb{Z}\,[K]/(1 + gh - g - h),$$

[(7)]It is worth noting that, since a is a sum of two squares, $c \in D(\langle 1, a\rangle)$ implies that c is actually a sum of *three* squares. By symmetry, the same holds for ac, so $(\sum F^2) \setminus \{0\} = D(3\langle 1\rangle)$. (In terminology to be introduced later, the field F has *Pythagoras number* 3.)

where K is the Klein 4-group generated by g and h.)

Case 2. $1 \notin D\langle a, b\rangle$. Here, $0 \neq \alpha\beta = \alpha + \beta + \delta$ implies that

$$W(F) = \mathbb{Z}\langle 1\rangle \oplus \mathbb{Z}_2\,\alpha \oplus \mathbb{Z}_2\,\beta \oplus \mathbb{Z}_2\,\delta.$$

For the ring C in 5.9, there is a surjective ring homomorphism $\varphi : C \to W(F)$ defined by $\varphi(t) = \alpha$, $\varphi(u) = \beta$ and $\varphi(v) = \delta$. In C, it is easy to see that not only $tu = t + u + v$, but also $tv = uv = t + u + v$. Therefore,

$$C = \mathbb{Z} \oplus \mathbb{Z}_2\,t \oplus \mathbb{Z}_2\,u \oplus \mathbb{Z}_2\,v.$$

From this, it follows that φ is a ring isomorphism, as desired.

It can be shown that the formally real fields F described in 5.4 and 5.7 do exist, for suitable choices of the elements a, b, c. To construct these fields, however, requires some future material. In the following constructions, we shall assume certain facts about real-closed fields and p-adic fields. Readers not yet familiar with these facts may postpone these constructions to a later reading.

To construct the field in 5.4, let K be the field of 3-adic numbers. According to VI.1.3 and VI.2.2, $\dot{K}/\dot{K}^2$ has a basis $\{2, 3\}$, and 3 is not a sum of two squares in K. These are the only properties of K to be used for our construction. In the algebraic closure $\overline{K}$ of K, let L be any real-closed subfield, and let $F := K \cap L$. We have a natural homomorphism

$$f:\ \dot{F}/\dot{F}^2 \to \dot{K}/\dot{K}^2 \times \dot{L}/\dot{L}^2 \tag{5.12}$$

defined by $f(x\dot{F}^2) = (x\dot{K}^2,\ x\dot{L}^2)$. It is easy to see that f is injective (Ch. I, Exer. 8), so $|\dot{F}/\dot{F}^2| \leq 8$. Now

$$f(2) = (2,\ 1), \qquad f(3) = (3,\ 1), \qquad f(-1) = (-1,\ -1)$$

form a basis of $\dot{K}/\dot{K}^2 \times \dot{L}/\dot{L}^2$. Therefore, $\{-1, 2, 3\}$ form a basis of $\dot{F}/\dot{F}^2$. Since $F \subseteq L$, F is formally real, and since $F \subseteq K$, 3 is not a sum of two squares in F. Thus, F fits the descriptions of 5.4, with $a = 2 = 1^2 + 1^2$ and $c = 3 = 1^2 + 1^2 + 1^2$. We have here

$$W(F) \cong \mathbb{Z} \oplus \mathbb{Z}_2 \oplus \mathbb{Z}_4. \tag{A}$$

To construct a field as in Case 1 of 5.7, we use the technique introduced in Remark 5.3. Starting with the independent square classes of $-1, 2, 3$ in $\mathbb{Q}$, we can construct a formally real algebraic extension $F/\mathbb{Q}$ such that $\dot{F}/\dot{F}^2$ has basis $\{-1, 2, 3\}$ and such that $5 \in \dot{F}^2$. Then $a = 2$ and $b = 3 = (1+5)/2$ are sums of two squares in F, and $\langle a, b\rangle$ represents $2+3 = 5 \in \dot{F}^2$, as desired. We have here

$$W(F) \cong \mathbb{Z} \oplus \mathbb{Z}_2 \oplus \mathbb{Z}_2. \tag{B}$$

To construct a field as in Case 2 of 5.7, let E be the field of 5-adic numbers and L be any real-closed subfield of $\overline{E}$. By VI.1.3, $\dot{E}/\dot{E}^2$ has basis $\{2,5\}$. As in our earlier construction, $F := E \cap L$ has square class basis $\{-1,a,b\}$, with $a = 2 = 1^2+1^2$ and $b = 5 = 1^2+2^2$. Here $1 \notin D_E(\langle 2,5\rangle)$ by VI.2.2, so we also have $1 \notin D_F(\langle 2,5\rangle)$. By Case 2 of 5.7, we have here

$$(C) \qquad W(F) \cong \mathbb{Z} \oplus \mathbb{Z}_2 \oplus \mathbb{Z}_2 \oplus \mathbb{Z}_2.$$

(The last three examples we constructed are all formally real fields. The Witt groups of *nonreal* fields F with $|\dot{F}/\dot{F}^2| = 8$ are rather different from those computed above. For instance, in case F is the field of 2-adic numbers, $\dot{F}/\dot{F}^2$ has also a basis $\{-1,2,5\}$ as in the last example, but the Witt group $W(F)$ is isomorphic to $\mathbb{Z}_8 \oplus \mathbb{Z}_2 \oplus \mathbb{Z}_2$, according to VI.2.29 below.)

After computing the Witt rings of fields in the two crucial examples 5.4 and 5.7, we are now in a position to tackle the classification of Witt rings of all nonpythagorean formally real fields F with $|\dot{F}/\dot{F}^2| = 8$.

Theorem 5.13. *Let F be a nonpythagorean formally real field with $|\dot{F}/\dot{F}^2| = 8$. Then $W(F)$ is isomorphic to one of the rings A, B, C (in 5.5, 5.8 and 5.9), or to the group ring $R[G]$ where $R = \mathbb{Z}[t]/(2t, t^2)$ and G is a two-element group $\{1, x\}$, or to the quotient ring $R[G]/(\bar{t}x - \bar{t})$.*

Proof. Let m (resp. n) be the number of square classes in $D(2\langle 1\rangle)$ (resp. $D(4\langle 1\rangle)$). If $m = 4$, we are in the situation of 5.7, where $W(F)$ is isomorphic to B or C. Since F is nonpythagorean (and $D(2\langle 1\rangle)$ is a group), we are left only with the case $m = 2$. If $n > 2$, then we are in the situation of 5.4, where $W(F)$ is isomorphic to A. The last case to consider is therefore $m = n = 2$. Here, any sum of squares in F is a sum of two squares. Take a basis $\{-1,a,x\}$ for $\dot{F}/\dot{F}^2$ with $a \in D(2\langle 1\rangle)$. Then $\pm x$ are not sums of squares, and $\alpha := \langle 1,-a\rangle \in W(F)$ satisfies the familiar relations $2\alpha = \alpha^2 = 0$. Let $\sigma := \langle x\rangle\alpha$, so that

$$W(F) = \mathbb{Z}\langle 1\rangle + \mathbb{Z}\langle x\rangle + \mathbb{Z}\alpha + \mathbb{Z}\sigma.$$

As in the proof of 5.1, $\mathbb{Z}\langle 1\rangle + \mathbb{Z}\langle x\rangle$ is a direct sum isomorphic to $\mathbb{Z} \oplus \mathbb{Z}$. Since $\mathbb{Z}\alpha + \mathbb{Z}\sigma$ is torsion, we have

$$(5.14) \qquad W(F) = \mathbb{Z}\langle 1\rangle \oplus \mathbb{Z}\langle x\rangle \oplus (\mathbb{Z}\alpha + \mathbb{Z}\sigma).$$

Noting that $D(\alpha) \supseteq \{\pm 1, \pm a\}\dot{F}^2$, we have the following two subcases.

Case 1. $D(\alpha) = \{\pm 1, \pm a\}\dot{F}^2$. Here $x \notin D(\alpha)$, and so $\sigma = \langle x\rangle\alpha \neq \alpha$. This implies that $\mathbb{Z}\alpha + \mathbb{Z}\sigma$ is a direct sum (isomorphic to $\mathbb{Z}_2 \oplus \mathbb{Z}_2$). Now, $\mathbb{Z}\langle 1\rangle \oplus \mathbb{Z}\alpha$ is just the ring $R = \mathbb{Z}[t]/(2t, t^2)$ (with α identified with $\bar{t}$), and

$$W(F) = (\mathbb{Z}\langle 1\rangle \oplus \mathbb{Z}\alpha) \oplus \langle x\rangle(\mathbb{Z}\langle 1\rangle \oplus \mathbb{Z}\alpha) = R \oplus R\langle x\rangle$$

is isomorphic to the group ring $R[G]$, where $G = \{\langle 1\rangle, \langle x\rangle\}$. As a group,

$$(D) \qquad W(F) \cong \mathbb{Z} \oplus \mathbb{Z} \oplus \mathbb{Z}_2 \oplus \mathbb{Z}_2.$$

Case 2. $D(\alpha) \supsetneqq \{\pm 1, \pm a\}\dot{F}^2$. Since $D(\alpha)$ is a group, this implies that $D(\alpha) = \dot{F}$. In particular, $x \in D(\alpha)$, so $\sigma = \langle x\rangle\, \alpha = \alpha$, and

$$W(F) = \mathbb{Z}\langle 1\rangle \oplus \mathbb{Z}\,\alpha \oplus \mathbb{Z}\,\langle x\rangle = R \oplus \mathbb{Z}\,\langle x\rangle.$$

It is easy to see that this is isomorphic to the quotient $R[G]/(\bar{t}x - \bar{t})$, where $G = \{\langle 1\rangle, \langle x\rangle\}$ as in Case 1. As a group,

$$(E) \qquad W(F) \cong \mathbb{Z} \oplus \mathbb{Z} \oplus \mathbb{Z}_2.$$

This completes the proof of 5.13. □

While we have constructed fields F whose Witt rings are of the type A, B, C, the classification work in 5.13 is complete only if we also *construct* fields F whose Witt rings are of the last two types in 5.13 (corresponding to Case 1 and Case 2 in the proof given above). Such constructions will require some techniques from the theory of ordered fields, so we shall postpone them to Chapter VIII: see the Appendix to VIII.4.

This concludes our discussion on formally real nonpythagorean fields F with $|\dot{F}/\dot{F}^2| = 8$. The pythagorean case can be analyzed too, by the method of VIII.4. It turns out that, for such fields, there are only two possible Witt rings (see VIII.4.13). Therefore, for formally real fields with eight square classes, there are altogether *seven* possible Witt rings. For nonreal fields, the corresponding number is *ten*. This makes for a total of *seventeen* Witt rings for fields F with $|\dot{F}/\dot{F}^2| = 8$. The subtlety of this classification problem can perhaps be seen from the fact that, in some early papers on this topic, the total number of possible Witt rings $W(F)$ (with $|\dot{F}/\dot{F}^2| = 8$) had sometimes been under-reported.

The enumeration of the ten types of Witt rings in the *nonreal* case is due to C. Cordes (and K. Szymiczek), with existence proofs for some of the types given later by L. Berman, M. Kula and T. L. Lee). We shall come back to give a partial discussion of this classification work in Ch. XII of this book; see XII.5 and XII.7. The classification of Witt rings of nonreal fields with $|\dot{F}/\dot{F}^2| = 4$ is, of course, easier; this will be taken up in an Appendix to VI.2, in conjunction with the study of quadratic forms over local fields.

Exercises for Chapter II

1. Let $G = \dot{F}/\dot{F}^2$, and let $\mathbb{Z}\,G$ denote the integral group ring of G. Let J be the ideal of $\mathbb{Z}\,G$ generated by the expressions $a\dot{F}^2 + b\dot{F}^2 - c\dot{F}^2 - d\dot{F}^2$, where $\langle a, b\rangle \cong \langle c, d\rangle$. Show that there is a natural ring isomorphism between $\mathbb{Z}\,G/J$ and $\widehat{W}(F)$. (Cf. Exercise 11(2) below.)

2. If $\langle a_1, \ldots, a_n \rangle \cong \langle b_1, \ldots, b_n \rangle$, show that

$$(\langle a_1 \rangle - 1) \cdots (\langle a_n \rangle - 1) = (\langle b_1 \rangle - 1) \cdots (\langle b_n \rangle - 1) \in \widehat{W}(F).$$

3. If we view $\widehat{I}F$ as an (abelian) group generated by the expressions $\langle a \rangle - 1$ ($a \in \dot{F}$), what are the relations among these generators?

4. Show that IF is the unique prime ideal of $W(F)$ that contains the element 2 ($= \langle 1, 1 \rangle$). [Do this exercise without looking at VIII.7.5.]

5. Prove the following converse to 3.5: *if* $I^2F = 0$, *then every binary form over* F *is universal.*

6. Find diagonalizations for the following quadratic forms over $\mathbb{Q}$, and compute their signatures over $\mathbb{R}$:
   (a) $q_1 = x^2 + y^2 + z^2 + xy + xz + yz$.
   (b) $q_2 = y^2 + 2z^2 + 4xy + 2xz$.
   (c) $q_3 = x_1x_2 + x_2x_3 + \cdots + x_{n-1}x_n$.
   (d) $q_4 = \sum_{i,j=1}^n ij\, x_ix_j$.
   (e) $q_5 = \sum_{i,j=1}^n (i+j)\, x_ix_j$.

7. Let $n \geq 2$. Over the rational field $\mathbb{Q}$, show that:
   (a) The form $\sum_{i,j=1}^n \min\{i,j\}\, x_ix_j$ is isometric to $\langle 1, 1, \ldots, 1 \rangle$.
   (b) The form $\sum_{i,j=1}^n \max\{i,j\}\, x_ix_j$ is isometric to $\langle n, -1, \ldots, -1 \rangle$.
   (c) The form $\sum_{i,j=1}^n |i-j|\, x_ix_j$ is isometric to $\langle 2 \rangle \langle n-1, -1, \ldots, -1 \rangle$.

   (In particular, the three quadratic forms above are all regular, and have real signatures n, $2-n$ and $2-n$ respectively. For more explicit information about these forms, see [L_5].)

8. (R. M. Robinson) Let (a_{ij}) be the symmetric matrix of the quadratic form in (c) above (i.e., $a_{ij} = |i-j|$ for $1 \leq i, j \leq n$). Show that $D_n := \det(a_{ij})$ is given by $(-1)^{n-1}2^{n-2}(n-1)$. (**Hint.** G. Szegö derived this expression for D_n by using the recurrence relation

$$D_{n+1} = -4(D_n + D_{n-1}).$$

   G. Marks found a simpler one: $D_{n+1} = (-1)^n 2^{n-1} - 2D_n$.)

9. (G. Marks) Show that the quadratic form $\sum_{i,j=1}^n |i-j|\, x_ix_j$ above has also the diagonalization $\langle -b_1, -b_2, \ldots, -b_n \rangle$ over $\mathbb{Q}$, where

$$b_1 = 1, \quad b_2 = -1, \quad \text{and} \quad b_r = 2(r-1)/(r-2) \ \text{ for } r \geq 3.$$

10. Show that $W(F)$ is finite iff -1 is a sum of squares in F and $\dot{F}/\dot{F}^2$ is finite. (**Hint.** Use the Pigeon Hole Principle.)

11. (1) Confirm that the "$\circ$" operation on $Q(F)$ in 2.5 is the correct multiplication that makes $Q(F)$ into a commutative ring isomorphic to $W(F)/I^2F$.

(2) Show that $Q(F)$ is isomorphic to the ring $\mathbb{Z}[G]/J$, where $G = \dot{F}/\dot{F}^2$ and J is the ideal of $\mathbb{Z}[G]$ generated by

$$[1]+[-1] \quad \text{and} \quad [1]+[ab]-[a]-[b] \quad (\text{for all } a,\, b \in G).$$

Also show that $\mathbb{Z}$ could have been replaced by $\mathbb{Z}_4$ in this statement.

12. In working out Example 5.4, we have shown that, if $c \in \dot{F}$ is a sum of four squares in F, then $\langle 1,1,1,1\rangle \cong \langle c,c,c,c\rangle$. Use this to show that the set of such elements c forms a subgroup of $\dot{F}$. (This generalizes easily to sums of 2^n squares, but perhaps we should not give away too much of the forthcoming theory of Pfister forms and round forms (Chapter X) at this point!)
13. List all anisotropic (quadratic) forms over a formally real field F with square class basis $\{-1, 2\}$.
14. Compute $W(F)$ for a formally real field F with square class basis $\{-1, 2, 7\}$, and again, list all anisotropic forms over F. (**Hint.** Show that 7 is a sum of two squares, and apply 5.7.)
15. Let K and E be subfields of a field such that K is quadratically closed and E is euclidean. Show that $K \cap E$ is euclidean. (For a related problem, see Ch. VIII, Exer. 4.)
16. Show that a field F is euclidean if and only if $W(F)$ is an infinite cyclic group. (For a related problem, see Ch. VIII, Exer. 3.)
17. For any algebraic extension F of a finite field, show that $|\dot{F}/\dot{F}^2| \leq 2$.
18. Supply a proof for the fact (mentioned in the text) that the field $\bigcup_{n \geq 1} \mathbb{F}_5(\sqrt[2^n]{2})$ is quadratically closed.
19. Let $a \in \dot{F}$, and let q be a form over F with $\dim(q) = 2m$. Show that $q \cong a \cdot q$ iff $q \cong q_1 \perp \cdots \perp q_m$, where each q_i is a binary form such that $q_i \cong a \cdot q_i$.
20. Let q be an even-dimensional form. If $q \cong -q$, show that the form $q \otimes q$ is hyperbolic. (In the language of Witt rings, $2q = 0 \Longrightarrow q^2 = 0$ in $W(F)$.) (**Hint.** Use Exercise 19 with $a = -1$.)
21. Does the converse hold in Exercise 20? (If you can't solve this exercise at this time, come back to solve it later after you have learned more about quadratic forms!)
22. For any formally real field F, and any integer $r \in \mathbb{Z}$, show that $r \cdot 1$ is a unit in the Witt ring $W(F)$ iff $r = \pm 1$. (**Note.** If F is *nonreal* instead, it will be clear from later material that $r \cdot 1$ is a unit in $W(F)$ iff r is odd; see, for instance, VIII.7.8.)

Chapter III

Quaternion Algebras and their Norm Forms

1. Construction of Quaternion Algebras

The construction of the usual quaternion algebra over the real numbers $\mathbb{R}$ has obvious analogues over an arbitrary field F (of characteristic not two). The generalized construction of quaternion algebras leads to the consideration of their norm forms, a class of 4-dimensional forms that plays a vital role in quadratic form theory. This chapter presents an introductory study of such 4-dimensional forms, taking full advantage of the fact that they arise as norm forms of the generalized quaternion algebras.

Let $a, b \in \dot{F}$. We define the quaternion algebra $A = \left(\frac{a,b}{F}\right)$ to be the F-algebra on two generators i, j with the defining relations

$$i^2 = a, \quad j^2 = b, \quad ij = -ji.$$

For $k := ij \in A$, we have $k^2 = (ij)(ij) = -i^2j^2 = -ab \in \dot{F}$, and

$$ik = -ki = aj, \qquad kj = -jk = bi.$$

Thus, *any two of the elements* $\{i, j, k\}$ *anticommute.* Note that, in the case where $F = \mathbb{R}$ and $a = b = -1$, $\left(\frac{-1,-1}{\mathbb{R}}\right)$ is the usual division ring of quaternions over the reals, which we will denote by $\mathcal{H}$. The quaternion algebra $\left(\frac{a,b}{F}\right)$ over F is a direct generalization of $\mathcal{H}$.

From the multiplication rules among $\{i, j, k\}$ derived above, it is clear that $A = \left(\frac{a,b}{F}\right)$ is spanned by $\{1, i, j, k\}$ over F.

Proposition 1.0. $\{1, i, j, k\}$ *form an F-basis for A (so that $\dim_F A = 4$).*

Proof. Fix α, β in the algebraic closure E of F such that $\alpha^2 = -a$, $\beta^2 = b$, and consider the two matrices: $i_0 = \begin{pmatrix} 0 & \alpha \\ -\alpha & 0 \end{pmatrix}$ and $j_0 = \begin{pmatrix} 0 & \beta \\ \beta & 0 \end{pmatrix}$ in $\mathbb{M}_2(E)$ (the algebra of 2×2 matrices over E). Direct computations show that

$$i_0^2 = aI, \quad j_0^2 = bI, \quad \text{and} \quad i_0 j_0 = \begin{pmatrix} \alpha\beta & 0 \\ 0 & -\alpha\beta \end{pmatrix} = -j_0 i_0.$$

Thus, there is an F-algebra homomorphism $\varphi : \left(\frac{a,b}{F}\right) \rightarrow \mathbb{M}_2(E)$ with $\varphi(i) = i_0$ and $\varphi(j) = j_0$. Since $\{I, i_0, j_0, i_0 j_0\}$ are (clearly) linearly independent over E, $\{1, i, j, ij\}$ are linearly independent over F. □

Note that the construction of the (generalized) quaternion algebra A is *symmetric* in a, b (that is, $\left(\frac{a,b}{F}\right) \cong \left(\frac{b,a}{F}\right)$ as F-algebras), and *functorial* in F; that is, if K is a field extension of F, then

$$K \otimes_F \left(\frac{a,\, b}{F}\right) \cong \left(\frac{a,\, b}{K}\right) \quad \text{(as } K\text{-algebras).}$$

We now proceed to prove some basic facts about quaternion algebras. In the following, it will be understood that, whenever we use the isomorphism symbol "$\cong$" in the context of algebras over a field L, we mean L-algebra isomorphisms.

Proposition 1.1.

(1) $\left(\frac{a,b}{F}\right) \cong \left(\frac{ax^2,by^2}{F}\right)$ *for any* $a, b, x, y \in \dot{F}$.

(2) $\left(\frac{-1,1}{F}\right) \cong \mathbb{M}_2(F)$ (*the algebra of* 2×2 *matrices over* F).

(3) *The center of* $\left(\frac{a,b}{F}\right)$ *is* F $(= F\cdot 1)$.

(4) $\left(\frac{a,b}{F}\right)$ *is a simple algebra* (*i.e. it has no nontrivial ideals*).

Proof. (1) Let $A = \left(\frac{a,b}{F}\right)$, with basis $\{1, i, j, k\}$ as in the general construction, and let $A' = \left(\frac{ax^2,by^2}{F}\right)$, with basis $\{1, i', j', k'\}$ such that $i'^2 = ax^2$, $j'^2 = by^2$, etc. Consider the elements xi and yj in A, for which we have

$$(xi)^2 = x^2 i^2 = ax^2, \quad (yj)^2 = y^2 j^2 = by^2, \quad \text{and}$$
$$(xi)(yj) = xy(ij) = -xy(ji) = -(yj)(xi).$$

Thus, $\varphi : A' \rightarrow A$ induced by sending $i' \mapsto xi$, $j' \mapsto yj$ furnishes an F-algebra isomorphism between A' and A.

(2) With $a = -1$ and $b = 1$, we can choose $\alpha = \beta = 1 \in F$ in the proof of 1.0. The work in that proof gives an F-algebra isomorphism $\varphi : \left(\frac{-1,1}{F}\right) \rightarrow \mathbb{M}_2(F)$ with $\varphi(i) = \begin{pmatrix} 0 & 1 \\ -1 & 0 \end{pmatrix}$ and $\varphi(j) = \begin{pmatrix} 0 & 1 \\ 1 & 0 \end{pmatrix}$.

(3), (4) Let E be the algebraic closure of F. By the functorial property, we have $E \otimes_F \left(\frac{a,b}{F}\right) \cong \left(\frac{a,b}{E}\right)$. Since $E = E^2 = -E^2$, (1) and (2) imply that $\left(\frac{a,b}{E}\right) \cong \mathbb{M}_2(E)$. Since the center of $\mathbb{M}_2(E)$ is $E \cdot 1$ (the scalar matrices), it follows that the center of $\left(\frac{a,b}{F}\right)$ is $F \cdot 1 = F$. Since $\mathbb{M}_2(E)$ is a simple E-algebra, it follows that $\left(\frac{a,b}{F}\right)$ is a simple F-algebra. □

Our next job is to introduce and characterize the so-called "pure quaternions". Let $A = \left(\frac{a,b}{F}\right)$, with basis $\{1, i, j, k\}$ as usual.

Definition 1.2. A quaternion $v = \alpha + \beta i + \gamma j + \delta k \in A$ is called a *pure quaternion* if $\alpha = 0$. The F-space of pure quaternions will be denoted by A_0. The following characterization shows that the notion of "purity" can be described independently of the choice of the basis $\{1, i, j, k\}$.

Proposition 1.3. *Let $0 \neq v \in A$. Then $v \in A_0$ iff $v \notin F$ and $v^2 \in F$.*

Proof. In general, if $v = \alpha + \beta i + \gamma j + \delta k$, then (by direct calculation)

$$(*) \qquad v^2 = (\alpha^2 + a\beta^2 + b\gamma^2 - ab\delta^2) + 2\alpha(\beta i + \gamma j + \delta k).$$

Thus, if v is pure, we have $v^2 = a\beta^2 + b\gamma^2 - ab\delta^2 \in F$. Conversely, if $v \notin F$ (so at least one of β, γ, δ is nonzero) and $v^2 \in F$, then the equation $(*)$ above implies that $\alpha = 0$; that is, v is pure. □

Corollary 1.4. *If $A = \left(\frac{a,b}{F}\right)$, $A' = \left(\frac{a',b'}{F}\right)$, and $\varphi : A \to A'$ is an F-algebra isomorphism, then $\varphi(A_0) = A'_0$. In particular, A_0 is stable under any algebra endomorphism of A.*

Proof. The characterization of pure quaternions in 1.3 yields the first conclusion. The second conclusion follows from the first, since any algebra endomorphism of A is an automorphism, in view of 1.1(4). □

We shall close this section by studying in more detail Hamilton's division ring of real quaternions $\mathcal{H} = \left(\frac{-1,-1}{\mathbb{R}}\right)$. The $\mathbb{R}$-subalgebra $\mathbb{R} + \mathbb{R}i$ is clearly isomorphic to the complex field $\mathbb{C}$, so we may write $\mathbb{C} = \mathbb{R} + \mathbb{R}i \subseteq \mathcal{H}$. Note that $\mathcal{H}$ is *not* a $\mathbb{C}$-algebra, but it is a right $\mathbb{C}$-vector space, with $\mathbb{C}$-basis $\{1, j\}$. In fact, any real quaternion $v = x + yi + zj + wk$ can be written as

$$(x + yi) + (zj - wji) = \alpha + j\beta,$$

where $\alpha = x + yi \in \mathbb{C}$, $\beta = z - wi \in \mathbb{C}$. We have a "left regular representation" of $\mathcal{H}$, constructed as follows. For $v \in \mathcal{H}$, let L_v denote the left multiplication by v (that is, $L_v(q) = vq$). Thanks to the associative law, L_v

is a linear endomorphism of the right $\mathbb{C}$-vector space $\mathcal{H}$, operating from the left. Since (clearly) $L_{vv'} = L_v \circ L_{v'}$, L defines an $\mathbb{R}$-algebra homomorphism

$$L : \mathcal{H} \longrightarrow \mathrm{End}_{\mathbb{C}}(\mathcal{H}) \cong \mathbb{M}_2(\mathbb{C})$$

(which is, of course, not $\mathbb{C}$-linear!). Let us compute L_i, L_j, L_k in matrix form (with respect to the $\mathbb{C}$-basis $\{1, j\}$ on $\mathcal{H}$). For instance, $i \cdot 1 = i$ and $i \cdot j = -ji = j(-i)$ lead to

$$L_i = \begin{pmatrix} i & 0 \\ 0 & -i \end{pmatrix} \in \mathbb{M}_2(\mathbb{C}),$$

and similarly,

$$L_j = \begin{pmatrix} 0 & -1 \\ 1 & 0 \end{pmatrix}, \qquad L_k = \begin{pmatrix} 0 & -i \\ -i & 0 \end{pmatrix}.$$

More generally, if $x, y \in \mathbb{R}$, we have

$$L_{x+yi} = \begin{pmatrix} x & 0 \\ 0 & x \end{pmatrix} + \begin{pmatrix} y & 0 \\ 0 & y \end{pmatrix}\begin{pmatrix} i & 0 \\ 0 & -i \end{pmatrix} = \begin{pmatrix} x+yi & 0 \\ 0 & \overline{x+yi} \end{pmatrix}.$$

Thus, for a general quaternion $v = \alpha + j\beta$ (where $\alpha,\ \beta \in \mathbb{C}$) we have

$$L_v = L_\alpha + L_j L_\beta = \begin{pmatrix} \alpha & 0 \\ 0 & \overline{\alpha} \end{pmatrix} + \begin{pmatrix} 0 & -1 \\ 1 & 0 \end{pmatrix}\begin{pmatrix} \beta & 0 \\ 0 & \overline{\beta} \end{pmatrix} = \begin{pmatrix} \alpha & -\overline{\beta} \\ \beta & \overline{\alpha} \end{pmatrix}.$$

Now L is a *faithful* representation of the $\mathbb{R}$-algebra $\mathcal{H}$, since

$$L_v = 0 \implies L_v(1) = 0 \implies v = v \cdot 1 = 0.$$

Consequently, $\mathcal{H}$ is isomorphic to the real subalgebra of $\mathbb{M}_2(\mathbb{C})$ consisting of all the matrices $\begin{pmatrix} \alpha & -\overline{\beta} \\ \beta & \overline{\alpha} \end{pmatrix}$, where $\alpha, \beta \in \mathbb{C}$.

Corollary 1.5. *The group of "unit quaternions"*

$$U_0 = \{x + yi + zj + wk \mid x^2 + y^2 + z^2 + w^2 = 1\}$$

(set-theoretically the 3-sphere S^3 in $\mathbb{R}^4$) is isomorphic to the special unitary group SU(2).

Proof. Under the faithful representation L above, the group U_0 maps isomorphically to

$$\left\{\sigma = \begin{pmatrix} \alpha & -\overline{\beta} \\ \beta & \overline{\alpha} \end{pmatrix} : \ \alpha, \beta \in \mathbb{C},\ \det\sigma = \alpha\overline{\alpha} + \beta\overline{\beta} = 1\right\},$$

which is precisely the group SU(2). □

For the real quaternions $\mathcal{H}$, there is a well-known notion of "conjugation," which we shall generalize in the next section. To close the present discussion, we wish to derive interpretations of the notions of "purity" and "conjugation" in the matrix model of the real quaternions.

For $v = x + yi + zj + wk \in \mathcal{H}$, the *conjugate* of v is defined to be $\overline{v} = x - yi - zj - wk$. If we resolve v in terms of the $\mathbb{C}$-basis $\{1, j\}$, we have $v = \alpha + j\beta$, where $\alpha = x + yi \in \mathbb{C}$, and $\beta = z - wi \in \mathbb{C}$. The conjugate $\overline{v}$ may then be rewritten as

$$\overline{\alpha + j\beta} = x - yi - zj + wji = \overline{\alpha} - j(z - wi) = \overline{\alpha} - j\beta,$$

where $\overline{\alpha}$ means the complex conjugate of α. In the matrix model $L(\mathcal{H})$,

$$\alpha + j\beta \leftrightarrow \begin{pmatrix} \alpha & -\overline{\beta} \\ \beta & \overline{\alpha} \end{pmatrix}, \quad \text{and} \quad \overline{\alpha} - j\beta \leftrightarrow \begin{pmatrix} \overline{\alpha} & \overline{\beta} \\ -\beta & \alpha \end{pmatrix}.$$

Thus, *conjugation in $\mathcal{H}$ corresponds precisely to "conjugate transpose" in* $\mathbb{M}_2(\mathbb{C})$! In particular,

$$(1.6) \qquad v \in \mathcal{H} \text{ is pure } \iff \overline{v} = -v \iff L_v \text{ is skew-Hermitian.}$$

Thus, the 3-dimensional space of pure quaternions is represented by the space of skew-Hermitian matrices in the model $L(\mathcal{H})$. On the other hand, the Hermitian matrices in $L(\mathcal{H})$ are just the scalar matrices over $\mathbb{R}$, and these correspond to the scalars in $\mathbb{R} \subseteq \mathcal{H}$.

2. Quaternion Algebras as Quadratic Spaces

Consider a quaternion algebra $A = \left(\frac{a, b}{F}\right)$ over F, with the usual basis $\{1, i, j, k\}$ $(a, b \in \dot{F})$. *We want to make A into a quadratic space.*

For an arbitrary quaternion $x = \alpha + \beta i + \gamma j + \delta k$, we define the *conjugate* of x to be $\overline{x} = \alpha - (\beta i + \gamma j + \delta k)$. A direct computation shows that

$$\overline{x + y} = \overline{x} + \overline{y}, \quad \overline{x\,y} = \overline{y}\,\overline{x}, \quad \overline{\overline{x}} = x,$$

and

$$\overline{r\,x} = r\,\overline{x} \qquad (r \in F).$$

If $x \in A_0$ (pure quaternions), then $\overline{x} = -x$, and conversely.

Definition 2.1. The map $x \mapsto \overline{x}$ is called the *bar involution* on A. For $x \in A$ as above, we define $\mathrm{N}\,x = x\,\overline{x}$ (norm of x), and $\mathrm{T}\,x = x + \overline{x}$ (trace of x).

Note that $\overline{\mathrm{T}\,x} = \overline{x} + x = \mathrm{T}\,x \implies \mathrm{T}\,x \in F$, and $\overline{\mathrm{N}\,x} = \overline{\overline{x}}\,\overline{x} = x\,\overline{x} = \mathrm{N}\,x \implies \mathrm{N}\,x \in F$; that is, the norm and the trace of x are both scalars.

Next, we define

$$B(x, y) := (x\,\overline{y} + y\,\overline{x})/2 = \mathrm{T}(x\,\overline{y})/2 \in F.$$

This is clearly a symmetric bilinear form on A, so (A, B) becomes a quadratic space over F. The quadratic form associated with this bilinear form B sends

$$x \mapsto B(x, x) = \mathrm{T}(x\,\overline{x})/2 = 2\,x\,\overline{x}/2 = x\,\overline{x} = \mathrm{N}\,x.$$

Thus, N is a quadratic form on A, which will be called the *norm form* of A in the sequel.

Consider *pure* quaternions $x, y \in A_0$. Their inner product under B becomes

$$B(x,\, y) = (x\,\overline{y} + y\,\overline{x})/2 = -(xy + yx)/2.$$

Consequently, x, y are *orthogonal* in the space (A_0, B) iff x, y *anticommute* in A_0. In particular, $\{i, j, k\}$ form an orthogonal basis for the quadratic subspace $A_0 \subseteq A$. Further, if x is pure, then $B(x,\, 1) = \mathrm{T}(x)/2 = 0$, so F $(= F \cdot 1)$ is orthogonal to the entire subspace A_0.

Corollary 2.2. *The quadratic space* $(A,\, B)$ *has orthogonal basis* $\{1, i, j, k\}$. *It is regular and isometric to*

$$\langle 1,\, -a,\, -b,\, ab\rangle \cong \langle 1,\, -a\rangle \otimes \langle 1,\, -b\rangle.$$

Proof. Along with $\mathrm{N}(1) = 1$, we have $\mathrm{N}\,i = i\,\overline{i} = -i^2 = -a$. Similarly, $\mathrm{N}\,j = -b$, and $\mathrm{N}\,k = ab$. □

We note, in passing, that $\langle 1, -a, -b, ab\rangle$ are precisely the 4-dimensional quadratic forms q over F such that $d(q) = 1$ and $1 \in D_F(q)$.

Corollary 2.3. *If* $x = \alpha + \beta i + \gamma j + \delta k \in A$, *then*

$$\mathrm{N}\,x = \alpha^2 - \beta^2 a - \gamma^2 b + \delta^2 ab.$$

Proof. This follows from the diagonalization given in 2.2. □

Remarks. (1) Of course, Corollary 2.3 also follows from a direct calculation of the product $x\,\overline{x}$. Note that such a direct determination of the norm form N gives another proof for the fact that $\{1, i, j, k\}$ is an orthogonal basis for the quadratic space $(A,\, B)$.

(2) From 2.3, we see that $\mathrm{N}\,\overline{x} = \mathrm{N}\,x$ for any quaternion $x \in A$. This means that $x \mapsto \overline{x}$ is an *isometry* of the quadratic space A, so that $B(\overline{x},\, \overline{y}) = B(x,\, y)$ for all $x,\, y \in A$. This amounts to the fact that $B(x,\, y) = \mathrm{T}(x\,\overline{y})/2$ is equal to $\mathrm{T}(\overline{x}\,y)/2$, which, of course, can also be checked directly (see Exercise 3).

(3) Any quaternion $x \in A$ satisfies a quadratic polynomial equation over the base field F. In fact,

$$x^2 - \mathrm{T}(x)\,x + \mathrm{N}(x) = x^2 - x^2 - \overline{x}\,x + x\,\overline{x} = 0.$$

(4) For the real quaternions $\mathcal{H} = \left(\frac{-1,\,-1}{\mathbb{R}}\right)$, the function N retrieves the (squared) euclidean norm:

$$\mathrm{N}(\alpha + \beta i + \gamma j + \delta k) = \alpha^2 + \beta^2 + \gamma^2 + \delta^2 \quad (\text{on } \mathbb{R}^4).$$

Incidentally, it is also worth pointing that, if we use the model $L(\mathcal{H})$ for $\mathcal{H}$ (constructed at the end of §1), the norm and the trace on $\mathcal{H}$ correspond *exactly* to the norm and the trace of 2×2 matrices over $\mathbb{C}$.

Proposition 2.4. (1) *For* $x, y \in A$, $\mathrm{N}(xy) = \mathrm{N}\,x \cdot \mathrm{N}\,y$.

(2) $x \in A$ *is invertible iff* $\mathrm{N}\,x \neq 0$ (*iff* x *is anisotropic in* (A, B)).

Proof. (1) follows from the following computation:

$$\mathrm{N}(xy) = x\,y\,\overline{x\,y} = x\,(y\,\overline{y})\,\overline{x} = (x\,\overline{x})\,(y\,\overline{y}) = \mathrm{N}\,x \cdot \mathrm{N}\,y.$$

For (2), note that if x^{-1} exists, then

$$\mathrm{N}\,x \cdot \mathrm{N}(x^{-1}) = \mathrm{N}(xx^{-1}) = \mathrm{N}(1) = 1,$$

so $\mathrm{N}\,x \neq 0$. Conversely, if $\mathrm{N}\,x \neq 0$, the equation $x\,\overline{x} = \mathrm{N}(x) \cdot 1$ implies that x^{-1} exists and is given by $\overline{x}/\mathrm{N}\,x \in A$. □

Corollary 2.4′. (1) N *is a group form over* F.

(2) *For any* $c \in \dot{F}$, *we have* $c \in D_F(\mathrm{N})$ *iff* $\langle c \rangle \cdot \mathrm{N} \cong \mathrm{N}$.

Proof. (1) is clear from 2.4(1). For (2), first note that

$$\langle c \rangle \cdot \mathrm{N} \cong \mathrm{N} \implies c \in D_F(\langle c \rangle \cdot \mathrm{N}) = D_F(\mathrm{N}),$$

so we need only prove the "only if" part. There are several ways to do this[(1)]; we shall give a proof which takes full advantage of 2.4 (and which also gives us a very good perspective on the result at hand). Let $c \in D_F(\mathrm{N})$, say $c = \mathrm{N}(x)$ where $x \in A$. Let $\varphi : A \to A$ be defined by $\varphi(y) = xy$. Since x is invertible by 2.4(2), φ is a vector space isomorphism. Let N' be the quadratic form $\langle c \rangle \cdot \mathrm{N}$ on A. Then $\varphi : (A, \mathrm{N}') \to (A, \mathrm{N})$ is an isometry, since

$$\mathrm{N}(\varphi(y)) = \mathrm{N}(xy) = \mathrm{N}\,x \cdot \mathrm{N}\,y = c \cdot \mathrm{N}(y) = \mathrm{N}'(y)$$

for any $y \in A$. This shows that $\mathrm{N} \cong \mathrm{N}' = \langle c \rangle \cdot N$. □

The important properties (1), (2) for N in 2.4′ will be generalized later to the so-called n-fold Pfister forms in Chapter X. What we dealt with in 2.4′ is the case $n = 2$; the case $n = 1$ was treated in II.3.7 (and Ch. I, Exer. 24(1)).

Returning now to quaternion algebras, we formulate the following important result on their classification by norm forms.

Theorem 2.5. *For* $A = \left(\frac{a,b}{F}\right)$, $A' = \left(\frac{a',b'}{F}\right)$, *the following statements are equivalent*:

(1) A *and* A' *are isomorphic as* F*-algebras.*

[(1)] One proof, for instance, can be obtained by generalizing the argument on the form $\langle 1,1,1,1 \rangle$ in II.5.4.

(2) *A and A' are isometric as quadratic spaces.*

(3) *A_0 and A_0' are isometric as quadratic spaces.*

Proof. The equivalence (2) $\Leftrightarrow$ (3) is clear from Witt's Cancellation Theorem (I.4.2). Let us show now (1) $\Rightarrow$ (2). Suppose $\varphi : A \to A'$ is an algebra isomorphism. Then, Corollary 1.4 implies that $\varphi(A_0) = A_0'$. If $x = \alpha + x_0$, where $\alpha \in F$ and $x_0 \in A_0$, then $\overline{x} = \alpha - x_0$, and hence $\varphi(x) = \alpha + \varphi(x_0)$ and $\varphi(\overline{x}) = \alpha - \varphi(x_0)$. Since $\varphi(x_0) \in A_0'$, we have $\overline{\varphi(x)} = \varphi(\overline{x})$. Therefore,

$$\mathrm{N}(\varphi(x)) = \varphi(x) \cdot \overline{\varphi(x)} = \varphi(x) \cdot \varphi(\overline{x}) = \varphi(x\,\overline{x}) = \varphi(\mathrm{N}\,x) = \mathrm{N}\,x,$$

so φ is an isometry from A to A'.

Finally, let us show that (3) $\Rightarrow$ (1). Start with an isometry $\sigma : A_0 \to A_0'$. Then,

$$\mathrm{N}(\sigma(i)) = \mathrm{N}(i) = -a, \quad \text{and also} \quad \mathrm{N}(\sigma(i)) = \sigma(i)\,\overline{\sigma(i)} = -\sigma(i)^2.$$

Therefore, $\sigma(i)^2 = a$, and similarly, $\sigma(j)^2 = b$. Lastly,

$$\begin{aligned} i \text{ orthogonal to } j &\implies \sigma(i) \text{ orthogonal to } \sigma(j) \\ &\implies \sigma(i)\sigma(j) = -\sigma(j)\sigma(i) \in A'. \end{aligned}$$

All of these put together imply that $A' \cong \left(\frac{a,\,b}{F}\right) = A$, proving (1). □

Corollary 2.6. $\left(\frac{a,\,a}{F}\right) \cong \left(\frac{a,\,-1}{F}\right)$.

Proof. These two quaternion algebras have norm forms respectively equal to $\langle 1, -a, -a, a^2\rangle$ and $\langle 1, -a, 1, -a\rangle$, which are evidently isometric. Now apply 2.5. □

Theorem 2.7. *For $A = \left(\frac{a,b}{F}\right)$, the following statements are equivalent:*

(1) $A \cong \left(\frac{1,-1}{F}\right)$ ($\cong \mathbb{M}_2(F)$ *by* 1.1(2)).

(2) *A is not a division algebra.*

(3) *A is isotropic as a quadratic space.*

(4) *A is hyperbolic as a quadratic space.*

(5) *A_0 is isotropic as a quadratic space.*

(6) $\left(\langle a\rangle - 1\right)\left(\langle b\rangle - 1\right) = 0$ *in* $\widehat{W}(F)$ *(or in* $W(F)$*).*

(7) *The binary form $\langle a, b\rangle$ represents* 1.

(8) *$a \in \mathrm{N}_{E/F}(E)$, where $E = F(\sqrt{b})$, and $\mathrm{N}_{E/F}$ is the field norm.*

If any of these conditions holds for A, we shall say that A is split, *or that A* splits over *F.*

Proof. Since the norm form for $\left(\frac{-1,1}{F}\right)$ is hyperbolic, the equivalence of (1), (4), (6) and (7) follows from 2.5. These are also equivalent to (3) since the norm form on A has determinant 1. Next, we have clearly (4) $\Rightarrow$ (5) $\Rightarrow$ (3), as well as (1) $\Rightarrow$ (2) $\Rightarrow$ (3) (in view of 2.4(2)). This shows the equivalence of (1) through (7).

We finish by proving (7) $\Leftrightarrow$ (8). We may assume that $b \notin \dot{F}^2$ (for otherwise (7), (8) are obviously both true). Let E be the quadratic extension $F(\sqrt{b})$. Since $\mathrm{N}_{E/F}(x + y\sqrt{b}) = x^2 - by^2$ (for $x, y \in F$), the norm form for E/F is $\langle 1, -b\rangle$. Thus, (8) amounts to $a \in D_F(\langle 1, -b\rangle)$, which in turn amounts to (7). $\square$

Remarks. (A) Note that the above proof made no use of Wedderburn's Theorem on the structure of simple algebras. Some of the implications can certainly be deduced directly from this theorem. For instance, to see that (2) $\Rightarrow$ (1), apply Wedderburn's Theorem to express the simple algebra A (see 1.1(4)) in the form $\mathbb{M}_m(D)$, where D is some division F-algebra. Under (2), we must have $m \geq 2$, and $\dim_F A = 4$ implies that $m = 2$ and $D = F$. Thus, $A \cong \mathbb{M}_2(F)$.

(B) Of all the splitting criteria for $A = \left(\frac{a,b}{F}\right)$ in 2.7, the criterion (7) is especially important. The equation $ax^2 + by^2 = 1$ over the field F is sometimes called the *Hilbert equation.* In elementary number theory, this equation is used to define the *Hilbert symbol* over $\mathbb{Q}$. We shall study Hilbert symbols later, both over $\mathbb{Q}$ and over $\mathbb{Q}_p$ (the p-adic fields). At this point, while we are still working over general fields, the splitting condition for $A = \left(\frac{a,b}{F}\right)$ is that the Hilbert equation $ax^2 + by^2 = 1$ be solvable in F. This may be called *Hilbert's Criterion* for the splitting of the quaternion algebra A. For instance, using this Criterion, along with Ch. I, Exercise 21, we see that $\left(\frac{17,\,13}{\mathbb{Q}}\right) \cong \mathbb{M}_2(\mathbb{Q})$.

Corollary 2.8. (1) *For any* $a \in \dot{F}$, $\left(\frac{1,a}{F}\right)$ *and* $\left(\frac{a,-a}{F}\right)$ *are both split.*

(2) *If* $a \neq 0, 1$, *then* $\left(\frac{a,1-a}{F}\right)$ *is also split.*

(3) $\left(\frac{-1,a}{F}\right)$ *splits iff* a *is a sum of two squares in* F.

Proof. The binary forms $\langle 1, a\rangle$, $\langle a, -a\rangle$ ($\cong \mathbb{H}$), and $\langle a, 1-a\rangle$ (in case $a \neq 0, 1$) all represent 1. Thus, (1) and (2) follow from Hilbert's Criterion for splitting. (3) follows similarly since $\langle -1, a\rangle$ represents 1 iff $\langle 1, 1\rangle$ represents a, iff a is a sum of two squares in F. $\square$

Corollary 2.9. *If F is a finite field or $F = k(t)$ where k is an algebraically closed field, then* $\left(\frac{a,b}{F}\right) \cong \mathbb{M}_2(F)$ *for all* $a, b \in \dot{F}$.

Proof. Any binary form is universal by II.3.4 (resp. II.3.8). Thus, the conclusion follows from 2.7(7). □

The perceptive reader might have noticed that, in 2.9, the conclusion in the finite field case could also be deduced from Wedderburn's Little Theorem, which says that any finite division ring is a field.

For one more application of 2.7, we note that the machinery of quaternion algebras also yields a satisfactory classification of binary forms, as follows.

Corollary 2.10 (Classification of Binary Forms). *The (non-singular) forms* $q = \langle a,\, b\rangle$ *and* $q' = \langle a',\, b'\rangle$ *are isometric iff* $d(q) = d(q')$ *and* $\left(\frac{a,b}{F}\right) \cong \left(\frac{a',b'}{F}\right)$.

Proof. (Necessity). Assume $q \cong q'$. Then $d(q) = d(q')$; that is, $ab = a'b' \in \dot{F}/\dot{F}^2$. It follows that

$$(*) \qquad \langle 1,\, -a,\, -b,\, ab\rangle \cong \langle 1,\, -a',\, -b',\, a'b'\rangle,$$

so $\left(\frac{a,b}{F}\right)$ and $\left(\frac{a',b'}{F}\right)$ have isometric norm forms. Now 2.5 implies that they are isomorphic.

(Sufficiency). If the two quaternion algebras are isomorphic, then $(*)$ holds. If, further, $d(q) = d(q')$, then cancellation of $\langle ab\rangle \cong \langle a'b'\rangle$ yields $\langle -a,\, -b\rangle \cong \langle -a',\, -b'\rangle$, which amounts to $q \cong q'$. □

Recall that, if B and C are F-algebras, the tensor product $B \otimes_F C$ becomes an F-algebra with a multiplication induced by the rule

$$(\beta \otimes \gamma)(\beta' \otimes \gamma') = \beta\beta' \otimes \gamma\gamma' \quad (\beta, \beta' \in B, \text{ and } \gamma, \gamma' \in C).$$

If B and C are quaternion algebras, the tensor product $B \otimes_F C$ (of dimension 16) is called a *biquaternion algebra.*[(2)] We now come to the last main result in this section, which computes the biquaternion algebra $B \otimes_F C$ in a special case.

Theorem 2.11 (Linearity). *For* $a, b, c \in \dot{F}$, *we have*

$$\left(\frac{a,\, b}{F}\right) \otimes \left(\frac{a,\, c}{F}\right) \cong \left(\frac{a,\, bc}{F}\right) \otimes \left(\frac{c,\, -a^2c}{F}\right) \cong \left(\frac{a,\, bc}{F}\right) \otimes \mathbb{M}_2(F).$$

(Here and in the following, all tensor products are over F*.)*

---

(2) This is a somewhat more modern usage of the term "biquaternion". In the older literature, "biquaternion algebra" used to mean the "algebra of complex quaternions" $\left(\frac{-1,-1}{\mathbb{C}}\right)$. For us, however, the latter algebra does not really need a name, since it is $\mathbb{C}$-isomorphic to $\mathbb{M}_2(\mathbb{C})$.

Proof. Let $\{1, i, j, k\}$ and $\{1, i', j', k'\}$ be the "standard" bases for $B = \left(\frac{a,b}{F}\right)$ and $C = \left(\frac{a,c}{F}\right)$. We wish to analyze the tensor product algebra $B \otimes C$. Consider the following span:

$$\begin{aligned} X &= F \cdot (1 \otimes 1) + F \cdot (i \otimes 1) + F \cdot (j \otimes j') + F \cdot (k \otimes j') \\ &= F \cdot 1 + F \cdot I + F \cdot J + F(I \cdot J), \end{aligned}$$

where we have set $I = i \otimes 1$, $J = j \otimes j'$ (with $IJ = k \otimes j'$). This is a four-dimensional subalgebra of $B \otimes C$. In fact, we have

$$\begin{gathered} I^2 = i^2 \otimes 1 = a, \qquad J^2 = j^2 \otimes j'^2 = bc, \\ -I \cdot J = -ij \otimes j' = ji \otimes j' = J \cdot I, \end{gathered}$$

so the subalgebra X is a "copy" of the quaternion algebra $\left(\frac{a,bc}{F}\right)$. Next, we look at another subalgebra:

$$\begin{aligned} Y &= F \cdot (1 \otimes 1) + F(1 \otimes j') + F(i \otimes k') + F(-ci \otimes i') \\ &= F \cdot 1 + F \cdot \widetilde{I} + F \cdot \widetilde{J} + F \cdot (\widetilde{I}\widetilde{J}), \end{aligned}$$

where $\widetilde{I} = 1 \otimes j'$, $\widetilde{J} = i \otimes k'$ (with $\widetilde{I}\widetilde{J} = i \otimes j'k' = -ci \otimes i'$). Here, we have

$$\begin{gathered} \widetilde{I}^2 = 1 \otimes j'^2 = c, \qquad \widetilde{J}^2 = i^2 \otimes k'^2 = -a^2c, \\ -\widetilde{J}\widetilde{I} = -i \otimes k'j' = i \otimes j'k' = \widetilde{I}\widetilde{J}. \end{gathered}$$

Therefore, Y is a copy of the quaternion algebra $\left(\frac{c,-a^2c}{F}\right)$. By 1.1(1) and 2.8(1), this is isomorphic to $\mathbb{M}_2(F)$. We now complete the proof of the theorem by showing that $B \otimes C \cong X \otimes Y$. First, by direct inspection, the set $\{I, J\}$ commutes elementwise with the set $\{\widetilde{I}, \widetilde{J}\}$. Thus, elements of X commute with elements of Y. Second, one can check easily that the subalgebras X and Y generate the entire algebra $B \otimes C$. From these two facts, it follows that

$$B \otimes C \cong X \otimes Y \cong \left(\frac{a,\, bc}{F}\right) \otimes \mathbb{M}_2(F), \qquad \square$$

To illustrate the material developed in this section, we offer a few simple worked examples over the field of the rational numbers. Note that, if $a,\, b < 0$ in $\mathbb{Q}$, then $\left(\frac{a,b}{\mathbb{Q}}\right)$ is *always* nonsplit (and hence a division algebra), since

$$\mathbb{R} \otimes_{\mathbb{Q}} \left(\frac{a,\, b}{\mathbb{Q}}\right) \cong \left(\frac{a,\, b}{\mathbb{R}}\right) \cong \left(\frac{-1,\, -1}{\mathbb{R}}\right)$$

is Hamilton's quaternion division algebra $\mathcal{H}$.

Example 2.12. Show that $\left(\frac{-1,-1}{\mathbb{Q}}\right) \cong \left(\frac{-2,-3}{\mathbb{Q}}\right)$.

To see this, we transform the norm form of the first algebra. Since

$$\langle 1,\ 1,\ 1,\ 1\rangle \cong \langle 1,\ 1,\ 2,\ 2\rangle \cong \langle 1,\ 3,\ 6,\ 2\rangle \cong \langle 1,\ 2,\ 3,\ 6\rangle,$$

we conclude from 2.5 that the two given quaternion algebras are isomorphic.

Example 2.13. Show that $\left(\frac{-1,\,-1}{\mathbb{Q}}\right) \not\cong \left(\frac{-2,\,-5}{\mathbb{Q}}\right)$.

To see this, note that the norm form of the second algebra is

$$\langle 1,\ 2,\ 5,\ 10\rangle \cong \langle 1,\ 7,\ 70, 10\rangle.$$

If the two given quaternion algebras were isomorphic, the above norm form would be isometric to $\langle 1,1,1,1\rangle$, and Witt cancellation would give $\langle 1,\ 1,\ 1\rangle \cong \langle 7,\ 70,\ 10\rangle$ (over $\mathbb{Q}$). This is impossible since 7 is not a sum of three squares over the rationals. Thus, the two given quaternion algebras are *not* isomorphic.

Example 2.14. *For a rational odd*[(3)] *prime p, show that $\left(\frac{-1,\,p}{\mathbb{Q}}\right)$ splits iff $p \equiv 1 \pmod 4$.*

To see this, first assume $\left(\frac{-1,\,p}{\mathbb{Q}}\right)$ splits. Then $\langle -1,\ p\rangle$ represents 1 over $\mathbb{Q}$, so there exist integers x, y, z with $z \neq 0$ and $\gcd(x,y,z) = 1$ such that $-x^2 + py^2 = z^2$. It is easy to see that $p \nmid x$, so -1 is a square modulo p. By the First Supplement to the Law of Quadratic Reciprocity, this means that $p \equiv 1 \pmod 4$. Conversely, if $p \equiv 1 \pmod 4$, then, according to Fermat, $p = x^2 + z^2$ for suitable $x,\ z \in \mathbb{Z}$. In particular, $\langle -1,\ p\rangle$ represents 1 over $\mathbb{Q}$, and hence $\left(\frac{-1,\,p}{\mathbb{Q}}\right)$ splits.

Example 2.15. *For a rational odd*[(4)] *prime p, show that $\left(\frac{-2,\,p}{\mathbb{Q}}\right)$ splits iff $p \equiv 1$ or $3 \pmod 8$.*

To see this, first assume $\left(\frac{-2,\,p}{\mathbb{Q}}\right)$ splits. Arguing as in 2.14, we see that -2 is a square modulo p. By the First and Second Supplements to the Law of Quadratic Reciprocity, we must have $p \equiv 1$ or $3 \pmod 8$. Conversely, if $p \equiv 1$ or $3 \pmod 8$, a classical result in number theory (also known to Fermat) says that p can be written in the form $2x^2 + z^2$ for suitable $x,\ z \in \mathbb{Z}$. Thus, $\langle -2,\ p\rangle$ represents 1 over $\mathbb{Q}$, and hence $\left(\frac{-2,\,p}{\mathbb{Q}}\right)$ splits.

Example 2.16. *Find a positive integer n such that $\left(\frac{5,7}{\mathbb{Q}}\right) \cong \left(\frac{13,n}{\mathbb{Q}}\right)$.*

To solve this problem, note that

$$\langle -5,\ -7\rangle \cong \langle -12,\ -12\cdot 35\,\rangle \cong \langle -3,\ -3\cdot 35\,\rangle,$$

---

(3) It suffices to consider the case where p is odd, since we already know that $\left(\frac{-1,2}{\mathbb{Q}}\right)$ splits.

(4) Again, if $p = 2$, $\left(\frac{-2,p}{\mathbb{Q}}\right)$ splits, so we may as well take p to be odd.

and that $\langle -3,\ 35\rangle \cong \langle -13,\ 3{\cdot}13{\cdot}35\rangle$ (since $-3\cdot 4^2+35=-13$). Thus, the norm form of the first algebra is

$$\begin{aligned}\langle 1,\ -5,\ -7,\ 35\rangle &\cong \langle 1,\ -3{\cdot}35,\ -3,\ 35\rangle\\ &\cong \langle 1,\ -3{\cdot}35,\ -13,\ 3{\cdot}13{\cdot}35\rangle.\end{aligned}$$

This implies (by 2.5) that

$$\left(\frac{5,\ 7}{\mathbb{Q}}\right)\cong\left(\frac{13,\ 3{\cdot}35}{\mathbb{Q}}\right),$$

so we can take n to be $3\cdot 35=105$. (Further food-for-thought: can you determine the *smallest* positive integer n that works?)

Example 2.17. *Show that* $A=\left(\frac{5,-3}{\mathbb{Q}}\right)$ *is a division algebra, but* $K\otimes_{\mathbb{Q}}A$ *is not a division algebra for* $K=\mathbb{Q}(\sqrt{17})$.

Indeed, if A splits over $\mathbb{Q}$, there would exist integers x,y,z with $z\neq 0$ and $\gcd(x,y,z)=1$ such that $5x^2-3y^2=z^2$. Then, as before, we must have $3\nmid x$, and $5x^2\equiv z^2 \pmod 3$ would imply that 2 is a square modulo 3, a contradiction. Thus, A is a division $\mathbb{Q}$-algebra. For $K=\mathbb{Q}(\sqrt{17})$, however,

$$K\otimes_{\mathbb{Q}}\left(\frac{5,-3}{\mathbb{Q}}\right)\cong\left(\frac{5,-3}{K}\right)$$

splits, since $5\cdot 2^2-3=17$ implies that $\langle 5,-3\rangle$ represents 1 over K.

For general methods for deciding if a rational quaternion algebra is a division algebra in terms of elementary number theory, see Ch. VI.

3. Coverings of the Orthogonal Groups

For a quadratic space (V,q) we write $\mathrm{O}(V)=\mathrm{O}_q(V)$ to denote the orthogonal group of V, consisting of all isometries of V onto itself. We define also the *special orthogonal group* $\mathrm{SO}(V)$ to be $\{\sigma\in\mathrm{O}(V)\mid \det\sigma=1\}$, which is clearly a (normal) subgroup of index 2 in $\mathrm{O}(V)$.

In this section, we shall consider the special orthogonal group of the space of pure quaternions A_0 in a quaternion algebra $A=\left(\frac{a,b}{F}\right)$. The quadratic form in question is the restriction of the norm form to A_0, which has a diagonalization $\langle -a,\ -b,\ ab\rangle$. Bear in mind that if $A=\mathcal{H}$ (over the real field $\mathbb{R}$), then $\mathrm{SO}(A_0)$ is no other than the group $\mathrm{SO}(3)$ in the standard notation of Lie groups.

Theorem 3.1. *Let* $A=\left(\frac{a,b}{F}\right)$, *and let* U *denote the group of invertible elements of* A. *Then there exists an exact sequence of groups*

$$\text{(A)}\qquad 1\longrightarrow \dot{F}\longrightarrow U\stackrel{c}{\longrightarrow}\mathrm{SO}(A_0)\longrightarrow 1$$

such that, for $y \in U \cap A_0$ *("pure and invertible"),* $c(y) = -\tau_y$ *(where* τ_y *denotes the hyperplane reflection in* A_0 *associated with the anisotropic vector* y*: see* I.4.5*). If* $\mathrm{N}(U) \subseteq \dot{F}^2$ *(every norm is a square), then there exists an exact sequence*

$$\text{(B)} \qquad 1 \longrightarrow \{\pm 1\} \longrightarrow U_0 = \{u \in U \mid \mathrm{N}\, u = 1\} \xrightarrow{\ c\ } \mathrm{SO}(A_0) \longrightarrow 1.$$

Proof. The (rather long) proof of this theorem will be given in a sequence of steps.

Step 1. Recall (from 2.4(2)) that U consists of all anisotropic vectors of the quadratic space A. For any such anisotropic vector $y \in U$, we define $c(y): A_0 \to A_0$ by the formula

$$c(y)(x) = yxy^{-1}, \quad \text{for any } x \in A_0.$$

The fact that $c(y)$ is a linear automorphism of A_0 follows from 1.4. Further, $c(y_1y_2) = c(y_1) \circ c(y_2)$.

Step 2. To show that $c(y)$ is an *isometry* of A_0, we compute as follows:

$$\begin{aligned} \mathrm{N}(c(y)(x)) &= \mathrm{N}(yxy^{-1}) = \mathrm{N}(y)\ \mathrm{N}(x)\ \mathrm{N}(y^{-1}) \\ &= \mathrm{N}(x)\ \mathrm{N}(yy^{-1}) = \mathrm{N}(x). \end{aligned}$$

It remains to show that $c(y)$ has determinant 1 as a linear automorphism of A. We first establish the following lemma, which is actually a part of the theorem.

Lemma 3.2. *If* $y \in U \cap A_0$ *(pure and invertible),* $c(y) = -\tau_y \in \mathrm{SO}(A_0)$.

Proof. Let us calculate $\tau_y(x)$, assuming only $y \in U$ and $x \in A$ (to begin with). In the quadratic space (A, B), we have y definition:

$$\begin{aligned} \tau_y(x) &= x - \frac{2B(x,y)}{\mathrm{N}(y)}\, y \\ &= x - (x\,\overline{y} + y\,\overline{x}) \cdot \frac{y}{\mathrm{N}(y)} \\ &= x - x\,\frac{\overline{y}\,y}{\mathrm{N}(y)} - y\,\overline{x} \cdot \frac{y}{\mathrm{N}(y)} \\ &= -y\,\overline{x}\,\overline{y}^{-1}. \end{aligned}$$

Specializing to *pure* quaternions $x \in A_0$, we have $\tau_y(x) = y\,x\,\overline{y}^{-1}$. If we require further that $y \in U \cap A_0$, then

$$\tau_y(x) = -yxy^{-1} = -c\,(y)(x).$$

This shows that $c(y) = -\tau_y$ on A_0. Finally, since $\dim A_0 = 3$, we have $\det(c(y)) = (-1)^3 \det(\tau_y) = 1$ (by I.4.5(4)). □

Step 3. To resume the proof of 3.1, we must show, in general, that $y \in U \implies \det(c(y)) = 1$. Assume, instead, that

$$\det(c(y)) = -1 \quad (\text{for some } y \in U). \tag{3.3}$$

By the theorem of Cartan-Dieudonné, $c(y) \in \mathrm{O}(A_0)$ is the product of three hyperplane reflections in A_0, say, $c(y) = \tau_{x_1}\tau_{x_2}\tau_{x_3}$ (where $x_i \in U \cap A_0$). Using the Lemma, we may write

$$c(y) = -(-\tau_{x_1})(-\tau_{x_2})(-\tau_{x_3}) = -c(x_1)\,c(x_2)\,c(x_3) = -c(z),$$

where $z = x_1x_2x_3 \in U$. (Presumably, z is no longer a pure quaternion.) We have now an equation

$$yxy^{-1} = -zxz^{-1} \quad \text{for all } x \in A_0.$$

If we write $w = z^{-1}y \in U$, then

$$wx = -xw \quad \text{for all } x \in A_0. \tag{3.4}$$

We claim this is impossible. In fact, decompose w into $d + w_0$, with $d \in F$ and $w_0 \in A_0$. In general, w commutes with its "pure part" w_0; namely,

$$w\,w_0 = (d + w_0)\,w_0 = w_0\,(d + w_0) = w_0\,w.$$

Thus, letting $x = w_0$ in (3.4), we get $2\,w_0w = 0$. But then $w_0 = 0$, since $2w \in U$. This means that $w = d \in \dot{F}$, which clearly contradicts (3.4). This contradiction stems from (3.3), so we have proved that $c(y) \in \mathrm{SO}(A_0)$, for all $y \in U$.

Step 4. *Let us now compute* $\ker(c)$. This subgroup is given by

$$\begin{aligned}\{y \in U \mid\ & yxy^{-1} = x \text{ for all } x \in A_0\}\\ &= \{y \in U \mid y \text{ commutes elementwise with } A_0\}\\ &= \{y \in U \mid y \text{ commutes elementwise with } A\}.\end{aligned}$$

Since A has center F (by 1.1(3)), it follows that $\ker(c) = \dot{F}$.

Step 5. *Here, we show that* $c(U) = \mathrm{SO}(A_0)$. By I.7.3, any $\sigma \in \mathrm{SO}(A_0)$ can be written as $\sigma = \tau_{x_1}\tau_{x_2}$, where $x_1, x_2 \in U \cap A_0$. Thus, by Lemma 3.2,

$$\sigma = (-\tau_{x_1})(-\tau_{x_2}) = c(x_1)\,c(x_2) = c(x_1x_2) \in c(U).$$

Step 6. We derive now the second sequence in 3.1, assuming that $\mathrm{N}(U) \subseteq \dot{F}^2$. This assumption implies that $c(U) = c(U_0)$, where U_0 is as defined in the statement of 3.1. In fact, for $y \in U$, let d be a scalar such that $d^2 = \mathrm{N}(y^{-1})$. Then $dy \in U_0$, and hence

$$c(y) = c(d)\,c(y) = c(dy) \in c(U_0).$$

To complete the proof of 3.1, we just observe that

$$\ker(c \,|\, U_0) = U_0 \cap \dot{F} = \{d \in \dot{F} \mid \mathrm{N}(d) = d^2 = 1\} = \{\pm 1\}.$$

Thus, U_0 provides a double covering of the special orthogonal group $\mathrm{SO}(A_0)$, as claimed. □

Of course, the hypothesis $\mathrm{N}(U) \subseteq \dot{F}^2$ needed for the exactness of the sequence (B) in 3.1 is a pretty strong one. Since the norm form of A is $\langle 1, -a, -b, ab\rangle$, this hypothesis amounts to $a, b \in -\dot{F}^2$ (so $A \cong \left(\frac{-1,-1}{F}\right)$) and F be a pythagorean field. Let us take a look, for instance, at the motivating case for the theorem, namely the case where $F = \mathbb{R}$. To apply the exact sequence 3.1(B), we take A to be $\mathcal{H}$, Hamilton's division ring of real quaternions. Here, we have

$$\mathrm{N}(x + yi + zj + wk) = x^2 + y^2 + z^2 + w^2 \in \mathbb{R}^2.$$

Identifying U_0 with $\mathrm{SU}(2)$ as in 1.5, the exact sequence 3.1(B) becomes

$$1 \longrightarrow \{\pm 1\} \longrightarrow \mathrm{SU}(2) \stackrel{c}{\longrightarrow} \mathrm{SO}(3) \longrightarrow 1.$$

It is not difficult to describe the covering c in purely matrix terms. As in Section 1, let us identify the quaternion $v = \alpha + j\beta$ $(\alpha, \beta \in \mathbb{C})$ with the complex matrix $\begin{pmatrix} \alpha & -\overline{\beta} \\ \beta & \overline{\alpha} \end{pmatrix}$. Under this identification, a pure quaternion $x_1 i + x_2 j + x_3 k$ goes over to the skew-Hermitian matrix

$$\begin{pmatrix} x_1 i & -x_2 - x_3 i \\ x_2 - x_3 i & -x_1 i \end{pmatrix} \quad \text{(see 1.6)}.$$

For a special unitary matrix $v = \begin{pmatrix} \alpha & -\overline{\beta} \\ \beta & \overline{\alpha} \end{pmatrix}$ (where $\alpha\overline{\alpha} + \beta\overline{\beta} = 1$), $c(v) \in \mathrm{SO}(3)$ is the proper rotation of $\mathbb{R}^3$ that takes (x_1, x_2, x_3) to (y_1, y_2, y_3), where

$$v \cdot \begin{pmatrix} x_1 i & -x_2 - x_3 i \\ x_2 - x_3 i & -x_1 i \end{pmatrix} \cdot v^{-1} = \begin{pmatrix} y_1 i & -y_2 - y_3 i \\ y_2 - y_3 i & -y_1 i \end{pmatrix}.$$

In this case, Lemma 3.2 also has the following nice interpretation. Let us visualize $\mathcal{H}$ as $\mathbb{R}^4$, and the pure quaternions $\mathcal{H}_0$ as $\mathbb{R}^3$. We have

$$U_0 = S^3 \text{ (the 3-sphere)}, \quad \text{and} \quad U_0 \cap \mathcal{H}_0 = S^3 \cap \mathbb{R}^3 = S^2.$$

Thus, Lemma 3.2 states that if $v \in S^2 \subseteq U_0 = \mathrm{SU}(2)$, then $c(v) \in \mathrm{SO}(3)$ is a $180°$-rotation of $\mathbb{R}^3$ around the axis determined by the vector v.

We shall now return to the case of an arbitrary ground field F, and try to apply 3.1 to a split quaternion algebra $A \cong \mathbb{M}_2(F)$. In this case, U corresponds to the group of invertible elements in $\mathbb{M}_2(F)$; that is, $U \cong \mathrm{GL}_2(F)$. To draw some corollaries from 3.1, let us consider specifically the quaternion algebra $A = \left(\frac{-1,1}{F}\right)$ ($\cong \mathbb{M}_2(F)$ by 1.1(2)). In this case, A_0 has diagonalization $\langle 1, -1, -1\rangle$. Thus, the first sequence in 3.1 gives the following interesting conclusion:

Corollary 3.5. $\mathrm{PGL}_2(F) := \mathrm{GL}_2(F)/\dot{F} \cong \mathrm{SO}(\langle 1, -1, -1\rangle)$.

The second sequence in 3.1 certainly applies to $A = \left(\frac{-1,1}{F}\right)$ when F is a quadratically closed field. In this case, the middle term of the sequence corresponds to $\mathrm{SL}_2(F)$ (see Exercise 8). Here, we have the following conclusion:

Corollary 3.6. *If F is quadratically closed, then*

$$\mathrm{PSL}_2(F) := \mathrm{SL}_2(F)/\{\pm 1\} \cong \mathrm{SO}(\langle 1,1,1\rangle).$$

Note that the group $\mathrm{PSL}_2(F)$ in question in 3.6 is a *simple group*, according to a classical theorem of Camille Jordan and E. H. Moore.

If A is not isomorphic to $\mathbb{M}_2(F)$, then the U in Theorem 3.1 is the multiplicative group of the division quaternion algebra A (see 2.7). In this case, 3.1, together with information about the special orthogonal group $\mathrm{SO}(A_0)$, may be used to study the commutator structure of the group U (see, e.g., Exercise 27).

4. Linkage of Quaternion Algebras

In this section, we return to the theme of §2, and study in more detail the notion of splitting for a quaternion algebra $A = \left(\frac{a,b}{F}\right)$. Recall that such a quaternion algebra is either a division algebra, or (isomorphic to) a matrix algebra $\mathbb{M}_2(F)$; in the latter case, A is said to be *split* over F. More generally, if $K \supseteq F$ is any field extension, we say that A *splits over* K if the scalar extension

$$K \otimes_F A = K \otimes_F \left(\frac{a,\, b}{F}\right) \cong \left(\frac{a,\, b}{K}\right)$$

is isomorphic to $\mathbb{M}_2(K)$. In this case, K is said to be a *splitting field* for A.

Such a splitting field always exists. For instance, we may take K to be $\overline{F}$ (the algebraic closure of F), or, if we prefer smaller extensions, $K = F(\sqrt{a}\,)$ or $K = F(\sqrt{b}\,)$. To give a somewhat less obvious example, recall from 2.16 that the rational quaternion algebra $A = \left(\frac{5,-3}{\mathbb{Q}}\right)$ does not split over $\mathbb{Q}$, but splits over $\mathbb{Q}(\sqrt{17}\,)$.

It is natural to try to describe the *quadratic* extensions of F that split a given quaternion F-algebra. The following theorem offers two such descriptions.

Theorem 4.1. *Let $A = \left(\frac{a,b}{F}\right)$, and let $K = F(\sqrt{c}\,)$ be a quadratic extension of F. Then the following are equivalent:*

(1) *A splits over K.*

(2) *$A \cong \left(\frac{c,d}{F}\right)$ for some $d \in \dot{F}$.*

(3) *K can be embedded (over F) in A.*

Proof. (2) $\Rightarrow$ (3). If (2) holds, there exists a basis $\{1, i, j, ij\}$ on A such that $i^2 = c$, $j^2 = d$, and $ij = -ji$. Now $F(i) \subseteq A$ is F-isomorphic to K.

(3) $\Rightarrow$ (1). To simplify notations, let us assume that $F \subseteq K \subseteq A$. Then

$$K \otimes_F A \supseteq K \otimes_F K \cong K \times K$$

shows that $K \otimes_F A$ is *not* a division algebra, proving (1).

(1) $\Rightarrow$ (2). Consider the quadratic space (A_0, q) supported by the space of pure quaternions A_0. Here, $q = \langle -a, -b, ab \rangle$. If q is isotropic, then it is universal, so there exists $d \in \dot{F}$ such that

$$\langle -a, -b, ab \rangle \cong \langle -c, -d, cd \rangle. \tag{4.2}$$

In this case, (2) holds. Now assume q is *anisotropic*. If (1) holds, there will be an equation

$$-a\big(x_1 + y_1\sqrt{c}\big)^2 - b\big(x_2 + y_2\sqrt{c}\big)^2 + ab\big(x_3 + y_3\sqrt{c}\big)^2 = 0, \tag{4.3}$$

where x_i, $y_i \in F$ are not all zero. Letting $x = (x_1, x_2, x_3)$ and $y = (y_1, y_2, y_3)$, expansion of (4.3) shows that $x \perp y$ in (A_0, q), and that $q(x) + c\,q(y) = 0$. If $y = 0$, then

$$q(x) = -c\,q(y) = 0 \implies x = 0,$$

a contradiction. Therefore, $y \neq 0$, and $x \perp y$ implies that $\{x, y\}$ can be completed into an orthogonal basis $\{x, y, z\}$ of the anisotropic space (A_0, q). Therefore,

$$q \cong \langle q(z),\, q(y),\, q(x) \rangle = \langle q(z),\, q(y),\, -c\,q(y) \rangle.$$

Since $d(q) = 1$, $q(z) \in -c \cdot \dot{F}^2$. Thus, (4.2) holds again with $d = -q(y) \in \dot{F}$, and we are done as before.[(5)] $\square$

In a manner of speaking, 4.1 identifies all quadratic splitting fields of a given quaternion F-algebra A. How about other splitting fields? It might be tempting to think that any splitting field E/F for A would contain a *quadratic* splitting field. However, the following example shows that this is not the case.

Example 4.4. Let $A = \left(\frac{-1,\,-1}{\mathbb{Q}}\right)$ and $E = \mathbb{Q}(\zeta)$, where $\zeta = e^{2\pi i/5}$. Then $E/\mathbb{Q}$ is a cyclic extension of degree 4. In E, we have

$$-1 = \zeta + \zeta^2 + \zeta^3 + \zeta^4 = (1 + \zeta^2)(\zeta + \zeta^2) = (1 + \zeta^2)((\zeta^3)^2 + \zeta^2),$$

which is a sum of two squares. Thus, by Hilbert's Criterion, E is a splitting field for A. Rewriting the cyclotomic equation $1 + \zeta + \zeta^2 + \zeta^3 + \zeta^4 = 0$ in

(5) The argument given here is very useful. In Ch. VII, we shall generalize it to give a computation of the kernel of $W(F) \to W(K)$ for a quadratic extension K/F.

terms of $\alpha := \zeta + \zeta^{-1}$, we get the equation $\alpha^2 + \alpha - 1 = 0$. Thus, we have $\alpha = (\sqrt{5}-1)/2$ (which is one less than the Golden Ratio).[(6)] It follows that the proper subfields of E are $\mathbb{Q}$ and $\mathbb{Q}(\sqrt{5})$. Neither of these splits A, since they are subfields of $\mathbb{R}$). Therefore, E is a *minimal* splitting field for A, with $[E : \mathbb{Q}] = 4$.

Later, we shall give another example where a rational quaternion division algebra A splits over a quartic extension $K/\mathbb{Q}$, but there is *no* quadratic extension of $\mathbb{Q}$ in K; see VII.2.11.

According to the criterion in 4.1(2), a quadratic extension $K = F(\sqrt{c})$ splits A iff "c is a slot for A", meaning that A can be expressed as a quaternion algebra with the element c as one of its two "slots". The idea of considering the possible slots in expressing A as a quaternion algebra leads to the following useful definition.

Definition 4.5. Two quaternion F-algebras B, C are said to be *linked* if they can be expressed "with a common slot"; that is, if there exist $x, y, z \in \dot{F}$ such that $B \cong \left(\frac{x,y}{F}\right)$ and $C \cong \left(\frac{x,z}{F}\right)$.

For instance, if B splits, then it is linked to *any* C. In fact, writing $C = \left(\frac{x,z}{F}\right)$, we can express B as $\left(\frac{x,1}{F}\right)$, so B, C are expressed with a common slot x. For a less trivial example, the nonisomorphic algebras $B = \left(\frac{-1,-1}{\mathbb{Q}}\right)$ and $C = \left(\frac{-2,-5}{\mathbb{Q}}\right)$ in Example 2.13 are linked, since B can also be expressed as $\left(\frac{-2,-3}{\mathbb{Q}}\right)$ by Example 2.12. It turns out that, in fact, *any pair of rational quaternion algebras are linked*! We shall prove this later, in Chapter VI.

Our next goal is to prove a theorem of A. A. Albert that gives some criteria for the linkage of a pair of quaternion algebras. As a preparation for this theorem, let us first introduce the notion of an Albert form, which is a certain 6-dimensional quadratic form associated with a pair of quaternion algebras.

Let $B = \left(\frac{b,b'}{F}\right)$ and $C = \left(\frac{c,c'}{F}\right)$ be given quaternion algebras over F. The respective spaces of pure quaternions, B_0 and C_0, carry the following ternary quadratic forms:

(4.6) $$q_B := \langle -b,\ -b',\ bb' \rangle, \quad \text{and} \quad q_C := \langle -c,\ -c',\ cc' \rangle.$$

We define the *Albert form* for the pair $\{B,\ C\}$ to be the 6-dimensional form

(4.7) $$q := q_B \perp \langle -1 \rangle\, q_C = \langle -b,\ -b',\ bb',\ c,\ c',\ -cc' \rangle.$$

(6) In terms of trigonometric functions, this calculation amounts to the fact that $\cos 72° = (\sqrt{5}-1)/4$. This, incidentally, implies the constructibility of the regular pentagon by ruler and compass. For a more thorough discussion on constructibility issues, see VII.7.

Of course, q is determined only "up to a sign", if we think of $\{B, C\}$ as an unordered pair.

Note that $d(q) = -1$, and $q \in I^2F$ (since q is the "difference" of the norm forms of B and C in the Witt ring). Conversely, if q_0 is any 6-dimensional form in I^2F, then $d(q_0) = -1$ (by I.2.2), and we see easily that q_0 "becomes" an Albert form (for a suitable pair of quaternion algebras) after a scaling by some unary form $\langle a \rangle$. Thus, up to such scalings, Albert forms are precisely the 6-dimensional forms in I^2F.

With the notion of Albert forms in place, we can now prove the following important result.

Albert's Theorem 4.8. *Let B, C be quaternion algebras over F, and let A be the biquaternion algebra $B \otimes_F C$. If q denotes the Albert form for the pair $\{B, C\}$, the following statements are equivalent:*

(1) *A is a division algebra.* (2) *B and C are not linked.*

(3) *The Albert form q is anisotropic over F.*

(4) *B and C are division algebras, and they do not have a common quadratic splitting field.*[7]

Proof. To simplify notations, let us write $\otimes$ for $\otimes_F$ throughout. As is customary in the theory of algebras, we shall identify B with $B \otimes 1$, and C with $1 \otimes C$, so that $B, C \subseteq A$ are elementwise-commuting subalgebras of the biquaternion algebra A.

(1) $\Rightarrow$ (2). If B and C are linked, then 2.11 implies that $B \otimes C$ is *not* a division algebra. (Alternatively, if B, C are as represented in 4.5, there exist pure quaternions $\beta \in B$, $\gamma \in C$ such that $\beta^2 = \gamma^2 \in \dot{F}$. Then

$$0 = \beta^2 - \gamma^2 = (\beta + \gamma)(\beta - \gamma) \in A$$

implies that A is not a division algebra.)

(2) $\Rightarrow$ (3). Assume the Albert form q in (4.7) is isotropic, and use the notations there. By I.3.6, q_B and q_C then represent a common nonzero element, say $-x$. We can then write

$$q_B \cong \langle -x, -y, xy \rangle \quad \text{and} \quad q_C \cong \langle -x, -z, xz \rangle,$$

for some $y, z \in \dot{F}$. This gives $B \cong \left(\frac{x,\, y}{F}\right)$, $C \cong \left(\frac{x,\, z}{F}\right)$, so B and C are linked.

(3) $\Rightarrow$ (4). If, say, B is not a division algebra, then q_B is isotropic by 2.7, so q is also isotropic. Next, assume that B and C have a common quadratic splitting field $F(\sqrt{a})$. By 4.1, $F(\sqrt{a})$ can be F-embedded into

---

[7] According to 4.1, the second condition here can also be paraphrased as follows: *no quadratic extension K/F can be F-embedded into both B and C.*

B and C. Therefore, there exist nonscalars $\beta \in B$ and $\gamma \in C$ such that $\beta^2 = a = \gamma^2$. By 1.3, β and γ are pure quaternions, and we have

$$q_B(\beta) = -\beta^2 = -a \quad \text{and} \quad q_C(\gamma) = -\gamma^2 = -a.$$

These clearly imply that $q = q_B \perp \langle -1 \rangle q_C$ is isotropic.

(4) $\Rightarrow$ (1). Assume (4), and let K (resp. L) denote a typical quadratic extension of F in B (resp. C). By 4.1, K splits B but not C, so $K \otimes C$ is a division algebra, and similarly, so is $B \otimes L$. To prove (1), we must show that *any nonzero* $\alpha \in A = B \otimes C$ *has a left inverse.* It suffices to show that there exists $\alpha^* \in A$ such that $\alpha^*\alpha$ is *nonzero* and lies in either some $K \otimes C$ or some $B \otimes L$ (for then $\alpha^*\alpha$ has a left-inverse, and so does α). Fix a quaternion basis $\{1, i, j, k\}$ for C, and let $L = F(j)$. We can then express α in the form

$$\alpha = (\beta_1 + \beta_2 j) + (\beta_3 + \beta_4 j)\, k, \quad \text{where } \beta_i \in B.$$

We may assume that $\gamma := \beta_3 + \beta_4 j \neq 0$ (for otherwise $\alpha \in B \otimes L$ already). Then γ^{-1} exists in $B \otimes L$, and, after left-multiplying α by γ^{-1}, we are reduced to the case where $\alpha = \beta_1 + \beta_2 j + k$. If $\beta_1\beta_2 = \beta_2\beta_1$, then $F(\beta_1, \beta_2)$ is either F or a quadratic extension K/F in B. In this case, $\alpha \in F \otimes C$ or $K \otimes C$, so we are done. Now assume $\beta_1\beta_2 \neq \beta_2\beta_1$. For $\alpha^* := \beta_1 - \beta_2 j - k$, we have

$$\begin{aligned} \alpha^*\alpha &= (\beta_1 - \beta_2 j - k)(\beta_1 + \beta_2 j + k) \\ &= (\beta_1 - \beta_2 j)(\beta_1 + \beta_2 j) - k^2 \\ &= \beta_1^2 - \beta_2^2 j^2 - k^2 + (\beta_1\beta_2 - \beta_2\beta_1) j, \end{aligned} \tag{4.9}$$

where the second equality holds since k commutes with β_1, β_2, and anti-commute with j. Recalling that $j^2, k^2 \in F$, and $\beta_1\beta_2 \neq \beta_2\beta_1$, we see from (4.9) that $\alpha^*\alpha \in (B \otimes L) \setminus \{0\}$, as desired. □

Albert's Theorem 4.8 has an interesting history. Its crucial part (4) $\Rightarrow$ (1) appeared in Albert's last paper [Al], published posthumously in the Proceedings of the A.M.S. This paper was received by the editors on January 6, 1972, exactly five months before Albert's death, and appeared in September of the same year. At the end of his short paper, Albert wrote: "I discovered this theorem some time ago. There appears to be some continuing interest in it, and I am therefore publishing it now." When exactly did Albert first prove the implication (4) $\Rightarrow$ (1) will perhaps remain a mystery. However, this crucial implication is also contained (if somewhat implicitly) in a result of Pfister in his paper [Pf3], published in 1966. While Albert's proof relied heavily on working with invertible elements in algebras, Pfister's proof made full use of the theory of quadratic forms. Pfister's approach to this problem will be presented later in Ch. XII: see the proof of (4) $\Rightarrow$ (1) in XII.2.7 for a good comparison.

Note that, in the notation of 4.8, if $A = B \otimes C$ fails to be a division algebra, then 2.11 and (1) $\Leftrightarrow$ (2) in 4.8 imply that

$$A \cong D \otimes \mathbb{M}_2(F) \cong \mathbb{M}_2(D)$$

for some quaternion algebra D. In the language of Brauer groups (to be introduced in IV.1), $B \otimes C$ is then "similar" to this quaternion algebra.

The beauty of Theorem 4.8 lies in the fact that one can decide if the biquaternion algebra $A = B \otimes C$ is a division algebra by a criterion (3) involving quadratic forms alone. Historically, A. A. Albert used this theorem to construct examples of "noncyclic" central division algebras of dimension 16. For instance, for the quaternion algebras

$$(4.10) \qquad B = \left(\frac{-1,\,-1}{F}\right) \quad \text{and} \quad C = \left(\frac{x,\; y}{F}\right) \quad \text{over } F = \mathbb{R}(x, y),$$

it can be shown that $B \otimes_F C$ is a division algebra by checking that the Albert form q for $\{B,\, C\}$ is *anisotropic* over F. Granted this fact, one can then try to show that the biquaternion division algebra $B \otimes_F C$ is *not* a cyclic algebra over $F = \mathbb{R}(x, y)$. Due to the elementary nature of this preliminary chapter, however, we shall not digress here into the technical work required for verifying the anisotropicity of the Albert form $q = \langle 1, 1, 1, x, y, -xy\rangle$, or checking the non-cyclicity of the tensor product algebra $B \otimes C$.

Actually, in Albert's original construction of a noncyclic biquaternion division algebra, he used the quaternion factors

$$(4.11) \qquad B = \left(\frac{x,\; -1}{F}\right) \quad \text{and} \quad C = \left(\frac{-x,\; y}{F}\right) \quad \text{over } F = \mathbb{R}(x, y).$$

Our choices in 4.10 seemed a little more "symmetrical", and had the advantage that the first factor B was the "usual" quaternion algebra (which is already defined over the prime field). The work in verifying $B \otimes_F C$ to be a division algebra turns out to be essentially the same, as we can check that the Albert forms for 4.10 and 4.11 are simply related to each other by a scaling and a change of variables. Detailed proofs for the anisotropicity of these Albert forms will be given later in VI.1.11 and VI.1.13.

In discussing the choices for B and C above, we should have mentioned perhaps that R. Brauer had constructed a noncyclic biquaternion division algebra a little bit earlier than Albert. Brauer used, instead, the smaller ground field $F_0 = \mathbb{Q}\,(x, y)$, and chose

$$(4.12) \qquad B = \left(\frac{b,\; x}{F_0}\right) \quad \text{and} \quad C = \left(\frac{c,\; y}{F_0}\right),$$

where $b,\, c \in \dot{\mathbb{Q}}$ represent two independent square classes in $\dot{\mathbb{Q}}/\dot{\mathbb{Q}}^2$. The proof of the anisotropicity of the Albert form for the pair $\{B,\, C\}$ in this case will be given later in VI.1.15.

It is of historical interest to note that Brauer approached the non-cyclic algebra problem from the viewpoint of group representations, while Albert approached the same problem from the viewpoint of quadratic forms. The formulation of the division algebra criteria in 4.8 for $A = B \otimes_F C$ in terms of quadratic forms comes directly from the work of Albert.

The consideration of the linkage of quaternion algebras leads naturally to the class of "linked fields"; that is, fields over which any two quaternion algebras are linked. We can already give some characterizations for such fields by using 4.8; however, we shall postpone this work to a later chapter (see X.4.20), where we can give a more complete list of characterizations. The theme of the linkage of quaternion algebras will occur a few more times later in this book, for instance, in the context of function fields (X.4), "u-invariants" (XI.6), and in the work on the axiomatic foundations of the theory of quadratic forms (XII.8).

To cap off this section, let us present one more result on the possible "slots" one can use for a given quaternion algebra. The following result was an exercise in the earlier versions of this book. It has been used rather frequently in the research literature in the theory of quaternion algebras. For ease of future reference, this exercise is elevated into a closing theorem for this section on linkage, with a short proof.

Common Slot Theorem 4.13. *Let* $B = \left(\frac{b,\, b'}{F}\right)$ *and* $C = \left(\frac{c,\, c'}{F}\right)$. *If* $B \cong C$, *then there exists* $x \in \dot{F}$ *such that* $B \cong \left(\frac{b,\, x}{F}\right)$ *and* $C \cong \left(\frac{c,\, x}{F}\right)$.

Proof. Since $B \cong C$, the Albert form q in 4.7 is hyperbolic. Therefore, its 4-dimensional subform $\langle -b',\, bb',\, c',\, -cc' \rangle$ is isotropic, according to Exercise 14 in Chapter I. This implies (by I.3.6) that $\langle c',\, -cc' \rangle$ and $\langle b',\, -bb' \rangle$ represent some common element $x \in \dot{F}$. It follows that

$$\langle b',\, -bb' \rangle \cong \langle x,\, -xb \rangle \quad \text{and} \quad \langle c',\, -cc' \rangle \cong \langle x,\, -xc \rangle,$$

so now $q_B \cong \langle -b,\, -x,\, xb \rangle$ and $q_C \cong \langle -c,\, -x,\, xc \rangle$, which lead directly to $B \cong \left(\frac{b,\, x}{F}\right)$ and $C \cong \left(\frac{c,\, x}{F}\right)$, as desired. □

5. Characterizations of Quaternion Algebras

In §1 of this chapter, quaternion algebras were defined and studied in a rather ad hoc (or matter-of-fact) fashion, and we have not yet tried to elucidate the more intrinsic nature of the family of quaternion algebras over a field. In this short section, we shall make up for this by mentioning a couple of characterization theorems for quaternion algebras that will, in particular, demonstrate the special role such algebras play in the general study of finite-dimensional algebras over fields.

The first characterization theorem says that quaternion algebras over a field F are precisely the simple algebras of dimension 4 with center equal to F. (The assumption that $\operatorname{char} F \neq 2$ will remain in force in the following, unless it is stated otherwise.)

Theorem 5.1. *Let $A \neq F$ be a simple F-algebra of dimension ≤ 4 with center F. Then A is isomorphic to a quaternion algebra over F.*

Proof. For a more efficient formulation of the proof, we'll invoke Wedderburn's theorem, to the effect that finite-dimensional simple algebras are matrix algebras over division algebras. Say $A \cong \mathbb{M}_n(D)$, where D is a division F-algebra. Then $n^2 \dim_F D \in \{2, 3, 4\}$ implies that either $n = 2$, $D = F$, or $n = 1$, $A \cong D$. In the former case, $A \cong \mathbb{M}_2(F) \cong \left(\frac{1,1}{F}\right)$. We may thus assume that A is a division algebra. Fix a noncentral element $i \in A$, and let K be the field $F(i) \subsetneq A$. Since $F \subsetneq K \subsetneq A$, we must have $\dim_F K = 2$ and $\dim_F A = 4$. Thus, K is a quadratic field extension of F, and we may assume $i \in K$ to have been chosen such that $i^2 = a \in \dot{F}$. (Here, we used the assumption that $\operatorname{char} F \neq 2$.) Let f be the inner automorphism on A defined by i. Then $f^2 = \mathrm{Id}$, and we have an eigenspace decomposition $A = A^+ \oplus A^-$, where

$$\begin{aligned} A^+ &= \{x \in A : \ f(x) = x\} = \{x \in A : \ ix = xi\}, \quad \text{and} \\ A^- &= \{x \in A : \ f(x) = -x\} = \{x \in A : \ ix = -xi\}. \end{aligned}$$

Fix a nonzero element $j \in A^-$. Since $K \subseteq A^+$ and $K \cdot j \in A^-$, we must have $K = A^+$ and $K \cdot j = A^-$. From $ij = -ji$, we have $j^2 i = ij^2$; that is, $j^2 \in A^+ = K$. On the other hand, $F(j)$ is also a quadratic field extension of F, so j satisfies a quadratic equation $j^2 + cj - b = 0$, for suitable b, $c \in F$. Now $cj = b - j^2 \in K$ implies that $c = 0$ and $j^2 = b \in \dot{F}$, so now

$$A = K \oplus K \cdot j = F \oplus Fi \oplus Fj \oplus Fij$$

is isomorphic to the quaternion algebra $\left(\frac{a,b}{F}\right)$, as desired. □

The second characterization of quaternion algebras is in terms of the canonical involution "bar" on such algebras. Note that this is an involution "of the first kind"; that is, it is an involution that restricts to the identity map on the center F of the algebra.

Theorem 5.2. *Let $B \neq F$ be a finite-dimensional simple F-algebra with center F equipped with an F-algebra involution of the first kind $x \mapsto \overline{x}$ such that $x + \overline{x} \in F$ and $x\overline{x} \in F$ (for all $x \in B$). Then B is isomorphic to a quaternion algebra over F.*

Although this is a very nice characterization of quaternion algebras, it will not be needed in the rest of this book. Therefore, we'll leave its proof as an exercise to the reader. The main idea of the proof is that, given the properties of the involution $x \mapsto \overline{x}$, every element $x \in B$ satisfies a quadratic equation over the center F; namely,

$$x^2 - x(x + \overline{x}) + x\,\overline{x} = 0 \quad (x + \overline{x},\ x\,\overline{x} \in F).$$

Exercises for Chapter III

1. Give another proof for the linear independence of $1, i, j, k$ in the quaternion algebra $A = \left(\frac{a,\,b}{F}\right)$ by finding an F-algebra homomorphism $\theta\colon A \to \mathbb{M}_2(\overline{F})$ ($\overline{F}$ = the algebraic closure of F) such that $1, \theta(i), \theta(j), \theta(k)$ are linearly independent over $\overline{F}$.

2. For any $b \in F$, show that -1 is a norm from $F(\sqrt{b}\,)$ iff b is a sum of two squares in F.

3. Show by direct computations that $\mathrm{T}(xy) = \mathrm{T}(yx)$ in any quaternion algebra, and that the bilinear form $B(x, y) = (x\,\overline{y} + y\,\overline{x})/2$ defined in the text is the same as the form $B'(x, y) = (\overline{x}\,y + \overline{y}\,x)/2$.

4. If one defines $\widetilde{\mathrm{N}}\,x = \mathrm{T}(x^2)/2$ on $A = \left(\frac{a,\,b}{F}\right)$, show that

   (1) $\widetilde{\mathrm{N}}$ is also a quadratic form on the quaternion algebra A, with a diagonalization $\langle 1, a, b, -ab\rangle$;
   (2) $\widetilde{\mathrm{N}}$ is a group form iff $-1 \in \dot{F}^2$, iff $\widetilde{\mathrm{N}} \cong \mathrm{N}$.

5. Given $A = \left(\frac{a,\,b}{F}\right)$, consider the following statements:

   (1) -1 is a square in A.
   (2) $A \cong \left(\frac{-1,\,c}{F}\right)$ for some $c \in \dot{F}$.
   (3) $\langle 1, a, b, -ab\rangle$ is isotropic.
   (4) There exists an $x \in A \setminus \{0\}$ such that x^2 is a pure quaternion.

   Show that (4) $\Leftrightarrow$ (3) $\Leftrightarrow$ (2) $\Rightarrow$ (1), and that (1) $\Rightarrow$ (2) if $-1 \notin \dot{F}^2$.

6. Show that $q = \langle a, b, c\rangle$ represents $-abc$ iff q is isotropic. In particular, a three-dimensional form is universal iff it is isotropic. (See Ch. X, Exercise 12 for a similar fact for five-dimensional forms.)

7. Show that the norm N for a quaternion algebra $A = \left(\frac{a,\,b}{F}\right)$ defined in Section 2 is the square root of the "algebra norm" $\mathrm{N}_{A/F}$. [For $x \in A$, the "algebra norm" $\mathrm{N}_{A/F}(x)$ is defined to be the determinant of the F-linear endomorphism of A given by left multiplication by x. The exercise is to show that $\mathrm{N}_{A/F}(x) = \mathrm{N}(x)^2$.]

8. Let A be a split quaternion algebra, given with a specific F-algebra isomorphism $\varphi\colon A \to \mathbb{M}_2(F)$. For any $x \in A$, show that

$$\mathrm{T}(x) = \mathrm{tr}\,(\varphi(x)) \quad \text{and} \quad \mathrm{N}(x) = \det\,(\varphi(x)),$$

where "tr" and "det" denote the usual trace and determinant functions on $\mathbb{M}_2(F)$.

9. Show that, for any quaternion algebra A over F, we have an F-algebra isomorphism $A \otimes_F A \cong \mathbb{M}_4(F)$. In particular, show that A embeds into $\mathbb{M}_4(F)$ as an F-algebra.
10. For elements x, y in a quaternion division algebra A, show that $x = u^{-1}yu$ for some unit $u \in A$ iff $\mathrm{T}\,x = \mathrm{T}\,y$ and $\mathrm{N}\,x = \mathrm{N}\,y$. (**Hint.** Use 3.1.) Is this also true if A is a *split* quaternion algebra?
11. For any element x in a quaternion algebra, show that x is nilpotent iff $x^2 = 0$, iff $\mathrm{T}\,x = \mathrm{N}\,x = 0$.
12. Let A be a quaternion algebra over F. If $I \in A \setminus F$ is an element such that $I^2 \in \dot{F}$, show that there exists $J \in A$ such that $\{1, I, J, K\}$ is a quaternion basis for A, where $K = IJ$.
13. Let $\varphi = \langle 1,\, a,\, b,\, ab\rangle$, $\gamma = \langle 1,\, c,\, d,\, cd\rangle$, and $r,\, s \in \dot{F}$. If $\langle r\rangle\varphi \cong \langle s\rangle\gamma$, show that $\varphi \cong \gamma$. (For a much more general result on n-fold Pfister forms, see X.5.4.)
14. (1) Use 2.11 and Wedderburn's Theorem on finite-dimensional simple algebras to show that if $\left(\frac{a,\, b}{F}\right)$ and $\left(\frac{a,\, c}{F}\right)$ are split quaternion algebras, then $\left(\frac{a,\, bc}{F}\right)$ is also a split quaternion algebra.
(2) Give a proof for (1) by using the theory of quadratic forms (especially II.3.7).
15. Use Hilbert's Criterion to show that $A = \left(\frac{3,\, -11}{\mathbb{Q}}\right)$ splits. Find a nonzero element $x \in A$ such that $x^2 = 0$.
16. Determine if the quaternion algebras
$$\left(\frac{2,\, \sqrt{7}}{F}\right) \quad \text{and} \quad \left(\frac{5 + 2\sqrt{7},\ 2 + \sqrt{7}}{F}\right)$$
split over the field $F = \mathbb{Q}\left(\sqrt{7}\right)$.
17. Determine the odd (rational) primes p for which
$$\left(\frac{-1,\, -1}{F}\right) \cong \left(\frac{-2,\, -p}{F}\right) \quad \text{over } F = \mathbb{Q}.$$
18. Let an odd prime p and $n \in \mathbb{Z}$ be such that $p - n$ is a square. Show that $\left(\frac{p,\, n}{\mathbb{Q}}\right)$ splits iff $p \equiv 1 \pmod 4$.
19. Let F be a formally real pythagorean field, and let $A = \left(\frac{-1,\, -1}{F}\right)$. Using Hilbert's Criterion, show that a quadratic extension K/F splits A iff $K = F(\sqrt{-1})$. [Note that, in view of 4.1, this means that all quadratic field extensions of F in A are isomorphic to $F(\sqrt{-1})$.]

20. Show that $\left(\frac{-2,-3}{\mathbb{Q}}\right)$ and $\left(\frac{-7,-23}{\mathbb{Q}}\right)$ are linked by exhibiting a common slot for the two quaternion algebras.
21. (Brauer-Noether) Let ζ be a primitive pth root of unity, where p is a prime divisor of 2^n+1 (for some $n \geq 1$). Show that $E = \mathbb{Q}(\zeta)$ is a splitting field for the rational quaternion algebra $\left(\frac{-1,-1}{\mathbb{Q}}\right)$. (**Hint.** Verify the identity
$$\prod_{k=0}^{n-1} (1+\zeta^{2^k}) = -\zeta^{2^n},$$
and note that the LHS is a sum of two squares in E.)
22. Show that $\left(\frac{a,b}{F}\right) \cong \left(\frac{a,c}{F}\right)$ iff $bc \in \mathrm{N}_{E/F}(\dot{E})$ for $E = F(\sqrt{a})$.
23. Let K be a finite extension of odd degree over F, and let $a, b \in \dot{F}$. Using the condition 2.7(8), show that $\left(\frac{a,b}{K}\right)$ splits iff $\left(\frac{a,b}{F}\right)$ splits.
24. Show that the conclusion of the above exercise also holds in case K is the rational function field $F(x)$.
25. For any formally real field F, show that $\left(\frac{-1,x}{K}\right)$ is a division algebra over the rational function field $K = F(x)$. (**Hint.** First try the case $K = F((x))$, the field of formal Laurent series in x over F.)
26. Let $A = \left(\frac{a,b}{F}\right)$. Show that each of the following fields
$$F(x)\left(\sqrt{b(x^2-a)}\right) \quad \text{and} \quad F(x)\left(\sqrt{ax^2+b}\right)$$
is a splitting field for A. (**Note.** As we shall see in X.3, these fields are, respectively, the function fields of $\langle 1, -b, ab\rangle$ and $\langle 1, -a, -b\rangle$, both of which are ternary subforms of the norm form of the quaternion algebra A.)
27. Let U be the multiplicative group of the real quaternions $\mathcal{H}$, and U_0 be the subgroup of unit quaternions (i.e. quaternions of norm 1). Show that
$$[U, U] = [U, U_0] = [U_0, U_0] = U_0.$$
28. Show that any element in a quaternion algebra can be written as a product of two pure quaternions.

Chapter IV

The Brauer-Wall Group

1. The Brauer Group

In this section, we shall recall briefly the idea of forming a Brauer group, since this idea will be generalized in Section 3. We shall also assemble a few properties of central simple algebras which will be needed later.

In the following, F will denote a field, and an F-algebra will always mean a *finite-dimensional* F-algebra. For any subset S of an F-algebra A, we shall write

$$C_A(S) = \{a \in A : \ as = sa \text{ for all } s \in S\},$$

which is called the *centralizer* of S in A. This is always a subalgebra of A. As a special case of this, we shall define $Z(A) = C_A(A)$, called the *center* of the algebra A.

Definition 1.1. (1) A is called *F-central* (or central over F) if $Z(A) = F$ $(= F \cdot 1)$. (2) A is called *simple* if A has no two-sided ideals other than (0) and A. (3) A is called a *central simple algebra* (CSA) over F if A satisfies both (1) and (2).

There are at least two kinds of basic examples. For any n-dimensional F-vector space V, the endomorphism algebra $A = \text{End}(V) \cong \mathbb{M}_n(F)$ is always a CSA over F. (In particular, $F = \mathbb{M}_1(F)$ is such a CSA.) Next, for nonzero elements a, b in F, the four-dimensional quaternion algebra $A = \left(\frac{a,b}{F}\right)$ is also a CSA over F (by III.1.1).

We begin our discussion by stating the following result on tensor products. (All unadorned tensor products will be over a fixed ground field F.)

Theorem 1.2. (1) *If A, B are F-algebras, and $A' \subseteq A, B' \subseteq B$ are subalgebras, then*

$$C_{A\otimes B}(A' \otimes B') = C_A(A') \otimes C_B(B').$$

In particular, if A, B are F-central, so is $A \otimes B$.

(2) *If A is a CSA over F, and B a simple algebra, then $A\otimes B$ is simple. In particular:*

(3) *If A, B are both CSA over F, so is $A \otimes B$.*

This theorem will be generalized to graded algebras later (see 2.3). In Section 2, we shall give the full proof of the generalized version; hence, it is unnecessary for us to prove the special case 1.2 here.

The idea of forming the Brauer group is to classify all CSAs over F by a suitable similarity relation, and then impose a group structure on the set of similarity classes by the tensor product. More formally, we proceed as follows.

Let A, A' both be CSAs over F. We shall say that A is *similar* to A' if there exist finite-dimensional vector spaces V, V' such that $A \otimes \operatorname{End} V \cong A' \otimes \operatorname{End} V'$ as F-algebras. It can easily be shown that similarity is an equivalence relation (for "transitivity," we have to know that

$$(\operatorname{End} V) \otimes (\operatorname{End} V'') \cong \operatorname{End}(V \otimes V''),$$

but that is easy). The equivalence class of A will be denoted by $[A]$, or sometimes just by A if no logicians are present. It is routine to check that the operation

$$[A_1] \cdot [A_2] = [A_1 \otimes A_2]$$

is well defined, and makes the set of similarity classes of CSAs into a commutative monoid, with $[F] = [\mathbb{M}_n(F)]$ as the identity. We denote this monoid by $B(F)$.

Proposition and Definition 1.3. *For any F-algebra A, let A^{op} denote the opposite algebra. If A is a CSA, so is A^{op}, and $A \otimes A^{\text{op}} \cong \operatorname{End}(A)$ (the algebra of linear endomorphisms of A). In particular, $B(F)$ is an (abelian) group, with $[A]^{-1} = [A^{\text{op}}]$ for any central simple algebra A. $B(F)$ will be called the Brauer group of F.*

Proof. Recall that $A^{\text{op}} = \{a^{\text{op}} : a \in A\}$ with $a^{\text{op}} \cdot b^{\text{op}} = (ba)^{\text{op}}$. Clearly, $Z(A^{\text{op}}) = \{a^{\text{op}} : a \in Z(A)\}$, so A central $\Rightarrow A^{\text{op}}$ central. If I is an ideal in A^{op}, then $\{a \in A : a^{\text{op}} \in I\}$ is an ideal in A. Thus, A simple $\Rightarrow A^{\text{op}}$ simple, and we have proved the first conclusion. Now define $\theta : A \otimes A^{\text{op}} \to \text{End}(A)$ by

$$\theta(a \otimes b^{\text{op}})(c) = acb \qquad (a,\, b,\, c \in A).$$

An easy calculation shows that θ is an algebra homomorphism. Since A is a CSA, so is $A \otimes A^{\text{op}}$ (by 1.2(3), and what we have just proved). Hence θ is injective, and it must be an isomorphism, by dimension count. The remaining conclusions now follow immediately. □

By the Wedderburn Theorem, every central simple algebra A over F is of the form $\mathbb{M}_n(D)$, where D is a central division algebra over F. We have then $[A] = [D]$ in the Brauer group, since

$$A \cong \mathbb{M}_n(D) \cong D \otimes \mathbb{M}_n(F).$$

Also, Wedderburn's Theorem says that A uniquely determines D (up to isomorphism). This implies that if D, D' are F-central division algebras,

$$\begin{aligned} [D] = [D'] &\implies \mathbb{M}_n(D) \cong \mathbb{M}_m(D') \quad \text{(for suitable } n,\, m) \\ &\implies D \cong D' \quad \text{(and } n = m). \end{aligned}$$

Hence we conclude that

Proposition 1.4. *The elements in the Brauer group $B(F)$ are in* 1-1 *correspondence with the isomorphism classes of F-central division algebras, by* $D \leftrightarrow [D]$.

As a corollary, distinct quaternion algebras will represent different elements of $B(F)$. In general, the classes of quaternion algebras need not form a subgroup of $B(F)$. However, each division quaternion algebra represents an element of order 2 in $B(F)$ (by III.2.11), so the subgroup generated by quaternion algebras in $B(F)$ will be of exponent 2 (or trivial). It is natural to ask if this subgroup contains *all* elements of order 2 in $B(F)$, and A. A. Albert had conjectured that this is always the case. Albert's conjecture remained unsolved for a long time, but was finally proved by A. Merkurjev in 1981. More information on this can be found in V.6.

We offer a few examples of Brauer groups, without proofs.

Example 1.5. (1) (Frobenius, 1877!) $B(\mathbb{R}) \cong \{\pm 1\}$, with $\left(\frac{-1,-1}{\mathbb{R}}\right)$ representing the unique nontrivial element.

(2) If F is a finite field, or F is an algebraic extension of the rational function field $\mathbb{C}(x)$, then $B(F) = 0$ (by the theorems of Wedderburn and Tsen).

(3) If F is the completion of a number field at a finite prime, then there exists an isomorphism $\text{inv} : B(F) \cong \mathbb{Q}/\mathbb{Z}$. This is one of the central facts in local class field theory.

We shall now state and prove a few facts about CSAs.

Proposition 1.6. *Let A be a CSA over F, and B be a simple subalgebra of A. Let $C = C_A(B)$. Then*

(1) *C is simple;*

(2) *$B = C_A(C)$;*

(3) *$\dim A = \dim B \cdot \dim C$.*

Proof. Let T denote the algebra $B \otimes A^{\text{op}}$; it is simple by 1.2(2). Let S be the unique irreducible left T-module, and $D = \text{End}_T(S)$, a division algebra by Schur's Lemma. We may view A as a left T-module, by the action $(b \otimes a^{\text{op}})(a') = ba'a$. The T-endomorphisms on A are precisely left multiplications on A by elements $c \in C_A(B) = C$, so we have $\text{End}_T A \cong C$. As a T-module, A is the direct sum of, say, m copies of S. Thus $C \cong \text{End}_T(A) \cong \mathbb{M}_m(D)$, a simple algebra as claimed. As a notational device, we shall write a, b, $c, \ldots$ etc. to denote the F-dimensions of A, B, $C, \ldots$. We also need another integer: $x = \dim_D S$. By Wedderburn's Theorem, $T \cong \mathbb{M}_x(D^{\text{op}})$, so $t = x^2 d$. Meanwhile, $s = xd$. Eliminating x, we get $s^2 = td$. Let us now calculate s, t, and d in terms of a, b, c, m. Clearly,

$$s = a/m, \qquad t = ba, \quad \text{and} \quad d = c/m^2$$

(the latter because $C \cong \mathbb{M}_m(D)$). Plugging these into $s^2 = td$, we get $(a/m)^2 = ba \cdot c/m^2$, so $a = bc$! It remains to show that $B = C_A(C)$. Let $B' = C_A(C) \supseteq B$, and $b' = \dim B'$. From (1) and (3), we see that $a = c \cdot b'$, so $b = b'$ and $B = B'$. □

Corollary 1.7. *Suppose $B \subseteq A$, and both are CSAs over F. If $C = C_A(B)$, then C is also a CSA, $B = C_A(C)$, and $B \otimes C \cong A$.*

Proof. Since B and C commute elementwise, there exists an algebra homomorphism $f : B \otimes C \to A$, induced by

$$f(b \otimes c) = bc \qquad (b \in B, \ c \in C).$$

But, $B \otimes C$ is simple by 1.2(2) and (1.6)(1). Hence, f is injective, and 1.6(3) implies that f is an isomorphism. In particular, B, C generate A as an algebra. It follows that $Z(C) \subseteq Z(A) = F$, so C is a CSA. □

We finish now with the Skolem-Noether Theorem.

Theorem 1.8. *Let A be a CSA over F, and B a simple algebra. If f, g are algebra homomorphisms from B to A, then there exists an invertible element $s \in A$, such that $f(b) = s^{-1}g(b)s$ for every $b \in B$ (i.e., f and g differ by an inner automorphism of A).*

Corollary 1.9. *If A is a CSA over F, every algebra endomorphism of A is an inner automorphism.*

Proof of 1.8. We first work with the special case where $A \cong \text{End}(V)$ for some vector space V (i.e., A is a matrix algebra over F). Using the homomorphisms f and g, we can make V into B-modules in two different ways. Let us denote these B-modules by V_f and V_g. Since these have the same F-dimension, and since B is simple, we must have $V_f \cong V_g$ as B-modules. Let $s : V_f \to V_g$ be a B-isomorphism. Then, s is an invertible element in $\text{End}(V) = A$, and

$$s(f(b)v) = g(b)s(v) \quad \text{for all } b \in B \text{ and } v \in V.$$

This implies the required equation $f(b) = s^{-1}g(b)s$, for all $b \in B$. To handle the general case, consider the homomorphisms $f \otimes 1$ and $g \otimes 1$ from $B \otimes A^{\text{op}}$ to $A \otimes A^{\text{op}}$. The latter is $\cong \text{End}(A)$ by the map θ of 1.3. Since $B \otimes A^{\text{op}}$ is simple by 1.2(2), the special case treated above shows that we can find an invertible $\overline{s} \in A \otimes A^{\text{op}}$, such that

$$\overline{s}^{-1}(g(b) \otimes a^{\text{op}})\overline{s} = f(b) \otimes a^{\text{op}} \qquad (b \in B,\ a \in A).$$

Setting $b = 1$, we see that $\overline{s}$ commutes elementwise with $1 \otimes A^{\text{op}}$. But, by 1.2(1), we have

$$C_{A \otimes A^{\text{op}}}(1 \otimes A^{\text{op}}) = A \otimes Z(A^{\text{op}}) = A \otimes F = A \otimes 1,$$

so $\overline{s} = s \otimes 1$ for some $s \in A$. Similarly, $\overline{s}^{-1} \in A \otimes 1$, so s is an invertible element of A. We have now

$$s^{-1}g(b)s \otimes a^{\text{op}} = f(b) \otimes a^{\text{op}},$$

for all $b \in B$ and $a \in A$. Setting $a = 1$, we obtain the desired equation $s^{-1}g(b)s = f(b)$, after using the identification $A \cong A \otimes 1 \subseteq A \otimes A^{\text{op}}$. □

2. Central Simple Graded Algebras (CSGA)

In this section, we shall extend the idea of a CSA to the setting of graded algebras, in preparation for the study of Clifford algebras in the next chapter.

Our major concern will be $\mathbb{Z}_2$-graded algebras over a field F. By definition, a $\mathbb{Z}_2$-graded F-algebra A is a finite-dimensional F-algebra given in the form $A = A_0 \oplus A_1$, such that $F = F \cdot 1 \subseteq A_0$ and that $A_iA_j \subseteq A_{i+j}$ when the subscripts are taken modulo 2. (In particular, A_0 is a subalgebra.) Since we shall only consider $\mathbb{Z}_2$-graded algebras, they will be called, from now on, just *graded algebras*.

For a graded algebra A as above, the elements in $h(A) = A_0 \cup A_1$ will be called the *homogeneous elements* of A. If $a \in h(A)$, we write $\partial(a) = i$ if $a \in A_i$ $(i = 0, 1)$. This "degree function" ∂ is not well defined at 0, but, in practice, this fact will not create any undue difficulty.

A subspace $S \subseteq A$ is called *graded* if it is the direct sum of the intersections $S_i = S \cap A_i$. This means that if $s \in S$ and $s = s_0 + s_1$ ($s_i \in A_i$), then each $s_i \in S$. We shall write $h(S) = S \cap h(A)$. "Graded subalgebra," "graded left ideal," etc. have obvious meanings.

For a graded subspace $S \subseteq A$, we define the *graded centralizer* $\widehat{C}_A(S)$ to be the graded subspace C such that

$$c \in h(C) \iff cs = (-1)^{\partial c \partial s} sc \quad \text{for all } s \in h(S).$$

This subspace can easily be checked to be a graded *subalgebra* of A. The ordinary centralizer $C_A(S)$, written without the cap, will continue to mean

$$\{c \in A : \ cs = sc \ \text{ for all } s \in S\}.$$

If $S \cap A_1 = 0$, then, of course, $\widehat{C}_A(S) = C_A(S)$.

Definition 2.1. $\widehat{Z}(A) = \widehat{C}_A(A)$ is called the *graded center* of A. We shall call A a *central graded algebra* (CGA) over F if $\widehat{Z}(A) = F$.

Note that $\widehat{Z}(A)$ is different from the ordinary center $Z(A)$ of A as an ungraded algebra. Nevertheless, $Z(A)$ is also a *graded* subalgebra, and (obviously) $Z(A)_0 = \widehat{Z}(A)_0$. In particular, A being CGA $\Longrightarrow Z(A)_0 = F$. The converse is true if A is a simple graded algebra (see 2.2), but is false in general (see Exercise 1).

Definition 2.2. A is called a *simple graded algebra* (SGA) over F if A has no proper ($\neq 0, \neq A$) graded (two-sided) ideals. If, in addition, A is a CGA, then we say A is a CSGA.

A graded algebra A is said to be *concentrated at degree* 0 if $A_1 = 0$. In this case, $A = A_0$ is just an ordinary F-algebra. Conversely, if B is any F-algebra, we may consider the associated graded algebra (B) that has components $(B)_0 = B$, $(B)_1 = 0$. The notation (B) will be used freely in the sequel.

If the graded algebra A is concentrated at degree 0, we clearly have $\widehat{Z}(A) = Z(A)$, so the notions of A being a CGA, SGA, or CSGA just boil down to the usual notions of A being, respectively, central, simple, or central simple. From now on, if we use any of these adjectives without the qualifier "graded," it will be understood to be in the ordinary (ungraded) sense.

We shall now introduce the *graded tensor product* of two graded algebras A, B, denoted by $A \hat{\otimes} B$. This, in turn, is a graded algebra, with i-component ($i = 0, 1$) defined to be $\sum A_j \otimes B_k$, summed over j, k with $j + k \equiv i$ (mod 2). The multiplication on $A \hat{\otimes} B$ is induced by

$$(a \otimes b)(a' \otimes b') = (-1)^{\partial b \, \partial a'} \cdot aa' \otimes bb' \qquad (a,\ a' \in h(A),\ \ b,\ b' \in h(B)).$$

A simple dimension count shows that

$$\dim_F(A \hat{\otimes} B) = \dim_F A \cdot \dim_F B.$$

(In fact, as vector spaces, $A \hat{\otimes} B$ is just the ordinary tensor product $A \otimes B$. The only point is that $A \hat{\otimes} B$ is not the same as the ordinary algebra tensor product $A \otimes B$.) It is also routine to verify that $\hat{\otimes}$ is associative.

We may identify A with $A \hat{\otimes} 1$ and B with $1 \hat{\otimes} B$ in $A \hat{\otimes} B$. This will be done automatically, without further mention. Note that, if either A or B is concentrated at degree zero, there is no distinction between $A \hat{\otimes} B$ and $A \otimes B$.

The following result is the graded version of 1.2.

Theorem 2.3. (1) *If A, B are graded F-algebras, and $A' \subseteq A$, $B' \subseteq B$ are graded subalgebras, then*

$$\widehat{C}_{A \hat{\otimes} B}(A' \hat{\otimes} B') = \widehat{C}_A(A') \hat{\otimes} \widehat{C}_B(B').$$

In particular, if A, B are both CGAs over F, so is $A \hat{\otimes} B$.

(2) *If A is a CSGA over F, and B an SGA, then $A \hat{\otimes} B$ is an SGA. In particular,*

(3) *If A, B are both CSGAs over F, so is $A \hat{\otimes} B$.*

Proof. We write $E = A \hat{\otimes} B$ in the following. In (1), the inclusion "$\supseteq$" follows quickly from a routine calculation. To establish the reverse inclusion, let $\{b_i\}$ be a homogeneous basis for B. Given a homogeneous element e of $\widehat{C}_E(A' \hat{\otimes} B')$, write $e = \sum a_i \otimes b_i$, where $a_i \in h(A)$. Writing $t = \partial(e)$, we have

$$\partial(a_i) + \partial(b_i) \equiv t \pmod 2, \quad \text{for every } i.$$

For any $a' \in h(A')$, the definition of $\widehat{C}_E(A' \hat{\otimes} B')$ yields an equation

$$(a' \otimes 1)e = (-1)^{t\,\partial(a')} e(a' \otimes 1).$$

The RHS equals

$$(-1)^{t\,\partial(a')} \sum (-1)^{\partial(a')\,\partial(b_i)} a_i a' \otimes b_i = \sum (-1)^{\partial(a')\,\partial(a_i)} a_i a' \otimes b_i.$$

Consequently, $a' a_i = (-1)^{\partial(a')\,\partial(a_i)} a_i a'$, which means that $a_i \in \widehat{C}_A(A')$, for all i. In particular, $e \in \widehat{C}_A(A') \hat{\otimes} B$. Now, let $\{\alpha_j\}$ be a homogeneous basis for $\widehat{C}_A(A')$, and express e as $e = \sum \alpha_j \otimes \beta_j$, $\beta_j \in h(B)$. Using the equation

$$(1 \otimes b')\, e = (-1)^{t\,\partial(b')}\, e\, (1 \otimes b') \qquad (b' \in B'),$$

and repeating the calculation above, we arrive at $\beta_j \in \widehat{C}_B(B')$, so e belongs to $\widehat{C}_A(A') \hat{\otimes} \widehat{C}_B(B')$, as claimed.

(2) Let $I \neq 0$ be a graded ideal in $E = A \hat{\otimes} B$. Our goal is to show that $1 \in I$. Each *homogeneous* element $z \in I$ can certainly be put in the form

$$z = \sum_{i=1}^{r} a_i \otimes b_i, \quad \text{where } a_i \in h(A),\ b_i \in h(B).$$

Among all nonzero *homogeneous* elements in I, let us pick z as above, such that r is as small as possible. Clearly, all a_i, b_i are nonzero, and, since z has a fixed degree in E, the sum $\partial(a_i) + \partial(b_i) \pmod 2$ is independent of i. Our first observation is that the a_i's (and, similarly, the b_i's) are linearly independent over F. Indeed, assume otherwise. Then, after renumbering, there exists a relation $a_1 = \sum_{i=2}^{s} e_i a_i$ ($e_i \in F$), where $a_1, \ldots, a_s$ are of the same degree. Then, the first s summands for z may be rewritten as $\sum_{i=2}^{s} a_i \otimes (e_i b_1 + b_i)$. But, $b_1, \ldots, b_s$ are also of the same degree, so each $e_i b_1 + b_i$ remains homogeneous, contradicting the choice of r. Our next step is to try to "make" a_1 equal to 1. The graded ideal $A a_1 A$ must be A, so there exists an equation

$$\sum_j c_j a_1 d_j = 1, \quad \text{with } c_j,\ d_j \in h(A).$$

By left and right multiplications, we obtain

$$c_j z d_j = \sum_i (-1)^{\partial(b_i)\,\partial(d_j)}\, c_j a_i d_j \otimes b_i,$$

for every j. Multiplying this by $(-1)^{\partial(b_1)\,\partial(d_j)}$, and then summing over j, we get a homogeneous element in I of the form

$$z_1 = 1 \otimes b_1 + \sum_{i=2}^{r} a_i' \otimes b_i,$$

where $a_i' = \sum_j \pm c_j a_i d_j$. Since $\partial(c_j) + \partial(d_j) \equiv \partial(a_1) \pmod 2$ independently of j, we see that each a_i' is still homogeneous. Further, z_1 is nonzero, since the b_i's are independent over F. Note that in the passage from z to z_1, the $\{b_i\}$ remain unchanged. Thus, pulling the same stunt on b_1, we can find a nonzero homogeneous element in I of the form

$$z' = 1 \otimes 1 + \sum_{i=2}^{r} a_i' \otimes b_i' \qquad (a_i',\ b_i' \text{ homogeneous}).$$

Taking the degree, we see that $\partial(a_i') \equiv \partial(b_i') \pmod 2$. For any homogeneous element $a \in h(A)$, calculate $az' - z'a \in I \cap h(E)$. The result is

$$\sum_{i=2}^{r} (a a_i' - (-1)^{\partial(b_i')\,\partial(a)} a_i' a) \otimes b_i'.$$

By the choice of r, we conclude that $aa'_i = (-1)^{\partial(a'_i)\partial(a)} a'_i a$, i.e., $a'_i \in \widehat{Z}(A)$ (see 2.1). Since A is a CGA, each a'_i must be a scalar. But $\{1, a'_2, \ldots, a'_r\}$ are linearly independent over F, as observed before, so we must have $r = 1$, i.e., $1 \in I$. □

We shall now look at some examples.

Example 2.4. Consider a quadratic extension $A = F(\sqrt{a}\,)$. We can make A into a graded F-algebra by declaring $A_0 = F$ and $A_1 = F \cdot \sqrt{a}$. We shall use the notation $A = F\langle\sqrt{a}\,\rangle$ to indicate the fact that A is made into a graded algebra in this way. Since A is commutative, and F has characteristic not 2, it is trivial to see that A is a CGA. Since A is a field, it follows that A is, in fact, a CSGA. Similarly, $B = F \oplus Fe$ subject to the relations $e^2 = 1$ and $\partial(e) = 1$ is also a CSGA. By analogy with the above, we shall write $F\langle\sqrt{1}\,\rangle$ for B.

Example 2.5. Consider a quaternion algebra $C = \left(\frac{a,b}{F}\right)$ with the usual basis $\{1, i, j, k\}$. We can make C into a graded F-algebra by setting $C_0 = F \oplus F \cdot k$, and $C_1 = Fi \oplus Fj$. We shall use the notation $C = \left\langle\frac{a,b}{F}\right\rangle$ to indicate this point of view. If we let

$$A = F \oplus Fi \cong F\langle\sqrt{a}\,\rangle \quad \text{and} \quad B = F \oplus Fj \cong F\langle\sqrt{b}\,\rangle$$

(graded subalgebras of C), then A, B are CSGAs by 2.4, and we clearly have $C \cong A \mathbin{\hat{\otimes}} B$. In particular, Theorem 2.3 implies that C itself is a CSGA. (This fact also follows from III.1.1).

Example 2.6. Let $V = V_0 \oplus V_1$ be a graded F-vector space. We may make the full endomorphism ring $E \cong \mathrm{End}(V)$ into a graded algebra by setting

$$E_i = \{f \in \mathrm{End}(V) \mid f(V_j) \subseteq V_{i+j}\},$$

where the subscripts are always taken modulo 2. From a matrix point of view, this means E_0 is the subalgebra of matrices of the form $\begin{pmatrix} * & 0 \\ 0 & * \end{pmatrix}$, and E_1 is the additive group of matrices of the form $\begin{pmatrix} 0 & * \\ * & 0 \end{pmatrix}$ (relative to a suitable basis). With this graded structure, E is a CSGA. The proof of this is trivial, in view of the fact that E is actually a CSA as an ungraded F-algebra.

Let us now look at matrix algebras in some more detail. There are usually many different ways of putting a grading on them. Consider, say, $\mathbb{M}_r(A)$, where A is a graded algebra. One natural way to grade $\mathbb{M}_r(A)$ is to assign a matrix degree i iff all its entries have degree i. This grading will be indicated by writing the matrix algebra as $\widetilde{\mathbb{M}}_r(A)$. It is trivial to see that

$\widetilde{\mathbb{M}}_r(F)$ is just $(\mathbb{M}_r(F))$, and

$$\widetilde{\mathbb{M}}_r(A) \cong \widetilde{\mathbb{M}}_r(F) \mathbin{\hat{\otimes}} A \cong \widetilde{\mathbb{M}}_r(F) \otimes A.$$

We can define *another* graded algebra $\widehat{\mathbb{M}}_r(A)$ out of $\mathbb{M}_r(A)$. In this new grading (the so-called "checkerboard grading"), we take:

$$\widehat{\mathbb{M}}_r(A)_0 = \begin{pmatrix} A_0 & A_1 & \\ A_1 & A_0 & \\ & & \ddots \end{pmatrix}, \qquad \widehat{M_r}(A)_1 = \begin{pmatrix} A_1 & A_0 & \\ A_0 & A_1 & \\ & & \ddots \end{pmatrix}.$$

It is easy to check that, indeed, $\widehat{\mathbb{M}}_r(A)_i \cdot \widehat{\mathbb{M}}_r(A)_j \subseteq \widehat{\mathbb{M}}_r(A)_{i+j}$ (subscripts mod 2). Note that, in case $A_1 = 0$ (i.e., A concentrated at degree 0), we will have simply

$$\widehat{\mathbb{M}}_r(A)_0 = \begin{pmatrix} A & 0 & \\ 0 & A & \\ & & \ddots \end{pmatrix}, \qquad \widehat{\mathbb{M}}_r(A)_1 = \begin{pmatrix} 0 & A & \\ A & 0 & \\ & & \ddots \end{pmatrix}.$$

In particular, consider $\widehat{\mathbb{M}}_r(F)$. View $\mathbb{M}_r(F)$ as the full endomorphism algebra of some vector space $V = Fe_1 \oplus \cdots \oplus Fe_r$. If we let

$$V_0 = Fe_1 \oplus Fe_3 \oplus \cdots \quad \text{and} \quad V_1 = Fe_2 \oplus Fe_4 \oplus \cdots,$$

then the i-component of $\widehat{\mathbb{M}}_r(F)$ consists precisely of

$$\{f \in \operatorname{End}(V) \mid f(V_j) \subseteq V_{i+j}\}.$$

In other words, the checkerboard grading on $\widehat{\mathbb{M}}_r(F)$ is consistent with the grading on $\operatorname{End}(V_0 \oplus V_1)$ defined in 2.6.

To show the existence of yet other gradings, let us specialize to $B = \mathbb{M}_2(F)$. Consider the matrices

$$i = \begin{pmatrix} 0 & 1 \\ -a & 0 \end{pmatrix}, \quad j = \begin{pmatrix} -1 & 0 \\ 0 & 1 \end{pmatrix}, \quad k = ij = \begin{pmatrix} 0 & 1 \\ a & 0 \end{pmatrix}.$$

These satisfy $i^2 = -aI_2$, $j^2 = I_2$, and $ij = -ji$, so we may identify $\mathbb{M}_2(F)$ with $\left(\frac{-a,1}{F}\right)$. The grading on $\left\langle\frac{-a,1}{F}\right\rangle$ therefore induces a grading on $\mathbb{M}_2(F)$. In this new grading, the zero component consists of $\begin{pmatrix} x & y \\ ay & x \end{pmatrix}$, and the 1-component consists of $\begin{pmatrix} x & y \\ -ay & -x \end{pmatrix}$. Since $k^2 = aI$, it is easy to see that, if a ranges over different square classes of F, we will get different gradings on $\mathbb{M}_2(F)$.

We shall close this section by examining more closely the relationship between $\hat{\otimes}$ and $\otimes$. The results we shall state are all tailored for specific applications in the sequel. Thus, if the reader so wishes, he/she may skip these results for a first reading, and come back to consult them when they arise in applications.

Computations involving the graded tensor product $\hat{\otimes}$ are invariably inconvenienced by the fact that one has to keep track of awkward sign changes. There exists no such difficulty with the ordinary tensor product $\otimes$. The ordinary tensor product of two graded algebras is again a graded algebra in the natural way—the disadvantage with $\otimes$ is only that (CSGA) $\otimes$ (CSGA) $\neq$ CSGA in general. This does not preclude the possibility that $A \hat{\otimes} B \cong A \otimes B$ (as graded algebras) for specific types of graded algebras A, B. Of course, a general criterion for such isomorphisms to exist will be extremely useful in practice.

Here is one such criterion:

Theorem 2.7. *Let A, B be graded algebras. Suppose there exists an element $z \in Z(A_0)$, such that $z^2 = 1$ and $za_1 = -a_1 z$ for every $a_1 \in A_1$. Then there exists a graded algebra isomorphism $A \hat{\otimes} B \cong A \otimes B$.*

Proof. Let B' be the subalgebra $1 \hat{\otimes} B_0 + z \hat{\otimes} B_1 \subseteq A \hat{\otimes} B$. Then, B' and $A = A \hat{\otimes} 1$ commute elementwise, since

$$(z \hat{\otimes} b_1)(a_1 \hat{\otimes} 1) = -za_1 \hat{\otimes} b_1 = (a_1 \hat{\otimes} 1)(z \hat{\otimes} b_1),$$

for $a_1 \in A_1$, $b_1 \in B_1$. Further, B' and A generate $A \hat{\otimes} B$ as an algebra. Thus, there exists a graded algebra isomorphism $A \otimes B' \cong A \hat{\otimes} B$. Lastly, the rule

$$b_0 + b_1 \mapsto 1 \otimes b_0 + z \otimes b_1 \qquad (b_i \in B_i)$$

clearly defines a graded algebra isomorphism from B to B', so we obtain $A \hat{\otimes} B \cong A \hat{\otimes} B$. □

Corollary 2.8. *Let B, C be graded algebras, where $C_1 = 0$. Then there exists a graded algebra isomorphism*

$$\widehat{\mathbb{M}}_r(C) \hat{\otimes} B \cong \widehat{\mathbb{M}}_r(C) \otimes B.$$

Proof. It suffices to show that $A = \widehat{\mathbb{M}}_r(C)$ satisfies the hypothesis of 2.7. Choose z to be the diagonal matrix $\operatorname{diag}(1, -1, 1, -1, \dots)$. For matrix units e_{ij}, we have

$$z \cdot e_{ij} = (-1)^{i+1} e_{ij} \quad \text{and} \quad e_{ij} \cdot z = (-1)^{j+1} e_{ij}.$$

Thus, $z \cdot e_{ij} = (-1)^{i+j} e_{ij} \cdot z$. Since A_k is spanned by $\{e_{ij} : i+j \equiv k \pmod 2\}$, the hypothesis of 2.7 follows. □

Corollary 2.9. *For any graded algebra B, there are graded algebra isomorphisms*

$$\widehat{\mathbb{M}}_r(F) \hat{\otimes} B \cong \widehat{\mathbb{M}}_r(F) \otimes B \cong \widehat{\mathbb{M}}_r(B).$$

Proof. The first isomorphism follows from 2.8 by letting $C = F$. The second isomorphism follows from the immediate observation, that the usual identification $\mathbb{M}_r(F) \otimes B \cong \mathbb{M}_r(B)$ is homogeneous of degree 0 relative to the gradings on $\widehat{\mathbb{M}}_r(F) \otimes B$ and on $\widehat{\mathbb{M}}_r(B)$. □

Corollary 2.10. *$\widehat{\mathbb{M}}_r(F) \hat{\otimes} \widehat{\mathbb{M}}_s(C) \cong \widehat{\mathbb{M}}_r(F) \otimes \widehat{\mathbb{M}}_s(C) \cong \widehat{\mathbb{M}}_{rs}(C)$ for any graded algebra C.*

Proof. Let $B = \widehat{\mathbb{M}}_s(C)$ in 2.9, and simply note that $\widehat{\mathbb{M}}_r(\widehat{\mathbb{M}}_s(C)) \cong \widehat{\mathbb{M}}_{rs}(C)$. □

There exists an "analog" for 2.7, as follows:

Theorem 2.11. *Let A, B be graded algebras. Suppose there exists an element $z \in A_1 \cap Z(A)$ such that $z^2 = -1$. Then there exists an (ungraded) algebra isomorphism $(A \hat{\otimes} B)_0 \cong A_0 \otimes B$.*

Proof. Proceeding as in the proof of 2.7, we define

$$B' = B_0 \oplus zB_1 \subseteq (A \hat{\otimes} B)_0.$$

Again, B' and A_0 commute elementwise, and they generate $(A \hat{\otimes} B)_0$ as an algebra. Hence $A_0 \otimes B' \cong (A \hat{\otimes} B)_0$. As before, the rule

$$b_0 + b_1 \mapsto b_0 + zb_1$$

is an algebra isomorphism from B to B', since

$$(z \otimes b_1)(z \otimes c_1) = -z^2 \otimes b_1 c_1 = 1 \otimes b_1 c_1$$

for $b_1, c_1 \in B_1$. Consequently, $A_0 \otimes B \cong (A \hat{\otimes} B)_0$. □

Corollary 2.12. *For any graded algebra B, there exists an algebra isomorphism $(F\langle\sqrt{-1}\,\rangle \hat{\otimes} B)_0 \cong B$.*

Remark 2.13. Let us assume the theory of CSGA to be covered in the next section. Then, 2.7 applies to any A which is a CSGA of even type having quadratic invariant 1, and 2.11 applies to any A which is a CSGA of odd type having quadratic invariant -1.

3. Structure Theory of CSGA

In Section 2, we have seen some typical examples of CSGAs. They are: (B) ($B =$ CSA over F), $F\langle\sqrt{a}\rangle$, $\left\langle\frac{a, b}{F}\right\rangle$, $\widehat{\mathbb{M}}_n(F)$, etc. The purpose of the present section is to classify all CSGAs, and to show that each CSGA is, essentially, a graded tensor product of the prototypes mentioned above.

We shall now develop this classification theory, following a paper of C. T. C. Wall [Wa].

Proposition 3.1. *Let A be an SGA over F, with $A_1 \neq 0$. Then $A_1^2 = A_0$. If $I \neq 0$ is any ideal of A_0, then $I + A_1 I A_1 = A_0$ and $A_1 I + I A_1 = A_1$.*

Proof. First, A_1^2 (the span of xy, where $x, y \in A_1$) must be the entire A_0, as otherwise $A_1^2 \oplus A_1$ would be a proper graded ideal in A. Similarly, consider the graded subspace

$$J = (I + A_1 I A_1) \oplus (A_1 I + I A_1).$$

This is easily checked to be an *ideal* in A. Since $J \neq 0$, we must have $J_0 = A_0$ and $J_1 = A_1$, which are the desired conclusions. □

Proposition 3.2. *Let A be an SGA, with $A_1 \neq 0$. Let J be a proper ideal in A (necessarily not graded). Then the projection maps $\pi_i : J \to A_i$ $(i = 0, 1)$ are isomorphisms.*

Proof. Set $I = J \cap A_0$, an ideal in A_0. By 3.1, we must have $I = 0$, for otherwise J would contain $I + A_1 I A_1 = A_0 \ni 1$. Next, $I' = \pi_0(J)$ is an ideal of A_0, and is nonzero since J cannot possibly be a graded subspace. Thus, 3.1 implies $A_1 I' A_1 + I' = A_0$. But, clearly, $A_1 I' A_1 \subseteq I'$, so we get $I' = A_0$. We have thus established the injectivity of $\pi_1 : J \to A_1$, and the surjectivity of $\pi_0 : J \to A_0$. But

$$J \cap A_1 = A_0 \cdot (J \cap A_1) = A_1 \cdot A_1 \cdot (J \cap A_1) \subseteq A_1 \cdot (J \cap A_0) = 0,$$

so $\pi_1 : J \to A_1$ is surjective. □

Proposition 3.3. *Let A be an SGA. Suppose A is not simple as an ordinary algebra. Then A_0 is a simple algebra, and $A_1 = A_0 \cdot u$, where $u \in Z(A) \cap A_1$ and $u^2 = 1$.*

Proof. Choose any proper ideal J in A (which exists by hypothesis). We have automatically $A_1 \neq 0$, so the two previous propositions apply. Since $\pi_0 : J \to A_0$ is an isomorphism, J contains a unique element of the form $1 + u$ $(u \in A_1)$. But J contains also $u(1 + u) = u^2 + u$, so we have $u^2 = 1$. We claim that $u \in Z(A)$. For $z \in A_0$, J contains both

$$z(1 + u) = z + zu \quad \text{and} \quad (1 + u)z = z + uz.$$

Applying the isomorphism $\pi_0 : J \to A_0$, we have $zu = uz$. Similarly, u commutes elementwise with A_1, so, indeed, $u \in Z(A)$. Next, for $x \in A_1$, we have $x = xu^2 \in A_0 u$, so $A_1 = A_0 u$. We finish by showing that A_0 is simple. Let $I \neq 0$ be an ideal of A_0. Then,

$$\begin{aligned} A_1 I A_1 &= A_0 u I A_0 u = A_0 u I u \\ &= A_0 I u^2 = A_0 I = I. \end{aligned}$$

By 3.1, we conclude that $I = A_0$. □

Theorem 3.4. *Let A be a CSGA, with $A_1 \neq 0$, and let $Z(A) = F \oplus Z_1$, where $Z_1 \subseteq A_1$. Then*

(1) $Z_1 = 0$ *iff A is a CSA over F (as an ungraded algebra!); and*

(2) $Z_1 \neq 0$ *iff A_0 is a CSA over F.*

Proof. (1) If $Z_1 = 0$, the preceding proposition implies that A must be a simple algebra, and therefore a CSA over F (in the sense of 1.1). The converse is trivial.

(2) Suppose $Z_1 \neq 0$. We claim that there exists $z_1 \in Z_1$, such that $z_1^2 \neq 0$. In fact, if we assume the contrary, then $Z(A)$ is certainly not a field, and hence A cannot be a simple algebra, in which case some $u \in Z_1$ satisfies $u^2 = 1$ by 3.3. Thus, there does exist $z_1 \in Z_1$ with $z_1^2 = a \in \dot{F}$. Since

$$A_1 = A_1 a = A_1 z_1^2 \subseteq A_0 z_1,$$

we have $A_1 = A_0 z_1$. This clearly implies that

$$Z(A_0) \subseteq Z(A) \cap A_0 = F,$$

so A_0 is a central F-algebra. Let $0 \neq I \subseteq A_0$ be an ideal of A_0. Then $0 \neq I + I z_1$ is a *graded* ideal in A, and so equals A. Thus $I = A_0$, proving that A_0 is a CSA over F. Conversely, suppose A_0 is central simple, and assume that $Z_1 = 0$. Then A itself is central simple by (1). Let $C = C_A(A_0)$, a graded subalgebra of A. Using 1.7, we know that C is also central simple, with $A \cong A_0 \otimes C$ as *ungraded* algebras. This forces $C_1 \neq 0$, so 3.1 applies to C, yielding $C_1^2 \neq 0$. Thus, there exist $u, v \in C_1$ with $0 \neq uv \in C_0 = F$. We have

$$C_1 = C_1 \cdot 1 = C_1 uv \subseteq C_0 v \subseteq Fv,$$

so $C = F \oplus Fv$. This implies that C is commutative, contradictory to the fact that C is a CSA over F. □

Motivated by the above result, we make the following:

Definition 3.5. Let A be a CSGA over F, and $Z(A) = F \otimes Z_1$ ($Z_1 \subseteq A_1$). If $Z_1 = 0$, we shall say that A is *of the even type.* If $Z_1 \neq 0$, we shall say that A is *of the odd type.* It follows immediately from the above theorem that A is of even type iff A is a CSA over F as an ungraded algebra. (And, A is of odd type iff $A_1 \neq 0$ and A_0 is a CSA over F, in which case A is not a CSA.)

We begin by analyzing the *odd type* case.

Theorem 3.6. *Let A be a CSGA of odd type. Then:*

(1) $Z(A) = C_A(A_0) = F \oplus Fz$, *where $z \in Z_1$ and $z^2 = a \in \dot{F}$. The square class of a does not depend on the choice of $z \in Z_1 \setminus \{0\}$, and $Z(A) \cong F\langle\sqrt{a}\rangle$ as graded algebras.*

(2) *There are graded algebra isomorphisms*

$$A \cong (A_0) \hat{\otimes} F\langle\sqrt{a}\rangle \cong (A_0) \otimes F\langle\sqrt{a}\,\rangle.$$

(3) *If* $a \notin \dot{F}^2$, *then* A *is a CSA over* $Z(A) \cong F(\sqrt{a})$. *If* $a \in \dot{F}^2$, *then* $Z(A) \cong F \times F$, *and* $A \cong A_0 \times A_0$.

In any case, A *is a semisimple (and in fact separable)* F*-algebra.*

Proof. In proving 3.4(2), we have shown that Z_1 contains some z with $z^2 = a \in \dot{F}$. We have also observed that $A_1 = A_0 z$, so, clearly,

$$Z(A) = C_A(A_0) \supseteq F \oplus Fz.$$

On the other hand, if $y \in Z_1$, then $yz^{-1} \in Z(A) \cap A_0 = F$, so $y \in F \cdot z$. If $y \neq 0$, and, say, $y = bz$ $(b \in \dot{F})$, then $y^2 = b^2 z^2 = b^2 a$. This establishes (1). Since $Z(A)$ and A_0 commute, and they generate A as an algebra, we have an algebra isomorphism $A \cong A_0 \otimes Z(A)$. This clearly implies the graded algebra isomorphisms in (2). If $a \notin \dot{F}^2$, $Z(A) \cong F\langle\sqrt{a}\,\rangle$ is the quadratic extension $F(\sqrt{a}\,)$ over F. Since A_0 is a CSA over F, 1.2 implies that $A \cong A_0 \otimes Z(A)$ is a CSA over $Z(A) \cong F(\sqrt{a}\,)$. Lastly, if $a \in \dot{F}^2$, we have $Z(A) \cong F \times F$ as an algebra, and hence

$$A \cong A_0 \otimes (F \times F) \cong A_0 \times A_0$$

as ungraded algebras. □

A trivial consequence of the above analysis of the odd type case is that A_0 *and* A_1 *must have the same dimension*, which is half of $\dim A$. In the even type case, this need not be true any more, as is shown by the trivial example, (F), or else by Example 2.6. Thus, the even case can be expected to be slightly harder.

Before we proceed to the even type case, it is convenient to introduce first the notion of the "main involution" for graded algebras.

Definition 3.7. Let $A = A_0 \oplus A_1$ be any $\mathbb{Z}_2$-graded algebra. For $x \in A_0$, $y \in A_1$, we define $\nu(x+y) = x - y$. This is an algebra automorphism of A, since

$$\begin{aligned}\nu((x+y)(x'+y')) &= \nu((xx'+yy') + (xy'+yx')) \\ &= (xx'+yy') - (xy'+yx') \\ &= (x-y)(x'-y') \\ &= \nu(x+y)\cdot\nu(x'+y').\end{aligned}$$

ν has order 2 (unless $A_1 = 0$), and will be called the *main involution* of A. It is uniquely determined by the properties that $\nu|A_0 =$ identity, and $\nu|A_1 =$ minus the identity.

Theorem 3.8. *Let A be a CSGA of even type, $A_1 \neq 0$. Suppose A is isomorphic, as an ungraded algebra, to $\mathbb{M}_n(D)$, where D is a central division algebra over F. Then the following statements hold:*

(1) *$Z(A_0) = C_A(A_0)$, and there exists $z \in Z(A_0)$, such that $Z(A_0) = F \oplus Fz$ and $z^2 = a \in \dot{F}$. The element z is determined up to a scalar multiple by these properties, and hence the square class of a is uniquely determined.*

(2) *Suppose $a \in \dot{F}^2$. Then $Z(A_0) \cong F \times F$, and there exists a graded F-vector space $V = V_0 \oplus V_1$, such that $A \cong \operatorname{End} V \hat{\otimes} (D)$ as graded algebras. (Here, $\operatorname{End} V$ is graded in the manner of 2.6.) Further, $A_0 \cong \mathbb{M}_r(D) \times \mathbb{M}_s(D)$, where $r = \dim V_0$, $s = \dim V_1$.*

(3) *Suppose $a \notin \dot{F}^2$, and the field $Z(A_0) \cong F(\sqrt{a})$ can be embedded into D. Then there exists a grading on D such that $A \cong \widetilde{\mathbb{M}}_n(D)$. In this case, $A_0 \cong \mathbb{M}_n(D_0)$ is a CSA over $Z(A_0)$.*

(4) *Suppose $a \notin \dot{F}^2$, and the field $Z(A_0) \cong F(\sqrt{a})$ cannot be embedded into D. Then $n = 2m$ is even, and $A \cong (\mathbb{M}_m(D)) \hat{\otimes} \left\langle \frac{-a, 1}{F} \right\rangle$ as graded algebras. In this case, $A_0 \cong \mathbb{M}_m(D) \otimes F(\sqrt{a})$ is a CSA over $Z(A_0)$.*

In any case, A_0 is a semisimple (and in fact separable) F-algebra.

Proof. Consider the main involution ν on A. Since A is a CSA over F (by 3.4(1)), the Skolem-Noether Theorem (1.9) implies that ν is an inner automorphism, i.e., there exists an invertible element $z \in A$, such that $\nu(x) = z^{-1}xz$ $(x \in A)$. Since $\nu(z) = z^{-1}zz = z$, we must have $z \in A_0$. By definition of ν, we also have $A_0 = C_A(z)$. Thus,

$$C_A(A_0) \subseteq C_A(z) = A_0.$$

Next, the fact that $\nu^2 =$ identity implies that $z^2 \in Z(A) = F$. But z is invertible, so z^2 is a nonzero scalar, say, $z^2 = a \in \dot{F}$. Let B denote the subalgebra $F \oplus Fz \subseteq A_0$. We must show $B = C_A(A_0)$, and show the uniqueness of z. Assume first $a \notin \dot{F}^2$. In this case, $B \cong F(\sqrt{a})$, and we have the simple algebra B contained in the F-central simple algebra A. We may then apply 1.6(2), which says that $B = C_A(C_A(B))$. But $C_A(B) = C_A(z) = A_0$, so $C_A(A_0) = B$, as required. To establish the "uniqueness" of z, let $y \in B$ be such that $\{1, y\}$ span B, and $y^2 \in F$. Write $y = \alpha + \beta z$, $\alpha, \beta \in F$, $\beta \neq 0$. We have

$$y^2 = (\alpha^2 + a\beta^2) + 2\alpha\beta z \in F,$$

which implies $\alpha = 0$, so $y = \beta z$. Also, A_0 is simple by 1.6(1). We shall now handle the case $a \in \dot{F}^2$, while trying to prove (2) at the same time. Here, changing z by a (nonzero) scalar multiple if necessary, we may assume that

$z^2 = 1$. The element $e = (1-z)/2$ is then a nontrivial idempotent of A. Fix any F-isomorphism $\theta : A \cong \mathbb{M}_n(D)$. In the matrix algebra $\mathbb{M}_n(D)$, the idempotent $\theta(e)$ must be conjugate to a standard idempotent of the form

$$e' = \begin{pmatrix} I_r & 0 \\ 0 & 0 \end{pmatrix}$$

(why?). Changing θ by an inner automorphism of $\mathbb{M}_n(D)$, we may assume that $\theta(e) = e'$. In particular,

$$\theta(z) = \theta(1-2e) = \begin{pmatrix} -I_r & 0 \\ 0 & I_s \end{pmatrix},$$

where $r + s = n$. With $\theta(z)$ in this form, it is now easy to determine the images of A_0 and A_1 under θ: $\theta(A_0)$ consists of all matrices which compute with $\begin{pmatrix} -I_r & 0 \\ 0 & I_s \end{pmatrix}$, i.e., matrices of the form $\begin{pmatrix} X & 0 \\ 0 & Y \end{pmatrix}$, and $\theta(A_1)$ consists of matrices M such that

$$M \cdot \begin{pmatrix} -I_r & 0 \\ 0 & I_s \end{pmatrix} = -\begin{pmatrix} -I_r & 0 \\ 0 & I_s \end{pmatrix} \cdot M,$$

i.e.,

$$\theta(A_1) = \left\{ M = \begin{pmatrix} 0 & P \\ Q & 0 \end{pmatrix} \right\}.$$

Let $V = V_0 \oplus V_1$, where $V_0 = F^r$ and $V_1 = F^s$. The above descriptions of $\theta(A_0)$ and $\theta(A_1)$ show that A is isomorphic, as a graded algebra, to $\operatorname{End} V \hat{\otimes} (D) = \operatorname{End} V \otimes (D)$, where $\operatorname{End} V$ is graded as in 2.6. We have also

$$A_0 = \left\{ \begin{pmatrix} X & 0 \\ 0 & Y \end{pmatrix} \right\} \cong \mathbb{M}_r(D) \times \mathbb{M}_s(D),$$

so $Z(A_0) \cong F \times F$, since D is F-central. This implies that $Z(A_0)$ is precisely $B = F \oplus Fz$ (both being two-dimensional). The uniqueness of z is trivial in this case, in view of $Z(A_0) \cong F \times F$. This establishes (2). In the following, we shall, therefore, suppose $a \notin \dot{F}^2$,

$$B = Z(A_0) \cong F(\sqrt{a}).$$

We first tackle (3), so assume that there exists $z_0 \in D$ such that $z_0^2 = a$. Take any F-isomorphism $\theta : A \cong \mathbb{M}_n(D)$. We shall identify D with the subalgebra of "scalar matrices" in $\mathbb{M}_n(D)$. The two subfields $F(z_0)$ and $F(\theta(z))$ of the matrix algebra are both $\cong F(\sqrt{a})$ and hence must be conjugate under an inner automorphism α of $\mathbb{M}_n(D)$, by the Skolem-Noether Theorem. Changing θ to $\alpha\theta$, we may assume that $\theta(z) = z_0$. Let us identify A with $\mathbb{M}_n(D)$, using this θ. Then, the main involution ν stabilizes D, since

$z_0^{-1} D z_0 = D$. We may thus put a grading on D, by setting

$$D_0 = \{d \in D \mid \nu(d) = d\} = C_D(z_0), \text{ and}$$
$$D_1 = \{d \in D \mid \nu(d) = -d\} = \{d \in D \mid z_0 d = -d z_0\}.$$

We have now

$$A_0 = C_A(z) = C_{\mathbb{M}_n(D)}(z_0) = \mathbb{M}_n(D_0),$$

and similarly, $A_1 = \mathbb{M}_n(D_1)$. Consequently, $A \cong \widetilde{\mathbb{M}}_n(D)$, as claimed in (3).

Finally, to handle (4), we assume that $B \cong F(\sqrt{a})$ *cannot* be embedded into D. Let $E = B \otimes D^{\mathrm{op}}$. We claim that

(3.9) *E is a central division algebra over $B = F(\sqrt{a})$.*

Let us first assume (3.9), and show how to finish. Let V be the simple right module over the simple algebra A. The ring of A-endomorphisms of V, written as left operators, is precisely the division algebra D. V itself is, therefore, a left D-vector space, or else a right D^{op}-vector space, of dimension n. Since A and D^{op} commute as right operators on V, we may, in particular, view V as a right module over $E = B \otimes D^{\mathrm{op}}$. But E is a division algebra by 3.9, so V is a right E-vector space, of some dimension, say, m. We therefore have

$$n = \dim_{D^{\mathrm{op}}} V = (\dim_E V) \cdot (\dim_{D^{\mathrm{op}}} E) = 2m = \text{even}.$$

Again, let us fix some F-isomorphism $\theta : A \cong \mathbb{M}_{2m}(D)$. Let z_0 denote the matrix with diagonal blocks of the form $\varepsilon = \begin{pmatrix} 0 & 1 \\ a & 0 \end{pmatrix}$. Then $z_0^2 = a$ (scalar matrix), and, as before, after changing θ, we may assume that $\theta(z) = z_0$. Using θ, we may identify A with $\mathbb{M}_{2m}(D)$. Then, A_0 consists of matrices in block form (M_{ij}) $(1 \le i, j \le m)$, where M_{ij} are 2×2 matrices, each commuting with ε. Similarly, A_1 consists of (N_{ij}), where $N_{ij} \cdot \varepsilon = -\varepsilon \cdot N_{ij}$. If we make the usual identification

$$(A =)\ \mathbb{M}_{2m}(D) = \mathbb{M}_m(D) \otimes \mathbb{M}_2(F) \quad (\text{as } F\text{-algebras}),$$

then $A_0 = \mathbb{M}_m(D) \otimes X$ and $A_1 = \mathbb{M}_m(D) \otimes Y$, where

$$X = \{x \in \mathbb{M}_2(F) \mid x\varepsilon = \varepsilon x\}, \text{ and}$$
$$Y = \{y \in \mathbb{M}_2(F) \mid y\varepsilon = -\varepsilon y\}.$$

(Note that, in deducing these expressions for A_0 and A_1, we don't need 1.2(1).) Let us calculate now X and Y in $\mathbb{M}_2(F)$. A matrix $\begin{pmatrix} p & q \\ r & s \end{pmatrix}$ lies in X iff $p = s$ and $r = aq$, i.e., $X = \left\{\begin{pmatrix} p & q \\ aq & p \end{pmatrix}\right\}$, so X has basis $\begin{pmatrix} 1 & 0 \\ 0 & 1 \end{pmatrix}$ and $\varepsilon = \begin{pmatrix} 0 & 1 \\ a & 0 \end{pmatrix}$. A matrix $\begin{pmatrix} p & q \\ r & s \end{pmatrix}$ lies in Y iff $p = -s$ and $r = -aq$, i.e.,

$Y = \left\{ \begin{pmatrix} p & q \\ -aq & -p \end{pmatrix} \right\}$, so Y has basis $i = \begin{pmatrix} 0 & 1 \\ -a & 0 \end{pmatrix}$ and $j = \begin{pmatrix} -1 & 0 \\ 0 & 1 \end{pmatrix}$. We have $i^2 = -a$, $j^2 = 1$, and $ij = \varepsilon = -ji$. Hence, we may identify $\mathbb{M}_2(F)$ with $\left\langle \frac{-a, 1}{F} \right\rangle$, thus making the former into a *graded* algebra. With this grading on $\mathbb{M}_2(F)$, we obtain, from above, a graded algebra isomorphism

$$A \cong (\mathbb{M}_m(D)) \mathbin{\hat{\otimes}} \mathbb{M}_2(F) \cong (\mathbb{M}_m(D)) \mathbin{\hat{\otimes}} \left\langle \frac{-a, 1}{F} \right\rangle.$$

Taking the zero components, we have $A_0 \cong \mathbb{M}_m(D) \otimes F(\sqrt{a})$, which is (by previous arguments) a CSA over $Z(A_0) \cong F(\sqrt{a})$. This completes the proof of (4).

It remains to prove the outstanding claim (3.9). Writing T for E^{op}, it suffices to show that $T = B \otimes D$ is a central division algebra over B. In any case, we know T is a CSA over B, by 1.2. Write T as $\mathbb{M}_r(S)$, where S is a central division algebra over B. We wish to show that $r = 1$. Let $s = \dim_B S$, $d = \dim_F D$, and let M be the irreducible left T-module. Then

$$\dim_B M = r \cdot s = r^2 s/r = d/r,$$

and so $\dim_F M = 2d/r$. M is a left T-module, so is also a left D-vector space. Consequently, $\dim_F M$ must be a multiple of $\dim_F D = d$. This shows r must be either 1 or 2. Assume, for the moment, that $r = 2$, so $\dim_F M = d$. Then $M \cong D$ as a left D-module. But $z = z \otimes 1 \in B \otimes D = T$ commutes with $1 \otimes D$, so z acts D-linearly on M. Since $z^2 = a$, this implies that D^{op} contains an element whose square is a. This contradicts the case assumption of (4). □

We shall now introduce some notations.

Definition 3.10. For any A which is a CSGA over F, write $\mathrm{type}(A) = 1$ if A is of *odd* type, and write $\mathrm{type}(A) = 0$ if A is of *even* type. If $\mathrm{type}(A) = 1$, let z_A denote the element z determined in 3.6, and write $\delta(A) = z_A^2$. Then z_A is well defined up to a scalar multiple, and $\delta(A)$ is well defined as an element in the square class group $\dot{F}/\dot{F}^2$. If $\mathrm{type}(A) = 0$, and $A_1 \neq 0$, let z_A denote the element z determined in 3.8, and, again, write $\delta(A) = z_A^2 \in \dot{F}/\dot{F}^2$. If $A_1 = 0$, we let z_A be 1, and write $\delta(A) = 1 \in \dot{F}/\dot{F}^2$. The square class $\delta(A)$ is called the *quadratic invariant* of A.

Note that in *all* cases, $\{1, z_A\}$ form a basis of $C_A(A_0) = \widehat{C}_A(A_0)$. We may now restate 3.6 and 3.8, as follows.

Classification Theorem 3.11. *Let A, B be CSGAs over F. If they are of odd type, then $A \cong B$ as graded algebras iff $A_0 \cong B_0$ and $\delta(A) = \delta(B)$. If A, B are of even type, and $\delta(A), \delta(B)$ are different from 1 in $\dot{F}/\dot{F}^2$, then $A \cong B$ as graded algebras iff $A \cong B$ as ungraded algebras and $\delta(A) = \delta(B)$.*

If A, B are of even type, $\delta(A) = \delta(B) = 1$, then $A \cong B$ as graded algebras iff $A \cong B$ as ungraded algebras and A, B determine the same unordered pair (r, s) as in 3.8(2).

We now finish with the

Product Theorem 3.12. *Let A, B be CSGAs over F, and $E = A \hat{\otimes} B$. Then, up to a scalar multiple, z_E may be taken as $z_A z_B$ $(= z_A \otimes z_B)$. Further, if* $\text{type}(A) = \varepsilon$ *and* $\text{type}(B) = \eta$, *then*

$$\delta(E) = (-1)^{\varepsilon\eta}\delta(A) \cdot \delta(B) \in \dot{F}/\dot{F}^2.$$

Proof. Note that $\partial z_A = \varepsilon$ and $\partial z_B = \eta$. If we may indeed take z_E to be $z_A z_B$, then

$$\delta(E) = (z_A \otimes z_B)^2 = (-1)^{\varepsilon\eta} z_A^2 \otimes z_B^2 = (-1)^{\varepsilon\eta}\delta(A) \cdot \delta(B).$$

It is, therefore, sufficient to prove the first conclusion in 3.12. If $A_1 = 0 = B_1$, then $z_A = z_B = z_E = 1$, and there is nothing to prove. Next, assume that $A_1 = 0$, but $B_1 \neq 0$. In this case, $E_0 = A_0 \otimes B_0$, so clearly, $1 \otimes z_B \in C_E(E_0)$. Further, $1 \otimes z_B$ is *not* a scalar, and $(1 \otimes z_B)^2 = 1 \otimes z_B^2$ *is* a scalar. Consequently, we may take z_E as $1 \otimes z_B = z_A z_B$. Finally, we treat the nontrivial case when $A_1 \neq 0$, $B_1 \neq 0$. We claim that $z_A z_B \in C_E(E_0)$. Indeed, let $a_i \in A_i$, $b_i \in B_i$ $(i = 0, 1)$. We have (by 3.6, 3.8):

	$\varepsilon = 0,\ i = 0$	$\varepsilon = 0,\ i = 1$	$\varepsilon = 1$
$z_A a_i$	$a_i z_A$	$-a_i z_A$	$a_i z_A$

Thus, $z_A a_i = (-1)^{(\varepsilon+1)i} a_j z_A$, and similarly for $z_B b_i$. Therefore,

$$\begin{aligned}(z_A \otimes z_B)(a_i \otimes b_i) &= (-1)^{i\eta}(z_A a_i \otimes z_B b_i)\\ &= (-1)^{i(\eta+\varepsilon+1+\eta+1)} a_i z_A \otimes b_i z_B\\ &= (-1)^{i\varepsilon} a_i z_A \otimes b_i z_B\\ &= (a_i \otimes b_i)(z_A \otimes z_B),\end{aligned}$$

which means $z_A z_B \in C_E(E_0)$. Since $z_A z_B$ is *not* a scalar, and since $(z_A z_B)^2 = (-1)^{\varepsilon\eta} z_A^2 \otimes z_B^2$ *is* a scalar, we may indeed choose z_E to be $z_A z_B$. □

4. The Brauer-Wall Group

In Section 1, we defined the Brauer group $B(F)$, whose elements are similarity classes of central simple F-algebras (CSAs). It was C. T. C. Wall [Wa] who first observed that it is possible (and expedient) to define a "graded Brauer group," using similarity classes of central simple graded F-algebras (CSGAs). Wall's "graded Brauer group" has since been known as the *Brauer-Wall group* (written $BW(F)$), and, in this book, we wish to perpetuate this terminology.

We proceed exactly as in Section 1. Let A, A' be CSGAs over F. We shall say that A is *similar* ("graded-similar"!) to A', if there exist graded vector spaces $V = V_0 \oplus V_1$, $V' = V_0' \oplus V_1'$, such that

$$A \hat{\otimes} \operatorname{End} V \cong A' \hat{\otimes} \operatorname{End} V'$$

as *graded* F-algebras. Here (as well as in the sequel), $\operatorname{End} V$ and $\operatorname{End} V'$ are made into graded algebras by the rule in 2.6. Again, it can be easily seen that similarity is an equivalence relation on CSGAs. *The equivalence class of A will be denoted by $\langle A \rangle$.* As before, it is routine to check (using 2.3(3)) that the operation

$$\langle A_1 \rangle \cdot \langle A_2 \rangle = \langle A_1 \hat{\otimes} A_2 \rangle$$

is well defined, and makes the set of similarity classes of CSGAs into a commutative monoid, $BW(F)$. The identity of this monoid is $\langle F \rangle$, where F is viewed as a CSGA over itself, concentrated at degree 0. The class $\langle F \rangle$ is, of course, the same as $\langle \operatorname{End}(V) \rangle$, where $V = V_0 \oplus V_1$.

Naturally, we expect that $BW(F)$ will be a *group*, and not just a monoid. To establish the existence of inverses, however, we must re-define an appropriate notion of the "opposite algebra," taking gradings into account. We proceed as follows.

Let A be a graded algebra. We define A^* to be $\{a^* : a \in A\}$, with the grading

$$(A^*)_0 = \{a^* : a \in A_0\}, \qquad (A^*)_1 = \{a^* : a \in A_1\},$$

and with multiplication induced by

$$a^* \cdot b^* = (-1)^{\partial a \cdot \partial b} (ba)^*, \quad \text{where } a,\, b \in h(A).$$

We call A^* the *graded opposite algebra* of A, to distinguish it from the usual opposite algebra A^{op}. One checks easily that

$$\widehat{Z}(A^*) = \{a^* : a \in \widehat{Z}(A)\};$$

in particular, if A is a CGA over F, so is A^*. It follows that A being CSGA over F implies the same for A^*.

Proposition and Definition 4.1. *If A is a CSGA, then $A \hat{\otimes} A^* \cong \operatorname{End} A$, as graded algebras. In particular, $BW(F)$ is an (abelian) group, with $\langle A \rangle^{-1} = \langle A^* \rangle$, for any A as above. $BW(F)$ will be called the Brauer-Wall group of F.*

Proof. Following the idea in the proof of 1.3, we define $\theta : A \hat{\otimes} A^* \to \operatorname{End}(A)$ to be the F-linear map induced by the rule

$$\theta(a \otimes b^*)(e) = (-1)^{\partial b\, \partial e} aeb,$$

where $a, b, e \in h(A)$. Clearly, θ "respects" the grading. The stake lies in checking that θ is a (graded-) algebra homomorphism. It suffices to check multiplicativity on homogeneous elements:

$$\begin{aligned}\theta\left((a \otimes b^*)(c \otimes d^*)\right)(e) &= \theta\left((-1)^{\partial b\,\partial c} \cdot (-1)^{\partial b\,\partial d} ac \otimes (db)^*\right)(e)\\ &= (-1)^{\partial b\,(\partial c+\partial d)} \cdot (-1)^{\partial e\,(\partial d+\partial b)}\, ac \cdot e \cdot db\\ &= (-1)^{(\partial c+\partial e+\partial d)\,\partial b} \cdot (-1)^{\partial d\,\partial e}\, a \cdot ced \cdot b\\ &= \theta\,(a \otimes b^*)((-1)^{\partial d\,\partial e}\, ced)\\ &= [\,\theta\,(a \otimes b^*)\,\theta\,(c \otimes d^*)\,](e).\end{aligned}$$

Since $A \hat{\otimes} A^*$ is a CSGA by 2.3(3), the graded ideal $\ker\theta$ is zero. Consequently, θ is an isomorphism by dimension count, and the remaining conclusions follow immediately. □

Our main task is to compare the Brauer-Wall group with the Brauer group. There exists a (clearly well defined) map $i : B(F) \to BW(F)$, which takes an algebra A (CSA over F) into the graded algebra (A). Not surprisingly, we have

Proposition 4.2. *$i : B(F) \to BW(F)$ is a monomorphism.*

Proof. i is clearly a homomorphism, so we need only show injectivity. Suppose A, B are CSAs over F. If $i([A]) = i([B]) \in BW(F)$, then, there exists a graded algebra isomorphism

$$(A) \hat{\otimes} \operatorname{End} V \cong (B) \hat{\otimes} \operatorname{End} W,$$

where V, W are graded vector spaces. But $(A) \hat{\otimes} \operatorname{End} V$ is identical with $A \otimes \operatorname{End} V$, and similarly for $(B) \hat{\otimes} \operatorname{End} W$. Thus, we get an algebra isomorphism

$$A \otimes \operatorname{End} V \cong B \otimes \operatorname{End} W,$$

which implies that $[A] = [B] \in B(F)$. □

Since i is a monomorphism, it is customary to view it as an inclusion map, i.e., we may think of $B(F)$ as a subgroup of $BW(F)$ via the "identification" i. We now wish to compute $BW(F)/B(F)$—the Structure Theory in Section 3 is tailor-made for this computation!

Recall the group $Q(F)$ constructed in II.2. It is the totality of elements (e, d), with $e \in \mathbb{Z}/2\mathbb{Z}$, $d \in \dot{F}/\dot{F}^2$, made into an (abelian) group, by the operation

$$(e,\, d)(e',\, d') = (e + e',\, (-1)^{ee'} dd').$$

In 3.10, we have seen that the isomorphism class of each CSGA gives rise to an element of $Q(F)$, via $A \mapsto (\text{type}(A),\, \delta(A))$. We are, therefore, led to the following statement.

Theorem 4.3. *The rule* $j : \langle A\rangle \to (\text{type}(A),\ \delta(A)) \in Q(F)$ *is a well-defined group homomorphism from* $BW(F)$ *to* $Q(F)$.

Proof. If K is a CSGA of the form $\operatorname{End} V$, where $V = V_0 \oplus V_1$, we clearly have $(\text{type}(K),\ \delta(K)) = (0, 1)$. To show that $(\text{type}(A),\ \delta(A)) \in Q(F)$ depends only on $\langle A\rangle \in BW(F)$, it suffices to prove that

$$(\text{type}(A),\ \delta(A)) = (\text{type}(A \mathbin{\hat\otimes} K),\ \delta(A \mathbin{\hat\otimes} K)),$$

for K as above. But, by 3.12,

$$\text{type}(A \mathbin{\hat\otimes} K) = \partial(z_{A\hat\otimes K}) = \partial(z_A z_K) = \partial(z_A) = \text{type}(A),\ \text{and}$$
$$\delta(A \mathbin{\hat\otimes} K) = (-1)^{\varepsilon\cdot 0}\delta(A)\cdot\delta(K) = \delta(A),$$

where $\varepsilon = \text{type}(A)$. So, we have verified that j is well defined. The fact that j is a homomorphism now follows from 3.12. □

Theorem 4.4. *The sequence*

$$0 \longrightarrow B(F) \xrightarrow{\ i\ } BW(F) \xrightarrow{\ j\ } Q(F) \longrightarrow 0$$

is exact. In fact, if A *is a CSGA such that* $\langle A\rangle \in \ker(j)$, *then* $\langle A\rangle = i[A]$. *(Here,* $[A] \in B(F)$ *is formed, of course, by viewing* A *as an ungraded* F*-algebra, which is a CSA.)*

Proof. If B is a CSA over F, then (B) is of type 0, and has quadratic invariant 1. This shows that the sequence is a zero sequence. To show the surjectivity of j, it suffices to show that $\operatorname{im}(j)$ catches all pairs of the form $(1, a)$, for then it also catches $(1, -a)(1, 1) = (0, a)$. Let $A = F\langle\sqrt{a}\rangle$, which is a CSGA (see 2.4). We have precisely $j(\langle A\rangle) = (1, a)$, so j is surjective. Finally, let A be any CSGA such that $\langle A\rangle \in \ker(j)$, i.e., A is of even type, having quadratic invariant 1. If $A_1 = 0$, then, $\langle A\rangle = i[A]$, and there is nothing to prove. Now, suppose $A_1 \neq 0$. According to 3.8(2), there exists a graded F-vector space V of dimension n, such that $A \cong \operatorname{End} V \mathbin{\hat\otimes} (D)$, where $A \cong \mathbb{M}_n(D)$, and D is an F-central division algebra. By the definition of (graded) similarity, we have

$$\langle A\rangle = \langle (D)\rangle = i\,[D] = i\,[A] \quad \text{in } BW(F). \qquad \square$$

Remark. By doing some additional work, it can be shown that $Q(F)$ is isomorphic to the "group of graded quadratic extensions" of F (with a suitable group law). If we replace $Q(F)$ by this new group in the exact sequence 4.4, then j will be replaced by the map $\langle A\rangle \mapsto C_A(A_0)$ (if $A_1 \neq 0$). The advantage of this alternative approach is that it is susceptible to a complete generalization to the case of graded algebras over arbitrary commutative rings. We have, however, no intention of going into such digressions.

A final remark. The exact sequence 4.4 defines an extension of the group $B(F)$ by the group $Q(F)$. To use 4.4 effectively in computing $BW(F)$, it is necessary to know the "factor set" of the extension class. In the next chapter, we shall determine this factor set, and then write down the multiplication table for $BW(F)$. We will also be able to scrutinize more examples, once we begin looking at $BW(F)$ in conjunction with Witt rings and Clifford algebras.

Exercises for Chapter IV

1. Let A be an SGA over F. Show that A is a CGA over F iff $Z(A)_0 = F$. Show that this statement is false if A is not assumed to be an SGA.
2. Construct two CSGAs, say, A and B, such that $A \otimes B$ is not a CSGA.
3. Show, in detail, that B and BW are (covariant) functors from fields to abelian groups.

Chapter V

Clifford Algebras

1. Construction of Clifford Algebras

To take a modern viewpoint, we shall define the Clifford algebra of a quadratic space by a universal property. For the definition, and for the development of some of the elementary properties of the Clifford algebra, we do not need the quadratic space to be regular. Hence, in this section, (V, q) will denote just an *arbitrary* quadratic space.

Definition 1.1. An F-algebra A containing (V, q) as a subspace is said to be *compatible with* q, if $x^2 = q(x) \cdot 1 \in A$ for any $x \in V$. If the role of q is clear from the context, we shall just say that A is compatible with V. Also, we shall identify $F \cdot 1$ with F, so the above equation becomes $v^2 = q(v)$.

Given the algebra A as above, the *quadratic structure* of (V, q) is closely tied in with the *algebra structure* of A. To make this point yet clearer, let us calculate $q(x + y) - q(x) - q(y)$ $(x, y \in V)$:

$$q(x + y) - q(x) - q(y) = (x + y)^2 - x^2 - y^2 = xy + yx.$$

Thus, if B denotes the bilinear form on V associated with q, one has the equation

$$2B(x, y) = xy + yx, \quad \text{for any } x,\ y \in V.$$

In particular, *x and y are orthogonal in V iff $xy = -yx$ in A*. We shall derive presently two more consequences of "compatibility" to illustrate its significance, as well as for later use.

Lemma 1.2. *For A as above, and $0 \neq x \in V$, x is invertible in A iff x is an anisotropic vector in V.*

Proof. If x is anisotropic, the equation $x^2 = q(x)$ implies that $x/q(x)$ is the inverse of x. Conversely, suppose $xy = 1$, $y \in A$. Then $q(x) \cdot y = x \cdot x \cdot y = x$, and, clearly, $x \neq 0 \Rightarrow q(x) \neq 0$. □

Lemma 1.3. *Let A be as above, and $u \in V$ a (nonzero) anisotropic vector. Then, the hyperplane reflection τ_u on V associated to u* (I.4.5) *is equal to* -1 *times the conjugation by u on V in the algebra A.*

Proof. The necessary calculations here are, of course, similar to those used in III.3.2. For any $x \in V$,

$$\begin{aligned}\tau_u(x) &= x - \frac{2B(x,u)}{q(u)} \cdot u = x - \frac{xu + ux}{q(u)} \cdot u \\ &= x - x - ux \cdot \frac{u}{q(u)} = -uxu^{-1}.\end{aligned}$$ □

Definition 1.4. An F-algebra $C \supseteq V$ compatible with q is said to be a *Clifford algebra* for (V, q) if it has the following universal property: given any F-algebra $A \supseteq V$ compatible with q, there exists a *unique* F-algebra homomorphism $\varphi : C \to A$, such that $\varphi(x) = x$ for any $x \in V$. By the usual arguments for universal constructions, one sees that, if a Clifford algebra exists for (V, q), then it is determined up to a canonical isomorphism.

To prove the existence, let $T(V)$ denote the tensor algebra of V, and let $I(q)$ be the two-sided ideal of $T(V)$ generated by elements of the form

$$x \otimes x - q(x) \cdot 1 \in T(V), \quad \text{where } x \in V.$$

We shall write $C(V) = C(V, q)$ to denote the quotient algebra $T(V)/I(q)$. It can be shown that $V = T^1(V)$ maps *injectively* into $C(V)$; we shall view this injection as an identification. It is, then, obvious that $C(V)$ is indeed a Clifford algebra for (V, q). From now on, multiplication in $C(V)$ will be expressed by juxtaposition in order to eliminate the tensor symbol $\otimes$. Note that V generates $C(V)$ as an F-algebra.

Now, $T(V)$ is a graded algebra (graded by $\{0, 1, 2, \dots\}$), and the generators $x \otimes x - q(x) \cdot 1$ of $I(q)$ all lie in the sum of homogeneous components of $T(V)$ of *even* degree. Thus, *$C(V)$ has the inherited structure of a $\mathbb{Z}_2$-graded algebra.* The "even part" of $C(V)$, which is the image of $\bigoplus T^i(V)$ ($i =$ even) under the quotient map, will be denoted by $C_0(V)$. Similarly, the "odd part" of $C(V)$, is the image of $\bigoplus T^j(V)$ ($j =$ odd), and will be denoted by $C_1(V)$. Since $V = T^1(V)$, we have $V \subseteq C_1(V)$ under the appropriate identifications. We have the usual relations $C_1 C_j \subseteq C_{i+j}$, where the subscripts are taken modulo 2 (as always). The subalgebra $C_0(V)$ is usually called the "even Clifford algebra" of (V, q).

Example 1.5. (1) Let $V = \langle a \rangle$ be the one-dimensional quadratic space with matrix (a), and basis $\{x\}$. In this case, we may identify the tensor algebra $T(V)$ with the polynomial ring $F[x]$, and $I(q)$ is, then, just the ideal generated by $x^2 - a$. Thus, $C(V)$ is the quotient $F[x]/(x^2 - a)$. If $a \neq 0$, $C(V)$ coincides with $F\langle\sqrt{a}\rangle$ (defined in IV.2.4). If $a = 0$, $C(V)$ is the graded algebra of "dual numbers" $F[x]/(x^2)$.

(2) If q is the zero quadratic form on V, $I(q)$ is just the two-sided ideal generated by $x \otimes x$ $(x \in V)$ in $T(V)$. Thus, $C(V, 0)$ coincides with the exterior algebra $\Lambda(V)$.

(3) Let V be a binary quadratic space with diagonalization $\langle a, b\rangle$, relative to an orthogonal basis $\{x, y\}$ $(a, b \in \dot{F})$. Let A be the *graded* quaternion algebra $\left\langle\frac{a, b}{F}\right\rangle$, with the usual generators $\{1, i, j, k\}$, $\partial(1) = \partial(k) = 0$ and $\partial(i) = \partial(j) = 1$ (see IV.2.5). We may embed V into A by identifying x with i and y with j. Then,

$$(\alpha x + \beta y)^2 = (\alpha i + \beta j)^2 = \alpha^2 a + \beta^2 b = q(\alpha x + \beta y),$$

which shows that A is compatible with (V, q). By checking generators and relations, it is easy to see that $A \supseteq V$ has the universal property of a Clifford algebra for (V, q). Thus, as graded algebras,

$$C(V, q) \cong A \cong \left\langle\frac{a, b}{F}\right\rangle.$$

Note, however, the quadratic form q on V is *not* the same as the restriction of the norm form of A to V: in fact, these two forms differ by a sign.

(4) By the above, we have $C(\mathbb{H}) \cong \left\langle\frac{-1, 1}{F}\right\rangle$, where $\mathbb{H}$ denotes the hyperbolic plane. Now review the proof of III.1.1(4). We showed there that there exists an isomorphism $\varphi : \left(\frac{-1, 1}{F}\right) \cong \mathbb{M}_2(F)$, with

$$\varphi(i) = \begin{pmatrix} 0 & 1 \\ -1 & 0 \end{pmatrix}, \quad \varphi(j) = \begin{pmatrix} 0 & 1 \\ 1 & 0 \end{pmatrix}, \quad \varphi(k) = \begin{pmatrix} 1 & 0 \\ 0 & -1 \end{pmatrix}.$$

Under this isomorphism φ, $F \oplus F \cdot k$ corresponds to the matrices $\begin{pmatrix} * & 0 \\ 0 & * \end{pmatrix}$, and $Fi \oplus Fj$ corresponds to the matrices $\begin{pmatrix} 0 & * \\ * & 0 \end{pmatrix}$. Thus, φ is a graded algebra isomorphism $C(\mathbb{H}) \cong \widehat{\mathbb{M}}_2(F)$ (cap = checkerboard grading).

We continue now with the general discussion of $C(V, q)$. Take an orthogonal basis $\{x_1, \ldots, x_n\}$ on V (this exists by I.2.4). By our earlier observations, $x_i x_j = -x_j x_i$ $(i \neq j)$, and $x_i^2 = q(x_i)$ in $C(V)$. Thus, as an F-space, $C(V)$ is spanned by products of the form $x_1^{e_1} \cdots x_n^{e_n}$, where $e_i = 0$ or 1. In particular, we may record the following:

Corollary 1.6. $\dim_F C(V) \leq 2^n$, *where* $n = \dim_F V$.

The main result in this section is precisely to show that the inequality above is actually an equality. The usual proofs of this fact (e.g., the one found in [O_1]) all involve rather complicated manipulations with the products $\{x_1^{e_1} \cdots x_n^{e_n}\}$. We shall, however, give a simple inductive proof, using the graded product $\hat{\otimes}$ set up in IV.2.

Lemma 1.7. *If $(V, q), (V', q')$ are quadratic spaces, there exists a surjection*

$$f : C(V \perp V') \to C(V) \hat{\otimes} C(V')$$

in the category of $\mathbb{Z}_2$-graded algebras. (It will be clear, from 1.8 *below, that f is in fact an isomorphism.)*

Proof. The rule $\varepsilon(x, x') = x \otimes 1 + 1 \otimes x'$ $(x \in V,\ x' \in V')$ clearly defines an injection of $V \perp V'$ into $C(V) \hat{\otimes} C(V')$. Calculating in the latter, we have

$$\begin{aligned}(x \otimes 1 + 1 \otimes x')^2 &= x^2 \otimes 1 + 1 \otimes x'^2 + (x \otimes 1)(1 \otimes x') + (1 \otimes x')(x \otimes 1) \\ &= q(x) + q'(x') + x \otimes x' + (-1) \cdot x \otimes x' \\ &= (q \perp q')(x,\, x').\end{aligned}$$

By the universal property of the Clifford algebra, we have a unique algebra homomorphism

$$f :\ C(V \perp V') \longrightarrow C(V) \hat{\otimes} C(V'),$$

which coincides with ε on $V \perp V'$. Easy inspection shows that f is homogeneous of degree 0. It remains only to show the *surjectivity* of f. As an F-algebra, $C(V) \hat{\otimes} C(V')$ is generated by the elements of the form $x \otimes 1$ and $1 \otimes x'$ $(x \in V,\ x' \in V')$. Since these all lie in the image of f, we conclude that f is onto. □

We can now prove our main result.

Theorem 1.8. *If (V, q) is an n-dimensional quadratic space, then* $\dim C(V) = 2^n$. *In particular, if $\{x_1, \ldots, x_n\}$ is an orthogonal basis for (V, q), then $\{x_1^{e_1} \cdots x_n^{e_n} :\ e_i = 0, 1\}$ constitutes an F-basis for $C(V)$.*

Proof. The second statement clearly follows from the first. We shall establish the first statement by induction on n. The induction is started off the ground by 1.5(1). For $n > 1$, take any orthogonal decomposition $V = U \perp U'$, where U' has dimension 1. Since $C(V) = C(U \perp U')$ maps onto $C(U) \hat{\otimes} C(U')$ by 1.7, we have

$$\dim C(V) \geq \dim C(U) \cdot \dim C(U').$$

By the inductive hypothesis, $\dim C(U') = 2^{n-1}$, and 1.5(1) implies that $\dim C(U') = 2$. Thus, $\dim C(V) \geq 2^n$, and Corollary 1.6 completes the proof. □

Corollary 1.9. $\dim C_0(V) = \dim C_1(V) = 2^{n-1}$.

Proof. Given any orthogonal basis $\{x_1, \ldots, x_n\}$ on (V, q), $C_0(V)$ is spanned by

$$\{x_1^{e_1} \cdots x_n^{e_n} \mid e_i = 0, 1;\ \sum e_i \in 2\mathbb{Z}\}.$$

Therefore, this set forms a basis for $C_0(V)$. The cardinality of this basis is clearly 2^{n-1}. Similar remarks hold for $C_1(V)$. □

We also obtain the following calculation of the Clifford algebra of a hyperbolic space $m\mathbb{H}$:

Corollary 1.10. *$C(m\,\mathbb{H}) \cong \widehat{\mathbb{M}}_{2^m}(F)$ (matrix algebra with checkerboard grading). In particular, $\langle C(m\,\mathbb{H})\rangle = 1 \in BW(F)$.*

Proof. This follows from 1.7, 1.5(4), and IV.2.10. □

We shall now make a few elementary observations about the Clifford algebra construction. Let V, V' be quadratic spaces, and σ an injective isometry of V into V'. The universal property of $C(V)$ shows that there exists a unique algebra homomorphism $C(\sigma)$ from $C(V)$ to $C(V')$ that "extends" σ. This remark accords the symbol C the prestigious role of a "functor." The domain category is the category of F-quadratic spaces (with injective isometries as morphisms), and the target is the category of $\mathbb{Z}_2$-graded algebras (with degree-preserving morphisms). It can be easily checked, also, that the functor C "behaves well" with respect to field extensions. Namely, if $F \subseteq K$ is an extension of fields, then, for any F-quadratic space (V, q), there exists a natural isomorphism

$$K \otimes_F C(V, q) \cong C(K \otimes_F V,\ K \otimes_F q).$$

This can be expressed by a square diagram (which commutes up to some natural equivalence) involving the functors $K \otimes_F -$ and C. The details are left to the meticulous reader.

We shall close this section by giving a passing mention of the "spinor norm." For this, we need two propositions.

Proposition 1.11. *There exists a unique algebra anti-automorphism ε on $C(V, q)$ that is the identity on V. This ε has order 2 (except when $\dim V = 1$), and stabilizes both $C_0(V)$ and $C_1(V)$.*

Proof. Let C^{op} denote the opposite algebra of $C = C(V)$. Then, clearly, $C^{\mathrm{op}} \supseteq V$ is still compatible with q. Thus, there exists a unique algebra homomorphism $\varepsilon : C \to C^{\mathrm{op}}$ that is the identity on V. ε is clearly an epimorphism, hence an isomorphism. If we identify C^{op} with C as a set, then ε amounts exactly to an anti-automorphism on C that is the identity on V. We have, therefore,

$$\varepsilon(u_1 \cdots u_m) = u_m \cdots u_1,$$

which implies the rest of the proposition. □

Proposition 1.12. *Suppose $u_1, \dots, u_r$ are anisotropic vectors in a regular quadratic space (V, q). If the product $\tau_{u_1}\tau_{u_2}\cdots\tau_{u_r}$ is the identity in $\mathrm{O}(V, q)$ (τ_{u_i} = hyperplane reflection associated with u_i), then the product $q(u_1)\cdots q(u_r)$ belongs to $\dot{F}^2$.*

Proof. For any anisotropic vector $u \in V$, let $c(u)$ denote the conjugation by u on the algebra $C(V, q)$: $c(u)z = uzu^{-1}$. We have shown in 1.3 that τ_u is equal to the restriction $-c(u)|V$. Since $c(u_i)c(u_j) = c(u_iu_j)$, our hypothesis implies that $(-1)^r c(u_1\cdots u_r)|V$ is the identity on V. But $(-1)^r = \det(\tau_{u_1}\cdots\tau_{u_r}) = 1$. Therefore, r is even, and the product

$$x = u_1\cdots u_r \in C_0(V, q)$$

commutes elementwise with V. Since V generates $C(V, q)$ as an algebra, we get $x \in Z(C(V, q))$. On the other hand, $C(V, q)$ is an F-central graded algebra (see 2.1 in the next section). This implies that

$$Z(C(V, q)) \cap C_0(V, q) = F,$$

so x is a nonzero scalar in F. Using the anti-automorphism ε in 1.11, we conclude that

$$\begin{aligned} \dot{F}^2 \ni x^2 &= x\cdot\varepsilon(x) = u_1\cdots u_r u_r\cdots u_1 \\ &= q(u_1)\cdots q(u_r). \end{aligned}$$

□

Consider any isometry $\sigma \in \mathrm{O}(V, q)$. By the Cartan-Dieudonné Theorem I.7.1, there exists a factorization $\sigma = \tau_{v_1}\cdots\tau_{v_m}$, where v_i are anisotropic vectors. We associate to σ the square class

$$\theta(\sigma) = q(v_1)\cdots q(v_m)\dot{F}^2 \in \dot{F}/\dot{F}^2.$$

By 1.12, it is clear that $\theta(\sigma)$ depends only on σ, and not on the factorization $\tau_{v_1}\cdots\tau_{v_m}$ chosen to define it. The square class $\theta(\sigma)$ is called the *spinor norm* of σ.

Theorem 1.13. *$\theta : \mathrm{O}(V, q) \to \dot{F}/\dot{F}^2$ is a group homomorphism. In fact, it is the unique such homomorphism satisfying the property that $\theta(\tau_u) = q(u)\dot{F}^2$ for all anisotropic vectors $u \in V$.*

Proof. Evident. □

2. Structure Theorems

We are now in a perfect position to view the Clifford algebra construction in the perspective of Chapter IV. Translating results of IV.3, we will get a complete structure theory for Clifford algebras. Of course, the results to be derived in this section have been the main motivation for the material in IV.3 (as will be clear to the reader on a second reading!).

From here on, *we restrict ourselves to regular quadratic spaces again*, and this will remain in force without further mention.

From IV.2.4 and 1.5(1), we have seen that the Clifford algebra of a (regular) 1-dimensional F-quadratic space is always a CSGA over F. By IV.2.3(3) and 1.7, we immediately obtain:

Theorem 2.1. *For a quadratic space* (V, q), $C(V)$ *is a CSGA over* F.

Since this is true, the Structure Theory of IV.3 applies! Let us first determine the "type" of $C(V)$ (see IV.3.10).

Theorem 2.2. *For a quadratic space* (V, q), $\operatorname{type} C(V) \equiv \dim V (\operatorname{mod} 2)$. *In other words,* $C(V)$ *is of odd type iff* $\dim V$ *is odd, and* $C(V)$ *is of even type iff* $\dim V$ *is even.*

Proof. Let $n = \dim V$, and let $e_1, \ldots, e_n$ be an orthogonal basis of V. We define z to be the element $e_1 \cdots e_n \in C(V)$, and write $Z(C(V)) = F \oplus Z_1$ $(Z_1 \subseteq C_1(V))$.

Case 1. $n =$ odd. Since $e_i e_j = -e_j e_i$ for $i \neq j$, z clearly commutes with all e_i, and hence $z \in Z_1$. This implies that $Z_1 \neq 0$, so $C(V)$ is of odd type.

Case 2. $n =$ even. Clearly, $e_i z = -z e_i$ for all i, and hence $e_i e_j z = z e_i e_j$ for all i, j. In particular, $z \in Z(C_0(V))$ and so $C_0(V)$ is not a central F-algebra. By IV.3.4(2), we infer that $Z_1 = 0$, i.e., $C(V)$ is of even type. □

The above shows, in both the odd and the even cases, that *the* z *here coincides* (up to a scalar multiple) *with the element* $z_{C(V)}$ *defined in* IV.3.10. Thus, if n is odd, we have $Z(C(V)) = F \oplus Fz$ (by IV.3.6), and if n is even, we have $Z(C_0(V)) = F \oplus Fz$ (by IV.3.8). Moreover, we can easily compute the "quadratic invariant" δ (see IV.3.10) for the Clifford algebra $C(V)$. Recall that, by definition, $\delta(C(V))$ is given by the square class of the scalar z^2. Keeping the notations in the proof of 2.2, we have

$$\begin{aligned} z^2 &= e_1 \cdots e_n e_1 \cdots e_n = (-1)^{n(n-1)/2} e_1^2 \cdots e_n^2 \\ &= (-1)^{n(n-1)/2} q(e_1) \cdots q(e_n). \end{aligned}$$

The square class of this element is (by definition) the signed determinant, $d_{\pm}(V)$, of the quadratic space V. We have thus proved that (for any dimension n, odd or even):

Theorem 2.3. $\delta(C(V)) = d_{\pm}(V) \in \dot{F}/\dot{F}^2$.

It is now a rather trivial matter to translate the Structure Theorems of IV.3 into the Clifford algebra language. The *odd* case (IV.3.6) transcribes as follows:

Theorem 2.4. *Suppose* $\dim V$ *is odd, and* $\delta = d_\pm(V)$ *denotes the signed determinant of* (V, q). *Then*:

(1) $C_0(V)$ *is a CSA over* F, *and* $C(V) \cong (C_0(V)) \hat{\otimes} F\langle\sqrt{\delta}\,\rangle$.

(2) *If* $\delta \notin \dot{F}^2$, *then* $Z(C(V)) \cong F(\sqrt{\delta}\,)$, *and* $C(V)$ *is a CSA over* $F(\sqrt{\delta}\,)$.

(3) *If* $\delta \in \dot{F}^2$, *then* $Z(C(V)) \cong F \times F$, *and* $C(V) \cong C_0(V) \times C_0(V)$.

In any case, $C(V)$ *is a semisimple (and, in fact, separable)* F*-algebra.*

The *even* case (IV.3.8) transcribes as follows:

Theorem 2.5. *Suppose* $\dim V = n$ *is even, and* $\delta = d_\pm V$ *denotes the signed determinant of* (V, q). *Then*:

(1) $C(V)$ *is a CSA over* F.

(2) *If* $\delta \notin \dot{F}^2$, *then* $Z(C_0(V)) \cong F(\sqrt{\delta}\,)$, *and* $C_0(V)$ *is a CSA over* $F(\sqrt{\delta}\,)$.

(3) *If* $\delta \in \dot{F}^2$, *then* $Z(C_0(V)) \cong F \times F$. *If* $C(V) \cong \mathbb{M}_t(D)$ *as algebras, where* D *is an* F*-central division algebra, then* $C(V) \cong \widehat{\mathbb{M}}_t((D))$ *as graded algebras. Further,* t *is a 2-power, and*

$$C_0(V) \cong \mathbb{M}_r(D) \times \mathbb{M}_r(D), \quad \textit{where } r = t/2.$$

In any case, $C_0(V)$ *is a semisimple (and, in fact, separable)* F*-algebra.*

Here, the statement (3) is slightly more specific than IV.3.8. Statement (2) of the latter gives only

$$C(V) \cong \operatorname{End} V \hat{\otimes} (D) \qquad (V = V_0 \oplus V_1,\ \dim V = t), \text{ and}$$
$$C_0(V) \cong \mathbb{M}_r(D) \times \mathbb{M}_s(D) \qquad (r = \dim V_0,\ s = \dim V_1).$$

We claim that $r = s$. Changing z $(= z_{C(V)})$ by a scalar multiple if necessary, we may assume that $z^2 = 1$. The elements

$$e = (1 + z)/2 \quad \text{and} \quad f = (1 - z)/2$$

are central orthogonal idempotents of $C_0(V)$, with $e + f = 1$, so the two simple components of $C_0(V)$ are, precisely, $C_0(V) \cdot e$ and $C_0(V) \cdot f$. The point is to prove that they are isomorphic (which implies that $r = s$). Let τ be the reflection which fixes the partial orthogonal basis $e_2, \ldots, e_n$, and flips e_1 to $-e_1$. The functorial map $C(\tau)$ (a graded automorphism of $C(V)$) takes $z = e_1 \cdots e_n$ to $(-e_1)e_2 \cdots e_n = -z$, so $C(\tau)$ interchanges e and f. In particular, $C(\tau)$ induces an isomorphism $C_0(V) \cdot e \cong C_0(V) \cdot f$. Now that we have proved $r = s$, it follows that

$$C(V) \cong \operatorname{End} V \mathbin{\hat{\otimes}} (D) \cong \widehat{\mathbb{M}}_t(F) \mathbin{\hat{\otimes}} (D) \cong \widehat{\mathbb{M}}_t((D)).$$

Finally, since t^2 divides $\dim C(V) = 2^n$, t is clearly a 2-power. ☐

We may now summarize all the key information in a chart:

<table>
<tr><th colspan="2">$\delta(C(V)) = d_{\pm}V = \delta$</th><th>$Z(C(V))$</th><th>$C(V) = $ CSGA</th><th>$Z(C_0(V))$</th><th>$C_0(V)$</th></tr>
<tr><td rowspan="2">$n = $ odd</td><td>$\delta \notin \dot{F}^2$</td><td>$F(\sqrt{\delta})$</td><td>CSA over $F(\sqrt{\delta})$</td><td rowspan="2">F</td><td rowspan="2">CSA over F</td></tr>
<tr><td>$\delta \in \dot{F}^2$</td><td>$F \times F$</td><td>product of two isomorphic CSAs over F</td></tr>
<tr><td rowspan="2">$n = $ even</td><td>$\delta \notin \dot{F}^2$</td><td rowspan="2">F</td><td rowspan="2">CSA over F</td><td>$F(\sqrt{\delta})$</td><td>CSA over $F(\sqrt{\delta})$</td></tr>
<tr><td>$\delta \in \dot{F}^2$</td><td>$F \times F$</td><td>product of two isomorphic CSAs over F</td></tr>
</table>

We shall now establish some more results relating $\hat{\otimes}$ and $\otimes$, in the same spirit as at the end of IV.2, but in the context of Clifford algebras. For a quadratic space (V, q), it will be convenient to write $C(q)$ for $C(V)$, since we shall have occasion to look at $C(\delta \cdot q)$, for $\delta \in \dot{F}$.

Theorem 2.6. *Let A be a graded algebra which has an element $z \in Z(A_0)$, such that $z^2 = \delta \in \dot{F}$, and $za_1 = -a_1 z$ for every $a_1 \in A_1$. Then, for any quadratic space (V, q), there exists a graded algebra isomorphism*

$$A \mathbin{\hat{\otimes}} C(q) \cong A \otimes C(\delta \cdot q).$$

Proof. Let B denote the graded subalgebra

$$C_0(q) \oplus zC_1(q) \subseteq A \mathbin{\hat{\otimes}} C(q).$$

Clearly, B commutes elementwise with $A = A \mathbin{\hat{\otimes}} 1$, while B and A span $A \mathbin{\hat{\otimes}} C(q)$ as an algebra. By dimension count, we have an isomorphism $A \otimes B \cong A \mathbin{\hat{\otimes}} C(q)$ in the category of graded algebras. If $v \in V$, the square of $zv = z \otimes v \in B$ is $z^2 \otimes v^2 = \delta \cdot q(v)$. Thus, the rule $v \mapsto zv \in B$ induces a graded algebra isomorphism $C(\delta \cdot q) \cong B$. ☐

Corollary 2.7. *If q' is an even-dimensional form, and q is any form, then*

$$C(q' \perp q) \cong C(q') \otimes C((d_{\pm}q') \cdot q) \quad (\textit{as graded algebras}).$$

The odd type analogue of 2.6 is as follows.

Theorem 2.8. *Let A be a graded algebra that has an element $z \in A_1 \cap Z(A)$, such that $z^2 = \delta \in \dot{F}$. Then, for any quadratic space (V, q), there exists an algebra isomorphism*

$$(A \hat{\otimes} C(q))_0 \cong A_0 \otimes C(-\delta \cdot q).$$

Proof. Let B denote the subalgebra

$$C_0(q) \oplus z \cdot C_1(q) \subseteq (A \hat{\otimes} C(q))_0.$$

Clearly, B commutes elementwise with $A_0 = A_0 \hat{\otimes} 1$, while B and A_0 span $(A \hat{\otimes} C(q))_0$ as an algebra. By dimension count, we have an algebra isomorphism $A_0 \otimes B \cong (A \hat{\otimes} C(q))_0$. If $v \in V$, the square of $zv = z \otimes v \in B$ is $-z^2 \otimes v^2 = -\delta \cdot q(v)$. Thus, the rule $v \mapsto zv \in B$ induces an algebra isomorphism $C(-\delta \cdot q) \cong B$. □

Corollary 2.9. *If q' is an odd-dimensional form, and q is any form, then*

$$C_0(q' \perp q) \cong C_0(q') \otimes C((-d_{\pm}q') \cdot q) \quad (\textit{as ungraded algebras}).$$

Taking $q' = \langle -d \rangle$, we obtain the very useful special case:

Corollary 2.10. *For any form q, and any $d \in \dot{F}$, there exists an algebra isomorphism $C_0(\langle -d \rangle \perp q) \cong C(d \cdot q)$. In particular, $C_0(\langle -1 \rangle \perp q) \cong C(q)$ (ungraded) for any q.*

Corollary 2.11. *For any form q, and any $a \in \dot{F}$, there exists an algebra isomorphism $C_0(a \cdot q) \cong C_0(q)$.*

Proof. Write $q = \langle b \rangle \perp q'$. Then $C_0(a \cdot q) = C_0(\langle ab \rangle \perp a \cdot q')$. By the preceding result, the latter is algebra-isomorphic to $C(-ab \cdot aq') \cong C(-b \cdot q')$. Using 2.10 again, $C_0(q) = C_0(\langle b \rangle \perp q')$ is also algebra-isomorphic to $C(-b \cdot q')$. □

We state one more similar fact which will be needed later.

Proposition 2.12. *If (V, q) is even-dimensional with $d_{\pm}q = \delta$, then there exists a graded algebra isomorphism $C(-\delta \cdot q) \cong C(q)$.*

Proof. Choose $z = z_{C(q)}$ such that $z^2 = \delta$, and consider the embedding $f : V \to C(V, q)$ defined by $f(v) = zv$. Since

$$f(v)^2 = zv \cdot zv = -z^2 v^2 = -\delta \cdot q(v),$$

f extends to a graded algebra homomorphism $f : C(V, -\delta \cdot q) \to C(V, q)$. One concludes easily that f must be an isomorphism (either by checking surjectivity directly, or else by recalling that $C(V, -\delta \cdot q)$ is a CSA). □

We observe, in closing, that classically the Structure Theorems (2.4, 2.5) for Clifford algebras are derived from 1.10 and 2.10. One proves first, by passing to the algebraic closure of F, that $C(q)$ is central simple over F when $\dim q$ is even. The structure for the odd-dimensional case will then follow from 2.10.

3. The Clifford Invariant, Witt Invariant, and Hasse Invariant

The idea of the Clifford algebra construction is that one associates, to the isometry class of each quadratic form, the isomorphism class of a graded algebra. Recalling the property that $C(V \perp V') \cong C(V) \hat{\otimes} C(V')$ (1.7), we see immediately that the rule

$$V \mapsto \langle C(V) \rangle \in BW(F)$$

induces a homomorphism from the Witt-Grothendieck group to the Brauer-Wall group. We shall denote this homomorphism by

$$\Gamma : \widehat{W}(F) \longrightarrow BW(F).$$

Since $\langle C(m\mathbb{H}) \rangle$ is the identity element in $BW(F)$ (by 1.10), Γ factors through $\widehat{W}(F)/(\mathbb{Z} \cdot \mathbb{H}) = W(F)$. The induced homomorphism $W(F) \to BW(F)$ will again be denoted by Γ, which will be called the "Clifford invariant."

Recall that IF denotes the ideal of all even-dimensional forms in $W(F)$. In II.2.1, we have shown that $f : W(F)/I^2F \to Q(F)$ defined by

$$f(V) = \big(\dim V \pmod 2,\ d_{\pm}(V)\big)$$

is an isomorphism. Using 2.2, 2.3 and (IV.4.4), we see that the Clifford invariant fits into a *commutative* diagram

$$(*) \qquad \begin{array}{ccccccccc} 0 & \longrightarrow & I^2F & \longrightarrow & W(F) & \longrightarrow & W(F)/I^2F & \longrightarrow & 0 \\ & & \Big\downarrow{\scriptstyle \gamma} & & \Big\downarrow{\scriptstyle \Gamma} & & \Big\downarrow{\scriptstyle f} & & \\ 0 & \longrightarrow & B(F) & \xrightarrow{\ i\ } & BW(F) & \xrightarrow{\ j\ } & Q(F) & \longrightarrow & 0. \end{array}$$

Thus Γ induces a homomorphism γ (dotted arrow), from I^2F to the Brauer group $B(F)$. Using the second statement of IV.4.4, we may record the following description of Γ:

Proposition 3.1. *If the class of the quadratic space V belongs to I^2F (i.e., $\dim V =$ even, and $d_{\pm}(V) = 1$), then $\gamma(V) = [C(V)] \in B(F)$.*

Note that the symbol $[C(V)]$ above denotes the similarity class of the ungraded algebra $C(V)$ in the Brauer group. It *does not* involve the graded structure of $C(V)$! To determine γ more explicitly, we shall compute γ on a four-dimensional form of (signed) determinant 1.

Lemma 3.2. *For any $a, b, c \in \dot{F}$, we have*

$$\gamma(\langle a,\, b,\, c, abc\rangle) = \left(\frac{-ab,\, -ac}{F}\right) \in B(F).$$

Proof. It suffices to compute $\Gamma(\langle a, b, c, abc\rangle)$, so let us work in the category of graded algebras. By 2.10,

$$C_0(\langle a, b, c\rangle) \cong C(\langle -ab, -ac\rangle) \cong \left(\frac{-ab, -ac}{F}\right),$$

as ungraded algebras. Thus, by 2.4(1), we have a graded algebra isomorphism

$$C(\langle a,\, b,\, c\rangle) \cong \left(\frac{-ab,\, -ac}{F}\right) \hat{\otimes}\, C(\langle -abc\rangle),$$

where he first factor is viewed as a graded algebra *concentrated at degree zero*. Using 1.7, and the associative law, we obtain

$$C(\langle a,\, b,\, c,\, abc\rangle) \cong \left(\frac{-ab,\, -ac}{F}\right) \hat{\otimes}\, C(\langle -abc,\, abc\rangle).$$

By 1.10, we conclude that $\Gamma(\langle a, b, c, abc\rangle)$ is precisely $i\left(\frac{-ab, -ac}{F}\right)$. □

Corollary 3.3. $\gamma(\langle 1, -a\rangle \otimes \langle 1, -b\rangle) = \left(\frac{a, b}{F}\right)$. *[Recall that $\langle 1, -a\rangle \otimes \langle 1, -b\rangle$ is the norm form of the quaternion algebra $\left(\frac{a, b}{F}\right)$.]*

Notation. We shall write $\mathrm{Quat}(F)$ for the subgroup of $B(F)$ generated by the classes of all quaternion algebras over F. Since each such class has order ≤ 2 in $B(F)$ (by III.2.11), $\mathrm{Quat}(F)$ is a group of exponent ≤ 2.

Corollary 3.4. $\gamma(I^3F) = \{1\}$, *and* $\gamma(I^2F) = \mathrm{Quat}(F)$.

Proof. Recall that IF is additively generated by $\langle 1, -a\rangle$ $(a \in \dot{F})$. Thus, I^2F is additively generated by $\langle 1, -a\rangle \otimes \langle 1, -b\rangle$ $(a, b \in \dot{F})$. The second conclusion thus follows from 3.3. Next, I^3F is additively generated by

$$\begin{aligned}\varphi &= \langle 1,\, -a\rangle \otimes \langle 1,\, -b\rangle \otimes \langle 1,\, -c\rangle \\ &= \langle 1,\, -a,\, -b,\, ab\rangle - \langle c,\, -ca,\, -cb,\, cab\rangle \in W(F).\end{aligned}$$

By 3.2, we have

$$\begin{aligned}\gamma(\langle c,\, -ca,\, -cb,\, cab\rangle) &= \left(\frac{c^2a,\, c^2b}{F}\right) = \left(\frac{a,\, b}{F}\right) \\ &= \gamma(\langle 1,\, -a,\, -b,\, ab\rangle).\end{aligned}$$

Thus, $\gamma(\varphi) = 1$. □

Corollary 3.5. *Let* $f = \langle 1, -a\rangle \otimes \langle 1, -b\rangle$ *and* $g = \langle 1, -c\rangle \otimes \langle 1, -d\rangle$. *If* $f \equiv g \pmod{I^3F}$, *then* $f \cong g$.

Proof. Applying γ to $f \equiv g \pmod{I^3F}$ and using 3.3, 3.4, we see that $\left(\frac{a,b}{F}\right) \cong \left(\frac{c,d}{F}\right)$, and hence $f \cong g$. ☐

Note that 3.5 would not have been easy to prove without the use of *some* invariants of quadratic forms.

Another consequence of 3.4 is that the earlier commutative diagram $(*)$ can be rewritten as

$$\begin{array}{ccccccccc} 0 & \longrightarrow & I^2F/I^3F & \longrightarrow & W(F)/I^3F & \longrightarrow & W(F)/I^3F & \longrightarrow & 0 \\ & & \downarrow{\scriptstyle \overline{\gamma}} & & \downarrow{\scriptstyle \overline{\Gamma}} & & \downarrow{\scriptstyle f} & & \\ 0 & \longrightarrow & B(F) & \longrightarrow & BW(F) & \longrightarrow & Q(F) & \longrightarrow & 0, \end{array}$$

where $\overline{\gamma}$ and $\overline{\Gamma}$ are the homomorphisms induced by γ and Γ. With respect to this new diagram, the following natural question arises:

(Q) *Is* $\overline{\Gamma}$ *a monomorphism? Or, equivalently (by diagram chase), is* $\overline{\gamma}$ *a monomorphism?*

This rather deep question has been answered affirmatively by A. Merkurjev in 1981. We will not be able to give Merkurjev's proof in this book; however, we shall return to this question in Section 6 of this chapter for a fuller discussion. At this time, let us just comment on a couple of special cases of the question (Q).

Example 3.6. (1) If every binary form over F is universal, then Γ and $\overline{\Gamma}$ are monomorphisms (γ and $\overline{\gamma}$ have domains equal to $\{0\}$!). See II.3.5.

(2) If $F = \mathbb{R}$, then $\overline{\Gamma}$ is an isomorphism, and $BW(\mathbb{R}) \cong \mathbb{Z}/8\mathbb{Z}$. This follows from the fact that

$$W(\mathbb{R}) \cong \mathbb{Z} \quad \text{and} \quad I^2\mathbb{R}/I^3\mathbb{R} \cong \mathbb{Z}/2\mathbb{Z} \cong B(\mathbb{R})$$

(see II.3.3), and that γ takes $\langle 1,1,1,1\rangle \in I^2\mathbb{R}$ to the generator $\left(\frac{-1,-1}{\mathbb{R}}\right) \in B(\mathbb{R})$. (See the next section for further development of this idea.)

We shall next determine the extension class of the bottom sequence of $(*)$ by explicitly calculating its factor set. To do this, we must first specify a "lifting" $t : Q(F) \to BW(F)$. This is supposed to be a *map* (not a homomorphism), such that $j \circ t =$ identity on $Q(F)$. Reviewing the proof of the surjectivity of j (IV.4.4), we are led to define

$$t(1, a) = \langle C(\langle a\rangle)\rangle, \quad t(0, a) = \langle C(\langle -a, 1\rangle)\rangle, \quad \text{for any } a \in \dot{F}.$$

We wish to compute the "factor set"

$$\Delta(x,\, y) = t(x)\, t(y)\, t(xy)^{-1} \in B(F) \subseteq BW(F),$$

where $x,\, y \in Q(F)$. The result is as follows:

Theorem 3.7. (1) $\Delta((1,a),\,(1,b)) = \Delta((0,a),\,(0,b)) = \left(\frac{a,\,b}{F}\right) \in \mathrm{Quat}(F)$.

(2) $\Delta((1,a),\,(0,b)) = \Delta((0,b),\,(1,a)) = \left(\frac{-a,\,b}{F}\right) \in \mathrm{Quat}(F)$.

Proof. By definition,

$$\begin{aligned}\Delta((1,a),\,(1,b)) &= C(\langle a\rangle) \mathbin{\hat{\otimes}} C(\langle b\rangle) \mathbin{\hat{\otimes}} C(\langle 1,\, ab\rangle)^{-1}\\ &= \Gamma(\langle a,\, b,\, -1,\, -ab\rangle) = \left(\frac{a,\, b}{F}\right) \in B(F)\end{aligned}$$

by (3.2). The calculations for the remaining equations are all similar. □

Now, the lifting t provides a complete system $t(Q(F))$ of coset representatives of the subgroup $B(F)$ in $BW(F)$. We may thus represent any element in $BW(F)$ by a "triple," as follows.

Definition 3.8. For $D \in B(F)$, and $(\varepsilon, a) \in Q(F)$, we write the triple (D, ε, a) to designate the element $D \cdot t(\varepsilon, a) \in BW(F)$. Every element in $BW(F)$ can be uniquely expressed by such a triple.

Using this triple representation, and using the factor set Δ already computed in 3.7, we can immediately write down the multiplication table for $BW(F)$, without further dealings with the twisted tensor product $\hat{\otimes}$.

Theorem 3.9. *For any* $a,\, b \in \dot{F}$:

(1) $(D,\, 1,\, a) \cdot (E,\, 1,\, b) = (D{\cdot}E{\cdot} \left(\frac{a,\,b}{F}\right),\, 0,\, -ab) \in BW(F)$.

(2) $(D,\, 0,\, a) \cdot (E,\, 0,\, b) = (D{\cdot}E{\cdot} \left(\frac{a,\,b}{F}\right),\, 0,\, ab) \in BW(F)$.

(3) $(D,\, 0,\, a) \cdot (E,\, 1,\, b) = (D{\cdot}E{\cdot} \left(\frac{a,\,-b}{F}\right),\, 1,\, ab) \in BW(F)$.

(4) $(D,\, 0,\, a)^{-1} = (D^{-1}{\cdot} \left(\frac{a,\,a}{F}\right),\, 0,\, a)$.

(5) $(D,\, 1,\, a)^{-1} = (D^{-1},\, 1,\, -a)$.

Next, we wish to explain how to express any $\langle A\rangle \in BW(F)$ in the triple form, where A is an arbitrary CSGA. Of course, the second and third coordinates are given by $j(A) = (\mathrm{type}\, A, \delta(A))$, so it is only a matter of determining the first coordinate.

Theorem 3.10.

(1) *If* A *has odd type,* $\langle A\rangle = ([A_0],\, 1,\, \delta(A)) \in BW(F)$.

(2) *If* B *has even type,* $\langle B\rangle = ([B],\, 0,\, \delta(B)) \in BW(F)$.

Proof. (1) In this case, we have $A \cong (A_0) \hat{\otimes} Z(A)$ as graded algebras. Further, $Z(A) \cong C(\langle\delta\rangle)$, where $\delta = \delta(A)$. Thus, we have $\langle A\rangle = [A_0]\cdot t(1,\delta)$, which implies the equation in (1).

(2) Writing $\langle B\rangle = (x, 0, \delta)$ ($\delta = \delta(B)$), we wish to determine x. Let us multiply $\langle B\rangle$ by $\langle C(-1)\rangle = (1, 1, -1)$, using 3.9(3) ($C(-1)$ means $C(\langle -1\rangle)$ for short). The result is

$$\langle C(-1) \hat{\otimes} B\rangle = \left(x \cdot \left(\frac{\delta,\, 1}{F}\right),\, 1,\, -\delta\right) = (x,\, 1,\, -\delta).$$

By the odd type case, x is the class of the zero component of $C(-1) \hat{\otimes} B$. Using IV.2.11, we have

$$x = [(C(-1) \hat{\otimes} B)_0] = [F \otimes B] = [B] \in B(F). \qquad \square$$

Using this result, and recalling the Classification Theorem, IV.3.11, we obtain

Corollary 3.11. *Let A, B be CSGAs over F. Assume either A, B are of odd type, or $\delta(A)$, $\delta(B)$ are different from 1. Then $A \cong B$ as graded algebras iff $\langle A\rangle = \langle B\rangle \in BW(F)$ and $\dim A = \dim B$.*

Continuing to use the triple notation, we may express the Clifford invariant of a quadratic space V in the fashion

$$\Gamma(V) = (c(V),\, \dim_0 V,\, d_\pm V),$$

where $c(V) \in B(F)$, and $\dim_0 V$ means $\dim V \bmod 2$ (as in II.1.6). By 3.10, we have

$$c(V) = \begin{cases} [C_0(V)] \in B(F) & \textit{if } \dim V \textit{ is odd,} \\ [C(V)] \in B(F) & \textit{if } \dim V \textit{ is even.} \end{cases} \tag{3.12}$$

This c is then a mapping $c : W(F) \to B(F)$, which is known as the *Witt invariant.* It is *not* a homomorphism on $W(F)$, although it is indeed a homomorphism on I^2F (on which $c = \gamma$). The formulas for $c(U \perp V)$ in the various cases can be read off immediately from 3.9, as follows:

(3.13)

$$c(U \perp V) = c(U) \cdot c(V) \cdot \left(\frac{d_\pm U,\, d_\pm V}{F}\right), \quad \textit{if } \dim U, \dim V \textit{ are both odd or both even;}$$

$$c(U \perp V) = c(U) \cdot c(V) \cdot \left(\frac{-d_\pm U,\, d_\pm V}{F}\right), \quad \textit{if } \dim U \textit{ is odd and } \dim V \textit{ is even.}$$

It follows by an obvious induction that

Corollary 3.14. *$c(V)$ always belongs to* Quat(F) *for every V.*

Using 3.12, we may also transcribe 2.7 and 2.9 into the language of the Witt invariant. They imply, respectively:

(3.15)
$$c(q' \perp q) = c(q') \cdot c((d_{\pm}q') \cdot q), \quad \textit{if both } \dim q', \dim q \textit{ are even;}$$
$$c(q' \perp q) = c(q') \cdot c((-d_{\pm}q') \cdot q), \quad \textit{if } \dim q' \textit{ is odd and } \dim q \textit{ is even.}$$

We can also write down the formulas for $c(a \cdot q)$ $(a \in \dot{F})$. They are:

(3.16)
$$c(a \cdot q) = c(q), \quad \textit{if } \dim q \textit{ is odd;}$$
$$c(a \cdot q) = c(q) \cdot \left(\frac{a,\, d_{\pm}q}{F}\right), \quad \textit{if } \dim q \textit{ is even.}$$

The first formula is just 2.11. To derive the second, simply let $(U, q') = \langle -a \rangle$, and eliminate $c(q')$ from the second formulas in 3.13 and 3.15.

If we take a diagonalization of V, say, $\langle a_1, \ldots, a_n \rangle$, the Witt invariant $c(V)$ can be calculated easily, with the aid of 3.13, or 3.15, or both. Predictably, $c(V)$ will be a product of the various quaternion algebras $\left(\frac{\pm a_i,\, \pm a_j}{F}\right)$, although the precise formula is somewhat hard to write down on the spot. In this connection, we may define another invariant, after Hasse:

Definition 3.17. If $\langle a_1, \ldots, a_n \rangle$ is a diagonalization of V, we define the *Hasse invariant*, $s(V)$, to be the class of

$$\prod_{i<j} \left(\frac{a_i,\, a_j}{F}\right)$$

in $B(F)$. [If $n = 1$, the product is taken to be 1.]

To see that this is a "good" invariant, the first task is to verify:

Proposition 3.18. *$s(V)$ depends only on the isometry type of V, but not on the particular diagonalization chosen to define it.*

Proof. Recall that any two diagonalizations of the same space V are chain-equivalent (I.5.2). Thus, it suffices to compare the two products defined by $\langle a, b, a_3, \ldots, a_n \rangle$ and $\langle c, d, a_3, \ldots, a_n \rangle$, where $\langle a, b \rangle \cong \langle c, d \rangle$. The last isometry implies that $ab = cd \in \dot{F}/\dot{F}^2$, and that $\left(\frac{a,b}{F}\right) \cong \left(\frac{c,d}{F}\right)$. The product formed from the diagonalization $\langle a, b, a_3, \ldots, a_n \rangle$ equals

$$\left(\frac{a,\, b}{F}\right)\left(\frac{a,\, a_3 \cdots a_n}{F}\right)\left(\frac{b,\, a_3 \cdots a_n}{F}\right)\prod_{3 \le i < j}\left(\frac{a_i,\, a_j}{F}\right)$$
$$= \left(\frac{c,\, d}{F}\right)\left(\frac{cd,\, a_3 \cdots a_n}{F}\right)\prod_{3 \le i < j}\left(\frac{a_i,\, a_j}{F}\right),$$

which is exactly the product formed from the second diagonalization $\langle c, d, a_3, \ldots, a_n\rangle$. □

The Hasse invariant is quite susceptible to computations, largely because of the (easily verified) fact that

$$s(U \perp V) = s(U)\,s(V) \cdot \left(\frac{d(U),\, d(V)}{F}\right) \qquad (d = \text{determinant}).$$

However, this very formula implies that s *does not* define a mapping on $W(F)$ as domain. This can be easily rectified by passing to the bigger Witt-Grothendieck ring $\widehat{W}(F)$. From 3.9(2) and the formula above, we immediately obtain:

Proposition 3.19. *The rule* $V \mapsto (s(V), 0, d(V)) \in BW(F)$ *defines a group homomorphism* $\widehat{W}(F) \to BW(F)$.

In practice, the Hasse invariant s seems to be slightly more convenient than the Witt invariant c. By working patiently enough, one may deduce without difficulty the exact relationship between the two. We will record the results, prove one of them, and leave the other (similar) proofs to the reader.

Proposition 3.20. *Let* $n = \dim V$.

(1) *If* $n \equiv 1, 2 \pmod 8$, *then* $c(V) = s(V)$.

(2) *If* $n \equiv 3, 4 \pmod 8$, *then* $c(V) = s(V) \cdot \left(\frac{-1,\, -d(V)}{F}\right)$.

(3) *If* $n \equiv 5, 6 \pmod 8$, *then* $c(V) = s(V) \cdot \left(\frac{-1,\, -1}{F}\right)$.

(4) *If* $n \equiv 7, 8 \pmod 8$, *then* $c(V) = s(V) \cdot \left(\frac{-1,\, d(V)}{F}\right)$.

Put in a single formula, we have

$$\text{(A)} \qquad c(V) = s(V) \cdot \left(\frac{-1,\, d(V)}{F}\right)^{\varepsilon} \cdot \left(\frac{-1,\, -1}{F}\right)^{\delta},$$

where $\varepsilon = (n-1)(n-2)/2$, $\delta = (n+1)n(n-1)(n-2)/24$. *And, in case* $[V] \in I^2F$ *with* $\dim V = 2m$, *we have*[(1)]

$$\text{(B)} \qquad c(V) = s(V) \left(\frac{-1,\, -1}{F}\right)^{m(m-1)/2}.$$

Proof. We induct on n, and will show how to do the inductive step in the case $n = 8r + 1$. Write $V \cong \langle a\rangle \perp U$, where $\dim U \equiv 0 \pmod 8$. By 3.13,

[(1)] In view of the formula (B), we may think of $c(V)$ as a "signed" Hasse invariant, for quadratic spaces V lying in I^2F.

we have

$$c(V) = c(U) \cdot \left(\frac{-a, \, d(U)}{F} \right),$$

and, by induction,

$$c(U) = s(U) \cdot \left(\frac{-1, \, d(U)}{F} \right).$$

Combining these, we get

$$c(V) = s(U) \cdot \left(\frac{a, \, d(U)}{F} \right) = s(V),$$

as in (1). The other (seven) cases are similar, and the formulas (A), (B) follow easily by inspection.

Caution. Our formula (A) does not agree with the one given by C. T. C. Wall in [Wa]. Wall's formula is, in fact, incorrect.

We shall now give a few applications of the invariants introduced. The first application is the classification of forms of dimension ≤ 3.

Theorem 3.21. *For forms q, q' such that $\dim q = \dim q' \leq 3$, the following three statements are equivalent*:

(1) *q, q' are isometric*;

(2) *$d(q) = d(q')$ and $c(q) = c(q')$*;

(3) *$d(q) = d(q')$ and $s(q) = s(q')$*.

Proof. We need only prove (3) $\Rightarrow$ (1), for the rest is trivial. The binary case has been settled in III.2.10, so let us assume that q, q' are ternary forms. Let d be the common determinant of q and q'. By a simple calculation, we have $s(\langle -d \rangle \cdot q) = s(q) \cdot \left(\frac{-d, -d}{F} \right)$. Replacing q and q' by $\langle -d \rangle\, q$ and $\langle -d \rangle\, q'$, we may assume that the common determinant is -1. Write

$$q = \langle x, \, y, \, -xy \rangle, \qquad q' = \langle x', y', -x'y' \rangle.$$

Then $s(q) = \left(\frac{x, \, y}{F} \right)$, and $s(q') = \left(\frac{x', \, y'}{F} \right)$. Since these are equal (under (3)), we have

$$\langle 1, \, -x, \, -y, \, xy \rangle \cong \langle 1, \, -x', \, -y', \, x'y' \rangle$$

by III.2.5. Cancelling $\langle 1 \rangle$, we deduce that $q \cong q'$. □

It should be noted, however, that 3.21 cannot be extended to higher-dimensional forms. In fact, from 3.2 (or else 3.16), it is clear that, if q is a 4-dimensional form of determinant 1, then q and $a \cdot q$ ($a \in \dot{F}$) have the same Witt invariant (and hence the same Hasse invariant by 3.20), but, of course, q need not be isometric to $a \cdot q$ (compare Exercise 11).

Next, we give a criterion for ternary forms to be isotropic.

Proposition 3.22. *A ternary form q is isotropic iff $c(q) = 1$, iff $s(q) = \left(\frac{-1,\,-d(q)}{F}\right)$.*

Proof. The last two conditions are equivalent by 3.20(2). Assume q is isotropic. Then $q \cong \langle 1, -1, a\rangle$ for some a, and, clearly,

$$s(q) = \left(\frac{-1,\ a}{F}\right) = \left(\frac{-1,\ -d(q)}{F}\right).$$

Conversely, suppose $d(q) = d$, and that $s(q) = \left(\frac{-1,-d}{F}\right)$. Then q and $\langle 1, -1, -d\rangle$ have the same dimension, same determinant ($= d$), and the same Hasse invariant. By 3.21, we have $q \cong \langle 1, -1, -d\rangle$, hence isotropic. □

There also exists a similar criterion for four-dimensional forms. We state it in 3.23 and 3.24.

Proposition 3.23. *Let q be a four-dimensional form of determinant* 1. *Then q is isotropic iff $c(q) = 1$, iff $s(q) = \left(\frac{-1,-1}{F}\right)$.*

Proof. The last two conditions are equivalent, again by 3.20(2). If q is isotropic, a determinant consideration shows q must be hyperbolic, so clearly $c(q) = 1$. Conversely, assume $c(q) = 1$. By the remark preceding 3.22, we have $c(a \cdot q) = 1$, for every $a \in \dot{F}$. We may, thus, assume that q represents 1, say, $q \cong \langle 1, -x, -y, xy\rangle$. By 3.2, $c(q) = \left(\frac{x,y}{F}\right)$. Since $c(q) = 1$, $\left(\frac{x,y}{F}\right)$ splits over F, and its norm form q must be hyperbolic. □

Remark 3.24. In general, let q be a four-dimensional form of determinant d over F, and let $K = F(\sqrt{d})$. From a later result (VII.3.1), one sees easily that q is isotropic over F iff q_K is isotropic over K. Since q_K has determinant 1 over K, 3.23 applies to q_K. Thus one can state: *q is isotropic over F iff $c_K(q_K) = 1 \in B(K)$, iff $s_K(q_K) = \left(\frac{-1,-1}{K}\right)$.*

We now conclude this section by proving a classification theorem for quadratic forms under a special hypothesis on the field.

Proposition 3.25. *Suppose every five-dimensional form over F is isotropic. Then two forms q, q' are isometric iff $\dim q = \dim q'$, $d(q) = d(q')$ and $s(q) = s(q')$.*

Proof (of sufficiency). If the common dimension n is ≤ 3, the result follows from 3.21. Now, suppose $n \geq 4$. The hypothesis then implies that q, q' represent 1. Say $q \cong \langle 1\rangle \perp \varphi$, $q' \cong \langle 1\rangle \perp \varphi'$. Clearly, φ, φ' have the same dimension, determinant, and Hasse invariant. By induction, we have $\varphi \cong \varphi'$, and hence $q \cong q'$. □

Actually, there is a better result in the literature. In 1972, R. Elman and the author [EL$_5$] proved that *quadratic forms over a field* F *are classified by dimension, determinant and the Hasse invariant iff* $I^3F = 0$. The "only if" part is easy. For any $q = \langle 1, a\rangle \otimes \langle 1, b\rangle$ and any $c \in \dot{F}$, the two forms q and $\langle c\rangle q$ have both dimension 4, determinant 1, and the same Hasse invariant (see Exercise 11). Thus, if quadratic forms over F are classified by dimension, determinant and the Hasse invariant, we must have $\langle c\rangle q \cong q$, which means that

$$\langle 1,\, a\rangle \otimes \langle 1,\, b\rangle \otimes \langle 1,\, -c\rangle = 0 \in W(F)$$

for all a, b, $c \in \dot{F}$. Since I^3F is additively generated by the forms on the LHS, it follows that $I^3F = 0$. The converse is harder, and will be proved in a more general form later (see XII.3). Note that this converse implies 3.25. For, if any 5-dimensional form over F is isotropic, then any form $q = \langle 1,\, a\rangle \otimes \langle 1,\, b\rangle$ represents any element $c \in \dot{F}$. By III.2.4$'$, we have $\langle c\rangle q \cong q$, and so $I^3F = 0$ as before, and the "if" part of the Elman-Lam Theorem applies to give 3.25.

Much more can be said about invariants on quadratic forms and the group I^2F/I^3F; we shall return to these topics in §6. This final section of the chapter is written independently of §§4 and 5, and can be read without difficulty at this point. Thus, if the reader so wishes, he/she can proceed directly to §6 before taking on §§4 and 5.

4. Real Periodicity and Clifford Modules

In 3.6(2), we have observed that $BW(\mathbb{R}) \cong W(\mathbb{R})/I^3\mathbb{R} \cong \mathbb{Z}/8\mathbb{Z}$. This fact turns out to have rather deep implications in topology, in relation to the phenomenon of "real periodicity" in K-theory. To explore this topological connection fully will take us too far afield. One may, however, give an easy account of the algebraic features involved, within the framework of Clifford algebras.

Let F be any field (in application, $F = \mathbb{R}$ or $\mathbb{C}$). Let $\varphi_{p,q}$ denote the form $p\langle -1\rangle \perp q\langle 1\rangle$ (p, $q \geq 0$) over F. Following Karoubi, we shall write $C^{p,q} = C(\varphi_{p,q})$, with the convention hat $C^{0,0} = F$. *We propose to calculate all* $C^{p,q}$. We first calculate these as *graded* algebras, and then specialize our information by passing to ungraded algebras.

We begin with the following substantial reduction of the problem:

Proposition 4.1. *There is a graded algebra isomorphism*

$$C^{p+n,q+n} \cong \widehat{\mathbb{M}}_{2^n}(C^{p,q})$$

(*cap* = *checkerboard grading*: *see* IV.2).

Proof. By the orthogonal decomposition $\varphi_{p+n,q+n} \cong \varphi_{p,q} \perp \varphi_{n,n}$, we have, by 1.7, $C^{p+n,q+n} \cong C^{p,q} \hat{\otimes} C^{n,n}$. But, by 1.10, $C^{n,n} \cong \widehat{\mathbb{M}}_{2^n}(F)$. Thus, IV.2.9 gives

$$C^{p+n,q+n} \cong C^{p,q} \hat{\otimes} \widehat{\mathbb{M}}_{2^n}(F) \cong \widehat{\mathbb{M}}_{2^n}(C^{p,q}).$$ □

In view of this, we need only calculate $C^{p,0}$ and $C^{0,q}$. A second reduction is achieved by the following:

Proposition 4.2 (**"Periodicity 8"**). $C^{p+8,q} \cong \widehat{\mathbb{M}}_{16}(C^{p,q}) \cong C^{p,q+8}$.

Proof. We need only prove the first isomorphism, because the second is similar. Consider, first, the case $p = q = 0$. By 2.12, we have $C^{4,0} \cong C^{0,4}$. Thus,

$$C^{8,0} \cong C^{4,0} \hat{\otimes} C^{4,0} \cong C^{4,0} \hat{\otimes} C^{0,4} \cong C^{4,4} \cong \widehat{\mathbb{M}}_{16}(F),$$

by 1.10. For the general case, we have

$$C^{p+8,q} \cong C^{p,q} \hat{\otimes} C^{8,0} \cong C^{p,q} \hat{\otimes} \widehat{\mathbb{M}}_{16}(F) \cong \widehat{\mathbb{M}}_{16}(C^{p,q}),$$

by IV.2.9. □

Thus, we are reduced to calculating only $C^{p,0}$ and $C^{0,q}$, $0 \leq p, q \leq 7$. We shall calculate these in terms of the four basic graded algebras $C^{1,0}, C^{2,0}$, $C^{0,1}$, and $C^{0,2}$, denoted respectively by X, Y, Z, and W. These are regarded as known, since

$$X \cong F\langle\sqrt{-1}\rangle, \quad Y \cong \left\langle \frac{-1,\ -1}{F} \right\rangle, \quad Z \cong F\langle\sqrt{1}\rangle, \quad \text{and} \quad W \cong \left\langle \frac{1, 1}{F} \right\rangle.$$

The other Clifford algebras are tabulated as follows:

n	0	1	2	3	4	5	6	7
$C^{n,0}$	F	X	Y	$Y \otimes Z$	$Y \otimes W$	$\widehat{\mathbb{M}}_2(X \otimes W)$	$\widehat{\mathbb{M}}_4(W)$	$\widehat{\mathbb{M}}_8(Z)$
$C^{0,n}$	F	Z	W	$X \otimes W$	$Y \otimes W$	$\widehat{\mathbb{M}}_2(Y \otimes Z)$	$\widehat{\mathbb{M}}_4(Y)$	$\widehat{\mathbb{M}}_8(X)$

We shall now explain the derivation of this chart. By 2.7,

$$C^{p+2,0} \cong C^{2,0} \hat{\otimes} C^{p,0} \cong C^{2,0} \otimes C^{0,p} \cong Y \otimes C^{0,p}.$$

This calculates $C^{3,0}$ and $C^{4,0}$, and we obtain $C^{0,3}$ and $C^{0,4}$ similarly. Finally, for $p \leq 4$,

$$C^{p+4,0} \cong C^{p,0} \hat{\otimes} C^{4,0} \cong C^{0,p} \hat{\otimes} C^{0,4} \cong C^{p,4} \cong \widehat{\mathbb{M}}_{2^p}(C^{0,4-p}),$$

by 4.1. Setting $p = 1, 2, 3$, we obtain the desired forms for $C^{5,0}, C^{6,0}$, and $C^{7,0}$. The three remaining ones are computed similarly.

We shall now specialize our information by descending to ungraded algebras. As such, $Z \cong F \times F$, and $W \cong \mathbb{M}_2(F)$, but the nature of X and Y

will depend largely on the properties of F. There are, namely, three cases to be considered:

Case 1. $-1 \in \dot{F}^2$ $(X \cong F \times F,\ Y \cong \mathbb{M}_2(F))$.

Case 2. $-1 \notin \dot{F}^2$, *but is a sum of two squares* ($X = F(\sqrt{-1}\,)$ *is a field, and* $Y \cong \mathbb{M}_2(F)$).

Case 3. -1 *is not a sum of two squares* ($X = F(\sqrt{-1}\,)$ *is a field, and* $Y = \left(\frac{-1,\,-1}{F}\right)$ *is a division algebra*).

In Case 1, we have, of course, $\varphi_{n,0} \cong \varphi_{0,n}$, so $C^{n,0} \cong C^{0,n}$. Furthermore, $\varphi_{2,0}$ is hyperbolic, so

$$C^{p+2,0} \cong C^{p,0} \mathbin{\hat{\otimes}} \widehat{\mathbb{M}}_2(F) \cong \widehat{\mathbb{M}}_2(C^{0,p}).$$

We have "periodicity 2" in this case:

Case 1	0	1	2	3
$C^{n,0} \cong C^{0,n}$	F	$F \times F$	$\mathbb{M}_2(F)$	$\mathbb{M}_2(F) \times \mathbb{M}_2(F)$
	4	5	6	7
$C^{n,0} \cong C^{0,n}$	$\mathbb{M}_4(F)$	$\mathbb{M}_4(F) \times \mathbb{M}_4(F)$	$\mathbb{M}_8(F)$	$\mathbb{M}_8(F) \times \mathbb{M}_8(F)$

In Case 2, we have $\langle 1,\, 1\rangle \cong \langle -1,\, -1\rangle$ (by I.5.1), so $\varphi_{4,0}$ and $\varphi_{0,4}$ are hyperbolic. Thus,

$$C^{p+4,0} \cong C^{p,0} \mathbin{\hat{\otimes}} \widehat{\mathbb{M}}_4(F) \cong \widehat{\mathbb{M}}_4(C^{p,0}),$$

and we have "periodicity 4" in this case:

Case 2	0	1	2	3
$C^{n,0}$	F	X	$\mathbb{M}_2(F)$	$\mathbb{M}_2(F) \times \mathbb{M}_2(F)$
$C^{0,n}$	F	$F \times F$	$\mathbb{M}_2(F)$	$\mathbb{M}_2(X)$
	4	5	6	7
$C^{n,0}$	$\mathbb{M}_4(F)$	$\mathbb{M}_4(X)$	$\mathbb{M}_8(F)$	$\mathbb{M}_8(F) \times \mathbb{M}_8(F)$
$C^{0,n}$	$\mathbb{M}_4(F)$	$\mathbb{M}_4(F) \times \mathbb{M}_4(F)$	$\mathbb{M}_8(F)$	$\mathbb{M}_8(X)$

In Case 3, we have no more "simplifications" other than the usual $X \otimes \mathbb{M}_n(F) \cong \mathbb{M}_n(X)$, $\mathbb{M}_r(\mathbb{M}_s(F)) \cong \mathbb{M}_{rs}(F), \ldots$, etc. So, the table reads:

Case 3	0	1	2	3
$C^{n,0}$	F	X	Y	$Y \times Y$
$C^{0,n}$	F	$F \times F$	$\mathbb{M}_2(F)$	$\mathbb{M}_2(X)$
	4	5	6	7
$C^{n,0}$	$\mathbb{M}_2(Y)$	$\mathbb{M}_4(X)$	$\mathbb{M}_8(F)$	$\mathbb{M}_8(F) \times \mathbb{M}_8(F)$
$C^{0,n}$	$\mathbb{M}_2(Y)$	$\mathbb{M}_2(Y) \times \mathbb{M}_2(Y)$	$\mathbb{M}_4(Y)$	$\mathbb{M}_8(X)$

Using these charts, we may now investigate the following question that arises naturally in geometry and topology: *Given an integer n, what is the biggest possible value $k = \rho_F(n)$, such that $C^{k-1,0}$ has an n-dimensional representation over F?* (The superscript $k-1$ is dictated by applications of this problem to the study of immersions of projective spaces, and to the study of vector fields on spheres. See also Section 5.)

Note that, if $m \le k = \rho_F(n)$, then $C^{m-1,0}$ may be viewed as a subalgebra of $C^{k-0,1}$; so $C^{m-1,0}$ has an n-dimensional representation over F, too. Therefore, $k = \rho_F(n)$ is "biggest" in the strong sense.

To compute the function $\rho_F(n)$, the key observation is that the Clifford algebras $C^{k-1,0}$ are "essentially" matrix algebras, and so the problem confronting us is, really, that of trying to map one matrix algebra into another. We begin with the simplest case, where the two matrix algebras are defined over the same ground field.

Lemma 4.3. *$\mathbb{M}_m(F)$ maps into $\mathbb{M}_n(F)$ (as an F-algebra) iff m divides n.*

Proof. The desired map exists iff F^n can be made into an $\mathbb{M}_m(F)$-module. But the unique irreducible $\mathbb{M}_m(F)$-module is m-dimensional. □

This is already sufficient for determining $\rho_F(n)$, when F is as in Case 1 $(-1 \in \dot{F}^2)$.

Theorem 4.4 (for Case 1). *Suppose $n = 2^a \cdot n_0$, where n_0 is odd. Then, $\rho_F(n) = 2a + 2$.*

Proof. When does $C^{m-1,0}$ possess an n-dimensional representation? Write $m - 1 = 2s + i$ ($i = 0$ or 1). By the table for Case 1, $C^{m-1,0} \cong \mathbb{M}_{2^s}(F)$ if $i = 0$, and

$$C^{m-1,0} \cong \mathbb{M}_{2^s}(F) \times \mathbb{M}_{2^s}(F)$$

if $i = 1$. Thus, by 4.3, $C^{m-1,0}$ maps into $\mathbb{M}_n(F)$ iff $2^s \mid n$, i.e., iff $s \le a$. The biggest possible value for m is, therefore, $\rho(n) = 2a + 2$. □

In the remaining cases, we assume that $-1 \notin \dot{F}^2$. (X, Y will continue to have their previous meanings.)

Lemma 4.5. $\mathbb{M}_m(X)$ *maps into* $\mathbb{M}_n(F)$ *(as an* F*-algebra) iff* $2m \mid n$.

Proof. $\mathbb{M}_m(X)$ is a simple algebra, and its unique irreducible module has F-dimension $2m$. Thus, F^n can be made into an $\mathbb{M}_m(X)$-module iff $2m$ divides n. □

Theorem 4.6 (for Case 2). *Suppose* $n = 2^{2a+b}n_0$, *where* n_0 *is odd and* $b = 0$ *or* 1. *Then* $\rho_F(n) = 4a + 4^b$.

Proof. Write $m - 1 = 4s + i$ $(0 \le i \le 3)$. By "4-periodicity," $C^{m-1,0}$ is

$$\mathbb{M}_{2^{2s}}(F), \text{ or } \mathbb{M}_{2^{2s}}(X), \text{ or } \mathbb{M}_{2^{2s+1}}(F), \text{ or } \mathbb{M}_{2^{2s+1}}(F) \times \mathbb{M}_{2^{2s+1}}(F),$$

according as $i = 0, 1, 2,$ or 3. If $b = 0$, the biggest m for which $C^{m-1,0}$ maps into $\mathbb{M}_n(F)$ is achieved by letting $s = a$ and $i = 0$, which give $m = 4a + 1$. Similarly, for $b = 1$, the biggest m is achieved by letting $s = a$ and $i = 3$, which give $m = 4a + 4$. □

For the rest, we assume that we are in Case 3.

Lemma 4.7. $\mathbb{M}_m(Y)$ *maps into* $\mathbb{M}_n(F)$ *(as an* F*-algebra) iff* $4m \mid n$.

Proof. $\mathbb{M}_m(Y)$ is, again, a simple algebra, so proceed as in 4.5. □

Theorem 4.8 (for Case 3). *Suppose* $n = 2^{4a+b}n_0$, *where* n_0 *is odd and* $0 \le b \le 3$. *Then,* $\rho_F(n) = 8a + 2^b$. *[This is known as the Hurwitz-Radon function.]*

Proof. Write $m - 1 = 8s + i$ $(0 \le i \le 7)$. By "8-periodicity," $C^{m-1,0}$ is given by:

	$C^{m-1,0}$
$i = 0$	$\mathbb{M}_{2^{4s}}(F)$
$i = 1$	$\mathbb{M}_{2^{4s}}(X)$
$i = 2$	$\mathbb{M}_{2^{4s}}(Y)$
$i = 3$	$\mathbb{M}_{2^{4s}}(Y) \times \mathbb{M}_{2^{4s}}(Y)$

	$C^{m-1,0}$
$i = 4$	$\mathbb{M}_{2^{4s+1}}(Y)$
$i = 5$	$\mathbb{M}_{2^{4s+2}}(X)$
$i = 6$	$\mathbb{M}_{2^{4s+3}}(F)$
$i = 7$	$\mathbb{M}_{2^{4s+3}}(F) \times \mathbb{M}_{2^{4s+3}}(F)$

When does $C^{m-1,0}$ map into $\mathbb{M}_n(F)$ $(n = 2^{4a+b}n_0)$? By 4.3, 4.5, and 4.7, the criterion is seen to be:

	Criterion
$i=0$	$4s \leq 4a+b$
$i=1$	$4s+1 \leq 4a+b$
$i=2$	$4s+2 \leq 4a+b$
$i=3$	$4s+2 \leq 4a+b$

	Criterion
$i=4$	$4s+3 \leq 4a+b$
$i=5$	$4s+3 \leq 4a+b$
$i=6$	$4s+3 \leq 4a+b$
$i=7$	$4s+3 \leq 4a+b$

For $b=0$, we achieve the maximum m by letting $s=a$, $i=0$, i.e., $m=8a+1$. For $b=1$, it is $s=a$, $i=1$, i.e., $m=8a+2$. For $b=2$, it is $s=a$, $i=3$, i.e., $m=8a+4$. For $b=3$, it is $s=a$, $i=7$, i.e., $m=8a+8$. This checks out the Hurwitz-Radon function in all cases, by easy inspection. □

One may also study, of course, the biggest value $k=\rho'_F(n)$ for which $C^{0,k-1}$ has an n-dimensional representation over F. Since the computations are completely parallel to the above, suffice it to record the results.

Theorem 4.4′ (for Case 1). *Suppose $n=2^a n_0$, where n_0 is odd. Then, $\rho'_F(n)=\rho_F(n)=2a+2$.*

Theorem 4.6′ (for Case 2). *Suppose $n=2^{2a+b}n_0$, where n_0 is odd and $b=0$ or 1. Then, $\rho'_F(n)=4a+b+2$.*

Theorem 4.8′ (for Case 3). *Suppose $n=2^{4a+b}n_0$, where n_0 is odd and $0\leq b\leq 3$. Then, $\rho'_F(n)=8a+b+[\frac{b}{3}]+2$. (Brackets denote the integral part of rational numbers.)*

Note that if F_1, F_2, F_3 are fields under Cases 1, 2, and 3, then, $\rho_{F_1}(n)\geq\rho_{F_2}(n)\geq\rho_{F_3}(n)$ (and the same for ρ') for all n.

5. Composition of Quadratic Forms

Historically, the problem of "composition of quadratic forms" over a field F asks the following: *For what values of m and n does there exist a formula*

$$(5.1) \qquad (x_1^2+\cdots+x_m^2)(y_1^2+\cdots+y_n^2)=z_1^2+\cdots+z_n^2,$$

where $z_1,\ldots,z_n$ are homogeneous bilinear forms in the two sets of variables $x_1,\ldots,x_m;\ y_1,\ldots,y_n$? This problem was first considered by Hurwitz, over the complex field, in the special case $m=n$. Subsequent contributions and generalizations were made by Radon, Albert, Eckmann, Jacobson, and others. In the late fifties, the problem received rejuvenated interest, as topologists began to apply it in studying projective spaces, Grassmannian manifolds, vector fields over spheres, etc.

In this section, we present an account for the algebraic side of the matter. We begin by generalizing and reformulating the composition problem posed in 5.1. Taking a more geometric point of view, we ask the following. Let (U, λ) and (V, φ) be F-quadratic spaces, of dimensions m and n. Does there exist a bilinear pairing $U \times V \to V$, denoted by $(x, y) \mapsto x \cdot y$, such that

$$\varphi(x \cdot y) = \lambda(x)\,\varphi(y) \quad \text{for all } x \in U \text{ and } y \in V? \tag{5.2}$$

It can be seen, easily, that 5.2 reduces to 5.1 in the special case where $\lambda \cong m\langle 1\rangle$, $\varphi \cong n\langle 1\rangle$.

Let us first comment on some affirmative cases. For $m = n = 1$, 5.2 is possible iff $\lambda \cong \langle 1\rangle$. For $m = n = 2$, 5.2 is possible for $\lambda \cong \varphi \cong \langle 1, a\rangle$, by virtue of the classical formula

$$(x_1^2 + ax_2^2)\,(y_1^2 + ay_2^2) = (x_1y_1 + ax_2y_2)^2 + a(x_1y_2 - x_2y_1)^2.$$

Similarly, for $m = n = 4$, 5.2 is possible for $\lambda \cong \varphi \cong \langle 1, -a, -b, ab\rangle$, by a formula constructed from the multiplication rule for the quaternion algebra $\left(\frac{a,\,b}{F}\right)$. (The precise formula can be found after the proof of X.2.11.)

Let us first make some deductions from 5.2. Let A and B denote the bilinear forms on U and V associated with the quadratic forms λ and φ. For $x \in U$, and $y, z \in V$, we have

$$\begin{aligned} B(x \cdot y,\, x \cdot z) &= \tfrac{1}{2}\left[\varphi(x \cdot (y+z)) - \varphi(x \cdot y) - \varphi(x \cdot z)\right] \\ &= \lambda(x) \cdot \tfrac{1}{2}\left[\varphi(y+z) - \varphi(y) - \varphi(z)\right] \\ &= \lambda(x) \cdot B(y,\, z). \end{aligned} \tag{5.3}$$

Doing the same thing for $B(x \cdot y,\ w \cdot y)$ $(x, w \in U,\ y \in V)$, we obtain likewise

$$B(x \cdot y,\ w \cdot y) = A(x,\, w) \cdot \varphi(y). \tag{5.4}$$

We shall now perform a normalization step which will greatly facilitate the subsequent computations. Let $u \in U$ be any vector, with $\lambda(u) = a \neq 0$. Then, the action of u by multiplication on V is a linear isomorphism. In fact, if $u \cdot y = 0$, then, for any $z \in V$,

$$0 = B(u \cdot y,\ u \cdot z) = a \cdot B(y,\, z) \Longrightarrow B(y,\, z) = 0.$$

Since B is regular, y must be zero. Let us now redefine the multiplication. For $x \in U$ and $y \in V$, let $x * y$ denote the unique vector in V such that $u \cdot (x * y) = x \cdot y$. Evaluating φ on both sides, we get $a\varphi(x * y) = \lambda(x)\varphi(y)$. If λ' denotes the quadratic form $a^{-1} \cdot \lambda$ on U, then we have $\varphi(x * y) = \lambda'(x)\varphi(y)$. Also, by easy nonsense, we see that the new multiplication $*$ remains bilinear in both variables. Finally, note that $\lambda'(u) = 1$, and that the definition of $*$ clearly forces $u * y = y$ for every $y \in V$.

The above normalization procedure enables us to assume (upon scaling λ by a multiple and modifying the multiplication) that $\lambda(u) = 1$, and that u acts as the identity on V by multiplication.

Theorem 5.5. *Suppose we have achieved the said normalization. Let U_0 be the orthogonal complement of $F \cdot u$ in U. Then, the Clifford algebra $C(U_0, -\lambda)$ admits an F-representation on V.*

Proof. For any $x \in U_0$, let L_x denote multiplication by x on V. If $y \in V$, we have

$$B(y, L_x(y)) = B(u \cdot y, x \cdot y) = A(u, x)\varphi(y) = 0.$$

Writing out $0 = B(y + z, L_x(y + z))$ in four terms, we obtain

$$B(y, L_x(z)) + B(z, L_x(y)) = 0 \qquad (y, z \in V). \tag{5.6}$$

If we replace z by $L_x(z)$, we obtain

$$B(y, L_x^2(z)) = -B(x \cdot z, x \cdot y) = -\lambda(x)B(y, z),$$

i.e.,

$$B\big(y, (L_x^2 + \lambda(x) \cdot 1_V)\, z\big) = 0,$$

where 1_V denotes the identity map on V. Consequently,

$$L_x^2 = -\lambda(x)\, 1_V \in \operatorname{End}_F V,$$

whenever $x \in U_0$. This shows that the rule $x \mapsto L_x$ defines an F-algebra homomorphism from $C(U_0, -\lambda)$ to $\operatorname{End}_F V$. $\square$

Remark 5.7. Conversely, let us suppose V is given to be a $C(U_0, -\lambda)$-module. Assume further that $x \in U_0$ acts by an endomorphism, L_x, such that $B(y, L_x(y)) = 0$ for all $y \in V$. We may then *create* a pairing $U \times V \to V$, by

$$(\alpha u + x) \cdot y = \alpha y + L_x(y) \qquad (\alpha \in F,\ x \in U_0,\ y \in V).$$

This is easily checked to be bilinear. As before, we derive 5.6. Replacing z by $L_x(z)$, and using $L_x^2 = -\lambda(x) \cdot 1_V$, we obtain

$$\lambda(x)\, B(y, z) = B(x \cdot y, x \cdot z), \quad \text{whenever } x \in U_0.$$

A trivial calculation shows that the same holds for any $x \in U$ (assuming, of course, $\lambda(u) = 1$). Putting $y = z$, we recapture the formula 5.2. Thus, *the composition 5.2 is possible if $C(U_0, -\lambda)$ can be represented on V in such a way that $B(y, L_x y) = 0$ for all $y \in V$.* (See Remark 5.13.)

Let us now apply 5.5 to the classical case when $\lambda \cong m\langle 1 \rangle$. Since λ represents 1, it remains unchanged in the normalization process. Starting with any $u \in U$ such that $\lambda(u) = 1$, the quadratic space $(U_0, -\lambda)$ involved in 5.5 is just $(m-1) \cdot \langle -1 \rangle$. In the notation of Section 4, $C(U_0, -\lambda) = C^{m-1,0}$. We conclude the following from 5.5.

Theorem 5.8. *If $\lambda \cong m\langle 1\rangle$, a necessary condition for the composition formula* 5.2 *to exist is that $m \le \rho_F(n)$, where $\rho_F(n)$ is determined as in* 4.4, 4.6, *and* 4.8. *Similarly, if $\lambda \cong \langle 1\rangle \perp (m-1)\langle -1\rangle$, then, a necessary condition for* 5.2 *to exist is that $m \le \rho'_F(n)$, where $\rho'_F(n)$ is determined as in* 4.4′, 4.6′, *and* 4.8′.

Corollary 5.9. *In general (no restrictions on λ), a necessary condition for the composition formula* 5.2 *to exist is that $m \le 2a+2$, where $n = 2^a n_0$ ($n_0 =$ odd).*

Proof. Suppose 5.2 holds. Let K be any extension field of F. It can be shown, by an easy but rather tedious calculation (which we leave to the reader, of course), that the K-quadratic spaces $(K \otimes U, K \otimes \lambda)$ and $(K \otimes V, K \otimes \varphi)$ have the same composition formula as U, V, under the scalar extension of the original pairing. Passing to the algebraic closure of F, we may assume that $\lambda \cong m\langle 1\rangle$. The function $\rho_F(n)$, in this case, gives $m \le 2a+2$, where $n = 2^a \cdot n_0$, according to 4.4. □

Theorem 5.10 (Hurwitz). *Let (V, φ) be any regular quadratic space over F. If there exists a bilinear pairing $V \times V \to V$, denoted by $(x, y) \mapsto x \cdot y$, such that $\varphi(x \cdot y) = \varphi(x)\varphi(y)$ for all $x, y \in V$, then $n = \dim V = 1, 2, 4$, or* 8.

Proof. We think of $U = V$, $m = n$, and $\lambda = \varphi$, for this special case. If $n = 2^a \cdot n_0$ ($n_0 =$ odd), 5.9 forces $2a+2 \ge 2^a \cdot n_0 \ge 2^a$. This is possible only for $0 \le a \le 3$ and $n_0 = 1$, i.e., n must be $1, 2, 4$, or 8. □

Let us begin to look for *sufficient conditions* for the existence of the formula 5.2. This is considerably harder, so *let us simplify matters by taking $\lambda \cong m\langle 1\rangle$, and also $\varphi \cong n\langle 1\rangle$. Further, we shall let F be the field of real numbers $\mathbb{R}$.*

Theorem 5.11 (Radon). *The formula* 5.1 *exists for the field $F = \mathbb{R}$ iff $m \le \rho_{\mathbb{R}}(n)$. The latter denotes the Hurwitz-Radon function (see* 4.8*), given by $\rho_{\mathbb{R}}(n) = 8a + 2^b$ for $n = 2^{4a+b} n_0$ ($n_0 =$ odd, $0 \le b \le 3$).*

Proof. Holding n fixed, let $m = \rho_{\mathbb{R}}(n)$. Our job is to show that we can construct the formula 5.1 with this specific m. Let $\{e_1, \dots, e_m\}$ be an orthonormal basis for (U, λ), and $U_0 = \sum_{i \ge 2} \mathbb{R} \cdot e_i$. Write $C = C(U_0, -\lambda) = C^{m-1,0}$. Since $\varphi \cong n\langle 1\rangle$, we may think of (V, B, φ) as $\mathbb{R}^n$ equipped with the usual inner product. By definition of $\rho_{\mathbb{R}}$, there exists a linear representation

$$L : C \longrightarrow \operatorname{End} V = \mathbb{M}_n(\mathbb{R}).$$

Let G be the multiplicative group generated by the invertible elements e_i $(2 \le i \le m)$ and -1 in C. Since we have the relations

$$e_i^2 = -1 \quad \text{and} \quad e_i e_j = -e_j e_i \quad (2 \le i \ne j \le m),$$

G is obviously a finite group. By the theory of group representations, $L : G \to \mathbb{M}_n(\mathbb{R})$ is equivalent to an orthogonal representation. Thus, after changing L by a conjugation on $\mathbb{M}_n(\mathbb{R})$, we may assume that each $L(e_i)$ $(2 \le i \le m)$ is an *isometry* on $\mathbb{R}^n$. The rule $(x, y) \mapsto x \cdot y = L(x)(y)$ clearly defines a bilinear pairing $U_0 \times V \to V$, and it satisfies

$$B(y,\, e_i \cdot y) = B(e_i \cdot y,\, e_i \cdot (e_i \cdot y)) = -B(e_i \cdot y,\, y) \quad (2 \le i \le m,\ y \in V).$$

This shows $B(y,\, e_i \cdot y) = 0$, and hence $B(y,\, x \cdot y) = 0$, for $x \in U_0$, $y \in V$. We are now done by Remark 5.7. □

Assuming a classical result of Schur in group representation theory, we can prove:

Corollary 5.12. *If* 5.1 *exists over the complex field* $\mathbb{C}$*, then it already exists over the real field* $\mathbb{R}$*.*

Proof. Keep all notations in the preceding proof. Thus, U and V denote $\mathbb{R}^m$ and $\mathbb{R}^n$, C denotes the *real* Clifford algebra $C^{m-1,0}$, etc. We use a "bar" to denote complexification $\mathbb{C} \otimes_{\mathbb{R}} -$. Assume 5.1 exists over $\mathbb{C}$. Then, by 5.5, we get a representation $\sigma : \overline{C} \to \operatorname{End}_{\mathbb{C}} \overline{V}$. Since

$$\overline{B}(e_i \cdot y,\, e_i \cdot z) = \overline{\lambda}(e_i) \cdot \overline{B}(y,\, z) = \overline{B}(y,\, z),$$

for all $y,\, x \in \overline{V}$, the $\sigma(e_i)$ are (complex) orthogonal matrices. In particular, σ restricted to G is a representation of G by (complex) orthogonal matrices. By the theorem of Schur, $\sigma \,|\, G$ is equivalent to a real representation, and hence equivalent to a suitable real *orthogonal* representation. Changing σ by a conjugation if necessary, we may suppose that $\sigma(G) \subseteq \mathrm{O}(V, \varphi)$. Since G spans C as a real algebra, we conclude that $\sigma(C) \subseteq \operatorname{End}_{\mathbb{R}} V$. From the proof of 5.11, we see that, with this information, we can construct the formula 5.1 over $\mathbb{R}$. □

Remark 5.13. Let n be such that $\rho_{\mathbb{R}}(n) < \rho_{\mathbb{C}}(n) = m$. Then, the *complex* Clifford algebra $C^{m-1,0}$ has an n-dimensional $\mathbb{C}$-representation, but 5.12 shows that the formula 5.1 does not exist for m, n over $\mathbb{C}$!

There is a large amount of beautiful and important mathematics associated with the problem of composition of quadratic forms and its various generalizations, done over a period of a hundred years. Our discussion in this section has barely scratched the surface of this voluminous work. For a complete and up-to-date discussion of this, see D. Shapiro's excellent monograph [Sh].

6. Steinberg Symbols and Milnor's Group k_2F

In this section, we return to the group I^2F/I^3F and the theme of invariants on quadratic forms over a field F. The work on Witt and Hasse invariants presented in §3 turns out to be closely related to work done by J. Milnor on the k_2-group of a field F. We shall include a discussion of the k_2-group here in order to round out our exposition on the beginning part of the theory of invariants. This section is written independently of §§4–5, and can be read right after §3. In fact, the only prerequisites for this section are a knowledge of the groups $I^nF/I^{n+1}F$ for $n \leq 2$, and the definitions of the Clifford, Witt, and Hasse invariants for quadratic introduced in §3.

In Quillen's algebraic K-theory, a series of K-groups denoted by $K_n^{\text{Q}}R$ $(n \geq 0)$ is associated with any ring R. For the purposes of studying the Witt ring $W(F)$, however, another series of higher K-groups $K_n^{\text{M}}(F)$, defined by Milnor [Mi] just for fields F, turns out to be more relevant. We shall postpone the discussion on $K_n^{\text{M}}(F)$ for higher n's to Chapter X, but will focus on the cases $n \in \{0, 1, 2\}$ in this section. For these three cases, Milnor's groups (for fields F) turn out to agree with Quillen's groups, so we may dispense with the superscripts "Q" and "M" in the K-group notations introduced above.

The fact that we are only interested in the cases $n = 0, 1, 2$ here enables us to make *ad hoc* definitions. For a ring R, K_0R is the Grothendieck group of finitely generated projective modules over R, and K_1R is the Bass-Whitehead group of R. If R is a field F (of any characteristic), it is well known that $K_0F \cong \mathbb{Z}$, and $K_1F \cong \dot{F}$. For the main purposes of this section, these may be taken to be the *definitions* for the first two K-groups of the field F. In particular, if we define $k_nF := K_nF/2K_nF$, then

$$k_0F \cong \mathbb{Z}_2 \cong W(F)/IF \quad \text{and} \quad k_1F \cong \dot{F}/\dot{F}^2 \cong IF/I^2F,$$

where the isomorphisms are defined, respectively, by dimension modulo 2 and by the signed determinant. This having been said, it would be natural to ask if the next filtration factor I^2F/I^3F of the Witt ring can also be described from the viewpoint of the algebraic K-theory of fields.

In this section we shall define the group K_2F following [Mi], and show that, indeed, I^2F/I^3F is isomorphic to the group $k_2F = K_2F/2K_2F$. To do this, we start with the definition of a Steinberg symbol on a field F (of general characteristic).

Definition 6.1. A pairing $f : \dot{F} \times \dot{F} \to G$ into a multiplicative abelian group G is said to be a *Steinberg symbol* if f is bimultiplicative, and $f(a, b) = 1$ whenever $a + b = 1$. (The latter property will be referred to as the *Steinberg property*.)

In view of the universal property of the tensor product $\dot{F} \otimes \dot{F}$ (formed over $\mathbb{Z}$), we can easily manufacture a *universal Steinberg symbol.* The recipient group for this universal symbol is taken to be the quotient

$$\dot{F} \otimes \dot{F}/(\text{subgroup generated by all } a \otimes b: \ a + b = 1), \tag{6.2}$$

which we define to be K_2F. The natural pairing $\dot{F} \times \dot{F} \to K_2F$ is then a Steinberg symbol with the universal property that an *arbitrary* Steinberg symbol $f : \dot{F} \times \dot{F} \to G$ (into an abelian group G) can be obtained from a unique group homomorphism $g : K_2F \to G$ by forming the composition in the following diagram:

$$\begin{array}{ccc} \dot{F} \times \dot{F} & \longrightarrow & K_2F \\ & {\scriptstyle f} \searrow & \Big\downarrow {\scriptstyle \exists!\, g} \\ & & G \end{array}$$

We should note that the definition given for K_2F above is *not* the original definition given by Milnor in 1968 for the K_2 of a general ring R in algebraic K-theory. This latter definition involves the consideration of universal central extensions of classical groups over the ring R, and is not directly relevant to quadratic form theory. Fortunately, according to a theorem of H. Matsumoto, the K_2F we defined above (via (6.2)) is isomorphic to Milnor's original K_2-group, in the case where F is a field. Therefore, although the two definitions are different, the end results are the same.

Writing $[a, b]$ for the image of $a \otimes b$ in K_2F, we record below some "bonus" relations in K_2F. The proofs of these relations also show us the various ways in which we can exploit the axioms for the universal Steinberg symbol $(a, b) \mapsto [a, b]$.

Lemma 6.3. *In K_2F, we have the following relations*:

(1) $[a, b] = 1$ *whenever* $a + b = 0$.

(2) $[a, b] = [b, a]^{-1}$.

(3) $[a, a] = [a, -1]$, *and this is an element of order* ≤ 2.

(4) $[a, b] = [a + b, -b/a]$ *whenever* $a + b \neq 0$.

Remark 6.4. (1) is a somewhat surprising analogue of the Steinberg property (the sum of the two entries here being 0 instead of 1). (2) shows the *skew-symmetry* of symbols. (4) is a self-strengthening of the Steinberg property. Of course, from (4) alone, we also get

$$[b, a] = [b + a, -a/b] = [a + b, -b/a]^{-1} = [a, b]^{-1},$$

so (4) $\Rightarrow$ (2). However, our proof of (4) below depends on (2).

Proof of 6.3. (1) We may assume that $a \neq 1$, so $1 - a^{-1} \neq 0$. Using $1 = [a^{-1},\, 1 - a^{-1}] = [a,\, 1 - a^{-1}]^{-1}$, we have

$$[a,\, -a] = [a,\, -a]\,[a,\, 1 - a^{-1}] = [a,\, -a + 1] = 1.$$

(2) Expanding $1 = [ab,\, -ab]$ (from (1)), we get

$$[a,\, -a]\,[a,\, b]\,[b,\, a]\,[b,\, -b] = 1.$$

Using (1) again to remove the end terms, we get the desired skew-symmetry.
(3) This follows from $[a,\, a]\,[a,\, -1] = [a,\, -a] = 1$ and the fact that $-1 \in \dot{F}$ has order ≤ 2.
(4) Letting $c = a + b \neq 0$, we have $ac^{-1} + bc^{-1} = 1$, so

$$1 = [ac^{-1},\, bc^{-1}] = [a,\, b]\,[c,\, b]^{-1}[a,\, c]^{-1}[c,\, c].$$

By transposition and (2), (3), we get

$$[a,\, b] = [c,\, b]\,[c,\, a]^{-1}[c,\, -1] = [c,\, -b/a].$$ □

Consistently with k_0 and k_1, we define k_2F to be the factor group $K_2F/(K_2F)^2$. Whenever there is no confusion possible, we shall write $[a,\, b]$ again for the image of $[a,\, b]$ in k_2F. In view of 6.3(2), the symbols $[a,\, b] \in k_2F$ are now *symmetric* in the two variables.

By an easy verification, we see that

$$(a,\, b) \mapsto [a,\, b] \in k_2F$$

is a "mod 2 universal" Steinberg symbol, in the sense that any Steinberg symbol into an abelian group G with $G^2 = \{1\}$ always factors uniquely through k_2F. Two Steinberg symbols of the latter type immediately come to mind, as in the proposition below. (From here on, the assumption that $\operatorname{char} F \neq 2$ will again be in force.)

Proposition 6.5. (1) *Let* ${}_2B(F) := \{x \in B(F) \mid x^2 = 1\}$.[(2)] *Then*

$$(a,\, b) \mapsto \left(\frac{a,\, b}{F}\right) \in {}_2B(F)$$

is a Steinberg symbol into ${}_2B(F)$. *This symbol is induced by a unique group homomorphism* $\beta : k_2F \to {}_2B(F)$ *given by*

$$\beta\,[a,\, b] = \left(\frac{a,\, b}{F}\right) \in {}_2B(F).$$

[(2)]We use the notation ${}_2B(F)$ for this, since $B(F)_2$ is usually reserved for the 2-primary part of the Brauer group.

(2) $(a,\, b) \mapsto \langle 1,\, -a\rangle\langle 1,\, -b\rangle + I^3F$ *is a Steinberg symbol into* I^2F/I^3F. *(Note, of course, that* $2 \cdot I^2F/I^3F = 0$.) *This symbol is induced by a unique group homomorphism* $\alpha : k_2F \to I^2F/I^3F$ *given by*

$$\alpha\,[a,\, b] = \langle 1,\, -a\rangle\,\langle 1,\, -b\rangle + I^3F.$$

Proof. (1) is clear since $\left(\frac{a,b}{F}\right) \in {}_2B(F)$ is bimultiplicative in a, b, and it is the identity of ${}_2B(F)$ when $a + b = 1$. Under this assumption, we also have $\langle 1, -a\rangle\langle 1, -b\rangle = 0 \in W(F)$, so to prove (2), it suffices to check that the pairing defined there is also "bimultiplicative" in a, b. This follows from the following routine calculation:

$$\begin{aligned}\langle 1, -a\rangle\,\langle 1, -b\rangle + \langle 1, -a'\rangle\,\langle 1, -b\rangle &= \langle 1, -a, 1, -a'\rangle\,\langle 1, -b\rangle \\ &= \langle 1, -a, -a', aa', 1, -aa'\rangle\,\langle 1, -b\rangle \\ &= (\langle 1, -a\rangle\,\langle 1, -a'\rangle + \langle 1, -aa'\rangle)\langle 1, -b\rangle \\ &\equiv \langle 1, -aa'\rangle\,\langle 1, -b\rangle \pmod{I^3F}.\end{aligned}$$ □

Combining 6.5 with 3.3, we see that there is a commutative diagram

(6.6)
$$\begin{array}{ccc} & k_2F & \\ {}^{\alpha}\swarrow & & \searrow^{\beta} \\ I^2F/I^3F & \xrightarrow{\quad\gamma\quad} & {}_2B(F)\end{array}$$

Here, γ is the homomorphism induced on I^2F/I^3F by the Clifford invariant Γ. We have denoted this homomorphism earlier by $\overline{\gamma}$; to simplify the notation (and to emphasize the parallel between the three maps in 6.6), we shall henceforth denote it by γ. Recall that γ coincides with the Witt invariant c on I^2F/I^3F. Also, ee should note that, for reasons that will soon be clear, we have used ${}_2B(F)$ for the target of γ, instead of the subgroup $\mathrm{Quat}(F)$ of $B(F)$ generated by the classes of the quaternion algebras.

Having set up the commutative diagram (6.6), it is natural to ask for further information on the three homomorphisms α, β and γ. In this connection, the first result is the following theorem, proved by J. Milnor [Mi] in 1970.

Theorem 6.7. *The map* α *is an isomorphism.*

To prove this, we shall try to construct an inverse for α. The idea for this construction comes from the description of γ via the Hasse invariant. Following Milnor, we define the ("second") *Stiefel-Whitney class* of a quadratic form $q \cong \langle a_1, \ldots, a_n\rangle$ to be

(6.8)
$$w(q) = \prod_{i<j}\,[a_i,\, a_j] \in k_2F.$$

Milnor's notation for this quantity is $w_2(q)$, but we shall write only $w(q)$ since we will not consider the higher Stiefel-Whitney classes.

The first essential task is to show that $w(q)$ is an invariant of the quadratic form, that is, it depends only on the isometry class of q. The proof of this is based on the Chain Equivalence Theorem I.5.2, in essentially the same way as in the proof of 3.18. In fact, this latter proof shows that it would suffice to check the invariance of $w(q)$ in the case of *binary* forms. In other words, all we need to check is that, if $\langle a,\, b\rangle \cong \langle c,\, d\rangle$, then $[a,\, b] = [c,\, d] \in k_2F$. This can be done as follows. Write $c = ax^2+by^2$, where $x,\, y \in F$. If (say) $y = 0$, then $c = ax^2$ and determinant consideration gives $d = bz^2$ for some $z \in \dot{F}$. In this case,

$$[c,\, d] = [ax^2,\, bz^2] = [a,\, b] \in k_2F.$$

If $x,\, y \neq 0$, we have in k_2F:

$$\begin{aligned} [a,\, b] &= [ax^2,\, by^2] = [ax^2 + by^2,\, -by^2/ax^2] && \text{(by 3.8(4))},\\ &= [c,\, -ab] = [c,\, -cd] = [c,\, d] && \text{(by 3.8(2))}. \end{aligned}$$

Note that, just as in the case of the Hasse invariant, we have

$$(6.9) \qquad w(q_1 \perp q_2) = w(q_1)\, w(q_2)\, [d(q_1),\, d(q_2)] \in k_2F.$$

Thus, although $w(\mathbb{H}) = [1, -1] = 1$, w does not "directly" define a homomorphism on the Witt group $W(F)$: there is an obstruction due to the determinant. If we take q_1, q_2 to be in I^2F, then $d(q_i) = (-1)^{m_i}$ for $m_i = (\dim q_i)/2$, and the "error term" for multiplicativity in 6.9 will be down to $[-1,\, -1]^{m_1 m_2}$.

In order to eliminate the error term altogether, we shall resort to a device similar to that used for defining the signed determinant. For any form q of even dimension $n = 2m$, we define the "signed" Stiefel-Whitney invariant by

$$w_\pm(q) = w(q)\, [-1,\, -1]^{m(m-1)/2} = \begin{cases} w(q) & \text{if } n \equiv 0,\, 2 \pmod 8,\\ w(q)\, [-1,\, -1] & \text{if } n \equiv 4,\, 6 \pmod 8. \end{cases}$$

Thus, for forms $q \in I^2F$, $w_\pm(q)$ bears exactly the same relationship to $w(q)$ as the Witt invariant $c(q)$ did to the Hasse invariant $s(q)$: see formula (B) in 3.20. In other words, for forms in I^2F, if we think of the Stiefel-Whitney invariant "w" as a "universal version" of the Hasse invariant "s", then signed Stiefel-Whitney invariant $w_\pm$ would be a corresponding universal version of the Witt invariant.

Some of the key properties of the new invariant $w_\pm(q)$ are summarized in the following result.

Proposition 6.10. (1) $w_\pm$ *gives a well-defined group homomorphism from* I^2F *to* k_2F.

(2) $w_\pm(\langle 1,\, a\rangle\,\langle 1,\, b\rangle) = [-a,\, -b] \in k_2F$.

(3) $w_\pm(I^3F) = 1$.

Proof. (1) For $q \in I^2F$, we must check that $w_\pm(q \perp \mathbb{H}) = w_\pm(q)$. If $\dim q = 2m$, we have

$$w(q \perp \mathbb{H}) = w(q)\,[1,\, -1]\,[d(q),\, d(\mathbb{H})] = w(q)\,[-1,\, -1]^m$$

by (6.9), since $d(q) = (-1)^m$. If we write $x = w(q)$ and $y = [-1,\, -1]$ in k_2F, the following simple chart checks the well-definition of $w_\pm$ on I^2F:

m (mod 4)	$w(q)$	$w_\pm(q)$	$w_\pm(q \perp \mathbb{H})$	$w(q \perp \mathbb{H})$	$m+1$ (mod 4)
0	x	x	x	x	1
1	x	x	$x \longleftarrow$	xy	2
2	$x \longrightarrow$	xy	$xy \longleftarrow$	x	3
3	$x \longrightarrow$	xy	xy	xy	4

same

Here, the arrows indicate the places where the extra factor $y = [-1, -1]$ is needed in going from w to $w_\pm$.

After checking the well-definition of $w_\pm$, we can show easily that $w_\pm$ is a homomorphism on I^2F; we shall leave this calculation to the reader since it is completely similar to the work done in the proof of II.2.1.

(2) Since $\langle 1, a\rangle\langle 1, b\rangle = \langle 1, a, b, ab\rangle$, we have

$$\begin{aligned} w(\langle 1,\, a\rangle\,\langle 1,\, b\rangle) &= [a,\, b]\,[a,\, ab]\,[b,\, ab] \\ &= [a,\, b]\,[ab,\, -1] \\ &= [a,\, b]\,[a,\, -1]\,[b,\, -1] \\ &= [a,\, -b]\,[b,\, -1]. \end{aligned}$$

Multiplying this by $[-1,\, -1]$, we get (2).

(3) For any 4-dimensional form q of determinant 1, a routine calculation shows that $w(\langle c\rangle q) = w(q)$ for any $c \in \dot{F}$. Thus,

$$\begin{aligned} w_\pm(\langle 1,\, a\rangle\langle 1, b\rangle\langle 1, c\rangle) &= w_\pm(\langle 1,\, a\rangle\langle 1,\, b\rangle \perp \langle c\rangle\langle 1,\, a\rangle\langle 1,\, b\rangle) \\ &= w_\pm(\langle 1,\, a\rangle\langle 1,\, b\rangle) \cdot w_\pm(\langle 1,\, a\rangle\langle 1,\, b\rangle) = 1. \end{aligned}$$ □

According to 6.10, $w_\pm$ defines a group homomorphism from I^2F/I^3F to k_2F. By checking on generators, we see that this homomorphism is an inverse to α. This finally completes the proof of Milnor's Theorem 6.7.

After Milnor proved that α is an isomorphism in 1970, it became clear that the next major project would be to deal with the maps β and γ in the commutative diagram (6.6), and to decide if these maps are *also* isomorphisms. This problem was completely solved in 1981 by A. Merkurjev [Me$_1$], who proved the following spectacular result.

Theorem 6.11. *In* (6.6), γ *is an isomorphism, and therefore, in view of* 6.7, β *is also an isomorphism.*

The proof of this theorem is too long for us to present here.(3) As a compromise, we shall just make some remarks on this result and on its significance. First, the *injectivity* of $\gamma : I^2F/I^3F \to {}_2B(F)$ (or, equivalently, the injectivity of the Clifford invariant homomorphism $\overline{\Gamma} : W(F)/I^3F \to BW(F)$) may be thought of as giving a characterization of the ideal I^3F:

A form q *is in* I^3F *iff* $\dim q = 2m$, $d(q) = (-1)^m$, *and* $c(q) = 1$.

Here, $c(q)$ denotes the Witt invariant of q. The last condition $c(q) = 1$ can also be expressed in terms of the Hasse invariant if we want. As we have pointed out in 3.20(B), for forms $q \in I^2F$, $c(q)$ is just the "signed" Hasse invariant $s(q)\left(\frac{-1,-1}{F}\right)^{m(m-1)/2}$, where $m = (\dim q)/2$. Therefore, the conditions for $q \in I^3F$ can also be expressed as

$$(6.12) \qquad \dim q = 2m, \quad d(q) = (-1)^m, \quad s(q) = \left(\frac{-1,\,-1}{F}\right)^{m(m-1)/2}.$$

Before 1981, this had been a much sought after result in the theory of quadratic forms, and was proved (by direct computations) only for even-dimensional forms of dimension ≤ 12 (see Ch. XII, Exer. 1).

The *surjectivity* of $\gamma : I^2F/I^3F \to {}_2B(F)$ is equally significant. Since by 3.4 $\mathrm{im}(\gamma)$ is $\mathrm{Quat}(F)$ (the subgroup of $B(F)$ generated by the classes of quaternion algebras), the surjectivity of γ means that *any element of order* ≤ 2 *in* $B(F)$ *is expressible as a product of quaternion algebras.* This result was conjectured by A. A. Albert in the 1930s. Albert proved that a central simple F-algebra A has order ≤ 2 in $B(F)$ iff A has an F-involution,(4) but was unable to show that such an algebra is similar to a tensor product of quaternion algebras. Merkurjev's result on the surjectivity of γ amounts precisely to an affirmative answer to Albert's question. Some years later, however, Amitsur, Rowen and Tignol showed that a central simple F-algebra

(3)Different approaches to the proof can be found in [Me$_1$], [Ara$_2$], and [Wad$_2$]. See also Ina Kersten's book [Ker].

(4)This means an involution ε whose restriction to F is the identity. In the theory of algebras, such an involution ε is said to be *of the first kind.* (If the restriction of ε to F is *not* the identity, the involution ε is said to be *of the second kind.*)

with an F-involution need not be *isomorphic* to a tensor product of quaternion algebras.

Another interpretation of Merkurjev's Theorem 6.11 is that the pairing

$$(a,\, b) \mapsto \left(\frac{a,\, b}{F}\right) \in {}_2B(F)$$

is a mod 2 *universal Steinberg symbol.* Granted this result, we can describe ${}_2B(F)$ as a group with generators and relations: the generators are the quaternion algebras, and the relations are exactly the bimultiplicativity relations and the Steinberg relations. Indeed, the power of 6.11 lies in the fact that it unifies three apparently disparate objects: k_2F (a group from the algebraic K-theory of fields), I^2F/I^3F (a group arising in quadratic form theory), and ${}_2B(F)$ (a group associated with central simple algebras with F-involutions).

Furthermore, the group ${}_2B(F)$ has an important cohomological interpretation. In fact, it turns out to be isomorphic to a certain second cohomology group of the field F. In Galois cohomology theory, there is a sequence of cohomology groups (with $\mathbb{Z}_2$-coefficients) associated with the Galois group of the separable closure of F. These groups are denoted by $H^n(F,\, \mathbb{Z}_2)$; for $n \leq 2$, one has

(6.13) $\quad H^0(F,\, \mathbb{Z}_2) \cong \mathbb{Z}_2, \quad H^1(F,\, \mathbb{Z}_2) \cong \dot{F}/\dot{F}^2, \quad H^2(F,\, \mathbb{Z}_2) \cong {}_2B(F).$

On the other hand, as we have pointed out at the beginning of this section, Milnor has defined a sequence of K-groups K_nF $(n \geq 0)$ for any field F. These groups will be introduced more formally in X.6, where we shall also return to comment on the relationship between the groups K_nF, $H^n(F,\, \mathbb{Z}_2)$, and $I^nF/I^{n+1}F$, etc.

Since this section is on Steinberg symbols and Milnor's group k_2F, let us finish by mentioning a few examples of K_2F. The first and the easiest example is the K_2 of a finite field.

Example 6.14 (Steinberg). *Let F be a finite field $\mathbb{F}_q$ (of general characteristic). Then $K_2F = \{1\}$.* To see this, we fix a generator a for the cyclic group $\dot{F}$. Then any generator $[b,\, c] \in K_2F$ has the form

$$[a^m,\, a^n] = [a,\, a]^{mn} = [a,\, -1]^{mn} \in K_2F.$$

From 6.3(3), K_2F is a cyclic group of order ≤ 2 on the generator $[a,\, a]$. We are done if we can show that $[a,\, a]^r = 1$ for some odd integer r. If $\operatorname{char} F = 2$, we can take r to be $q - 1$. Now assume $\operatorname{char} F \neq 2$. Take a nonsquare element $b \in \dot{F}$. Since $\langle b,\, b\rangle$ is universal, there exists an equation $bx^2 + by^2 = 1$, where x, y are necessarily nonzero. Writing $bx^2 = a^s$ and $by^2 = a^t$, we see that s, t must be odd integers. Now the Steinberg relation gives

$$1 = [bx^2,\, by^2] = [a^s,\, a^t] = [a,\, a]^{st} \in K_2F,$$

as desired.

Since we have just computed $K_2(\mathbb{F}_q)$, a remark on the higher K-groups of a finite field would be appropriate. In Quillen's algebraic K-theory, in fact $K^{\mathrm{Q}}_{2i}(\mathbb{F}_q) = 0$ for all $i \geq 1$. On the other hand,

$$K^{\mathrm{Q}}_{2i-1}(\mathbb{F}_q) \cong \dot{\mathbb{F}}_{q^i} \cong \mathbb{Z}_{q^i-1},$$

which, of course, recaptures the fact that $K_1(\mathbb{F}_q) \cong \dot{\mathbb{F}}_q$ when $i = 1$.

Later in Chapter X, we shall introduce Milnor's higher K-groups $K_n(F) = K^M_n(F)$ (for fields) in direct generalization of the group $K_2(F)$ studied in this section. It will be seen (in X.6.11) that, for any finite field $\mathbb{F}_q$, $K_n(\mathbb{F}_q) = 0$ for any $n \geq 2$. Through this example, we see that Milnor's K-groups are not the same as Quillen's K-groups in general, making for a "different" algebraic K-theory — just for fields.

Example 6.15. *If F is an algebraically closed field, then K_2F is a divisible group.* In fact, for any element $\alpha = \prod_i [x_i,\, y_i] \in K_2F$ and any positive integer m, we can write y_i in the form z_i^m for some z_i's, thus getting

$$\alpha = \prod_i\, [x_i,\, z_i^m] = \prod_i\, [x_i,\, z_i]^m \in \big(K_2F\big)^m.$$

Example 6.16. *It is known that $K_2(\mathbb{R}) \cong \{\pm 1\} \times A$, where A is a divisible group.*[(5)] This information on $K_2(\mathbb{R})$ implies that

$$k_2(\mathbb{R}) \cong (\{\pm 1\} \times A)/A \cong \{\pm 1\}.$$

This is consistent with 6.7, since we know that $W(\mathbb{R}) \cong \mathbb{Z}$, and this implies that

$$I^2(\mathbb{R})/I^3(\mathbb{R}) \cong 4\,\mathbb{Z}/8\,\mathbb{Z} \cong \mathbb{Z}_2.$$

The divisible group A above turns out to be*uncountable*. In general, K_2 *of any uncountable field is always an uncountable group*, although we will not be able to prove it here.

Exercises for Chapter V

1. Show that the functorial map $C(-1) : C(V) \to C(V)$ associated with the isometry -1_V coincides with the main involution ν (see IV.3.7) on the (graded) Clifford algebra $C(V)$.
2. Show that, for $\alpha \in \mathrm{O}(V)$, the maps $C(\alpha)$, ν and ε (see 1.11) commute pairwise.

---

(5) We shall prove this later, more generally, for $K_n(\mathbb{R})$ $(n \geq 1)$; see X.6.6.

3. For any $z \in C(V, q)$, define $\overline{z} = \varepsilon\nu(z) = \nu\varepsilon(z)$, and $\mathrm{N}(z) = z\overline{z} \in C(V)$.
   (1) Show that "bar" is an anti-automorphism of $C(V, q)$.
   (2) If $V = \langle a,\, b\rangle$, show that the bar map on $C(V)$ coincides with the bar involution on the quaternion algebra $\left(\frac{a,\,b}{F}\right)$ (defined in III.2.1), and N coincides with the quaternionic norm.
   (3) Give an example to show that, if $\dim V \geq 3$, the map N need not be scalar-valued.
   (4) Show that the restriction of N to V coincides with the form $-q$.
   (5) If $\mathrm{N}(z)$ is a scalar, show that $\mathrm{N}(yz) = \mathrm{N}(y)\,\mathrm{N}(z)$.
4. Show that, if an element $z \in C_1(V)$ anti-commutes with every $x \in V$ (i.e., $zx = -xz$), then $z = 0$.
5. Let $\{e_1, \ldots, e_n\}$ be an orthogonal basis for $(V,\, q)$. Let α be a linear endomorphism of V. Show that if we resolve $\alpha(e_1)\cdots\alpha(e_n) \in C(V)$ into components relative to the basis $\{e_1^{i_1}\cdots e_n^{i_n} : i_j = 0, 1\}$ on $C(V)$, then the coordinate associated with $e_1\cdots e_n$ is precisely $\det(\alpha)$. In particular, if $\alpha \in \mathrm{O}(V, q)$, show that
$$C(\alpha)(e_1\cdots e_n) = \pm e_1\cdots e_n\,,$$
according as $\det(\alpha) = \pm 1$.
6. Let x be the element of $BW(F)$ represented by the Clifford algebra of the form $\langle 1\rangle$. Using the multiplication formulas in 3.9, compute the eight powers x^i $(1 \leq i \leq 8)$ in the triple notation. Show that x has order 8 iff $\left(\frac{-1,\,-1}{F}\right)$ is nonsplit. Otherwise, x has order 2 or 4 according as -1 is or is not a square in F.
7. Let q be a (regular) quadratic form of dimension < 8 over a field F. Using 3.5, show that $q \in I^3F$ iff q is hyperbolic. (This has an important generalization: see X.3.1.)
8. Show that if q has dimension n and determinant d, the following formula holds for the Stiefel-Whitney invariant:
$$w(\langle a\rangle\, q) = w(q)\,[a,\,(-1)^{n(n-1)/2}d^{n-1}] \in k_2F \quad (\forall\, a \in \dot{F}).$$
(In view of the commutative diagram in 6.6, the same formula holds for the Hasse invariant, with the symbol on the RHS replaced by the corresponding quaternion algebra.)
9. (Dieudonné) Show that $\langle a\rangle \cdot q \cong q$ implies that $a \in D\langle 1, -d_{\pm}(q)\rangle$.
10. Let f, g be forms of the same dimension over F, and let $d = d(f)\,d(g)$. For any $a \in \dot{F}$, show that
$$f \cdot \langle 1,\, a\rangle \equiv g \cdot \langle 1,\, a\rangle \pmod{I^3F}$$

iff $d \in D_F\langle 1,\ a\rangle$. (In particular, if $f \cdot \langle 1,\ a\rangle \cong g \cdot \langle 1,\ a\rangle$, then $\langle 1,\ a\rangle$ represents d. This special case is due to A. Wadsworth. For a partial converse, see Exercise 31 in Chapter VIII.)

11. Suppose f, g are four-dimensional forms of determinant 1. Show that f, g have the same Hasse invariant iff f is isometric to a scalar multiple of g.
12. Show that a quaternion F-algebra with a norm form q splits over a quadratic extension $F(\sqrt{c}\,)$ iff $q \in \langle 1,\ -c\rangle \cdot W(F)$.
13. Prove the following special case of the injectivity of the homomorphism γ in 6.6: if q_i are norm forms of quaternion algebras over F, then $\gamma(q_1 + q_2 + q_3) = 1 \in B(F) \Rightarrow q_1 + q_2 + q_3 \in I^3F$. (**Hint.** First handle the case of two norm forms; then apply III.4.8.)
14. Let F be a field, and V be a quadratic space with an anisotropic form q not representing 1. Show that if $a \in F$ and $v \in V$ are not both zero, then $a+v$ is an invertible element in the Clifford algebra $C(V, q)$. (In other words, $(F \otimes V) \setminus \{0\}$ is in the group of units of $C(V, q)$.)
15. Let F be a field in which -1 is not a sum of squares. Let n be a given positive integer, and $k = \rho_F(n)$ (see 4.8). Show that the matrix algebra $\mathbb{M}_n(F)$ contains a k-dimensional linear subspace X such that $X \setminus \{0\} \subseteq \mathrm{GL}_n(F)$. (For example, $\mathbb{M}_8(F)$ contains an 8-dimensional subspace X such that $X \setminus \{0\} \subseteq \mathrm{GL}_8(F)$.)

Chapter VI

Local Fields and Global Fields

1. Springer's Theorem for C.D.V. Fields

In this chapter, we present an elementary exposition on the theory of quadratic forms over local fields and global fields. Much of this theory has provided the motivation and techniques for the investigation of the general algebraic theory of quadratic forms over the years.

We begin by recalling the basic terminology of nonarchimedean valuated fields. By a *discretely valuated field* (or a *d.v. field* for short), we mean a field F equipped with a discrete (rank 1) Krull valuation; that is, a surjective map $v\colon \dot{K} \to \mathbb{Z}$ such that

(1) $v(ab) = v(a) + v(b)$ (for all $a, b \in \dot{F}$); and

(2) $v(a+b) \geq \min\{v(a), v(b)\}$ (for $a, b, a+b \in \dot{F}$).

Following conventional wisdom, we take $v(0) = \infty$ (if necessary). In view of (1), (2), the set

$$A = \{x \in F\colon\ v(x) \geq 0\} \quad \text{(including } 0)$$

is a subring of F, called the *valuation ring* of F. The following properties of A are easily verified:

(3) The quotient field of A is F.

(4) A has a unique maximal ideal

$$\mathfrak{p} = \{x \in F\colon\ v(x) \geq 1\} \quad \text{(including } 0),$$

which is generated by any element π such that $v(\pi) = 1$. Such an element π is determined up to a unit in A, and is called a *uniformizer* of A (or of F).

(5) The group of units of the ring A is given by

$$\begin{aligned} U = \mathrm{U}(A) &= \{x \in A\colon\ x \notin \mathfrak{p}\} \\ &= \{x \in \dot{F}\colon\ v(x) = 0\}, \end{aligned}$$

and every element $y \in \dot{F}$ can be written uniquely in the form $y = u\pi^{v(y)}$, where $u \in U$ (and π is a fixed uniformizer).

(6) The lattice of ideals of A is just

$$A \supsetneq \mathfrak{p} \supsetneq \mathfrak{p}^2 \supsetneq \cdots \supsetneq (0), \quad \text{with } \bigcap_i \mathfrak{p}^i = 0.$$

(7) (terminology only) The field $\overline{F} := A/\mathfrak{p}$ is called the *residue class field* of A (relative to the valuation v), and the projection of A onto $\overline{F}$ is expressed as a "bar" map: $a \mapsto \overline{a} = a + \mathfrak{p}$.

In commutative algebra, a local Dedekind domain is called a *discrete valuation ring* (or DVR for short). The ring A obtained from the valuation v above is such a DVR. Conversely, if $(A, \mathfrak{p})$ is a DVR (with unique maximal ideal $\mathfrak{p}$), then the ("exponential") $\mathfrak{p}$-adic valuation v on the quotient field of A satisfies the properties (1), (2) above, and A is precisely the associated valuation ring of v. Thus, the study of discretely valuated fields is essentially equivalent to the study of discrete valuation rings and their quotient fields.

Returning to the pair (F, v) introduced at the beginning of this section, we say that (F, v) is a *complete discretely valuated field* (or *c.d.v. field* for short) if F is complete with respect to v. This means that, with respect to the metric

$$d(x, y) = e^{-v(x-y)} \qquad (x, y \in F),$$

every Cauchy sequence in F converges. (Here, e is the basis of the natural logarithm, although we could have replaced it by a real number > 1.)

Note that in the above metric, two elements x and y are "close" iff $x - y$ lies in a "high power" $\mathfrak{p}^n$. A typical way of using the completeness property is the following. Suppose $x_1, x_2, \ldots$ is a sequence in A such that $x_{i+1} \equiv x_i \pmod{\mathfrak{p}^i}$ for all i. Then A contains an element x $(= \lim x_i)$ such that $x \equiv x_i \pmod{\mathfrak{p}^i}$ for all i. This can, in fact, be taken as the definition of completeness. A somewhat more abstract way of expressing the completeness property is to say that the natural map from A to the projective limit $\varprojlim A/\mathfrak{p}^i$ is a surjection (and hence an isomorphism of rings).

Let us now give some examples of c.d.v. fields. For any field k, let $F = k((t))$ be the field of formal Laurent series (in the variable t) over k.

This consists of all series of the type

$$f = \sum_{i=m}^{\infty} a_i t^i \qquad (a_i \in k \text{ and } m \in \mathbb{Z}),$$

under formal addition and multiplication. (Notice that the integer m here can be negative.) If $a_m \neq 0$ in the above, we can define $v(f)$ to be m.

It is easy to check that (F, v) satisfies the properties (1), (2) for a Krull valuation. Obviously, the valuation ring of v is $A = k[[t]]$ (the power series ring), and the residue class field $\overline{F}$ of v is isomorphic to k.

For another example, let K be any *global field*; that is, either a finite extension of $\mathbb{Q}$ (called a number field), or of $\mathbb{F}_q(t)$ (called a function field in one variable over a finite field $\mathbb{F}_q$). The completion of K under any non-archimedean valuation is a c.d.v. field that has a finite residue class field.

The purpose of the present section is to prove a fundamental result of Springer about quadratic forms over *nondyadic* c.d.v. fields (that is, c.d.v. fields (F, v) for which $\text{char}(\overline{F}) \neq 2$).[(1)] The significance of the nondyadic assumption will be seen from the following result, which is a special case of the forthcoming Hensel's Lemma.

Lemma 1.1. *Let (F, v) be a c.d.v. field, with* $\text{char}(\overline{F}) \neq 2$. *For any $u \in U$, u is a square in F (or in U) iff $\overline{u}$ is a square in $\overline{F}$.*

Proof. (Sufficiency) We shall construct a sequence $\{b_i\}$ in U such that

$$b_i^2 \equiv u \pmod{\mathfrak{p}^i} \quad \text{and} \quad b_{i+1} \equiv b_i \pmod{\mathfrak{p}^i}$$

for all $i \geq 1$. If we have such a sequence, its limit b will satisfy $b^2 - u = \lim\,(b_i^2 - u) = 0$.

The existence of b_1 is guaranteed by the hypothesis that $\overline{u} \in \dot{\overline{F}}^2$. Inductively, suppose we have constructed the element b_i as required. Let $b_{i+1} = b_i + \pi^i z$, where $z \in A$ is to be determined. If we write $b_i^2 - u = \pi^i c$ (for some $c \in A$), then

$$b_{i+1}^2 - u \equiv (b_i + \pi^i z)^2 - u \equiv \pi^i(c + 2b_i z) \pmod{\mathfrak{p}^{i+1}}.$$

Since $\text{char}(\overline{F}) \neq 2$, we have $2b_i \in U$. Choosing z such that $c + 2b_i z = \pi$, we will have $b_{i+1} \in U$, and

$$b_{i+1}^2 \equiv u \pmod{\mathfrak{p}^{i+1}}, \qquad b_{i+1} \equiv b_i \pmod{\mathfrak{p}^i},$$

as desired. □

Later, in §2, we will extend the above argument to cover the case of *dyadic* c.d.v. fields; see 2.19.

---

(1) Whenever we consider a d.v. field (F, v), the notations A, $\mathfrak{p}$, π, U, $\overline{F}$ etc. (introduced earlier) will be fixed.

Corollary 1.2. *Under the hypothesis of* 1.1, *a nonzero element* $u\pi^n$ $(u \in U,$ $n \in \mathbb{Z})$ *is a square in* F *iff* n *is even and* $\overline{u} \in \dot{\overline{F}}^2$. *In particular,* $\overline{u} = 1 \implies u \in U^2$.

It is useful to state this in the form of a short exact sequence. First, let us define a group homomorphism i from $\dot{\overline{F}}/\dot{\overline{F}}$ to $\dot{F}/\dot{F}^2$. Given $\overline{u} \in \dot{\overline{F}}$, let $u \in U$ be any lifting of $\overline{u}$. If v is another lifting, then $\overline{u/v} = 1$, so 1.2 implies that $u \in v\dot{F}^2$. The rule $i(\overline{u}) = u\dot{F}^2$ is therefore a well-defined homomorphism from $\dot{\overline{F}}/\dot{\overline{F}}^2$ to $\dot{F}/\dot{F}^2$.

Corollary 1.3. *Under the hypothesis of* 1.1, *the sequence*

$$1 \longrightarrow \dot{\overline{F}}/\dot{\overline{F}}^2 \xrightarrow{\ i\ } \dot{F}/\dot{F}^2 \xrightarrow{\ v\ } \mathbb{Z}/2\mathbb{Z} \longrightarrow 0$$

is split exact (with a splitting depending on the choice of π*).*

Proof. The exactness is clear. For the splitting, just define $\mathbb{Z}/2\mathbb{Z} \to \dot{F}/\dot{F}^2$ by sending 1 to $\pi\dot{F}^2$. □

Assuming $\operatorname{char}(\overline{F}) \neq 2$ (through the rest of this section), we shall "compute" the Witt-Grothendieck ring $\widehat{W}(F)$. First, the rule $\langle\overline{u}\rangle \mapsto \langle u\rangle$ $(u \in U)$ extends, by linearity, to a well-defined ring homomorphism from $\widehat{W}(\overline{F})$ to $\widehat{W}(F)$. This is clear by checking the relations in II.4.1 for $\widehat{W}(\overline{F})$. By abuse of notation, we denote this homomorphism again by i. Next, we define a second homomorphism $j\colon \widehat{W}(\overline{F}) \to \widehat{W}(F)$ by taking the composite

$$\widehat{W}(\overline{F}) \xrightarrow{\ i\ } \widehat{W}(F) \xrightarrow{\langle\pi\rangle\cdot} \widehat{W}(F).$$

Since $j(\mathbb{H}_{\overline{F}}) = \langle\pi\rangle\cdot\mathbb{H}_F = \mathbb{H}_F$, the pair (i, j) induces a group homomorphism

$$f\colon\ \widehat{W}(\overline{F}) \oplus \widehat{W}(\overline{F})/\mathbb{Z}\cdot(\mathbb{H}, -\mathbb{H}) \longrightarrow \widehat{W}(F).$$

The main theorem of this section is the following result from [Sp2].

Theorem 1.4 (Springer). *For any nondyadic c.d.v. field* F, *the map* f *is a group isomorphism.*

Before we prove 1.4, let us first give some of its more natural interpretations. First, by factoring out the subgroup generated by $(\mathbb{H}, 0)$ in the domain of f, we conclude the following from 1.4.

Corollary 1.5. (i, j) *induces a group isomorphism from* $W(\overline{F}) \oplus W(\overline{F})$ *to* $W(F)$.

Clearly, any 1-dimensional form over F can be written as $\langle u\rangle$ or $\langle\pi u\rangle$, where $u \in U$. Thus, an arbitrary form q over F can be written as $q_1 \perp \langle\pi\rangle q_2$, where

$$q_1 = \langle u_1, \ldots, u_r\rangle, \quad q_2 = \langle u_{r+1}, \ldots, u_n\rangle \qquad (u_i \in U).$$

Since $q = i\,\langle \overline{u}_1, \ldots, \overline{u}_r\rangle + j\,\langle \overline{u}_{r+1}, \ldots, \overline{u}_n\rangle$ in $W(F)$, the isomorphism in 1.5 implies that the classes

$$\overline{q}_1 = \langle \overline{u}_1, \ldots, \overline{u}_r\rangle \quad \text{and} \quad \overline{q}_2 = \langle \overline{u}_{r+1}, \ldots, \overline{u}_n\rangle$$

are uniquely determined in $W(\overline{F})$. These are called, respectively, the *first and second residue forms of* q. Using this terminology, we may restate 1.5 in the following form.

Corollary 1.6. *We have a group isomorphism*

$$(\partial_1, \partial_2)\colon\ W(F) \longrightarrow W(\overline{F}) \oplus W(\overline{F}),$$

where $\partial_i(q) = \overline{q}_i$ $(i = 1, 2)$. *(∂_1, ∂_2 are called the first and second residue homomorphisms.) Note that*

$$\partial_1\langle u\rangle = \langle \overline{u}\rangle, \quad \partial_1\langle \pi u\rangle = 0, \qquad \textit{and} \qquad \partial_2\langle u\rangle = 0, \quad \partial_2\langle \pi u\rangle = \langle \overline{u}\rangle,$$

for any $u \in U$. *Here,* ∂_2 *depends on the choice of* π, *but* ∂_1 *does not.*

Using the map i, we may identify $W(\overline{F})$ with the subring $i(W(\overline{F})) \subseteq W(F)$. The subgroup $(\langle\pi\rangle - 1)\cdot W(\overline{F})$ is an ideal in $W(F)$. In fact, if $u \in U$, then

$$\langle u\rangle(\langle\pi\rangle - 1)\cdot W(\overline{F}) = (\langle\pi\rangle - 1)\cdot W(F)$$

and

$$\langle\pi\rangle(\langle\pi\rangle - 1)\cdot W(\overline{F}) = (1 - \langle\pi\rangle)\cdot W(F).$$

Stating 1.5 in yet another way, we have

Corollary 1.7. $W(F)$ *is the direct sum of the ideal* $(\langle\pi\rangle - 1)\cdot W(\overline{F})$ *and the subring* $W(\overline{F})$. *In particular,* $W(\overline{F})$ *is a retract of* $W(F)$, *and* $W(F)$ *is isomorphic to the group ring* $W(\overline{F})[C]$, *where* C *is a group of order* 2.

We shall now begin the proof of 1.4. To facilitate the notations, let us write M for the quotient $\widehat{W}(\overline{F}) \oplus \widehat{W}(\overline{F})/\mathbb{Z}\cdot(\mathbb{H}, -\mathbb{H})$. Elements of the first and second copy of $\widehat{W}(F)$ will be distinguished by the subscripts 1 and 2. These elements will be taken automatically modulo the subgroup $\mathbb{Z}\cdot(\mathbb{H}, -\mathbb{H})$.

The strategy is to define a homomorphism $g\colon \widehat{W}(F) \to M$ that will be inverse to f. We shall define g on 1-dimensional forms $\langle x\rangle$ $(x \in \dot{F})$, and then use II.4.3 to check that, by linearity, g yields a group homomorphism from $\widehat{W}(F)$ to M.

Writing $x = \pi u^m$, where $u \in U$ and $m \in \mathbb{Z}$, we set $g(\langle x\rangle) = \langle \overline{u}\rangle_1 \in M$ if m is even, and $g(\langle x\rangle) = \langle \overline{u}\rangle_2 \in M$ if m is odd. Once we see that g respects the relations $(R'1)$ and $(R'2)$ in II.4.3, it is clear that f and g are mutually inverse isomorphisms, thus proving 1.4.

Suppose we change x to xy^2, where $y = \pi^n z$ $(z \in U)$. Then $xy^2 = uz^2 \cdot \pi^{m+2n}$, so clearly $g(\langle x \rangle) = g(\langle xy^2 \rangle)$. Next, assume that $x + y \neq 0$, where x, y are as above. We must verify that

$$g(\langle x \rangle) + g(\langle y \rangle) = g(\langle x+y \rangle) + g(\langle xy(x+y) \rangle). \tag{1.8}$$

Assume, without loss of generality, that $m \leq n$. Then

$$0 \neq x + y = \pi^m(u + \pi^{n-m} z).$$

First, suppose $m < n$, so that $t = u + \pi^{n-m} z \in U$, with $\overline{t} = \overline{u} \in \dot{\overline{F}}$. If m, n are both even, then the LHS of (1.8) is $\langle \overline{u} \rangle_1 + \langle \overline{z} \rangle_1$, while the RHS is

$$\langle \overline{u} \rangle_1 + \langle \overline{uzt} \rangle_1 = \langle \overline{u} \rangle_1 + \langle \overline{z} \rangle_1$$

too. Similarly, if m, n are both odd, the two sides of 1.8 are both $\langle \overline{u} \rangle_2 + \langle \overline{z} \rangle_2$. If m is even and n is odd, then the two sides of 1.8 are both $\langle \overline{u} \rangle_1 + \langle \overline{z} \rangle_2$. In the remaining case, the two sides are both $\langle \overline{u} \rangle_2 + \langle \overline{z} \rangle_1$.

Finally, suppose $m = n$. In this case, $0 \neq x + y = \pi^m(u + z)$. Write $u + z = \pi^r w$, where $w \in U$ and $r \geq 0$. Suppose m and n are both odd. Then the LHS of 1.8 is $\langle \overline{u} \rangle_2 + \langle \overline{z} \rangle_2$. If $r = 0$, we have $\overline{u} + \overline{z} = \overline{w} \neq \overline{0}$, so the RHS of 1.8 is

$$\begin{aligned}\langle \overline{w} \rangle_2 + \langle \overline{uzw} \rangle_2 &= \langle \overline{u} + \overline{z} \rangle_2 + \langle \overline{u} \cdot \overline{z}(\overline{u} + \overline{z}) \rangle_2 \\ &= \langle \overline{u} \rangle_2 + \langle \overline{z} \rangle_2.\end{aligned}$$

Assume now $r \geq 1$. Then $\overline{u} + \overline{z} = \overline{0}$, so the LHS of 1.8 is $\mathbb{H}_2$. Since $x + y = \pi^{m+r} w$, the RHS of 1.8 is $\langle \overline{w} \rangle_i + \langle \overline{uzw} \rangle_i = \mathbb{H}_i$, where $i = 1$ if $m + r$ is even, and $i = 2$ if $m + r$ is odd. But $\mathbb{H}_1 = \mathbb{H}_2 \in M$, so the two sides of 1.8 are again equal. If m and n are even, a similar argument works. $\square$

Proposition 1.9. *Let F be a nondyadic c.d.v. field.*

(1) *If $\varphi = \langle u_1, \ldots, u_r \rangle$, where $u_i \in U$, then φ is anisotropic over F iff $\overline{\varphi} = \langle \overline{u}_1, \ldots, \overline{u}_r \rangle$ is anisotropic over $\overline{F}$.*

(2) *Suppose that $\varphi = q_1 \perp \langle \pi \rangle q_2$, where $q_1 = \langle u_1, \ldots, u_r \rangle$ and $q_2 = \langle u_{r+1}, \ldots, u_n \rangle$ $(u_i \in U)$. Then q is anisotropic over F iff $\overline{q}_1$ and $\overline{q}_2$ are anisotropic over $\overline{F}$.*

Proof. (1) If $\overline{\varphi}$ is isotropic, we have

$$\langle \overline{u}_1, \ldots, \overline{u}_r \rangle \cong \langle \overline{1}, -\overline{1}, \overline{w}_3, \ldots, \overline{w}_r \rangle$$

for suitable $w_i \in U$. Clearly, φ and $\langle 1, -1, w_3, \ldots, w_r \rangle$ have the same image under the map (∂_1, ∂_2) in 1.6, so $\varphi \cong \langle 1, -1, w_3, \ldots, w_r \rangle$ is isotropic over F. Conversely, if φ is isotropic, let $\varphi \cong \varphi_a \perp \varphi_h$ be a Witt decomposition of φ over F. Applying the first residue homomorphism ∂_1, we see that

$$\overline{\varphi} = \partial_1(\varphi) = \partial_1(\varphi_a) \in W(\overline{F}),$$

so $\overline{\varphi}$ is Witt-similar to a form of smaller dimension, a contradiction.

(2) If q is anisotropic, then so are q_1 and q_2. Hence $\overline{q}_1$, $\overline{q}_2$ are anisotropic over $\overline{F}$, by (1). Conversely, assume that $\overline{q}_1$, $\overline{q}_2$ are anisotropic over $\overline{F}$, but q is isotropic over F. Writing down a Witt decomposition for q as before, we see that one of $\overline{q}_1$, $\overline{q}_2$ is Witt-similar to a form of smaller dimension—a contradiction. □

Corollary 1.10 (Hypothesis as in 1.9). *If every quadratic form of dimension $n+1$ over $\overline{F}$ is isotropic, then every quadratic form of dimension $2n+1$ over F is isotropic. If $\overline{F}$ has an anisotropic form of dimension n, then F has an anisotropic form of dimension $2n$.*

After we have introduced the notion of the *u-invariant* of a field (see Ch. XI), we will see that 1.10 above has a very natural interpretation in terms of the u-invariant; namely, it implies that the u-invariant of the nondyadic c.d.v. field F is given by two times the u-invariant of its residue class field $\overline{F}$. In particular, the foregoing conclusions apply very well to the Laurent series field $F = k((x))$ with the x-adic valuation (that is trivial on k), where the residue class field $\overline{F}$ is just isomorphic to the field k.

To show some applications of 1.9, let us work out some explicit examples of biquaternion division algebras using the result 1.9. Recall from III.4 that, for any biquaternion algebra A over a field F represented in the form $A = B \otimes_C F$, where B and C are quaternion algebras over F, there is an associated *Albert form* $q := q_B \perp \langle -1 \rangle q_C$, where q_B, q_C are respectively the norm forms on the spaces of the pure quaternions in B and C. According to Albert's Theorem III.4.8, A is a division algebra iff the Albert form q above is anisotropic over F. The following covers the various examples of biquaternion division algebras mentioned in III.4.10, III.4.11, and III.4.12.

Example 1.11. Let $F = \mathbb{R}(x,y)$, and let $B = \left(\frac{-1,-1}{F}\right)$, $C = \left(\frac{x,y}{F}\right)$. To see that $A = B \otimes_F C$ is a division algebra, we are free to replace F by the bigger field $\mathbb{R}((x))((y))$. (This is an iterated Laurent field, not to be confused with $\mathbb{R}((x,y))$, which is usually taken to mean the quotient field of the power series ring $\mathbb{R}[[x,y]]$.) It suffices to show that the Albert form q for $B \otimes C$ is anisotropic over $\mathbb{R}((x))((y))$. Viewing the latter as $K((y))$, where $K = \mathbb{R}((x))$, the Albert form

$$q_1 \cong \langle 1, 1, 1, x, y, -xy \rangle \tag{1.12}$$

has first and second residue class forms $\langle 1, 1, 1, x \rangle$ and $\langle 1, -x \rangle$, both of which are anisotropic over $K = \mathbb{R}((x))$ (again by 1.9 if we wish). Therefore, q is indeed anisotropic over $\mathbb{R}((x))((y))$.

Example 1.13. If we take, instead, $B = \left(\frac{x,-1}{F}\right)$ and $C = \left(\frac{-x,y}{F}\right)$ over $F = \mathbb{R}(x,y)$ (as in III.4.11), the corresponding Albert form will be

$$q_2 = \langle 1, -x, -x, -x, y, xy\rangle. \tag{1.14}$$

This can be shown to be anisotropic over $\mathbb{R}((x))((y))$ by the same method. Indeed, upon a scaling

$$\langle -x\rangle q_2 \cong \langle -x, 1, 1, 1, -xy, -y\rangle.$$

This form is "essentially" the form q_1 in (1.12), upon a change of variables $(x,y) \mapsto (-x,-y)$.

Example 1.15. Here, we take $B = \left(\frac{b,x}{F_0}\right)$ and $C = \left(\frac{c,y}{F_0}\right)$ over $F_0 = \mathbb{Q}(x,y)$, where $b, c \in \dot{\mathbb{Q}}$ represent two independent square classes in $\dot{\mathbb{Q}}/\dot{\mathbb{Q}}^2$ (as in III.4.12). Here the Albert form is

$$q_3 \cong \langle -b, -x, bx, c, y, -cy\rangle. \tag{1.16}$$

Over the field $K_0((y))$, where $K_0 = \mathbb{Q}((x))$, q has first and second residue class forms $\langle c, -b, -x, bx\rangle$ and $\langle 1, -c\rangle$, both of which are anisotropic over $K_0 = \mathbb{Q}\left((x)\right)$ (noting that $\langle c, -b, -x, bx\rangle$ itself has first and second residue class forms $\langle c, -b\rangle$ and $\langle -1, b\rangle$, both of which are anisotropic over $\mathbb{Q}$). It follows that q_3 is anisotropic over $\mathbb{Q}\left((x)\right)((y))$, and hence $B \otimes_{F_0} C$ is a division algebra over F_0.

Recall that, for B, C as in 1.13 and 1.15, $B \otimes C$ were the examples of biquaternion division algebras, furnished by Albert and Brauer respectively, which are *noncyclic* algebras. A general treatment of the question of which biquaternion algebras are cyclic (making full use of quadratic form theory) can be found in a paper by Leep, Tignol, and the author ([LLT]).

To conclude this section, we note that 1.11, 1.13, and 1.15 gave examples of pairs of quaternion algebras $\{B, C\}$ that are not linked. In particular, $\mathbb{R}(x,y)$, $\mathbb{Q}(x,y)$ (as well as $\mathbb{R}\left((x)\right)((y))$ and $\mathbb{Q}\left((x)\right)((y))$) are examples of *non-linked fields*. Some nontrivial examples of linked fields will be given later in this chapter.

2. Quadratic Forms over Local Fields

In this section, we shall specialize our study of c.d.v. fields to the case where the residue class fields are *finite*. The general notations

$$(F, v, A, \mathfrak{p}, \pi, U, \overline{F})$$

for a c.d.v. field F will remain in force, unless it is explicitly stated otherwise.

Definition 2.1. A c.d.v. field F (as above) is called a *local field* if the residue class field $\overline{F}$ is finite. In this case, the cardinality $|\overline{F}|$ of the residue class field will be denoted by $\mathbb{N}\,\mathfrak{p}$.

Fortuitously, local fields in the sense of 2.1 have been "classified": it is known that they are either a formal Laurent series field $k((x))$ (with the x-adic valuation that is trivial on a finite field k), or a finite algebraic extension of $\mathbb{Q}_p$ (the field of the p-adic numbers), carrying the unique extension of the canonical valuation on $\mathbb{Q}_p$. The former kind, $k((x))$, is the completion of $k(x)$ with respect to the x-adic valuation (trivial on k), and the latter kind is the completion of some number field with respect to a nonarchimedean valuation of residue characteristic p. The former kind is the "equi-characteristic" case, in that the field itself and its residue class field *both* have characteristic p (a prime), while the latter kind is the "non-equi-characteristic" case, in that the field itself has characteristic 0, whereas the residue class field has characteristic p (a prime). Local fields of the latter kind are also known as $\mathfrak{p}$-*adic fields* in the literature.

The classification of local fields described above is quite well-known. A proof of it can be found in many standard texts in algebra and number theory; we recommend, for instance, the excellent exposition in Jacobson's "Basic Algebra II" (Ch. 9, §12), where this classification is worked out in considerable detail. On the other hand, local fields, together with the fields of real numbers and complex numbers, can also be characterized as the nondiscrete locally compact (commutative) topological fields; see, for instance, W. Więsław's "Topological Fields" (Ch. 6, §2). For a very interesting sketch of the history of local field methods, see the preface to Cassels' book "Local Fields" [Ca$_2$].

The topological classification mentioned above is nice to know, but it will not be really essential for our purposes. The first classification, however, is important for us, since it clearly delineates the two specific classes of local fields we will be dealing with; namely, $k((x))$ for finite fields k, and the $\mathfrak{p}$-adic fields (that is, finite extensions of $\mathbb{Q}_p$ for prime numbers p). While the proof for this classification of local fields is not given in this book, a reasonable compromise would be to assume that the term "local fields" simply means these two specific classes of c.d.v. fields. (The fact that these fields have finite residue class fields is, of course, easily verified.) This will, then, be the position taken in this book, as far as local fields are concerned.

In treating the theory of quadratic forms over local fields, extra difficulties are often present in the "dyadic" case, namely the case where $2 \in \mathfrak{p}$, or equivalently, where the residue class field $\overline{F}$ has characteristic 2. In the case of Laurent series fields, the dyadic case corresponds to the fields $k((x))$, where k is a finite field of characteristic 2. The consideration of such local fields will have to be bypassed, on the grounds that we have not developed the theory of quadratic forms over fields of characteristic 2 in this text. As a result, the dyadic case of interest to us will be finite extensions of the field of 2-adic numbers, $\mathbb{Q}_2$.

We start our considerations with the *nondyadic* case. For nondyadic local fields, the Springer theory of quadratic forms developed in Section 1 applies. Upon recalling the simple structure theory of quadratic forms over a finite field of characteristic $\neq 2$ (II.3.5), we may specialize the results 1.3, 1.4 and 1.9 as follows.

Theorem 2.2. *Let F be a nondyadic local field, and let $u \in U$ be such that $\overline{u} \notin \dot{\overline{F}}^2$. Then,*

(1) *$\dot{F}/\dot{F}^2$ consists of four cosets, represented by 1, u, π, $u\pi$.*

(2) *Every quadratic form over F of dimension ≥ 5 is isotropic.*

(3) *There is a unique 4-dimensional anisotropic form over F, namely, the form*

$$\varphi_F := \langle 1, -u, -\pi, u\pi \rangle = \langle 1, -u \rangle \otimes \langle 1, -\pi \rangle.$$

(4) *There is a unique quaternion division algebra over F, namely, $\left(\frac{\pi, u}{F}\right)$. (In particular, F is a linked field in the sense of* III.4.*)*

(5) *Assume that $\mathbb{N}\,\mathfrak{p} \equiv 1 \pmod 4$. Then*

$$\widehat{W}(F) \cong \mathbb{Z} \oplus \mathbb{Z}_2^3,$$

and we have a ring isomorphism $W(F) \cong \mathbb{Z}_2[K]$, where K is the Klein 4-group.

(6) *Assume that $\mathbb{N}\,\mathfrak{p} \equiv 3 \pmod 4$. Then*

$$\widehat{W}(F) \cong \mathbb{Z} \oplus \mathbb{Z}_4 \oplus \mathbb{Z}_2,$$

and we have a ring isomorphism $W(F) \cong \mathbb{Z}_4[C]$, where C is the group of order 2.

Proof. (1) follows from 1.3, and (2) follows from 1.10 (since any ternary form is isotropic over $\overline{F}$). The unique anisotropic binary form over $\overline{F}$ is $\langle 1, -\overline{u} \rangle$, so (3) follows from 1.9(2). (4) is a consequence of (3), since quaternion (division) algebras are in one-one correspondence with their (anisotropic) norm forms.

For (5), (6), recall (from II.3.5) that the Witt-Grothendieck group of $\overline{F}$ is

$$\widehat{W}(\overline{F}) = \mathbb{Z} \cdot \langle \overline{1} \rangle \oplus \mathbb{Z}_2 \cdot (\langle \overline{u} \rangle - \langle \overline{1} \rangle) \cong \mathbb{Z} \oplus \mathbb{Z}_2. \tag{2.3}$$

First assume that $\mathbb{N}\mathfrak{p} \equiv 1 \pmod 4$. Then $-1 \in \dot{F}^2$, so the form $\mathbb{H}$ corresponds to the element $2\langle\overline{1}\rangle$ in the first summand. Using 1.4 to calculate $\widehat{W}(F)$, we see that $\widehat{W}(F) \cong \mathbb{Z} \oplus \mathbb{Z}_2^3$, with (direct) summands generated by

$$\langle 1\rangle, \quad \langle u\rangle - \langle 1\rangle, \quad \langle \pi\rangle - \langle 1\rangle, \quad \text{and} \quad \langle \pi u\rangle - \langle \pi\rangle.$$

To form $W(F)$, we "quotient out" $\mathbb{H} = 2\langle 1\rangle$ from the first summand, so the result is $W(F) \cong \mathbb{Z}_2^4$. Note that the elementary 2-group $W(F)$ has $\mathbb{Z}_2$-basis $\langle 1\rangle$, $\langle u\rangle$, $\langle \pi\rangle$ and $\langle \pi u\rangle$; so as a ring, $W(F)$ is isomorphic to the $\mathbb{Z}_2$-group algebra of the Klein 4-group.

Finally, assume that $\mathbb{N}\mathfrak{p} \equiv 3 \pmod 4$, in which case we may pick $u = -1$. Write $a = \langle\overline{1}\rangle$ and $b = \langle -\overline{1}\rangle - \langle\overline{1}\rangle$ for the generators of the summands in 2.3, so

$$\mathbb{H} = \langle\overline{1}\rangle + \langle -\overline{1}\rangle = 2a + b \in \widehat{W}(\overline{F}).$$

Using 1.4, we obtain, after a simple calculation

$$\widehat{W}(F) \cong \mathbb{Z} \oplus \mathbb{Z}_4 \oplus \mathbb{Z}_2,$$

with summands generated by

$$\langle 1\rangle, \quad \langle \pi\rangle - \langle 1\rangle, \quad \text{and} \quad \langle -1\rangle - \langle 1\rangle.$$

To pass to $W(F)$, we factor out

$$\langle 1\rangle + \langle -1\rangle = 2\langle 1\rangle + (\langle -1\rangle - \langle 1\rangle),$$

getting $W(F) \cong \mathbb{Z}_4^2$, with the summands generated by $\langle 1\rangle$ and $\langle \pi\rangle - \langle 1\rangle$. As a $\mathbb{Z}_4$-module, $W(F)$ has basis $\langle 1\rangle$ and $\langle \pi\rangle$. Hence, as a ring, $W(F)$ is isomorphic to the $\mathbb{Z}_4$-group algebra of the cyclic group of order 2. □

The Witt rings of nondyadic local fields described in (5), (6) in the theorem above turn out to be two of the four possible Witt rings of "nonreal" fields F with $|\dot{F}/\dot{F}^2| = 4$. (Nonreal fields are fields in which -1 is a sum of squares.) In fact, they are precisely the two with the property that $I^2F \neq 0$. This shows the special role played by the fields studied in 2.2. However, to present this characterization result here would interrupt the flow of our discussion on local fields. Therefore, we shall postpone the characterization result to an Appendix of this section, and proceed with our main line of investigation on local fields here.

Having proved the basic result 2.2, we shall record a few of its immediate corollaries. The notations in 2.2 will be fixed in the following.

Corollary 2.4. *F has exactly 16 anisotropic quadratic forms (including the 0-form).*

Corollary 2.5. (1) *If $x, y \in U$, then $\left(\frac{x,y}{F}\right)$ splits.*

(2) *If $x, y, z \in U$, then $\langle x, y, z\rangle$ is isotropic.*

(3) *If φ is any anisotropic ternary form over F, then φ does not represent $-d(\varphi)$, but represents all other (three) square classes.*

Proof. 1.9(1) implies (2), and (2) clearly implies (1). For (3), let $\varphi = \langle a,b,c\rangle$ be anisotropic. If φ represents $-abc$, then

$$(*) \qquad \langle a,b,c,abc\rangle \cong \langle 1,-1,d,e\rangle \qquad (d,e\in \dot{F}).$$

Computing determinants, we see that

$$\langle d,e\rangle \cong \langle 1,-1\rangle \cong \langle abc,-abc\rangle.$$

Cancelling $\langle abc\rangle$ in $(*)$, we get $\varphi \cong \langle 1,-1,-abc\rangle$, a contradiction.

Finally, let w be a square class of F different from $-abc$. Then

$$d(\varphi \perp \langle -w\rangle) = (-abc)w \notin \dot{F}^2.$$

From 2.2(3), we see that $\varphi \perp \langle -w\rangle$ must be isotropic, which means that φ represents w. □

Corollary 2.6. (1) *If $\mathbb{N}\mathfrak{p} \equiv 1 \pmod 4$, then $-1 \in \dot{F}^2$.*

(2) *If $\mathbb{N}\mathfrak{p} \equiv 3 \pmod 4$, then $-1 \notin \dot{F}^2$, but -1 is a sum of two squares in F.*

Proof. (1) and the first part of (2) both follow from 1.1. To see that -1 is a sum of two squares in F, it suffices to show that $\langle 1,1,1\rangle$ is isotropic, but this is a special case of 2.5(2). □

Remark 2.7. For u as in 2.2, $K_1 = F(\sqrt{u})$ is the unique unramified quadratic extension of F. The (unique) quaternion division algebra $D = \left(\frac{\pi,u}{F}\right)$ clearly splits over K_1. The two other quadratic extensions

$$K_2 = F(\sqrt{\pi}) \quad \text{and} \quad K_3 = F(\sqrt{\pi u})$$

are both ramified over F. Again, they both split D_1 too, by an easy check. Thus, *any quadratic extension $K \supset F$ splits D*. For instance, in the case of an equi-characteristic local field $F = k((x))$, where k is a finite field, u may be taken to be a nonsquare in k. Then the unramified quadratic extension of F is $K_1 = k'((x))$, where k' is the finite field $k(\sqrt{u})$, and the two other quadratic extensions are $K_2 = k((y))$ and $K_3 = k((z))$, where $y^2 = x$ and $z^2 = ux$.

The above results, essentially an offspring of the Springer theory in Section 1, provide a satisfactory and rather complete picture for the structure of quadratic forms over nondyadic local fields. Unfortunately, the arguments no longer work if $\operatorname{char}(\overline{F}) = 2$. In the dyadic case, for instance, the results 2.2(1), (5), (6), and 2.4, 2.5, 2.6 all need to be modified.

In order to cover the dyadic case as well, we must therefore develop some new tools that can be applied equally well to *both* cases. For nondyadic fields, we expect nothing new. Nevertheless, the techniques to be introduced now will reprove 2.2, 2.4, and 2.5, while these results will also serve as a guide to (or a motivation for) the following unified approach.

The key turns out to be the property 2.2(4), namely, the existence of a unique quaternion division algebra. Indeed, our first step, 2.10 below, will be precisely to establish this property for *all* local fields.

For the balance of this section, we assume that the reader is mildly familiar with the elementary notions in local number theory. We begin with two basic observations (2.8 and 2.9).

Proposition 2.8. *A local field F has a unique unramified quadratic extension K.*

(Handwaving) Proof. K has to be the splitting field of the polynomial $X^m - X$ (where $m = (\mathrm{N}\,\mathfrak{p})^2$), since the unique quadratic extension of $\overline{F}$ is the splitting field of $X^m - X$ over $\overline{F}$. □

Proposition 2.9. *Let F and K be as above. Then,*

(1) *there exists $u \in U$ such that $K = F(\sqrt{u})$ (the square class of u in F is uniquely determined).*

(2) *Any unit $x \in U$ is a norm from K.*

(3) *The quotient group $\dot{F}/\mathrm{N}_{K/F}(\dot{K})$ consists of two elements, represented by 1 and u.*

Proof. Say $K = F(\alpha)$, where $\alpha^2 = \pi^r u$ ($r \in \mathbb{Z}$, $u \in U$). Since π remains a uniformizer for K, and u remains a unit, the equation $\alpha^2 = \pi^r u \in K$ clearly shows that r is even. But then $K = F(\sqrt{\pi^r u}) = F(\sqrt{u})$, proving (1).

For (2) and (3), note that the valuation function $v'\colon \dot{K} \to \mathbb{Z}$ extends the original v, since K is unramified over F. Let σ be the F-automorphism of K (written exponentially) that takes $\sqrt{u}$ to $-\sqrt{u}$. For any $z \in \dot{K}$, we have $v'(z) = v'(z^\sigma)$. In particular,

$$v(\mathrm{N}_{K/F}(z)) = v(zz^\sigma) = v'(z) + v'(z^\sigma) = 2v'(z)$$

is always even. This implies that $\pi \notin \mathrm{N}_{K/F}(\dot{K})$. Therefore, if we prove (2), (3) follows automatically.

To prove (2), let A' be the valuation ring of the local field K, and let $\mathfrak{p}' = \mathfrak{p}A'$ be its maximal ideal. Since there is no ramification, $\overline{K} = A'/\mathfrak{p}'$ is a quadratic extension of $\overline{F}$. Let

$$\overline{\mathrm{N}}\colon \dot{\overline{K}} \longrightarrow \dot{\overline{F}} \quad \text{and} \quad \overline{\mathrm{T}}\colon \overline{K} \longrightarrow \overline{F}$$

denote, respectively, the norm map and the trace map. Both maps are surjective since $\overline{F}$ is a finite field. Given $x \in U$, we shall construct a sequence $\{b_n\}$ $(n \geq 0)$ in A', such that

$$x \equiv \mathrm{N}_{K/F}\left(\sum_{i=0}^{n-1} b_i \pi^i\right) \pmod{\mathfrak{p}^n}$$

for every $n \geq 1$. Once we have constructed $\{b_n\}$, then, setting

$$b = \lim_{n\to\infty} \sum_{i=0}^{n-1} b_i \in A',$$

we get $x = \mathrm{N}_{K/F}(b)$, as desired.

To construct $\{b_n\}$, we proceed by induction on n. For $n = 0$, there exists $b_0 \in A'$ solving the congruence $x \equiv \mathrm{N}_{K/F}(b_0) \pmod{\mathfrak{p}}$, since $\mathrm{N}_{K/F}$ induces $\overline{\mathrm{N}}$, and $\overline{\mathrm{N}}$ is onto. Inductively, suppose we have constructed

$$a_n = b_0 + b_1\pi + \cdots + b_{n-1}\pi^{n-1}$$

satisfying $x \equiv \mathrm{N}_{K/F}(a_n) \pmod{\mathfrak{p}^n}$. We must determine some $b_n \in A'$ such that

$$x \equiv \mathrm{N}_{K/F}(a_n + b_n\pi^n) \pmod{\mathfrak{p}^{n+1}}.$$

For any $b_n \in A'$, we have

$$\begin{aligned}\mathrm{N}_{K/F}(a_n + b_n\pi^n) &= (a_n + b_n\pi^n)(a_n^\sigma + b_n^\sigma\pi^n)\\ &\equiv \mathrm{N}_{K/F}(a_n) + \pi^n(b_n a_n^\sigma + a_n b_n^\sigma) \pmod{\mathfrak{p}^{n+1}}.\end{aligned}$$

Say $\mathrm{N}_{K/F}(a_n) - x = \pi^n y$, where $y \in A$. Then the required condition on b_n reads

$$\pi^n(y + \mathrm{T}_{K/F}(b_n a_n^\sigma)) \in \mathfrak{p}^{n+1},$$

where $\mathrm{T}_{K/F}$ is the trace map from K to F. Consequently, it suffices to pick $b_n \in A'$ so that

$$\mathrm{T}_{K/F}(b_n a_n^\sigma) \equiv -y \pmod{\mathfrak{p}}.$$

Such b_n exists, since $\mathrm{T}_{K/F}$ induces $\overline{\mathrm{T}}$, and $\overline{\mathrm{T}}$: $\overline{K} \to \overline{F}$ is onto. □

Having proved the proposition above, we are now in a position to determine the family of quaternion algebras over a local field. This work, however, requires a sequence of steps.

Theorem 2.10. *For an arbitrary local field F, let u be as in* 2.9(1). *Then F has a unique quaternion division algebra, namely,* $D = \left(\frac{\pi, u}{F}\right)$.

Proof. Since we have observed that π is not a norm from $K = F(\sqrt{u})$, it follows from III.2.7(8) that D is indeed a division algebra. Conversely, let E be any quaternion division algebra over F. We will show that $E \cong D$ in a number of steps.

Step 1. Define a homomorphism $w'\colon E \setminus \{0\} \to \mathbb{Z}$ by $w'(x) = v(\mathrm{N}(x))$, where N denotes the (anisotropic) norm form of E. Let d be the unique positive integer such that $w'(E \setminus \{0\}) = d\mathbb{Z}$. Since

$$w'(\pi) = v(\mathrm{N}(\pi)) = v(\pi^2) = 2,$$

d must be either 2 or 1. We may "normalize" w' by setting $w(x) = w'(x)/d$ for $x \in E \setminus \{0\}$, and (by convention) $w(0) = \infty$. Let $B = \{x \in E\colon w(x) \geq 0\}$, which is a subring of E, and let $\mathfrak{P} = \{x \in E\colon w(x) \geq 1\}$, which is a 2-sided ideal in B. Clearly, $B \cap F = A$, and

$$\begin{aligned} \mathfrak{P} \cap A &= \{x \in A\colon\ v(\mathrm{N}(x)) \geq d\} \\ &= \{x \in A\colon\ v(x) \geq d/2\} = \mathfrak{p}. \end{aligned}$$

The factor ring $B/\mathfrak{P}$ is a division ring, since $w(x) = 0 \implies w(x^{-1}) = 0$.

Step 2. We have $E = F \cdot B$. In fact, if $x \in E$, then $\mathrm{N}(\pi^m x) = \pi^{2m}\mathrm{N}(x)$. If m is sufficiently large, then $\mathrm{N}(\pi^m x) \in A$, which implies that $\pi^m x \in B$, or $x \in (1/\pi^m) \cdot B$.

Step 3. Since N maps B into A, the inner product associated with N clearly maps $B \times B$ into A. By a familiar dual basis argument, we see that B is contained in a finitely generated A-submodule of E. Since A is a discrete valuation ring, this shows that B must be a free (left) A-module (of rank 4, by Step 2). Setting $e = w(\pi) = 2/d$, we have $\mathfrak{p}B = \pi B = \mathfrak{P}^e$. Let f be the $\overline{F}$-dimension of $B/\mathfrak{P}$. Then

$$\begin{aligned} 4 &= \dim_{\overline{F}}(B/\mathfrak{p}B) = \dim_{\overline{F}}(B/\mathfrak{P}^e) \\ &= e \cdot \dim_{\overline{F}}(B/\mathfrak{P}) = ef. \end{aligned}$$

Since $d \leq 2$, we have $e \leq 2$, so f is either 2 or 4.

Step 4. We now invoke the finiteness of $\overline{F}$. By Wedderburn's Little Theorem, the finite division ring $B/\mathfrak{P}$ must be a commutative field, and therefore a simple extension of $\overline{F}$. Choose $s \in B$ so that the residue class of s generates $B/\mathfrak{P}$ over $\overline{F}$. The field $L = F(s)$ is a quadratic extension of F (by degree consideration). The quadratic minimal polynomial of s over F necessarily has coefficients in A, since $\mathrm{N}(s) = \mathrm{N}_{L/F}(s) \in A$. Thus, $e = \dim_{\overline{F}}(B/\mathfrak{P})$ must be 2 (so $e = 2$ and $d = 1$), L is a local field, with valuation ring $L \cap B$, and residue class field $\overline{L} = (L \cap B)/(L \cap \mathfrak{P})$. Since $\overline{L}$ contains the residue class of s, we conclude that $\dim_{\overline{F}} \overline{L} = 2$; i.e., L is unramified over F. From here on, we shall identify L with $F(\beta)$, where $\beta = \sqrt{u}$, and u is as in 2.9.

Step 5. Again, let σ denote the F-automorphism of L taking β to $\beta^\sigma = -\beta$. By the Skolem-Noether Theorem (IV.1.8), σ is induced by a suitable inner automorphism of E. In other words, there exists $\alpha \in E$ such that

$z^\sigma = \alpha^{-1}z\alpha$ for every $z \in L$. Since $\sigma^2 = \mathrm{Id}_L$, α^2 commutes with L. But $E = L \oplus L \cdot \alpha$, so α^2 lies in the center of E; that is, $\alpha^2 \in \dot{F}$.

Step 6. Write $\alpha^2 = \pi^m x$, where $m \in \mathbb{Z}$ and $x \in U$. In the algebra E, we have the two quantities α, β, which satisfy

$$\alpha^2 = \pi^m x \in \dot{F}, \quad \beta^2 = u \in \dot{F}, \quad \alpha\beta = -\beta\alpha.$$

Therefore, E is precisely the quaternion algebra $\left(\frac{\pi^m x, u}{F}\right)$. But x is a norm from $L = F(\sqrt{u})$ (by 2.9(2)), so, using III.2.11 and III.2.7(8),

$$E = \left(\frac{\pi^m x,\, u}{F}\right) = \left(\frac{\pi^m,\, u}{F}\right)\left(\frac{x,\, u}{F}\right) = \left(\frac{\pi^m,\, u}{F}\right) \in B(F),$$

where $B(F)$ denotes the Brauer group of F. Since E is a division algebra, we see that m must be odd, and so $E \cong \left(\frac{\pi, u}{F}\right) = D$. $\square$

Remark. The above six steps, in fact, contain essentially all the arguments needed in local class field theory to classify *arbitrary* central division algebras over local fields, due to H. Hasse.

Let us now reap the fruits of 2.9 and 2.10.

Corollary 2.11. *Keep the notations in* 2.10. *Then the (anisotropic) norm form* $\varphi_F := \langle 1, -\pi, -u, \pi u\rangle$ *of* D *is universal.*

Proof. The group $\dot{F}$ is generated by $-\pi$ and all $x \in U$. Since x is a norm from $L = F(\sqrt{u})$, $\langle 1, -u\rangle$ represents x and hence φ_F represents x. But the values of φ_F form a subgroup of $\dot{F}$, so φ_F is clearly universal. $\square$

We now come to the structure of the quadratic forms over F. Recall that "s" denotes the Hasse invariant of (quadratic) forms.

Theorem 2.12. *Let F be any local field. Then any five-dimensional form f over F is isotropic. Two forms q, q' over F are isometric iff* $\dim q = \dim q'$, $d(q) = d(q')$, *and* $s(q) = s(q')$.

Proof. Assume that f is anisotropic. "Scaling" f by an element that it represents, we may assume that $f \cong \langle 1, a, b, c, d\rangle$. Since $\langle 1, a, b\rangle$ is anisotropic, so is $\langle 1, a, b, ab\rangle$. We claim that $\langle 1, -c, -d\rangle$ must be isotropic. For, if otherwise, then $\langle 1, -c, -d, cd\rangle$ is also anisotropic, and 2.10 would imply

$$\langle 1, a, b, ab\rangle \cong \langle 1, -c, -d,\, cd\rangle,$$

which then gives the contradiction $\langle a, b, c, d\rangle \cong \langle 1, -1, -ab, cd\rangle$. Consequently, we must have $\langle c, d\rangle \cong \langle 1, cd\rangle$. Repeating this argument (twice more), we obtain

$$f \cong \langle 1, a, b, 1, cd\rangle \cong \langle 1, 1, ab, 1, cd\rangle \cong \langle 1, 1, 1, 1, abcd\rangle.$$

Again using 2.10, we have $\langle 1,1,1,1\rangle \cong \varphi_F$. But 2.11 says that φ_F is universal, so f is isotropic, a contradiction. The last assertion of the theorem follows now from V.3.25. □

Remark 2.13. Adding just one more step (explained in Exercise 1) to the above argument, one can prove the following theorem of Kaplansky (cf. XII.6.11(2)). *If F is any nonreal field (that is, $-1 \in \Sigma F^2$) that has a unique quaternion division algebra, then the conclusions of* 2.12 *hold for F.* This theorem will be further generalized in XI.6.21 and XI.6.22.

Recall that $\text{Quat}(F)$ denotes the subgroup of the Brauer group $B(F)$ generated by the classes of quaternion algebras. From 2.10, we saw that, if F is a local field, $\text{Quat}(F)$ may be identified with the multiplicative group $\{\pm 1\}$. Thus, we may write $\left(\frac{a,b}{F}\right) = -1$ if $\langle a, b\rangle$ fails to represent 1, and write $\left(\frac{a,b}{F}\right) = 1$ otherwise. This rule defines a *symmetric and bilinear* pairing

$$\dot{F}/\dot{F}^2 \times \dot{F}/\dot{F}^2 \longrightarrow \{\pm 1\},$$

the so-called *Hilbert symbol* over the local field F. Changing notations slightly, we shall write $(\ ,\ )_F$, or sometimes $(\ ,\ )_{\mathfrak{p}}$, to denote this Hilbert symbol. Our next goal is to show that this Hilbert symbol is nondegenerate.(2)

Lemma 2.14. *Keep the notations of* 2.10. *Any quadratic extension $F' \supseteq F$ splits the quaternion algebra D. In particular, the natural map* $\text{Quat}(F) \to \text{Quat}(F')$ *is the trivial homomorphism.*

Proof. If F' is unramified over F, then u has a square root in F', so F' clearly splits D. Next, suppose F' is ramified over F. Let $L' = F' \cdot L$, where $L = F(\sqrt{u})$ is the unramified quadratic extension of F. Then $L' = F'(\sqrt{u})$. We have

$$[\,\overline{L'}\colon \overline{F}\,] = [\,\overline{L'}\colon \overline{L}\,]\,[\,\overline{L}\colon \overline{F}\,] = 2\,[\,\overline{L'}\colon \overline{L}\,] \geq 2,$$

while $[\,\overline{F'}\colon \overline{F}\,] = 1$. Thus, $[\,\overline{L'}\colon \overline{F'}\,] = 2$, and L' is *unramified over* F'. Let π' be a uniformizer for F'. Then $\pi = \pi'^2 x'$, where x' is a unit in the valuation ring of F'. We have

$$(\pi,\, u)_{F'} = (\pi'^2 x',\, u)_{F'} = (x',\, u)_{F'}.$$

The latter is 1, since, by 2.9(2), x' is a norm from $F'(\sqrt{u}) = L'$. This proves that D splits over F'. □

Corollary 2.15. (1) *The form φ_F in* 2.11 *is the unique* 4*-dimensional anisotropic form over F.*

(2) In the terminology of XII.6, F is a *Hilbert field.*

(2) *If f is any anisotropic ternary form over F, then it does not represent $-d(f)$, but represents all other square classes.*

(3) $I^3F = 0$, *where IF is the ideal of even-dimensional forms in $W(F)$.*

Proof. Once we establish (1), then (2) follows by repeating the earlier proofs for 2.5(3). Let q be any 4-dimensional anisotropic form over F. *We claim that $d(q) = 1 \in \dot{F}/\dot{F}^2$.* Suppose we have proved this. By 2.12, q is universal, so we may write $q \cong \langle 1, -y, -z, yz\rangle$. Using 2.10, we conclude that $q \cong \varphi_F$. Let us now prove the claim. Suppose, on the contrary, that $d(q) = d \notin \dot{F}^2$. Let F' be the quadratic extension $F(\sqrt{d})$. The form $F' \otimes q$ has Witt invariant equal to 1 in $B(F')$, since $\mathrm{Quat}(F) \to \mathrm{Quat}(F')$ is the trivial map. By V.3.24, q must be isotropic over F, a contradiction. The last conclusion (3) follows from (1), since I^3F is generated by $\langle 1, a\rangle \cdot \varphi_F \cong \varphi_F \perp \langle -1\rangle\varphi_F \cong 4\mathbb{H}$. □

Theorem 2.16 (Nondegeneracy of the Hilbert Symbol). *If $0 \neq y \notin \dot{F}^2$, then there exists $z \in \dot{F}$ such that $(y, z)_F = -1$.*

Proof. By 2.15(2), $f = \langle -u, -\pi, u\pi\rangle$ represents $-y$, since $-y \neq -d(f) = -1$ in $\dot{F}/\dot{F}^2$. Thus there exists $z \in \dot{F}$ such that $f \cong \langle -y, -z, yz\rangle$. Adding $\langle 1\rangle$ and passing to the corresponding quaternion algebras, we obtain $(y, z)_F = (u, \pi)_F = -1$. □

The final goal of this section will be to calculate the size of $\dot{F}/\dot{F}^2$, and to determine the structure of $W(F)$ for a local field F. The nondyadic case is much easier, and has been completed in the beginning part of this section. What follows is specially tailored for the dyadic case, although it holds word for word for the nondyadic case as well. The basic idea is to refine the earlier Lemma 1.1, which was a procedure for lifting roots of polynomial equations from a residue class field to a (complete) valuation ring.

Lemma 2.17. *Let $(F, v, A, U, \dots)$ denote a c.d.v. field, its valuation, etc. Let $x \in A$, $f \in A[t]$ (polynomial ring over A), and $\mathfrak{A} \subseteq \mathfrak{p}$ an ideal in A. Suppose $f'(x) = e$ (f' = derivative of f), and $f(x) \equiv 0 \pmod{e^2\,\mathfrak{A}}$. Then there exists $y \in A$ with $y \equiv x \pmod{e\,\mathfrak{A}}$ such that*

$$f'(y) \in e \cdot U \quad \textit{and} \quad f(y) \equiv 0 \pmod{e^2\,\mathfrak{A}^2}.$$

Proof. By hypothesis, $f(x) = e^2 z$, where $z \in \mathfrak{A}$. Set $y = x - ez \equiv x \pmod{e\,\mathfrak{A}}$. By Taylor's formula,

$$\begin{aligned} f(y) &= f(x - ez) = f(x) - ezf'(x) + e^2z^2a \qquad (a \in A) \\ &= e^2z - e^2z + e^2z^2a = e^2z^2a \\ &\equiv 0 \pmod{e^2\,\mathfrak{A}^2}. \end{aligned}$$

On the other hand,

$$\begin{aligned} f'(y) &= f'(x-ez) = f'(x) - ezb \qquad (b \in A) \\ &= e(1-zb) \in e \cdot U. \end{aligned}$$
□

Let us write y_1 for the y above. Repeating the same process for y_1 (with $\mathfrak{A}^2$ replacing $\mathfrak{A}$), we obtain a certain $y_2 \in A$ with $y_2 \equiv y_1 \pmod{e\,\mathfrak{A}^2}$, such that

$$f'(y_2) \in e \cdot U \quad \text{and} \quad f(y_2) \equiv 0 \pmod{e^2\,\mathfrak{A}^4}.$$

Thus, we may construct, inductively, a sequence y_n $(n \geq 1)$ in A with $y_n \equiv y_{n-1} \pmod{e\,\mathfrak{A}^{2^{n-1}}}$, such that

$$f'(y_n) \in e \cdot U \quad \text{and} \quad f(y_n) \equiv 0 \pmod{e^2\,\mathfrak{A}^{2^n}}.$$

Now recall that $\mathfrak{A}$ is a proper ideal, i.e., $\mathfrak{A} \subseteq \mathfrak{p}$. Thus, $\{y_n\}$ is a Cauchy sequence, and has a limit $x_0 \in A$. Clearly, $x_0 \equiv x \pmod{e\,\mathfrak{A}}$, and $f(x_0) = 0$; i.e., x_0 is a root of f. Further, since U is closed in the metric topology of A, we have $f'(x_0) \in e \cdot U$. Thus, we have proved the following result, which is a version of what is called "Hensel's Lemma".

Theorem 2.18. *Keep the c.d.v. field notations in* 2.17. *Let* $x \in A$, $f \in A[t]$, *and* $\mathfrak{A} \subseteq \mathfrak{p}$. *Suppose* $f'(x) = e$, *and* $f(x) \equiv 0 \pmod{e^2\mathfrak{A}}$. *Then there exists* $x_0 \in A$ *with* $x_0 \equiv x \pmod{e\,\mathfrak{A}}$, *such that* $f(x_0) = 0$ *and* $f'(x_0) \in e \cdot U$.

Let us now deduce some consequences of 2.18. The first one is the following.

Local Square Theorem 2.19. *Keep the usual c.d.v. field notations. Given any* $a \in A$, *there exists* $b \in A$ *such that* $1 + 4\pi a = (1 + 2\pi b)^2$. *In particular,* $1 + 4\,\mathfrak{p} \subseteq U^2$. *(The reader can see easily that this generalizes* 1.1.*)*

Proof. Let $f(t) = t^2 - (1 + 4\pi a)$, and apply the above theorem with $x = 1$, $e = f'(x) = 2$, and $\mathfrak{A} = \mathfrak{p}$. We obtain a square root x_0 of $1 + 4\pi a$, such that $x_0 \equiv 1 \pmod{2\,\mathfrak{p}}$ and $2x_0 \in 2 \cdot U$. Thus, $x_0 \in U$, and it can be written in the form $1 + 2\pi b$, for some $b \in A$. □

Corollary 2.20. *Let* $c, d \in U$. *If* $c \equiv d \pmod{4\,\mathfrak{p}}$, *then* $c \in d \cdot U^2$. *In particular,* $\dot{F}^2$ *is an open subgroup of* $\dot{F}$.

Proof. We have $c^{-1}d \equiv 1 \pmod{4\,\mathfrak{p}}$. Now apply 2.19. □

For the rest of the section, we will assume that F *is a local field.* We shall also fix the two integers $q = \mathbb{N}\,\mathfrak{p}$, and $s = v(2)$ (so that $(2) = \mathfrak{p}^s$). In case F is nondyadic, then, of course, $s = 0$.

For $r \geq 1$, we shall write $U_r = 1 + \mathfrak{p}^r$. These subgroups provide a filtration

$$U \supseteq U_1 \supseteq \cdots \supseteq U_r \supseteq \cdots .$$

Clearly, $U/U_1 \cong \dot{\overline{F}}$, and each $U_i/U_{i+1} \cong$ (the additive group) $\overline{F}$. In particular, each $|U/U_r| < \infty$.

Lemma 2.21. *If $r > s$, the squaring map $\sigma(a) = a^2$ sends U_r onto U_{r+s}.*

Proof. If $x \in \mathfrak{p}^r$, we have

$$(1+x)^2 = 1 + 2x + x^2 \in 1 + \mathfrak{p}^{r+s} + \mathfrak{p}^{2r} = 1 + \mathfrak{p}^{r+s},$$

since $r > s$. Conversely, take any $z \in U_{r+s}$, i.e., $z \equiv 1 \pmod{2\,\mathfrak{p}^r}$. We want to solve for a root for $f(t) = t^2 - z$. Apply 2.19 with $x = 1$, $e = f'(x) = 2$, and $\mathfrak{A} = \mathfrak{p}^{r-s}$. Note that

$$e^2\,\mathfrak{A} = 2\,\mathfrak{p}^s \cdot \mathfrak{p}^{r-s} = 2\,\mathfrak{p}^r.$$

so the initial condition $f(x) \equiv 0 \pmod{e^2\,\mathfrak{A}}$ is fulfilled. Thus, 2.19 gives a square root y for z such that $y \equiv 1 \pmod{e\,\mathfrak{A}}$. But $e\,\mathfrak{A} = 2\,\mathfrak{p}^{r-s} = \mathfrak{p}^r$, so $y \in U_r$. □

Theorem 2.22. *Let F be a local field, with $q = \mathbb{N}\,\mathfrak{p} = |\overline{F}|$, and $s = v(2)$. Then $|\dot{F}/\dot{F}^2| = 4 \cdot q^s$. (In particular, if F is nondyadic, we have $s = 0$ and $|\dot{F}/\dot{F}^2| = 4$, as obtained previously in* 1.3.*)*

Proof. It suffices to show that $|U/U^2| = 2 \cdot q^s$, since we have an exact sequence

$$1 \longrightarrow U/U^2 \longrightarrow \dot{F}/\dot{F}^2 \longrightarrow \mathbb{Z}/2\mathbb{Z} \longrightarrow 0.$$

Pick any integer $r > s$. Then, $2 \notin \mathfrak{p}^r$, and so $-1 \notin U_r$. By 2.21, the squaring map σ defines an exact sequence

$$1 \longrightarrow \{\pm 1\} \longrightarrow U/U_r \xrightarrow{\ \sigma\ } U^2/U_{r+s} \longrightarrow 1,$$

where all three terms are finite groups. We have therefore

$$\begin{aligned}[U : U^2] &= \frac{[U : U_{r+s}]}{[U^2 : U_{r+s}]} = \frac{[U : U_{r+s}]}{[U : U_r]/2} \\ &= 2\,[U_r : U_{r+s}] = 2q^s,\end{aligned}$$

where the last equation follows from $U_r \supseteq U_{r+1} \supseteq \cdots \supseteq U_{r+s}$, whose filtration factors are all $\cong \overline{F}$. □

Corollary 2.23. *If F is a finite extension of degree n over $\mathbb{Q}_2$ (= the 2-adic numbers), then $|\dot{F}/\dot{F}^2| = 2^{n+2}$.*

Proof. Since 2 is a uniformizer for $\mathbb{Q}_2$, $s = v(2)$ is precisely the ramification index e of F over $\mathbb{Q}_2$. Further, $q = |\overline{F}| = 2^f$, where $f = [\overline{F} : \overline{\mathbb{Q}}_2]$. From elementary field theory, we have $e \cdot f = n$. Consequently,

$$|\dot{F}/\dot{F}^2| = 4 \cdot (2^f)^e = 2^{2+ef} = 2^{2+n}.$$

□

Corollary 2.24. *Let $F = \mathbb{Q}_2$. Then,*

(1) $|\dot{F}/\dot{F}^2| = 8$ *and* $U^2 = U_3$. *In particular, a 2-adic unit* $x \in U$ *is a square in* $\mathbb{Q}_2$ *iff* $x \equiv 1 \pmod{8A}$.

(2) *The set* $\{-1, 2, 5\}$ *forms a* $\mathbb{Z}_2$*-basis for* $\dot{F}/\dot{F}^2$.

(3) *The unique unramified quadratic extension over* F *is* $\mathbb{Q}_2(\sqrt{5})$.

(4) *The unique quaternion division algebra corresponds to* $(2,5)_2 = (-1,-1)_2 = -1$.

Proof. That $|\dot{F}/\dot{F}^2| = 8$ follows from 2.23. Since $\overline{F} = \mathbb{F}_2$, clearly, $U = U_1$. Thus, we have $U = \{\pm 1\} \times U_2$ and so $U^2 = U_2^2$. By 2.21 (with $r = 2$), we have $U^2 = U_3$, proving (1).

Next, note that U_1/U_2 and U_2/U_3 are both (cyclic) of order 2, generated by -1 and 5. Thus, U_1/U_3 is the Klein 4-group, with $\mathbb{Z}_2$-basis $\{-1, 5\}$. Throwing in the uniformizer, we get the desired basis for $\dot{F}/\dot{F}^2$.

To prove (3), note that the "golden ratio" $(1 + \sqrt{5})/2$ satisfies the integral equation $x^2 - x - 1 = 0$ over A. Thus the residue class field of $L = \mathbb{Q}_2(\sqrt{5})$ has degree 2 over $\mathbb{F}_2$, i.e., L is *the* unramified quadratic extension of $\mathbb{Q}_2$. Hence we have $(2, 5)_F = -1$, by 2.10. By (1), we know that -3 and 5 are in the same square class, so $(2,5)_2 = (2,-3)_2$. To get $(-1,-1) = -1$, it suffices to show that $\langle 1, -2, 3, -6\rangle \cong 4\langle 1\rangle$. This follows easily, since $\langle -2, 3\rangle \cong \langle 1, -6\rangle$, and

$$\langle -6, -6\rangle \cong \langle 10, -6\rangle \cong \langle 4, 4\rangle \cong \langle 1, 1\rangle. \qquad \square$$

We shall now calculate the 2-adic Hilbert symbol $(a,b)_2$ for all $a, b \in \dot{\mathbb{Q}}_2$. First, let us introduce the two important arithmetic functions $\varepsilon(x)$ and $\omega(x)$, defined for $x \in U$ (= 2-adic units). The definitions are

(2.25)
$$\varepsilon(x) = \overline{(x-1)/2} \in \mathbb{F}_2 \quad \text{and} \quad \omega(x) = \overline{(x^2-1)/8} \in \mathbb{F}_2 \quad (\text{for } x \in U).$$

These are meaningful since $x \equiv 1 \pmod{2A}$ and $x^2 \equiv 1 \pmod{8A}$ for all $x \in U$. For $x, y \in U$, we have

$$(xy - 1) - (x - 1) - (y - 1) = (x - 1)(y - 1) \in 4A$$

and

$$(x^2y^2 - 1) - (x^2 - 1) - (y^2 - 1) = (x^2 - 1)(y^2 - 1) \in 64A.$$

So $\varepsilon(xy) = \varepsilon(x) + \varepsilon(y)$, and $\omega(xy) = \omega(x) + \omega(y)$ in $\mathbb{F}_2$; in other words, ε and ω are homomorphisms $U \to \mathbb{F}_2$. They clearly vanish on U_3, so we obtain a homomorphism

$$(\varepsilon,\, \omega)\colon\ U/U_3 \longrightarrow \mathbb{F}_2 \oplus \mathbb{F}_2.$$

Since -1 goes to $(1,0)$ and 5 goes to $(0,1)$, we have proved the following.

Corollary 2.26. (ε, ω): $U/U_3 \cong \mathbb{F}_2 \oplus \mathbb{F}_2$ *(for* $F = \mathbb{Q}_2$*).*

Now we can compute $(a,b)_2$ quite explicitly.

Theorem 2.27. *Let* $F = \mathbb{Q}_2$*, and* $a, b \in \dot{F}$*. Write* $a = 2^j x$*,* $b = 2^J y$ $(x, y \in U;\ j, J \in \mathbb{Z})$*. Further, write* $x = (-1)^i 5^k x_0$*,* $y = (-1)^I 5^K y_0$ *(where* $x_0, y_0 \in U^2$*, and* $i, I, k, K \in \mathbb{Z}$*). In other words,* $a = (-1)^i 2^j 5^k x_0$*, and* $b = (-1)^I 2^J 5^K y_0$*. Then*

$$(a,b)_2 = (-1)^{iI+jK+kJ} = (-1)^{\varepsilon(x)\varepsilon(y)+j\omega(y)+J\omega(x)}.$$

Proof. First, x_0 and y_0 may be ignored, since they are squares. Next, note that $(-1,2)_2$ and $(-1,5)_2$ are both 1, since the corresponding norm forms are clearly isotropic. By the bilinearity, we obtain immediately

$$(a,\ b)_2 = (-1)^{iI}(2^j 5^k,\ 2^J 5^K)_2,$$

since we have seen already (in 2.24) that $(-1,-1)_2 = -1$. By III.2.6, we have $(2,2)_2 = (-1,2)_2 = 1$ and $(5,5) = (-1,5)_2 = 1$, so $(a,b)_2$ boils down to

$$(-1)^{iI}(2,\ 5)_2^{jK+kJ} = (-1)^{iI+jK+kJ},$$

proving the first equality in 2.27. The second follows immediately, since the isomorphism in 2.26 shows that $\varepsilon(x) = i$, $\varepsilon(y) = I$, and $\omega(x) = k$, $\omega(y) = K$. □

Corollary 2.28. *If* x, y *are 2-adic units* $(x, y \in U)$*, then*

$$(x,\ y)_2 = (-1)^{\varepsilon(x)\varepsilon(y)} \quad \textit{and} \quad (2,\ y)_2 = (-1)^{\omega(y)}.$$

Example. Calculate $(-120, 38)_2$. We have $a = -120 = 8 \cdot (-15)$, and $b = 38 = 2 \cdot 19$. Further, $\varepsilon(-15) = \overline{-8} = 0$, $\varepsilon(19) = \overline{9} = 1$; $j = v(a) = 3$, $J = v(b) = 1$; and finally, $\omega(-15) = \overline{28} = 0$, $\omega(19) = \overline{45} = 1$. Thus $(a,b)_2 = (-1)^{0+3+0} = -1$. This means that $-120x^2 + 38y^2 = 1$ *cannot be solved in* $\mathbb{Q}_2$!

Remark. A similar (but much simpler) calculation yields the Hilbert symbol for nondyadic fields in terms of the usual "Legendre symbols" (see Exercise 10).

We conclude this section by determining the additive structure of the Witt ring $W(F)$, for F a local field. We may restrict ourselves to the *dyadic* case, since the nondyadic case was already settled in 2.2(6).

It turns out that, again, we have to consider three different cases, as in V.4; namely,

Case 1. $-1 \in \dot{F}^2$;

Case 2. $-1 \notin \dot{F}^2$, but -1 is a sum of two squares;

Case 3. -1 is not a sum of two squares.

In the following computation of $W(F)$ in these cases, $\mathbb{Z}_n^k$ denotes the direct sum of k copies of $\mathbb{Z}_n = \mathbb{Z}/n\mathbb{Z}$.

Theorem 2.29. *Let F be a dyadic local field, with $|\dot{F}/\dot{F}^2| = 2^m$ (m can be determined by 2.22, which shows that $m \geq 3$). Then:*

Case 1 $\implies W(F) \cong \mathbb{Z}_2^{m+2}$.

Case 2 $\implies W(F) \cong \mathbb{Z}_4^2 \oplus \mathbb{Z}_2^{m-2}$.

Case 3 $\implies W(F) \cong \mathbb{Z}_8 \oplus \mathbb{Z}_2^{m-1}$.

Proof. Recall that IF, the ideal of even-dimensional forms in $W(F)$, is generated as a group by all $\langle 1, a\rangle$ ($a \in \dot{F}$). Its square I^2F is therefore generated by the norm forms of quaternion division algebras. Thus, $I^2F = \mathbb{Z} \cdot \varphi_F$, where φ_F is as in 2.11. But by 2.15, $\varphi_F \cong -\varphi_F$, so 2 annihilates φ_F in $W(F)$. This yields $I^2F \cong \mathbb{Z}_2$. The filtration $0 \subseteq I^2F \subseteq IF \subseteq W(F)$ together with II.2.3 implies that $|W(F)| = 2^{m+2}$.

Suppose, first, $-1 \in \dot{F}^2$. Then $\langle 1, 1\rangle \cong \mathbb{H}$, so $W(F)$ is a ring of characteristic 2, and hence $W(F) \cong \mathbb{Z}_2^{m+2}$. Next, suppose we are in Case 2. Then, $\langle 1, 1\rangle$ represents -1, and so $\langle 1, 1\rangle \cong \langle -1, -1\rangle$, showing that $W(F)$ has characteristic 4. Write

$$W(F) \cong \mathbb{Z}_4^a \oplus \mathbb{Z}_2^b.$$

A cardinality count shows that $m + 2 = 2a + b$. Multiplying by 2, we get $2W(F) \cong \mathbb{Z}_2^a$. But $W(F) = (IF) \cup (1 + IF)$, so

$$2W(F) = (2IF) \cup (2 + 2IF).$$

Now $2IF \subseteq I^2F$, and, by the nondegeneracy of the Hilbert symbol (2.16), the quaternion division algebra over F can be written as $\left(\frac{-1, y}{F}\right)$ (for some $y \in \dot{F}$). The latter shows that $\varphi_F \cong \langle 1, 1, y, y\rangle \in 2IF$, so we conclude that $2IF = I^2F$. Consequently, $2W(F) = \mathbb{Z}_2 \cdot \varphi_F \oplus \mathbb{Z}_2 \cdot 2$, so $a = 2$. Solving $m + 2 = 2a + b$, we get $b = m - 2$.

Finally, suppose are are in Case 3. Then, $4\langle 1\rangle$ must be anisotropic (otherwise, $\langle 1, 1\rangle \cong \langle -1, -1\rangle$), so $4\langle 1\rangle \cong \varphi_F$, and $W(F)$ has characteristic 8. Write

$$W(F) \cong \mathbb{Z}_8^a \oplus \mathbb{Z}_4^b \oplus \mathbb{Z}_2^c,$$

where $m + 2 = 3a + 2b + c$. We have here

$$2W(F) = (I^2F) \cup (2 + I^2F) = \mathbb{Z}_4 \cdot 2 \cong \mathbb{Z}_4.$$

This yields $a = 1$ and $b = 0$, by inspection. Solving for c, we get $c = m - 1$. □

Corollary 2.30. *If* $[F: \mathbb{Q}_2] = n$, *then* F *has* 2^{n+4} *anisotropic forms (including the* 0*-form).*

Remark 2.31. Consider the case $F = \mathbb{Q}_2$. Here, $(-1,-1)_2 = -1$ (by 2.27), so we are in "Case 3", and $W(F) \cong \mathbb{Z}_8 \oplus \mathbb{Z}_2^2$. We claim that *the generators can be taken to be*

$$\varepsilon = \langle 1 \rangle, \quad \alpha = \langle 1, -2 \rangle, \quad \text{and} \quad \beta = \langle 1, -5 \rangle.$$

Indeed, since $\langle 1, -2 \rangle \cong \langle -1, 2 \rangle$ and $\langle 1, -5 \rangle \cong \langle -1, 5 \rangle$, we know that α, β are of order 2 in $W(\mathbb{Q}_2)$, and, of course, ε has order 8. Thus, it suffices to show that ε, α, β generate $W(F)$ as a group. The span W' of ε, α, β already contains $\pm\langle 1 \rangle$, $\pm\langle 2 \rangle$ and $\pm\langle 5 \rangle$, so it suffices to see that $\langle 10 \rangle \in W'$. But 2.24(4) implies that $4\langle 1 \rangle \cong \langle 1, -2, -5, 10 \rangle$, so transposition yields the desired result. The ring structure on $W(\mathbb{Q}_2)$ can be easily determined too. Indeed, the last isometry gives $\alpha\beta = 4\langle 1 \rangle$, and, of course,

$$\alpha^2 = 2\alpha = 0 \quad \text{and} \quad \beta^2 = 2\beta = 0.$$

These rules completely determine the multiplication in $W(\mathbb{Q}_2)$, and consequently $W(\mathbb{Q}_2)$ is isomorphic to the following ring:

$$\mathbb{Z}_8\,[s,t]/(2s,\ 2t,\ s^2,\ t^2,\ st-4).$$

In a similar spirit, one can try to determine the multiplicative structure of $W(F)$ for a general dyadic field F. We shall leave the computations to the reader.

Let us now offer some worked examples on quadratic forms over $\mathbb{Q}_p$.

Example 2.32. *Over which p-adic fields* $\mathbb{Q}_p$ *is the form* $f = \langle 5, -1, -3 \rangle$ *isotropic?* If q is a prime $\neq 2$, 3, 5, then f is isotropic over $\mathbb{Q}_q$, by 2.5(2). Over $\mathbb{Q}_3$, f has first and second residue forms $\langle 5, -1 \rangle$ and $\langle -1 \rangle$ in $W(\mathbb{F}_3)$, both of which are anisotropic. Thus, f is anisotropic over $\mathbb{Q}_3$, by 1.9(2). Similarly, it can be checked that f is anisotropic over $\mathbb{Q}_5$. Over $\mathbb{Q}_2$, f has Hasse invariant

$$s(f) = (5,-1)_2(5,-3)_2(-1,-3)_2 = 1,$$

while $(-1, -d(f))_2 = (-1, -15)_2 = 1$, also. Thus, by V.3.22, f is isotropic over $\mathbb{Q}_2$. [Alternatively, note that $\langle 5, -1 \rangle$ represents $5 \cdot 2^2 - (\sqrt{17})^2 = 3$ over $\mathbb{Q}_2$.]

Example 2.33. *Over which p-adic fields* $\mathbb{Q}_p$ *are the forms*

$$f = \langle 1, -6, 15 \rangle \quad \text{and} \quad g = \langle 3, -10, 3 \rangle$$

isometric? To begin with, they both have determinant $-2 \cdot 3^2 \cdot 5$. If q is a prime $\neq 2$, 3, 5, then f, g are both isotropic over $\mathbb{Q}_q$, and hence isometric. Over $\mathbb{Q}_3$, f has second residue form $\langle -2, 2 \rangle \cong \mathbb{H}$ over $\mathbb{F}_3$, while g has second residue form $\langle 1, 1\rangle$. Thus $f \not\cong g$ over $\mathbb{Q}_3$ because f is isotropic but g is not. Over $\mathbb{Q}_5$, f, g both have first residue form $\cong \mathbb{H}$ over $\mathbb{F}_5$, so f, g are isotropic over $\mathbb{Q}_5$, hence isometric. Over $\mathbb{Q}_2$, $s(f) = 1$ and $s(g) = -1$ by a simple calculation, so $f \not\cong g$ over $\mathbb{Q}_2$.

Example 2.34. *Over which p-adic fields $\mathbb{Q}_p$ does $f = \langle 2, 3, 21\rangle$ represent 1?* Certainly over those $\mathbb{Q}_q$ where $q \neq 2, 3, 7$, for f becomes isotropic. Over $\mathbb{Q}_3$, f is anisotropic, with determinant -1, hence f cannot represent 1 (by 2.5(3)). Over $\mathbb{Q}_7$, f is isotropic and certainly represents 1. Over $\mathbb{Q}_2$, $s(f) = 1$ which is the same as $(-1, -d(f))_2$, so f is isotropic over $\mathbb{Q}_2$ (by V.3.22) and represents 1.

Appendix: Nonreal Fields with Four Square Classes

As an afterthought on this section, we shall solve a small classification problem for the Witt ring. In Ch. 2, §5, we mentioned the problem of classifying Witt rings of nonreal fields with four square classes. Having studied the Witt rings of $\mathfrak{p}$-adic fields in this section, we are now in a good position to complete the solution of this particular classification problem.

Theorem 2.35. *Let F be a nonreal field with $|\dot{F}/\dot{F}^2| = 4$. Then $W(F)$ is isomorphic to one of the following rings:*

$$\frac{\mathbb{Z}_2[s,t]}{(s^2,\, t^2,\, st)},\quad \frac{\mathbb{Z}_4[t]}{(2t,\, t^2)},\quad \mathbb{Z}_2[K],\quad \mathbb{Z}_4[C],$$

where C is the cyclic group of order 2, and K is the Klein 4-group. (And all four possibilities arise.)

Proof. To classify $W(F)$, we consider separately the cases $I^2F = 0$ and $I^2F \neq 0$.

Case 1. $I^2F = 0$. In this case, we have (by II.2) a ring isomorphism $W(F) \cong Q(F)$, where $Q(F)$ is the extended square class group, made into a (commutative) ring as in II.2. Two possible rings arise in this way, depending on whether $-1 \in \dot{F}^2$. Actually, even without using the isomorphism $W(F) \cong Q(F)$, it is quite easy to work out the two possibilities directly. *First assume* $-1 \in \dot{F}^2$. Let $\{a, b\}$ be a $\mathbb{Z}_2$-basis for $\dot{F}/\dot{F}^2$. Since $\langle 1, a, b, ab\rangle = 0 \in W(F)$, $W(F)$ is additively generated by $\langle 1\rangle$, $\alpha = \langle 1, -a\rangle$, and $\beta = \langle 1, -b\rangle$. We have

$$2 = 0 = \alpha^2 = \beta^2 = \alpha\beta \in W(F),$$

so clearly $W(F)$ is isomorphic to the ring $\mathbb{Z}_2[s,t]/(s^2, t^2, st)$. *Next, assume* $-1 \notin \dot{F}^2$. Let $\{-1, b\}$ be a $\mathbb{Z}_2$-basis for $\dot{F}/\dot{F}^2$. We have $2 \neq 0 \in W(F)$, but

$4 \in I^2F = 0$. Let $\beta = \langle 1, -b \rangle$, which has order 2 since $2\beta \in I^2F = 0$. Since $\beta \notin \mathbb{Z} \cdot \langle 1 \rangle = \mathbb{Z}_4 \cdot \langle 1 \rangle$, we have

$$W(F) = \mathbb{Z}_4 \cdot \langle 1 \rangle \oplus \mathbb{Z}_2 \cdot \beta.$$

Using also the information $\beta^2 \in I^2F = 0$, we see that $W(F)$ is isomorphic as a ring to $\mathbb{Z}_4\,[t]/(2t,\, t^2)$.

Case 2. $I^2F \neq 0$. We have again two cases, depending on whether $-1 \in \dot{F}^2$.

Subcase A. $-1 \in \dot{F}^2$. Fix a nonsplit quaternion algebra $\left(\frac{a,b}{F}\right)$. Clearly, $\{a, b\}$ forms a basis for $\dot{F}/\dot{F}^2$ (since $\left(\frac{a,\,a}{F}\right) \cong \left(\frac{a,-1}{F}\right)$ splits). We have $\langle 1, a, b, ab \rangle \neq 0 \in W(F)$, so

$$W(F) = \mathbb{Z}_2\langle 1 \rangle \oplus \mathbb{Z}_2\langle a \rangle \oplus \mathbb{Z}_2\langle b \rangle \oplus \mathbb{Z}_2\langle ab \rangle.$$

As a ring, $W(F)$ is clearly isomorphic to $\mathbb{Z}_2[K]$, where K is the Klein 4-group $\{1, \langle a \rangle, \langle b \rangle, \langle ab \rangle\}$.

Subcase B. $-1 \notin \dot{F}^2$. Fix an integer s such that $-1 \in D(s\langle 1 \rangle)$. Then $s\langle 1 \rangle \perp \langle 1 \rangle$ is isotropic, so $D((s+1)\langle 1 \rangle) = \dot{F}$. If $D(2\langle 1 \rangle) = \dot{F}$ already, we can take a basis $\{-1, a\}$ for $\dot{F}/\dot{F}^2$ and argue easily that every quaternion algebra splits (e.g. $\left(\frac{a,\,a}{F}\right) \cong \left(\frac{a,-1}{F}\right) \cong \mathbb{M}_2(F)$). Since this is not the case, we must have $D(2\langle 1 \rangle) \subsetneq \dot{F}$. Therefore, $D(2\langle 1 \rangle) \subsetneq D(4\langle 1 \rangle)$ (for otherwise $D(2\langle 1 \rangle) = D((s+1)\langle 1 \rangle) = \dot{F}$), and of course $D(2\langle 1 \rangle) \neq \dot{F}^2$. Since $D(4\langle 1 \rangle)$ is a group, we must have $D(4\langle 1 \rangle) = \dot{F}$. *We claim that* $-1 \in D(2\langle 1 \rangle)$. Indeed, write $-1 = x^2 + y^2 + z^2 + w^2$. We may assume that $x^2 + y^2 \neq 0 \neq z^2 + w^2$. Since $D(2\langle 1 \rangle)$ has only one nontrivial square class, we must have $z^2 + w^2 = r^2(x^2 + y^2)$ for some r, and hence

$$-1 = (x^2 + y^2)(1 + r^2) \in D(2\langle 1 \rangle),$$

as claimed. Now let $\{-1, c\}$ be a basis for $\dot{F}/\dot{F}^2$. Then $W(F) = \mathbb{Z}\langle 1 \rangle + \mathbb{Z}\langle c \rangle$. Since $\langle 1, 1 \rangle \cong \langle -1, -1 \rangle$, $\langle 1 \rangle$ has additive order 4 in $W(F)$. This implies that $\langle c \rangle$ also has order 4 (as $-1 \notin \dot{F}^2$). Now the sum $\mathbb{Z}\langle 1 \rangle + \mathbb{Z}\langle c \rangle$ must be direct, for otherwise we would have $\langle 1, 1 \rangle \cong \langle c, c \rangle$, which is not the case. The decomposition

$$W(F) = \mathbb{Z}_4\langle 1 \rangle \oplus \mathbb{Z}_4\langle c \rangle$$

shows now that $W(F) \cong \mathbb{Z}_4\,[C]$ where $C = \{\langle 1 \rangle, \langle c \rangle\}$ is a group of order 2.

To complete the proof of 2.35, we must show that all four subcases considered above do arise. The two subcases under Case 2 present no problems, since we can take F to be a nondyadic $\mathfrak{p}$-adic field, with $-1 \in \dot{F}^2$ if $\mathbb{N}\mathfrak{p} \equiv 1 \pmod 4$, and $-1 \notin \dot{F}^2$ if $\mathbb{N}\mathfrak{p} \equiv 3 \pmod 4$. (Alternatively, we could also take $F = \mathbb{F}_5((x))$ and $F = \mathbb{F}_3((x))$ respectively.) It remains then only to realize the two subcases in Case 1.

To produce an F with $-1 \in \dot{F}^2$, we start with $\mathbb{Q}$ and the basis -1, 2, 3, 5, 7, ... for $\dot{\mathbb{Q}}/\dot{\mathbb{Q}}^2$. We go to $K = \mathbb{Q}(\sqrt{-1}, \sqrt{5}, \sqrt{7}, \dots)$ whose square class group has basis 2, 3, ..., as in the construction in II.5.3. Iterating this construction and taking the ascending union F of the fields obtained, we get $\sqrt{-1} \in F$, and $\dot{F}/\dot{F}^2$ has basis $\{2, 3\}$. Since $2 + 3 = 5 \in \dot{K}^2 \subseteq \dot{F}^2$, it is easy to see that all quaternion algebras split over F, so $I^2F = 0$. To produce an F with $-1 \notin \dot{F}^2$, we start instead with the basis $\{-1, 2, -3, -5, -7, \dots\}$ for $\dot{\mathbb{Q}}/\dot{\mathbb{Q}}^2$ and form $L = \mathbb{Q}(\sqrt{-3}, \sqrt{-5}, \dots)$. Repeating this construction and taking the ascending union (again as in II.5.3), we get a field F with a square class basis $\{2, -1\}$ (so $-1 \notin \dot{F}^2$), and F is nonreal since $-1 = 2^2 + (\sqrt{-5})^2$. This equation implies that $\left(\frac{-1,-1}{F}\right)$ splits. Since $\left(\frac{2,2}{F}\right)$ also splits, it follows easily that all quaternion algebras over F split, and therefore $I^2F = 0$, as desired. $\square$

To summarize the above information, we have the following chart for the four possible Witt groups of nonreal fields F with 4 square classes. To complete the information, we also record, in each case, the level s, the u-invariant u (see XI.4), and the number m of distinct quaternion algebras (including the split one) over the field. As before, $\mathbb{Z}_n^k$ denotes the direct sum of k copies of $\mathbb{Z}_n = \mathbb{Z}/n\mathbb{Z}$.

	Witt Group $W(F)$	$\lvert W(F)\rvert$	s	u	m
(1)	$\mathbb{Z}_2^3$	2^3	1	2	1
(2)	$\mathbb{Z}_4 \oplus \mathbb{Z}_2$	2^3	2	2	1
(3)	$\mathbb{Z}_2^4$	2^4	1	4	2
(4)	$\mathbb{Z}_4^2$	2^4	2	4	2

The next case to consider is, of course, that of nonreal fields with 8 square classes. In this case, the complete classification of Witt rings is also known, thanks to the work of Cordes [Co$_1$] and Szymiczek [Sz$_1$]. At this point, we do not go into the detailed classification work in this case. However, later in XII.5, we will give the complete information on the 10 possible Witt rings, in a chart similar to the one above. The construction of these ten Witt rings will be described in detail in XII.5 and XII.7.

3. Hasse-Minkowski Principle

In the two previous sections, we have achieved a fairly complete understanding of the behavior of quadratic forms over local fields. For the rest of this chapter, our main focus will be shifted to the class of global fields. Following the standard terminology in number theory, by a *global field*, we mean a number field (i.e. a finite extension field of $\mathbb{Q}$), or a function field in one variable over a finite field (i.e. a finite extension field of $\mathbb{F}_q(t)$, where $\mathbb{F}_q$ is a finite field of q elements). The theory of quadratic forms over global fields is considerably harder, and requires a deeper analysis.

Throughout this section, F will denote a global field, and Ω the set of places on F. The latter means the totality of the archimedean and non-archimedean absolute values on F. For $\mathfrak{p} \in \Omega$, $F_\mathfrak{p}$ will denote the completion of F at $\mathfrak{p}$. For the archimedean places $\mathfrak{p}$, $F_\mathfrak{p}$ is either $\mathbb{R}$ or $\mathbb{C}$, while for the finite (non-archimedean) $\mathfrak{p}$'s, $F_\mathfrak{p}$ is a local field in the sense of Section 2. (Of course, in the case where $\operatorname{char} F = p$ (a prime), there are no archimedean places on F, so all completions $F_\mathfrak{p}$ are local fields of the type $k((t))$ for finite fields $k \supseteq \mathbb{F}_p$.) We assume, in this section, that the reader is mildly familiar with the elementary facts in number theory.

The central pillar of the global theory of quadratic forms is the following celebrated result.

Hasse-Minkowski Principle 3.1. *If q is a quadratic form over a global field F, then q is isotropic over F iff q is isotropic over all $F_\mathfrak{p}$ $(\mathfrak{p} \in \Omega)$.*

The striking thing about this famous "principle" is that it is a result holding *only* for quadratic forms, but not for forms of higher degrees (in general), over global fields. For instance, it is shown in [Ca$_2$, p. 256] that the following rational ternary form of degree 4:

$$(3x^2 - yz)^2 + 5(y^2 - zx)^2 - 2(z^2 - 3xy)^2 \in \mathbb{Q}[x, y, z]$$

is isotropic over all of the completions of $\mathbb{Q}$ (namely, $\mathbb{R}$ and all $\mathbb{Q}_p$), although it is *anisotropic over* $\mathbb{Q}$. But of course, the Hasse-Minkowski Principle may still hold for some *specific types* (or families) of forms of degree > 2.

To appreciate the usefulness of the Hasse-Minkowski Principle, let us first record some of its well-known consequences (for quadratic forms) below.

Corollary 3.2. *For F as above, let q be a quadratic form over F, and let $a \in \dot{F}$. Then q represents a in F iff q represents a in all $F_\mathfrak{p}$ $(\mathfrak{p} \in \Omega)$.*

Proof. Apply 3.1 to $q \perp \langle -a \rangle$, and use I.3.5. $\square$

Corollary 3.3 (Weak Hasse-Minkowski Principle). *If q, q' are quadratic forms over F, then $q \cong q'$ over F if $q \cong q'$ over all $F_\mathfrak{p}$ $(\mathfrak{p} \in \Omega)$, iff $\dim q = \dim q'$, $d(q) = d(q')$, $s(q) = s(q')$ over all non-archimedean $F_\mathfrak{p}$, and q, q' have the same signature over all $F_\mathfrak{p} \cong \mathbb{R}$. In particular, the ring homomorphism $\theta \colon W(F) \to \prod_{\mathfrak{p} \in \Omega} W(F_\mathfrak{p})$ is a monomorphism.*

Proof. We need only show the "if" parts. Let $\varphi = q' \perp \langle -1 \rangle q$. By hypothesis, φ is hyperbolic over all $F_\mathfrak{p}$. From 3.1 and induction, it follows that φ must be hyperbolic over F. $\square$

Remark 3.4. It is easy to see that 3.3 actually implies 3.1 if dim $q \leq 3$ in 3.1.

Corollary 3.5. *Suppose* dim $q \geq 5$. *If q is isotropic over all $F_{\mathfrak{p}} \cong \mathbb{R}$, then q is isotropic over F. The latter is automatic if none of $F_{\mathfrak{p}}$ is $\mathbb{R}$. This applies, in particular, in the function field case, and in the case of totally imaginary number fields.*

Proof. If $F_{\mathfrak{p}} \ncong \mathbb{R}$, then $F_{\mathfrak{p}}$ is either $\mathbb{C}$, or a local field in the sense of Section 2. By 2.12, q is automatically isotropic over such $F_{\mathfrak{p}}$. Now apply 3.1. □

Corollary 3.6. *If F is a local field or a global field, then F is a linked field; that is, any two quaternion algebras over F are linked (in the sense of* III.4.5).

Proof. If F is a local field, this is clear since there is only one quaternion division algebra over F. Now assume that F is a global field. By III.4.8, it is sufficient to prove that any 6-dimensional F-form q with $d(q) = -1$ is isotropic. This is clear from 3.5 since $d(q) = -1$ implies that q is indefinite (and hence isotropic) over any real completion of F. □

Unfortunately, it is not easy to give a *completely* self-contained proof of the Hasse-Minkowski Principle here — a full proof would usually invoke some class field theory, or else some deep arithmetic fact such as Dirichlet's theorem on primes in an arithmetic progression. As a compromise, *we will present a proof of this principle, modulo two specific results.* In this way, the reader will hopefully be able to see the gist of the proof, yet not having to wander too far afield.

The two facts we wish to assume are the following.

Theorem 3.7 (Global Square Theorem). *If $a \in \dot{F}$, then $a \in \dot{F}^2$ iff $a \in \dot{F}_{\mathfrak{p}}^2$ for all $\mathfrak{p} \in \Omega$.*

Theorem 3.8. *Let A be a quaternion algebra over F. Then A splits over F iff A splits over $F_{\mathfrak{p}}$ for all $\mathfrak{p} \in \Omega$.*

Theorem 3.8 can either be viewed as a special case of the Brauer-Hasse-Noether Theorem (with A replaced by an arbitrary central simple F-algebra), or as a special case of the Hasse Norm Theorem for cyclic (even quadratic) extensions. (A full proof of 3.8 for $F = \mathbb{Q}$ can be found in 4.5.)

We shall now commence the proof of (the sufficiency part) of 3.1. If dim $q = 1$, 3.1 is vacuous. Assume dim $q = 2$. After a scaling, we may take $q = \langle 1, -a \rangle$ $(a \in \dot{F})$. If q is isotropic over every $F_{\mathfrak{p}}$, then $a \in \dot{F}_{\mathfrak{p}}^2$ for all $\mathfrak{p} \in \Omega$. Therefore, 3.7 gives $a \in \dot{F}^2$, so $q \cong \mathbb{H}$. Assume now dim $q = 3$, and, as before, we may take $q = \langle 1, a, b\rangle$. Since this is isotropic over each $F_\mathfrak{p}$, the quaternion algebra $A = \left(\frac{-a,-b}{F}\right)$ splits over $F_\mathfrak{p}$ (by III.2.7). By 3.8, A splits over F, so its norm form $\langle 1, a, b, ab\rangle$ is isometric to the hyperbolic form $\langle 1, -1, -ab, ab\rangle$. Cancellation of $\langle ab\rangle$ yields $q \cong \langle 1, -1, -ab\rangle$, so q is isotropic. For $\dim q = 4$, take $q = \langle 1, a, b, c\rangle$, and let $K = F(\sqrt{abc})$. Each completion of K is an extension of some completion of F. Thus, our hypothesis guarantees that q is isotropic over each completion of K. But $q \cong \langle 1, a, b, ab\rangle$ over K since $abc \in \dot{K}^2$. Applying 3.8 to $A = \left(\frac{-a,-b}{F}\right)$, we conclude that q is isotropic over K. But then V.3.24 implies that q is already isotropic over F. Finally, suppose $\dim q \geq 5$. Write

$$q = q_1 \perp q_2, \quad \text{with } q_1 = \langle a, b\rangle \text{ and } \dim q_2 \geq 3.$$

By 2.5(2), there exists a finite subset $T \subseteq \Omega$ such that q_2 is isotropic at every $F_\mathfrak{p}$, $\mathfrak{p} \in \Omega \setminus T$. For $\mathfrak{p} \in T$, fix a value $z_\mathfrak{p} \in \dot{F}_\mathfrak{p}$ such that q_1 represents $z_\mathfrak{p}$ and q_2 represents $-z_\mathfrak{p}$. (This is possible since q is isotropic over $F_\mathfrak{p}$.) Write

$$x_\mathfrak{p}^2 a + y_\mathfrak{p}^2 b = z_\mathfrak{p}, \quad \text{where } x_\mathfrak{p}, y_\mathfrak{p} \in F_\mathfrak{p}.$$

Pick $x, y \in F$ that approximate respectively $x_\mathfrak{p}$ and $y_\mathfrak{p}$ at the finite number of places $\mathfrak{p} \in T$. Then $z = x^2 a + y^2 b$ is close to $z_\mathfrak{p}$ at all $\mathfrak{p} \in T$. Approximating closely enough, $z, z_\mathfrak{p}$ will belong to the same square class in $F_\mathfrak{p}$. Since q_1 represents z (which is clearly nonzero), q contains the subform $q' = \langle z\rangle \perp q_2$. The latter is surely isotropic at all $F_\mathfrak{p}$, $\mathfrak{p} \in \Omega \setminus T$ (since q_2 is). For $\mathfrak{p} \in T$, q_2 represents $-z_\mathfrak{p}$ and hence also $-z$ over $F_\mathfrak{p}$. Consequently, q' is isotropic over *all* completions. Invoking an inductive hypothesis, we may conclude that q' is isotropic over F. But then so is q. □

We shall close this section by making some observations about the group structure and ring structure of $W(F)$. For simplicity, we label the real completions of F by $F_1, \ldots, F_r$ (if any). By the approximation theorem for absolute values, there exist e_i ($1 \leq i \leq r$) that is negative in F_i but positive in all other F_j ($j \neq i$). Recall that IF denotes the ideal of even-dimensional forms in $W(F)$.

Corollary 3.9. (1) *The subgroup G in $W(F)$ generated by $\langle 1\rangle, \langle e_2\rangle, \ldots, \langle e_r\rangle$ is free abelian of rank r, and $W(F) = G \oplus W_t(F)$, where $W_t(F)$ denotes the torsion subgroup of $W(F)$.*

(2) $8W_t(F) = 0$.

(3) *For $n \geq 3$, I^nF is free abelian, with basis $2^{n-1}\langle 1, -e_i\rangle$ ($1 \leq i \leq n$). In particular, $I^nF = 2^{n-1}IF$.*

Proof. Let $\sigma_i\colon W(F) \to W(F_i) = \mathbb{Z}$ be the signature map induced by the real completion F_i, and let σ be the map

$$(\sigma_1, \ldots, \sigma_r)\colon\ W(F) \to \mathbb{Z}^r.$$

Since $W(F_\mathfrak{p})$ is torsion if $F_\mathfrak{p} \not\cong \mathbb{R}$, we clearly have $\ker \sigma = W_t(F)$, in view of the embedding θ in 3.3. By a straightforward computation, $\sigma(G)$ is generated by

$$(1, 1, \ldots, 1),\ (0, 2, \ldots, 0),\ \ldots\ ,\ (0, 0, \ldots, 2).$$

This is precisely the subgroup G' of $\mathbb{Z}^r$ consisting of r-tuples $(a_1, \ldots, a_r)$ with $a_i \equiv a_j \pmod 2$ for all i, j. Since $\sigma\langle a\rangle \in G'$ for any $a \in \dot{F}$, we have $\sigma(W(F)) \subseteq G' = \sigma(G)$ and hence $\sigma(W(F)) = \sigma(G)$. This proves both statements in (1).

(2) is clear since 8 kills $W(\mathbb{F}_\mathfrak{p})$ whenever $F_\mathfrak{p} \not\cong \mathbb{R}$.

For (3), recall that for $n \geq 3$, $I^n F_\mathfrak{p} = 0$ if $F_\mathfrak{p} \not\cong \mathbb{R}$, so the embedding θ in 3.3 restricted to $I^n F$ becomes essentially an embedding $\sigma\colon I^n F \to \bigoplus_i I^n F_i$. Since $\{\sigma(2^{n-1}\langle 1, -e_i\rangle)\}$ clearly forms a $\mathbb{Z}$-basis for $\bigoplus_i I^n F_i$, the assertion (3) follows. $\square$

Corollary 3.10. *The ring homomorphism*

$$\overline{\theta}\colon\ W(F)/I^3F \longrightarrow \prod_{\mathfrak{p}} W(F_\mathfrak{p})/I^3F_\mathfrak{p}$$

is a monomorphism. The Clifford invariant $\overline{\Gamma}\colon W(F)/I^3F \to BW(F)$ *is a monomorphism (where* $BW(F)$ *denotes the Brauer-Wall group of* F*).*

Proof. At each completion $F_\mathfrak{p}$, the Clifford invariant $\overline{\Gamma}_\mathfrak{p}\colon W(F_\mathfrak{p})/I^3F_\mathfrak{p} \to BW(F_\mathfrak{p})$ is certainly a monomorphism (see 2.12). If we know that $\overline{\theta}$ is a monomorphism, then the commutative diagram

$$\begin{array}{ccc} W(F)/I^3F & \xrightarrow{\ \overline{\theta}\ } & \prod_\mathfrak{p} W(F_\mathfrak{p})/I^3F_\mathfrak{p} \\ \Big\downarrow{\scriptstyle \overline{\Gamma}} & & \Big\downarrow{\scriptstyle \prod_\mathfrak{p} \overline{\Gamma}_\mathfrak{p}} \\ BW(F) & \longrightarrow & \prod_\mathfrak{p} BW(F_\mathfrak{p}) \end{array}$$

implies that $\overline{\Gamma}$ is also a monomorphism. To show that $\overline{\theta}$ is a monomorphism, take an anisotropic form q over F such that $q \in I^3F_\mathfrak{p}$ at each completion $F_\mathfrak{p}$. Then q is hyperbolic over all the $F_\mathfrak{p} \not\cong \mathbb{R}$, and, at each F_i, q has signature $\sigma_i(q) = 8r_i$. If we set $z = \sum_i 4r_i\langle 1, -e_i\rangle \in I^3F$, then clearly $\theta(q) = \theta(z)$, and so $q = z \in I^3F$. $\square$

Corollary 3.11. $\overline{\theta}$ *induces a monomorphism* $I^2F/I^3F \to \bigoplus_\mathfrak{p} I^2F_\mathfrak{p}/I^3F_\mathfrak{p}$.

Proof. In view of 3.10, we need only show that $\theta(I^2F)$ lies in the direct sum $\bigoplus_{\mathfrak{p}} I^2F_{\mathfrak{p}}$. Consider a typical additive generator $\langle 1,-a\rangle \otimes \langle 1,-b\rangle = \langle 1,-a,-b,ab\rangle$ for I^2F. This is the norm form of the quaternion algebra $A = \left(\frac{a,b}{F}\right)$. If we choose a big finite set $T \subseteq \Omega$ containing all the dyadic and archimedean places such that a, b are both $\mathfrak{p}$-adic units for $\mathfrak{p} \notin T$, then A splits over all $F_{\mathfrak{p}}$ except possibly for $\mathfrak{p} \in T$ (by 2.5(1)). Consequently, the "coordinates" of $\theta(\langle 1,-a,-b,ab\rangle)$ are almost all zero. □

The monomorphism in 3.11 is of great arithmetic interest. Its cokernel is computed by the existence and the uniqueness of Hilbert's Reciprocity Law, and turns out to be $\mathbb{Z}/2\mathbb{Z}$. For details, we refer the reader to O'Meara's text [O'M]. For the special case $F = \mathbb{Q}$, we shall offer a complete treatment in 5.8.

Example 3.12 (Gauss). *What (positive) integers n are sums of three squares in $\mathbb{Q}$?* Since $n > 0$, $f = \langle 1,1,1\rangle$ represents n at all completions of $\mathbb{Q}$ except perhaps at $\mathbb{Q}_2$. Thus, f represents n over $\mathbb{Q}$ iff n is distinct from the square class $-\det f = -1$ in $\mathbb{Q}_2$. Now n and -1 are in the same square class in $\mathbb{Q}_2$ iff n is of the form $4^a(8b-1)$ $(a \geq 0,\ b \in \mathbb{Z})$. Thus a positive integer n is a sum of three squares in $\mathbb{Q}$ iff n is not of the form $4^a(8b-1)$. [Such n's are automatically sums of squares of three *integers*. See IX.Exercise 3.]

Example 3.13. *Determine the square classes of $\mathbb{Q}$ represented by the form $f = \langle 2,-6,15\rangle$.* Note that f is isotropic over all completions of $\mathbb{Q}$ except at $\mathbb{Q}_2$ and $\mathbb{Q}_5$ (by the methods of 2.32). Thus, for $n \in \mathbb{Z}$, f represents n over $\mathbb{Q}$ iff f represents n over $\mathbb{Q}_2$ and $\mathbb{Q}_5$, iff n is different from the square class 5 over both $\mathbb{Q}_2$ and $\mathbb{Q}_5$. We may assume that n is square free. Thus, f *fails* to represent n over $\mathbb{Q}$ iff n is either $\equiv 5 \pmod 8$, or of the form $5m$, $m \equiv \pm 1 \pmod 5$. In particular, if p is a prime, f represents p iff $p \not\equiv 5 \pmod 8$.

For more results about quadratic forms over global fields, see Chapter XI.

4. Witt Ring of $\mathbb{Q}$

According to 3.3, we have an embedding

$$\theta\colon\ W(\mathbb{Q}) \to W(\mathbb{R}) \oplus \prod_p W(\mathbb{Q}_p),$$

where p ranges over all (finite) primes, and $\mathbb{Q}_p$ denotes the completion of $\mathbb{Q}$ at p (that is, the field of p-adic numbers). To some extent, this "computes" $W(\mathbb{Q})$. However, it does not yield the explicit structure of $W(\mathbb{Q})$, since nothing is said about the image of the embedding θ.

We shall try to rectify this by presenting in this section an explicit computation of $W(\mathbb{Q})$. The idea of this computation is supposed to have originated from Gauss' first proof of quadratic reciprocity—and is rediscovered by Milnor and Tate. Most significant is the fact that *this computation does not presuppose the Hasse-Minkowski Principle* 3.1, *thus rendering the results of this section completely independent of those in Section* 3.

To begin with, let $i\colon \mathbb{Z} \to W(\mathbb{Q})$ be the unique homomorphism that sends 1 to $\langle 1\rangle$. This is a split monomorphism, because the functorial map $W(\mathbb{Q}) \to W(\mathbb{R}) \cong \mathbb{Z}$ (the "signature") clearly splits i. Our main task is that of calculating coker(i). For any prime p, the residue class field $\overline{\mathbb{Q}}_p$ of the local field $\mathbb{Q}_p$ is just $\mathbb{Z}/p\mathbb{Z} = \mathbb{F}_p$. According to 1.6, for $p \neq 2$, there exists a "second residue homomorphism" $W(\mathbb{Q}_p) \to W(\mathbb{F}_p)$ (relative to the uniformizer p). Composing this with the functorial map $W(\mathbb{Q}) \to W(\mathbb{Q}_p)$, we obtain a group homomorphism $W(\mathbb{Q}) \to W(\mathbb{F}_p)$, which will be denoted by ∂_p $(p \neq 2)$. Note that if a is prime to p (i.e., a p-adic unit), then

$$\partial_p\langle a\rangle = 0 \quad \text{and} \quad \partial_p\langle pa\rangle = \langle\, \overline{a}\,\rangle\,,$$

where the bar here denotes the projection of the valuation ring of $\mathbb{Q}_p$ to $\overline{\mathbb{Q}}_p$. Since Springer's Theorem does not apply to $\mathbb{Q}_2$, we need a separate definition for ∂_2. We simply define $\partial_2\colon\ W(\mathbb{Q}) \to \mathbb{Z}/2\mathbb{Z}$ by

$$\partial_2(q) \equiv \varphi_2(\det(q)) \pmod 2,$$

where φ_2 denotes the 2-adic valuation. Since

$$\begin{aligned}\partial_2(q_1 \perp q_2) &\equiv \varphi_2((\det q_1)(\det q_2))\\ &\equiv \varphi_2(\det q_1) + \varphi_2(\det q_2) \pmod 2\end{aligned}$$

and $\partial_2(\mathbb{H}) \equiv \varphi_2(-1) \equiv 0 \pmod 2$, ∂_2 is a well-defined group homomorphism on $W(\mathbb{Q})$.

Theorem 4.1. $0 \to \mathbb{Z} \xrightarrow{i} W(\mathbb{Q}) \xrightarrow{\oplus\, \partial_p} \mathbb{Z}/2\mathbb{Z} \oplus \bigoplus_{p\neq 2} W(\mathbb{F}_p) \to 0$ *is split exact.* (*This calculates* $W(\mathbb{Q})$ *explicitly as an additive group.*)

Proof. For $d \geq 1$, define L_d to be the subring of $W(\mathbb{Q})$ generated by $\langle a\rangle$, where $|a| \leq d$. For example, if $b = a_1 \cdots a_r$, where $|a_i| \leq d$, then $\langle b\rangle \in L_d$, and L_d is additively spanned by such $\langle b\rangle$'s. Clearly, L_1 is just $i(\mathbb{Z}) = \mathbb{Z}\cdot\langle 1\rangle$, and L_d coincides with L_{d-1} unless d is a prime. For the ascending chain of subrings

$$L_1 \subseteq L_2 \subseteq L_3 \subseteq L_5 \subseteq \cdots \subseteq W(\mathbb{Q}),$$

we wish to determine the filtration quotients L_p/L_{p-1} $(p = \text{prime})$.

The quotient L_2/L_1 is clearly cyclic on the generator $\langle 2\rangle$. Further, since $\langle -1,2\rangle \cong \langle 1,-2\rangle$, $\langle 2\rangle$ has order 2 in L_2/L_1. Consequently, ∂_2 induces an isomorphism from L_2/L_1 to $\mathbb{Z}/2\mathbb{Z}$. We have thus an inverse isomorphism $v_2\colon \mathbb{Z}/2\mathbb{Z} \to L_2/L_1$. For each *odd* prime p, we shall try to define a suitable $v_p\colon W(\mathbb{F}_p) \to L_p/L_{p-1}$. We first establish

Lemma 4.2. *For any prime p, the quotient L_p/L_{p-1} is additively generated by $\langle pa_1 \cdots a_s\rangle$, where $|a_i| < p$. If $|a| < p$, and $a \equiv a_1 \cdots a_s \pmod{p}$, then $\langle pa_1 \cdots a_s\rangle \equiv \langle pa\rangle \pmod{L_{p-1}}$.*

Proof. The first statement is obvious, since repeated factors of p can be eliminated in pairs. To show the second statement, we calculate $\langle pa_1 \cdots a_s\rangle$ modulo L_{p-1} by "shrinking" s. Say $a_1a_2 = pk + h$, where $|h| < p$. By the isometry $\langle h,\, pk\rangle \cong \langle a_1a_2,\, a_1a_2phk\rangle$, we have an equation

$$\langle a_1a_2\rangle = \langle h\rangle + \langle pk\rangle - \langle a_1a_2phk\rangle \in W(\mathbb{Q}).$$

Multiplying this by $\langle pa_3 \cdots a_s\rangle$, we obtain

$$\begin{aligned}\langle pa_1 \cdots a_s\rangle &= \langle pha_3 \cdots a_s\rangle + \langle ka_3 \cdots a_s\rangle - \langle hka_1 \cdots a_s\rangle \\ &\equiv \langle pha_3 \cdots a_s\rangle \pmod{L_{p-1}},\end{aligned}$$

since clearly $|k| < p$ also. Shrinking the a 's in this way, the second assertion of 4.2 follows by induction on s. □

Let $p \neq 2$. For each residue class $\overline{a} \in \dot{\overline{\mathbb{Q}}}_p = \dot{\mathbb{F}}_p$, let a denote the unique integer in $(0,p)$ that lifts $\overline{a}$. *We claim that the rule*

$$(*) \qquad \langle \overline{a}\rangle \mapsto \langle pa\rangle + L_{p-1}$$

gives a well-defined homomorphism $v_p\colon W(\mathbb{F}_p) \to L_p/L_{p-1}$. Now, the additive relations among the generators $\langle \overline{a}\rangle \in W(\mathbb{F}_p)$ were set forth in II.4. To show that v_p is well-defined, we must check that the rule $(*)$ respects the relations $(R'1)$, $(R'2)$ and $(R'3)$ given in the paragraph following II.4.3.

First, suppose $bc^2 \equiv a \pmod{p}$, where $a,\, b,\, c \in (0,p)$. By 4.2, we have

$$\langle pb\rangle = \langle pbc^2\rangle \equiv \langle pa\rangle \pmod{L_{p-1}}.$$

This checks the relation $(R'1)$ in II.4.3. Next, suppose $\overline{a} + \overline{b} \neq 0$ in $\mathbb{F}_p$. Choose $a, b, c, d \in (0,\, p)$ lifting $\overline{a}$, $\overline{b}$, $\overline{a} + \overline{b}$, and $\overline{a}\,\overline{b}\,(\overline{a} + \overline{b})$ respectively. To check that $(*)$ respects $(R'2)$, we must show that

$$\text{(A)} \qquad \langle pa\rangle + \langle pb\rangle \equiv \langle pc\rangle + \langle pd\rangle \pmod{L_{p-1}}.$$

To see this, choose $e \in (0,p)$ such that $ae \equiv b \pmod{p}$. We must have $1 + e < p$ (since $-\overline{a} \neq \overline{b}$). In $W(\mathbb{Q})$, we have the usual isometry:

$$\text{(B)} \qquad \langle pa\rangle + \langle pae\rangle = \langle pa(1+e)\rangle + \langle pae(1+e)\rangle \in L_p.$$

Modulo L_{p-1}, $\langle pae\rangle$ may be replaced by $\langle pb\rangle$, by 4.2. Similarly, since

$$a(1+e) \equiv a+b \equiv c \ (\mathrm{mod}\ p),$$

$\langle pa(1+e)\rangle$ may be replaced by $\langle pc\rangle$. Finally, we have

$$d \equiv ab(a+b) \equiv a^3e(1+e) \pmod{p},$$

so another application of 4.2 gives $\langle pd\rangle \equiv \langle pae(1+e)\rangle \ (\mathrm{mod}\ L_{p-1})$. Therefore, (B) yields the desired congruence (A).

Finally, by $(*)$, $\langle\overline{1}\rangle \mapsto \langle p\rangle$ and $\langle -\overline{1}\rangle \mapsto \langle p(p-1)\rangle$ (modulo L_{p-1}). Since

$$\langle p\rangle + \langle p(p-1)\rangle = \langle p^2\rangle + \langle p^2(p-1)\rangle = \langle 1\rangle + \langle p-1\rangle \in L_{p-1},$$

we see that the relation $(R'3)$ is also respected, thus proving the well-definition of v_p $(p \neq 2)$.

It follows from 4.2 that v_p is *onto*. On the other hand, ∂_p induces a homomorphism $\overline{\partial}_p \colon L_p/L_{p-1} \to W(\mathbb{F}_p)$, with

$$\overline{\partial}_p v_p \langle\, \overline{a}\,\rangle = \overline{\partial}_p\langle pa\rangle = \langle\, \overline{a}\,\rangle \in W(\mathbb{F}_p).$$

Consequently, v_p and $\overline{\partial}_p$ are inverse isomorphisms.

The sequence in 4.1 is clearly a zero sequence. The exactness will then follow if we can show that

$$\bigoplus_{p\le q} \partial_p\colon \quad L_q/L_1 \longrightarrow \mathbb{Z}_2 \oplus \bigoplus_{p \le q} W(\mathbb{F}_p)$$

is an isomorphism, for every prime q. For $q=2$, this has been observed before. Inductively, if we have established this isomorphism for a prime q, and if r is the next prime following q, then, in the commutative diagram (4.3)

$$\begin{array}{ccccccccc}
0 & \longrightarrow & L_{r-1}/L_1 & \longrightarrow & L_r/L_1 & \longrightarrow & L_r/L_{r-1} & \longrightarrow & 0 \\
 & & \downarrow \cong & & \downarrow & & \cong \downarrow \overline{\partial}_r & & \\
0 & \longrightarrow & \dfrac{\mathbb{Z}}{2\mathbb{Z}} \oplus \bigoplus\limits_{2<p\le q} W(\mathbb{F}_p) & \longrightarrow & \dfrac{\mathbb{Z}}{2\mathbb{Z}} \oplus \bigoplus\limits_{2<p\le r} W(\mathbb{F}_p) & \longrightarrow & W(\mathbb{F}_r) & \longrightarrow & 0
\end{array}$$

(where $L_{r-1} = L_q$), the flanking maps are isomorphisms. From the five-lemma, it follows that the middle map is also an isomorphism. □

Remark. In a later chapter, we shall apply the same technique to calculate the Witt group of a rational function field $E = F(x)$. If the reader so wishes, he/she can read IX.3.1 at this point to compare the two parallel proofs.

Since the monomorphism i in 4.1 is split by the signature map $\partial_\infty \colon W(\mathbb{Q}) \to W(\mathbb{R}) = \mathbb{Z}$, another (equivalent) way to state 4.1 is that $\partial = \partial_\infty \oplus \bigoplus_p \partial_p$

defines an *isomorphism*:

$$\partial\colon\ W(\mathbb{Q}) \cong \mathbb{Z} \oplus \mathbb{Z}/2\mathbb{Z} \oplus \bigoplus_{p\neq 2} W(\mathbb{F}_p). \tag{4.4}$$

Since the maps ∂_p are uniquely "determined" by the maps $W(\mathbb{Q}) \to W(\mathbb{Q}_p)$, it is clear that 4.4 *implies the Weak Hasse-Minkowski Principle* 3.3 *for* $\mathbb{Q}$, independently of Section 3. (In particular, as remarked in 3.4, 4.4 actually implies the Hasse-Minkowski Principle 3.1 for forms of dimension ≤ 3 over $\mathbb{Q}$.)

Corollary 4.5. *Let* $A = \left(\frac{a,b}{\mathbb{Q}}\right)$. *Then,* A *splits over* $\mathbb{Q}$ *iff* A *splits over all* $\mathbb{Q}_p$ *and over* $\mathbb{R}$ *(proving* 3.8 *in the special case* $F = \mathbb{Q}$*).*

Proof. Apply the Weak Hasse-Minkowski Principle over $\mathbb{Q}$ (which we have just proved) to the two forms $\langle 1, -a, -b, ab\rangle$ and $\langle 1, -1, 1, -1\rangle$. □

5. Hilbert Reciprocity and Quadratic Reciprocity

The goal of this section is to establish the existence and uniqueness of Hilbert's reciprocity law for $\mathbb{Q}$, and then deduce from it Gauss' quadratic reciprocity law. There are two main tools to be used in the forthcoming proof. The first is the direct sum decomposition of $W(\mathbb{Q})$ given in 4.4, and the second is an "inductive" device which, in some sense, enables us to induct on primes. We shall first discuss the latter, which is due to Gauss.

Gauss Lemma 5.1. *If* p *is a prime* $\equiv 1 \pmod 8$, *then there exists an odd prime* $q < \sqrt{p}$ *such that* p *is not a square modulo* q.

Note. The lemma is false (in general) if $p \not\equiv 1 \pmod 8$. For instance, $109 \equiv 2^2 \pmod{3 \cdot 5 \cdot 7}$ is a counterexample.

Proof. (From notes of Bass, who ascribed it to Tate.) Let $f_n(X)$ denote the (binomial) polynomial function

$$X(X-1)\cdots(X-n+1)/n!.$$

If x_0 is any real number such that $n-1 < x_0 < n$, then $f_n(x_0)$ is the product of n real numbers in $(0,1)$, so also $f_n(x_0) \in (0,1)$. Let $2m-1$ be the unique odd integer such that

$$(2m-1)^2 < p < (2m+1)^2.$$

This inequality can be transformed into $m-1 < \alpha < m$, where $\alpha = (\sqrt{p}-1)/2$. Adding m to both sides, we have $2m-1 < \alpha + m < 2m$, so by our earlier observation, $f_{2m}(\alpha+m) \in (0,1)$. We claim that the real number $f_{2m}(\alpha+m) \in \mathbb{Q}$. For this, it suffices to show that

$$[(\alpha+m)(\alpha+m-1)\cdots][(\alpha)(\alpha-1)\cdots(\alpha-m+1)] \in \mathbb{Z}.$$

To handle this product, let us multiply out pairs of factors that are equidistant from the "middle point". The result is

$$(\alpha+k+1)(\alpha-k)=\alpha^2+\alpha-k^2-k \qquad (0\le k\le m-1).$$

This is indeed an integer, since α satisfies the minimal equation

$$\alpha^2+\alpha+(1-p)/4=0.$$

We have thus shown that $f_{2m}(\alpha+m)$ can be written as a/b, where a, b are relatively prime positive integers, with $a<b$. Let q be any prime factor of b. Since q divides $(2m)!$, we clearly have $q\le 2m-1<\sqrt{p}$. The final step is to show:

(5.2) $\qquad \sqrt{p}$ does not exist in $\mathbb{Q}_q$ (q-adic numbers).

If this is true, then $q\neq 2$, since $p\equiv 1 \pmod 8$ implies that $\sqrt{p}$ exists in $\mathbb{Q}_2$ (by 2.24). Further, p cannot be a square mod q, otherwise $\sqrt{p}$ again exists in $\mathbb{Q}_q$ (by 1.1). Thus, q satisfies all conclusions of 5.1. To show 5.2, let us assume that $\mathbb{Q}_q$ contains $\sqrt{p}$, and therefore also $\alpha=(\sqrt{p}-1)/2$. Since α is integral over $\mathbb{Z}$, it must lie in A, the ring of q-adic integers. Consider the polynomial map $x\mapsto f_{2m}(x)$ from $\mathbb{Q}_q$ to itself. Since $f_{2m}(\mathbb{Z})\subseteq\mathbb{Z}$, we see, by continuity, that $f_{2m}(A)\subseteq A$ (A being the closure of $\mathbb{Z}$ in the q-adic topology of $\mathbb{Q}_q$). In particular, $a/b=f_{2m}(\alpha+m)\in A\cap\mathbb{Q}$. But $A\cap\mathbb{Q}$ consists of rational numbers with q-free denominators, which contradicts the fact that $q\,|\,b$. $\square$

We shall now use the notations in Section 4, and, in particular, invoke the isomorphism ∂ in 4.4.

(5.3) $$\partial=\bigoplus_{p\in\Omega}\partial_p\colon\ W(\mathbb{Q})\cong W(\mathbb{R})\oplus\frac{\mathbb{Z}}{2\mathbb{Z}}\oplus\bigoplus_{p\neq 2,\infty}W(\mathbb{F}_p).$$

Here, $\Omega=\{\infty,2,3,5,\dots\}$. Recall that $\partial_\infty\colon W(\mathbb{Q})\to W(\mathbb{R})=\mathbb{Z}$ corresponds to the signature, and $\partial_2(q):=\varphi_2(\det q) \pmod 2$, where φ_2 is the 2-adic valuation. The other ∂_p's are defined by second residue homomorphisms.

Let A be any abelian group. By the universal property of $\oplus$, a homomorphism $\chi\colon W(\mathbb{Q})\to A$ is uniquely determined by a family of homomorphisms

$$\chi_\infty\colon W(\mathbb{R})\to A,\quad \chi_2\colon \mathbb{Z}/2\mathbb{Z}\to A,\quad\text{and}\quad \chi_p\colon W(\mathbb{F}_p)\to A,$$

such that $\chi=\sum_{p\in\Omega}\chi_p\partial_p$. We shall indicate this correspondence by writing $\chi\leftrightarrow(\chi_p)$.

Recall that $W(\mathbb{Q}_2)$ is generated as a direct sum by $\alpha=\langle 1\rangle$, $\beta=\langle 1,-2\rangle$, and $\gamma=\langle 1,-5\rangle$, whose orders are, respectively, 8, 2, and 2 (see 2.31). Define

$\eta\colon W(\mathbb{Q}_2) \to \mathbb{Z}/8\mathbb{Z}$ by

$$\eta\,(\alpha) \equiv 1, \quad \eta(\beta) \equiv 0, \quad \text{and} \quad \eta\,(\gamma) \equiv 4 \pmod 8.$$

We shall write χ for the composition $W(\mathbb{Q}) \to W(\mathbb{Q}_2) \xrightarrow{\eta} \mathbb{Z}/8\mathbb{Z}$, and determine the family of maps $(\chi_p) \leftrightarrow \chi$.

Theorem 5.4. *For the χ defined above, we have*

$$\chi_\infty(n\langle 1\rangle) \equiv n \pmod 8 \ \ (n \in \mathbb{Z}), \qquad \chi_2 = 0,$$
$$\chi_p\langle 1\rangle \equiv p - 1 \pmod 8, \quad \chi_p(\psi_p) \equiv 4 \pmod 8 \qquad (p \neq \infty, 2),$$

where χ_p denotes the unique binary anisotropic form over $\mathbb{F}_p$.

The significance of this result lies in the fact that it implies Hilbert's reciprocity law. Let us first explain this to get the necessary motivation. Recall that $(\ ,\ )_p$ denotes the Hilbert symbol over $\mathbb{Q}_p$. For $p = \infty$, we may similarly define a Hilbert symbol

$$(\ ,\ )_\infty : \ \dot{\mathbb{R}}/\dot{\mathbb{R}}^2 \times \dot{\mathbb{R}}/\dot{\mathbb{R}}^2 \to \text{Quat}(\mathbb{R}) = \{\pm 1\}.$$

(The behavior of this symbol is quite simple; namely, $(a, b)_\infty = -1$ iff a, b are both negative.) The beautiful fact connecting all symbols $(\ ,\ )_p$ $(p \in \Omega)$ is the following

Hilbert's Reciprocity Law 5.5. *If $a, b \in \dot{\mathbb{Q}}$, then the set*

$$\Lambda = \{p \in \Omega\colon\ (a,\, b)_p = -1\}$$

is finite with even cardinality. In other words, $\prod_{p\in\Omega} (a,\, b)_p = 1$.

Proof. Write q for the norm form $\langle 1, -a, -b, ab\rangle$. Let $p \in \Omega$, $p \neq 2, \infty$. Then, $p \in \Lambda$ iff q is the unique 4-dimensional anisotropic form over $\mathbb{Q}_p$, iff $\partial_p(q) = \psi_p$, iff $\chi_p\partial_p(q) \equiv 4 \pmod 8$ (by 5.4). Similarly, $\infty \in \Lambda$ iff $q \cong 4\langle -1\rangle$ over $\mathbb{R}$, iff $\chi_\infty\partial_\infty(q) \equiv 4 \pmod 8$. Using $\chi = \sum_{p\in\Omega} \chi_p\partial_p$, and the fact that $\chi_2 = 0$, we get

$$\chi(q) \equiv |\Lambda \setminus \{2\}| \cdot 4 \pmod 8.$$

If $\mathbb{Q}_2$ splits q, then $\chi(q) = \eta\,(\mathbb{Q}_2 \otimes q) = 0$, so the above congruence yields $|\Lambda| = |\Lambda \setminus \{2\}| =$ even. If $\mathbb{Q}_2$ does not split q, then $\chi(q) = \eta\,(4\langle 1\rangle) \equiv 4 \pmod 8$ and $2 \in \Lambda$. The congruence again implies that $|\Lambda|$ is even. ☐

Proof of 5.4 (following Scharlau [Sc$_3$]). First, the definition of χ implies that χ_∞ is reduction mod 8. Secondly, $\langle -1, 2\rangle$ has image 0 under all ∂_p $(p \neq 2)$, and $\partial_2\langle -1, 2\rangle = \varphi_2(-2) = 1$. Since $\chi\langle -1, 2\rangle = \eta\,(-\beta) = 0$, we conclude that $\chi_2 = 0$. Consider now $p \neq \infty, 2$. We have $\chi_p\langle 1\rangle = \chi\langle -1, p\rangle$ since $\langle -1, p\rangle$ maps to zero under all ∂_r $(r \neq p)$, and maps to $\langle 1\rangle$ under ∂_p. Noting that $\eta\,\langle 5\rangle = \eta\,(1 - \gamma) \equiv 5 \pmod 8$, we have

$$\chi\langle p\rangle = \eta\,(\mathbb{Q}_2 \otimes \langle p\rangle) \equiv p \pmod 8$$

for any odd prime p (using 2.2). Thus,

$$\chi_p\langle 1\rangle = \chi\langle p\rangle - 1 \equiv p - 1 \pmod 8.$$

It remains only to calculate $\chi_p(\psi_p)$. First, suppose $p \equiv 3, 7 \pmod 8$. Then we may pick ψ_p to be $\langle 1, 1\rangle$ and hence

$$\chi_p(\psi_p) = 2\chi_p\langle 1\rangle \equiv 2p - 2 \equiv 4 \pmod 8.$$

Next, suppose $p \equiv 5 \pmod 8$, and consider the form $\langle -p, 2p\rangle \in W(\mathbb{Q})$. This maps to zero under all ∂_r $(r \neq p)$, and maps to $\langle -1, 2\rangle$ under ∂_p. Thus,

$$\chi_p\langle -1,\, 2\rangle = \chi\langle -p,\, 2p\rangle = \eta\,\langle -5,\, 10\rangle.$$

Evaluating η on $4\langle 1\rangle \cong \langle 1, -2, -5, 10\rangle$ over $\mathbb{Q}_2$ (see 2.24(4)), we get $\chi_p\langle -1, 2\rangle \equiv 4 \pmod 8$. This *implies* that $\langle -1, 2\rangle \neq 0 \in W(\mathbb{F}_p)$, and thus

$$\chi_p(\psi_p) = \chi_p\langle -1, 2\rangle \equiv 4 \pmod 8,$$

as required. Finally, we handle the case $p \equiv 1 \pmod 8$. Choose an odd prime q as in Gauss' Lemma (we only need $q < p$ for the following). Let ψ be the form $\langle 1, -p, -q, pq\rangle$. Since p is a square in $\mathbb{Q}_2$ (by 2.24), we clearly have $\chi(\psi) = 0$. On the other hand, for $r \in \Omega \setminus \{p, q\}$, $\partial_r(\psi) = 0$ by 2.5(1). Consequently, we get

$$0 = \chi(\psi) = \chi_p\,\partial_p(\psi) + \chi_q\,\partial_q(\psi) = \chi_p\,\langle -1, q\rangle + \chi_q\,\langle -1,\, p\,\rangle.$$

Since p is not a square modulo q, we have $\psi_q \cong \langle -1, p\rangle$. Invoking an inductive hypothesis, we may assume that $\chi_q(\psi_q) \equiv 4 \pmod 8$. But then $\chi_p\langle -1, q\rangle \equiv 4 \pmod 8$, and à fortiori $\langle -1, q\rangle \equiv \psi_p$ over $\mathbb{F}_p$. In particular, this yields $\chi_p(\psi_p) \equiv 4 \pmod 8$, as desired. □

We may now deduce the Quadratic Reciprocity Law with bravado. For odd prime p, and $b \in \mathbb{Z}$, the classical "Legendre symbol" $\left(\frac{b}{p}\right)$ is defined to be 1 or -1 according as b is or is not a square modulo the prime p. Recall also the two special functions ε and ω defined in 2.25, which will be used below.

Quadratic Reciprocity Law 5.6. *For odd primes p and q,*

$$\left(\frac{p}{q}\right)\left(\frac{q}{p}\right) = (-1)^{\varepsilon(p)\varepsilon(q)}.$$

We have also the following "First and Second Supplements" to Quadratic Reciprocity:

$$\left(\frac{-1}{p}\right) = (-1)^{\varepsilon(p)} \qquad \textit{and} \qquad \left(\frac{2}{p}\right) = (-1)^{\omega(p)}.$$

Proof. By 1.1 and 2.2(4), we have $\left(\frac{q}{p}\right) = (p,\, q)_p$. But $(p,\, q)_r = 1$ for $r \in \Omega \setminus \{2, p, q\}$ (by 2.5(1)). Thus the Hilbert Reciprocity Law for the pair $\{p,\, q\}$ boils down to

$$(p,\, q)_p\,(p,\, q)_q = (p,\, q)_2 = (-1)^{\varepsilon(p)\varepsilon(q)}$$

(using 2.28), which is the desired quadratic reciprocity formula. Similarly, using the pair $\{p,\, 2\}$, we deduce $(p,\, 2)_p\,(p,\, 2)_2 = 1$, and hence $\left(\frac{2}{p}\right) = (p,\, 2)_2 = (-1)^{\omega(p)}$ by 2.28. The "First Supplement" $\left(\frac{-1}{p}\right) = (-1)^{\varepsilon(p)}$ follows likewise from the Hilbert Reciprocity Law for the pair $\{p,\, -1\}$. (Note that the "First Supplement" amounts to saying that -1 *is a square modulo* p *iff* $p \equiv 1 \pmod 4$, a fact we have been using freely all along. Similarly, we observe that the "Second Supplement" amounts to: 2 *is a square modulo* p *iff* $p \equiv \pm 1 \pmod 8$.) $\square$

Let us now take the isomorphism ∂ in 5.3, and reduce it modulo 8. On the right side of 5.3, this changes $W(\mathbb{R})$ into $W(\mathbb{R})/8W(\mathbb{R})$, but does not affect other summands. On the other hand, $8W(\mathbb{Q})$ equals the ideal power $I^3\mathbb{Q}$ by 3.9(3). [Note: The latter depends only on the Weak Hasse-Minkowski Principle, and is thus applicable to this section.] Consequently, ∂ induces a monomorphism $\overline{\partial}$ in the following diagram:

(5.7)
$$\begin{array}{ccc} I^2\mathbb{Q}/I^3\mathbb{Q} & \xrightarrow{\quad h \quad} & I^2\mathbb{R}/I^3\mathbb{R} \oplus \bigoplus_{p\neq\infty} I^2\mathbb{Q}_p \\ & \overline{\partial} \searrow \qquad \swarrow g & \\ & (I^2\mathbb{R}/I^3\mathbb{R}) \oplus (\mathbb{Z}/2\mathbb{Z}) \oplus \bigoplus_{p\neq\infty,2} W(\mathbb{F}_p) & \end{array}$$

(noting that $I^3\mathbb{Q}_p = 0$). Here, h is the functorial map (essentially giving the Hilbert symbols), g is the identity on the first component, zero on the second component $I^2\mathbb{Q}_2$, and is the second residue homomorphism on the other components. All components are $\cong \mathbb{Z}/2\mathbb{Z}$, so we may think of the domain of g as a $\mathbb{Z}_2$-vector space, say, G. The kernel of g is the second component $I^2\mathbb{Q}_2$, since, for $p \neq 2$, g takes the generator of $I^2\mathbb{Q}_p$ to ψ_p (= the anisotropic binary form over $\mathbb{F}_p$). Let V be the hyperplane in G consisting of "vectors" $(a_\infty, a_2, a_3, \dots)$ such that $\sum_{p\in\Omega} a_p = 0$. Clearly, we have a vector space direct sum $G = I^2\mathbb{Q}_2 \oplus V$. Let $V' = \mathrm{im}(h)$. By Hilbert's reciprocity law, V' lies in V. *We claim that* $V = V'$.

To see this, it suffices to show that $\mathrm{im}(\overline{\partial}) = \mathrm{im}(g)$. The inclusion $\subseteq$ is clear. For the reverse inclusion, we must show that each $\psi_p \in \mathrm{im}(\overline{\partial})$ $(p \neq 2, \infty)$. If $p \equiv 3, 7 \pmod 8$, we have $\psi_p = \langle 1, 1\rangle$, so

$$\psi_p = \overline{\partial}\,\langle -1,\, -1,\, p,\, p\rangle \in \overline{\partial}\,(I^2\mathbb{Q}).$$

If $p \equiv 5 \pmod 8$, we have $\psi_p = \langle 1, -2 \rangle$ (since 2 is a nonsquare mod p), hence

$$\psi_p = \overline{\partial}\,\langle 1,\, -2,\, p,\, -2p \rangle \in \overline{\partial}\,(I^2\mathbb{Q}).$$

Finally, for $p \equiv 1 \pmod 8$, pick q as in Gauss' Lemma. Again, consider $\psi = \langle 1, -p, -q, pq \rangle \in I^2\mathbb{Q}$. We have

$$\overline{\partial}\,(\psi) = (\langle -1,\, q \rangle \text{ in } W(\mathbb{F}_p)) \oplus (\langle -1,\, p \rangle \text{ in } W(\mathbb{F}_q)).$$

By the choice of q, we have $\langle -1, p \rangle = \psi_q$. But by the quadratic reciprocity law, q is also a nonsquare mod p, so $\langle -1, q \rangle = \psi_p$. We obtain therefore $\overline{\partial}\,(\psi) = \psi_p \oplus \psi_q$. By induction, we may assume $\psi_q \in \overline{\partial}\,(I^2\mathbb{Q})$, and hence $\psi_p \in \overline{\partial}\,(I^2\mathbb{Q})$.

The meaning of $V = V'$ can be interpreted as follows:

Uniqueness of Hilbert's Reciprocity Law 5.8. *We have an exact sequence*

$$0 \longrightarrow I^2\mathbb{Q}/I^3\mathbb{Q} \xrightarrow{\ h\ } I^2\mathbb{R}/I^3\mathbb{R} \oplus \bigoplus_{p \neq \infty} I^2\mathbb{Q}_p \xrightarrow{\ f\ } \mathbb{Z}/2\mathbb{Z} \longrightarrow 0,$$

where $f(a_\infty, a_2, a_3, \dots) = \sum_{p \in \Omega} a_p$. *If* $\{i_p \colon p \in \Omega\}$ *is a family of integers chosen from* $\{0, 1\}$ *such that*

$$\prod_{p \in \Omega} (a, b)^{i_p} = 1 \quad \textit{for all } a, b \in \dot{\mathbb{Q}}\,,$$

then $\{i_p\}$ *are either all equal to* 0 *or else all equal to* 1.

(Note that the injectivity of $\overline{\partial}$ implies that of h. Alternatively, one may use 3.11.)

Exercises for Chapter VI

1. Prove the theorem of Kaplansky stated in 2.13. (**Hint.** Repeating the proof of 2.12, we may assume that $4\langle 1 \rangle$ is anisotropic. In this case, show that, given any $a \in \dot{F}$, one of $\pm a$ is a sum of two squares. Next, show that -1 is a sum of four squares, and therefore, any a is a sum of four squares.)
2. Let F be a c.d.v. field whose residue class field $\overline{F}$ has characteristic different from 2. Using the decomposition

$$W(F) = W(\overline{F}) \oplus (\langle \pi \rangle - 1) \cdot W(\overline{F})$$

in 1.7, show that $I^nF = I^n\overline{F} \oplus (\langle \pi \rangle - 1) \cdot I^{n-1}\overline{F}$.

3. Let $F = k((t))$, where $\mathrm{char}(k) \neq 2$. If q_1, q_2 are anisotropic forms over k, show that

$$D_F(q_1 \perp \langle t \rangle q_2) = D_k(q_1)\dot{F}^2 \cup D_k(q_2) \cdot t\dot{F}^2.$$

[In particular, the binary form $\langle 1,\, t\rangle$ represents exactly two square classes over F, namely those of 1 and t.]

4. Compute the 16 square classes of $\mathbb{Q}_2(\sqrt{5}\,)$, and determine its unique unramified quadratic extension.
5. Determine the p-adic fields $\mathbb{Q}_p$ over which the form $\langle 3, 7, -15\rangle$ is isotropic.
6. Over which p-adic fields $\mathbb{Q}_p$ are the forms $\langle 6, 7, 1\rangle$ and $\langle 21, -2, -1\rangle$ isometric?
7. Over which p-adic fields $\mathbb{Q}_p$ does $\langle 3, 3, 11\rangle$ represent 2?
8. Using 2.16, show that, if F is a local field, and $[K\colon F] = 2$, then $[\dot{F}\colon \mathrm{N}_{K/F}(\dot{K})] = 2$. From this, show that any anisotropic binary form over F represents exactly half of the elements of $\dot{F}/\dot{F}^2$.
9. Which square classes of $\mathbb{Q}_2$ are represented by the form $\langle 10, 13\rangle$?
10. Verify the following formula for the Hilbert symbol $(a,b)_p$ over $\mathbb{Q}_p$, where $p \neq 2$. If $a = p^\alpha x$, $b = p^\beta y$ $(\alpha, \beta \in \mathbb{Z};\ x, y = p\text{-adic units})$, then
$$(a,\, b)_p = (-1)^{\alpha\beta\cdot\varepsilon(p)}\left(\frac{x}{p}\right)^{\beta}\left(\frac{y}{p}\right)^{\alpha},$$
where $\varepsilon(p) = (p-1)/2 \pmod 2$, as in (2.25).
11. (Conway) Let p be a given prime, or ∞. Let C be a cyclic group of order 4, written multiplicatively as $\{\pm 1, \pm i\}$, where $i^2 = -1$. Show that there exists a mapping $[\ ]_p\colon \dot{\mathbb{Q}}_p/\dot{\mathbb{Q}}_p^2 \to C$ such that
$$(a,\, b)_p = [\,a\,]_p\,[\,b\,]_p/[\,ab\,]_p \quad \text{for all } a,\, b \in \dot{\mathbb{Q}}_p\,.$$
If $f = \langle a_1, \ldots, a_n\rangle$ is any form over $\mathbb{Q}_p$, show by induction on n that the Hasse invariant of f is given by
$$s(f) = [\,a_1\,]_p \cdots [\,a_n\,]_p/[\,a_1 \cdots a_n\,]_p\,.$$
(**Hint.** For $p = \infty$, define $[a]_\infty = 1$ if $a > 0$, and $[a]_\infty = i$ if $a < 0$. For $p \equiv -1 \pmod 4$, define $[1]_p = [-1]_p = 1$, $[p]_p = i$, and $[-p]_p = -i$. For $p \equiv 1 \pmod 4$, and u a nonsquare mod p, define $[1]_p = [u]_p = [p]_p = 1$, and $[up]_p = -1$. Finally, define $[\ ]_2$ by the chart:

a	1	-1	2	-2	5	-5	10	-10
$[a]_2$	1	$-i$	1	$-i$	1	$-i$	-1	i

and check that these definitions work over $\mathbb{Q}_p$ for all $p \le \infty$.)
12. If a, b, c are 2-adic units in $\mathbb{Q}_2$, show that $(b^2 - 4ac,\, 2a)_2 = -1$.

13. For the eight square classes of $\mathbb{Q}_2$ represented by $\{\pm 1,\ \pm 2,\ \pm 5,\ \pm 10\}$, verify the following (symmetrical) table for the Hilbert symbol $(\ ,\ )_2$ over $\mathbb{Q}_2$:

	1	−1	2	−2	5	−5	10	−10
1	+	+	+	+	+	+	+	+
−1	+	−	+	−	+	−	+	−
2	+	+	+	+	−	−	−	−
−2	+	−	+	−	−	+	−	+
5	+	+	−	−	+	+	−	−
−5	+	−	−	+	+	−	−	+
10	+	+	−	−	−	−	+	+
−10	+	−	−	+	−	+	+	−

where "+" means the symbol value 1 (trivial), and "−" means the symbol value -1 (nontrivial).

14. (1) Is 383 a square modulo 443?
(2) For any odd prime p, show that 3 is a square modulo p iff $p \equiv \pm 1 \pmod{12}$.

15. Show that $\langle -1, 3, 5\rangle$ is isometric to $\langle 1, 7, -105\rangle$ over $\mathbb{Q}$.

16. Which of the forms $\langle 1, -2, 5\rangle$, $\langle 1, -1, 10\rangle$ and $\langle 3, -1, 30\rangle$ are isometric over $\mathbb{Q}$?

17. Show that $\langle 1, 2, 5, -10\rangle$ is anisotropic and universal over $\mathbb{Q}$.

18. If p, q, r, s are distinct odd primes such that $pqrs \not\equiv 1 \pmod 8$, show that $\langle p, q, -r, -s\rangle$ is isotropic over $\mathbb{Q}$.

19. (Legendre) Let a, b, c be square-free nonzero integers that are pairwise relatively prime, and not all of the same sign. Show that $\langle a, b, c\rangle$ is isotropic over $\mathbb{Q}$ iff $-bc$ is a square mod a, $-ac$ is a square mod b, and $-ab$ is a square mod c.

20. Let a, b be square-free nonzero integers. If $\left(\frac{a,b}{\mathbb{Q}}\right)$ splits, show that:

(1) a, b are not both negative,
(2) a is a square modulo b, and
(3) b is a square modulo a.

If a and b are relatively prime, show that the converse also holds, by using the previous exercise, or by applying an induction on the positive integer $|a| + |b|$.

21. Let σ be a five-dimensional form over a local field or a global field F. Show that there exist $a, b, c, d \in \dot{F}$ such that $c\cdot\sigma \cong \langle 1, a\rangle \otimes \langle 1, b\rangle \perp \langle d\rangle$.

(*Remark.* Recall that F is a linked field, by 3.6. The property of 5-dimensional forms above is actually characteristic for linked fields in general. For more details on this, see X.4.20 and X.4.21.)

22. Let F be a c.d.v. field for which $\text{char}(\overline{F}) \neq 2$. Let $f(X) = \sum_{i,j} a_{ij} X_i X_j$, where (a_{ij}) is a symmetric matrix in $\text{GL}_n(A)$ (A = valuation ring of F). Show that f has a nontrivial zero in F iff $\overline{f}$ has a nontrivial zero in $\overline{F}$. (**Hint.** Use 2.18.)

23. Let f be a quadratic form over $\mathbb{Q}$. Assume either that $\dim f = 3$, or that $\dim f = 4$ and $\det(f) = 1$. Using Hilbert's Reciprocity Law, show that the number of completions of $\mathbb{Q}$ at which f is anisotropic is finite and even. In particular, if f is isotropic over all but perhaps one completion of $\mathbb{Q}$, then f is isotropic over $\mathbb{Q}$.

24. Let $f(X) = \sum_{i,j} a_{ij} X_i X_j$, where (a_{ij}) is a symmetric matrix in $\text{GL}_n(\mathbb{Z})$. Show that f has a zero over $\mathbb{Z}$ iff f has a zero over $\mathbb{R}$. (**Hint.** Use 3.1, 3.5, and the two preceding exercises.)

25. If f is any quadratic form over $\mathbb{Q}$, show that the Hasse invariant of $\mathbb{Q}_p \otimes f$ is equal to 1 except for a finite, even number of completions $\mathbb{Q}_p$ $(p \leq \infty)$.

Chapter VII

Quadratic Forms Under Algebraic Extensions

1. Scharlau's Transfer

Let K be an extension of a field F. From a given F-quadratic space (V, B, q), we can construct a K-quadratic space (V_K, B_K, q_K). The underlying space V_K is taken to be $K \otimes_F V$, and B_K is taken to be the unique symmetric bilinear form on V_K satisfying

$$B_K(k \otimes v,\, k' \otimes v') = kk'B(v,\, v') \qquad (k, k' \in K;\ v, v' \in V).$$

Correspondingly, the K-quadratic form q_K associated with B_K is uniquely given by

$$q_K(k \otimes v) = k^2 q(v) \qquad (k \in K,\ v \in V).$$

Note that the symmetric matrix of q with respect to an F-basis $\{v_1, \ldots, v_n\}$ on V is the same as that of q_K with respect to the K-basis $\{1 \otimes v_1, \ldots, 1 \otimes v_n\}$ on V_K. In particular, if q is regular, so is q_K. By abuse of notation, we shall sometimes express B_K and q_K just by B and q (again).

Recall that $M(F)$ (resp. $M(K)$) denotes the monoid of all isometry classes of quadratic forms over F (resp. over K), under the operation $\perp$. The construction $q \mapsto q_K$ above induces a monoid homomorphism $M(F) \to M(K) \subseteq \widehat{W}(K)$. If we give a name, say r, to the inclusion map $F \subseteq K$, then we write $\hat{r}^*$ to denote the map $\widehat{W}(F) \to \widehat{W}(K)$ given by $\hat{r}^*(V) = V_K$. It is easy to check that $\hat{r}^*$ is actually a ring homomorphism. More ceremoniously, we proclaim that the rules $F \mapsto \widehat{W}(F)$, $r \mapsto \hat{r}^*$ define a *functor* from the category of fields (of characteristic not 2) to the category of commutative rings.

Note that, under $\hat{r}^*$, the F-hyperbolic plane $\mathbb{H}_F = \langle 1, -1\rangle_F$ goes over to the K-hyperbolic plane $\mathbb{H}_K = \langle 1, -1\rangle_K$. Thus, $\hat{r}^*\colon \widehat{W}(F) \to \widehat{W}(K)$ induces a ring homomorphism $W(F) \to W(K)$, which will be denoted by the symbol r^*. Again, $F \mapsto W(F)$ and $r \mapsto r^*$ define a functor just as in the last paragraph.

In general, the functorial maps $\hat{r}^*$ and r^* cannot be expected to be monomorphisms. For example, for $r\colon \mathbb{R} \hookrightarrow \mathbb{C}$, neither

$$\hat{r}^*\colon \widehat{W}(\mathbb{R})\,(\cong \mathbb{Z}\oplus\mathbb{Z}) \to \widehat{W}(\mathbb{C}) \cong \mathbb{Z} \quad\text{nor}\quad r^*\colon W(\mathbb{R})\,(\cong \mathbb{Z}) \to W(\mathbb{C}) \cong \mathbb{Z}_2$$

is a monomorphism. In fact, the binary anisotropic form $\langle 1, 1\rangle_{\mathbb{R}}$ goes over to $\langle 1, 1\rangle_{\mathbb{C}} \cong \mathbb{H}_{\mathbb{C}}$. More generally, if $a \in \dot{F}$ is a nonsquare, and $K = F(\sqrt{a}\,)$, then the binary anisotropic form $\langle 1, -a\rangle$ over F goes over to the K-hyperbolic form $\mathbb{H}_K$; hence $\widehat{W}(F) \to \widehat{W}(K)$ and $W(F) \to W(K)$ both fail to be monomorphisms. We shall see, however, that (for $[K : F] < \infty$) this phenomenon can occur only with *even* degree field extensions (see Springer's Theorem below).

The most powerful tool in studying the functorial maps $\hat{r}^*$ and r^* is Scharlau's "transfer map" (from [Sc$_1$]), which we shall now define.

Let $r\colon F \subseteq K$ be a field extension of finite degree. Let $s\colon K \to F$ be a nonzero F-linear functional on the F-vector space K. (Note that such a functional is automatically surjective.) For any K-quadratic space (U, B), we may compose the pairing $B\colon U \times U \to K$ with the functional s to get an F-bilinear pairing

$$sB\colon\ U \times U \longrightarrow F.$$

Thus, the K-quadratic space (U, B) gives rise to an F-quadratic space $(U,\, sB)$. A very nice observation here is the following.

Proposition 1.1. *If $(U,\, B)$ is a regular K-quadratic space, then $(U,\, sB)$ is a regular F-quadratic space.*

Proof. If otherwise, there would exist a nonzero vector $x_0 \in U$ such that $(sB)(x_0,\, U) = 0$. On the other hand, by the regularity of (U, B), there exists $y_0 \in U$ such that $B(x_0,\, y_0) \neq 0$. For any scalar $c \in K$, we have

$$B\left(x_0,\, \frac{c}{B(x_0, y_0)}\, y_0\right) = \frac{c}{B(x_0,\, y_0)} \cdot B(x_0,\, y_0) = c.$$

Applying the functional s to this equation, we get

$$s(c) \in (sB)(x_0,\, U) = 0.$$

This contradicts the fact that $s \not\equiv 0$. □

Notation 1.2. Let $s: K \to F$ be as above. If U denotes a quadratic space over K with bilinear form B, then we write $s_*(U)$ to denote the quadratic space U over F with the bilinear form sB. We call $s_*(U)$ the "transfer" of U, noting that

$$\dim_F\, s_*(U) = [K\colon F] \cdot \dim_K U.$$

This construction applies, in particular, to the 1-dimensional K-space K that carries the natural bilinear form $(x, y) \mapsto xy$ $(x, y \in K)$. As a K-quadratic space, this is just $\langle 1 \rangle_K$. The transfer $s_*(\langle 1 \rangle_K)$ will have underlying space K, with the F-bilinear form $(x, y) \mapsto s(xy) \in F$. This transfer space $s_*(\langle 1 \rangle_K)$ will be calculated in detail in Theorem 2.2 below, for a specific choice of s (and for a simple algebraic extension K/F).

It is tempting to take the s above to be the field trace

$$\mathrm{tr} = \mathrm{tr}_{K/F}\colon\ K \longrightarrow F,$$

since this is an F-linear functional on K. However, we need s to be nonzero, so we should focus on finite extensions K/F with $\mathrm{tr} \not\equiv 0$ only. By Ch.I, Exer. 30, these are precisely the finite *separable* extensions K/F. For any such extension, the transfer $\mathrm{tr}_*(\langle 1 \rangle_K)$ obtained is the *trace form* on the F-algebra K (defined in Ch.I, Exer. 29). More generally, if we transfer a unary form $\langle \alpha \rangle$ $(\alpha \in \dot{K})$, we get a quadratic F-form on K given by $x \mapsto \mathrm{tr}(\alpha x^2)$, which is called a "scaled" trace form for the extension K/F. This notion is a very important special case of the transfer construction.

We now come to an interesting formal relationship between $\hat{r}^*$ and s_* which reminds us of similar relationships in group representations, group cohomology, etc.

Theorem 1.3 (Frobenius Reciprocity). *Let r, s, F and K be as above. Let V be a quadratic space over F and U a quadratic space over K. Then there exists an F-isometry*

$$s_*((\hat{r}^* V) \otimes_K U) \cong V \otimes_F s_*(U).$$

In particular, setting $U = \langle 1 \rangle_K$, we have

$$s_*(\hat{r}^* V) \cong V \otimes_F s_*(\langle 1 \rangle_K).$$

Proof. To simplify notations, we shall write $\langle\ ,\ \rangle$ for *all* inner products. But, wherever necessary, we shall write subscripts to indicate the quadratic spaces being considered. Define a map

$$f\colon\ s_*((K \otimes_F V) \otimes_K U) \longrightarrow V \otimes_F s_*(U)$$

by $(k \otimes_F v) \otimes_K u \mapsto v \otimes_F (ku)$. This is (easily seen to be) an F-isomorphism, with inverse provided by $v \otimes u \mapsto (1 \otimes v) \otimes u$. *We claim that f is the required*

F-*isometry.* To check this, take $k, k' \in K$, $v, v' \in V$, and $u, u' \in U$. In $V \otimes_F s_*(U)$, we have

$$\begin{aligned}\langle f((k \otimes v) \otimes u),\, f((k' \otimes v') \otimes u')\rangle &= \langle v \otimes (ku),\, v' \otimes (k'u')\rangle \\ &= \langle v,\, v'\rangle_V \cdot \langle ku,\, k'u'\rangle_{s_*U} \\ &= \langle v,\, v'\rangle_V \cdot s(kk'\langle u,\, u'\rangle_U).\end{aligned}$$

On the other hand, calculating in $s_*((K \otimes_F V) \otimes_K U)$, we get

$$\begin{aligned}\langle (k \otimes v) \otimes u,\, (k' \otimes v') \otimes u'\rangle_{s_*(\cdots)} &= s\big(\langle k \otimes v,\, k' \otimes v'\rangle_{K\otimes V}\langle u,\, u'\rangle_U\big) \\ &= s\big(kk'\langle v,\, v'\rangle_V \cdot \langle u,\, u'\rangle_U\big) \\ &= \langle v,\, v'\rangle_V \cdot s\big(kk'\langle u,\, u'\rangle_U\big),\end{aligned}$$

since $\langle v, v'\rangle_V \in F$, and s is F-linear. This checks that f is an F-isometry, as desired. □

Corollary 1.4. *If U is a hyperbolic space over K, then $s_*(U)$ is a hyperbolic space over F.*

Proof. Since $s_*(U_1 \perp U_2) \cong s_*(U_1) \perp s_*(U_2)$, it suffices for us to check the case $U = \mathbb{H}_K$. Using the last statement in 1.3, we have

$$s_*(U) = s_*(\mathbb{H}_K) = s_*(\hat{r}^*(\mathbb{H}_F)) \cong \mathbb{H}_F \otimes_F s_*(\langle 1\rangle_K).$$

The last quadratic space is $\cong [K\colon F] \cdot \mathbb{H}_F$, as desired. □

Corollary 1.5. (1) *$U \mapsto s_*(U)$ defines group homomorphisms*

$$s_*\colon \widehat{W}(K) \longrightarrow \widehat{W}(F) \quad \text{and} \quad s_*\colon W(K) \longrightarrow W(F).$$

(2) *The composition*

$$\widehat{W}(F) \xrightarrow{\hat{r}^*} \widehat{W}(K) \xrightarrow{s_*} \widehat{W}(F)$$

coincides with the multiplication by $s_*(\langle 1\rangle_K)$. (Same for W.)

(3) $\mathrm{im}(s_*)$ *is an ideal in $\widehat{W}(F)$. (Same for W.)*

Proof. All statements follow from 1.3 and 1.4. □

Remarks 1.6.

(A) To put Theorem 1.3 in a better perspective, we note that $s_*\colon \widehat{W}(K) \to \widehat{W}(F)$ is a morphism of $\widehat{W}(F)$-modules, if we view $\widehat{W}(K)$ as a $\widehat{W}(F)$-module via the ring homomorphism $\hat{r}^*\colon \widehat{W}(F) \to \widehat{W}(K)$. (And again, same for W.) This is actually a very nice way to interpret the Frobenius Reciprocity Law in 1.3.

(B) Although s_* is not a ring homomorphism, it is nevertheless "functorial". Namely, if $F \subseteq K \subseteq L$ is a tower of finite extensions, and if $s\colon K \to F$ is F-linear and $t\colon L \to K$ is K-linear (both nonzero), then $(s \circ t)_* = s_* \circ t_*$.

(C) It is natural to ask to what extent does $s_*\colon \widehat{W}(K) \to \widehat{W}(F)$ depend on the choice of the (nonzero) F-linear functional s. Since $s_*(\langle 1\rangle_K)$ is a regular F-quadratic space by 1.1, *every* F-linear functional on K is of the form $z \mapsto s(kz)$ for some $k \in K$. Thus, for any (nonzero) $t\colon K \to F$, there exists a commutative diagram

$$\begin{array}{ccc} \widehat{W}(K) & \xrightarrow{\quad\langle k\rangle\cdot\quad} & \widehat{W}(K) \\ & {\scriptstyle t_*}\searrow \qquad \swarrow{\scriptstyle s_*} & \\ & \widehat{W}(F) & \end{array}$$

where $k \in \dot{K}$ is such that $t(z) = s(kz)$ for all $z \in K$. This says that s_* and t_* are the same up to a group automorphism of $\widehat{W}(K)$ (given by multiplication by the unary form $\langle k\rangle_K$). In particular, the ideal $s_*(\widehat{W}(K))$ is independent of the choice of s; it may be called the *transfer ideal* with respect to K/F. (Again, the same for W.)

(D) From the above, it follows that s_* *is onto iff the form* $\langle 1\rangle_F$ *lies in the image of* s_*.

We conclude with the following

Corollary 1.7. *Let* $T \subseteq W(F)$ *be the transfer ideal for a finite field extension* K/F*, and let* $W(K/F)$ *denote the kernel of* $r^*\colon W(F) \to W(K)$*. Then* $T \cdot W(K/F) = 0$.

Proof. Fixing any nonzero F-linear functional $s\colon K \to F$, we can take T to be $\operatorname{im}(s_*\colon W(K) \to W(F))$. For any K-quadratic space U and F-quadratic space V, Frobenius Reciprocity gives

$$s_*(V_K \otimes_K U) \cong V \otimes_F s_*(U). \tag{1.8}$$

If V_K is hyperbolic, the LHS is hyperbolic by 1.4. In this case, 1.8 shows that V is annihilated by T in $W(F)$. □

2. Simple Extensions and Springer's Theorem

Our first goal in this section is to look at a *simple* algebraic extension $K = F(x)$, and compute the important transfer space $s_*(\langle 1\rangle_K)$ with respect to a strategically chosen F-linear functional $s\colon K \to F$. This computation will be very useful for the purposes of understanding the natural Witt ring map $r^*\colon W(F) \to W(K)$.

Let $n = [K\colon F]$, where $K = F(x)$. An F-basis on K is given by $\{1, x, \ldots, x^{n-1}\}$, so we have a unique F-linear functional $s\colon K \to F$ given by

(2.1) $$s(1) = 1 \quad \text{and} \quad s(x) = s(x^2) = \cdots = s(x^{n-1}) = 0.$$

Theorem 2.2 (Scharlau [Sc$_1$]). *With respect to the notations above:*

(1) *If $n = 2m+1$, then $s_*(\langle 1\rangle_K) \cong m\,\mathbb{H}_F \perp \langle 1\rangle$.*

(2) *If $n = 2m$, then $s_*(\langle 1\rangle_K) \cong (m-1)\,\mathbb{H}_F \perp \langle 1,\, -\mathrm{N}_{K/F}(x)\rangle$.*

Proof. As in §1, we think of $s_*(\langle 1\rangle_K)$ as supported by the F-space K, with the F-bilinear form $(y,z) \mapsto s(yz)$. In this bilinear structure, the subspaces $F\cdot 1$ and $K_0 := \sum_{i=1}^{n-1} F\cdot x^i$ are clearly orthogonal to each other, so $s_*(\langle 1\rangle_K) \cong \langle 1\rangle_F \perp K_0$.

Case 1. $n = 2m+1$. The subspace K_0 has F-dimension $n-1 = 2m$, and $x, x^2, \dots, x^m$ clearly span a totally isotropic subspace of K_0. Therefore, $K_0 \cong m\,\mathbb{H}_F$ by I.3.4, and we are done.

Case 2. $n = 2m$. Here, $x, x^2, \dots, x^{m-1}$ span a totally isotropic subspace of K_0, so by I.3.4 again, there exists a splitting

$$K_0 \cong (m-1)\,\mathbb{H}_F \perp K',$$

where

$$\dim_F K' = \dim_F K_0 - 2(m-1) = 1.$$

It is easy to determine the structure of K' by using the determinant invariant. If $t^n + a_{n-1}t^{n-1} + \cdots + a_0$ is the minimal polynomial of x over F, the symmetric matrix associated with the F-quadratic space K_0 with respect to the basis $\{x, x^2, \dots, x^{n-1}\}$ is

$$\begin{pmatrix} 0 & 0 & \cdots & 0 & -a_0 \\ 0 & 0 & \cdots & -a_0 & * \\ \vdots & \vdots & \ddots & \vdots & \vdots \\ -a_0 & * & \cdots & * & * \end{pmatrix} \in \mathbb{M}_{2m-1}(F),$$

which has determinant $(-1)^{m-1}(-a_0)^{2m-1} = (-1)^m a_0^{2m-1}$. On the other hand,

$$d((m-1)\,\mathbb{H}_F \perp K') = (-1)^{m-1} d(K').$$

Thus, $d(K') = -a_0^{2m-1}\dot{F}^2 = -a_0\dot{F}^2$, which means that $K' \cong \langle -a_0\rangle$. Observing that

$$\mathrm{N}_{K/F}(x) = (-1)^n a_0 = (-1)^{2m} a_0 = a_0,$$

we deduce that

$$s_*(\langle 1\rangle_K) \cong \langle 1\rangle_F \perp K_0 \cong (m-1)\mathbb{H}_F \perp \langle 1,\, -\mathrm{N}_{K/F}(x)\rangle. \qquad \square$$

For the same functional s as defined in 2.1, there exists an entirely analogous calculation for the transfer space $s_*(\langle x\rangle_K)$. Here, we think of $s_*(\langle x\rangle_K)$ as supported again by the F-space K, but the F-bilinear form to be considered now is $(y, z) \mapsto s(xyz)$. The computation of $s_*(\langle x\rangle_K)$ is as follows.

Theorem 2.3. *With respect to the notations in* 2.2:

(1) *If* $n = 2m + 1$, *then* $s_*(\langle x\rangle_K) \cong m\,\mathbb{H}_F \perp \langle \mathrm{N}_{K/F}(x)\rangle$.

(2) *If* $n = 2m$, *then* $s_*(\langle x\rangle_K) \cong m\,\mathbb{H}_F$.

Proof. (2) $n = 2m$. Here, the vectors $1, x, \ldots, x^{m-1}$ span a totally isotropic subspace of half the dimension of K. Thus, the desired conclusion follows directly from I.3.4.

(1) $n = 2m + 1$. The vectors $1, x, \ldots, x^{m-1}$ span a totally isotropic subspace of K, so I.3.4 implies

$$s_*(\langle x\rangle_K) \cong m\,\mathbb{H}_F \perp \langle d\rangle$$

for some $d \in \dot{F}$. By a determinant calculation similar to that in Case (2) of 2.2, we can check that $\langle d\rangle \cong \langle \mathrm{N}_{K/F}(x)\rangle$. □

Combining 2.2 and 2.3, we get the following.

Corollary 2.4. *For any simple algebraic extension* $K = F(x)$, *we have*

$$s_*\langle 1, -x\rangle = \langle 1, -\mathrm{N}_{K/F}(x)\rangle \in W(F).$$

Applying 2.2 to the study of the functorial map r^*, we obtain the following.

Theorem 2.5 (Scharlau). *Let* $K = F(x)$ *be a simple algebraic extension of degree* n.

(1) *If* $n = 2m + 1$, *then* $r^* \colon W(F) \to W(K)$ *is a split monomorphism in the category of* $W(F)$*-modules, and the transfer ideal for* K/F *is the unit ideal* $W(F)$.

(2) *If* $n = 2m$, *then the kernel* $W(K/F)$ *of* $r^* \colon W(F) \to W(K)$ *is annihilated by the form* $\langle 1, -\mathrm{N}_{K/F}(x)\rangle$.

Proof. We use again the functional $s \colon K \to F$ constructed in 1.7.

Case 1. $n = 2m + 1$. By 1.5, the composition

$$W(F) \xrightarrow{r^*} W(K) \xrightarrow{s_*} W(F)$$

coincides with the multiplication by $s_*(\langle 1\rangle_K)$, and this is just $\langle 1\rangle_F$ in $W(F)$ by 2.2(1). Thus, $s_* r^* = \mathrm{Id}_{W(F)}$. Since s_* is a morphism of $W(F)$-modules, the first conclusion of (1) follows. The fact that $s(\langle 1\rangle_K) = \langle 1\rangle_F \in W(F)$ also implies that $\operatorname{im}(s_*) = W(F)$.

Case 2. By 2.2(2), $\langle 1, -\mathrm{N}_{K/F}(x)\rangle$ belongs to the transfer ideal of K/F, so it annihilates $W(K/F)$ by 1.7. $\square$

Corollary 2.6. *Let* $r\colon F \hookrightarrow K$ *be an extension of odd degree. Then*

$$r^*\colon W(F) \to W(K)$$

is a split monomorphism in the category of $W(F)$*-modules.*

Proof. K can be obtained from F by successive simple extensions, so the conclusion follows by repeated applications of 2.5(1). $\square$

An obvious consequence of 2.6 is the fact that, if K/F is an odd degree extension, then the natural map $\dot F/\dot F^2 \to \dot K/\dot K^2$ on the square class groups is injective. This is clear in any case, since the existence of an element $a \in (\dot F \cap \dot K^2) \setminus \dot F^2$ would have yielded a quadratic subextension $F(\sqrt{a})/F$ in K, in contradiction to $[K\colon F]$ being odd. For an arbitrary field extension K/F, however, the injectivity of $W(F) \to W(K)$ is generally a stronger condition than the injectivity of $\dot F/\dot F^2 \to \dot K/\dot K^2$; for more discussions on this issue, see Examples 2.10–2.14 below.

Another relevant remark is that the injectivity of $r^*\colon W(F) \to W(K)$ for odd extensions could have been proved without using the method of transfer. Historically, the first proof of the injectivity of r^* in 2.6 was found by T. A. Springer some fifteen years before the formulation of the transfer technique. We shall now present Springer's result (from [Sp₁]), which is, in fact, considerably stronger than the injectivity of r^*. This result will be useful in §4 below, as well as later in Chapter VIII.

Theorem 2.7 (Springer). *Let* $F \subseteq K$ *be an extension of odd degree. If an* F*-quadratic form* q *is anisotropic over* F*, then* q_K *is anisotropic over* K*. (This implies the injectivity part of* 2.6*, in view of* II.1.4.)

Proof. Suppose $(K/F, q)$ is a counterexample with $n = [K\colon F]$ minimal. Clearly, $n > 1$, and $K = F(x)$ for some x. Let $p(t) \in F[t]$ be the minimal polynomial of x over F. Since q_K is isotropic, there exists an equation

$$q(g_1(t), \dots, g_d(t)) = p(t)h(t) \in F[t], \tag{2.8}$$

where $d = \dim q$, $m = \max_j(\deg g_j) \le n-1$, and the g_j's are not all zero. We may assume that no irreducible polynomial $f(t)$ divides all g_j (for otherwise $f^2|h$ and we could have cancelled out f^2 from 2.8). This condition means that $\sum_j F[t] \cdot g_j(t) = F[t]$, so in particular, the g_j's cannot have a common zero in the algebraic closure $\overline{F}$ of F. Since q itself is anisotropic, the LHS of 2.8 has degree $2m \le 2n-2$, so $h(t)$ has *odd* degree $\le n-2$.

Now pick any root $y \in \overline{F}$ of an irreducible odd degree factor of h in $F[t]$. Plugging y into (2.8), we see that $(g_1(y), \dots, g_d(y))$ is an isotropic vector for $q_{F(y)}$. But by choice, $[F(y): F]$ is odd and $\leq n-2$, which contradicts the minimal choice of n. □

We note in passing that Springer's theorem 2.7 was originally a conjecture of Witt. The following is essentially an equivalent version of it.

Corollary 2.9. *Let K/F be as in 2.7, and let $a \in \dot{F}$. For any quadratic form q_0 over F, q_0 represents a over F iff q_0 represents a over K.*

Proof. The "only if" part is clear, and the "if" part follows by applying 2.7 to the form $q := q_0 \perp \langle -a \rangle$. □

We close this section with a discussion on the relationship between the following conditions, where K/F is any finite field extension:

(A) $\dot{F}/\dot{F}^2 \to \dot{K}/\dot{K}^2$ is injective.

(B) $r^*: W(F) \to W(K)$ is injective.

(C) Anisotropic F-forms remain anisotropic over K.

(D) $[K: F]$ is odd.

In general, we have (D) $\Rightarrow$ (C) $\Rightarrow$ (B) $\Rightarrow$ (A), but the discussions below show that the reverse implications are not true.

Example 2.10. Let K be a finite extension over a finite field F. In this case, (A), (B), (C), (D) turn out to be equivalent, and each of them is equivalent to $r^*: W(F) \to W(K)$ being an isomorphism. The proof of this is left as an exercise.

The next example, showing that (A) $\not\Rightarrow$ (B) in general, requires some familiarity with Galois theory.

Example 2.11. To produce an extension K/F satisfying (A) but not (B), we must work in the case where $n = [K: F]$ is even and ≥ 4. Let us construct an explicit example with $n = 4$. Let L be the splitting field of the irreducible polynomial

$$f(x) = 4x^4 + 12x + 9 = 4(x^4 + 3x + 9/4) \tag{2.12}$$

over $F = \mathbb{Q}$. The resolvent cubic[(1)] for $f(x)$ is $g(x) = x^3 - 9x - 9$, which has discriminant 3^6. This being a square in $\mathbb{Q}$, the Galois group of g is A_3. From this, it follows that the Galois group of f is A_4. Now let θ be a root of f in L, and let $K = F(\theta)$. The subgroup $H \subseteq A_4$ corresponding to K has index 4, so H is a maximal subgroup. This means that there are no fields

[(1)] For the explicit computation of the resolvent cubic needed here, see Hungerford's "Algebra", p. 273, GTM Vol. 73, Springer-Verlag, 1974.

properly between F and K, so $\dot{F}/\dot{F}^2 \to \dot{K}/\dot{K}^2$ *is injective.* To construct a nonzero element in the kernel of r^*, note that, by completing squares,

$$0 = 4\theta^4 + 12\theta + 9 = 4\,(\theta^2 + 1)^2 - 2\,(2\theta - 3/2)^2 + 19/2.$$

Thus, $\langle 1, -2, 38\rangle$ is isotropic over K, so the F-form

$$q = \langle 1, -2, -19, 38\rangle = \langle 1, -2\rangle\,\langle 1, -19\rangle$$

is hyperbolic over K. However, q is the norm form of $\left(\frac{2,19}{\mathbb{Q}}\right)$: this is a division algebra since 2 is not a quadratic residue modulo 19 (by VI.5.6). Thus, q is an *anisotropic* form lying in the kernel of $W(F) \to W(K)$.

Example 2.13. W. Waterhouse has constructed a finite extension K/F of algebraic number fields for which (B) holds but (C) does not. This construction requires a considerable amount of number theory; for the details, we refer the reader to the exposition given by R. Ware in [War$_2$].

Example 2.14. It is much easier to construct an example K/F for which (A), (B), (C) hold but (D) does not. Indeed, consider any finite extension K/F where F is quadratically closed. Since $\dot{F}/\dot{F}^2$ is the trivial group and $W(F) = \{0, \langle 1\rangle_F\}$, (A), (B) and (C) obviously all hold. However, $[K\colon F]$ may not be odd. To see this, let us produce an extension of degree 4 over $F = \widetilde{\mathbb{Q}}$, the field of constructible numbers. (For properties of the constructible numbers needed here, and a more detailed discussion on the field $\widetilde{\mathbb{Q}}$, see §7 below.) Let $f(x) \in \mathbb{Q}[x]$ be the irreducible polynomial in 2.12. Then $f(x)$ has no root in $\widetilde{\mathbb{Q}}$. (For, if $\theta \in \widetilde{\mathbb{Q}}$ is a root, then $\mathbb{Q}(\theta)$ would be contained in a finite 2-extension of $\mathbb{Q}$. This is impossible by Galois theory since there are no fields strictly between $\mathbb{Q}$ and $\mathbb{Q}(\theta)$.) On the other hand, there are no irreducible quadratic polynomials over $\widetilde{\mathbb{Q}}$, since $\widetilde{\mathbb{Q}}$ is quadratically closed. Therefore, $f(x)$ *remains irreducible over* $\widetilde{\mathbb{Q}}$, so if θ is any root of $f(x)$, then $\widetilde{\mathbb{Q}}(\theta)$ is a degree 4 extension of $\widetilde{\mathbb{Q}}$. (For other constructions of quartic extensions of $\widetilde{\mathbb{Q}}$, see Exercise 24.) Later in §7, we shall show that $\widetilde{\mathbb{Q}}$ has, in fact, finite extensions of *any* prescribed degree > 2.

3. Quadratic Extensions

In this section, we shall investigate the behavior of r^* in the case of an even (mainly quadratic) extension. As we have observed at the beginning of §1, 2.6 and 2.7 break down completely if $[K\colon F]$ is even. To see more clearly how things happen, we focus our attention here on the case of quadratic extensions and formulate several basic results in this case. Throughout this section, $K = F(\sqrt{a})$ *denotes a fixed quadratic field extension of* F, *and we write* $\alpha := \langle 1, -a\rangle$, which is an anisotropic F-form that becomes the hyperbolic plane over K. As it turns out, the form α "controls" much of what happens under the quadratic extension $r\colon F \to K$.

Theorem 3.1. *Let q be an anisotropic form over F. Then q_K is isotropic over K iff q contains a binary subform isometric to $\langle b\rangle\cdot\alpha$ for some $b\in\dot{F}$.*

Proof. The "if" part is clear, since $\alpha_K\cong\mathbb{H}_K$. Conversely, let $\langle b_1,\ldots,b_n\rangle$ be a diagonalization of q, and assume that q_K is isotropic. Then there exists an equation

$$\sum_i b_i\big(x_i+y_i\sqrt{a}\,\big)^2=0,$$

where $x_i,\,y_i\in F$ are not all zero. Considering the "rational" and the "irrational" parts of this equation, we get

$$\sum b_ix_i^2+a\sum b_iy_i^2=0,\quad\text{and}\quad\sum b_ix_iy_i=0.$$

The latter says that the vectors $x=(x_1,\ldots,x_n)$ and $y=(y_1,\ldots,y_n)$ are orthogonal in the quadratic space (F^n,q). The other equation says $q(x)=-a\,q(y)$, which implies that x and y must *both* be nonzero (since q is anisotropic). Therefore, q contains the binary form

$$\langle q(x),\,q(y)\rangle=\langle -a\,q(y),\,q(y)\rangle\cong q(y)\cdot\alpha,$$

as desired. □

By specializing 3.1, we can now determine the (anisotropic) F-forms that become hyperbolic over K. Both 3.1 and 3.2 are due to A. Pfister ([Pf$_3$]).

Theorem 3.2. *An anisotropic F-form q becomes hyperbolic over K iff $q\cong\theta\otimes\alpha$ for some F-form θ. In particular, the kernel of $r^*\colon W(F)\to W(K)$ is given by the principal ideal $W(F)\cdot\alpha$.*

Proof. The sufficiency part is clear, again because $\alpha_K\cong\mathbb{H}_K$. For the necessity part, we induct on $m=(\dim q)/2$. The case $m=0$ is trivial, and starts the induction. If $m>0$, the lemma gives an isometry $q\cong\langle b\rangle\alpha\perp q'$, where $b\in\dot{F}$ and $(\dim q')/2=m-1$. By Witt's Cancellation Theorem I.4.2, $(q')_K$ is hyperbolic over K. Our inductive hypothesis then gives a form θ' such that $q'\cong\theta'\otimes\alpha$. We now have

$$q\cong\langle b\rangle\,\alpha\perp(\theta'\otimes\alpha)\cong\theta\otimes\alpha,\quad\text{where }\theta=\langle b\rangle\perp\theta',$$

as desired. □

Corollary 3.3. *Let q be an F-form of dimension $2m$ that becomes hyperbolic over $K=F(\sqrt{a}\,)$. Then:*

(1) *$-a\cdot q\cong q$ over F.*

(2) *If q is anisotropic over F, then $d(q)=(-a)^m$ and $d_\pm(q)=a^m$.*

(3) *If q also becomes hyperbolic over $F(\sqrt{-a}\,)$, then $2q=0\in W(F)$.*

Proof. By 3.2, we can write $q \cong r \cdot \mathbb{H}_F \perp \theta \otimes \alpha$ for some F-form θ. Since $-a \cdot \mathbb{H}_F \cong \mathbb{H}_F$ and $-a \cdot \alpha \cong \alpha$, it follows that $-a \cdot q \cong q$. If q is anisotropic over F, we have $r = 0$, so $\dim \theta = m$, and computing determinants from $q \cong \theta \otimes \alpha$ shows that

$$d(q) = (-a)^m, \quad d_\pm(q) = (-1)^m d(q) = a^m \quad \text{in } \dot{F}/\dot{F}^2.$$

Finally, for (3), assume that q is also hyperbolic over $F(\sqrt{-a}\,)$. If $F(\sqrt{-a}\,) = F$, then $q = 0 \in W(F)$. If $F(\sqrt{-a}\,) \neq F$, then by (1) (applied to the quadratic extension $F(\sqrt{-a}\,)/F$), we have $a \cdot q \cong q$, along with $-a \cdot q \cong q$. Adding these, we get $2q = 0 \in W(F)$. □

The next step is to compute the transfer ideal (defined in 1.6(C)) for the quadratic extension K/F. For this computation, the most convenient F-linear functional $s \colon K \to F$ turns out to be the one defined by $s(1) = 0$ and $s(\sqrt{a}\,) = 1$ (which is *not* the one in 2.1 for $x = \sqrt{a}\,$).

Theorem 3.4. *Let $s \colon K \to F$ be defined as above. Then:*

(1) *For any $x \in \dot{K}$, $s_*\langle x\rangle \cong d\langle 1, -\mathrm{N}_{K/F}(x)\rangle$ for some $d \in \dot{F}$. (In particular, $s_*(\langle b\rangle_K) \cong \mathbb{H}_F$ for every $b \in \dot{F}$.)*

(2) *The transfer ideal T for K/F is generated by the binary forms $\langle 1, -\mathrm{N}_{K/F}(x)\rangle$, where $x \in \dot{K}$. (In particular, $T \subseteq IF$.)*

(3) *$T = \operatorname{ann}(\alpha)$ (the annihilator of α in $W(F)$).*

Proof. (1) The transfer $s_*\langle x\rangle$ has underlying space K, and F-bilinear form $(u, v) \mapsto s(xuv)$. With respect to the F-basis $\{1, \sqrt{a}\,\}$ on K, this bilinear form has the symmetric matrix $\begin{pmatrix} s(x) & s(x\sqrt{a}\,) \\ s(x\sqrt{a}\,) & s(ax) \end{pmatrix}$. If $x = b + c\sqrt{a}$ (where $b, c \in F$), this matrix has determinant

$$\begin{vmatrix} c & b \\ b & ac \end{vmatrix} = -(b^2 - ac^2) = -\mathrm{N}_{K/F}(x).$$

Thus, if $d \in \dot{F}$ is any value represented by $s_*\langle x\rangle$, we have[(2)] $s_*\langle x\rangle \cong d\,\langle 1, -\mathrm{N}_{K/F}(x)\rangle$.

(2) This follows from (1), since T is an ideal in $W(F)$, and $W(K)$ is additively generated by the unary forms $\langle x\rangle$ $(x \in \dot{K})$.

(3) Since $\alpha_K \cong \mathbb{H}_K$, Frobenius Reciprocity 1.3 implies that $T \subseteq \operatorname{ann}(\alpha)$. (We could equally well get this information from (1) too. Since the norm form for K/F is precisely $\langle 1, -a\rangle = \alpha$, we have $\mathrm{N}_{K/F}(x) \cdot \alpha \cong \alpha$ for any $x \in \dot{K}$. This shows that $\alpha \cdot s_*\langle x\rangle = 0$, and so $\alpha \cdot T = 0$.) To see that

[(2)] In fact, it will be essentially correct to write $s_*\langle x\rangle = c\langle 1, -\mathrm{N}_{K/F}(x)\rangle \in W(F)$. For, if $c \neq 0$, then $c = s(x) \in D_F(\langle x\rangle)$, so we could have taken d to be c. On the other hand, if $c = 0$, then $s_*\langle x\rangle = 0 \in W(F)$ and the equation would still be true!

$\operatorname{ann}(\alpha) \subseteq T$, consider any F-form $q \in \operatorname{ann}(\alpha)$. This means that $q \cong a \cdot q$. Since $a \notin \dot{F}^2$, taking determinants shows that $\dim q$ is even. By Chapter II, Exercise 19, we can then decompose q into $q_1 \perp \cdots \perp q_m$, where each q_i is a binary F-form such that $q_i \cong a \cdot q_i$. Let $q_i = c_i\langle 1, -b_i\rangle$. Then $\langle 1, -b_i\rangle \cong a\langle 1, -b_i\rangle$, so $b_i \in D_F(\alpha)$. Thus, $b_i = \mathrm{N}_{K/F}(x_i)$ for some $x_i \in \dot{K}$. By (1), $q_i \in c_i\dot{F} \cdot T$. Since T is an ideal, it follows that $q \in \sum_i c_i\dot{F} \cdot T \subseteq T$, as desired. □

The above result is from [EL$_6$], which also contained the first full statement of the following fact.

Exact Triangle Theorem 3.5. *Let $r\colon F \hookrightarrow K$ be the inclusion map, and let $s\colon K \to F$ be as in 3.4. If t denotes the multiplication by α on $W(F)$, then we have an exact triangle:*

(3.6)
$$\begin{array}{ccccc} & & W(K) & & \\ & \overset{r^*}{\nearrow} & & \overset{s_*}{\searrow} & \\ W(F) & & \xleftarrow{\quad t \quad} & & W(F) \end{array}$$

Alternatively, we have a symmetric exact hexagon

(3.7)
$$\begin{array}{ccccc} & W(F)\cdot\alpha & \longrightarrow & W(F) & \\ \nearrow & & & & \overset{r^*}{\searrow} \\ 0 & & & & W(K) \\ \nwarrow & & & & \overset{s_*}{\swarrow} \\ & W(F)\cdot\alpha & \xleftarrow[t]{} & W(F) & \end{array}$$

Proof. In (3.6), exactness at $W(F)$ on the left is just a restatement of the last part of 3.2. Exactness at $W(F)$ on the right is a restatement of 3.4(3). Thus, it only remains to check the exactness at $W(K)$.

That $\operatorname{im}(r^*) \subseteq \ker(s_*)$ follows from the last part of 3.4(1). For the reverse inclusion, let (V, B) be an anisotropic K-quadratic space such that $s_*(V)$ is a hyperbolic F-space. In particular, $s_*(V)$ is isotropic, so $s(B(v,v)) = 0$ for some nonzero $v \in V$. The definition of s shows clearly that $b := B(v,v) \in \dot{F}$. Therefore, we have a K-splitting $V \cong \langle b\rangle_K \perp V'$. Since $s_*(\langle b\rangle_K)$ is hyperbolic (by 3.4(1)), the Cancellation Theorem implies that $s_*(V')$ is also hyperbolic. Now an obvious induction on $\dim_K V$ shows that *the K-quadratic space V is defined over F*; in particular, $[V] \in \operatorname{im}(r^*)$. □

By specializing 3.5, we obtain the following result on square class groups for a quadratic extension.

Theorem 3.8. *Let ε be the homomorphism from $\dot{F}/\dot{F}^2$ to the Brauer group $B(F)$ defined by $b\dot{F}^2 \mapsto \left(\frac{a,b}{F}\right)$. Then there is an exact sequence*

$$1 \longrightarrow \{\dot{F}^2, a\dot{F}^2\} \longrightarrow \dot{F}/\dot{F}^2 \xrightarrow{r^*} \dot{K}/\dot{K}^2 \xrightarrow{\mathrm{N}} \dot{F}/\dot{F}^2 \xrightarrow{\varepsilon} B(F).$$

Here, $r^*(c\dot{F}^2) = c\dot{K}^2$, and N is induced by $\mathrm{N}_{K/F}$.

Proof. If $r^*(c\dot{F}^2) = \dot{K}^2$, then $r^*\langle 1, -c\rangle \cong \mathbb{H}_K$. By 3.1, we have either $c \in \dot{F}^2$ or $\langle 1, -c\rangle \cong b\langle 1, -a\rangle$ for some $b \in \dot{F}$. The latter implies that $c \in a\dot{F}^2$.

Next, we have clearly $\mathrm{N} \circ r^* = 1$. If $x \in \dot{K}$ is such that $\mathrm{N}_{K/F}(x) \in \dot{F}^2$, then $s_*\langle x\rangle \cong \mathbb{H}_F$ by 3.4(1). The proof of 3.5 shows that the K-form $\langle x\rangle$ is defined over F; that is, $x\dot{K}^2 \subseteq \mathrm{im}(r^*)$.

Finally, for $c \in \dot{F}$, $\varepsilon(c) = 1$ iff $1 \in D_F\langle a, c\rangle$, and this is the case iff $c \in D_F\langle 1, -a\rangle = \mathrm{N}_{K/F}(\dot{K})$. □

Remark 3.9. We proved 3.8 by using 3.5 mainly to illustrate the viewpoint that the former is the "special case" of the latter for binary forms. In classical algebra, the proof of 3.8 would have proceeded as follows. To compute $\ker(r^*)$, suppose $c \in \dot{F} \cap \dot{K}^2$. Then $c = (r + s\sqrt{a}\,)^2$ for some $r, s \in F$. This implies that $2rs = 0$ and $c = r^2 + as^2$. If $s = 0$, then $c = r^2 \in \dot{F}^2$; if $r = 0$, then $c = as^2 \in a\dot{F}^2$. To compute $\ker(\mathrm{N})$, suppose $\mathrm{N}_{K/F}(x) = b^2$ for some $b \in \dot{F}$. Then $\mathrm{N}_{K/F}(xb^{-1}) = 1$, so by Hilbert's Theorem 90, $xb^{-1} = \overline{y}/y$ for some $y \in \dot{K}$ (where $\overline{y}$ denotes the "conjugate" of y in the quadratic extension K/F). Then $xy^2 = by\overline{y} \in \dot{F}$, so $x\dot{K}^2 \in \mathrm{im}(r^*)$. For an even more elementary proof (without Hilbert's Theorem), let $z = b - \overline{x}$. Then

$$xz^2 = x(x\,\overline{x} - 2b\overline{x} + \overline{x}^2) = (x\,\overline{x})\,(x + \overline{x} - 2b) \in F.$$

If $z \neq 0$, we are done. Otherwise, $\overline{x} = b$, so $x = \overline{b} = b \in \dot{F}$, and we are also done.

Let us now record some consequences of 3.8 and 3.5.

Corollary 3.10. *For a quadratic extension K/F, we have $|\dot{F}/\dot{F}^2| < \infty$ iff $|\dot{K}/\dot{K}^2| < \infty$. More precisely, we have the following inequalities:*

$$\frac{1}{2}\,|\dot{F}/\dot{F}^2| \leq |\dot{K}/\dot{K}^2| \leq \frac{1}{2}\,|\dot{F}/\dot{F}^2|^2.$$

Proof. This is immediate from 3.8. (For some refinements of the estimates here, see Exercise 4.) □

The complete analogue of the first statement in 3.10 does not work for Witt rings. For instance, for the quadratic extension $\mathbb{C}/\mathbb{R}$, $W(\mathbb{C}) \cong \mathbb{Z}_2$ is finite but $W(\mathbb{R}) \cong \mathbb{Z}$ is not. Nevertheless, we can state the following result (from [EL$_6$]).

Corollary 3.11. (1) *$W(K)$ is finite iff the endomorphism t (multiplication by $\langle 1, -a\rangle$) on $W(F)$ in 3.6 has finite kernel and finite cokernel.*

(2) *If $W(F)$ has finite rank (as an abelian group), then* $\operatorname{rank} W(K) = 2 \cdot \operatorname{rank}(\operatorname{coker}(t))$.

(3) *If $|W(F)| < \infty$, then $|W(K)| = |\operatorname{coker}(t)|^2$.*

Proof. (1) follows from the short exact sequence

$$0 \longrightarrow \operatorname{coker}(t) \longrightarrow W(K) \longrightarrow \ker(t) \longrightarrow 0,$$

which can be gleaned from the exact triangle in 3.6.

(2) If $W(F)$ has finite rank, then all groups in the exact hexagon 3.7 also do, and we have

$$\begin{aligned} 0 &= \operatorname{rk}(W(F)\alpha) - \operatorname{rk}(W(F)) + \operatorname{rk}(W(K)) - \operatorname{rk}(W(F)) + \operatorname{rk}(W(F)\alpha) \\ &= \operatorname{rk}(W(K)) - 2\,[\operatorname{rk}(W(F)) - \operatorname{rk}(W(F)\alpha)\,]. \end{aligned}$$

Since $\operatorname{rk}(W(F)) - \operatorname{rk}(W(F)\alpha) = \operatorname{rk}(\operatorname{coker}(t))$, (2) follows.

(3) is proved by the same calculation, with the ranks replaced by the (finite) cardinalities, and with addition replaced by multiplication. □

Note that the formula in (3) does not work (in general) if the group $W(F)$ is only finitely generated (instead of finite). For instance, for $F = \mathbb{R}$ and $K = \mathbb{C}$ again, $|\operatorname{coker}(t)|^2 = |\mathbb{Z}/2\mathbb{Z}|^2 = 4$, but $|W(K)| = 2$.

Using 3.10, we do get some information on the square class group under arbitrary *finite* field extensions.

Theorem 3.12. *Let K/F be any finite extension. If $|\dot{K}/\dot{K}^2| < \infty$, then $|\dot{F}/\dot{F}^2| < \infty$. The converse holds also, provided K/F is Galois and $[K : F]$ is a power of 2.*

Proof. Among all subfields of K that can be obtained from F by successive quadratic extensions, let E be one with $[E\colon F]$ maximal. Then E has no quadratic extensions within K, so $\dot{E}/\dot{E}^2 \to \dot{K}/\dot{K}^2$ is injective. If $|\dot{K}/\dot{K}^2| < \infty$, then $|\dot{E}/\dot{E}^2| < \infty$. Applying the "if" part of 3.10 a finite number of times, we get $|\dot{F}/\dot{F}^2| < \infty$.

If K/F is Galois and $[K\colon F] = 2^n$, then $G = \operatorname{Gal}(K/F)$ is a finite 2-group. Thus, there is a subgroup series

$$G = G_0 \supset G_1 \supset \cdots \supset G_n = \{1\}$$

such that $[G_i\colon G_{i+1}] = 2$ for all i. Taking $F_i = K^{G_i}$, we get a tower of fields

$$F = F_0 \subset F_1 \subset \cdots \subset F_n = K$$

where $[F_{i+1}\colon F_i] = 2$ for all i. Thus, if $|\dot{F}/\dot{F}^2| < \infty$, the "only if" part of 3.10 implies that $|\dot{K}/\dot{K}^2| < \infty$. □

For the first part of 3.12, some estimates on $|\dot{F}/\dot{F}^2|$ are possible (in terms of $\dot{K}/\dot{K}^2$). For more details, we refer the reader to Exercise 6. As for the question of whether $|\dot{F}/\dot{F}^2| < \infty$ implies $|\dot{K}/\dot{K}^2| < \infty$, the second part of 3.12 only dealt with the special case of a Galois extension of a 2-power degree. We shall return to consider the more general case of this question in §7 below.

For later reference, it will be useful for us to record here some further results on the behavior of the transfer map s_* under a quadratic extension. Thus, we go back to the notation $K = F(\sqrt{a}\,)$ for such an extension, and let $s: K \to F$ be, again, the F-linear functional defined by $s(1) = 0$ and $s(\sqrt{a}\,) = 1$. To simplify notations, we'll use (ahead of Chapter X) the symbol $\langle\langle z_1, \ldots, z_n \rangle\rangle$ $(z_i \in \dot{K})$ for the "n-fold Pfister form" $\langle 1, z_1\rangle \otimes \cdots \otimes \langle 1, z_n\rangle$ over K. Note that these forms generate I^nK as an additive group. We begin with a lemma on the ideal $I^2K \subseteq W(K)$.

Lemma 3.13. *Let $K = F(\sqrt{a}\,)$ be a quadratic extension of F. Then I^2K coincides with the $W(F)$-module J in $W(K)$ generated by $\langle\langle e, z\rangle\rangle$, where $e \in \dot{F}$ and $z \in \dot{K}$.*

Proof. Let $x, y \in \dot{K}$ and $b \in \dot{F}$. From the equation

$$\langle\langle xb, y\rangle\rangle = \langle b\rangle \langle\langle x, y\rangle\rangle + \langle\langle -b, y\rangle\rangle \in W(K),$$

it follows that $\langle\langle x, y\rangle\rangle \in J \Longrightarrow \langle\langle xb, y\rangle\rangle \in J$. Thus, we need only show that $\varphi = \langle\langle c + \sqrt{a}, d - \sqrt{a}\,\rangle\rangle \in J$ for every $c, d \in F$. If $c = -d$, φ is hyperbolic. If $c \neq -d$, we see easily that

$$\varphi \cong \langle\langle c + d,\ (c + \sqrt{a}\,)(d - \sqrt{a}\,)\rangle\rangle \in J,$$

so we are done. □

Corollary 3.14. *For $s: K \to F$ as above, $s_*(I^nK) \subseteq I^nF$ for all $n \geq 1$.*

Proof. We may assume $n \geq 2$ (the case $n = 1$ being clear). By repeated applications of 3.13, we see that I^nK is generated as a $W(F)$-module by the Pfister forms $\langle\langle e_1, \ldots, e_{n-1}, z\rangle\rangle$, where $e_i \in \dot{F}$ and $z \in \dot{K}$. By Frobenius Reciprocity (1.3), we have

$$s_*\langle\langle e_1, \ldots, e_{n-1}, z\rangle\rangle = \langle\langle e_1, \ldots, e_{n-1}\rangle\rangle\, s_*\langle\langle z\rangle\rangle \in I^nF,$$

so the desired conclusion follows. □

Proposition 3.15. *For K/F as above, let r^* denote the functorial map from $W(F)$ to $W(K)$. For $n \geq 1$, we have a 0-sequence*

$$0 \longrightarrow \langle\langle -a\rangle\rangle I^{n-1}F \longrightarrow I^nF \xrightarrow{r^*} I^nK \xrightarrow{s_*} I^nF. \tag{3.16}$$

(1) *For $n = 1, 2$, this sequence is exact.*

(2) *For $n = 3$, this sequence is exact except possibly at I^3K.*

Proof. From 3.5 and 3.14 above we know that (3.16) is a 0-sequence for all n, and that it is exact for $n = 1$. Now let $n = 2$, and suppose q is an anisotropic form in I^2F with $r^*(q) = 0$. By 3.2, $q \cong \langle\langle -a\rangle\rangle\, q_1$ for some form q_1 over F. If $q_1 \notin IF$, then $d(q) = -a$, which contradicts $q \in I^2F$. Thus, $q_1 \in IF$, and $q \in \langle\langle -a\rangle\rangle IF$. Next, suppose $\gamma \in I^2K$ with $s_*(\gamma) = 0$. Then $\gamma \cong \langle a_1, \ldots, a_{2m}\rangle_K$ for suitable $a_i \in \dot{F}$. Since $d(\gamma) = (-1)^m$ over K, we must have $(-1)^m a_1 \cdots a_{2m} = 1$ or a, up to a square in $\dot{F}$. In the first case, $\gamma \in r^*(I^2F)$, as desired. In the second case,

$$\gamma \cong r^*(\langle aa_1, a_2, \ldots, a_{2m}\rangle) \in r^*(I^2F)$$

also. Finally, let $n = 3$ and let q be an anisotropic form in I^3F such that $r^*(q) = 0$. Then $q \cong \langle\langle -a\rangle\rangle q_1$, where $\dim q_1 = 2m$ for some m. Write

$$q_1 = \langle\langle (-1)^m d\rangle\rangle + q_2 \in W(F), \quad \text{where } d = d(q_1) \text{ and } q_2 \in I^2F.$$

Then $q = \langle\langle -a,\, (-1)^m d\,\rangle\rangle + \langle\langle -a\rangle\rangle\, q_2 \in I^3F$ implies that $\langle\langle -a,\, (-1)^m d\,\rangle\rangle \in I^3F$. Taking Witt invariants, we see that $\langle\langle -a,\, (-1)^m d\rangle\rangle = 0$, and thus $q = \langle\langle -a\rangle\rangle\, q_2 \in \langle\langle -a\rangle\rangle\, I^2F$. □

We close this section by discussing a technique of field construction that goes back to H. Gross and H. R. Fischer [GF]. In its original conception, this technique served to prove the existence of fields with certain prescribed square class groups. Later, this construction technique was generalized in other contexts to produce fields with other prescribed quadratic invariants (e.g. u-invariants and Pythagoras numbers of fields). We shall have occasions to apply the more general versions of the Gross-Fischer technique too. Here, since we have just presented the 5-term exact sequence (in 3.8) for square class groups under a quadratic extension, it is fitting to give a formal description of the original Gross-Fischer construction for fields with prescribed square class groups. (Versions of this construction technique have, in fact, been used somewhat informally in earlier parts of this book; see, e.g., II.5.)

Gross-Fischer Theorem 3.17. *Let $\{a_i\}$ be a subset of $\dot{F}$ representing a set of $\mathbb{Z}_2$-independent square classes in $\dot{F}/\dot{F}^2$. Then there exists a field K, obtainable from F by successive quadratic extensions, such that $\{a_i\}$ represents a $\mathbb{Z}_2$-basis of square classes for the group $\dot{K}/\dot{K}^2$.*

Proof. To begin with, take another subset $\{b_j\} \subseteq \dot{F}$ such that $\{a_i, b_j\}$ give a $\mathbb{Z}_2$-basis of square classes for $\dot{F}/\dot{F}^2$. Let $F_1 = F(\{\sqrt{b_j}\})$, the field obtained by adjoining the square roots of the elements b_j to F. By a judicious application of the beginning part of the exact sequence in 3.8, we see that the elements $\{a_i\}$ remain to represent $\mathbb{Z}_2$-independent square classes in $\dot{F}_1/\dot{F}_1^2$. We can thus add a set of elements $\{c_k\} \subseteq \dot{F}_1$ to $\{a_i\}$ so that $\{a_i, c_k\}$

form a $\mathbb{Z}_2$-basis of square classes for $\dot{F}_1/\dot{F}_1^2$, and construct $F_2 = F_1(\{\sqrt{c_k}\})$. Continuing this construction, we obtain a field tower

$$F \subseteq F_1 \subseteq \cdots \subseteq F_n \subseteq \cdots .$$

Taking $K = \bigcup_{n\geq 1} F_n$, we see easily that $\{a_i\}$ remain $\mathbb{Z}_2$-independent in $\dot{K}/\dot{K}^2$. We finish by showing that the square classes of $\{a_i\}$ in fact *generate* $\dot{K}/\dot{K}^2$. Given $x \in \dot{K}$, we have $x \in F_n$ for some $n \geq 1$. In the construction of F_{n+1} from F_n, we have used a square class basis of the form $\{a_i, d_\ell\}$. Therefore, up to a square in F_n, x can be written as $\alpha\gamma$, where α is a product of the a_i's, and γ is a product of the d_ℓ's. In F_{n+1}, the d_ℓ's all become squares, so up to a square factor, x is equal to α. Since this remains true in K, the square class of x in K is represented by a product of the elements in $\{a_i\}$, as desired. □

In applying the Gross-Fischer construction, it is sometimes necessary to fine-tune the construction steps so as to guarantee that the field K obtained above has certain additional properties. This can be done by choosing the sets $\{b_j\}$, $\{c_k\}$, etc. more judiciously in the construction. We will not dwell on this point here, but will give an example of this in the form of an exercise (see Exercise 15).

4. Scharlau's Norm Principle

In this and the next section, we shall present a couple of norm principles for quadratic forms with respect to finite extensions of fields. The first one, due to W. Scharlau, concerns the groups of similarity factors of quadratic forms.

The idea of "similarity factors" has been used implicitly in earlier parts of this book. Let us now introduce this notion more formally, and give it a permanent notation. Recall that, for a quadratic form q over a field F and any scalar $a \in \dot{F}$, $a \cdot q$ means the tensor product of $\langle a \rangle$ with q. This operation makes $W(F)$ into an $\dot{F}$-module, in the usual sense of modules over multiplicative groups. The *group of similarity factors* of a form q is defined to be the isotropy subgroup of q under the above action; that is,

$$\begin{aligned} G(q) = G_F(q) &:= \{a \in \dot{F} \mid a \cdot q \cong q\} \\ &= \{a \in \dot{F} \mid a \cdot q = q \in W(F)\} \qquad (4.1) \\ &= \{a \in \dot{F} \mid \langle 1, -a \rangle \cdot q = 0 \in W(F)\}. \end{aligned}$$

Remarks 4.2. (1) Since $\dot{F}^2$ acts as the identity, $G(q)$ is a union of cosets of $\dot{F}$ modulo $\dot{F}^2$. Therefore, we may also think of $G(q)$ as a group of square classes (that is, we can work instead with the factor group $G(q)/\dot{F}^2$).

(2) If q is a hyperbolic form, then $G(q) = \dot{F}$.

(3) If $\dim q$ is odd, then $G(q) = \dot{F}^2$.

(4) If $b \in D_F(q)$, then $b \cdot G(q) \subseteq D_F(q)$; in other words, $D_F(q)$ is a union of cosets of $G_F(q)$. (To see this, let $q \cong \langle b \rangle \perp q'$. If $a \in G(q)$, then $q \cong a \cdot q \cong \langle ab \rangle \perp aq'$ implies that $ba \in D_F(q)$.) From the observation above, we see, in particular, that $G(q) \subseteq D_F(q)$ iff $1 \in D_F(q)$.

(5) If $q = \langle 1, a \rangle$, then $G(q) \subseteq D_F(q)$ is an equality. (For, if $b \in D_F(q)$, then $q \cong \langle b, ab \rangle \cong b\,q$.) This fact will be generalized later to n-fold Pfister forms for any n.

We now come to the main result in this section.

Theorem 4.3 (Scharlau's Norm Principle). *Suppose K/F is a finite field extension, and q is a (regular) quadratic form over F. Then, for any $x \in \dot{K}$,*

$$x \in G_K(q_K) \implies \mathrm{N}_{K/F}(x) \in G_F(q).$$

Proof. *Step* 1. $[K\colon F(x)] = 2m$. In this case

$$\begin{aligned}\mathrm{N}_{K/F}(x) &= \mathrm{N}_{F(x)/F}(\mathrm{N}_{K/F(x)}(x)) \\ &= \mathrm{N}_{F(x)/F}(x^{2m}) \\ &= [\mathrm{N}_{F(x)/F}(x)^m]^2 \in \dot{F}^2 \subseteq G_F(q).\end{aligned}$$

Step 2. $[K\colon F(x)] = 2m+1$. In this case, $W(F(x)) \to W(K)$ is injective (by 2.5(1) or 2.7). Thus, $x \cdot q_K \cong q_K$ implies that $x \cdot q_{F(x)} \cong q_{F(x)}$. As in Step 1, $\mathrm{N}_{K/F}(x)$ can be checked to be in the same square class as $\mathrm{N}_{F(x)/F}(x)$. Thus, we are done if we can show that $\mathrm{N}_{F(x)/F}(x) \in G_F(q)$.

Step 3. We are now reduced to the case where $K = F(x)$ (a simple algebraic extension). Let $s\colon K \to F$ be the F-linear functional defined in 2.1. Applying the transfer s_* to the equation $\langle 1, -x \rangle \cdot q = 0 \in W(K)$ and using Frobenius Reciprocity, we get $s_*\langle 1, -x \rangle \cdot q = 0 \in W(F)$. But by 2.4, $s_*\langle 1, -x \rangle = \langle 1, -\mathrm{N}_{K/F}(x) \rangle$ in $W(F)$. Therefore, $\mathrm{N}_{K/F}(x) \in G_F(q)$, as desired. □

An interesting consequence of 4.3 is the following.

Corollary 4.4. *In the notation of 4.3, if q_K is hyperbolic over K, then $\mathrm{N}_{K/F}(\dot{K}) \subseteq G_F(q)$. If, moreover, $b \in D_F(q)$, then $b \cdot \mathrm{N}_{K/F}(\dot{K}) \subseteq D_F(q)$.*

Proof. This follows from 4.3, in view of (2) and (4) in Remark 4.2. □

In the case of a *quadratic* extension, Scharlau's Norm Principle can be appropriately sharpened into an "iff" statement. The following sharpening, for instance, was given in [EL$_6$].

Theorem 4.5. *Let $K = F(\sqrt{a}\,)$ be a quadratic extension of F, and let q be any F-form. For any $x \in \dot{K}$, we have $\mathrm{N}_{K/F}(x) \in G_F(q)$ iff $x \cdot q_K \cong q'_K$ for some F-form q'.*

Proof. The "if" part certainly implies Scharlau's Norm Principle (in the case of quadratic extensions), since we can take q' to be q if $x \in G_K(q_K)$. The "iff" statement can be proved in one stroke as follows.

Let $s\colon K \to F$ be the F-linear map defined by $s(1) = 0$ and $s(\sqrt{a}\,) = 1$. By 3.4(1) and Frobenius Reciprocity,

$$s_*(x \cdot q_K) \cong s_*\langle x\rangle \cdot q \cong d \cdot \big(q \perp \langle -\mathrm{N}_{K/F}(x)\rangle q\big)$$

for some $d \in \dot{F}$. Therefore,

$$\begin{aligned} \mathrm{N}_{K/F}(x) \in G_F(q) &\iff s_*(x \cdot q_K) \text{ is } F\text{-hyperbolic} \\ &\iff x \cdot q_K \cong q'_K \text{ for some } F\text{-form } q', \end{aligned}$$

where the last equivalence follows from the proof of the exactness at $W(K)$ in the Exact Triangle Theorem 3.5. (See the last statement in that proof.) □

5. Knebusch's Norm Principle

In §4, we have investigated the behavior of the group of similarity factors $G_F(q)$ with respect to the norm map for a finite field extension. It will be of interest to establish a parallel result for $D_F(q)$ (the set of values of $\dot{F}$ represented by the form q). However, the fact that $D_F(q)$ is not a group (in general) causes some technical difficulties. To rectify this, we allow ourselves to replace $D_F(q)$ by $[D_F(q)]$, the subgroup of $\dot{F}$ generated by $D_F(q)$. Then, everything works out nicely, and we have the following result, due to M. Knebusch [Kn$_1$].

Theorem 5.1 (Knebusch's Norm Principle). *Suppose K/F is a finite extension of degree n, and q is a (regular) quadratic form over F. Let $x \in \dot{K}$. If $x \in D_K(q_K)$, then $\mathrm{N}_{K/F}(x)$ is a product of n elements of $D_F(q)$. (In particular, $\mathrm{N}_{K/F}(x) \in [D_F(q)]$.)*

Proof. We may assume q is anisotropic (for otherwise $D_F(q) = \dot{F}$). For $r \geq 1$, let us write $D_F^r(q)$ for the set of products of r elements of $D_F(q)$. Note that

$$\dot{F}^{2r} \subseteq D_F^{2r}(q). \tag{5.2}$$

To see this, it suffices to show that $\dot{F}^2 \subseteq D_F^2(q)$. Fix any $a \in D_F(q)$. Then, for any $b \in \dot{F}$,

$$b^2 = a \cdot [(b/a)^2 a] \in D_F(q)\, D_F(q) = D_F^2(q).$$

The proof of 5.1 proceeds by induction on $n = [K : F]$. If $n = 1$, there is nothing to prove, so we assume $n > 1$ in the following.

Step 1. We first handle the case where $x \in F$. If $n = 2t$ (an even integer), then $\mathrm{N}_{K/F}(x) = x^{2t} \in \dot{F}^{2t} \subseteq D_F^n(q)$ by (5.2). Now assume $n = 2t + 1$ (an odd integer). By Springer's Theorem (in the form 2.9), $x \in D_K(q_K)$ implies $x \in D_F(q)$. Thus,

$$\mathrm{N}_{K/F}(x) = x^{2t+1} \in x \cdot \dot{F}^{2t} \subseteq D_F(q) \cdot D_F^{2t}(q) = D_F^n(q),$$

as desired.

Step 2. $x \notin F$. Consider the intermediate field $E = F(x) \supsetneq F$, and let $r = [K\colon E]$, $r' = [E\colon F]$. In this step, we assume that $E \subsetneq K$ (that is, $r > 1$), and will finish by induction. Note that now *both* r and r' are $< n$. By Step 1 (applied to q_E over the ground field E), we have $\mathrm{N}_{K/E}(x) \in D_E^r(q_E)$. Applying $\mathrm{N}_{E/F}$, we have $\mathrm{N}_{K/F}(x) \in \mathrm{N}_{E/F}(D_E^r(q_E))$. Using the inductive hypothesis on E/F, we have $\mathrm{N}_{E/F}(D_E(q_E)) \subseteq D_F^{r'}(q)$. By the multiplicativity of the norm map $\mathrm{N}_{E/F}$, we deduce that

$$\mathrm{N}_{K/F}(x) \in D_F^{rr'}(q) = D_F^n(q),$$

which is the desired conclusion.

Step 3. We are now reduced to the case where $K = F(x)$, a simple (algebraic) extension of F. Let $p(t)$ be the minimal polynomial of x over F. The hypothesis $x \in D_K(q_K)$, interpreted as $x^{-1} \in D_K(q_K)$, can be expressed by an equation

$$t \cdot q(g_1(t), \dots, g_d(t)) = 1 + p(t)h(t) \in F[t], \tag{5.3}$$

where $d = \dim q$ and h, $g_i \in F[t]$, with $m := \max\{\deg(g_i)\} \leq n - 1$. Since q is anisotropic, 5.3 shows that

$$n_0 := \deg h = 2m + 1 - n \leq 2(n-1) + 1 - n = n - 1. \tag{5.4}$$

At this point, the proof looks suspiciously similar to the earlier one given for Springer's Theorem! Write down a complete factorization of $h(t)$, say

$$h(t) = c \cdot h_1(t) \cdots h_s(t) \qquad (c \in \dot{F}), \tag{5.5}$$

where the h_i's are monic irreducible polynomials in $F[t]$. Then c is the leading coefficient of $1 + p(t)h(t)$. Since q is anisotropic, (5.3) shows that $c \in D_F(q)$.

Assume for the moment that, in (5.5), $s \geq 1$; that is, h is not a constant polynomial. Let x_i be a root of h_i in the algebraic closure of F. Plugging x_i into (5.3), we get

$$x_i^{-1} = q(g_1(x_i), \dots, g_d(x_i)) \in D_{F(x_i)}(q_{F(x_i)}).$$

Since $[F(x_i)\colon F] \le \deg h \le n-1$ (by (5.4)), our inductive hypothesis, applied to $F(x_i)/F$, yields

$$\mathrm{N}_{F(x_i)/F}(x_i) = (-1)^{\deg h_i} h_i(0) \in D_F^{\deg h_i}(q).$$

Taking the product of these over i, we get

$$(-1)^{n_0} h_1(0) \cdots h_s(0) = (-1)^{n_0} h(0)\, c^{-1} \in D_F^{n_0}(q).$$

Since $c \in D_F(q)$, we have

$$(5.6) \qquad (-1)^{n_0} h(0) \in D_F^{(n_0+1)}(q).$$

In proving this, we have assumed that h was nonconstant. Should h be a constant polynomial, $(-1)^{n_0} h(0)$ would be c, and (5.6) is true anyway. Now by (5.4), $n_0 + 1 \equiv n \pmod 2$, so we may use (5.2) to rewrite (5.6) as

$$(5.7) \qquad (-1)^{n+1} h(0) \in D_F^n(q).$$

Therefore, $\mathrm{N}_{K/F}(x) = (-1)^n p(0) = (-1)^{n+1} h(0)^{-1} \in D_F^n(q)$, which is what we want! □

Corollary 5.8. *(1) If $[K\colon F] = n$, and q is an F-form such that q_K is isotropic (or just universal), then $\mathrm{N}_{K/F}(\dot{K}) \subseteq D_F^n(q)$.*

(2) An anisotropic F-form q remains anisotropic over $F(\sqrt{a})$ unless $-a \in D_F^2(q)$.

Proof. (1) is clear, since the assumption(s) on q give $D_K(q_K) = \dot{K}$. For (2), we may assume that $K = F(\sqrt{a})$ is a quadratic extension of F (for otherwise there is nothing to prove). If q becomes isotropic over K, then (1) implies $-a = \mathrm{N}_{K/F}(\sqrt{a}) \in D_F^2(q)$. (This conclusion could also have been proved directly from 3.1.) □

As another application of Theorem 5.1, taking q to be the F-form $\langle 1, \ldots, 1\rangle$ yields the following memorable result.

Corollary 5.9. *For any finite extension K/F, we have*

$$\mathrm{N}_{K/F}\left(\sum K^2\right) \subseteq \sum F^2.$$

A particularly nice thing about this Corollary is that we have proved a result on sums of squares in fields, but without ever writing down *any* sums of squares in any field!

As was the case for Scharlau's Norm Principle, for quadratic extensions, Theorem 5.1 can be sharpened into an "iff" statement, following [EL$_6$].

Theorem 5.10. *Let $K = F(\sqrt{a})$ be a quadratic extension of F, and let q be any F-form. For any $x \in \dot{K}$,*

$$\mathrm{N}_{K/F}(x) \in D_F^2(q) \iff x \in \dot{F} \cdot D_K(q_K).$$

Proof. Again, the implication "$\Longleftarrow$" here contains Knebusch's Norm Principle for quadratic extensions. To prove "$\Longleftrightarrow$" in one stroke, we need only modify slightly the earlier proof given for 4.5, as follows:

$$\begin{aligned} \mathrm{N}_{K/F}(x) \in D_F^2(q) &\iff q \perp \langle -\mathrm{N}_{K/F}(x)\rangle\, q \text{ is } F\text{-isotropic} \\ &\iff s_*(x\cdot q_K) \text{ is } F\text{-isotropic (see 3.4)} \\ &\iff D_K(x\cdot q_K)\cap \dot F \neq \varnothing \\ &\iff x \in \dot F \cdot D_K(q_K). \end{aligned}$$

Here, as in the proof of 4.5, $s\colon K \to F$ is the F-linear functional defined by $s(1)=0$ and $s(\sqrt{a}\,)=1$. □

Remark 5.11. Of course, Knebusch's Norm Principle 5.1 would have looked simpler if its conclusion was

$$x \in D_K(q_K) \implies \mathrm{N}_{K/F}(x) \in D_F(q). \tag{5.12}$$

Unfortunately, this implication is not true in general. For instance, if $K = F(\sqrt{a}\,)$ is a quadratic extension and q is the unary form $\langle a\rangle$ over F, then $1 \in D_K(q_K)$ (since $q_K \cong \langle 1\rangle_K$), but $\mathrm{N}_{K/F}(1) = 1 \notin D_F(q)$ (since $a \notin \dot F^2$). Of course, in the special case where q is a "group form" (i.e., $D_F(q)$ is a group), (5.12) will be true for any finite extension K/F, by 5.1.

Remark 5.13. Note that Knebusch's Norm Principle implies that $\mathrm{N}_{K/F}$ maps the group $[D_K(q_K)]$ to the group $[D_F(q)]$ (for any F-form q). After reviewing the definition of the "spinor norm" of isometries in V.1, we may make the following statement. If σ is an isometry of the form q_K (for q above), then the spinor norm of $\sigma \in \mathrm{O}(q_K)$ goes under the norm map $\mathrm{N}_{K/F}$ to the spinor norm of a suitable isometry $\sigma_0 \in \mathrm{O}(q)$.

6. Galois Extensions and Trace Forms

This section, and to some extent also the next section, represent our first attempts at studying quadratic forms and Witt rings under Galois field extensions. More connections to Galois theory will be given later in VIII.5.

Let K be a field, and σ an automorphism of K. For any K-quadratic space (V, B), we can define a new K-quadratic space (V^σ, B^σ) as follows. The K-space V^σ is taken to have the same underlying abelian group as V, but with a new K-action

$$(k,\, v) \mapsto k * v = \sigma(k)v \qquad (k \in K,\ v \in V).$$

The new form B^σ on V^σ is defined by $B^\sigma(u,v) = \sigma^{-1}(B(u,v))$, for any $u, v \in V$. The K-linearity of B^σ can be checked as follows:

$$\begin{aligned} B^\sigma(k * u,\, v) &= \sigma^{-1}(B(\sigma(k)u,\, v)) = \sigma^{-1}(\sigma(k)B(u,\, v)) \\ &= k\sigma^{-1}(B(u,\, v)) = kB^\sigma(u,\, v), \end{aligned}$$

where $k \in K$ and $u,\, v \in V$.

The above construction induces a right action of the group $\operatorname{Aut}(K)$ on the Witt-Grothendieck ring $\widehat{W}(K)$. In terms of 1-dimensional forms, we simply have $\langle a\rangle^\sigma \cong \langle \sigma^{-1}(a)\rangle$, for any $a \in \dot{K}$. In light of this observation, we can quickly check that $\operatorname{Aut}(K)$ also acts on the Witt ring $W(K)$, via

$$(\langle a\rangle,\, \sigma) \mapsto \langle a\rangle^\sigma.$$

Let K/F be a finite field extension. If $s\colon K \to F$ is a nonzero F-linear functional on K, then it defines a Scharlau transfer $q \mapsto s_*(q)$ for K-quadratic forms q. We have observed before that, in case K is separable over F, we may take s to be the trace form $\mathrm{tr} = \mathrm{tr}_{K/F}$, since this trace is not identically zero. A transferred form $s_*(q)$ in this case will be denoted by $\mathrm{tr}_*(q)$. For Galois extensions K/F, we have the following basic result on $\mathrm{tr}_*(q)$ (from [KS]).

Theorem 6.1 (Knebusch-Scharlau). *Let K/F be a finite Galois extension, with Galois group $G = \operatorname{Gal}(K/F)$. Then, for any K-quadratic form q, there is a K-isometry*

$$K \otimes_F \mathrm{tr}_*(q) \cong \perp_{\sigma\in G} q^\sigma.$$

In particular, the form $\perp_{\sigma\in G} q^\sigma$ lies in $\operatorname{im}(\widehat{W}(F) \to \widehat{W}(K))$.

Proof. Let (V, B, q) be the K-quadratic space in question. The quadratic form $\perp_{\sigma\in G} q^\sigma$ is supported by the K-space $\perp_{\sigma\in G} V^\sigma$, while $\mathrm{tr}_*(q)$ is supported by V (as an F-space), with the transferred form $(x, y) \mapsto \mathrm{tr}(B(x, y))$. Thus, we need a K-isometry

$$f\colon K \otimes_F V \longrightarrow \perp_{\sigma\in G} V^\sigma.$$

We define f in the "obvious" way:

$$f(k \otimes_F v) = \sum_{\sigma\in G} k * v,$$

with the cautioning note that the summands $k*v$ here are formed in *different* V^σ's. To see that f is well-defined, we must show that it associates the same image to $ka \otimes v$ and $k \otimes av$, for any $a \in F$. This can be checked as follows (where all summations are over $\sigma \in G$):

$$\begin{aligned} f(ka \otimes v) &= \sum ka * v = \sum \sigma(ka)\, v \\ &= \sum \sigma(k) \cdot av = \sum k * (av) \\ &= f(k \otimes av). \end{aligned}$$

Next, f is K-linear. In fact,

$$\begin{aligned} f(k'(k \otimes v)) = f(k'k \otimes v) &= \sum k'k * v \\ &= \sum k' * (k * v) = k' * f(k \otimes v), \end{aligned}$$

bearing in mind that K acts "diagonally" on $\perp_{\sigma\in G} V^\sigma$. To show that f preserves inner products, we proceed as follows (where the first inner product $\widetilde{B}$ is for the space $\perp_{\sigma\in G} V^\sigma$, and all sums are again over $\sigma \in G$):

$$\begin{aligned} \widetilde{B}(f(k \otimes v), f(k' \otimes v')) &= \sum B^\sigma(k * v,\, k' * v') \\ &= \sum \sigma^{-1} B(\sigma(k)v,\, \sigma(k')v') \\ &= \sum \sigma^{-1} \left[\sigma(k)\sigma(k') B(v,\, v')\right] \\ &= kk' \sum \sigma(B(v,\, v')) \\ &= kk' \operatorname{tr}(B(v,\, v')) \\ &= (K \otimes \operatorname{tr}_*(B))\,(k \otimes v,\, k' \otimes v'). \end{aligned}$$

It remains only to show that f is an isomorphism. Since both $K \otimes_F V$ and $\perp_{\sigma\in G} V^\sigma$ have K-dimension $[K\colon F] \cdot \dim_K V$, it suffices to show that f is injective. Suppose $f(z) = 0$, where $z \in K \otimes_F V$. Since f preserves inner products, it follows that z lies in the radical of the K-quadratic space $(K \otimes_F V,\ K \otimes \operatorname{tr}_*(B))$. But this quadratic space is regular (by 1.1), so $z = 0$, as desired. □

Corollary 6.2. *Keep the notations in* 6.1, *and let* $W(K)^G$ *denote the subring of fixed points of* $W(K)$ *under the right action of* G. *Then the kernel of the transfer map* $\operatorname{tr}_*\colon W(K)^G \to W(F)$ *and the cokernel of the functorial map* $W(F) \to W(K)^G$ *are both killed by the field extension degree* $n = [K\colon F]$.

Proof. As in §1, let us write r for the inclusion map $F \hookrightarrow K$, so that we may write r^* for the functorial map $W(F) \to W(K)$. The image of r^* clearly lies in the subring $W(K)^G$. The composition

$$W(K)^G \xrightarrow{\operatorname{tr}_*} W(F) \xrightarrow{r^*} W(K)^G \tag{6.3}$$

can be computed as follows. Take any K-form q with the property that $q^\sigma \cong q$ for all $\sigma \in G$. By 6.1,

$$r^*(\mathrm{tr}_*(q)) = \sum q^\sigma = n \cdot q. \tag{6.4}$$

Thus, the composition $r^* \circ \mathrm{tr}_*$ in 6.3 is just multiplication by n. Both conclusions in the Corollary follow immediately from this remark. □

Corollary 6.5. *In the notations in* 6.1 *and* 6.2, *if the map* r^* *is injective (this is the case, for instance, when* n *is odd), then* $\mathrm{tr}_* \langle 1\rangle_K \cong n \cdot \langle 1\rangle_F$.

Proof. Applying (6.4) to the form $q = \langle 1\rangle_K \in W(K)^G$, we have

$$r^*\big(\mathrm{tr}_* \langle 1\rangle_K\big) = n\,\langle 1\rangle_K = r^*\big(n\,\langle 1\rangle_F\big) \in W(K)^G.$$

Since r^* is injective, the desired conclusion follows. □

In the case where $n = [K\colon F]$ is odd, 6.2 may be further refined into the results 6.6 and 6.8 below. These finer results, esentially from [RW], are made possible by the use of a *different* transfer s_* in place of tr_*.

Theorem 6.6 (Rosenberg-Ware). *Let* $r\colon F \hookrightarrow K$ *be a Galois extension of odd degree* $2m+1$, *with Galois group* $G = \mathrm{Gal}(K/F)$. *Then*

$$r^*\colon\ W(F) \longrightarrow W(K)^G$$

is an isomorphism of rings.

Proof. Since K/F is separable, we may represent K as a simple extension $K = F(x)$. We shall use again the F-linear functional $s\colon K \to F$ defined in 2.1 by

$$s(1) = 1, \qquad s(x) = s(x^2) = \cdots = s(x^{2m}) = 0.$$

First, let us relate this functional s to the trace map. Since K/F is separable, we know that the F-bilinear pairing

$$K \times K \to F \quad \text{defined by} \quad (x, y) \mapsto \mathrm{tr}(xy)$$

is nonsingular. Thus, there exists an element $a \in \dot{K}$ such that $s(y) = \mathrm{tr}(ay)$ for all $y \in K$. Now, consider any K-quadratic space $(V,\,q)$. The supporting space of $s_*(q)$ is the same V, with the form

$$(s_*q)(v) = s(q(v)) = \mathrm{tr}(a \cdot q(v)).$$

Thus, $s_*(q) = \mathrm{tr}_*(a \cdot q)$. Assume now $(V,\,q) \in W(K)^G$. Then, 6.1 gives

$$r^*s_*(q) = K \otimes_F \mathrm{tr}_*(a\cdot q) = \perp_{\sigma\in G} (a\cdot q)^\sigma = \Big(\perp_{\sigma\in G} \langle a\rangle^\sigma\Big)\cdot q,$$

since $q^\sigma \cong q$. This means that the composition in the sequence

$$W(K)^G \xrightarrow{s_*} W(F) \xrightarrow{r^*} W(K)^G \tag{6.7}$$

is the multiplication by a *fixed* form $\tau = \perp_{\sigma \in G} \langle a \rangle^{\sigma}$ over K. To determine τ explicitly, we set q equal to $\langle 1 \rangle_K \in W(K)^G$. This yields $\tau = r^* s_*(\langle 1 \rangle_K) \in W(K)$. By 2.2(1), $s_*(\langle 1 \rangle_K) = \langle 1 \rangle_F \in W(F)$. Thus,

$$\tau = r^*(\langle 1 \rangle_F) = \langle 1 \rangle_K \in W(K).$$

This says that the composition in (6.7) is the identity map on $W(K)^G$. In particular, the second map in (6.7) is onto. But by 2.5 (or by Springer's Theorem 2.7), this map is also one-one. Therefore, it is an isomorphism (with inverse given by $s_*\colon W(K)^G \to W(F)$). $\square$

Assuming the result that $W(F)$ has no odd torsion (which will be proved in the next chapter), we record the following consequence of 6.6.

Corollary 6.8. *Under the hypotheses of* 6.6, *the composition*

(6.9) $$W(F) \xrightarrow{r^*} W(K)^G \xrightarrow{\mathrm{tr}_*} W(F)$$

is given by multiplication by $n = [K\colon F]$. *The second map* tr_* *here is injective, and has a cokernel isomorphic to* $W(F)/n \cdot W(F)$.

Proof. For any F-form q_0, we have by 6.4:

$$r^*(\mathrm{tr}_*(r^*(q_0))) = n \cdot r^*(q_0) = r^*(n \cdot q_0).$$

Since r^* is injective, it follows that $\mathrm{tr}_*(r^*(q_0)) = n \cdot q_0$. This proves the first conclusion in 6.8. By a later result (see VIII.3.2), multiplication by an *odd* integer n on $W(F)$ is an injective endomorphism. The last two conclusions in 6.8 follow immediately from this, since $r^*\colon W(F) \to W(K)^G$ is an isomorphism. $\square$

To conclude this section, we shall prove a few results on (and give some examples of) trace forms on algebraic field extensions. Recall that, for any finite-dimensional algebra K over a field F, we can view K as an F-quadratic space with the pairing

(6.10) $$(x, y) \mapsto \mathrm{tr}_{K/F}(xy) \in F,$$

where $\mathrm{tr}_{K/F}$ denotes the "algebra trace" on K. We call (6.10) the *trace form* on K (see Ch. I, Exer. 29). If this F-form happens to be *regular*, we shall denote it by the symbol $\langle K \rangle$. In particular, if K/F is a finite separable field extension, then $\langle K \rangle$ is defined, and it is in fact just the transferred form $\mathrm{tr}_* \langle 1 \rangle_K$. Corollary 6.5 gives us a surprisingly simple computation of this trace form in case K/F is a Galois extension of odd degree n: namely, it is just isometric to the "sum of squares" form $n \langle 1 \rangle_F$! It behooves us to double-check this nice result in a concrete example.

Example 6.11. Let $F = \mathbb{Q}$ and $K = \mathbb{Q}(\alpha)$, where α has (irreducible) minimal polynomial $g(t) = t^3 - 9t - 9$ over $\mathbb{Q}$. In Example 2.11, we have observed that K is the splitting field for $g(t)$, so K/F is a cubic Galois extension. To determine the trace form $\langle K\rangle$ ("by bare hands"), we first compute that

$$\mathrm{tr}(\alpha) = 0, \quad \mathrm{tr}(\alpha^2) = 18, \quad \mathrm{tr}(\alpha^3) = 27, \quad \mathrm{tr}(\alpha^4) = 162,$$

where $\mathrm{tr} = \mathrm{tr}_{K/F}$. Thus, in the F-basis $\{1, \alpha, \alpha^2\}$ on K, the quadratic form in question is

$$\begin{aligned}\mathrm{tr}((x + y\alpha + z\alpha^2)^2) &= \mathrm{tr}(x^2 + y^2\alpha^2 + z^2\alpha^4 + 2xy\alpha + 2xz\alpha^2 + 2yz\alpha^3)\\ &= 3x^2 + 18y^2 + 162z^2 + 36xz + 54yz\\ &= 3\left[(x + 6z)^2 + 9z^2/2 + 6(y + 3z/2)^2\right].\end{aligned}$$

Therefore, we have the following diagonalizations over $\mathbb{Q}$:

$$\begin{aligned}\langle K\rangle &\cong \langle 3\rangle\,\langle 1,\,2,\,6\rangle \cong \langle 3,\,6,\,2\rangle \cong \langle 9,\,2,\,2\rangle\\ &\cong \langle 9,\,4,\,1\rangle \cong \langle 1,\,1,\,1\rangle,\end{aligned}$$

just as predicted by 6.5!

It turns out that, by applying a somewhat different method, we can generalize 6.5 considerably. To this end, let us first prove the following nice fact about trace forms on separable extensions.

Theorem 6.12. *Let E/F be a finite Galois extension, and let $K \subseteq E$ be any subextension, with degree n over F. Under these assumptions, $\langle K\rangle$ and $\langle E\rangle$ are both defined as regular quadratic spaces over F.*

(1) *The tensor product $\langle E\rangle \otimes \langle K\rangle$ is isometric to $n\,\langle E\rangle$.*

(2) *$\langle K\rangle_E \cong n \cdot \langle 1\rangle_E$. (Here, $\langle K\rangle_E$ denotes the scalar extension of the F-form $\langle K\rangle$ to E.)*

Proof. Since K/F is separable, $K = F(\alpha)$ for an element α with a separable minimal polynomial $g(t) \in F[t]$ of degree n. Over E, $g(t)$ factors completely into $(t - \alpha_1)\cdots(t - \alpha_n)$ for *distinct* $\alpha_1, \ldots, \alpha_n$ in E. Thus, we have E-algebra isomorphisms

$$E \otimes_F K \cong E \otimes_F \frac{F[t]}{(g(t))} \cong \frac{E[t]}{(g(t))} \cong E \times \cdots \times E \quad (n \text{ copies}).$$

Applying Ch.I, Exercise 29 (to algebras over E), we get (2). On the other hand, viewing the above as F-algebra isomorphisms, and applying the same exercise (to algebras over F), we get (1). □

The theorem above leads directly to the following generalization of 6.5.

Corollary 6.13. *Let $F \subseteq K \subseteq E$ be as in 6.12, with $n = [K\colon F]$. If the functorial map $r^*\colon W(F) \to W(E)$ is injective (this is the case, for instance, when $[E\colon F]$ is odd), then $\langle K\rangle \cong n \cdot \langle 1\rangle_F$; in other words, the trace form on K admits an orthonormal basis over F.*

Proof. Using 6.12(2), we have

$$r^*\big(\langle K\rangle - n\cdot\langle 1\rangle_F\big) = \langle K\rangle_E - n\cdot\langle 1\rangle_E = 0 \in W(E).$$

If $r^*\colon W(F) \to W(E)$ is injective, this gives $\langle K\rangle \cong n\cdot\langle 1\rangle_F$, since both F-forms here have dimension n.. □

Another generalization (or self-strengthening) of 6.5 is the following "reduction" result in computing trace forms.

Theorem 6.14. *Let E/F be a finite Galois extension, and let K be any field between F and E, with $[E\colon K] = 2r+1$. Then $\langle E\rangle \cong (2r+1)\cdot\langle K\rangle$ (as F-quadratic spaces).*

Proof. Applying 6.5 to the odd-degree Galois extension E/K, we have $\big(\mathrm{tr}_{E/K}\big)_*(\langle 1\rangle_E) \cong (2r+1)\cdot\langle 1\rangle_K$ (as K-quadratic spaces). By the functoriality of Scharlau's transfer, it follows that

$$\begin{aligned}\langle E\rangle &= \big(\mathrm{tr}_{E/F}\big)_*(\langle 1\rangle_E) = \big(\mathrm{tr}_{K/F}\big)_*\big(\mathrm{tr}_{E/K}\big)_*(\langle 1\rangle_E)\\ &= \big(\mathrm{tr}_{K/F}\big)_*\big((2r+1)\cdot\langle 1\rangle_K\big) \cong (2r+1)\cdot\langle K\rangle,\end{aligned}$$

as claimed. □

To show how 6.14 can be used toward concrete computations, let us now determine the trace form on a Galois extension whose degree is twice an odd integer (the case of Galois extensions of odd degree having been handled already in 6.5).

Theorem 6.15. *Let E/F be a finite Galois extension with $[E\colon F] = 2m$ where m is odd, and let $d \in \dot F/\dot F^2$ be the field discriminant of E/F. Then $\langle E\rangle \cong m\,\langle 2,\, 2d\,\rangle$.*

Proof. Let $G = \mathrm{Gal}\,(E/F)$. Under the regular representation of G (where every $g \in G$ acts on G by left multiplication), an element of order 2 acts as an odd permutation. Therefore, the set H of elements of G acting as even permutations constitute a subgroup of index 2. Thus, the fixed field K of H is a quadratic extension of F, say $K = F(\sqrt{a}\,)$, where $a \in \dot F \setminus \dot F^2$. Since $[E : K] = m$ is odd, 6.14 yields $\langle E\rangle \cong m\,\langle K\rangle$. A special case of the lemma below shows that $\langle K\rangle \cong \langle\, 2,\, 2a\,\rangle$. Therefore, $\langle E\rangle \cong m\,\langle 2,\, 2a\rangle$. Computing determinants, we see that $a = \det\langle E\rangle \in \dot F/\dot F^2$. But by definition, $\det\langle E\rangle$ is the discriminant d of E/F, so we are done. □

In the proof above, we needed the computation for the trace form $\langle K\rangle$ on a quadratic extension K/F. This is covered by the following more general result for "scaled" trace forms

$$(x,y)\mapsto \mathrm{tr}_{K/F}(\alpha\, xy)\quad (x,\, y\in K;\ \ \alpha\in \dot{K}) \tag{6.16}$$

on such an extension, which we could have included in §3.

Lemma 6.17. *Let $K = F(\sqrt{a}\,)$ be a quadratic extension of F, and let $\alpha = b + c\sqrt{a} \in \dot{K}$ (where one of $b,\, c\in F$ is possibly zero). Then the scaled trace form in* (6.16) *above is given by the transferred form*

$$\mathrm{tr}_*\,\langle\alpha\rangle \cong \langle 2b\rangle\,\langle 1,\, a\,\mathrm{N}_{K/F}(\alpha))\,\rangle \in W(F).$$

In particular, for $\alpha = 1$, the trace form $\langle K\rangle$ on K is given by $\langle 2,\, 2a\rangle$.

Proof. With respect to the F-basis $\{1,\, \sqrt{a}\,\}$ on K, the Scharlau transfer $\varphi := \mathrm{tr}_*\langle\alpha\rangle$ has the symmetric matrix

$$\begin{pmatrix} \mathrm{tr}\,(\alpha) & \mathrm{tr}\,(\alpha\sqrt{a}\,) \\ \mathrm{tr}\,(\alpha\sqrt{a}\,) & \mathrm{tr}\,(a\,\alpha) \end{pmatrix} = \begin{pmatrix} 2b & 2ac \\ 2ac & 2ab \end{pmatrix}$$

If $b = 0$, this is the hyperbolic plane, so the claimed equation formally holds. If $b \neq 0$, the matrix of φ has determinant $4a(b^2 - ac^2)$. Since φ represents $2b$, we have

$$\varphi \cong \langle 2b\rangle\,\langle 1,\ 4a(b^2 - ac^2)\rangle \cong \langle 2b\rangle\,\langle 1,\ a\,\mathrm{N}_{K/F}(\alpha)\rangle. \qquad \square$$

The formula in 6.17 can be used to handle the trace forms on repeated quadratic extensions. As a final illustration for this section, let us apply 6.17 to work out the case of a double quadratic extension.

Theorem 6.18. *Let $L = F(\sqrt{a}\,)$ be a quadratic extension, and assume that $\alpha = b + c\sqrt{a} \in \dot{L}\setminus \dot{L}^2$ (where one of $b,\, c\in F$ is possibly zero). Then $K = L(\sqrt{\alpha}\,)$ has the trace form*

$$\langle K\rangle \cong \langle 1,\, a,\, b,\, ab\,\mathrm{N}_{L/F}(\alpha)\rangle \in W(F).$$

(Here, the RHS is to be interpreted as $\langle 1,\, a\rangle$ if $b = 0$.)

Proof. Since $\left(\mathrm{tr}_{K/L}\right)_*\langle 1\rangle_K \cong \langle 2,\, 2\alpha\rangle_L$ by 6.17, we have

$$\begin{aligned} \langle K\rangle &\cong \left(\mathrm{tr}_{L/F}\right)_*\langle 2\rangle_L + \left(\mathrm{tr}_{L/F}\right)_*\langle 2\alpha\rangle_L \\ &\cong \langle 4\rangle\,\langle 1,\, a\,\mathrm{N}_{L/F}(2)\rangle + \langle 4b\rangle\,\langle 1,\, a\,\mathrm{N}_{L/F}(2\alpha)\rangle \\ &\cong \langle 1,\, a\rangle + \langle b\rangle\,\langle 1,\, a\,\mathrm{N}_{L/F}(\alpha)\rangle \\ &\cong \langle 1,\, a,\, b,\, ab\,\mathrm{N}_{L/F}(\alpha)\rangle \in W(F), \end{aligned}$$

as desired. $\square$

We'll record an important special case of 6.18, where we take $a = r^2+s^2$ in $\dot{F} \setminus \dot{F}^2$, and $\alpha = t(a + r\sqrt{a}\,) \in \dot{L}$ $(t \in \dot{F})$. Note that, for these special choices of parameters,

$$\mathrm{N}_{L/F}(\alpha) = t^2(a^2 - r^2 a) = t^2\left(a^2 - (a - s^2)a\right) = t^2 s^2 a \notin \dot{F}^2.$$

Therefore, $\alpha \notin \dot{L}^2$, so $K = L(\sqrt{\alpha}\,)$ is a quadratic extension over L.

Corollary 6.19. *With $a \in \dot{F}$ and $\alpha \in \dot{L}$ chosen as above, the quartic extension field $K = F\big(\sqrt{t\,(a + r\sqrt{a}\,)}\big)$ over F has the trace form*

$$\langle K \rangle \cong \langle 1,\, a,\, t,\, t \rangle.$$

Here a is the discriminant of the extension K/F in $\dot{F}/\dot{F}^2$.

Proof. Applying 6.18 and the above computation of $\mathrm{N}_{L/F}(\alpha)$, we get

$$(6.20) \qquad \langle K \rangle \cong \langle 1,\, a,\, ta,\, a(ta)(t^2 s^2 a) \rangle \cong \langle 1,\, a,\, ta,\, ta \rangle \cong \langle 1,\, a,\, t,\, t \rangle,$$

since $a = r^2 + s^2$ implies that $\langle a,\, a \rangle \cong \langle 1,\, 1 \rangle$ over F. Computing determinants from (6.20), we see that, up to squares, a is $\det\big(\langle K \rangle\big)$, which is the discriminant of K/F in $\dot{F}/\dot{F}^2$. □

The reason we singled out the particular quartic field K/F for study in 6.20 is that the extension K/F is *cyclic*, and in fact, the field K there is the *most general* cyclic extension of degree 4 over F. To avoid an unnecessary digression, we refrain from proving this fact here. Anyway, a special case of this will be dealt with later in VIII.5.1, and the proof we'll give for VIII.5.1 can be generalized without much difficulty to verify the claim above. The upshot of these remarks is that, in the result 6.19, we have, in fact, computed the trace form on *any* cyclic quartic extension K/F.

The methods used in the proofs of 6.12–6.19 have been generalized in various ways to get information on the trace forms of finite separable field extensions. We have included a short discussion on trace forms here, since the investigation of such forms and the Witt classes they represent in $W(F)$ (known as the *algebraic elements* of $W(F)$) has been an active topic for research in quadratic form theory in recent years. Serre's computation of the Witt invariants of the trace forms of separable field extensions in [Se$_2$] has especially heightened the interest in the general theory of trace forms (and reduced trace forms). For a comprehensive introduction to this area of investigation (with special emphasis on the case of algebraic number fields), see the book of Conner and Perlis [CP]. For examples of many explicit computations of trace forms on separable field extensions, see the article of Drees, Epkenhans, and Krüskemper [DEK]. Our short exposition on trace forms above was partly inspired by the material presented in these two sources.

7. Quadratic Closures of Fields

For any finite field extension K/F, we have shown in 3.12 that if $\dot{K}/\dot{K}^2$ is finite, then so is $\dot{F}/\dot{F}^2$. A statement of this sort may be called a "Going-Down Theorem". In this section, we shall address the corresponding "Going-Up" question, namely the following:

Question 7.1. *Let K be a finite extension of F. If $\dot{F}/\dot{F}^2$ is finite, is $\dot{K}/\dot{K}^2$ necessarily finite?*

In the second part of Theorem 3.12, we were also able to answer this question affirmatively, in case K/F is a Galois extension with $[K\colon F] = 2^n$ for some n. We shall show, however, that the answer to 7.1 is "no" in general, even if $[K\colon F]$ is a power of 2. In fact, we shall construct quadratically closed fields F with the property that $\dot{K}/\dot{K}^2$ is infinite for *every* finite extension $K \supsetneqq F$ (see Corollary 7.11). A part of the exposition here grew out of several discussions I had with David Goldschmidt some years ago; I wish to thank him for suggesting several of the key ideas used in the second half of this section.

For this section, we assume the reader has some familiarity with number theory and Galois theory. All fields considered below will be assumed to be in a common ambient field. We begin with the following crucial result in Galois theory.

Theorem 7.2. *Let F/F_1 be a Galois extension, not necessarily of finite degree. Let K_1 be an algebraic extension of F_1 such that $K_1 \cap F = F_1$. If K denotes the field compositum $K_1 \cdot F$, then the natural map*

$$\dot{K}_1/\dot{K}_1^2\dot{F}_1 \to \dot{K}/\dot{K}^2\dot{F}_1$$

is an injection.

Proof. We may clearly assume that $[F\colon F_1] < \infty$ and $[K_1\colon F_1] < \infty$. Suppose $b \in \dot{K}_1 \cap \dot{K}^2\dot{F}_1$. To show that $b \in \dot{K}_1^2\dot{F}_1$, it suffices to work in the case where $b \notin \dot{K}_1^2$ and $b \in \dot{K}_1 \cap \dot{K}^2$. Consider the quadratic extension

$$K_2 = K_1(\sqrt{b}) \subseteq K,$$

and let $F_2 = K_2 \cap F$. Since F/F_1 is Galois and $K_1 \cap F = F_1$, the Theorem on Natural Irrationalities[3] implies that

$$[K\colon K_1] = [F\colon F_1] \quad \text{and} \quad [K\colon K_2] = [F\colon F_2].$$

---

[3] See, e.g., Artin's "Galois Theory", or Theorem 1.12 in the chapter on Galois theory in Lang's "Algebra".

From these, it follows that $[F_2 : F_1] = [K_2 : K_1] = 2$.

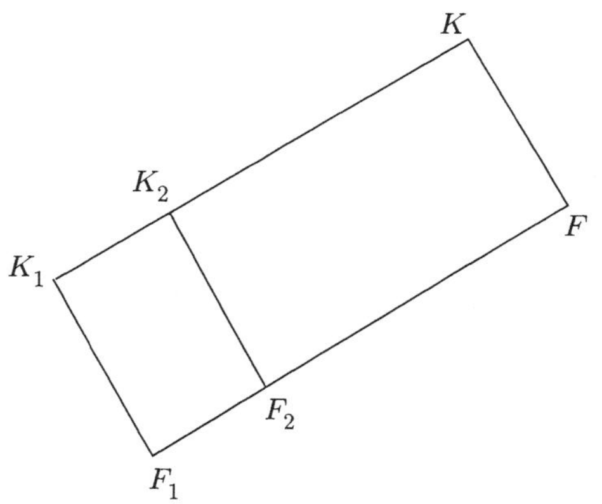

Say $F_2 = F_1(\sqrt{a})$ (where $a \in \dot{F}_1 \setminus \dot{F}_1^2$). Then a is a nonsquare in K_1 that becomes a square in $K_1(\sqrt{b})$. Theorem 3.8 then implies that $a \in \dot{K}_1^2 b$, so now $b \in \dot{K}_1^2 \dot{F}_1$. $\square$

Remark 7.3. Readers with expertise in field theory will have recognized that the lemma above was essentially an exercise on linear disjointness. The hypotheses that F/F_1 is Galois and $K_1 \cap F = F_1$ were used to show that, for any field K_3 between K_1 and K, K_3 and F are *linearly disjoint* over $K_3 \cap F$. This fact was the crux of the argument used above to prove 7.2.

The study of the Going-Down Question 7.1 is a good excuse for us to formally introduce the notion of the quadratic closure of a field. Recall that a field F is *quadratically closed* if $F = F^2$. Given any field E with algebraic closure $\overline{E}$, there clearly exists a smallest quadratically closed field $\widetilde{E} \subseteq \overline{E}$ that contains E, namely, the intersection of all quadratically closed subfields of $\overline{E}$ containing E. (See Chapter I, Exercise 8.) The field $\widetilde{E}$ is called the *quadratic closure* of E. Note that, if K/E is *any* field extension that is quadratically closed, then the algebraic closure K_0 of E in K is easily seen to be quadratically closed, and so $K \supseteq K_0 \supseteq \widetilde{E}$. Thus, $\widetilde{E}$ is the "smallest" quadratically closed extension of E in the strongest possible sense. From the way $\widetilde{E}$ is defined, it is clear that $\widetilde{E}/E$ is a Galois extension—possibly of infinite degree.

Of course, the quadratic closure $\widetilde{E}$ can also be constructed "from below", starting from E. In fact, the set of elements of $\overline{E}$ that lie in *some* subfield obtained from E by a finite number of successive quadratic extensions clearly forms a quadratically closed subfield of $\widetilde{E}$ — thus necessarily equal to $\widetilde{E}$. This leads quickly to the following important Galois-theoretic characterization of $\widetilde{E}$.

Theorem 7.4. *$\widetilde{E}$ is the union of all finite 2-extensions of E in $\overline{E}$. (A "finite 2-extension" of E means a Galois extension L/E with $[L\colon E] = 2^n$ for some $n \geq 0$.)*

Proof. By Galois theory, any finite 2-extension L/E can be obtained from E by a finite number of successive quadratic extensions. By what we said in the paragraph preceding the theorem, $L \subseteq \widetilde{E}$. Conversely, consider any $a \in \widetilde{E}$. Let $a_1, \ldots, a_r$ be all the conjugates of a in $\overline{E}$. Since $\widetilde{E}/E$ is Galois, all $a_i \in \widetilde{E}$. Therefore,

$$L := E(a_1, \ldots, a_r) \subseteq \widetilde{E}$$

is Galois over E. Now L is contained in a field L' obtainable from E by a finite number of successive quadratic extensions (since each a_i is). Therefore, $[L\colon E]$ divides $[L'\colon E]$, which is a power of 2. Thus, L/E is a finite 2-extension (in $\widetilde{E}$) that contains a. $\square$

In view of the conclusion of 7.4, $\widetilde{E}$ is often called the "maximal 2-extension" of E. Loosely speaking, the elements of $\widetilde{E}$ are those in $\overline{E}$ that are "constructible" from the elements of E. In particular, for $E = \mathbb{Q}$, *the quadratic closure $\widetilde{\mathbb{Q}}$ is called the field of all constructible numbers.*[(4)]

We note the following *necessary* condition on the so-called constructible elements.

Corollary 7.5. *For any $a \in \widetilde{E}$, $[E(a)\colon E] = 2^k$ for some $k \geq 0$.*

The condition that $[E(a)\colon E] = 2^k$ for some k is, however, not a sufficient condition for a to lie in $\widetilde{E}$, as we have seen in Example 2.14. (The cases $k = 0, 1$ constitute the only trivial exceptions.)

We offer below a few standard examples of $\widetilde{E}$.

Example 7.6. If $E = \mathbb{R}$, the quadratic closure is $\widetilde{E} = \mathbb{C}$. More generally, if E is any euclidean field (that is, a formally real field with $|\dot{E}/\dot{E}^2| = 2$), an easy application of 3.8 shows that $E(\sqrt{-1}\,)$ is quadratically closed. (For more details on this, see VIII.1.7.) Thus, $\widetilde{E} = E(\sqrt{-1}\,)$.

Example 7.7. If $E = \mathbb{F}_5$ (the finite field of 5 elements), then $\widetilde{E}$ is the field $\bigcup_{n\geq 1} \mathbb{F}_5(\sqrt[2^n]{2})$ in Chapter II, Exercise 18. In this example, the description of $\widetilde{E}$ given in 7.4 is especially transparent. The 2-extensions of $E = \mathbb{F}_5$ form a chain, since $\mathbb{F}_5(\sqrt[2^n]{2})$ is the only extension of degree 2^n over $\mathbb{F}_5$. The

(4) Of course, "constructible" here means "constructible by using straightedge and compass alone."

maximal 2-extension of E is simply their ascending union. The quadratic closure of *any* finite field can be described in a similar way.

Example 7.8. The proof or disproof that certain numbers are constructible is a time-honored part of classical mathematics. For instance, we know that:

- $\sqrt{\pi}$ is not constructible ("you can't square the circle");
- $\sqrt[3]{2}$ is not constructible ("you can't double the cube"); and
- $\cos 20^\circ$ is not constructible ("you can't trisect a 60° angle").

And the young Gauss became justly famous (at the tender age of 19) by proving that, for an odd prime p, *the side* s_p *of a regular* p*-gon inscribed in the unit circle is constructible iff* p *is a Fermat prime* (that is, $p = 2^{2^r}+1$ for some r). The proof of the "if" part of this result appeared in Article 365 of Gauss's tour de force *Disquisitiones Arithmeticae* (c. 1801). However, Gauss did not give a proof for the converse, stating only that he can demonstrate it "with all rigor", but that "the limits of this work do not allow including the demonstration here."

From the modern viewpoint of Galois theory, a proof of Gauss's theorem can be given very easily via 7.4 and 7.5, following largely van der Waerden's "Algebra". To "construct" s_p is equivalent to constructing the primitive p-th root of unity $\zeta_p = e^{2\pi i/p}$. A necessary condition, according to 7.5, is that $[\mathbb{Q}(\zeta_p) \colon \mathbb{Q}] = p - 1$ be of the form 2^k. But then $p = 2^k + 1$ being prime forces k to be of the form 2^r, so $p = 2^{2^r} + 1$ is a Fermat prime. Conversely, if p is a Fermat prime $2^{2^r} + 1$, then $\mathbb{Q}(\zeta_p)/\mathbb{Q}$ is Galois of degree 2^{2^r}, so $\mathbb{Q}(\zeta_p) \subseteq \widetilde{\mathbb{Q}}$ by 7.4. Of course, this proof depends on knowing that the cyclotomic polynomial

$$\Phi_p(x) = x^{p-1} + \cdots + p + 1 \in \mathbb{Q}[x]$$

is irreducible. Luckily, this can be proved easily by applying Eisenstein's Criterion to $\Phi_p(x+1)$.

For the Fermat prime $p = 5 = 2^2 + 1$, a quick method for constructing the regular pentagon can be gleaned from our earlier example III.4.4, where we showed that

$$\cos 72^\circ = \left(\zeta_5 + \zeta_5^{-1}\right)/2 = \left(\sqrt{5} - 1\right)/4.$$

For the "next" Fermat prime $p = 17 = 2^{2^2} + 1$, Gauss also gave (what amounts to) an explicit construction for the regular 17-gon. Indeed, it would have been uncharacteristic of Gauss if he did not! As in the $p = 5$ case, a construction amounts to writing down the real number

$$\cos \frac{2\pi}{17} = \left(\zeta_{17} + \zeta_{17}^{-1}\right)/2$$

in terms of iterated square roots of rationals. Miraculously, Gauss expressed this algebraic number in the form

$$\frac{1}{16}\Big(-1+\sqrt{17}+\sqrt{\alpha}+2\,\big(\,17+3\sqrt{17}-\sqrt{\alpha}-2\sqrt{\beta}\,\big)^{1/2}\Big), \tag{7.9}$$

where $\alpha = 2\left(17-\sqrt{17}\right)$ and $\beta = 2\left(17+\sqrt{17}\right)$. This expression appeared explicitly in Article 365 of *Disquisitiones Arithmeticae*. In decimal expansion, this number is 0.93247... .

Gauss took pleasure in pointing out that the regular 17-gon was the first "truly new" regular n-gon to have been constructed since Euclid's era, so it is certainly fitting that, after Gauss's death, a regular 17-gon was carved on his tombstone (apparently by his own desire). But would Gauss turn in his grave to know that, even more than 200 years after his monumental discovery, no new Fermat primes have ever been found (beyond $3, 5, 17, 257$, and 65537)?

To conclude our discussion of Example 7.8, let us add a few words on the constructibility of a regular n-gon for any integer n. With some additional work, this case can be settled as well: the criterion for constructibility is that n be a product of a power of 2 and *distinct* Fermat primes. This criterion was also claimed by Gauss, in Article 366 of *Disquisitiones Arithmeticae*, but again, no proof was given. An explicit proof of this criterion appeared only much later in 1837, in a paper of Pierre Wantzel. But according to my colleague Professor Robin Hartshorne, even this proof contained a small gap. Fortunately, this gap can be fixed without too much difficulty.

After giving the above historical background on the field $\widetilde{\mathbb{Q}}$ of constructible numbers, we now return to our quest for finite extensions of quadratically closed fields that have infinitely many square classes. From here until the result 7.14, *we shall work within* $\overline{\mathbb{Q}}$, the field of all algebraic numbers. By a "number field", we shall mean a finite extension of the rational number field $\mathbb{Q}$. Our goal is to prove the following result on the infinitude of square classes for certain subfields of $\overline{\mathbb{Q}}$.

Theorem 7.10. *Let F be a (possibly infinite) Galois extension of a number field F_0. If K is any proper finite extension of F, then $\dot{K}/\dot{K}^2$ is infinite.*

With this result, we can settle Question 7.1 in a rather satisfactory way, as follows.

Corollary 7.11. (1) *Let F be the quadratic closure of a number field F_0. Then any finite extension $K \supsetneq F$ has an infinite square class group.*

(2) *The field $\widetilde{\mathbb{Q}}$ of constructible numbers has extensions of any given finite degree > 2, and all of these finite extensions have infinitely many square classes.*

Proof. (1) is a special case of 7.10, since $\widetilde{F}_0/F_0$ is Galois. For (2), take $F_0 = \mathbb{Q}$, so $\widetilde{F}_0 = \widetilde{\mathbb{Q}}$. Given any integer $n > 2$, write $n = 2^r n_0$, where n_0 is odd. First assume $n_0 > 1$. The field $\mathbb{Q}$ certainly has an extension K_0 of degree n_0, and we have $K_0 \cap \widetilde{\mathbb{Q}} = \mathbb{Q}$ since n_0 is odd. Then the Theorem on Natural Irrationalities implies that the compositum $K = K_0 \widetilde{\mathbb{Q}}$ has degree $n_0 > 1$ over $\widetilde{\mathbb{Q}}$. Then $\dot{K}/\dot{K}^2$ is infinite by (1), so by taking successive quadratic extensions of K, we can produce an extension $L/\widetilde{\mathbb{Q}}$ with degree $n = 2^r n_0$. Finally, assume $n = 2^r \geq 4$. In Example 2.14, we have seen that $\widetilde{\mathbb{Q}}$ has a quartic extension, say K, of degree 4. Repeating the above construction, we get again an extension $L/\widetilde{\mathbb{Q}}$ with degree $n = 2^r$. □

Proof of Theorem 7.10. Keeping the notations in 7.10, let $K = F(\alpha)$, and choose a number field L between F_0 and F such that the minimal polynomial of α over F has all coefficients in L. Then $K_1 := L(\alpha)$ is a number field, with $K_1 \cdot F = F(\alpha) = K$. Let $F_1 := K_1 \cap F$. Since $\alpha \notin F$, we have $K_1 \supsetneq F_1$. Applying Theorem 7.2, we see that $\dot{K}_1/\dot{K}_1^2\dot{F}_1$ maps injectively into $\dot{K}/\dot{K}^2\dot{F}_1$. To prove 7.10, it is thus sufficient to establish the following fact.

Theorem 7.12. *Let $K_1 \supsetneq F_1$ be a proper extension of number fields. Then $\dot{K}_1/\dot{K}_1\dot{F}_1$ is infinite.*

Note that the conclusion implies that $\dot{K}_1/\dot{K}_1^2$ is infinite, which is, of course, well-known. In fact, 7.12 should be regarded as a generalization of this. The special case of 7.12 for quadratic extensions is also noteworthy. For $K_1 = F_1(\sqrt{a}\,)$, the 5-term exact sequence in 3.8 shows that $\dot{K}_1/\dot{K}_1\dot{F}_1$ is isomorphic to $\mathrm{N}_{K_1/F_1}(\dot{K}_1)/\dot{F}_1^2$. Thus, in this case, 7.12 says that the norm form $\langle 1, -a\rangle_{F_1}$ represents infinitely many square classes in F_1. This statement does not seem to be obvious either.

To find a proof for 7.12, we go back to the arena of classical algebraic number theory. In what follows, let A and B be, respectively, the full rings of algebraic integers in the number fields F_1 and K_1. To prove Theorem 7.12, we shall use

Lemma 7.13. *In the setting of 7.12 there exist infinitely many primes $\mathfrak{p} \subset A$ such that $\mathfrak{p}B$ is not prime in B.*

Let us show that this Lemma implies Theorem 7.12. Indeed, suppose $\dot{K}_1/\dot{K}_1^2\dot{F}_1$ is finite, and let $a_i \in \dot{K}_1$ $(1 \leq i \leq r)$ be a finite set representing all the elements of $\dot{K}_1/\dot{K}_1^2\dot{F}_1$. We may assume that all $a_i \in B$. Let P be a finite set of primes of A such that

(1) *any prime in A that ramifies in B belongs to P, and*

(2) *for any prime divisor $\mathfrak{P} \subset B$ of any a_i, $\mathfrak{P} \cap A \in P$.*

By Lemma 7.13, there exists a prime $\mathfrak{p}$ in A such that $\mathfrak{p} \notin P$ and $\mathfrak{p}B$ is not prime in B. Since $\mathfrak{p}$ is unramified, we have

$$\mathfrak{p}B = \mathfrak{P}_1 \cdots \mathfrak{P}_g$$

with $\mathfrak{P}_1, \ldots, \mathfrak{P}_g$ distinct primes in B and $g \geq 2$. By the Approximation Theorem in number theory, we can find an element $x \in \dot{K}_1$ such that $v_{\mathfrak{P}_1}(x) = 1$ and $v_{\mathfrak{P}_2}(x) = 0$ (where $v_{\mathfrak{P}_i}$ is the valuation associated with the finite prime $\mathfrak{P}_i$). Write $x = a_j y^2 z$, where $1 \leq j \leq r$, $y \in \dot{K}_1$, and $z \in \dot{F}_1$. We have, for any i,

$$v_{\mathfrak{P}_i}(x) \equiv v_{\mathfrak{P}_i}(a_j) + v_{\mathfrak{P}_i}(z) \equiv v_{\mathfrak{P}_i}(z) \pmod 2,$$

since each a_j is a $\mathfrak{P}_i$-adic unit. On the other hand, $v_{\mathfrak{P}_i}(z)$ depends only on $v_{\mathfrak{p}}(z)$ and not on i. The equation above, therefore, implies that

$$v_{\mathfrak{P}_1}(x) \equiv v_{\mathfrak{P}_2}(x) \pmod 2,$$

which is a contradiction. □

It remains only to prove Lemma 7.13. Write $K_1 = F_1(\theta)$, where $\theta \in B$. The monic minimal polynomial h of θ over F_1 belongs then to $A[X]$. We need yet another number-theoretic fact.

Lemma 7.14. *There exist infinitely many primes* $\mathfrak{p} \subset A$ *such that* $\overline{h} \in (A/\mathfrak{p})[X]$ *has a root in the (finite) field* $A/\mathfrak{p}$.

Suppose this is proven. Consider any prime $\mathfrak{p} \subset A$ with the property in the lemma, and such that $\mathfrak{p}$ is prime to the index $[B\colon A[\theta]]$ (there are infinitely many such primes). Clearly, $\mathfrak{p}B$ cannot be a prime ideal in B, in view of the familiar formula $\sum e_i f_i = [K_1\colon F_1]$ for prime ideal decompositions. We have thus established the conclusion of Lemma 7.13. □

We now finish by proving 7.14. Assume the contrary. Then the set of prime ideals

$$\{\mathfrak{p} \subset A \mid \mathfrak{p} \text{ divides some } h(a)\ (a \in A)\}$$

is a finite set, say $\{\mathfrak{p}_1, \ldots, \mathfrak{p}_s\}$. Write the polynomial $h(X+1)$ as $X \cdot g(X) + h(1)$, where $g \in A[X]$. Let $\mathfrak{p}_1^{n_1} \cdots \mathfrak{p}_s^{n_s}$ ($n_i \geq 0$) be the prime factorization of the principal ideal $A \cdot h(1)$. For any i ($1 \leq i \leq s$), fix an element $a_i \in A$ such that $v_{\mathfrak{p}_i}(a_i) = 1$. For any integer $t \in \mathbb{Z}$, evaluate the polynomial $h(X+1)$ at $X = x_t = a_1^{n_1+1} \cdots a_s^{n_s+1} \cdot t$. We get an algebraic integer

$$b_t = h(x_t + 1) = x_t \cdot g(x_t) + h(1).$$

Since $v_{\mathfrak{p}_i}(h(1)) = n_i$ and $v_{\mathfrak{p}_i}(x_t) \geq n_i + 1$, we have $v_{\mathfrak{p}_i}(b_t) = n_i$, independently of t. But the prime divisors of b_t are among $\mathfrak{p}_1, \ldots, \mathfrak{p}_s$. Therefore,

$$A \cdot b_t = \mathfrak{p}_1^{n_1} \cdots \mathfrak{p}_s^{n_s} = A \cdot h(1).$$

Taking the norm $\mathrm{N} = \mathrm{N}_{F_1/\mathbb{Q}}$ into $\mathbb{Q}$, we get

$$|\,\mathrm{N}(b_t)\,| = |\,\mathrm{N}(h(1))\,| = \text{ constant.}$$

This contradicts the obvious fact that $|\,\mathrm{N}(b_t)\,| \to \infty$ as $t \to \infty$. $\square$

At this point, we have successfully carried out our investigation on the number of square classes of finite extensions of the quadratic closure of a number field F_0. For the sake of completeness, we should give a similar discussion on the case where F_0 is a local field. This discussion will conclude our study of the "Going-Up Question" for square class groups raised in 7.1. I thank David Leep for pointing out that the techniques of exploiting linear disjointness used for number fields earlier in this section can also be adapted to the case of local fields.

As it turns out, the behavior of the quadratic closure of a local field F_0 depends critically on the characteristic of its residue field. We start with the case of a *nondyadic* field F_0. The result here will be deduced from the following general field-theoretic observation of David Leep.

Theorem 7.15. *Let F_0 be a field with the property that $|\dot{F}_0/\dot{F}_0^2| = |\dot{K}_0/\dot{K}_0^2| < \infty$ for any finite extension K_0/F_0, and let F be the quadratic closure of F_0. Then F is* hereditarily quadratically closed, *in the sense that every finite (or algebraic) extension K of F is a quadratically closed field.*

Proof. Let $\alpha \in K$, where K is a finite field extension of F. Let

$$t^n + a_{n-1}t^{n-1} + \cdots + a_0 \in F[t]$$

be the minimal polynomial of α over F, and let F_1 be any finite extension of F_0 such that

$$(7.16) \qquad F_0(a_0, \ldots, a_{n-1}) \subseteq F_1 \subseteq F.$$

Then $K_1 := F_1(\alpha)$ has degree n over F_1, and $K_1 \cap F = F_1$. (If $\beta \in K_1 \cap F$, say

$$\beta = b_0 + b_1\alpha + \cdots + b_{n-1}\alpha^{n-1} \in F \quad \text{where } b_0, \ldots, b_{n-1} \in F_1,$$

then $b_1, \ldots, b_{n-1} = 0$, so $\beta = b_0 \in F_1$.) Now, for any $a \in \dot{F}_1 \setminus \dot{F}_1^2$, we have $F_1 \subsetneq F_1(\sqrt{a}\,) \subseteq F$. Therefore, $a \notin \dot{K}_1^2$ (for otherwise $K_1 \cap F \supseteq F_1(\sqrt{a}\,) \supsetneq F_1$). This shows that the natural map $i : \dot{F}_1/\dot{F}_1^2 \to \dot{K}_1/\dot{K}_1^2$ is injective. By hypothesis, these two groups have the same finite cardinality, so i is also surjective. Thus, $\alpha = b\,\beta^2$, for some $b \in \dot{F}_1$ and $\beta \in \dot{K}_1$. Since $b \in \dot{F}^2 \subseteq \dot{K}^2$, it follows that $\alpha \in \dot{K}^2$. This shows that $\dot{K} = \dot{K}^2$. $\square$

From 7.15, we see that the nature of the quadratic closure of a *nondyadic* local field is quite different from that of the quadratic closure of a number field, as follows.

Theorem 7.17. *Let F_0 be a nondyadic local field. Then its quadratic closure F is hereditarily quadratically closed.*

Proof. This follows from 7.15 since, according to VI.2.2(1), $|\dot{K}_0/\dot{K}_0^2| = 4$ for any finite extension K_0 of F_0. □

Finally, let us consider the case of *dyadic* local fields. Since our base field F_0 is assumed to have characteristic $\neq 2$, the dyadic case can arise only when F_0 has characteristic 0, in which case F_0 is a finite extension of $\mathbb{Q}_2$ (see the discussion preceding VI.2.2). The analysis in this case is a combination of those in the proofs of 7.2 and 7.15. It turns out that the conclusion here is similar to that in the case of number fields.

Theorem 7.18. *Let F_0 be a dyadic local field (that is, a finite extension of $\mathbb{Q}_2$), and let F be its quadratic closure. Then $\dot{K}/\dot{K}^2$ is infinite for any finite extension $K \supsetneq F$.*

Proof. Let $K = F(\alpha)$ be of degree n over F, and define F_1 and $K_1 = F_1(\alpha)$ as in the proof of 7.15. As before, we have $K_1 \cap F = F_1$, so 7.2 applies to give the injectivity of the map

$$\dot{K}_1/\dot{K}_1^2 \cdot \dot{F}_1 \longrightarrow \dot{K}/\dot{K}^2 \cdot \dot{F}_1 = \dot{K}/\dot{K}^2.$$

Now $|\dot{K}_1/\dot{K}_1^2 \cdot \dot{F}_1| \geq |\dot{K}_1/\dot{K}_1^2|/|\dot{F}_1/\dot{F}_1^2|$. Thus, if $m := [F_1 : \mathbb{Q}_2]$, an application of VI.2.23 gives

$$|\dot{K}/\dot{K}^2| \geq |\dot{K}_1/\dot{K}_1^2 \cdot \dot{F}_1| \geq 2^{mn+2}/2^{m+2} = (2^{n-1})^m. \tag{7.19}$$

Recalling (from (7.16)) that F_1 is *any* finite extension of $F_0(a_0, \ldots, a_{n-1})$ within F, we see that $m = [F_1 : \mathbb{Q}_2]$ can be made as large as we please. (Note that F has infinite degree over any dyadic field it contains.) Since $n > 1$, it follows from (7.19) that $\dot{K}/\dot{K}^2$ is an *infinite* group, as desired. □

Exercises for Chapter VII

1. For finite fields $F \subseteq K$, prove the equivalence of the conditions (A), (B), (C), (D) claimed in 2.10.

   (1) If $[K\colon F]$ is odd, show that $r^*\colon W(F) \to W(K)$ is an isomorphism.
   (2) If $[K\colon F]$ is even, show that r^* has kernel IF and image $\{0, \langle 1\rangle\}$.
   (3) If φ_F and φ_K denote the unique anisotropic binary forms over F and K, show that $s_*(\varphi_K) = \varphi_F \in W(F)$ for any nonzero linear functional $s\colon K \to F$.

2. Let $r\colon F \subseteq K$ be a finite extension of degree n, and let $s\colon K \to F$ be a nonzero F-linear map. Show that there exists an F-basis $\{\alpha_1, \ldots, \alpha_n\}$ on K such that $s(\alpha_i^2) \neq 0$ for all i, while $s(\alpha_i\alpha_j) = 0$ for all $i \neq j$.

Suppose (U, q) is an arbitrary K-quadratic space with orthogonal K-basis $x_1, \ldots, x_m$ such that $q(x_j) \in F$ for all j. Show that $\{\alpha_i x_j\}$ is an orthogonal F-basis for the transfer $s_*(U)$, and use this to give an alternative proof for the last statement in Theorem 1.3.

3. In Theorem 3.8, for a quadratic extension K/F, show that the exactness of the sequence at $\dot{K}/\dot{K}^2$ is *equivalent* to Hilbert's Theorem 90 for the cyclic extension K/F. (One implication was already proved in the text.)

4. In the notation of Theorem 3.8, show that $\mathrm{N}\colon \dot{K}/\dot{K}^2 \to \dot{F}/\dot{F}^2$ is the trivial homomorphism iff F is pythagorean (that is, $F^2 + F^2 \subseteq F^2$) and $K = F(\sqrt{-1})$. Excluding this case, show that Corollary 3.10 can be improved into

$$|\dot{F}/\dot{F}^2| \leq |\dot{K}/\dot{K}^2| \leq |\dot{F}/\dot{F}^2|^2/2.$$

Show that this always holds for a quadratic extension K/F if F is a nonreal field.

5. If K/F is a quadratic extension, use Exercise 4 to show that K cannot be a euclidean field.

6. (Cf. 3.12) Let K be a finite extension of F. Use Exercise 4 to show that $|\dot{F}/\dot{F}^2| \leq 2\,|\dot{K}/\dot{K}^2|$. If F is nonreal, show that in fact $|\dot{F}/\dot{F}^2| \leq |\dot{K}/\dot{K}^2|$; deduce from this that, if K is quadratically closed, then so is F. (For a completely different approach to the last statement, see VIII.5.11.)

7. Let $K = F(\sqrt{a})$ be a quadratic extension of F, and let q be an F-form in I^2F with $\dim q \equiv 2 \pmod 4$. If q_K is hyperbolic over K, show that q is isotropic over F.

8. Show that two quaternion algebras over F are linked if they become isomorphic over a quadratic extension K/F.

9. ([Sc$_2$]) Let a, b be $\mathbb{Z}_2$-independent square classes in $\dot{F}/\dot{F}^2$. If an F-form q becomes hyperbolic over $F(\sqrt{a})$, $F(\sqrt{b})$ and $F(\sqrt{ab})$, show that $2q = 0 \in W(F)$.

10. Let $K = F(\sqrt{a})$ be a quadratic extension of F. If $I^2K = 0$, show that $I^2F = \langle 1, -a\rangle\, IF$. Using this, deduce that, for any field F of transcendence degree 1 over $\mathbb{R}$, $I^2F = 2 \cdot IF$. (**Hint.** For the second part, apply II.3.8.)

11. (1) Let $K = F(\sqrt{a})$ be a quadratic extension of F. Show that

$$\sqrt{a} \in \dot{K}^2 \iff -4a \in \dot{F}^4.$$

In this case, show that we must have $K = F(\sqrt{-1})$.
(2) Let $K = F(\sqrt{-1})$ be a quadratic extension of F. Show that

$$\sqrt{-1} \in \dot{K}^2 \iff 2 \in \dot{F}^2 \cup (-\dot{F}^2).$$

12. If F is a pythagorean field with a pythagorean quadratic extension $F(\sqrt{a})$, show that (up to squares) the form $\langle 1, a\rangle$ represents only 1 and a over F.

13. For $a, b, c \in \dot{F}$, show that the following are equivalent:
(1) $c \in D_{F(\sqrt{ab})}\langle 1, -a\rangle$.
(2) $c \in D_F\langle 1, -a\rangle \cdot D_F\langle 1, -b\rangle$.
(3) $\langle 1, -b, -c, ac\rangle$ is isotropic over F.
(4) $\langle 1, -b, -c\rangle$ is isotropic over $F(\sqrt{ab})$.
(**Hint.** First observe that (2) $\Leftrightarrow$ (3) and (1) $\Leftrightarrow$ (4). Then prove (3) $\Rightarrow$ (4) by cancellation, and (4) $\Rightarrow$ (3) by 3.8.)

14. If $-1 \notin \dot{F}^2$, use Theorem 3.8 to show that all solutions of the equation $x^2 + y^2 = z^2$ in F can be expressed parametrically in the form

$$x = \lambda(a^2 - b^2), \quad y = 2\lambda ab, \quad z = \lambda(a^2 + b^2)$$

for suitable $\lambda, a, b \in F$. What more can be said if $F = \mathbb{Q}$ and x, y, z are integers?

15. In the Gross-Fischer Construction 3.17, suppose the $\mathbb{Z}_2$-span of the a_i's in $\dot{F}/\dot{F}^2$ contains the square class of -1. If F is a formally real field, show that the field K constructed in 3.17 may be chosen to be also formally real.

16. For any F-form q and any $a \in \dot{F}$, show that $G_F(a \cdot q) = G_F(q)$.

17. Let N be any positive integer, and F be any field (of characteristic $\neq 2$). Show that $|\dot{F}/\dot{F}^2| \geq N$ iff F has a Galois extension K/F with a Galois group G of exponent 2 such that $|G| \geq N$.

18. If K/F is a finite extension of even degree, show that

$$\dot{F}/\dot{F}^2 \xrightarrow{r^*} \dot{K}/\dot{K}^2 \xrightarrow{\mathrm{N}_{K/F}} \dot{F}/\dot{F}^2$$

is a 0-sequence. Is the sequence exact at $\dot{K}/\dot{K}^2$?

19. Let K/F be a field extension of odd degree, and let φ, ψ be quadratic forms over F. If ψ_K is isometric to a subform of φ_K, show that ψ is isometric to a subform of φ.

20. (Ware) Let E/F be a finite Galois extension with Galois group G, and let $W(E/F) = \ker(W(F) \to W(E))$.
(1) If G has a subgroup of index 2, show that $W(E/F) \neq 0$.
(2) If $|G| = 2m$ where m is odd, show that $W(E/F) = W(F)\langle 1, -a\rangle$, where $a \in \dot{F}/\dot{F}^2$ is the discriminant of the field extension E/F.

(**Hint.** See the proof of 6.15.)

(3) If G is a nilpotent group, show that $W(E/F) = 0$ iff $|G|$ is odd.

21. (Knebusch) Let G be a finite group having no subgroup of index 2. Show that there exists a Galois extension E/F with Galois group G such that $W(E/F) = 0$.

22. Let $E = F(\theta)$ be a simple algebraic extension of F, where θ has a minimal polynomial $f(x)$ over F. Show that the signed determinant of the trace form on E is given by $\mathrm{N}_{E/F}(f'(\theta))$, where f' denotes the formal derivative of f. Re-confirm this formula in the case of a quadratic extension $E = F(\sqrt{a}\,)$ by using the trace form computation in Lemma 6.17.

23. Let $E = \mathbb{Q}\,(\theta)$, where $\theta^3 + \theta^2 - 2\theta - 1 = 0$. Show that $E/\mathbb{Q}$ is a Galois extension, and find an *orthonormal* basis for the trace form on E.

24. Let $f(x) \in \mathbb{Q}\,[x]$ be an irreducible quartic polynomial with no real roots.

    (1) If the Galois group of f over $\mathbb{Q}$ is A_4 or S_4, show that f remains irreducible over the quadratic closure $\widetilde{\mathbb{Q}}$.

    (2) Using (1), show that $\widetilde{\mathbb{Q}}\,[x]/(x^4 + x + 1)$ is a quartic field extension of $\widetilde{\mathbb{Q}}$.

Chapter VIII

Formally Real Fields, Real-Closed Fields, and Pythagorean Fields

1. Structure of Formally Real Fields

In Chapter II, we have already made use of the notion of formally real fields in our work on classifying Witt rings of fields with a small number of square classes. And in Chapter VI, we saw that the $\mathfrak{p}$-adic fields are all nonreal fields. In this chapter, we shall study more systematically the theory of formally real fields.

Let us first recall the basic definitions. For any field F, the following two conditions are easily seen to be equivalent:

(1) *-1 is not a sum of squares in F.*

(2) *For any natural number n, the quadratic form $n\langle 1\rangle = \langle 1, \ldots, 1\rangle$ is anisotropic over F.*

If F satisfies one (and hence both) of these conditions, we say that F is *formally real.* Otherwise, we say that F is *nonreal.* In general, a formally real field must have characteristic zero, although a field of characteristic zero may not necessarily be formally real (e.g. the complex field $\mathbb{C}$, or any of its extension fields).

Permanent Notation. For an arbitrary field F, let $\sigma(F)$ denote the set of elements of F that can be expressed as a sum of squares in F. We shall also write $\dot\sigma(F)$ for $\sigma(F) \setminus \{0\}$.

Proposition 1.1. (1) *$\dot{\sigma}(F)$ is a subgroup of the multiplicative group $\dot{F}$.*

(2) *If F is nonreal and* $\operatorname{char}(F) \neq 2$, *then $\sigma(F) = F$. (Of course, if* $\operatorname{char}(F) = 2$, *$\sigma(F) = F^2$.)*

Proof. (1) Clearly, $\dot{\sigma}(F)$ is closed under multiplication. If $0 \neq x = x_1^2 + \cdots + x_n^2$ in F, then

$$x^{-1} = x/x^2 = (x_1/x)^2 + \cdots + (x_n/x)^2 \in \dot{\sigma}(F),$$

so $\dot{\sigma}(F)$ is a subgroup of $\dot{F}$.

(2) Let $x \in F$. Since the hyperbolic plane $\langle 1, -1\rangle$ is universal (assuming $\operatorname{char}(F) \neq 2$), there exist $y, z \in F$ such that $x = y^2 - z^2$. (More explicitly, we could just take $y = (x+1)/2,\ z = (x-1)/2$.) If $-1 \in \sigma(F)$, we get

$$x = y^2 + (-1)z^2 \in \sigma(F) + \sigma(F) \cdot \sigma(F) \subseteq \sigma(F).$$

Therefore, $\sigma(F) = F$. □

In the 1920s, Artin and Schreier [AS] developed the algebraic theory of formally real fields and formulated the basic relationship between formally real fields and fields with orderings. This will be our main topic of discussion in the present section. We begin by introducing the notion of orderings on a field, which is probably familiar to the reader from an earlier course in abstract algebra.

Definition 1.2. An *ordering* on a field F is the assignment of a set $P \subsetneq F$ (called the *positive cone* of the ordering) which possesses the following properties:

(P1) $P + P \subseteq P$.

(P2) $P \cdot P \subseteq P$.

(P3) $P \cup (-P) = F$.

Given such a set P, we shall say briefly that F is *ordered by* P, or that (F, P) is an *ordered field.*

The following proposition collects some of the basic properties of an ordered field.

Proposition 1.3. *Let (F, P) be any ordered field. Then:*

(1) *$\sigma(F) \subseteq P$. (In particular, $0, 1 \in P$.)*

(2) *$-1 \notin P$, and $P \cap (-P) = \{0\}$.*

(3) *F is formally real (and so* $\operatorname{char}(F) = 0$*).*

(4) *$\dot{P} := P \setminus \{0\}$ is a subgroup of index 2 in $\dot{F}$.*

(5) *If $P' \subsetneq F$ gives another ordering on F, then $P \subseteq P' \Rightarrow P = P'$.*

Proof. (1) Since $P + P \subseteq P$, it suffices to prove that $F^2 \subseteq P$. Let $x \in F$. By (P3), we have $x \in P$ or $-x \in P$. If $x \in P$, then $x^2 \in P \cdot P \subseteq P$. If $-x \in P$, then $x^2 = (-x)(-x) \in P \cdot P \subseteq P$.

(2) Since $P \subsetneq F$, (P3) implies that $\mathrm{char}(F) \neq 2$. Assume that $-1 \in P$. For any $a \in F$, we have

$$a = \left(\frac{a+1}{2}\right)^2 + (-1)\left(\frac{a-1}{2}\right)^2 \in P + P \cdot P \subseteq P,$$

contradicting $P \subsetneq F$. Therefore $-1 \notin P$. Next, consider $x \in P \cap (-P)$. If $x \neq 0$, we would have $-1 = (x^{-1})^2 x(-x) \in P$, a contradiction. This shows that $P \cap (-P) = \{0\}$.

(3) Since $-1 \notin P$ and $\sigma(F) \subseteq P$, we have $-1 \notin \sigma(F)$, so F is formally real.

(4) For $x \in \dot{P}$, we have $x^{-1} = (x^{-1})^2 x \in P$. Hence $\dot{P}$ is a subgroup of $\dot{F}$. Since $\dot{F} = \dot{P} \cup (-\dot{P})$, we have $[\dot{F} : \dot{P}] = 2$.

(5) This is clear since $[\dot{F} : \dot{P}] = [\dot{F} : \dot{P}'] = 2$. □

In view of (P3) and (2) in the above proposition, we see that F is the *disjoint* union of $\{0\}$, $\dot{P}$ and $-\dot{P}$. This is "the law of trichotomy" in an ordered field (F, P). As usual, we may introduce the notation $x \leq_P y$ to mean that $y - x \in P$ (and $x <_P y$ to mean $y - x \in \dot{P}$). With respect to this notation, the law of trichotomy is expressed by the fact that, for any $x, y \in F$, we have *exactly* one of the three possibilities: $x = y$, $x <_P y$, or $y <_P x$. Otherwise put, $\leq_P$ is a *total ordering* on the elements of F. If P is given and fixed, we shall often write "$\leq$" and "$<$" instead of "$\leq_P$" and "$<_P$".

Let (F, P) be an ordered field. For any subfield F_0 of F, we may order F_0 by taking $P_0 := F_0 \cap P$ to be its positive cone. Clearly, P_0 satisfies all the axioms of an ordering on F_0; this ordering is said to be *induced* (on F_0) by the ordering P on F.

The quintessential example of an ordered field is $F = \mathbb{R}$, which has a (unique) ordering given by the positive cone $P = \mathbb{R}^2$. By what we said in the last paragraph, any subfield $F_0 \subseteq \mathbb{R}$ inherits an ordering $F_0 \cap \mathbb{R}^2$ from $\mathbb{R}$. Thus, the rational field $\mathbb{Q}$, all real quadratic fields, the field $\mathbb{Q}(\sqrt[3]{2})$, and the field of all real algebraic numbers, etc. are all equipped with natural orderings. More examples of orderings on fields will be given below in 1.13.

In order to better understand orderings and existence questions about them, let us develop a few more preliminary results on formally real fields.

Basic Lemma 1.4. *Let F be formally real and $K = F(\sqrt{a})$ be a quadratic extension of F. Then K is nonreal iff $-a \in \dot{\sigma}(F)$.*

Proof. If $-a \in \dot{\sigma}(F)$, the equation $(\sqrt{a})^2 + (-a) = 0$ shows that K is nonreal. Conversely, if K is nonreal, there exists an equation

$$-1 = \sum (b_i + c_i\sqrt{a}\,)^2, \quad \text{where } b_i, c_i \in F.$$

In particular, $-1 = \sum b_i^2 + a \sum c_i^2$. Now $\sum c_i^2 \neq 0$ (lest $-1 = \sum b_i^2 \in \sigma(F)$). Therefore,

$$-a = \left(1 + \sum b_i^2\right)\left(\sum c_i^2\right)^{-1}.$$

Since $\dot{\sigma}(F)$ is a group (by 1.1(1)), we conclude that $-a \in \dot{\sigma}(F)$. (Alternatively, we can also deduce the "only if" part above from our earlier result VII.5.8(2).) □

For a more quantitative version of this Basic Lemma, we refer the reader to Exercise 4 in Chapter XI.

Our next goal is to introduce the notion of real-closed fields. Before we do this, it is convenient to discuss first the weaker notions of euclidean fields and pythagorean fields. We have already briefly encountered these notions in Chapter II; let us recall them here.

Definition 1.5. (1) A field F is called *euclidean* if F is formally real and $|\dot{F}/\dot{F}^2| = 2$. (In such a field, $\dot{F} = \dot{F}^2 \cup (-\dot{F}^2)$.)

(2) A field F is called *pythagorean* if the sum of two squares (or any number of squares) in F is always a square. (It is, of course, sufficient to require that $1 + y^2$ be a square for every $y \in F$.) For such F, we have $\sigma(F) = F^2$.

A relationship between "euclidean" and "pythagorean" is given in the following easy result.

Proposition 1.6. *If F is euclidean, then F is pythagorean with a unique ordering.*

Proof. We claim that $P := F^2$ is an ordering. For this P, we clearly have

$$P \neq F, \quad P \cdot P \subseteq P, \quad \text{and} \quad P \cup (-P) = F.$$

Thus, we need only prove that $P + P \subseteq P$, that is, F is pythagorean. Consider a sum $1+y^2$, where $y \in F$. If $1+y^2 \in -F^2$, F would be nonreal, which is not the case. Since F is euclidean, we must have then $1+y^2 \in F^2$, as desired. Once we know $P = F^2$ is an ordering, it is clear (from 1.3(5)) that it is the unique ordering (on F). □

Remark. The *converse* of 1.6 is true too: we shall prove it later in 4.2.

The next result offers several important characterizations of euclidean fields.

Theorem 1.7. *For any field F (of any characteristic), the following are equivalent:*

(1) *F is euclidean.*

(2) *F is formally real, but every quadratic extension of F is nonreal.*

(3) *$i := \sqrt{-1} \notin F$, and $K := F(i)$ is quadratically closed (that is, $K^2 = K$).*

(4) *$\operatorname{char}(F) \neq 2$ and there exists a quadratic extension $L \supseteq F$ that is quadratically closed.*

Proof. (2) $\Rightarrow$ (1). Assume (2), and consider any nonsquare $a \in F$. Then $F(\sqrt{a}\,)$ is nonreal, and so by 1.4,

$$-a = a_1^2 + \cdots + a_n^2 \quad \text{for some } a_i \in F.$$

We may assume this equation is taken with n minimal. (In particular, each $a_i \neq 0$.) If $n \geq 2$, $a_1^2 + a_2^2 \notin F^2$ would imply as above that

$$-(a_1^2 + a_2^2) = b_1^2 + \cdots + b_m^2 \quad (\text{for some } b_j \in F).$$

This contradicts the formal reality of F. Hence we must have $n = 1$, and so $a = -a_1^2 \in -F^2$.

(1) $\Rightarrow$ (3). Under (1), certainly $i \notin F$. For $K = F(i)$, the homomorphism $\dot{K}/\dot{K}^2 \to \dot{F}/\dot{F}^2$ induced by the norm map is trivial, since F is pythagorean (by 1.6). Therefore, by VII.3.8, $\dot{K}/\dot{K}^2$ is isomorphic to $\dot{F}/\{\pm\dot{F}^2\} = \{1\}$, so $\dot{K} = \dot{K}^2$.

(3) $\Rightarrow$ (4) is clear.

(4) $\Rightarrow$ (2). This implication is essentially a special case of Ch. VII, Exer. 4. To keep our exposition self-contained, we offer a direct proof. Since $\operatorname{char}(F) \neq 2$, we may represent the field L in (4) in the form $F(\sqrt{a}\,)$, where $a \notin \dot{F}^2$. The norm map $\dot{L}/\dot{L}^2 \to \dot{F}/\dot{F}^2$ has image $D_F(\langle 1, -a\rangle)/\dot{F}^2$. Since $\dot{L} = \dot{L}^2$, we have $D_F(\langle 1, -a\rangle) = \dot{F}^2$. In particular, $-a \in \dot{F}^2$. Thus, we may assume that $a = -1$, and the equation $D_F(\langle 1, -a\rangle) = \dot{F}^2$ now says that F is pythagorean. If F is nonreal, then $-1 \in \sigma(F) = F^2$, a contradiction. Thus, F is formally real. Also, the exact sequence in VII.3.8 implies that $\dot{F}/\dot{F}^2 = \{\dot{F}, -\dot{F}^2\}$, so L/F is the *only* quadratic extension of F, and (2) follows. □

Remark. In condition (4) above, the assumption that $\operatorname{char}(F) \neq 2$ is essential. For instance, $F = \mathbb{F}_2$ has a quadratically closed quadratic extension $\mathbb{F}_4$, but F certainly does not satisfy (1), (2), or (3).

Examples. Two obvious examples of euclidean fields are the real field $\mathbb{R}$, and the field A of real algebraic numbers. Here, $A = \overline{\mathbb{Q}} \cap \mathbb{R}$, where $\overline{\mathbb{Q}}$ denotes the algebraic closure of $\mathbb{Q}$ (taken within the complex field $\mathbb{C}$). We can give a smaller example of a euclidean field, as follows. Let $\widetilde{\mathbb{Q}}$ be the field of constructible numbers; that is, $\widetilde{\mathbb{Q}}$ is the quadratic closure of $\mathbb{Q}$ within $\overline{\mathbb{Q}}$ (see VII.7). It is easy to show that $F := \widetilde{\mathbb{Q}} \cap \mathbb{R}$ (the field of *real* constructible numbers) is a euclidean field, with $F(\sqrt{-1}) = \widetilde{\mathbb{Q}}$; for more details on this, see Exercise 4.

We are now well-prepared to introduce and to study the all-important notion of a real-closed field, due to Artin and Schreier.

Definition. A field F is called *real-closed* if F is formally real, but no proper algebraic extension of F is formally real.

As an immediate consequence of 1.7, we have

Corollary 1.8. *Let F be a real-closed field. Then F is euclidean (with a unique ordering F^2), and $F(\sqrt{-1})$ is quadratically closed.*

Next we shall prove the following result, which shows that there is an abundant supply of real-closed fields.

Proposition 1.9. *Let F be any formally real field, and $\overline{F}$ be its algebraic closure. Then there exists a real-closed field R between F and $\overline{F}$.*

Proof. Consider the collection $\mathcal{S}$ of all formally real subfields of $\overline{F}$ containing F. If $\{F_\alpha\}$ is an inductive family (relative to inclusion) of such fields, then $F_0 = \bigcup_\alpha F_\alpha$ clearly belongs to the same family $\mathcal{S}$. By Zorn's Lemma, there exists $R \in \mathcal{S}$ that is a maximal member of $\mathcal{S}$ with respect to inclusion. Such a field R is clearly real-closed! □

From this proposition, we can easily deduce the following famous Artin-Schreier Criterion ([AS], ca. 1927) for formally real fields.

Theorem 1.10. *F is formally real iff F possesses at least one ordering.*

Proof. The "if" part is just 1.3(2). Conversely, assume that F is formally real. By the above proposition, there exists an algebraic extension $R \supseteq F$ that is real-closed. The unique ordering on R (see 1.8) therefore induces an ordering on F. □

Definition 1.11. An element $b \in \dot{F}$ is said to be *totally positive* if it is positive with respect to all orderings on F. (Note that if F has no orderings, this condition is vacuously satisfied.)

Artin's Theorem 1.12. *For a field F of characteristic $\neq 2$, an element $b \in \dot{F}$ is totally positive iff $b \in \sigma(F)$.*

Proof. If F is nonreal (and $\mathrm{char}(F) \neq 2$), every $b \in \dot{F}$ belongs to $\sigma(F)$ (by 1.1) and every $b \in \dot{F}$ is totally positive (since F admits no orderings). In this case, the desired result is a tautology. For F formally real, we need only show the "only if" part. Suppose $b \notin \sigma(F)$. By the Basic Lemma 1.4, $K = F(\alpha)$ is formally real for $\alpha = \sqrt{-b}$. Pick an ordering on K (which is possible by 1.10). It induces an ordering on F in which $b = -\alpha^2$ is *negative*, so b is not totally positive in F. □

The characteristic assumption in 1.12 is necessary (for the "only if" part) since, in the case where F is a field of characteristic 2, every element $b \in \dot{F}$ is (vacuously) totally positive, but $\dot{\sigma}(F)$ is just the group $\dot{F}^2$. In this case, the conclusion of the theorem would hold only for *perfect* fields (of characteristic 2).

The criterion for total positivity in 1.12 is often attributed to Artin and Schreier in the literature. However, the result 1.12 appeared *not* in Artin-Schreier's inaugural paper [AS] on formally real fields, but rather in Artin's solo paper [Ar] (in which he solved Hilbert's 17th Problem[(1)]). Therefore, it seems that the correct attribution of 1.12 should be to Emil Artin alone.

Let us now give some examples of ordered fields.

Examples 1.13. (A) In relation to II.5.4 and II.5.7, we have constructed several formally real fields F with $|\dot{F}/\dot{F}^2| = 8$. Each of the fields constructed has the property that $\dot{F}/\dot{F}^2$ has a $\mathbb{Z}_2$-basis consisting of $\{-1, x, y\}$ where x, y are sums of squares in F. Therefore, we have $[\dot{F} : \dot{\sigma}(F)] = 2$. Since F must have some orderings by 1.10, it follows that $\sigma(F)$ gives the unique ordering on F for these fields. (Of course, these are all *nonpythagorean* fields.) On the other hand, for the formally real fields F with 8 square classes discussed in Case 1 and Case 2 in the proof of II.5.13, we have $[\dot{F} : \dot{\sigma}(F)] = 4$. Here, F must have exactly two orderings (and F is nonpythagorean). (However, a model field for Case 2 there still awaits construction.)

(B) Let $F = \mathbb{Q}(\alpha)$ where $\alpha^2 = 2$. We can define an ordering P on F by using the embedding $\varphi : F \to \mathbb{R}$ with $\varphi(\alpha) = \sqrt{2}$. Similarly, we can define another ordering $P' \neq P$ on F by using the embedding $\varphi' : F \to \mathbb{R}$ with $\varphi'(\alpha) = -\sqrt{2}$. It can be shown that P, P' are the only two orderings on F (see 2.20 below). Granted this fact, Artin's Theorem 1.12 leads to the interesting equation $\sigma(F) = P \cap P'$. (In particular, we have $[\dot{F} : \dot{\sigma}(F)] = 4$.) Consider, for instance, the element $\theta = 5 + 3\alpha \in \dot{F}$. Since

$$\varphi(\theta) = 5 + 3\sqrt{2} > 0 \quad \text{and} \quad \varphi'(\theta) = 5 - 3\sqrt{2} > 0,$$

[(1)] In fact, the result 1.12 was used as one of the key tools by Artin in his solution of this Hilbert Problem (on the structure of positive semidefinite rational functions).

the equation $\sigma(F) = P \cap P'$ predicts that $\theta \in \sigma(F)$. This can be verified directly if we want. Indeed,

$$2\theta = 10 + 6\alpha = 1 + (\alpha+1)^2 + (\alpha+1)^2 + (\alpha+1)^2 \in \sigma(F)$$

implies that $\theta \in \sigma(F)$. On the other hand, for an element such as $\theta' = 4 + 3\alpha \in \dot{F}$, we may conclude that $\theta' \notin \sigma(F)$, since $\varphi'(\theta') = 4 - 3\sqrt{2} < 0$.

(C) Let $F = k(x)$, where k is a field given with an ordering P_0. We can extend this ordering to F in several ways. First, we declare a polynomial

$$f(x) = a_0 + a_1x + \cdots + a_nx^n \in k[x]$$

positive if $a_n \in \dot{P}_0$. Then we declare a rational function $g(x)/f(x)$ positive if the polynomial $f(x)g(x)$ is positive. It is easy to check that the set of positive elements in F defined in this way, together with 0, gives an ordering P_1 on F. Note that in this ordering, we have

$$0 < \cdots < x^{-2} < x^{-1} < a < x < x^2 < \cdots$$

for any $a \in \dot{P}_0$, as we can readily check. We can also get a second extension of P_0 as follows. Declare a polynomial

$$f(x) = a_rx^r + a_{r+1}x^{r+1} + \cdots + a_nx^n \in k[x]$$

positive if $a_r \in \dot{P}_0$, and extend this positivity notion to $F = k(x)$ as before. This results in a second ordering P_2 extending P_0. With respect to this ordering P_2, we have instead

$$0 < \cdots < x^2 < x < a < x^{-1} < x^{-2} < \cdots$$

for any $a \in \dot{P}_0$. It is easily seen that the two orderings P_1, P_2 are actually "conjugate" under the k-automorphism of F defined by $x \mapsto x^{-1}$. In general, for any automorphism σ of a field K ordered by P, $\sigma(P)$ is another ordering on K, said to be *conjugate to* P. Of course $\sigma(P)$ may be equal to P, even with $\sigma \neq \mathrm{Id}_K$. For instance, in the example $F = k(x)$ above, if σ is the k-automorphism on F given by $\sigma(x) = x + 1$, we can show easily that $\sigma(P_1) = P_1$. However, if we take τ to be the k-automorphism on F given by $\tau(x) = -x$, then $P_3 = \tau(P_1)$ and $P_4 = \tau(P_2)$ are two new orderings on F with respect to which x is negative. Note that the P_i's obtained above are examples of *nonarchimedean* orderings on F: these are orderings with respect to which there are elements that are larger than all integers (and hence all rational numbers) in F.

In view of 1.10, the extendibility of orderings from k to $k(x)$ implies immediately that $k(x)$ is formally real if (and only if) k is formally real. This fact can also be verified directly without using orderings: for the details, see IX.1.2.

(D) In the above example, take $k = \mathbb{R}$, and take P_0 to be the usual ordering on $\mathbb{R}$. Let C be any subset of $\mathbb{R}$ with the property that

$$(*) \qquad \text{For any pair } a < b \in \mathbb{R}: \quad b \in C \Longrightarrow a \in C.$$

We can define an ordering P_C on $F = \mathbb{R}(x)$ as follows. For any nonzero polynomial $f(x) \in \mathbb{R}[x]$, write down the factorization of f into irreducible factors:

$$f(x) = r(x - a_1) \cdots (x - a_n) q_1(x) \cdots q_m(x),$$

where $r, a_1, \ldots, a_n \in \mathbb{R}$, and the $q_i(x)$'s are monic irreducible quadratics. We shall take $f(x) \in \dot{P}_C$ iff $r \in \dot{P}_0$ and the number of $a_i \notin C$ is even, or $r \notin \dot{P}_0$ and the number of $a_i \notin C$ is odd. For nonzero rational functions $g(x)/f(x)$, we take (as before) $g/f \in \dot{P}_C$ iff $gf \in \dot{P}_C$. It can be shown that the P_C obtained in this manner is an ordering on $\mathbb{R}(x)$, and is, in fact, the unique ordering P on $\mathbb{R}(x)$ with respect to which $C = \{b \in \mathbb{R} : b <_P x\}$ (see Exercise 26). Note that the cases $C = \mathbb{R}$ and $C = \varnothing$ are certainly allowed in this construction, since these choices of C do have the required property $(*)$. For these choices, we simply have $P_{\mathbb{R}} = P_1$ and $P_{\varnothing} = P_3$ in the notation of (C) above.

According to Exercise 26, every ordering on $\mathbb{R}(x)$ has the form P_C for some $C \subseteq \mathbb{R}$ satisfying $(*)$, so the next question is: what are the possible choices of C? The answer to this goes back to Dedekind: if $C \neq \mathbb{R}, \varnothing$, then C must be $(-\infty, a]$ or $(-\infty, a)$ for some $a \in \mathbb{R}$ (the real number system $\mathbb{R}$ is "cut complete"). Thus, each real number $a \in \mathbb{R}$ gives rise to two orderings on $\mathbb{R}(x)$: for the choice $C = (-\infty, a]$, P_C is an ordering in which x is between a and s for any real number $s > a$; for the choice $C = (-\infty, a)$, P_C is an ordering in which x is between t and a for any real number $t < a$. For instance, for the real number $a = 0$, these two orderings are precisely what we called P_2 and P_4 in (C) above.

(E) Here we take $k = \mathbb{Q}$, and P_0 to be the usual ordering on $\mathbb{Q}$. It can be shown that the orderings on $\mathbb{Q}(x)$ are of the following kinds (see Exercises 27, 28).

(1) The two orderings P_1 and P_3 defined in (C).

(2) The restrictions of the orderings P_C on $\mathbb{R}(x)$ for $C = (-\infty, a]$ and $C = (-\infty, a)$, where a is any *real algebraic number*. It can be shown that, for such a number a, the two restrictions to $\mathbb{Q}(x)$ are really different orderings. The reader should try to supply a proof for this fact. Here, we shall just give an illustrative example, for the real algebraic number $a = \sqrt{2}$. In the ordering P_C for $C = (-\infty, \sqrt{2}\,]$, we have $x > \sqrt{2}$, so

$$f(x) = x^2 - 2 = \left(x + \sqrt{2}\right)\left(x - \sqrt{2}\right)$$

is positive; in the ordering P_C where $C = (-\infty, \sqrt{2})$, however, we have $x < \sqrt{2}$, so $f(x) = x^2 - 2$ is negative instead. Since $x^2 - 2 \in \mathbb{Q}(x)$, the two orderings P_C on $\mathbb{R}(x)$ restrict to different ones on $\mathbb{Q}(x)$. For more details, see Exercises 26–28.

(3) The common restriction to $\mathbb{Q}(x)$ of the orderings P_C on $\mathbb{R}(x)$ for $C = (-\infty, a]$ and $C = (-\infty, a)$, where a is *any real transcendental number* (such as π or e). The fact that the two restrictions to $\mathbb{Q}(x)$ are the same stems from the observation that, when we factor a monic polynomial $f(x) \in \mathbb{Q}[x]$ into irreducible factors over $\mathbb{R}[x]$, the linear factors are of the form $x - a_i$ where the a_i's are real *algebraic* numbers. (The positivity or negativity of $f(x)$ in the ordering P_C depends solely on how many a_i's are in C. This number is the same for $C = (-\infty, a]$ and for $C = (-\infty, a)$ if a is a real transcendental number.)

We finish with two more observations. The first is that the ordering in (3) above is exactly the one obtained on $\mathbb{Q}(x)$ by identifying $\mathbb{Q}(x)$ with $\mathbb{Q}(a) \subseteq \mathbb{R}$ (a = transcendental) and restricting the usual ordering on $\mathbb{R}$. The second observation is that (2), (3) lead to *archimedean* orderings on $\mathbb{Q}(x)$, while (1) leads to (two) *nonarchimedean* ones. According to Exercise 30, these are all the possible orderings on the rational function field $\mathbb{Q}(x)$.

2. Characterizations of Real-Closed Fields

In the last section, we have obtained some characterizations of euclidean fields. In the present section, we shall derive the analogues of these results for real-closed fields.

We begin with a simple field-theoretic fact.

Proposition 2.1. *For any field F, the following are equivalent:*

(1) *Any odd-degree polynomial $f \in F[x]$ has a root in F.*

(2) *F has no proper odd-degree extension.*

Proof. (2) $\Rightarrow$ (1). We induct on $n = \deg f$. The result is trivial for $n = 1$, so assume $n > 1$. If $f(x)$ is irreducible, $F[x]/(f(x))$ would be a proper odd-degree field extension of F, contradicting (2). Therefore, there exists a proper factorization $f = f_1 f_2$, where, say, deg f_1 is odd. By the inductive hypothesis, f_1 has a root in F, so f also has a root in F.

(1) $\Rightarrow$ (2). Say K/F has odd degree $n > 1$. Take $\theta \in K \setminus F$ and let $f(x)$ be the minimal polynomial of θ over F. Then $\deg f = [F(\theta) : F]$ is an odd integer > 1. Since f is irreducible in $F[x]$, it has no root in F. $\square$

Next we note the following nice property of a formally real field.

Proposition 2.2. *If F is formally real, so is every odd-degree extension K of F.*

This is an immediate consequence of Springer's Theorem VII.2.3, which says that if $q = n\langle 1\rangle$ is anisotropic over F, then $q_K = n\langle 1\rangle_K$ remains anisotropic over K.

Corollary 2.3. *If F is real-closed, then any odd-degree polynomial $f \in F[x]$ has a root in F.*

Proof. Since any odd-degree extension of F is formally real by 2.2, F cannot have any proper odd-degree extension. Therefore, the desired conclusion follows from 2.1. □

Remark 2.4. The conclusion of the corollary above need not hold over a euclidean field. For instance, over the euclidean field F of all real constructible numbers, the cubic polynomial $f(x) = x^3 - 2$ has no root (since $\sqrt[3]{2}$ is not constructible). Here F admits a cubic extension $F[x]/(x^3 - 2)$.

We now come to the following main characterization theorem for real-closed fields, which is due to Artin and Schreier. This result is to be compared with 1.7.

Theorem 2.5. *For any field F, the following are equivalent:*

(1) *F is real-closed.*

(2) *F is euclidean, and every odd-degree polynomial in $F[x]$ has a root in F.*

(3) *$i := \sqrt{-1} \notin F$ and $K = F(i)$ is algebraically closed.*

This result may be viewed as an "algebraic version" of the Fundamental Theorem of Algebra. To justify this viewpoint, let us deduce the "usual version" of the Fundamental Theorem of Algebra as an immediate consequence of 2.5.

Corollary 2.6. *The real field $\mathbb{R}$ is real-closed, and the complex field $\mathbb{C} = \mathbb{R}(i)$ is algebraically closed.*

Proof. To begin with, $\mathbb{R}$ is a euclidean field. By the usual continuity argument in calculus, every real polynomial of odd degree has a real root. Therefore, (2) in the above theorem is satisfied for $F = \mathbb{R}$, and we get the desired conclusions from (1) and (3) respectively! □

Proof of 2.5. (3) $\Rightarrow$ (1). By 1.7, F must be euclidean. In particular, F is formally real. Since the only proper algebraic extension of F is K (which is nonreal), F is real-closed.

(1) $\Rightarrow$ (2) follows from 1.8 and 2.3.

(2) $\Rightarrow$ (3). (This implication is sometimes called the "Euler-Lagrange-Gauss Theorem".) First, we know from 1.7 that K is quadratically closed. Let $\alpha \mapsto \overline{\alpha}$ denote the complex conjugation on K. If $f(x) \in K[x] \setminus K$, then $f(x)\overline{f}(x) \in F[x]$. If $f(x)\overline{f}(x)$ has a root in K, then f itself has a root in K. It is thus sufficient to show that any $g(x) \in F[x] \setminus F$ has a root in K. Consider a splitting field E of $(x^2+1)g(x)$ over F, which is a Galois extension[(2)] of F containing K. By 2.1, the second hypothesis in (2) means that F has no proper odd-degree extensions. By Galois theory (and the existence of a 2-Sylow group in $\text{Gal}(E/F)$), this implies that $[E : F] = 2^n$ for some n. But K has no quadratic extensions, so by Galois theory again, $E = K$. Since E is the splitting field of $(x^2+1)g(x)$, we have shown that $g(x)$ has a root in K. □

Remark. If K is an algebraically closed field, then there exists a (real-closed) subfield $F \subsetneq K$ such that $K = F(\sqrt{-1}\,)$ if and only if $\text{char}(K) = 0$. We shall leave this as an exercise for the reader (see Exercise 10). It will soon be clear, however, that the real-closed field F mentioned in the last sentence need not be unique.

We shall next introduce the notion of a "real-closure", which will be used to relate ordered fields to real-closed fields.

Definition 2.7. Let F be a field ordered by a positive cone P. An extension field $R \supseteq F$ is called a *real-closure* of F (relative to P) if it satisfies the following three conditions:

(1) R is real-closed;

(2) R is algebraic over F; and

(3) the given ordering on F is induced by the unique ordering on R (in other words, $P = R^2 \cap F$).

We have the following existence and uniqueness result.

Theorem 2.8. (1) *Every ordered field (F, P) possesses a real-closure.*

(2) *If (F_1, P_1), (F_2, P_2) are ordered fields, and R_1, R_2 are their real-closures, then any order isomorphism $f : F_1 \to F_2$ (isomorphism such that $f(P_1) = P_2$) extends uniquely to an isomorphism $\tilde{f} : R_1 \to R_2$, which is automatically an order isomorphism.*

This theorem means that the possible orderings which can be put on F are in 1-1 correspondence with the F-isomorphism classes of the real-closed algebraic extensions of F.

[(2)]Note that $\text{char}(F) = 0$ here.

In the following, we shall present only the proof of the *existence* part of 2.8. The proof of the *uniqueness* part will be given later in Appendix A to this section. Our proof for the existence part of 2.8 is based on the following lemma.

Lemma 2.9. *Let (F, P) be an ordered field, and $\overline{F}$ be the algebraic closure of F. Then $E = F(\{\sqrt{b} : b \in \dot{P}\}) \subseteq \overline{F}$ is formally real.*

Proof. It suffices to prove the following statement:

$$\text{(2.10)} \qquad \begin{array}{l} \textit{For any } c_1, \ldots, c_n \textit{ and } b_1, \ldots, b_r \textit{ in } \dot{P}, \textit{ the form} \\ \langle c_1, \ldots, c_n \rangle \textit{ is anisotropic over } F(\sqrt{b_1}, \ldots, \sqrt{b_r}). \end{array}$$

Once we know this, then, for any n, $n\langle 1 \rangle$ is anisotropic over $F(\sqrt{b_1}, \ldots, \sqrt{b_r})$ as above, and therefore over E. The proof of (2.10) (for all n) is by induction on r, the case $r = 0$ being trivial. Suppose (2.10) is already true over

$$F(\sqrt{b_1}, \ldots, \sqrt{b_{r-1}}) \subsetneq F(\sqrt{b_1}, \ldots, \sqrt{b_r}).$$

If $\langle c_1, \ldots, c_n \rangle$ has a zero $(z_1, \ldots, z_n)$ over $F(\sqrt{b_1}, \ldots, \sqrt{b_r})$, write $z_i = x_i + y_i\sqrt{b_r}$ where $x_i, y_i \in F(\sqrt{b_1}, \ldots, \sqrt{b_{r-1}})$. Then

$$0 = \sum c_i z_i^2 = \sum c_i x_i^2 + \sum b_r c_i y_i^2 + 2 \sum c_i x_i y_i \sqrt{b_r}$$

implies that $(x_1, \ldots, x_n, y_1, \ldots, y_n)$ is a zero of $\langle c_1, \ldots, c_n, b_r c_1, \ldots, b_r c_n \rangle$ over $F(\sqrt{b_1}, \ldots, \sqrt{b_{r-1}})$. Hence $x_i = y_i = 0$ for all i, as desired. □

Existence Proof for 2.8. For an ordered field (F, P), let E be defined as in the above lemma. Since E is formally real, there exists (by 1.9) a real-closed field $R \subseteq \overline{F}$ containing E. Since each $b \in \dot{P}$ has a square root in R, we have $P \subseteq F \cap R^2$. According to 1.3(5), $P = F \cap R^2$, so R is a real-closure of (F, P). □

Another proof for the existence part can also be given as follows. Consider the family of ordered fields (F', P') where $F' \supseteq F$ is any algebraic extension and $P' \cap F = P$. We can order this family in an obvious way (by inclusion and compatibility of the orderings), and apply Zorn's Lemma to prove the existence of a maximal member (R, P_1). One can then prove that R is real-closed, and hence a real-closure of (F, P). We shall leave the checking of the details to the reader, with the remark that this proof is actually not very different in substance from the proof we gave above using Lemma 2.9.

Note that in general, an ordered field (F, P) may have more than one real-closure (though different real-closures must be isomorphic over F). To construct an example, let $F = \mathbb{Q}$ and let P_1, P_2 be the two orderings on $\mathbb{Q}(\sqrt{2})$ such that $\sqrt{2} \in P_1$ and $-\sqrt{2} \in P_2$. Let R_i be a real-closure of

$(\mathbb{Q}(\sqrt{2}), P_i)$ in $\overline{\mathbb{Q}}$ for $i = 1, 2$. Then $R_1 \neq R_2$, since $\sqrt{2} \in R_1^2$ but $\sqrt{2} \notin R_2^2$. However, since P_1 and P_2 induce the same ordering P (the unique one) on $F = \mathbb{Q}$, R_1, R_2 are clearly both real-closures of (F, P).

The above construction can also be used to show that there are "unusual" copies of the real number field $\mathbb{R}$ sitting inside the field $\mathbb{C}$ of complex numbers! In fact, let $\{x_i : i \in I\}$ be a transcendence basis of $\mathbb{R}$ over $\mathbb{Q}$, and let P_1 and P be, respectively, the orderings induced by $\mathbb{R}^2$ on $K = \mathbb{Q}(\{x_i\})$ and on $F = \mathbb{Q}(x_1^2, \{x_i\}_{i \neq 1})$, where we take "1" to be a given element in the indexing set I. Let α be the generator of $\text{Gal}(K/F)$, and let $P_2 = \alpha(P_1)$. Then

$$P_2 \cap F = P_1 \cap F = P,$$

but $\alpha(x_1) = -x_1$ implies that $P_2 \neq P_1$. Let $R \subseteq \mathbb{C}$ be a real-closure of (K, P_2). Then R has codimension 2 in $\mathbb{C}$, with $R(\sqrt{-1}) = \mathbb{C}$. Here, $R \neq \mathbb{R}$, since

$$R^2 \cap K = P_2 \neq P_1 = \mathbb{R}^2 \cap K.$$

However, R and $\mathbb{R}$ are both real-closures of (F, P), so by 2.8, R is isomorphic to $\mathbb{R}$ (over F).

If the above example was not "scary" enough, let us now point out that there are also real-closed fields of codimension 2 in $\mathbb{C}$ that are not even isomorphic to the real field $\mathbb{R}$! To prove the existence of such a real-closed field, keep the notation $K = \mathbb{Q}(\{x_i\})$ in the previous paragraph. Fix any ordering P_3 on K with respect to which $x_1 > \mathbb{Q}$. (Such an ordering exists since K can be represented as a rational function field in x_1 over $\mathbb{Q}(\{x_i\}_{i \neq 1})$.) Let R' be a real-closure of (K, P_3) in $\mathbb{C}$, so again $\mathbb{C} = R'(\sqrt{-1})$. However, (R', R'^2) is a *nonarchimedean* ordered field, so clearly $R' \not\cong \mathbb{R}$.

Of course, the situation is different with $\overline{\mathbb{Q}}$, the field of all algebraic numbers. Any real-closed subfield R of $\overline{\mathbb{Q}}$ is a real-closure of $\mathbb{Q}$ with respect to its usual ordering. Therefore, by the uniqueness part of 2.8, R is isomorphic to the field of real algebraic numbers — the "usual" real-closure of $\mathbb{Q}$.

We shall finish this section by including a discussion of the notion of euclidean closures of a formally real field, which is a natural generalization of the notion of real closures. To facilitate this discussion, let us first recall, from VII.7, the notion of the *quadratic closure* $\widetilde{F}$ of a field F. By definition, $\widetilde{F}$ is the smallest quadratically closed extension of F; it may also be thought of as the "maximal 2-extension" of F, namely, the compositum of all Galois extensions K/F where $[K : F]$ is a power of 2. For formally real fields F, we introduce the following notion of "euclidean closures" of F, due to Eberhard Becker [Be$_1$].

Definition 2.11. Let F be a formally real field. By a *euclidean closure* of F, we mean a field $E \supseteq F$ that is euclidean and is minimal with respect to this property.

Note that such a field E must be algebraic over F. For, let E_0 be the algebraic closure of F within E. Clearly E_0 is formally real. For any element $a \in E_0$, we have $\pm a = x^2$ (for a suitable choice of sign) for some $x \in E$. Then x is algebraic over F, so $x \in E_0$, and hence $a \in \pm E_0^2$. This shows that E_0 is also euclidean, and hence $E = E_0$.

The following result, together with 1.9, implies that any formally real field has a euclidean closure.

Proposition 2.12. *Let $F \subseteq L$ be a field extension, where L is euclidean. Then L contains a unique euclidean closure of F.*

Proof. Let P $(= L^2)$ be the unique ordering on L. Define fields $\{F_i\}$ by putting $F_0 = F$, and inductively

$$F_{i+1} = F_i(\{\sqrt{a}: \ a \in F_i \cap P\}).$$

We see easily that $E := \bigcup_{i \geq 0} F_i \subseteq L$ is euclidean. Now if K is *any* euclidean subfield of L, then P induces the unique ordering on K. Thus, if $a \in K$ and $a \in P$, we will have $\sqrt{a} \in K$. By induction on i, we see that $F_i \subseteq K$ and therefore $E \subseteq K$. This shows that E is the unique euclidean closure of F in L. □

We can now describe *all* the euclidean closures of a formally real field.

Theorem 2.13. *Let F be a formally real field. Then a field $E \supseteq F$ is a euclidean closure of F iff E is a subfield of codimension 2 in $\widetilde{F}$ (the quadratic closure of F).*

Proof. First assume $E \subseteq \widetilde{F}$ is of codimension 2. Using the characterization (4) in 1.7, we see that E is euclidean. To see that E is a euclidean closure of F, let $E \supseteq L \supseteq F$ where L is also euclidean. Then by 1.7, $L(\sqrt{-1}) \subseteq \widetilde{F}$ is quadratically closed, and therefore $L(\sqrt{-1}) = \widetilde{F}$, which implies that $L = E$.

Conversely, let E be *any* euclidean closure of F. The proof of 2.12 shows that $E \subseteq \widetilde{F}$. Since $E(\sqrt{-1})$ is quadratically closed and $\widetilde{F}$ is the smallest quadratically closed field containing F, we have $E(\sqrt{-1}) = \widetilde{F}$, as desired. □

If E is a euclidean closure of F, then the unique ordering on E induces an ordering P on F. We shall say that E is a euclidean closure of the ordered field (F, P).

Theorem 2.14. *Let (F, P) be an ordered field. Then the euclidean closures of (F, P) are exactly the fields $R \cap \widetilde{F}$ where R is any real closure of (F, P). Any two euclidean closures of (F, P) are isomorphic over F.*

Proof. First, let $E := R \cap \widetilde{F}$, where R is a real-closure of (F, P). By Exercise 15 in Chapter II, E is a euclidean subfield of $\widetilde{F}$. Since the unique ordering on E has to be $R^2 \cap E$, E is a euclidean closure of (F, P).

Conversely, let $E \subseteq \widetilde{F}$ be a euclidean closure of (F, P). Let R be a real-closure of (E, E^2). Then R is also a real-closure of F with respect to $E^2 \cap F = P$. Since $\sqrt{-1} \notin R$, we have clearly $R \cap \widetilde{F} = E$.

To prove the last part of this theorem, let E_1, E_2 be euclidean closures of (F, P). As above, we can write $E_i = R_i \cap \widetilde{F}$ where R_i is a suitable real-closure of (F, P). By 2.8, there exists an F-isomorphism $\varphi : R_1 \to R_2$. Since $\overline{F} = R_i(\sqrt{-1}\,)$, φ can be extended to an automorphism ψ of $\overline{F}$. We must have $\psi(\widetilde{F}) = \widetilde{F}$, as $\widetilde{F}$ is the smallest quadratically closed field containing F. Therefore,

$$\varphi(E_1) = \psi(R_1 \cap \widetilde{F}) = \psi(R_1) \cap \psi(\widetilde{F}) = R_2 \cap \widetilde{F} = E_2,$$

so $\varphi : E_1 \to E_2$ is the desired isomorphism over F. $\square$

Actually, concerning the last part of the theorem, a much sharper statement can be made: *If E_1 is a given euclidean closure of (F, P), then for any euclidean field $E \supseteq F$ whose unique ordering extends P, there exists a unique F-embedding of E_1 into E.* The proof of this will be left as an exercise.

In the case where $F = \mathbb{Q}$, one euclidean closure of $\mathbb{Q}$ (with its usual ordering P) is the field of all real constructible numbers. However, the same construction used before for showing the nonuniqueness of the real-closures of $(\mathbb{Q}, P)$ shows clearly that $(\mathbb{Q}, P)$ has other euclidean closures.

Appendix A: Uniqueness of Real-Closure

In this Appendix, we shall give a proof for the uniqueness of the real-closure of an ordered field (F, P), which was left out in the text above. Classically, this uniqueness proof is based on the use of a technical theorem of Sturm. Here we shall give a modern version of this proof, in which Sturm's Theorem is replaced by the use of a certain result on trace forms.

Let $f(t) \in F[t]$ be a polynomial that is without multiple roots in the algebraic closure of a field F, and let $A = F[t]/(f(t))$. As a finite-dimensional F-algebra, A carries a trace function $\mathrm{tr} : A \to F$, so we may define on A the associated trace form:

$$T_f(x, y) = \mathrm{tr}(xy) \qquad (x, y \in A).$$

This is a symmetric bilinear form over F, which we shall identify, as usual, with its depolarized quadratic form $x \mapsto \mathrm{tr}(x^2)$. (See Ch.I, Exercise 27.)

Let P be a *fixed* ordering on F and let R be any real-closed field containing F such that $R^2 \cap F = P$. Then the signature $\mathrm{sgn}(T_f) \in \mathbb{Z}$ is defined by thinking of T_f (by scalar extension) as a quadratic form over R. Note that since a diagonalization of the form T_f can be taken in F, the integer $\mathrm{sgn}(T_f)$ depends only on f and on P, but not on the choice of the real-closed field R. To emphasize this particular fact, let us denote this signature by $\mathrm{sgn}(T_f)_P$.

The following important interpretation of $\mathrm{sgn}(T_f)_P$ appeared in a paper of Olga Taussky in 1968 (in the case $F = \mathbb{Q}$ and A a number field), but the idea of the result goes as far back as to the work of J. J. Sylvester.

Theorem 2.15. *In the above notations, the signature* $\mathrm{sgn}(T_f)_P$ *is equal to the number* r *of roots of* f *in the real-closed field* R.

Proof. (Cf. the proof of VII.6.12) We can think of the scalar extension of T_f, to R as the trace form of the R-algebra $R \otimes_F A$. Over the real-closed field R, we have a factorization

$$f = g_1 \cdots g_r \cdot h_1 \cdots h_s \in R\,[t],$$

where the factors are irreducible and distinct, $\deg g_i = 1$ and $\deg h_j = 2$. Then by the Chinese Remainder Theorem:

$$\begin{aligned} R \otimes_F A &\cong \frac{R\,[t]}{(g_1 \cdots g_r \cdot h_1 \cdots h_s)} \\ &\cong \prod_{i=1}^{r} \frac{R\,[t]}{(g_i)} \times \prod_{j=1}^{s} \frac{R\,[t]}{(h_j)} \\ &\cong R \times \cdots \times R \times \overline{R} \times \cdots \times \overline{R}, \end{aligned}$$

where $\overline{R} = R(i)$, $i^2 = -1$. Here, different factors have products equal to zero, so they are orthogonal under the trace form. On a factor R, the trace form is $\langle 1 \rangle$, but on a factor $\overline{R}$, the trace form has matrix $\begin{pmatrix} 2 & 0 \\ 0 & -2 \end{pmatrix}$, so it is $\mathbb{H}$. From this, it follows immediately that[(3)] $\mathrm{sgn}(T_f)_P = r$. □

A main point of 2.15 is that the number r of roots of f in R depends only on the ordering P induced by R on F, and not at all on the choice of R itself. We shall now use this fact to prove the following special embedding property of a real-closed field.

---

(3)There seems to be no a priori reason for the signature $\mathrm{sgn}(T_f)_P$ to be a *nonnegative* integer. But that turns out to be the case.

Proposition 2.16. *Suppose $(F, P) \subseteq (K, P')$ is a finite extension of ordered fields, with $P' \cap F = P$. Let S be any real-closed field. Then any order-embedding $\varepsilon : (F, P) \to S$ can be extended to an order-embedding $\varepsilon' : (K, P') \to S$.*

Proof. For convenience let us think of F as embedded in S with $S^2 \cap F = P$. Let R be a fixed real-closure of (K, P'), hence of (F, P). Since $\operatorname{char} F = 0$, we may write $K = F(b)$ for some $b \in K$, say with minimal polynomial $f(t)$ over F. Since f has at least one root in R, it must have at least one root in S, by 2.15. Thus, there certainly exists an F-embedding of K into S. However, we would like to get an *order-embedding*, i.e. one that is order-preserving. This can be accomplished by the following trick. Let $\sigma_1, \ldots, \sigma_n$ be all the F-embeddings of K into S, and suppose *none* of them is order-preserving. Then, for every i, there exists $a_i \in \dot{P}'$ such that $\sigma_i(a_i) \notin S^2$. Consider

$$L = K\big(\sqrt{a_1}, \ldots, \sqrt{a_n}\big) \subseteq R,$$

which inherits an ordering, say P'', from R. By what we have said before, there exists an F-embedding $\sigma : L \to S$. Suppose the restriction of σ to K is σ_j $(1 \le j \le n)$. Then

$$\sigma_j(a_j) = \sigma(a_j) = \sigma\big((\sqrt{a_j}\,)^2\big) = \big(\sigma\sqrt{a_j}\,\big)^2 \in S^2,$$

a contradiction. This shows that at least one of $\sigma_1, \ldots, \sigma_n$ is order-preserving, as desired. □

We arrive now at the following universal property of a real-closure of an ordered field.

Theorem 2.17. *Let R be a real-closure of (F, P) and S be any real-closed field. If $\varepsilon : (F, P) \to S$ is an order-embedding, then ε extends uniquely to R (necessarily to an order-embedding).*

Proof. The existence of an extension follows easily from Zorn's Lemma and 2.16. For the uniqueness, suppose σ, σ' are two extensions of ε to R. Both of these are necessarily order-embeddings of R into S. Now consider any $b \in R$, with minimal polynomial f over F. Let $b_1 < \cdots < b_r$ be all roots of f in R, arranged according to the unique ordering of R. Then $\sigma b_1 < \cdots < \sigma b_r$ are *all* the roots of f in S (by 2.15!), and the same holds for $\sigma' b_1 < \cdots < \sigma' b_r$. It follows that $\sigma b_i = \sigma' b_i$ for every i, and, in particular, $\sigma b = \sigma' b$. This shows that $\sigma = \sigma'$. □

The theorem above implies immediately the uniqueness of the real-closure, by the usual proof of the uniqueness of universal objects. A moment of thought shows that the theorem also implies each of the following three results.

Corollary 2.18. *If R is a real-closure of (F, P), then any F-endomorphism of the field R must be the identity.*

Corollary 2.19. *Proposition* 2.16 *remains true if K/F is assumed to be an algebraic (instead of finite) extension, and the order-embedding $\varepsilon' : (K, P') \to S$ extending the given $\varepsilon : (F, P) \to S$ is unique.*

Corollary 2.20. *Let (F, P) be an ordered field with a real-closure R, and K/F be an algebraic extension. Then the orderings on K extending P are in one-one correspondence with the F-embeddings σ of K into R. (An F-embedding σ corresponds to the ordering $\sigma^{-1}(R^2)$ on K.)*

For instance, if K is any number field, that is, an algebraic extension of the rational field $F = \mathbb{Q}$, then the orderings on K are to be obtained by taking various embeddings σ of K into the field R of real algebraic numbers (the real-closure of $\mathbb{Q}$ with respect to its unique ordering), and pulling back the ordering on R. Of course, the σ's are essentially embeddings of K into $\mathbb{R}$; they are called the "real embeddings" of K in number theory. In the case when $[K : \mathbb{Q}] < \infty$, we can take K as $\mathbb{Q}[x]/(f(x))$ for an irreducible polynomial $f(x) \in \mathbb{Q}[x]$. Here, the number of orderings on K is given by the number of real roots of $f(x)$, as well as by the signature of the trace form T_f on K, according to 2.15. In particular, K is a formally real field iff $f(x) \in \mathbb{Q}[x]$ has at least one real root. If K is nonreal, it has no embeddings into $\mathbb{R}$; in number theory, such a number field K is said to be *totally imaginary*.

Let us close this Appendix by making some remarks on the automorphisms of the real field $\mathbb{R}$ and the complex field $\mathbb{C}$. It is well-known that $\mathrm{Aut}(\mathbb{R})$ *is trivial*: this is easily proved by first checking that any automorphism (or even endomorphism) of $\mathbb{R}$ is continuous, and then observing that $\mathrm{Aut}(\mathbb{Q}) = \{\mathrm{Id}\}$. It was a claim of Dedekind that, once we know $\mathrm{Aut}(\mathbb{R}) = \{\mathrm{Id}\}$, then $\mathrm{Aut}(\mathbb{C}) = \{\mathrm{Id}, \tau\}$, where τ denotes the complex conjugation automorphism of $\mathbb{C}$. This claim[(4)] is, however, *incorrect*.

Dedekind, the well-acknowledged father of modern abstract algebra, made few mistakes in his long and illustrious mathematical career. In fact, the claim he made above on $\mathrm{Aut}(\mathbb{C})$ may very well have been the only blemish in his writings. Thus, it is of interest to find out exactly where Dedekind went astray. Dedekind's determination of a nonidentity $\sigma \in \mathrm{Aut}(\mathbb{C})$ apparently went as follows. Look at $\sigma \mid \mathbb{R}$. Given that $\mathrm{Aut}(\mathbb{R})$ is trivial, σ is the identity on $\mathbb{R}$. Now, for $i = \sqrt{-1} \in \mathbb{C}$,

$$\sigma(i)^2 = \sigma(i^2) = \sigma(-1) = -1 \Longrightarrow \sigma(i) = \pm i.$$

---

(4)See *Ges. Math. Werke*, Vol. II, p. 277, Braunschweig, 1930.

Since $\sigma \neq \text{Id}$, one has $\sigma(i) = -i$. Hence $\sigma(a+bi) = a - bi$ for all $a, b \in \mathbb{R}$, so $\sigma = \tau$.

The "bug" in this argument lies, of course, in the assumption that σ must take the real numbers to themselves! This would have been true if σ is assumed to be *continuous* (for then $\sigma(\mathbb{Q}) = \mathbb{Q} \Rightarrow \sigma(\mathbb{R}) = \mathbb{R}$). However, the continuity of σ is far from automatic.

To see that $\text{Aut}(\mathbb{C}) \supsetneq \{\text{Id}, \tau\}$, we can proceed as follows. Prior to 2.11, we have seen that $\mathbb{C}$ contains real-closed subfields $R \neq \mathbb{R}$ such that $R(i) = \mathbb{C}$. Fixing such a field R, we see that the generator σ of the Galois group $\text{Gal}(\mathbb{C}/R)$ is an automorphism of $\mathbb{C}$ that is different from Id and τ. By Dedekind's own argument, σ *cannot* take $\mathbb{R}$ to itself. Instead, $\sigma(\mathbb{R})$ is a codimension 2 real-closed subfield of $\mathbb{C}$ that is isomorphic, but not equal, to the real field $\mathbb{R}$.

We will not dwell on the problem of computing the automorphism group of $\mathbb{C}$. Let us just point out that, by Galois theory and another theorem of Artin and Schreier to be proved in Appendix B below, any nonidentity torsion element of $\text{Aut}(\mathbb{C})$ is an involution, and there are infinitely many involutions in $\text{Aut}(\mathbb{C})$. (The cardinality of $\text{Aut}(\mathbb{C})$ turns out to be $2^{|\mathbb{C}|}$.) General information on the automorphism groups of algebraically closed fields of characteristic zero is available from Reinhold Baer's paper [Ba].

Appendix B: Another Artin-Schreier Theorem

In the literature on formally real fields, one of the most famous classical results is that of Artin-Schreier (ca. 1926–27) characterizing real-closed fields as proper subfields of finite codimension in algebraically closed fields. Since Section 2 is supposed to be on the characterizations of real-closed fields, we would be remiss if we did not include a discussion of this beautiful result of Artin and Schreier.

We start by making a full statement of this result.

Artin-Schreier Theorem 2.21. *Let C be any algebraically closed field, and $F \subsetneq C$ be a subfield such that $[C : F] < \infty$. Then* $\text{char}(F) = 0$*, F is real-closed, and $C = F(\sqrt{-1})$.*

Part of the depth of this theorem lies in the fact that the assumptions $1 < [C : F] < \infty$ actually *imply* that $\text{char}(F) = 0$. If we assume that $\text{char}(F) = 0$ to begin with, the proof becomes considerably easier. In fact, Artin and Schreier first proved their result under the assumption that $\text{char}(F) = 0$. In the following, we shall present a proof of 2.21 for fields of characteristic zero, following the original approach of Artin and Schreier.(5) In writing up this proof, I have profited from reading a paper of Guralnick and Miller, and a related exposition by Keith Conrad.

Proof of 2.21 ($\operatorname{char} F = 0$). *We first claim that* $[C : F]$ *is a power of* 2. To see this, assume, on the contrary, that there is an *odd* prime $p \mid [C : F]$. Since C/F is a finite Galois extension, Cauchy's Theorem implies that C has a subfield K ($K \supseteq F$) of codimension p. Fix a primitive pth root of unity $\zeta \in C$. Since ζ has degree $\le p-1$ over K, we must have $\zeta \in K$. By Kummer theory, $C = K(x)$ where $x^p = a \in K$. Let $\operatorname{Gal}(C/K) = \langle \sigma \rangle$, and let $y \in C$ be such that $y^p = x$ (so that $y^{p^2} = a$). Clearly, $\sigma(y) = \alpha y$ where $\alpha^{p^2} = 1$. If $\alpha^p = 1$, then $\sigma(x) = \sigma(y)^p = y^p = x$, which is impossible. Therefore, α is a *primitive* p^2th root of unity, so $\sigma(\alpha) = \alpha^r$ for some r (necessarily prime to p). Therefore, $\sigma^2(y) = \alpha^{r+1}y$, $\sigma^3(y) = \alpha^{r^2+r+1}y$, etc., leading to

$$y = \sigma^p(y) = \alpha^{r^{p-1}+\cdots+r+1}y.$$

Thus, $r^{p-1} + \cdots + r + 1 \equiv 0 \pmod{p^2}$. This implies that $r^p \equiv 1 \pmod{p^2}$. Specializing this to $r^p \equiv 1 \pmod{p}$ and using Fermat's Little Theorem, we see that $r = 1 + kp$ for some integer k. But then

$$\begin{aligned} r^{p-1} + \cdots + r + 1 &= [(1+kp)^p - 1] \cdot (kp)^{-1} \\ &= \left[kp^2 + \binom{p}{2}(kp)^2 + \binom{p}{3}(kp)^3 + \cdots \right] \cdot (kp)^{-1} \\ &= p + \left(\frac{p-1}{2}\right) kp^2 + \binom{p}{3}(kp)^2 + \cdots \\ &\equiv p \pmod{p^2}, \end{aligned}$$

since p is odd. This is a contradiction.

Having now proved that $[C : F] = 2^n$ for some n, we claim that $n = 1$. Indeed, assume that $n \ge 2$. By Galois theory again, there exist subfields $E \subsetneq L \subsetneq C$ with $[C : L] = [L : E] = 2$. By 1.7 ((4) $\Rightarrow$ (1)), L is euclidean. In particular, $\sqrt{-1} \notin L$. But then $E(\sqrt{-1})$ is of codimension 2 in C, and 1.7 ((4) $\Rightarrow$ (1)) would imply again that $E(\sqrt{-1})$ is euclidean, a contradiction. Therefore, we must have $[C : F] = 2$. As above, we have $\sqrt{-1} \notin F$, so $C = F(\sqrt{-1})$. By 2.5, F is a real-closed field. □

Note that the argument in the last paragraph above actually only made use of the fact that C is the *quadratic closure* of F, except in the very last step. Therefore, the argument can be used to give the following analogue of 2.21 for quadratic closures (in the case $\operatorname{char}(F) \ne 2$), due to G. Whaples [Wh].

---

(5)We suppress the proof of 2.21 in nonzero characteristic only because it has essentially no bearing on the theory of formally real fields.

Theorem 2.22. *Let F be a field and $\widetilde{F}$ be its quadratic closure* (*see* VII.7). *If* $1 < [\widetilde{F} : F] < \infty$, *then* $[\widetilde{F} : F] = 2$ *and F is a euclidean field.*

For an arbitrary prime p, we can define the "maximal p-extension" F_p of a field F to be the compositum of all finite Galois extensions K/F for which $\text{Gal}(K/F)$ is a p-group. This is a generalization of the formation of the quadratic closure $\widetilde{F}$ of F, since $\widetilde{F}$ is just the maximal 2-extension F_2 of F. Whaples had also proved the following result, which shows the special role played by the prime 2: *Let F be a field such that* $1 < [F_p : F] < \infty$. *Then we must have* $p = 2$, $[F_p : F] = 2$, *and F is euclidean.* See [Wh].

A note of caution is in order. In Whaples' Theorem 2.22, $\widetilde{F}$ is assumed to be the quadratic closure of F. The theorem does not tell us anything if we are just given a subfield $K \subseteq \widetilde{F}$ with $1 < [\widetilde{F} : K] < \infty$. Here we cannot apply 2.22 since $\widetilde{F}$ may not be the quadratic closure[6] of K. One of our later results implies that K must be either euclidean or quadratically closed (see 5.11); however, there is no control on the codimension $[\widetilde{F} : K]$.

We close with the following consequence of the Artin-Schreier Theorem 2.21. (Of course, we have proved 2.21 only in characteristic zero, so our proof below is complete in this case only.)

Corollary 2.23. *Let C be an algebraically closed field, and σ be an automorphism of finite order on C. If* $\text{char}(C) \neq 0$, *then* $\sigma = \text{Id}$. *If* $\text{char}(C) = 0$, *then σ has order* ≤ 2.

Proof. By Artin's Theorem in Galois theory, the fixed field F of σ is such that C/F is Galois with Galois group generated by σ. In particular, $[C : F]$ is given by the (finite) order of σ. Now apply 2.21. □

3. Pfister's Local-Global Principle

In this section, we shall apply the theory of ordered fields to the theory of quadratic forms. For a field F, let us write X_F (or sometimes $X(F)$) for the (possibly empty) set of orderings on F. In order to work with X_F as a set of "points", we shall write α for a typical element in X_F, and write "$\leq_\alpha$" for the total ordering given by α on F. To reconcile this with our earlier notation, we shall write $P_\alpha = \{a \in F : a \geq_\alpha 0\}$ for the positive cone of the ordering α.

For each $\alpha \in X_F$, let us fix a real-closure F_α of F with respect to α. (We could equally well have taken F_α to be a euclidean closure of F with respect to α throughout the following.) Letting $r_\alpha : F \to F_\alpha$ be the inclusion map,

---

[6]We can conclude that $\widetilde{F}$ is the quadratic closure of K only if more information is given on K, for instance $K \supseteq F$. Otherwise, it won't even follow that $\widetilde{F}/K$ is a Galois extension.

we have a functorial homomorphism $r_\alpha^* : W(F) \to W(F_\alpha)$. Now, just as in II.3.2, we have a canonical isomorphism $W(F_\alpha) \cong \mathbb{Z}$. The composition of these two maps gives a surjection $\text{sgn}_\alpha : W(F) \to \mathbb{Z}$, which sends an F-quadratic form q to its signature $\text{sgn}_\alpha(q)$ with respect to α. (As we have noted before, the map sgn_α *does not* depend on the choice of F_α, or on the fact that F_α is uniquely determined up to an F-isomorphism.)

Letting α range over the set of orderings X_F, we get a "total signature" map

$$\text{sgn} : W(F) \longrightarrow \prod_\alpha W(F_\alpha) \cong \prod_\alpha \mathbb{Z}, \tag{3.1}$$

which sends a form q to $(\text{sgn}_\alpha(q))_\alpha$ on the RHS. The first main goal of this section is to compute the kernel of this total signature map. The result is given in the following fundamental theorem, first published in Pfister's paper [Pf3].

Theorem 3.2. (Pfister's Local-Global Principle) *For any field F,* $\ker(\text{sgn}) = W_t(F)$*, the torsion subgroup of the Witt group $W(F)$. Moreover, every element in $W_t(F)$ is 2-primary torsion.*

An equivalent way to state the first part of the theorem is that *two quadratic forms q_1, q_2 over F have the same signature relative to all orderings on F iff $n \cdot q_1 = n \cdot q_2 \in W(F)$ for some integer $n \geq 1$.* The second part of the theorem says that, in this case, we could have taken n to be of the form 2^r for some r. Yet another way to express 3.2 is to say that, if a form q is hyperbolic in all real-closures of F, then for some integer $r \geq 0$, $2^r \cdot q$ is hyperbolic over F. These alternative formulations of 3.2 explain why this result is called a Local-Global Principle.

Example 3.3. The simplest example of a formally real field F for which the total signature map "sgn" has a nontrivial kernel is the field with a square-class basis $\{-1, 2\}$ constructed in II.5.3. Such a field F has clearly a unique ordering $P_\alpha = F^2 \cup 2F^2$. By the computations in II.5.1, $W(F) \cong \mathbb{Z} \oplus \mathbb{Z}_2$. Thus, $W_t(F) \cong \mathbb{Z}_2$, and in fact, $W_t(F) = \{0, \varphi\}$, where $\varphi = \langle 1, -2 \rangle$. (By listing all anisotropic forms over F, it is also easy to verify directly that, up to isometry, the only nonzero anisotropic form with a zero α-signature is φ.) Similar examples with somewhat larger $\ker(\text{sgn})$ can be found in II.5.4 and II.5.7.

Our strategy for proving 3.2 is as follows. We first check the truth of 3.2 in two special cases, and then give the general proof by making a reduction to these special cases.

The first special case is when F is a euclidean field. In this case, F has a unique ordering α with $P_\alpha = F^2$, and the total signature map

$$\text{sgn}: W(F) \longrightarrow W(F_\alpha) \cong \mathbb{Z}$$

is obviously an isomorphism. Here, $W_t(F) = \{0\}$, so 3.2 is certainly true.

The second special case of 3.2 is when F is a *nonreal* field, for which X_F is the empty space. Here, $\prod_\alpha W(F_\alpha)$ is an "empty" direct product, which is, as usual, taken to be $\{0\}$. *In this case,* 3.2 *asserts that* $W(F)$ *is a 2-primary torsion group.* This is equivalent to saying that the ring $W(F)$ has characteristic 2^r for some integer r, so the proof of 3.2 boils down to checking this statement for any nonreal field F. We shall do this by appealing to the following observation on the prime ideals of $W(F)$ for any field F.

Lemma 3.4. *Let* $\mathfrak{p}$ *be any prime ideal in* $W(F)$ *(where* F *is any field).*

(1) *If* $2 \in \mathfrak{p}$, *then* $\mathfrak{p} = IF$.

(2) *If* $2 \notin \mathfrak{p}$, *then* $P := \{0\} \cup \{a \in \dot{F} : \langle a \rangle \equiv 1 \pmod{\mathfrak{p}}\}$ *is an ordering on* F.

Proof. Note that for any $a \in \dot{F}$, $\langle a \rangle^2 = 1 \in W(F)$ implies that $\langle a \rangle \equiv \pm 1 \pmod{\mathfrak{p}}$ (since $W(F)/\mathfrak{p}$ is an integral domain). If $2 \in \mathfrak{p}$, then $\langle a \rangle \equiv 1 \pmod{\mathfrak{p}}$, so for any $2n$-dimensional form q, we have $q \equiv 2n \equiv 0 \pmod{\mathfrak{p}}$. Thus, $IF \subseteq \mathfrak{p}$, and equality must hold. (See Ch. II, Exer. 4.)

Now assume $2 \notin \mathfrak{p}$, and define P as in the statement of (2). Clearly $P \cdot P \subseteq P$, $P \cup (-P) = F$, and $\langle -1 \rangle \not\equiv \langle 1 \rangle \pmod{\mathfrak{p}}$ yields $-1 \notin P$. We finish by checking that $a, b \in P$, $c := a + b \neq 0$ imply that $c \in P$. From the isometry $\langle a, b \rangle \cong \langle c \rangle \langle 1, ab \rangle$, we have $2 \equiv 2\langle c \rangle \pmod{\mathfrak{p}}$. Since $2 \notin \mathfrak{p}$, we have $\langle c \rangle \equiv 1 \pmod{\mathfrak{p}}$, as desired. □

(This lemma is actually the first step toward the classification of the prime ideals of $W(F)$. We shall return to this matter a little later in §7.)

For a *nonreal* field F, the above lemma implies that IF *is the unique prime ideal of* $W(F)$. But then, by a standard theorem in commutative algebra, IF must be the nilradical of $W(F)$. In particular, for the element $2 \in IF$, we have $2^r = 0 \in W(F)$ for some $r \geq 1$. This proves 3.2 for *nonreal* fields F. (Another proof using the quadratic closure of F will be given later in §4: see the discussion following Theorem 4.10.)

We are now in a good position to prove 3.2. For any F, we have clearly $W_t(F) \subseteq \ker(\text{sgn})$ (since $\prod_\alpha \mathbb{Z}$ is torsionfree). The main job is to show that, if a form $q \in W(F)$ is not 2-primary torsion, then $\text{sgn}_\alpha q \neq 0$ for some ordering $\alpha \in X_F$. By Zorn's Lemma, there exists a field $K \supseteq F$ within the algebraic closure of F that is maximal with respect to the property that $q_K \in W(K)$ is not 2-primary torsion. *We claim that* K *is euclidean.*

Surely, K is formally real (for otherwise $2^r W(K) = 0$ for some r). Assume, for the moment, that K has an element $a \notin \pm K^2$. By the "maximality" of K, q_K must become 2-primary in $K(\sqrt{a}\,)$ and $K(\sqrt{-a}\,)$, and so for a large integer N, $2^N q_K$ is hyperbolic over both $K(\sqrt{a}\,)$ and $K(\sqrt{-a}\,)$. But then by VII.3.3(3), $2 \cdot 2^N q_K = 0 \in W(K)$, a contradiction. This shows that K is *euclidean*, and we have $\operatorname{sgn}_\alpha(q) \neq 0$ for the ordering $\alpha \in X_F$ induced on F by the unique ordering on K. □

This proof illustrates in a remarkable way the role played by euclidean fields in quadratic form theory. While there are several other known proofs for 3.2, we believe the proof given above is the best and the most interesting.

Having proved the Local-Global Principle 3.2, let us note one of its interesting consequences. Since this principle yields the fact that the only torsion in the Witt ring $W(F)$ is 2-primary torsion, we can deduce the following fact, which we try to couch in the elementary language of matrices:

For two (nonsingular) symmetric $n \times n$ matrices M, N over a field of characteristic not 2, if $\begin{pmatrix} M & & \\ & \ddots & \\ & & M \end{pmatrix}$ is congruent to $\begin{pmatrix} N & & \\ & \ddots & \\ & & N \end{pmatrix}$ for a certain odd number of blocks, then M is congruent to N.

It is not clear how such a fact can be proved by matrix-theoretic techniques, without using the powerful tools of Witt rings and quadratic forms. Note that the proof we gave for 3.2 depended very much on *changing the base field* in the argument; ordinarily one would not be applying such a technique when dealing with a specific matrix-theoretic problem over a given field.

4. Pythagorean Fields

By definition, a field F is *pythagorean* if $F^2 + F^2 \subseteq F^2$, or equivalently, if $\sigma(F) = F^2$. We have dealt with such fields on several previous occasions, for instance, in II.5 and in 1.6. In this section, we shall develop the theory of pythagorean fields in more detail. Before we begin, let us point out that several other characterizations of pythagorean fields are available in Exercise 13 in this chapter.

Note that, if $\operatorname{char}(F) = 2$, we have $a^2 + b^2 = (a+b)^2$, so F is always pythagorean. However, we usually assume $\operatorname{char}(F) \neq 2$ in this book. In this case, the main case of interest is that of *formally real pythagorean* fields. For, if a pythagorean field F is nonreal, then $-1 = a_1^2 + \cdots + a_n^2$ for some a_i, and for any $a \in F$, we can write $a = x^2 - y^2$ for some $x, y \in F$ to get

$$a = x^2 + (a_1^2 + \cdots + a_n^2)\, y^2 \in \sigma(F) = F^2.$$

This means that F is in fact a quadratically closed field.

The distinction between real and nonreal pythagorean fields can also be formulated in terms of the Witt group.

Theorem 4.1. (1) *F is formally real pythagorean iff $W(F)$ is torsionfree.*

(2) *F is nonreal pythagorean iff $W(F) \cong \mathbb{Z}/2\mathbb{Z}$.*

Proof. (2) is clear in view of what we have said about nonreal pythagorean fields, so it suffices to prove (1). First assume F is formally real pythagorean. For any anisotropic form $q = \langle a_1, \ldots, a_n \rangle \in W(F)$, note that *$r \cdot q$ is also anisotropic for any natural number r*. For, if $r \cdot q$ vanishes on a vector $(e_{11}, \ldots, e_{1r}, \ldots, e_{n1}, \ldots, e_{nr})$, i.e.

$$a_1e_{11}^2 + \cdots + a_1e_{1r}^2 + \cdots + a_ne_{n1}^2 + \cdots + a_ne_{nr}^2 = 0,$$

we can write $e_{i1}^2 + \cdots + e_{ir}^2 = e_i^2$ (for suitable $e_i \in F$) to get

$$a_1e_1^2 + \cdots + a_ne_n^2 = 0.$$

This implies that $e_i = 0$ for all i, and therefore $e_{ij} = 0$ for all i, j (by formal reality). This implies, in particular, that $W(F)$ is torsionfree. Conversely, if $W(F)$ is torsionfree (or just free of 2-primary torsion), we have for any $c = a^2 + b^2 \neq 0$ an isometry $\langle 1, 1 \rangle \cong \langle c, c \rangle$, which implies that $\langle c \rangle = \langle 1 \rangle \in W(F)$, so $c \in F^2$. This shows that F is pythagorean, and F must be formally real according to (2)! □

It is clear from the above proof that one could also recognize pythagorean fields via properties of the Witt-Grothendieck group $\widehat{W}(F)$. We leave this analogous formulation to the reader (see, in part, Exercise 34).

The following easy result characterizes the euclidean fields among all pythagorean fields.

Proposition 4.2. *A field F is euclidean iff F is a pythagorean field with a unique ordering.*

Proof. The "only if" part has already appeared in 1.6. For the converse, assume that F is pythagorean with a unique ordering P. By 1.12, we have $P = \sigma(F)$ $(= F^2)$. In particular,

$$F = P \cup (-P) = F^2 \cup (-F^2),$$

so F is euclidean. □

It turns out that, quite generally, the formally real pythagorean fields "arise" from euclidean fields in a certain fashion. To formulate this more precisely, we first make the following crucial observation.

Proposition 4.3. *If $\{F_i\}$ is a family of pythagorean subfields of a field K, then their intersection $F = \bigcap_i F_i$ is also a pythagorean field.*

Proof. Consider $a^2 + b^2$, where $a, b \in F$. For each i, there exists $c_i \in F_i$ such that $c_i^2 = a^2 + b^2$. For any pair of indices i, j, we have $c_i = \pm c_j$. For a fixed index j, we have then $c_j \in \bigcap_i F_i = F$ and $a^2 + b^2 = c_j^2 \in F^2$. □

This proposition leads us to the following "construction" of formally real pythagorean fields, due to E. Becker [Be_1].

Theorem 4.4. *A field F is formally real pythagorean iff F is the intersection of a nonempty family of euclidean subfields of $\overline{F}$ (the algebraic closure of F).*

Proof. The "if" part follows from 4.2 and 4.3. Conversely, assume F is formally real pythagorean. Let $\{E_i\}$ be the family of all euclidean closures of F within $\overline{F}$. We finish by showing that $L := \bigcap_i E_i$ is equal to F. Recall that each E_i lies in the quadratic closure of F. If $F \subsetneq L$, we would have $F \subsetneq F(\sqrt{a}) \subseteq L$ for some $a \in \dot{F}$. Let P be any ordering on F and pick i such that E_i is a euclidean closure of (F, P). Then

$$a = (\sqrt{a})^2 \in F \cap L^2 \subseteq F \cap E_i^2 = P.$$

Thus, a is totally positive in F, so Artin's Theorem 1.12 implies that $a \in \sigma(F)$. But then $a \in F^2$ (since F is pythagorean), a contradiction. □

In the proof of the "only if" part above, it is not surprising that we have only used the euclidean closures of F to form the intersection, since *any* euclidean field containing F must contain one of these euclidean closures, according to 2.12.

The theorem we just proved in 4.4 portrays the class of formally real pythagorean fields in good light. If we work only with the class $\mathcal{E}$ of euclidean fields, this class would not be closed with respect to the formation of intersections. By forming the class of all intersections of the members of $\mathcal{E}$, we get the smallest class $\mathcal{P} \supseteq \mathcal{E}$ that is closed with respect to intersections. This $\mathcal{P}$ is precisely the class of all formally real pythagorean fields.

Given what is said above, it is now natural to introduce the notion of a pythagorean closure.

Definition 4.5. For any field F, there exists a smallest pythagorean subfield of $\overline{F}$ that contains F, namely, the intersection of all pythagorean subfields of $\overline{F}$ containing F. This field is called the *pythagorean closure* (or *hull*) of F, and is denoted by F_{py}.

Of course, we could have also given a construction of F_{py} "from below". Let $\mathcal{F}$ be the family of extensions K/F for which there exists a tower

$$F = K_0 \subseteq K_1 \subseteq \cdots \subseteq K_n = K$$

such that $K_{i+1} = K_i((1+a_i^2)^{1/2})$, where $a_i \in K_i$. Clearly, each such K lies in F_{py}, and the compositum of all the fields in $\mathcal{F}$ is easily seen to be F_{py}. In this construction of F_{py}, we obtain a pythagorean field from F by doing what is "absolutely necessary", namely, we "keep adjoining" the square roots of $1+a^2$ whenever $1+a^2$ is not yet a square in a field containing F. This constructive description of F_{py} turns out to be quite useful for getting information about the relationship between a field F and its pythagorean closure. It shows right away, for instance, that *F_{py} is a Galois extension of F.*

If F is nonreal, then of course F_{py} is just $\widetilde{F}$, the quadratic closure of F. In the formally real case, we have $F \subseteq F_{\text{py}} \subseteq \widetilde{F}$, and the proof of 4.4 yields the following description of F_{py}.

Corollary 4.6. *For a formally real field F, F_{py} is the intersection of the euclidean closures of F. The mapping $X(F_{\text{py}}) \to X(F)$ (defined by the restriction of orderings from F_{py} to F) is surjective.*

As an explicit example, let us take a look at the pythagorean hull $\mathbb{Q}_{\text{py}}$ of the field $\mathbb{Q}$ of rational numbers. This field deserves to be listed in the Guinness Book of Records, as it is the world's smallest formally real pythagorean field. We claim that $\mathbb{Q}_{\text{py}}$ is an *infinite* (algebraic) extension of $\mathbb{Q}$. In fact, for any prime $p \equiv 1 \pmod 4$, we have $p = a^2 + b^2$ in $\mathbb{Z}$ (by the theorem of Fermat), so $\sqrt{p} \in \mathbb{Q}_{\text{py}}$. This shows that $\mathbb{Q}_{\text{py}}$ contains

$$E = \mathbb{Q}(\{\sqrt{p} : \ p \equiv 1 \pmod 4\}),$$

which has infinite degree over $\mathbb{Q}$. (This argument implies, in particular, that *a finite number field can never be pythagorean.*) We note further that, from 2.20, the field E above has infinitely many (in fact uncountably many) orderings. Since $\mathbb{Q}_{\text{py}} = E_{\text{py}}$, it follows that $\mathbb{Q}_{\text{py}}$ has uncountably many orderings, even though $\mathbb{Q}$ itself has a *unique* ordering. Of course, $\mathbb{Q}_{py}$ is a subfield of the field K of all constructible real numbers, and K (being euclidean) has also a unique ordering.

In order to understand, in general, the structure of the "fibers" of the map $X(F_{\text{py}}) \to X(F)$ in 4.6, we prove the following general result for order-preserving automorphisms of fields. Recall that, for a G-set S over a group G, the G-action on S is said to be *semiregular* if any 1-point stabilizer is trivial, and *regular* if it is transitive and semiregular.

Theorem 4.7. *Let K/F be an algebraic extension, and G be the group of F-automorphisms of K. Let $\rho : X_K \to X_F$ be the map given by the restriction of orderings, and let $P_0 \in \text{im}(\rho)$.*

(1) *(Geyer) The fiber $\rho^{-1}(P_0)$ is a semiregular G-set.*

(2) *If K/F is Galois, $\rho^{-1}(P_0)$ is a regular G-set.*

Proof. (1) If $P \in X_K$ extends P_0 and $\sigma \in G$, clearly $\sigma(P) \in X_K$ also extends P_0. Thus, the rule $(\sigma, P) \mapsto \sigma(P)$ makes $\rho^{-1}(P_0)$ into a G-set. To show that this G-set is semiregular means showing that $\sigma(P) = P \Rightarrow \sigma = 1 \in G$. Let $x \in K$. Clearly, $\sigma^i(x)$ is a conjugate of x over F for any i. Since there are only finitely many such conjugates, we have $\sigma^n(x) = x$ for some $n \geq 1$. If $x < \sigma x$ with respect to the ordering P, then, applying successively the order-preserving map σ, we get

$$x < \sigma x < \sigma^2 x < \cdots < \sigma^n x = x,$$

a contradiction. Similarly, $x > \sigma x$ would also lead to a contradiction, so we must have $\sigma(x) = x$.

(2) Assume now K/F is Galois, and let $P, P' \in \rho^{-1}(P_0)$. Fix a real-closure R for (F, P_0). By 2.20, there exist F-embeddings φ, $\varphi' : K \to R$ such that

$$\varphi(P) = R^2 \cap \varphi(K) \quad \text{and} \quad \varphi'(P') = R^2 \cap \varphi'(K).$$

Since K/F is Galois, we have $\varphi(K) = \varphi'(K)$. Therefore, it makes sense to define $\sigma \in \mathrm{Gal}(K/F)$ by $\sigma = \varphi^{-1}\varphi'$. Since $\varphi(P) = \varphi'(P')$, it follows that $P = \varphi^{-1}\varphi'(P') = \sigma(P')$. □

Remark 4.8. (1) above can also be formulated without the field F as follows: *If $\sigma \in \mathrm{Aut}(K)$ is "algebraic" in the sense that K is algebraic over the field of fixed points K^σ, then for any $P \in X_K$, $\sigma(P) = P \Rightarrow \sigma = \mathrm{Id}$.* (For instance, this implication is valid for any $\sigma \in \mathrm{Aut}(K)$ of finite order, since in this case $[K : K^\sigma] < \infty$ by a theorem of Artin in Galois theory.) But if K/K^σ is not algebraic, the implication may no longer hold; for an example to this effect, see Exercise 25.

In the case when $K = F_{\mathrm{py}}$ where F is a formally real field, 4.7(2) shows that *each fiber of the map*

$$\rho : \ X(F_{\mathrm{py}}) \to X(F)$$

is isomorphic to $G = \mathrm{Gal}(F_{\mathrm{py}}/F)$ as a G-set. If F is *not* pythagorean, it will be shown later (in 5.7) that F_{py}/F is always an *infinite* Galois extension. Therefore, the above statement shows that each ordering $P_0 \in X_F$ extends to infinitely many orderings on F_{py}.

Having discussed the ordering structure on F_{py}, let us now return to the consideration of quadratic forms. In §3, we have proved Pfister's Local-Global Principle for the Witt ring. In the context of formally real fields and their pythagorean closures,[7] this principle has the following two remarkable consequences.

---

[7] We focus on the formally real case, since a nonreal pythagorean field is quadratically closed, as we have observed earlier.

Theorem 4.9. *Let F be any formally real pythagorean field. Then the total signature map*

$$\mathrm{sgn}:\ W(F) \longrightarrow \prod_{\alpha \in X_F} \mathbb{Z}$$

is injective. Equivalently, two quadratic forms q_1, q_2 are isometric over F iff $\dim q_1 = \dim q_2$ and $\mathrm{sgn}_\alpha q_1 = \mathrm{sgn}_\alpha q_2$ for any ordering $\alpha \in X_F$.

This follows from 3.2 since $W_t(F) = 0$ by 4.1(1). In the language of ring theory, this theorem may be expressed by the statement that, for any formally real pythagorean field F, $W(F)$ is a subdirect product of $|X_F|$ copies of the ring of integers $\mathbb{Z}$.

Theorem 4.10. *For any formally real field F, the functorial map $W(F) \to W(F_{\mathrm{py}})$ has kernel $W_t(F)$, which is a 2-primary torsion group.*

(This follows from 3.2, 4.1(1) and the last part of 4.6.)

It is worth noting that Theorem 4.10 can also be proved *directly*, by using the description of F_{py} as the field obtained by "successive adjunction" of square roots of elements of the form $1 + a^2$. Using earlier results on quadratic extensions, it is easy to see that, for any field K (formally real or nonreal), *the kernel of*

$$W(K) \to W\big(K(\sqrt{1+a^2}\,)\big) \qquad (1 + a^2 \notin K^2)$$

is always killed by 2. (Indeed, by VII.3.2, the kernel is $W(K) \cdot \alpha$ where $\varphi = \langle 1, -(1+a^2)\rangle$, and we have $2\,\alpha = 0 \in W(K)$.) By repeated use of this, we see that $\ker(W(F) \to W(F_{\mathrm{py}}))$ *is always 2-primary torsion.* In case F is formally real, this recaptures 4.10. In case F is nonreal, $\ker(W(F) \to W(F_{\mathrm{py}}))$ is just IF (since F_{py} is quadratically closed); in this case, we conclude that IF is 2-primary torsion, so we get another proof for the fact that $W(F)$ is of characteristic 2^r for some r.

As Scharlau noted in [Sc$_2$], in the formally real case, the above considerations can be used to give a second proof for Pfister's Local-Global Principle 3.2. In fact, 4.10 reduces the proof to the case when the field F is *pythagorean*, and in this case, a proof of 3.2 can be obtained by considering quadratic extensions and applying Exer. 9 in Ch. VII. Indeed, this was the proof given for 3.2 in the earlier versions of this book. We believe, however, the proof given in §3 is better (and more efficient), as it makes a direct reduction to euclidean fields without going through the pythagorean closure.

Before we move on to give more results on pythagorean fields, we would like to devote some time to looking at examples.

We begin by reviewing the structure of a Laurent series field $F_1 = F((t))$ over a given field F. This consists of all formal Laurent series of the form

$$f = a_m t^m + a_{m+1}t^{m+1} + \cdots \qquad (m \in \mathbb{Z},\ a_i \in F),$$

under the usual series addition and multiplication. The field F_1 is a *local field* in the sense of VI.1, with valuation ring $F[[t]]$ (the ring of all power series in t). The information on the orderings on F_1 is given in the following result.

Proposition 4.11. *Let $F_1 = F((t))$ as above.*

(1) Given an ordering ">" on F, there exist precisely two orderings on F_1 extending it, one making t positive, and the other making t negative. (In particular, F is formally real iff F_1 is.)

(2) F is formally real pythagorean iff F_1 is.

(3) If F is euclidean, the iterated Laurent series field

$$F_n = F((t_1))\cdots((t_n))$$

is a pythagorean field with 2^{n+1} square classes and 2^n orderings.

Proof. (1) If $f = a_m t^m + \cdots \in F_1$ with $a_m \neq 0$, declare $f > 0$ iff $a_m > 0$. It is easy to see that this satisfies all the axioms of an ordering. Further, this ordering is the unique one (extending the original ">" on F) in which t is positive. This is because any power series $1 + a_1 t + a_2 t^2 + \cdots$ is a square in $F[[t]]$ (and hence in F_1), by VI.1.1. Observing that $t \mapsto -t$ defines an F-automorphism of F_1, we see that F_1 has also a unique ordering extending ">", in which t is negative. (In this ordering, $f = a_m t^m + \cdots$ is positive iff $(-1)^m a_m > 0$ in F.)

Note that in the first ordering above, we have $0 < t < a$ for any $a > 0$ in F; in the second ordering, we have $b < t < 0$ for any $b < 0$ in F.

(2) This follows from the isomorphism $W(F_1) \cong W(F) \oplus W(F)$ (from VI.1.5), applied in conjunction with 4.1(1). However, it is perhaps more illuminating to give an *ad hoc* proof. First assume F_1 is formally real pythagorean. Certainly F is also formally real. For $a, b \in \dot{F}$, we have

$$a^2 + b^2 = (c_m t^m + \cdots)^2 = c_m^2 t^{2m} + \cdots$$

for some $c_i \in F$ with $c_m \neq 0$. Clearly, m must be 0 and we have $a^2 + b^2 = c_0^2$. Conversely, assume F is formally real pythagorean. Consider $f^2 + g^2$ in F_1, where

$$f = a_m t^m + \cdots, \quad g = b_n t^n + \cdots, \qquad \text{with } a_m \neq 0 \neq b_n.$$

If, say, $m < n$, then $f^2 + g^2 = (a_m t^m)^2(1 + \cdots)$ is a square in F_1, by VI.1.1. Now assume $m = n$, and write

$$a_m^2 + b_m^2 = c_m^2 \quad (c_m \in \dot{F}).$$

We then have $f^2 + g^2 = (c_m t^m)^2(1 + \cdots)$, which is again a square in F_1, as before. The fact that F_1 is formally real follows from (1) (or a direct argument).

(3) This part follows from (2) and VI.1.3. □

Remark 4.12. There is a new phenomenon exhibited by the orderings on the fields F_n in 4.11(3) above that is worth mentioning. In general, if P_1, P_2, P_3 are three different orderings on a field, an easy combinatorial argument (Exercise 7) shows that there always exists an element which is positive in P_2, P_3, but negative in P_1; in other words, we always have $P_2 \cap P_3 \not\subseteq P_1$. However, for *four* different orderings $P_1, \ldots, P_4$, it is possible that $P_2 \cap P_3 \cap P_4 \subseteq P_1$. For instance, let $K = \mathbb{R}((x))((y))$, and let the four orderings $P_1, \ldots, P_4$‘, on K be labelled such that the signs of $(x,\, y)$ are

$$(+,\, +), \quad (+,\, -), \quad (-,\, +), \quad (-,\, -)$$

under $P_1, \ldots, P_4$ respectively. Then $P_1, \ldots, P_4$ consist respectively of the following square classes:

$$\{1,\, x,\, y,\, xy\}, \quad \{1,\, x,\, -y,\, -xy\}, \quad \{1,\, -x,\, y,\, -xy\}, \quad \{1,\, -x,\, -y,\, xy\}.$$

By inspection, we see that $P_i \cap P_j \cap P_k = K^2$ for any three distinct indices i, j, k, so we have in particular $P_2 \cap P_3 \cap P_4 \subseteq P_1$, as desired. Another way to see this is to note that each ordering P_i defines a character $\chi_i : \dot{K}/\dot{K}^2 \to \{\pm 1\}$. In the character group of $\dot{K}/\dot{K}^2$, we have $\chi_1\chi_2\chi_3\chi_4 = 1$ (since this holds on $-1, x$ and y). From this, it follows that if $\chi_i(a) = 1$ for $i = 2, 3, 4$, then we also have $\chi_1(a) = 1$. This shows that

$$P_2 \cap P_3 \cap P_4 \subseteq P_1.$$

In the literature, four distinct orderings $P_1, \ldots, P_4$ on a field are said to form a "(4-element) fan" if their characters $\chi_1, \ldots, \chi_4$ satisfy the relation $\chi_1\chi_2\chi_3\chi_4 = 1$. The study of these fans is of great importance in the modern theory of ordered fields.

As an illustration, let us now classify the Witt rings of formally real pythagorean fields F with number of square classes up to (and including) 8. This was, in fact, a project we started already in II.5. In that section, we have dealt with the nonpythagorean (formally real) fields F with $|\dot{F}/\dot{F}^2| \leq 8$, but the case of pythagorean F was settled with only $|\dot{F}/\dot{F}^2| \leq 4$. We shall now complete this classification.

Theorem 4.13. *Let F be formally real pythagorean, with $|\dot{F}/\dot{F}^2| \leq 8$.*

(1) *If $|\dot{F}/\dot{F}^2| = 2$, then (F is euclidean and) $W(F) \cong \mathbb{Z}$.*

(2) *If $|\dot{F}/\dot{F}^2| = 4$, then $W(F) \cong \mathbb{Z}[G]$, where G is a group of order 2.*

(3) *For $|\dot{F}/\dot{F}^2| = 8$, there are two cases.*

(a) *If* $|X_F| = 4$, *then* $W(F) \cong \mathbb{Z}[K]$ *where* K *is the Klein* 4*-group.*

(b) *If* $|X_F| = 3$, *then* $W(F)$ *is isomorphic to the subring of* $\mathbb{Z} \times \mathbb{Z} \times \mathbb{Z}$ *consisting of* (a, b, c) *where* a, b, c *are integers of the same parity.*

Proof. (1) is clear, and (2) has been settled before in II.5.1. So let us now assume $|\dot{F}/\dot{F}^2| = 8$, with a square class basis $\{-1, x, y\}$. Since $\dot{\sigma}(F) = \dot{F}^2$ has index 8 in $\dot{F}$, $|X_F|$ is either 3 or 4. We consider these two cases separately.

(3a) $|X_F| = 4$. Let P_i $(1 \le i \le 4)$ be the four orderings. With a suitable labelling of these, we may assume that the signs of the ordered pair (x, y) exhaust the following four possibilities:

$$(+,+), \quad (+,-), \quad (-,+), \quad (-,-)$$

under P_1, P_2, P_3, P_4 respectively. Consider the total signature map

$$\text{sgn}: \ W(F) \longrightarrow \mathbb{Z} \times \mathbb{Z} \times \mathbb{Z} \times \mathbb{Z},$$

which is injective by 4.7. Here, $W(F)$ is additively generated by $\langle 1 \rangle, \langle x \rangle, \langle y \rangle, \langle xy \rangle$, and

$$\text{sgn}\langle 1 \rangle = (1,1,1,1), \quad \text{sgn}\langle x \rangle = (1,1,-1,-1),$$
$$\text{sgn}\langle y \rangle = (1,-1,1,-1), \quad \text{sgn}\langle xy \rangle = (1,-1,-1,1)$$

are easily seen to be linearly independent in $\mathbb{Z}^4$. Therefore, we have

$$(4.14) \qquad W(F) = \mathbb{Z}\langle 1 \rangle \oplus \mathbb{Z}\langle x \rangle \oplus \mathbb{Z}\langle y \rangle \oplus \mathbb{Z}\langle xy \rangle \cong \mathbb{Z}[K],$$

where K is the Klein 4-group in $\dot{F}/\dot{F}^2$ generated by x, y.

(3b) $|X_F| = 3$. Let P_i $(1 \le i \le 3)$ be the three orderings. By replacing x, y by $\pm x$, $\pm y$ in a suitable way, we may assume that the signs of (x, y) are

$$(+,+), \qquad (+,-), \qquad (-,+)$$

under P_1, P_2, P_3 respectively. Just as in (3a), "sgn" identifies $W(F)$ with the subring of $\mathbb{Z}^3$ additively spanned by

$$(1,1,1), \quad (1,1,-1), \quad (1,-1,1), \quad (1,-1,-1).$$

An easy calculation shows that this is precisely the subring consisting of (a, b, c) where $a, b, c \in \mathbb{Z}$ have the same parity. $\square$

The existence question in connection with the classification result 4.13 can be settled as well. For (1), we can take any euclidean field; for (2), we can take the Laurent series field $\mathbb{R}((x))$; and for (3a), we can take the iterated Laurent series field $\mathbb{R}((x))((y))$.[8] In the following, we shall construct a pythagorean field for (3b).

Start with a field E with exactly three orderings, say P_1, P_2, P_3. By Exercise 7, there exist elements

$$a \in (P_2 \cap P_3) \setminus P_1, \quad b \in (P_1 \cap P_3) \setminus P_2, \quad c \in (P_1 \cap P_2) \setminus P_3.$$

Let R_i be a real-closure of (E, P_i), and let $F = R_1 \cap R_2 \cap R_3$. In the commutative diagram

$$\text{(4.15)} \qquad \begin{array}{ccc} \dot{E}/\dot{E}^2 & \xrightarrow{\ \alpha\ } & \dot{F}/\dot{F}^2 \\ & {\scriptstyle\gamma}\searrow & \downarrow{\scriptstyle\beta} \\ & & \prod_i \dot{R}_i/\dot{R}_i^2 \end{array}$$

we have $\gamma(\overline{a}) = (-1, 1, 1)$, $\gamma(\overline{b}) = (1, -1, 1)$ and $\gamma(\overline{c}) = (1, 1, -1)$, so γ is surjective. Since β is injective by Ch. I, Exer. 8, it follows that β is bijective and α is surjective. In particular, $|\dot{F}/\dot{F}^2| = 8$ (and F is pythagorean by 4.3). Clearly, $\{R_i^2 \cap F\}$ give three different orderings on F. Now by Exercise 8, the surjectivity of α implies the injectivity of $X_F \to X_E$. Since $|X_E| = 3$, it follows that $|X_F| = 3$, as desired.

As it turns out, the two cases (3a) and (3b) in Theorem 4.13 are special cases of two important families of (formally real) pythagorean fields. Without going into all the details, let us try to give an idea of what these two families are.

For simplicity, we shall restrict ourselves to the case of *finitely many square classes*. Let F be a formally real pythagorean field, with $|\dot{F}/\dot{F}^2| = 2^n$, and let $r = |X_F|$ be the number of orderings on F. It is an elementary exercise to show that we have always $n \leq r \leq 2^{n-1}$ (see Exercise 16). The two cases $r = 2^{n-1}$ and $r = n$ are therefore "extreme" in the sense that they realize the upper and lower bounds in the inequalities. The case $r = 2^{n-1}$ is where we have "very many" orderings, and the case $r = n$ is where we have "very few". In the first case, if $\{-1, x_2, \ldots, x_n\}$ is a basis of square classes, essentially the same proof as given for (3a) will show that $W(F) \cong \mathbb{Z}[K]$ where K is the subgroup of $\dot{F}/\dot{F}^2$ generated by $x_2, \ldots, x_n$. Here, F is called a *superpythagorean field* (following [EL$_2$]). In the second case, the proof for (3b) will show that $W(F)$ is given by the subring of $\mathbb{Z}^n$ consisting of integer n-tuples $(a_1, \ldots, a_n)$ where the a_i's have the same parity. Here, F is called a *pythagorean* SAP *field*.[9] The Witt ring $W(F)$ here can also be described

[8] This is not to be confused with $\mathbb{R}((x, y))$, the field of fractions of the power series ring $\mathbb{R}[[x, y]]$. The latter field is formally real, but not pythagorean.

[9] "SAP field" here is an acronym, often used in the literature, for a "field with the Strong Approximation Property". See, e.g. [KRW], and [EL$_2$].

as the "pullback" of n copies of the ring epimorphism $\mathbb{Z} \to \mathbb{Z}/2\mathbb{Z}$. In the language of abstract category theory, $W(F)$ is just the "direct product of n copies of $\mathbb{Z}$ in the category of Witt rings"; see XII.7.

In the special cases $n = 1, 2$, the two extreme cases coincide, so the pythagorean field in question is *both* superpythagorean *and* SAP. In the case $n = 3$, the pythagorean field is *either* superpythagorean *or* SAP, depending on whether there are 4 or 3 orderings. These were the two cases (3a) and (3b) in 4.13. For a general n, again both extreme cases ($|X_F| = 2^{n-1}$ and $|X_F| = n$) occur for pythagorean fields F: for the former case, we can take $F = \mathbb{R}((x_2))\cdots((x_n))$, and for the latter case, we can take F to be a "suitable" intersection of n real-closed fields, generalizing the construction we gave above for $n = 3$. (Starting with a *number field* E with exactly n orderings will be sufficient.)

Superpythagorean fields and SAP fields have been much studied in the recent literature on ordered fields. For an elementary exposition on these two classes of fields, the reader may consult my CBMS volume [L3]. In this connection, it is also relevant to mention that, if F is a formally real field with $|\dot{F}/\dot{\sigma}(F)| = 2^n$ $(n < \infty)$, the possible number of orderings on F has been explicitly determined; see Bröcker's paper [Br].

Appendix: Fields with 8 Square Classes and 2 Orderings

Let F be a formally real field with $|\dot{F}/\dot{F}^2| = 8$. If F happens to be a pythagorean field, then F has either three or four orderings, and the Witt ring of F is classified in 4.13(3). If F is *not* pythagorean, the group $\dot{\sigma}(F)$ of nonzero sums of squares has index 2 or 4 in $\dot{F}$, and the number of orderings on F will be 1 or 2 accordingly. The Witt ring $W(F)$ has also been determined in II.5.13: we showed there that if $[\dot{F} : \dot{\sigma}(F)] = 2$, there are three possible Witt rings, and if $[\dot{F} : \dot{\sigma}(F)] = 4$, there are two possible Witt rings. Model fields of the former kind have been constructed in II.5 (before II.5.13), but the construction of model fields of the latter kind was lacking in II.5. In this Appendix, we propose to complete this construction. After this work is done, we would have accounted for the Witt rings of all formally real fields F with $|\dot{F}/\dot{F}^2| \leq 8$.

The fields F in question are formally real with exactly two orderings and with $|\dot{F}/\dot{F}^2| = 8$. From the analysis in II.5.13, we know that if $a \in \dot{F}$ is a sum of two squares but not a square, then $\dot{\sigma}(F) = \{1, a\}\dot{F}^2$, and $W(F)$ is determined uniquely in each of the following two cases, where α denotes the binary form $\langle 1, -a\rangle$:

Case 1. $D(\alpha) = \{\pm 1, \pm a\}\dot{F}^2$.

Case 2. $D(\alpha) = \dot{F}$.

Our job is to construct a field for each of these cases. For Case 1, this is easy. Indeed, let F_0 be a formally real field with square class basis $\{-1, 2\}$, and let $F = F_0((x))$. Then $|\dot{F}/\dot{F}^2| = 8$, and $|X_F| = 2$ by 4.11. For $a = 2$ and $\alpha = \langle 1, -a\rangle$, an easy application of Springer's Theorem shows that $D_F(\alpha) = \{\pm 1, \pm a\}\dot{F}^2$. Therefore, F is as in Case 1. Indeed, by VI.1.7, $W(F) \cong R[G]$ where G is a group of order 2 and

$$R = W(F_0) \cong \mathbb{Z}\,[t]/(2t,\, t^2),$$

as predicted by II.5.13. As a group,

$$W(F) \cong \mathbb{Z} \oplus \mathbb{Z} \oplus \mathbb{Z}_2 \oplus \mathbb{Z}_2.$$

To construct a field F corresponding to Case 2, we follow an idea of J. Mináč. Start with the field $F_0 = \mathbb{Q}(\sqrt{3}\,)$. In this field, $a := 7 = 4 + (\sqrt{3}\,)^2 \in D_{F_0}(\langle 1, 1\rangle)$, and

$$(2 + \sqrt{3}\,)^2 = 4 + 3 + 4\sqrt{3} = a + 4\sqrt{3} \tag{4.16}$$

shows that $\alpha := \langle 1, -a\rangle$ represents $\sqrt{3}$ (as well as -1 and a). But of course, $\dot{F}_0/\dot{F}_0^2$ is infinite, so we need to do something to F_0 to "cut down" its number of square classes. Our plan is to use a suitable modification of the Gross-Fischer method in VII.3.17 to construct an extension F/F_0 with $|X_F| = 2$ such that $\dot{F}/\dot{F}^2$ has $\mathbb{F}_2$-basis $\{-1, a, \sqrt{3}\}$. By what we said above about α, we see that $D_F(\alpha) = \dot{F}$, as required in Case 2.

To construct F, let $P_{0,1}, P_{0,2}$ be the two orderings on F_0, labelled such that $\sqrt{3} \in P_{0,1} \setminus P_{0,2}$. Here, $\dot{F}_0/\dot{P}_{0,1} \cap \dot{P}_{0,2}$ has $\mathbb{F}_2$-basis $-1, \sqrt{3}$. Let $\{7, b_i\}$ be an $\mathbb{F}_2$-basis for $\dot{P}_{0,1} \cap \dot{P}_{0,2}/\dot{F}^2$, so that $\dot{F}_0/\dot{F}_0^2$ has $\mathbb{F}_2$-basis $\{-1, \sqrt{3}, 7, b_i\}$. Let $F_1 = F_0(\{\sqrt{b_i}\})$, and let $P_{1,1}$ and $P_{1,2}$ be orderings on F_1 extending, respectively, $P_{0,1}$ and $P_{0,2}$. Then $\sqrt{3} \in P_{1,1} \setminus P_{1,2}$, and $-1, \sqrt{3}$ remain different elements in $\dot{F}_1/\dot{P}_{1,1} \cap \dot{P}_{1,2}$, with $7 \in P_{1,1} \cap P_{1,2}$. As before, we can pick $c_j \in \dot{P}_{1,1} \cap \dot{P}_{1,2}$ such that $\{-1, \sqrt{3}, 7, c_j\}$ is an $\mathbb{F}_2$-basis for F_1. Let $F_2 = F_1(\{\sqrt{c_j}\})$ and continue the process. For the direct limit $F = \bigcup_{i=0}^{\infty} F_i$, $\{-1, \sqrt{3}, 7\}$ forms an $\mathbb{F}_2$-basis for $\dot{F}/\dot{F}^2$ (according to Gross and Fischer), and

$$P_1 = \bigcup_i P_{i,1}, \qquad P_2 = \bigcup_i P_{i,2}$$

give different orderings on F (since $\sqrt{3} \in P_1 \setminus P_2$). From $7 \notin \dot{F}^2$, we see that F is not pythagorean. Hence $|X_F| \le 2$ and we must have $|X_F| = 2$. By the last paragraph, we have automatically $D_F(\alpha) = \dot{F}$ for $\alpha = \langle 1, -a\rangle = \langle 1, -7\rangle$, so F is the field we want. The Witt ring $W(F)$ was determined in II.5.13; as a group,

$$W(F) \cong \mathbb{Z} \oplus \mathbb{Z} \oplus \mathbb{Z}_2.$$

In summary, we list below the seven types of possible Witt groups for formally real fields F with eight square classes. The last two types (which are free abelian) are the ones arising from pythagorean fields.

	Witt Group $W(F)$	# (Orderings)	Reference
(A)	$\mathbb{Z} \oplus \mathbb{Z}_2 \oplus \mathbb{Z}_4$	1	II.5.4
(B)	$\mathbb{Z} \oplus \mathbb{Z}_2 \oplus \mathbb{Z}_2$	1	II.5.8
(C)	$\mathbb{Z} \oplus \mathbb{Z}_2 \oplus \mathbb{Z}_2 \oplus \mathbb{Z}_2$	1	II.5.9
(D)	$\mathbb{Z} \oplus \mathbb{Z} \oplus \mathbb{Z}_2 \oplus \mathbb{Z}_2$	2	this Appendix
(E)	$\mathbb{Z} \oplus \mathbb{Z} \oplus \mathbb{Z}_2$	2	this Appendix
(F)	$\mathbb{Z} \oplus \mathbb{Z} \oplus \mathbb{Z}$	3	VIII.4.13(3)
(G)	$\mathbb{Z} \oplus \mathbb{Z} \oplus \mathbb{Z} \oplus \mathbb{Z}$	4	VIII.4.13(3)

A corresponding chart for the *nonreal* fields F with $|\dot{F}/\dot{F}^2| = 8$ listing the 10 possible Witt groups appears later in XII.5.

5. Connections with Galois Theory

Having discussed pythagorean fields in general in §4, we shall explore in this section some interesting and possibly surprising connections between pythagorean fields and Galois theory. These connections were found by G. Whaples, J. Diller and A. Dress.

For the reader's convenience, we first review some terminology in basic Galois theory. Whenever K/F is a Galois extension, say with Galois group $G = \text{Gal}(K/F)$, we shall say that K/F is a *G-extension.* If G happens to be cyclic (or abelian, or ...), we also say that K/F is a cyclic (or abelian, ...) extension. An interesting problem in Galois theory is the so-called

Embedding Problem. *Given a group G and a field extension L/F, when is L embeddable in a G-extension of F?*

The first Galois-theoretic result related to the notion of pythagorean fields is the following. (We remind the reader that all fields considered are assumed to be of characteristic $\neq 2$.)

Theorem 5.1. *Let $L = F(\sqrt{a})$ be a quadratic extension of F. Then L can be embedded in a $\mathbb{Z}_4$-extension (cyclic extension of degree 4 over F) iff a is a sum of two squares in F.*

Proof. Suppose $L \subseteq K$, where K/F is a $\mathbb{Z}_4$-extension, with Galois group $\langle\sigma\rangle$ of order 4. Write $K = L(\sqrt{v})$ where $v = x + y\sqrt{a} \in L$, with $x, y \in F$, and let $u := \sqrt{v}$. Since $\text{Gal}(K/L) = \langle\sigma^2\rangle$, we have $\sigma^2(u) = -u$. Thus,

$$\sigma(u\sigma(u)) = \sigma(u)\sigma^2(u) = -u\sigma(u) \neq u\sigma(u),$$

so $u\sigma(u) \notin F$. On the other hand, $\mathrm{Gal}(L/F) = \langle \overline{\sigma} \rangle$, so $\sigma(v) = x - y\sqrt{a}$, and

$$u\sigma(u)]^2 = u^2\sigma(u^2) = v\sigma(v) = x^2 - ay^2 \in F.$$

Therefore, $F(u\sigma(u))$ must be L (the unique quadratic extension of F in K). Now $x^2 - ay^2$ is a square in L, but not a square in F, so VII.3.8 implies that $x^2 - ay^2 = az^2$ for some $z \in F$. If $x \neq 0$, then $0 \neq x^2 = a(y^2 + z^2)$ shows that a is a sum of two squares in F. If $x = 0$, then $y \neq 0$ and we get $-1 \in F^2$. In this case, *every* element in F is a sum of two squares in F.

Conversely, assume $a = b^2 + c^2$ in F. We may assume (after changing a to $b^{-2}a$) that $b = 1$. Let $v = a + \sqrt{a} \in L$, which is not in L^2 since

$$\mathrm{N}_{L/F}(v) = a^2 - a = a(a-1) = ac^2 \notin F^2. \tag{5.2}$$

For $K := L(u)$ where $u = \sqrt{v}$, we have then $[K : F] = 4$. Note that we have also $K = F(u)$ since $F(u)$ contains $u^2 - a = \sqrt{a}$. Let $\varphi : K \to \overline{F}$ be a typical F-embedding of K into $\overline{F}$. Since $\varphi(\sqrt{a}) = \pm\sqrt{a}$, we see that $\varphi(u) \in \{\pm u, \pm w\}$ where $w := \sqrt{a - \sqrt{a}} \in \overline{F}$. But

$$uw = \sqrt{a^2 - a} = \pm c\sqrt{a} \in \dot{L} \tag{5.3}$$

shows that $w \in K$, so all conjugates of u are in K and K/F is Galois.[10] Let $\sigma \in \mathrm{Gal}(K/F)$ be that particular embedding of K with $\sigma(u) = w$. Then

$$\sigma(u^2) = w^2 \Longrightarrow \sigma(a + \sqrt{a}) = a - \sqrt{a} \Longrightarrow \sigma(\sqrt{a}) = -\sqrt{a}.$$

From (5.3), we see therefore that $\sigma(uw) = -uw$, and cancellation of $\sigma(u) = w$ yields $\sigma^2(u) = -u$. Thus, $\sigma^2 \neq \mathrm{Id}$, and we must have $\mathrm{Gal}(K/F) = \langle \sigma \rangle$. The field L is now embedded in the $\mathbb{Z}_4$-extension K/F. □

Example 5.4. Note that 5.1 can be used to give a proof for the fact that *any prime number* $p \equiv 1 \pmod 4$ *is a sum of two squares in* $\mathbb{Q}$ *(and hence also in* $\mathbb{Z}$*).* Indeed, let $E = \mathbb{Q}(\zeta)$, where ζ is a primitive p-th root of unity, and $p = 4r + 1$. Then $E/\mathbb{Q}$ is a cyclic extension of degree $4r$. It is well-known that E contains a unique quadratic extension of $\mathbb{Q}$, namely $L = \mathbb{Q}(\sqrt{p})$. By Galois theory, L is contained in a unique quartic extension $K/\mathbb{Q}$ in E. Then $K/\mathbb{Q}$ is a $\mathbb{Z}_4$-extension, and hence p is a sum of two squares in $\mathbb{Q}$ by 5.1!

An interesting consequence of 5.1 is the following Galois-theoretic characterization of pythagorean fields, which was proved by G. Whaples [Wh] in 1957, and rediscovered by Diller and Dress [DD] in 1965. (The second part of this result is due to Diller and Dress.)

---

[10] Of course, the argument on conjugates showed only that K/F is normal. We then invoke the assumption that $\mathrm{char}\, F \neq 2$ to deduce that K/F is Galois.

Theorem 5.5. *F is pythagorean iff F does not admit a $\mathbb{Z}_4$-extension. If F is not pythagorean, it admits a $\mathbb{Z}_4$-extension in F_{py}.*

Proof. First suppose there is a $\mathbb{Z}_4$-extension K/F. Let $L = F(\sqrt{a}\,)$ be the unique quadratic extension of F in K. By 5.1, $a = b^2 + c^2 \in F \setminus F^2$, so F is not pythagorean. Conversely, assume F is not pythagorean, say $a = 1 + c^2 \in F \setminus F^2$. In the construction of the $\mathbb{Z}_4$-extension K/F in the proof of 5.1, we have

$$2v = 2\big(a + \sqrt{a}\,\big) = c^2 + 1 + a + 2\sqrt{a} = c^2 + \big(1 + \sqrt{a}\,\big)^2. \tag{5.6}$$

Since $\sqrt{a} \in F_{\text{py}}$, this shows that $v \in F_{\text{py}}^2 + F_{\text{py}}^2 = F_{\text{py}}^2$, and so $K = F(\sqrt{a}\,)(\sqrt{v}\,) \subseteq F_{\text{py}}$, as desired. □

Remark. (1) The following special case of the theorem is note-worthy: *every nonreal field F with $F \neq F^2$ has a $\mathbb{Z}_4$-extension.*

(2) In the second part of the proof above, we have to make a *judicious choice* of the $\mathbb{Z}_4$-extension K/F in order to have $K \subseteq F_{\text{py}}$. It is not difficult to give examples of $\mathbb{Z}_4$-extensions E/F that are *not* contained in F_{py}. For instance, take $p = 5$ in 5.4. In the notation there, $E = \mathbb{Q}(\zeta)$ is a $\mathbb{Z}_4$-extension of $\mathbb{Q}$ (for ζ a primitive 5th root of unity). However, E is *nonreal*, since -1 is a sum of two squares in E, by III.4.4. Therefore, E is not contained in the (formally real) field F_{py}. (We have here $E \cap F_{\text{py}} = \mathbb{Q}(\sqrt{5}\,)$.)

A small subset of the considerations in the proof of 5.5 leads to the following remarkable "going down" theorem for pythagorean fields.

Diller-Dress Theorem 5.7. *Let E/F be a finite extension, where E is pythagorean. Then F is also pythagorean.*

Proof. We induct on $[E : F]$. If F is not pythagorean, it would have an element $a = 1 + c^2 \notin F^2$, and we would have $L = F(\sqrt{a}\,) \subseteq E$. By the inductive hypothesis, L is pythagorean. Then, for the element

$$v := a + \sqrt{a} \in L,$$

(5.6) shows that $v \in L^2$. But, by (5.2), $\mathrm{N}_{L/F}(v) \notin F^2$, a contradiction. □

Example 5.8. (1) It is of interest to mention an explicit example for the situation in 5.7. The simplest such example is when $E = \mathbb{R}((t))$ and $F = \mathbb{R}((t^n))$. Here, $[E : F] = n$, and $E \cong F$ are both formally real pythagorean fields.

(2) As a source of "non-examples", we might mention that finite number fields E are never pythagorean. This follows from 5.7 since $[E : \mathbb{Q}] < \infty$, and $\mathbb{Q}$ is not pythagorean.

Next, we shall record some nice consequences of the Diller-Dress Theorem.

Corollary 5.9. *For any nonpythagorean field F, F_{py}/F is an infinite Galois extension, with a torsionfree Galois group.*

Proof. If $[F_{py} : F] < \infty$, 5.7 would imply that F is pythagorean, which is not the case. Hence F_{py}/F is an infinite extension. Let $G = \text{Gal}(F_{py}/F)$, and let $\sigma \in G$ be of finite order. For the fixed field E of σ, Artin's Theorem (in Galois theory) implies that F_{py}/E is a *finite* (Galois) extension. Hence by 5.7, E is pythagorean. This yields $E = F_{py}$, so $\sigma = \text{Id}$. □

Corollary 5.10. *Let E/F be a finite Galois extension, with $G = \text{Gal}(E/F)$. If E is pythagorean, a 2-Sylow group H of G must be an elementary abelian 2-group.*

Proof. If H is not elementary abelian, it would have a cyclic subgroup C of order 4. Let K be the fixed field of C. Then E/K is a $\mathbb{Z}_4$-extension, so by 5.5, K is not pythagorean. But this contradicts 5.7. Hence H is elementary abelian. □

From the Diller-Dress Theorem 5.7, we can easily deduce similar "going down" theorems for quadratically closed fields and euclidean fields.

Corollary 5.11. *Let E/F be a finite extension, where E is quadratically closed. If $\sqrt{-1} \in F$, then F is quadratically closed. If $\sqrt{-1} \notin F$, then F is euclidean.*

Proof. By 5.7, F is pythagorean. If $\sqrt{-1} \in F$ (or more generally, F is nonreal), then F is quadratically closed. Now assume $\sqrt{-1} \notin F$. Then by the case just settled, $F(\sqrt{-1})$ is quadratically closed, and 1.7 implies that F is euclidean. □

Corollary 5.12. *Let E/F be a finite extension, where E is euclidean. Then F is also euclidean.*

Proof. $E(\sqrt{-1})$ is quadratically closed by 1.7, and is a finite extension of F. Since $\sqrt{-1} \notin F$, 5.11 implies that F is euclidean. □

Note that, in the above, we derived 5.11 and 5.12 from the Diller-Dress Theorem mainly in order to illustrate the utility of the latter result. It is not difficult to give direct proofs for 5.11 and 5.12 using only the square-class exact sequence of a quadratic extension (without using Diller-Dress). In fact, the first case in 5.11 was already given as Exercise 5 in Ch. VII, and the rest follows as above. Yet another proof for 5.12 not going through quadratically closed fields can be found in the hint to Exer. 35.

6. Harrison Topology on X_F

The main goal of this section is to introduce the Harrison topology on the space X_F of orderings of a field F, and to explore the consequences of the presence of such a topology. I am not sure if the empty set can legally carry a topology, so let us assume throughout that F is a *formally real* field, so that $X_F \neq \varnothing$ by 1.10. The exposition in this section follows largely my Queen's University Lecture Notes [L1].

To set up the Harrison topology on X_F, first note that each ordering $\alpha \in X_F$ determines a map (actually a group epimorphism) $\dot{F} \to \{\pm 1\}$, in the obvious way. Thus, we have an embedding

$$(6.1) \qquad X_F \hookrightarrow \{\pm 1\}^{\dot{F}} := \text{Maps}(\dot{F} \to \{\pm 1\}).$$

The function space $\{\pm 1\}^{\dot{F}}$ has a natural product topology, if $\{\pm 1\}$ is given the discrete topology. Thus, there is a subspace topology induced on X_F; this is, by definition, the *Harrison topology*, named after David Harrison, who first pointed out its existence in his work.

To get a better view of this topology, let us first write down the defining subbase of the product topology on $\{\pm 1\}^{\dot{F}}$:

$$(6.2) \qquad H_{a,\varepsilon} = \{f : \dot{F} \to \{\pm 1\} \mid f(a) = \varepsilon\} \qquad (a \in \dot{F},\ \varepsilon = \pm 1).$$

This is a *clopen* (= closed and open) set, since its complement is $H_{a,-\varepsilon}$. Thus, $\{\pm 1\}^{\dot{F}}$ is a *Boolean space*; that is, it is compact, Hausdorff, and totally disconnected.(11) Here, of course, the Tychonoff Theorem is needed to guarantee the *compactness* of the space $\{\pm 1\}^{\dot{F}}$.

Theorem 6.3. *X_F, with the Harrison topology, is also a Boolean space.*

Proof. It is sufficient to show that X_F is a *closed* subspace of $\{\pm 1\}^{\dot{F}}$ (since any closed subspace of a Boolean space remains Boolean). Take any map $s : \dot{F} \to \{\pm 1\}$ that *does not* yield an ordering. If s is identically 1 (or -1), the subbasic open set $H_{-1,1}$ (resp. $H_{1,1}$) clearly separates s from X_F. We may thus assume that s is *surjective*. Using this s, we can thus talk about "positive" elements and "negative" elements in $\dot{F}$ in an obvious way; however, there must exist some "positive" a, b such that c will be "negative", where c stands for either $a + b$ or ab. But then the basic open set

$$H_{a,1} \cap H_{b,1} \cap H_{c,-1}$$

clearly separates s from X_F. $\square$

(11) A topological space is called *totally disconnected* if each of its connected components is a singleton set.

To get a subbasis (of open sets) for X_F, we need only take the following intersections:

$$(6.4)\qquad H(a) := H_{a,1} \cap X_F = \{\alpha \in X_F : a >_\alpha 0\} \qquad (a \in \dot{F}).$$

The reason we can restrict our attention to $\varepsilon = 1$ is, of course, that $H_{a,-1} \cap X_F$ is given by $H(-a)$. The family $\{H(a) : a \in \dot{F}\}$ may be called the *Harrison subbasis* for the Boolean space X_F.

Corollary. *Let K/F be a field extension. Then the map $\rho : X_K \to X_F$ obtained by the restriction of orderings is continuous and closed (with respect to the Harrison topologies on X_K and X_F). ("Closed" means that ρ takes closed sets to closed sets.)*

Proof. For any $a \in \dot{F}$, clearly $\rho^{-1}(H_F(a)) = H_K(a)$, where the subscripts refer to the respective fields. Since $\{H_F(a) : a \in \dot{F}\}$ is a subbasis for X_F, the continuity of ρ follows. If C is a closed subset of X_K, then C is compact (since X_K is), and therefore $\rho(C)$ is also compact. It follows that $\rho(C)$ is closed in X_F. □

Elman, Lam, and Wadsworth have proved that, if K/F is a finitely generated field extension, then the map $\rho : X_K \to X_F$ is also *open* [ELW₂]. However, we will not need this harder result below.

Next, we note that each quadratic form q over F defines a map

$$(6.5)\qquad \hat{q} : X_F \to \mathbb{Z}, \quad \text{where } \hat{q}(\alpha) := \operatorname{sgn}_\alpha(q).$$

The significance of the Harrison topology is largely clarified by the following observation.

Proposition 6.6. *For each quadratic form q, the signature map $\hat{q}$ in 6.5 is continuous with respect to the Harrison topology on X_F and the discrete topology on $\mathbb{Z}$. In fact, with the latter topology fixed, the Harrison topology is the coarsest topology on X_F that makes all the maps $\hat{q}$ continuous.*

Proof. To prove the continuity of $\hat{q}$, it is sufficient to treat the case $q = \langle a \rangle$ (since the sum of continuous functions into an additive topological group is continuous). In this case, we note that

$$\hat{q}^{-1}(i) = \{\alpha \in X_F : \operatorname{sgn}_\alpha \langle \alpha \rangle = i\} = \begin{cases} \varnothing & \text{if } i \neq \pm 1, \\ H(a) & \text{if } i = 1, \\ H(-a) & \text{if } i = -1. \end{cases}$$

From these calculations, the desired conclusions in the Proposition follow immediately. □

Previously, we have written the total signature map in the form

$$\text{(6.7)} \qquad \operatorname{sgn} : W(F) \longrightarrow \prod_{\alpha \in X_F} \mathbb{Z},$$

where the RHS is just $\operatorname{Maps}(X_F, \mathbb{Z})$. Since each $\hat{q}$ in 6.5 is *continuous*, we may as well use a smaller target set for "sgn", and re-express the latter as a ring homomorphism

$$\text{(6.8)} \qquad \operatorname{sgn} : W(F) \longrightarrow C(X_F, \mathbb{Z}) \quad (\text{given by } q \mapsto \hat{q}),$$

where $C(X_F, \mathbb{Z})$ denotes the ring of continuous functions from X_F to $\mathbb{Z}$. (The topologies on X_F and $\mathbb{Z}$ will henceforth be *fixed*.)

The advantage of using the smaller target group $C(X_F, \mathbb{Z})$ is that we can now more meaningfully study the cokernel of the map "sgn" in 6.8. In 3.2, we have shown that ker(sgn) is a 2-primary torsion group. Our first main result in this section is the following "dual" statement.

Theorem 6.9. *For the map* "sgn" *in* 6.8, coker(sgn) *is also a* 2-*primary torsion group.*

The proof of this is based on the lemma below concerning the existence of quadratic forms in I^nF with certain prescribed signature properties. Here, I^nF denotes the nth power of the "fundamental ideal" IF.

Lemma 6.10. *For any clopen set $C \subseteq X_F$, there exists a form $q \in I^nF$ (for some $n \geq 0$) such that $2^n\chi_C = \operatorname{sgn}(q)$, where χ_C denotes the characteristic function on X_F associated with the subset $C \subseteq X_F$.*

Proof. *Step* 1. If the lemma holds for two clopen sets C_1, C_2, then it holds for $C_1 \cup C_2$. Indeed, suppose $q_i \in I^{m_i}F$ are such that $2^{m_i}\chi_{C_i} = \operatorname{sgn}(q_i)$, $i = 1, 2$. After multiplying these equations by powers of 2 if necessary, we may assume that $m_1 = m_2 = m$ (say). Now take the well-known equation

$$\chi_{C_1 \cup C_2} = \chi_{C_1} + \chi_{C_2} - \chi_{C_1}\chi_{C_2},$$

and multiply it by 2^{2m} to get

$$2^{2m}\chi_{C_1 \cup C_2} = 2^m \operatorname{sgn}(q_1 + q_2) - \operatorname{sgn}(q_1\, q_2) = \operatorname{sgn}(q),$$

where $q = 2^m(q_1 + q_2) - q_1\, q_2 \in I^{2m}F$.

Step 2. A basis of open sets in X_F is given by the sets

$$\text{(6.11)} \qquad H(a_1, \ldots, a_n) := H(a_1) \cap \cdots \cap H(a_n) \qquad (a_i \in \dot{F}).$$

Since X_F is compact, so is the given clopen set C. Thus, C can be written as a *finite* union of sets of the form $H(a_1, \ldots, a_n)$. By Step 1, the proof of the lemma is now reduced to the case where $C = H(a_1, \ldots, a_n)$.

Step 3. Let $q_i = \langle 1, a_i \rangle \in IF$. Clearly, $\operatorname{sgn}(q_i) = 2\chi_{H(a_i)}$. Therefore, for $q = q_1 \cdots q_n \in I^n F$, we have

(6.12) $$\operatorname{sgn}(q) = 2^n \chi_{H(a_1)} \cdots \chi_{H(a_n)} = 2^n \chi_{H(a_1,\ldots,a_n)}.$$

This proves the lemma for the case $C = H(a_1, \ldots, a_n)$. ☐

The quadratic form $q = \langle 1, a_1 \rangle \cdots \langle 1, a_n \rangle$ above is called *an n-fold Pfister form.* Such forms will be the principal objects for study in a future chapter (see Ch. X). Given some hindsight, it would perhaps be relatively easy to anticipate the significant role played by these special forms in quadratic form theory. First, these forms have the very nice signature property in 6.12. Second, since the forms $\langle 1, a \rangle$ provide a set of additive generators for IF, the n-fold Pfister forms likewise provide a set of additive generators for $I^n F$. Last but not least, $\langle 1, 1 \rangle \cdots \langle 1, 1 \rangle$ is the "sum of 2^n squares" quadratic form over F.

We are now in a good position to supply the following.

Proof of 6.9. Given a continuous function $f \in C(X_F, \mathbb{Z})$, let $C_i = f^{-1}(i)$ ($i \in \mathbb{Z}$). These sets are clopen, and form a partition of X_F. Since X_F is compact, all but a finite number of the C_i's must be empty. This means that f is a bounded function on X_F, so we can express f as a *finite* sum $\sum_i i \cdot \chi_{C_i}$. By 6.10,

$$2^{n_i} \chi_{C_i} \in \operatorname{im}(\operatorname{sgn}) \quad \text{for suitable } n_i \geq 0.$$

From this, it follows immediately that $2^n f \in \operatorname{im}(\operatorname{sgn})$ for some n, as desired. ☐

Returning to 6.10, we note that there is a further self-strengthening of this result that can be stated in the form of a "separation theorem." We shall call this "Urysohn's Lemma", in view of its obvious resemblance to the familiar topological result about separation in normal spaces. The results in the balance of this section come from my joint work with R. Elman [EL$_2$].

Urysohn's Lemma 6.13. *For any two disjoint closed sets A, B in X_F, there exists $q \in I^n F$ (for some n) such that* $\operatorname{sgn}(q) \equiv 0$ *on B, and* $\operatorname{sgn}(q) \equiv 2^n$ *on A.*

Note that if $C \subseteq X_F$ is clopen, and we apply 6.13 to $A = C$ and $B = X_F \setminus C$, then we get back 6.10. However, we shall *use* 6.10 to prove 6.13, which is why we called 6.13 a self-strengthening of 6.10.

Proof of 6.13. The complement of B is a union of sets of the form 6.11. Since A is compact, a finite number of these, say $C_1, \ldots, C_r$, will cover A, and $C_i \cap B = \varnothing$. If we apply 6.10 to the clopen set $C = C_1 \cup \cdots \cup C_r$, the conclusion in Urysohn's Lemma follows immediately. ☐

As another application of 6.10, we shall give a characterization for quadratic forms q with the property that $\operatorname{sgn}_\alpha(q)$ is divisible by 2^n for every $\alpha \in X_F$, where n is a given integer. Note that these are precisely the forms q such that $\operatorname{sgn}(q) \in C(X_F, 2^n\mathbb{Z})$.

Theorem 6.14. *For $n \geq 0$ and any quadratic form q, the following are equivalent:*

(1) $\operatorname{sgn}(q) \in C(X_F, 2^n\mathbb{Z})$.

(2) $2^t \cdot q \in I^{t+n}F$ *for some integer* $t \geq 0$.

Proof. (2) $\Rightarrow$ (1). Since even-dimensional forms have even signatures (at any ordering), $\operatorname{sgn}(IF) \subseteq C(X_F, 2\mathbb{Z})$. Recalling that "sgn" is a ring homomorphism, we have $\operatorname{sgn}(I^mF) \subseteq C(X_F, 2^m\mathbb{Z})$ for any m. Thus, if $2^t \cdot q \in I^{t+n}F$, we get

$$2^t \cdot \operatorname{sgn}(q) \in C(X_F, 2^{t+n}\mathbb{Z}).$$

Now cancellation of 2^t yields (1).

(1) $\Rightarrow$ (2). Assuming (1), let $D_i = (\operatorname{sgn}(q))^{-1}(2^n i)$: these clopen sets form a partition of X_F. As before, at most a finite number of these clopen sets can be nonempty, so we can resolve $\operatorname{sgn}(q)$ in a finite sum

$$\operatorname{sgn}(q) = \sum_i 2^n i \cdot \chi_{D_i}.$$

For each $D_i \neq \varnothing$, apply 6.10 to find a form $q_i \in I^{m_i}F$ such that $2^{m_i}\chi_{D_i} = \operatorname{sgn}(q_i)$. Since only a finite number of these are involved, we may again arrange that all m_i's be equal (say $= m$). Thus,

$$2^m \cdot \operatorname{sgn}(q) = \sum_i 2^n i \cdot 2^m \chi_{D_i} = \sum_i 2^n i \cdot \operatorname{sgn}(q_i).$$

By Pfister's Local-Global Principle 3.2, it follows that

$$2^m q \in \sum_i 2^n i \cdot q_i + W_t(F).$$

Multiplying this by a sufficiently large power of 2, say 2^k, we can eliminate the torsion error term, and arrive at

$$2^{k+m} q = 2^{n+k} \sum_i i \cdot q_i \in I^{n+k+m}F,$$

which proves (2) with $t = k + m$. $\square$

There is another condition on the form q that is obviously related to the two conditions in 6.14, namely:

(3) $q \in I^nF + W_t(F)$.

The argument above on the elimination of torsion error terms shows clearly that (3) implies (2), and in any case, an application of the map "sgn" shows that (3) implies (1). In my Queen's University Lecture Notes [L_1] (ca. 1976), I asked if (3) is *equivalent* to (1) and (2). At around the same time, M. Marshall had raised the same question for formally real pythagorean fields F (for which $W_t(F) = 0$), and answered it affirmatively in the case where $|X_F| < \infty$. Later, a possible "yes" answer for (1) $\Leftrightarrow$ (3) (for general fields F) became known as "Lam's Conjecture". Recently, this "Conjecture" has been proved by Dickmann-Miraglia [DM_1, DM_2] (1998, 2003) and Arason-Elman [AE] (2001), using the solution of Milnor's Conjectures due to Voevodsky [Vo] and Orlov-Vishik-Voevodsky [OVV_1].(12) For a related work on "Lam's Conjecture" (still assuming the Milnor Conjectures), see the paper [Mon] of Monnier.

Note that the equivalence (1) $\Leftrightarrow$ (3) may be thought of as a characterization of the forms in I^nF "up to a torsion summand". In passing, we should note that "Lam's Conjecture" also has an analogue in the "reduced theory" of quadratic forms with respect to a preordering T (in the sense of §9 below). A sketch of a proof of the Conjecture in that case has been given by M. Marshall in his review of Monnier's paper (*loc. cit.*); see MR 2001g:11056.

To conclude this section, let us record the following interesting consequence of 3.2 and 6.9 on the structure of the additive group $W(F)$.

Corollary 6.15. *$W(F) = W_t(F) \oplus G$, where G is a free abelian group with* rank G = rank $C(X_F, \mathbb{Z})$. *If $|X_F| = r < \infty$, then* rank $G = r$.

Proof. By 3.2, we may view $W(F)/W_t(F)$ as a subgroup of $C(X_F, \mathbb{Z})$, which is in turn a subgroup of $\mathrm{Bdd}(X_F, \mathbb{Z})$, the group of all bounded functions from X_F to $\mathbb{Z}$. By a classical theorem of Specker and Nöbeling, $\mathrm{Bdd}(X_F, \mathbb{Z})$ is a free abelian group. Since any subgroup of a free abelian group is free, it follows that the short exact sequence

$$0 \to W_t(F) \to W(F) \xrightarrow{\text{sgn}} \mathrm{im(sgn)} \to 0$$

splits. Also, we have $\mathrm{rank}\,(\mathrm{im}\,(\mathrm{sgn})) = \mathrm{rank}\; C(X_F, \mathbb{Z})$ since coker(sgn) has rank 0. This proves the first part of the theorem. The second part follows upon noting that, in case $|X_F| = r < \infty$,

$$C(X_F, \mathbb{Z}) = \mathrm{Maps}(X_F, \mathbb{Z}) \cong \mathbb{Z}^r. \qquad \square$$

In conclusion, we have seen in this section the significance of having a topology on the space X_F of orderings. We have shown (in 6.3) that, for any formally real field F, X_F (with the Harrison topology) is a Boolean

(12) A quick description of Milnor's Conjectures will be given later in X.6.

space. It is appropriate to mention, therefore, that T. Craven has also proved the converse of this statement; namely, *any (nonempty) Boolean space is homeomorphic to* X_F *for some formally real field* F. Interested readers can find this result in [Cr].

7. Prime Spectrum of $W(F)$

In this section, we shall determine the prime ideal spectrum of the Witt ring $W(F)$. Recall that, for any commutative ring A, the set of prime ideals of A, denoted by $\mathrm{Spec}(A)$, is called the *prime spectrum* of A. This is a topological space carrying the *Zariski topology*, in which the closed sets are of the form

$$V(I) = \{ \mathfrak{p} \in \mathrm{Spec}(A) : \mathfrak{p} \supseteq I \}, \tag{7.0}$$

where I is any ideal in A. This prime spectrum is usually not Hausdorff, but it is always compact, and a subbasis of its topology is given by the sets

$$D(f) = \{ \mathfrak{p} \in \mathrm{Spec}(A) : f \notin \mathfrak{p} \}, \tag{7.1}$$

where f is any element of A.

If $\mathfrak{p} \in \mathrm{Spec}(A)$, then $A/\mathfrak{p}$ is an integral domain, so it has characteristic p, where p is a prime number, or 0. We shall say, for short, that $\mathfrak{p}$ is a prime ideal of characteristic p (or, symbolically, $\mathrm{char}(\mathfrak{p}) = p$).

The main idea needed for determining the prime spectrum of the Witt ring $W(F)$ is already implicit in Lemma 3.4. Let us recall its two-part statement here. First, the only prime ideal of characteristic 2 in $W(F)$ is IF; and second, if $\mathfrak{p} \in \mathrm{Spec}(W(F))$ has characteristic $\neq 2$, then

$$\alpha_{\mathfrak{p}} := \{0\} \cup \{ a \in \dot{F} : \langle a \rangle \equiv 1 \pmod{\mathfrak{p}} \} \tag{7.2}$$

is an ordering on F. In the case where F is nonreal, therefore, we have $\mathrm{Spec}(W(F)) = \{IF\}$. We may dismiss this case in the following, and shall assume henceforth, until further notice, that F is formally real.

We start out by defining some prime ideals in $W(F)$. (Eventually, these will be shown to be *all* of the prime ideals.) For any ordering $\alpha \in X_F$, we fix a real-closure F_α for (F, α), and define

$$\begin{aligned} \mathfrak{p}_\alpha &= \ker(\mathrm{sgn}_\alpha : W(F) \to W(F_\alpha) \cong \mathbb{Z}), \\ \mathfrak{p}_{\alpha,p} &= \{ \varphi \in W(F) : \mathrm{sgn}_\alpha(\varphi) \equiv 0 \pmod{p} \} \quad (p = \text{prime}). \end{aligned} \tag{7.3}$$

Clearly, both $\mathfrak{p}_\alpha \subsetneq \mathfrak{p}_{\alpha,p}$ are prime ideals of $W(F)$, with

$$\mathrm{char}(\mathfrak{p}_\alpha) = 0, \quad \text{and} \quad \mathrm{char}(\mathfrak{p}_{\alpha,p}) = p \quad (\text{for any prime } p).$$

In fact, we have

$$W(F)/\mathfrak{p}_\alpha \cong \mathbb{Z} \quad \text{and} \quad W(F)/\mathfrak{p}_{\alpha,p} \cong \mathbb{Z}/p\mathbb{Z} \quad (\text{for any prime } p).$$

Also, from our earlier remark about prime ideals of characteristic 2, we see that $\mathfrak{p}_{\alpha,2} = IF$ for *every* $\alpha \in X_F$. This means that the $\mathfrak{p}_{\alpha,2}$'s only give one prime ideal (namely IF). But other than this, there is no more "collapsing" among the prime ideals defined in 7.3. First, the $\mathfrak{p}_\alpha$'s are distinct from one another and from the $\mathfrak{p}_{\alpha,p}$'s. Second, if $\mathfrak{p}_{\alpha,p} = \mathfrak{p}_{\beta,q}$, clearly $p = q$ (by considering characteristics). Third, if $\mathfrak{p}_{\alpha,p} = \mathfrak{p}_{\beta,p}$ where $p \neq 2$, then

$$\begin{aligned} a >_\alpha 0 &\implies \langle a \rangle \equiv 1 \pmod{\mathfrak{p}_{\alpha,p}} \\ &\implies \langle a \rangle \equiv 1 \pmod{\mathfrak{p}_{\beta,p}} \\ &\implies a >_\beta 0 \quad (\text{since } \operatorname{char}(\mathfrak{p}_{\beta,p}) = p \neq 2). \end{aligned}$$

This shows that $\alpha = \beta \in X_F$.

Proposition 7.4. *The map $\alpha \mapsto \mathfrak{p}_\alpha$ gives a one-one correspondence between X_F and the set Y_F of prime ideals of characteristic 0 in $W(F)$.*

Proof. If $\mathfrak{p} \in Y_F$, 7.2 defines an ordering $\alpha_\mathfrak{p} \in X_F$. It is straightforward to check that $\alpha \mapsto \mathfrak{p}_\alpha$ and $\mathfrak{p} \mapsto \alpha_\mathfrak{p}$ are mutually inverse maps between X_F and Y_F. (The key idea behind this check is the one used already in the proof of 3.4; namely, for any $a \in \dot{F}$, $\langle a \rangle^2 = 1 \in W(F)$ implies that $\langle a \rangle \equiv \pm 1 \pmod{\mathfrak{p}}$.) □

From this partial result is it now easy to reach the complete classification of prime ideals in $W(F)$.

Theorem 7.5. (Harrison [Ha], Lorenz-Leicht [LL]) $\operatorname{Spec}(W(F))$ *consists of three types of prime ideals*:

(1) $\mathfrak{p}_\alpha$ $(\alpha \in X_F)$; *these are all prime ideals of characteristic* 0.

(2) $\mathfrak{p}_{\alpha,p}$ $(\alpha \in X_F$, *p an odd prime*); *these are all prime ideals of prime characteristic* $\neq 2$.

(3) $IF = \mathfrak{p}_{\alpha,2}$ $(\alpha \in X_F)$; *this is the unique prime ideal of characteristic* 2.

Proof. It only remains to analyze the prime ideals of odd prime characteristic p. Let $\mathfrak{p} \subseteq W(F)$ be such a prime ideal. Then the construction in 7.2 produces an ordering $\alpha = \alpha_\mathfrak{p}$. Consider any form q with $\operatorname{sgn}_\alpha(q) = 0$. Then, in a diagonalization of q, half of the entries are positive (and the other half negative) in the ordering α. The positive entries are $\equiv 1 \pmod{\mathfrak{p}}$, and the others are $\equiv -1 \pmod{\mathfrak{p}}$, by definition. Thus, $q \equiv 0 \pmod{\mathfrak{p}}$, which shows that $\mathfrak{p}_\alpha \subseteq \mathfrak{p}$. Since $\mathfrak{p}_{\alpha,p}$ is the unique prime ideal of characteristic p that contains $\mathfrak{p}_\alpha$, we conclude that $\mathfrak{p} = \mathfrak{p}_{\alpha,p}$. This completes the classification of prime ideals. □

A somewhat impressionistic picture of the prime spectrum of $W(F)$ is as follows:

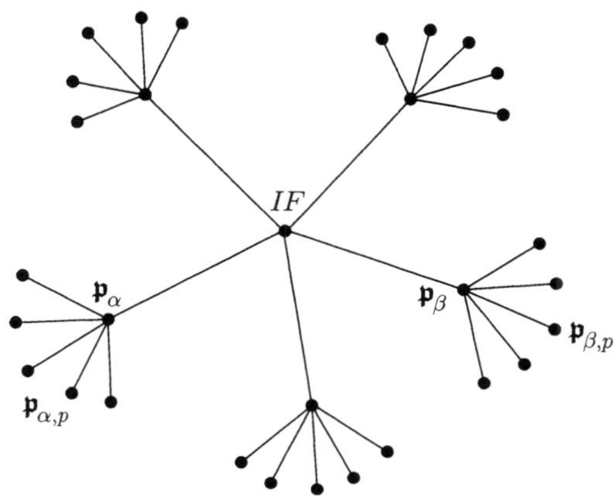

Here, p denotes any odd prime.

The corollary below (especially its last statement) serves to show that the Harrison topology is indeed the most reasonable topology to be put on the space X_F of orderings for a formally real field F.

Corollary 7.6. $\mathrm{Max}(W(F))$ *(the maximal ideal spectrum of* $W(F)$*) consists of the height one primes* $\{\mathfrak{p}_{\alpha,p}\}$*. On the other hand,* $\mathrm{MinSpec}(W(F))$ *(the minimal prime ideal spectrum) is just the space* Y_F *in* 7.4 *consisting of the* $\mathfrak{p}_\alpha$ *'s. The one-one correspondence in* 7.4 *is a homeomorphism between* X_F *(with the Harrison topology) and* $\mathrm{MinSpec}(W(F))$ *(with the induced Zariski topology).*

Proof. Only the last statement needs a verification. Using the notation in 7.1, consider a subbasic open set in $\mathrm{MinSpec}(W(F))$, which has the form

$$\begin{aligned} D(q) \cap \mathrm{MinSpec}(W(F)) &= \{\, \mathfrak{p}_\alpha : \ q \notin \mathfrak{p}_\alpha\} \\ &= \{\, \mathfrak{p}_\alpha : \ \mathrm{sgn}_\alpha(q) \neq 0\}, \end{aligned}$$

where $q \in W(F)$. Under the one-one correspondence in 7.4, this corresponds to the following subset in X_F:

$$\{\alpha \in X_F \ : \ \ \mathrm{sgn}_\alpha(q) \neq 0\},$$

which is open in X_F since $\alpha \mapsto \mathrm{sgn}_\alpha(q)$ is a continuous mapping from X_F to $\mathbb{Z}$. Conversely, specializing the above information to the form $q = \langle 1, a\rangle$

where $a \in \dot{F}$, we have

$$\{\alpha \in X_F : \operatorname{sgn}_\alpha\langle 1, a\rangle \neq 0\} = \{\alpha \in X_F : a >_\alpha 0\},$$

which is the Harrison subbasic set $H(a) \subseteq X_F$. Therefore, under the one-one correspondence in 7.4, $H(a) \subseteq X_F$ also corresponds to a (subbasic) open set in MinSpec($W(F)$). This shows that the one-one correspondence in question is a *homeomorphism*. □

Corollary 7.7. *The Witt ring $W(F)$ has Krull dimension one if F is formally real, and Krull dimension zero if F is nonreal.*

Proof. This follows directly from the enumeration of prime ideals in $W(F)$. For a *second* proof, consider the subring $R = \mathbb{Z}\langle 1\rangle$ of $W(F)$. Every unary form $\langle a\rangle$ is integral over R, as $\langle a\rangle^2 = 1$. Since the sum of integral elements is integral, it follows that $W(F)$ is an integral extension of R. By the Cohen-Seidenberg Theorem, we see that $W(F)$ has the same Krull dimension as R. Clearly, this latter Krull dimension is one if F is formally real, and zero if F is nonreal. □

The third corollary below is now also self-evident.

Corollary 7.8. *The following three statements are equivalent*:

(1) *F is nonreal.*

(2) *$W(F)$ has a unique prime ideal (which must be IF).*

(3) *$W(F)$ is a local ring (with maximal ideal IF).*

Note that, in this case, $W(F)$ is a *noetherian* local ring iff the square class group $\dot{F}/\dot{F}^2$ is finite, according to II.2.4.

To close this section, we use the results above to prove a theorem on the extension of orderings from one field to another. Again, the Cohen-Seidenberg theorem is needed for this application.

Theorem 7.9. *For any field extension K/F, let $W(K/F)$ denote the kernel of the functorial map $\varepsilon : W(F) \to W(K)$. Then an ordering $\alpha \in X_F$ can be extended to an ordering on K iff* $\operatorname{sgn}_\alpha(W(K/F)) = 0$.

Proof. Let $q \in W(K/F)$. If $\beta \in X_K$ extends α, then $\operatorname{sgn}_\alpha(q) = \operatorname{sgn}_\beta(q_K)$ $= 0$ since q_K is hyperbolic. This proves the "only if" part. For the converse, assume that $\operatorname{sgn}_\alpha(W(K/F)) = 0$. Then the prime ideal $\mathfrak{p}_\alpha$ contains $W(K/F)$, so $\varepsilon(\mathfrak{p}_\alpha)$ is a prime ideal of the subring $\varepsilon(W(F)) \subseteq W(K)$. Since this is an integral extension, the Cohen-Seidenberg theorem implies that there is a prime ideal $\mathfrak{P} \subseteq W(K)$ that contracts to $\mathfrak{p}_\alpha$. We see easily that $\mathfrak{P}$ is of characteristic 0, so it gives rise to some ordering $\beta \in X_K$. For any $a \in \dot{F}$, we have

$$a >_\alpha 0 \iff \langle a \rangle - 1 \in \mathfrak{p}_\alpha \iff \langle a \rangle_K - 1 \in \mathfrak{P} \iff a >_\beta 0.$$

Therefore, $\beta \in X_K$ is an extension of α from F to K. □

Corollary 7.10. (1) (Artin-Schreier) *If K/F is a finite extension of odd degree, then any ordering on F extends to an ordering on K.*

(2) *If $K = F(\sqrt{a})$ where $a \in \dot{F} \setminus \dot{F}^2$, then $\alpha \in X_F$ extends to an ordering on K iff $a >_\alpha 0$.*

Proof. (1) Here, $W(K/F) = 0$ by VII.2.2. Therefore, $\text{sgn}_\alpha(W(K/F)) = 0$ is satisfied by every $\alpha \in X_F$.

(2) Here, $W(K/F) = \langle 1, -a \rangle\, W(F)$ by VII.3.2. Thus, by 7.9, $\alpha \in X_F$ extends to K iff $\text{sgn}_\alpha \langle 1, -a \rangle = 0$, which amounts to $a >_\alpha 0$. □

For a somewhat different derivation of 7.10, see Exercise 6. For another criterion on the extendibility of an ordering from one field to another, see 9.8.

8. Applications to the Structure of $W(F)$

In this section, using results in §§3, 6, and 7, we shall determine the following objects which are of interest for the structure of the Witt ring:

(A) $\text{nil}\,(W(F))$: this is the nil radical, consisting of all nilpotent elements in $W(F)$. By commutative ring theory, we know that $\text{nil}\,(W(F))$ is the intersection of all prime ideals in $W(F)$.

(B) $\text{rad}\,(W(F))$: this is the Jacobson radical of $W(F)$, i.e., the intersection of all maximal ideals of $W(F)$.

(C) $\text{zd}\,(W(F))$: the set of zero-divisors in $W(F)$ (including 0).

(D) The set of idempotents in $W(F)$.

(E) $\text{U}(W(F))$: the multiplicative group of units in $W(F)$.

The results in the first half of this section are mostly due to Pfister (see [Pf3]).

The prime ideal notations set up in §7 will remain in force throughout this section. We begin with the following calculations of (A) and (B).

Theorem 8.1. (1) *If F is nonreal, then* $\text{nil}\,(W(F)) = \text{rad}\,(W(F)) = IF$.

(2) *If F is formally real, then* $\text{nil}\,(W(F)) = \text{rad}\,(W(F)) = W_t(F)$.

Proof. (1) is clear in view of 7.8 and the remark made in (A) above. For (2), note that, for any ordering α on F, the intersection $\bigcap \mathfrak{p}_{\alpha,p}$ (p ranging over all primes $\neq 0$) is just $\mathfrak{p}_\alpha$. Thus,

$$\text{(8.2)} \qquad \operatorname{rad}(W(F)) = \bigcap_{\alpha,p} \mathfrak{p}_{\alpha,p} = \bigcap_{\alpha} \mathfrak{p}_\alpha .$$

This is clearly just the intersection of all the prime ideals in $W(F)$. Consequently, $\operatorname{rad}(W(F)) = \operatorname{nil}(W(F))$, again by the remark made in (A) above. Further, 8.2 says that $\operatorname{rad}(W(F))$ is the intersection of the kernels of $W(F) \to W(F_\alpha)$, where F_α ranges over all the real-closures of F. By Pfister's Local-Global Principle 3.2, we conclude that $\operatorname{rad}(W(F)) = W_t(F)$. □

Next, we try to determine the set of zero-divisors $\operatorname{zd}(W(F))$. We need the following general observation about $\operatorname{zd}(R)$ for any commutative ring R.

Lemma 8.3. *If R is a commutative ring, $\operatorname{zd}(R)$ is the union of a certain set of prime ideals in R.*

Proof. It suffices to show that any 0-divisor z is contained in a prime ideal $\mathfrak{p} \subseteq \operatorname{zd}(R)$. Let S be the multiplicative set of all non 0-divisors (i.e., the complement of $\operatorname{zd}(R)$). By Zorn's Lemma, there exists an ideal $\mathfrak{p}$ maximal with respect to the properties $\mathfrak{p} \cap S = \varnothing$ and $z \in \mathfrak{p}$. We finish by showing that $\mathfrak{p}$ is prime. Indeed, suppose $xy \in \mathfrak{p}$, but $x, y \notin \mathfrak{p}$. By the maximality property of $\mathfrak{p}$, there exist , $s, s' \in S$ such that $s \in \mathfrak{p} + xR$ and $s' \in \mathfrak{p} + yR$. Multiplying these equations, we get $ss' \in \mathfrak{p} \cap S$, which contradicts the choice of $\mathfrak{p}$. □

Theorem 8.4. (1) *If F is nonreal, $\operatorname{zd}(W(F)) = IF$.*

(2) *If F is formally real but not pythagorean, then also $\operatorname{zd}(W(F)) = IF$.*

(3) *If F is formally real and pythagorean, then $\operatorname{zd}(W(F))$ is the union of the minimal prime ideals $\mathfrak{p}_\alpha$ $(\alpha \in X_F)$.*

Proof. (1) is clear since, in this case,

$$IF = \operatorname{nil}(W(F)), \quad \text{and} \quad W(F) \setminus IF = \mathrm{U}(W(F)).$$

For (2), suppose F is formally real, but *not* pythagorean. Then there exists $0 \neq q \in W_t(F)$, by 4.1. Since the additive order of q is a power of 2, we see that $2 \in \operatorname{zd}(W(F))$. Now, IF is the unique prime ideal of characteristic 2, so 8.3 implies that $IF \subseteq \operatorname{zd}(W(F))$. On the other hand, any prime ideal of the form $\mathfrak{p}_{\alpha,p}$ $(\alpha \in X_F,\ p \neq 2)$ cannot be contained in $\operatorname{zd}(W(F))$, since $p \cdot 1 \in \mathfrak{p}_{\alpha,p}$ is not a 0-divisor. The remaining primes $\mathfrak{p}_\alpha$ are already contained in $IF \subseteq \operatorname{zd}(W(F))$. Thus, 8.3 yields $IF = \operatorname{zd}(W(F))$.

Finally, suppose F is formally real and pythagorean. The prime ideals $\mathfrak{p}_{\alpha,p}$ ($\alpha \in X_F$, p any prime) cannot be contained in $\mathrm{zd}\,(W(F))$, since $W(F)$ is now torsion-free. Therefore, 8.3 implies that $\mathrm{zd}\,(W(F)) \subseteq \bigcup \mathfrak{p}_\alpha$ (α ranging over X_F). We finish by proving that, for each $\alpha \in X_F$, $\mathfrak{p}_\alpha$ consists entirely of zero-divisors. If a form $q \in \mathfrak{p}_\alpha$, then

$$q = \langle a_1, \ldots, a_m, -b_1, \ldots, -b_m \rangle$$

with all a_i, b_j positive at α. Letting q' be the product of the binary forms $\langle a_i, b_i \rangle$ ($1 \le i \le m$), we have clearly $q' \neq 0$ in $W(F)$ (since $\mathrm{sgn}_\alpha(q') = 2^m\langle 1 \rangle \in W(F_\alpha)$), and $q \cdot q' = 0 \in W(F)$. Therefore, $q \in \mathrm{zd}\,(W(F))$, as desired. □

It is perhaps a little surprising that, in the formally real case, the determination of $\mathrm{zd}\,(W(F))$ depends on whether or not F is pythagorean. But, as we saw from the proof above, this distinction of cases is necessary since we need to know whether or not 2 is a 0-divisor. There is, however, a nice piece of information that is common to all three cases in Theorem 8.4; we record this below.

Corollary 8.5. *For any field F, $q \in \mathrm{zd}\,(W(F))$ only if $\dim(q)$ is even. (In other words, odd-dimensional forms cannot be 0-divisors in $W(F)$.)*

Proof. In the first two cases in 8.4, we know that even the "if and only if" statement holds. But, in the formally real case, each prime ideal $\mathfrak{p}_\alpha$ ($\alpha \in X_F$) obviously lies in IF, so even in Case (3) in 8.4, we have

$$\mathrm{zd}\,(W(F)) \subseteq \bigcup_\alpha \mathfrak{p}_\alpha \subseteq IF. \qquad \square$$

Remark. It is also possible to give a *uniform* proof for 8.5 (for any field) F), independently of the computation of the set $\mathrm{zd}\,(W(F))$ in 8.4. Suppose $q \cdot q' = 0 \in W(F)$, where q is odd-dimensional. Then q' must have total signature 0, so by the Local-Global Principle $2^k \cdot q' = 0$ for some $k \geq 1$. Let

$$J = q \cdot W(F) + 2^k \cdot W(F).$$

If this is not the unit ideal, then it is contained in some maximal ideal $\mathfrak{m} \subseteq W(F)$. Then $2^k \in \mathfrak{m}$ implies that $2 \in \mathfrak{m}$. Therefore, $\mathfrak{m} = IF$. But then $q \in IF$, a contradiction. Thus, we must have $J = W(F)$. Since $J \cdot q' = 0$, it follows that $q' = 0 \in W(F)$. (For yet other approaches to 8.5, see the end of this section, as well as Ex. 33.)

We come now to the determination of the idempotents in $W(F)$. It turns out, however, that there are no interesting ones!

Theorem 8.6. *The only idempotents in $W(F)$ are 0 and 1 (i.e., $W(F)$ is a "connected" ring).*

Proof. Suppose we have an equation $1 = e_1 + e_2 \in W(F)$, where e_1, e_2 are mutually orthogonal idempotents, other than 0, 1. Then, $e_1, e_2 \in \text{zd}\,(W(F)) \subseteq IF$ by 8.5, and $1 = e_1 + e_2 \in IF$ gives the desired contradiction! □

Our final task is that of describing $\text{U}(W(F))$, the group of units of the Witt ring $W(F)$. A preliminary result, first obtained by Pfister, is the following.

Theorem 8.7. (1) *If F is nonreal,* $\text{U}(W(F))$ *consists of all odd-dimensional forms.*

(2) *If F is formally real, a form q lies in* $\text{U}(W(F))$ *iff* $\text{sgn}_\alpha(q) = \pm 1$ *for every $\alpha \in X_F$.*

Proof. (1) follows directly from 7.8(3). For (2), assume that F is formally real. The "only if" part in (2) is trivial, since $\text{U}(\mathbb{Z}) = \{\pm 1\}$. For the "if" part, consider any form q with the given signature property. Then $\text{sgn}_\alpha(q^2) = 1$ for every $\alpha \in X_F$, and hence, by 3.2 and 8.1(2),

$$q^2 - 1 \in W_t(F) = \text{nil}\,(W(F)).$$

We then have $q^2 \in 1 + \text{nil}\,(W(F)) \subseteq \text{U}((F))$, so certainly $q \in \text{U}(W(F))$. □

A more explicit computation of $\text{U}(W(F))$ has been given by M. Marshall [Ma$_2$] and D. W. Lewis [Le$_3$], as follows.

Theorem 8.8. *Let $I_t^2F := I^2F \cap W_t(F)$. Then $1 + I_t^2F$ is a multiplicative group, and*

$$\text{U}(W(F)) = (\dot{F}/\dot{F}^2) \times (1 + I_t^2F).$$

(Here, $\dot{F}/\dot{F}^2$ is identified with the subgroup of $W(F)$ consisting of the unary forms.) In particular, I^2F is torsionfree iff $\text{U}(W(F)) = \dot{F}/\dot{F}^2$.

Proof. First note that, by 8.1, I_t^2F is a nil ideal. Thus, $1 + I_t^2F$ is a subgroup of $\text{U}(W(F))$. This subgroup has trivial intersection with $\dot{F}/\dot{F}^2$, for, if $\langle a\rangle \in 1 + I_t^2F$, then $1 - \langle a\rangle \in I^2F$ implies that $\langle a\rangle = 1 \in W(F)$. It only remains to check that $\dot{F}/\dot{F}^2$ *and* $1 + I_t^2F$ *generate* $\text{U}(W(F))$. If $q \in \text{U}(W(F))$, then $\dim(q)$ must be odd, and we'll have $q_0 := q \perp \langle -a\rangle \in I^2F$ for some $a \in \dot{F}$. This gives $a \cdot q = 1 + q_1$, where $q_1 = a \cdot q_0 \in I^2F$. We are done if we can show that $q_1 \in W_t(F)$. We may assume that F is formally real (for otherwise $W(F) = W_t(F)$). Taking signatures with respect to any $\alpha \in X_F$, we have

$$\text{sgn}_\alpha(a \cdot q) = 1 + \text{sgn}_\alpha(q_1) \equiv 1 \pmod 4.$$

By 8.7(2), the LHS can only be ± 1, so it must be 1, which implies that $\text{sgn}_\alpha(q_1) = 0$. Now 3.2 implies that $q_1 \in W_t(F)$, as desired. □

Marshall and Lewis have also proved the following interesting result on the structure of the group $\mathrm{U}(W(F))$.

Theorem 8.9. $\mathrm{U}(W(F))$ *is a 2-primary torsion group.*

In the following, we shall try to give an elementary proof for this result using solely the fact that $W_t(F)$ is a 2-primary torsion group. This will be done with the help of the following ring-theoretic lemma, taken verbatim from Exercise 1.34 from the author's "Exercises in Classical Ring Theory" (2nd ed.), Springer-Verlag, 2003.

Lemma 8.10. *Let x be an element in any ring (with 1) such that $mx = 0 = x^{2^r}$, where $m \geq 1$ and $r \geq 0$ are given integers. Then $(1+x)^{m^r} = 1$.*

Proof. The proof is by induction on r. The case $r = 0$ being clear, we assume $r > 0$. Since $mx = 0$, the binomial theorem gives $(1+x)^m = 1+x^2y$, where y is a polynomial in x with integer coefficients. Since

$$m(x^2y) = 0 \quad \text{and} \quad (x^2y)^{2^{r-1}} = x^{2^r}y^{2^{r-1}} = 0,$$

the inductive hypothesis (applied to the element x^2y) implies that

$$1 = (1+x^2y)^{m^{r-1}} = [(1+x)^m]^{m^{r-1}} = (1+x)^{m^r}. \qquad \square$$

We now return to

Proof of 8.9. Let $q \in I_t^2F$. Since $W_t(F)$ is 2-primary torsion, we have $mq = 0$ for some $m = 2^k$. By 8.1, q is also nilpotent, so $q^{2^r} = 0$ for some r. Applying 8.10, we see that $(1+q)^{2^{kr}} = 1$. Thus, 8.9 follows from 8.8. $\square$

In the proof above, it can be shown that $(1+q)^{2^k}$ is already equal to 1, by using some more sophisticated techniques. We'll come back to prove this in a later chapter (see XI.3.9). Without knowing this for now, we can nevertheless obtain the following refinement of 8.9 in a special case.

Corollary 8.11. *If F is a field such that I^3F is torsionfree, then $\mathrm{U}(W(F))$ is a group of exponent ≤ 2.*

Proof. In view of 8.8, it suffices to show that $(1+q)^2 = 1$ for every $q \in I_t^2F$. Now

$$2q \in 2 \cdot I_t^2F \subseteq I^3F \cap W_t(F) = 0, \text{ and}$$
$$q^2 \in q \cdot I_t^2F \subseteq I^3F \cap W_t(F) = 0,$$

so indeed $(1+q)^2 = 1 + 2q + q^2 = 1$, as desired. $\square$

In particular, the conclusion of 8.11 applies to local fields and global fields. It behooves us to give an explicit example.

Example. Let $F=\mathbb{Q}$. By III.2.14, $\left(\frac{-2,5}{F}\right)$ is nonsplit, so its norm form

$$q=\langle 1,2,-5,-10\rangle=\langle 1,2\rangle\,\langle 1,-5\rangle$$

is anisotropic. Since 5 is a sum of two squares, $\langle 1,-5\rangle$ is killed by 2, so $q\in I_t^2F$. The anisotropic form $q_0=\langle 1,2,-5\rangle$ is therefore a unit in $W(F)$, since

$$q_0=\langle 10\rangle+q\in\langle 10\rangle\,(1+I_t^2F).$$

The reader can quickly check that $q_0^2=1\in W(F)$, as is predicted by 8.11.

While "nontrivial" units may exist in a Witt ring, the situation with the "reduced" Witt ring $W(F)/W_t(F)$ is much simpler. Indeed, since units of $W(F)/W_t(F)$ lift to units of $W(F)$, the following is an immediate consequence of 8.8.

Corollary 8.12. *For any field F, the units of $W(F)/W_t(F)$ are given by the images of the* 1*-dimensional forms.*

To complete our discussions on the structure of the Witt ring, let us also present some ideas of D. W. Lewis on the annihilating polynomials of quadratic forms.

In §7, we have observed that $W(F)$ is an integral extension of its subring $\mathbb{Z}\cdot\langle 1\rangle$. In [Le$_1$], Lewis constructed an *explicit* universal monic polynomial equation of degree $n+1$ over $\mathbb{Z}$ that is satisfied (in the Witt ring) by any (regular) n-dimensional quadratic form over any field. More explicitly, Lewis's result is as follows.

Theorem 8.13. *For any $n\geq 1$, there exists a universal monic polynomial $\sigma_n(x)\in\mathbb{Z}[x]$ of degree $n+1$ such that, for any field F and any n-ary (regular) quadratic form q over F, one has $\sigma_n(q)=0\in W(F)$. Moreover, such a polynomial is unique, and is given as follows:*

(1) *For n odd, $\sigma_n(x)=(x^2-1^2)(x^2-3^2)\cdots(x^2-n^2)$.*

(2) *For n even, $\sigma_n(x)=x(x^2-2^2)(x^2-4^2)\cdots(x^2-n^2)$.*

Proof. We first prove the *uniqueness* of $\sigma_n(x)$. Suppose $\sigma_n(x)$ has the stated properties, and consider the Witt ring $W(\mathbb{R})$, which we *identify* with $\mathbb{Z}$ (by the signature map). Under this identification, an n-ary (regular) form over $\mathbb{R}$ corresponds to one of the following $n+1$ integers:

$$(n=\text{odd}):\qquad \pm 1,\ \pm 3,\ \ldots,\ \pm n;$$

$$(n=\text{even}):\qquad 0,\ \pm 2,\ \ldots,\ \pm n.$$

Since $\sigma_n(x)$ vanishes on these, and is monic of degree $n+1$, $\sigma_n(x)$ must be precisely as given in (1) and (2) respectively!

For the *existence* part, we must show that the given $\sigma_n(x)$ has the desired properties. Of course, $\sigma_n(x)$ is monic of degree $n+1$. It is also useful to point out that $\sigma_n(x)$ is an even polynomial when n is odd, and an odd polynomial when n is even. The following quick argument for showing that

(8.14) $\qquad \sigma_n(q) = 0 \in W(F)$ for any n-ary form q over F

is due to K. H. Leung.

By factoring the polynomials $\sigma_n(x)$, we see easily that the following recursion formula holds:

(8.15) $\qquad \sigma_n(x) = \sigma_{n-1}(x-1)\cdot(x+n)$ (for any $n \geq 2$).

Now consider any n-ary form q over a field F, and any element $a \in \dot{F}$. Since $(\langle a\rangle\, q)^2 = q^2 \in W(F)$, to check $\sigma_n(q) = 0$ is the same as checking $\sigma_n(\langle a\rangle\, q) = 0$ (thanks to the fact that $\sigma_n(x)$ is either even or odd). Thus, after a scaling, we may assume that $q \cong \langle 1\rangle \perp q'$, where $\dim(q') = n-1$. Now 8.15 gives

$$\sigma_n(q) = \sigma_{n-1}(q')\cdot(q+n) \in W(F),$$

so 8.14 follows by induction on n. (The case $n=1$ is trivial, since every $\langle a\rangle$ is a root of $\sigma_1(x) = x^2 - 1$.) □

Remark. Of course, the formula 8.10 shows that the $\sigma_n(x)$'s could have been constructed recursively, starting with $\sigma_1(x) = x^2 - 1$. However, with such a recursive construction, it would not be immediately clear that $\sigma_n(x)$ is either odd or even (depending on whether n is even or odd).

In [Le$_2$], Lewis observed that his result 8.13 can be used to give quick proofs for some of the structural theorems on $W(F)$ obtained earlier in this section. These alternative proofs are so nice and so short that they bear repeating here. The only fact we shall use below is that the torsion subgroup $W_t(F)$ is 2-primary, which was proved before in 3.2.[13]

Proof of 8.5. Suppose $q \cdot q' = 0 \in W(F)$, where $\dim(q) = n$ is odd. Multiplying the equation

$$(q^2 - 1^2)(q^2 - 3^2)\cdots(q^2 - n^2) = 0 \in W(F)$$

by q', we get $(1^2\cdot 3^2 \cdots n^2)q' = 0$. Thus. $q' = 0 \in W(F)$, since $W_t(F)$ is a 2-primary torsion group. □

[13] Actually, even this can be proved by using 8.13. To keep matters simple, however, we won't do it here.

Proof of (1), (2) in 8.4. We need only show that $IF \subseteq \mathrm{zd}\,(W(F))$ in case F is *not* a pythagorean field. Say $a^2 + b^2 \notin F^2$, and let

$$f = \langle 1,\ -(a^2 + b^2)\rangle,$$

which satisfies $2f = 0$ in $W(F)$. For any n-ary form $q \in IF$, we have

$$q(q^2 - 2^2)(q^2 - 4^2)\cdots(q^2 - n^2) = 0 \in W(F).$$

Multiplying this by f, we get $f \cdot q^{n+1} = 0$ (since $2f = 0$). If $q \notin \mathrm{zd}\,(W(F))$, we would have $f = 0$, which is not the case. Therefore, we must have $q \in \mathrm{zd}\,(W(F))$. □

Finally, let us also reprove the following:

(8.16) *If* $q \in \mathrm{nil}\,(W(F))$, *then* $q \in W_t(F)$, *and the converse holds if* $\dim(q)$ *is even.*

To see this, let $n = \dim(q)$. If $q^r = 0 \in W(F)$ where $r \geq 1$, then n must be even, and 8.13(2) gives an equation

$$q^{n+1} + \cdots + k \cdot q = 0 \in W(F),$$

where k is a nonzero integer. Thus, $k \cdot q \in q^2 \cdot W(F)$. By induction, we have then $k^{r-1} \cdot q \in q^r \cdot W(F) = 0$, so $q \in W_t(F)$. Conversely, if $q \in W_t(F)$ and $n = \dim(q)$ is even (which would be automatic if F is formally real), then $\sigma_n(q) = 0$ leads to $q^{n+1} \in 2q \cdot W(F)$. If, say, $2^s q = 0$, then $(q^{n+1})^s = 2^s q^s = 0$, so $q \in \mathrm{nil}\,(W(F))$, completing the proof of 8.16. □

The idea of finding monic annihilating polynomials for quadratic forms seems to have some good potential. For instance, some work has been done toward finding such polynomials for quadratic forms that arise as trace forms. For a recent survey on this circle of ideas, see [Le$_4$].

9. An Introduction to Preorderings

In Section 1 of this chapter, we have presented the basic Artin-Schreier theory of formally real fields by using the notion of real-closed fields. In particular, Artin's characterization of totally positive elements in 1.12 was proved by using the fact that a formally real field has a real-closed algebraic extension. In this section, we shall present a different approach to these basic results that is independent of the use of real-closed fields. The central notion in this new approach is that of a *preordering*. In the modern theory of ordered fields, preorderings have played an increasingly important role. Thus, it seems particularly appropriate that we end this chapter with a brief coverage of the basic definitions and results in the theory of preorderings in a formally real field.

Definition 9.1. A *preordering* on a field F is a set $T \subseteq F$ such that

$$T + T \subseteq T, \quad T \cdot T \subseteq T, \quad F^2 \subseteq T, \quad \text{and} \quad -1 \notin T.$$

Note that, *in case* $\text{char}(F) \neq 2$, the last axiom $-1 \notin T$ can be replaced by the ostensibly weaker axiom $T \subsetneq F$ (in the presence of the other axioms). In fact, if $-1 \in T$, then any $a \in F$ can be written as $x^2 - y^2$ for some x, y, and hence $a = x^2 + (-1)y^2 \in T$. This would show that $T = F$. In case $\text{char}(F) = 2$, however, the axiom $T \subsetneq F$ would be too weak. In fact, if F is a nonperfect field of characteristic 2, then $T := F^2$ would satisfy $T + T \subseteq T$, $T \cdot T \subseteq T$, $F^2 \subseteq T$ and $T \subsetneq F$; but we have $-1 = 1 \in T$, so T fails to be a preordering in the sense of 7.1.

The following proposition is the analogue of 1.3 for fields F with a preordering T. The proof is analogous to that of 1.3, so it can safely be omitted.

Proposition 9.2. *Let (F, T) be a "preordered field", that is, a field given with a preordering T. Then:*

(1) $\sigma(F) \subseteq T$.

(2) $T \cap (-T) = \{0\}$.

(3) *F is formally real.*

(4) *$\dot{T} := T \setminus \{0\}$ is a subgroup of $\dot{F}$.*

The main difference between a preordering T and an ordering is that we may no longer have $F = T \cup (-T)$. In fact, T *will be an ordering iff* $[\dot{F} : \dot{T}] = 2$. In general, however, the index $[\dot{F} : \dot{T}]$ may very well be infinite.

Obviously, the intersection of any nonempty family of preorderings on F is itself a preordering. In particular, we can produce examples of preorderings by taking any (nonempty) family of orderings and forming their intersections. It turns out that all preorderings on a field actually arise in this way. To see this, we shall first prove the following preliminary result.

Extension Lemma 9.3. *Let T be a preordering on F, and let $a \in F \setminus T$. Then $T' := T - T \cdot a$ is a preordering on F containing both T and $-a$.*

Proof. Since $0, 1 \in T$, T' clearly contains T and $-a$. For two elements $x = t_1 - t_2 a$ and $y = t_3 - t_4 a$ in T' (where $t_i \in T$), we have $x + y = (t_1 + t_3) - (t_2 + t_4)a \in T'$ and

$$xy = (t_1 t_3 + t_2 t_4 a^2) - (t_1 t_4 + t_2 t_3)a \in T'.$$

Also, $F^2 \subseteq T \subseteq T'$, so it only remains to check that $-1 \notin T'$. If $-1 \in T'$, we would have $-1 = t_1 - t_2 a$ for some $t_i \in T$. If $t_2 \neq 0$, then $a = t_2^{-1}(1 + t_1) \in T$, contrary to our assumption. Therefore, $t_2 = 0$, and we get $-1 = t_1 \in T$, a contradiction. □

Corollary 9.4. *Let T be a preordering on F. Then*

(1) *T is an ordering iff T is a maximal preordering.*

(2) *$T \subseteq P$ for some ordering P.*

Proof. (1) If T is an ordering, then $[\dot{F} : \dot{T}] = 2$, so T is clearly a maximal preordering. Conversely, suppose T is a maximal preordering. For any $a \notin T$, $T' := T - T \cdot a$ is a preordering containing T by 9.3, so $T' = T$, and hence $-a \in T' = T$. This shows that T is an ordering.

(2) By an easy application of Zorn's Lemma, we see that T can be enlarged into a maximal preordering P, which is then an ordering by (1). $\square$

From the above Corollary, we can easily recapture the Artin-Schreier Criterion for formally real fields.

Theorem 9.5. *A field F is formally real iff $\sigma(F)$ is a preordering, iff F has at least one ordering.*

Proof. The first "iff" is clear: note that if F is formally real, $\sigma(F)$ is in fact the *smallest* preordering on F. (In the following, we shall call $\sigma(F)$ the *weak preordering* of F). In this case, 9.4(2) above implies that F has an ordering. Conversely, if F has an ordering, F is clearly formally real. $\square$

We also get the following description of a preordering, which may be viewed as a sharpening of Artin's Criterion 1.12 for the elements in $\sigma(F)$.

Theorem 9.6. *For any preordering T on a field F, we have $T = \bigcap_P P$, where P ranges over X/T, the space of all orderings of F containing T.*

Proof. We need only prove the inclusion "$\supseteq$". Consider any element $a \notin T$. By the Extension Lemma, $T' = T - T \cdot a$ is a preordering containing T and $-a$. Let $P \in X/T'$, which exists by 9.4(2). We have then $P \in X/T$, and $-a \in \dot{P}$ implies that $a \notin P$, as desired. $\square$

Next, we apply 9.4(2) to give a criterion for the existence of orderings satisfying certain prescribed conditions.

Theorem 9.7. *Let G be a submonoid of $\dot{F}$, and let $y \in \dot{F}$.*

(1) *F admits an ordering containing G iff $\langle a_1, \ldots, a_n \rangle$ is anisotropic for any finite collection of elements $a_1, \ldots, a_n \in G$.*

(2) *$y \in P$ for all orderings P on F containing G iff $y = a_1x_1^2 + \ldots + a_nx_n^2$ for some $a_i \in G$, $x_i \in F$, and $n \geq 1$.*

Proof. (1) The "only if" part is clear. For the "if" part, assume the anisotropicity condition in the theorem. The set T consisting of finite sums of the form $a_1x_1^2 + \ldots + a_nx_n^2$ (where $a_i \in G$, $x_i \in F$, and $n \geq 1$) is readily checked to be a preordering, since $G \cdot G \subseteq G$, and $1 \in G$. (Note that $-1 \in T$ would have violated the given anisotropicity condition.) Therefore, by 9.4(2), $T \subseteq P$ for some ordering P, which clearly contains G.

(2) Here, it suffices to prove the "only if" part. Suppose $y \in G$ for all orderings $P \supseteq G$. If there is no ordering containing G, then, by (1), $\langle a_1, \ldots, a_n \rangle$ is isotropic (and hence universal) for some $a_i \in G$. In this case, y certainly has the desired expression. Now assume there exist orderings of F containing G. Let T be the preordering constructed in the proof of (1) above. By 9.6, T is the intersection of all orderings of F containing T (or equivalently, containing G). By the assumption on y, we have thus $y \in T$, as desired. □

We have given earlier in 7.9 a criterion for the extendibility of an ordering P from one field to another. Using the theorem above (with $G = \dot{P}$), we get now a second criterion, as in (1) below, along with a bonus part (2).

Corollary 9.8. *Let (F, P) be an ordered field, and K/F be a field extension, with a given element $y \in \dot{K}$.*

(1) *P can be extended to an ordering on K iff, for any $a_1, \ldots, a_n \in \dot{P}$, the form $\langle a_1, \ldots, a_n \rangle$ is anisotropic over K.*

(2) *y is positive with respect to all orderings of K extending P iff $y = a_1 x_1^2 + \ldots + a_n x_n^2$ for some $a_i \in \dot{P}$, and $x_i \in K$.*

The idea of using preorderings to prove the basic results of the Artin-Schreier Theory (and to derive other existence results on orderings) is due to J.-P. Serre. He reported this approach to the theory of orderings in a short note [Se$_1$] in C.R. Acad. Sci. Paris in 1949. In my survey article [L$_2$] on ordered fields, I observed that this was apparently Serre's first paper. Upon receiving my survey paper, Serre wrote to me on June 4, 1980: "... this CR note is indeed the first paper I wrote. But I was not interested in "preorderings" at the time: I just wanted to catch the total orderings! It does not matter ... ". Nowadays, preorderings are as important as orderings, and are used extensively in the theory of ordered fields and rings, real algebra, and real algebraic geometry.

It turns out that, for any preordering T, we can define a Witt ring $W_T(F)$ relative to T, and also develop an associated theory of "quadratic forms over T." This theory (due to Becker and Köpping [BK]) shares many of the formal properties of the quadratic form theory developed so far in this book, but in many ways the new theory seems to be simpler and more manageable. And it turns out that the new Witt ring $W_T(F)$ is isomorphic to the factor ring of the ordinary Witt ring $W(F)$ modulo the ideal generated by the binary forms $\langle 1, -t\rangle$, where t ranges over $\dot{T}$. For a detailed introduction to this "reduced theory of quadratic forms" (modulo the preordering T), see §1 of my CBMS Lecture Notes [L3].

In the special case where T is the weak preordering $\sigma(F)$ of a formally real field F, it will be shown in Ch. XI that the ideal of $W(F)$ generated by $\langle 1, -t\rangle$ $(t \in \dot{T})$ is exactly $W_t(F)$, the torsion ideal of $W(F)$. Since $W_t(F)$ is precisely the kernel of the total signature map

$$\text{(9.9)} \qquad \text{sgn} : \ W(F) \longrightarrow C(X_F, \mathbb{Z})$$

(by Pfister's Local-Global Principle), it follows that

$$\text{(9.10)} \qquad W_T(F) \cong W(F)/W_t(F) \cong \text{im}(\text{sgn}),$$

which is a subring of the ring of continuous functions $C(X_F, \mathbb{Z})$. On the other hand, since (by (8.1(2)) $W_t(F) = \text{rad}\,(W(F)) = \text{nil}\,(W(F))$, we can also think of $W(F)/W_t(F)$ as the maximal Jacobson-semisimple factor ring or the maximal reduced factor ring of $W(F)$ (hence the name "reduced Witt ring" for the three rings listed in (9.10)).

For a general preordering T, the Witt ring $W_T(F)$ can likewise be embedded as a subring of $C(X/T, \mathbb{Z})$, the ring of $\mathbb{Z}$-valued continuous functions on X/T (the space of orderings P on F such that $P \supseteq T$). The remarkable thing here is, of course, that there is a ("reduced") theory of quadratic forms associated with the arbitrary preordering T, which is nontrivial even in the special case where $T = \sigma(F)$.

Exercises for Chapter VIII

1. Show that a subset $P \subseteq F$ gives an ordering on the field F iff it satisfies

$$P + P \subseteq P, \quad P \cdot P \subseteq P, \quad P \cup (-P) = F, \quad P \cap (-P) = \{0\}.$$

(The point is that we have replaced the axiom $P \subsetneq F$ by $P \cap (-P) = \{0\}$.)

2. If P_1, P_2, P_3 are three different orderings on F, show that $P_2 \cap P_3 \not\subseteq P_1$.
3. Show that a field F is euclidean iff $W(F)$ is an integral domain. (Cf. Ch. II, Exercise 16.)
4. (1) Exercise 15 in Chapter II stated that, if K and E are subfields of a field Ω such that K is quadratically closed and E is euclidean, then $F := K \cap E$ is euclidean. Is it true that $F(\sqrt{-1}) = K$?
(2) In (1) above, let $\Omega = \mathbb{C}$, $E = \mathbb{R}$, and $K = \widetilde{\mathbb{Q}}$ (the field of constructible numbers). Here, $F := K \cap E$ is the field of real constructible numbers. Show that F is euclidean, with $F(\sqrt{-1}) = K$.

5. (Knebusch) Let K/F be a finite extension. Suppose F admits an ordering P that cannot be extended to K. Show that, for any nonzero F-linear functional $s : K \to F$, and any K-quadratic space V, the transferred F-space $s_*(V)$ has zero signature relative to the ordering P. (**Hint.** Use Frobenius's Reciprocity Law VII.1.3, and the Cohen-Seidenberg Theorem.)

6. As an alternative approach to 7.10, deduce the following conclusions from Exercise 5:

   (A) If $[K : F]$ is odd, every $P \in X_F$ extends to an ordering on K.
   (B) (Ware) Suppose $[K : F] < \infty$ and that $P \in X_F$ does not extend to an ordering on K. Then $\mathrm{N}_{K/F}(\dot{K}) \subseteq \dot{P}$. (**Hint.** If $\mathrm{N}_{K/F}(a) \in -\dot{P}$, then $[K : F(a)]$ must be odd. Now apply Exercise 5, part (A) above, and VII.2.2.)
   (C) Let $K = F(\sqrt{a}\,)$ where $a \notin \dot{F}^2$. Then $P \in X_F$ extends to K iff $a \in P$.
   (D) If $a \in P$ in (C), there are exactly two orderings on K extending P. Construct these two extensions of P explicitly.

7. Let K/F be a Galois extension of degree n, and $\langle K \rangle$ be the trace form on K. Show that, for any ordering P on F, $\mathrm{sgn}_P(\langle K \rangle)$ is either 0 or n. (**Hint.** If P extends to K, use VII.6.12. Otherwise, use Exercise 5.)

8. Let $E \subseteq F$ be fields such that the natural map $\dot{E}/\dot{E}^2 \to \dot{F}/\dot{F}^2$ is onto. Show that the map $X_F \to X_E$ defined by restriction of orderings is one-one.

9. Let K/F be a field extension, and let E be the algebraic closure of F in K. If K is real-closed (resp. euclidean, pythagorean), show that so is E.

10. For any algebraically closed field K of characteristic 0, show that there exists a real-closed field $F \subseteq K$ such that $K = F(\sqrt{-1}\,)$.

11. Show that the number of orderings on F is finite iff $\dot{\sigma}(F)$ has finite index in $\dot{F}$. (For more quantitative results, see Exercise 16.)

12. If F is a formally real pythagorean field, show that a form over F is universal iff it is isotropic.

13. Show that the following conditions are each equivalent to F being pythagorean.

    (1) If a, b are different square classes in $\dot{F}/\dot{F}^2$, then F has an ordering at which a, b have different signs.
    (2) If c is a square class different from -1, then there exists an ordering for F at which c is positive.
    (3) Given any form $\varphi = \langle 1, a, b, ab \rangle$ where a, b represent distinct

square classes, there exists a real-closure F_α (for some ordering α) over which φ is hyperbolic.

(4) For any quaternion algebra $A = \left(\frac{a,\,b}{F}\right)$, where a, b represent distinct square classes, there exists a real-closure F_α over which A splits.

14. Let F be a pythagorean field, and let q_1, q_2 be forms of dimension n. If $n \leq 2$, show that $q_1 \cong q_2$ iff $D(q_1) = D(q_2)$. If $n = 3$, show that $q_1 \cong q_2$ iff $d(q_1) = d(q_2)$ and $D(q_1) = D(q_2)$. If $n \leq 3$, show that q_1 is isotropic over F iff q_1 is isotropic over all real-closures F_α.

15. For any ordering $\alpha \in X_F$, let χ_α denote the group homomorphism $\dot{F}/\dot{\sigma}(F) \to \{\pm 1\}$ induced by $\chi_\alpha(a) = 1$ or -1 according as a is positive or negative at α. Let $\alpha_1, \ldots, \alpha_m$ be a set of m different orderings of F, and let χ_i $(1 \leq i \leq m)$ denote the corresponding homomorphisms induced on $\dot{F}/\dot{\sigma}(F)$. Show that $\chi_1, \ldots, \chi_m$ are $\mathbb{F}_2$-linearly independent in $\mathrm{Hom}(\dot{F}/\dot{\sigma}(F),\ \{\pm 1\})$ iff, for each i, there exists $a_i \in \dot{F}$ that is negative at α_i and positive at every α_j, $j \neq i$. In this case, show that $a_1, \ldots, a_m$ are $\mathbb{F}_2$-linearly independent in $\dot{F}/\dot{\sigma}(F)$.

16. Suppose $|\dot{F}/\dot{\sigma}(F)| = 2^n$ and that F has exactly r (≥ 1) orderings $\alpha_1, \ldots, \alpha_r$. Let $\chi_1, \ldots, \chi_r$ be the functionals (in $\mathrm{Hom}\,(\dot{F}/\dot{\sigma}(F),\ \{\pm 1\})$ associated with $\alpha_1, \ldots, \alpha_r$, as in the last exercise.

    (1) Show that $\chi_1, \ldots, \chi_r$ span the dual space $\mathrm{Hom}(\dot{F}/\dot{\sigma}(F),\ \{\pm 1\})$.
    (2) Deduce from (1) that $n \leq r \leq 2^{n-1}$.
    (3) Show that $r = n$ iff, for each i $(1 \leq i \leq r)$, there exists $a \in \dot{F}$ that is negative at α_i and positive at every other α_j.
    (4) Show that $r = 2^{n-1}$ iff, for every subgroup $G \subseteq \dot{F}$ containing $\dot{\sigma}(F)$ with $[\dot{F} : G] = 2$ and $-1 \notin G$, $\{0\} \cup G$ is an ordering on F.
    (5) For any given integer $n > 3$, construct a formally real pythagorean field F with $|\dot{F}/\dot{F}^2| = 2^n$ such that the number of orderings on F is strictly between n and 2^{n-1}.

17. Show that Artin's Theorem 1.12 about elements in $\sigma(F)$ can be formally deduced from Pfister's Local-Global Principle 3.2.

18. For a formally real field F, show that $\mathrm{rad}\,(W(F))$ is a prime ideal iff F has a unique ordering.

19. Using Krull's Intersection Theorem in commutative algebra (and II.2.4), show that if $\dot{F}/\dot{F}^2$ is finite, then $\bigcap_n I^n F = 0$. (Compare X.5.2.)

20. Let $(F_0, P_0) \subseteq (F, P)$ be an extension of ordered fields (that is, with $P \cap F_0 = P_0$). If F/F_0 is algebraic, show that (F, P) is *archimedean over* (F_0, P_0) in the sense that, for any $a \in F$, there exists $a_0 \in F_0$ such that $a <_P a_0$. (**Hint.** If $b \in F$ satisfies a monic equation $b^n + a_1 b^{n-1} + \cdots + a_n = 0$ over F_0, show that $|b| \leq 1 + \sum_i |a_i|$, where the "absolute values" are formed with respect to P_0 in the usual way.)

21. Let $(F_0, P_0) \subseteq (F, P)$ be a finite extension of ordered fields. *True or False*: For any $a <_P b$ in F, the open interval
$$(a,\, b) = \{x \in F:\ 0 <_P x <_P b\}$$
in F contains an element of F_0? (**Hint.** Consider the fields
$$F_0 = \mathbb{Q}\,(t^2) \subseteq F = \mathbb{Q}\,(t),$$
with P an ordering on F such that $\mathbb{Q} <_P t$, and look at the interval $(t,\, 2t) \subseteq F$.)

22. Show that any field with an archimedean ordering is order-isomorphic to a subfield of $\mathbb{R}$.

23. Let $K = \mathbb{Q}\,(\alpha)$, where $\alpha^4 = \alpha + 1$. Can K be ordered, and if so, in how many ways? (You should solve this problem by both computing the number of real roots of $x^4 - x - 1$, and computing the signature of the trace form of K via a diagonalization over $\mathbb{Q}$.)

24. Let K be the intersection of all the real-closures of $\mathbb{Q}$ (with respect to its unique ordering). Show that K consists of all algebraic numbers whose minimal polynomials over $\mathbb{Q}$ have only real roots.

25. In Remark 4.8, it was pointed out that, if P is an ordering on a field K and if σ is an algebraic automorphism of K, then $\sigma(P) = P$ implies that $\sigma = \text{Id}$. Give an example to show that this implication need not hold if σ is not assumed to be algebraic. (**Hint.** Take $K = \mathbb{Q}\,(x)$ and let P be the ordering on K in which $x > \mathbb{Q}$. Consider the $\mathbb{Q}$-automorphism σ on K defined by $\sigma(x) = x + 1$.)

26. (Bourbaki Exercise) Let k be a real-closed field, with unique ordering $P_0 = \dot{k}^2$. Show that there is a one-one correspondence between the orderings on $F = k(x)$ and subsets $C \subseteq k$ with the property that, whenever $a \leq b$ in k, $b \in C \Rightarrow a \in C$. (Note that the two valid choices, $C = \varnothing$ and $C = k$, are certainly allowed.) Here, an ordering P on F corresponds to the set $C = \{b \in k\,:\, b <_P x\}$.

27. In 1.13(D), we have described how, using the above exercise, we can determine all the orderings on $\mathbb{R}(x)$. Now determine all the orderings on $k(x)$ for k the field of all real algebraic numbers. (**Hint.** Let $a \in \mathbb{R}$. If a is an algebraic number, show that the orderings $P_{(-\infty,a]}$ and $P_{(-\infty,a)}$ restrict to different orderings on $k(x)$. If a is a transcendental number, show that $P_{(-\infty,a]}$ and $P_{(-\infty,a)}$ both restrict to the ordering on $k(x)$ obtained by identifying $k(x)$ as $k(a) \subseteq \mathbb{R}$ and restricting the order of $\mathbb{R}$ to $k(a)$.)

28. Let k be a field with an ordering P, and let K be a real-closure of (k, P). Show that there is a one-one correspondence (by restriction) between the orderings of $K(x)$ and the orderings of $k(x)$ extending P.

(Note that this result, together with Exercise 26, leads to a description of *all* orderings on $k(x)$ for any formally real field k.)

29. It follows from Exercise 20 that an algebraic extension of $\mathbb{Q}$ has only archimedean orderings. In contrast to this, show that the field

$$\mathbb{Q}(x)\left(\{\sqrt{x-n}:\ n=1,2,3,\dots\}\right)$$

(of transcendence degree 1 over $\mathbb{Q}$) has uncountably many orderings, all of which are nonarchimedean.

30. Let F be a field such that I^2F is torsionfree. For $q \in W(F)$, show that $q^2 = 2q$ in $W(F)$ iff $q = \langle 1,\, a\rangle \in W(F)$ for some $a \in \dot{F}$.

31. (Cf. Exercise 10 in Ch. V.) Let $a \in \dot{F}$, where F is a field such that I^3F is torsionfree, and let f, g be forms of the same dimension. If a is a sum of squares or $\operatorname{sgn}(f) = \operatorname{sgn}(g)$, show that $f \cdot \langle 1,\, a\rangle \cong g \cdot \langle 1,\, a\rangle$ iff $\langle 1,\, a\rangle$ represents $d(f)\,d(g)$.

32. Let $q_1, \dots, q_n$ be forms over a formally real pythagorean field F. If $q_1^2 + \cdots + q_n^2 = 1 \in W(F)$, show that, after a re-indexing, $q_2 = \cdots = q_n = 0$, and $q_1 = \langle a\rangle$ for some $a \in \dot{F}$.

33. Devise a proof to show that $W_t(F)$ being 2-primary torsion is equivalent to the statement that odd-dimensional forms are not 0-divisors in $W(F)$. (**Hint.** If $m \cdot q = 0$, factor m in the form $2^k r$ where r is odd. If odd-dimensional forms are not 0-divisors, then $2^k\, q = 0$. Conversely, assume $W_t(F)$ is 2-primary torsion, and let $q \cdot q' = 0$, where $q = \langle 1, a_1, \dots, a_{2r}\rangle$. Multiply $q \cdot q' = 0$ by $\langle 1, -a_i\rangle$ to get $\langle 1, -a_i\rangle \cdot q' = 0$ by induction. Then $0 = q \cdot q' = (2r+1) \cdot q'$.)

34. Assuming that odd-dimensional forms are not 0-divisors in $W(F)$, give a direct proof for the fact that $q^{2n+1} = 1 \Rightarrow q = 1$ in $W(F)$. (**Hint.** Note that $q^{2n} + \cdots + q + 1 \notin IF$.) The result in this exercise is *weaker* than that in 8.9: why?

35. In the notations of §8, calculate the objects (A), (B), (C), (D), (E) for the Witt-Grothendieck ring $\widehat{W}(F)$, instead of the Witt ring.

36. Let K/F be a finite extension, where K is euclidean. Show that F is euclidean (Cor. 5.12) without using the Diller-Dress Theorem. (**Hint.** Let $P_0 = P \cap F$, where P is the unique ordering on K. It suffices to show that $a \in P_0 \Longrightarrow a \in F^2$. For this, invoke an inductive hypothesis, and use Ch. VII, Exer. 5.)

37. (Whaples) For any field F, show that the following statements are equivalent:

   (1) F is euclidean.
   (2) $F \neq F^2$, and F has no Galois extension of degree 4.

(2) $F \neq F^2$, and F has no Galois extension of degree 2^n for any $n \geq 2$.

38. (1) Let $q \in W(F)$. If $q^n - 1 \in \mathrm{U}(W(F))$ for some $n \geq 2$, show that $q \in W_t(F)$.
(2) For any ring R, the additive group

$$\mathrm{rad}_0\,(R) := \{r \in R \mid r + \mathrm{U}(R) \subseteq \mathrm{U}(R)\}$$

contains (but is in general not equal to) $\mathrm{rad}\,(R)$, the Jacobson radical of R. If $R = W(F)$, use (1) (in the case $n = 2$) to show that $\mathrm{rad}_0\,(W(F)) = \mathrm{rad}\,(W(F))$.

39. Prove *Weierstrass' Nullstellensatz*: if f is a polynomial over a real-closed field F such that $f(a) < 0$ and $f(b) > 0$, then there exists between a and b at least one element $c \in F$ such that $f(c) = 0$.

Chapter IX

Quadratic Forms under Transcendental Extensions

1. Cassels-Pfister Theorem

In his famous paper [Ar] on the solution of Hilbert's 17th Problem (ca. 1927), Artin also proved the following remarkable fact: *if* $p(x) \in F[x]$ *is a sum of squares in the rational function field* $F(x)$ *(where* F *is a field), then* $p(x)$ *is already a sum of squares in* $F[x]$. Artin's result, however, was not of a quantitative nature, so in the end, there was no control on the number of polynomial squares needed to express $p(x)$. In 1964, J. W. S. Cassels [Ca$_1$] obtained the sharp quantitative version, by proving that, if $p(x)$ is a sum of n squares in $F(x)$, then it is a sum of n squares in $F[x]$. Shortly after Cassels' proof appeared, A. Pfister extended this result further to the case of *arbitrary* quadratic forms over F (rather than just sums of squares). Pfister's generalization (now called the Cassels-Pfister theorem) will occupy the entirety of this section.

Using his quantitative improvement of Artin's Theorem, Cassels went on to show that *the polynomial* $x_0^2 + x_1^2 + \cdots + x_n^2$ *is not a sum of* n *squares in the field* $\mathbb{R}(x_0, \ldots, x_n)$. This also generalizes in a suitable way to quadratic forms over formally real fields. These results marked the beginning of the study of quadratic forms under transcendental extensions, and have led to very fruitful results in quadratic form theory. The application of such transcendental methods will ultimately culminate in the study of the function fields of quadratic forms in the next chapter.

Recall (from VII.1.7) that, for any field extension K/F, $W(K/F)$ denotes the kernel of the functorial Witt ring map $W(F) \to W(K)$. To begin our study of rational function fields, we first make the following simple observation.

Lemma 1.1. *Let γ be a quadratic form over a field F. If γ is anisotropic over F, then γ remains anisotropic over the rational function field $F(x)$. In particular, the Witt kernel $W(F(x)/F)$ is the zero ideal in $W(F)$.*

Proof. We may assume that γ is a diagonal form $\langle a_1, \ldots, a_n \rangle$, where $a_i \in \dot{F}$. Assume that γ is isotropic over $F(x)$. After clearing denominators, we obtain an equation $\sum a_i f_i(x)^2 = 0$, where $f_i(x) \in F[x]$ are not all zero. Changing the f_i's if necessary, we may further assume that x does not divide all of the polynomials $f_i(x)$. Setting $x = 0$, we get $\sum_i a_i f_i(0)^2 = 0$, where the $f_i(0) \in F$ are not all zero. This says that γ is isotropic over F. The last part of the theorem now follows immediately. □

For a quadratic form φ over F, recall that $D_F(\varphi)$ denotes the set of nonzero values of F represented by φ. If K is an extension field of F, we usually write $D_K(\varphi)$ for the value set $D_K(K \otimes_F \varphi)$.

Corollary 1.2. (1) *For a quadratic form φ over F,*

$$D_F(\varphi) = \dot{F} \cap D_{F(x)}(\varphi).$$

(2) *-1 is a sum of n squares in F iff -1 is a sum of n squares in $F(x)$.*

Proof. For (1), we apply 1.1 to the form $\gamma = \varphi \perp \langle -d \rangle$, where $d \in \dot{F} \cap D_{F(x)}(\varphi)$. (2) follows from (1) by letting $\varphi = n\langle 1 \rangle$. □

Note that (2) above implies, in particular, that $F(x)$ *is formally real iff F is formally real.* We have already given a proof for this by using the extendibility of orderings from F to $F(x)$: see the last paragraph of VIII.1.13(C).

We come now to the main result in this section.

Cassels-Pfister Theorem 1.3. *Let γ be a (regular) quadratic form over F, and let $p(x) \in F[x] \cap D_{F(x)}(\gamma)$. Then,*

(1) *$p(x)$ is already represented by γ over $F[x]$, and*

(2) *if $e \in F$ is such that $p(e) \neq 0$, then $p(e) \in D_F(\gamma)$.*

Proof. We need only prove (1), since (1) $\Rightarrow$ (2) by substitution.

We may assume that γ is a diagonal form $\langle a_1, \ldots, a_n \rangle$ $(a_i \in \dot{F})$. Also, we may assume that γ is anisotropic over F. (For, if otherwise, γ contains a subform $\langle 1, -1\rangle$, and from the identity

$$p(x) = [(p(x)+1)/2]^2 - [(p(x)-1)/2]^2 \in F[x],$$

the desired conclusion (1) follows.)

By the hypothesis, we have an equation

$$p(x) = a_1(f_1(x)/f_0(x))^2 + \cdots + a_n(f_n(x)/f_0(x))^2, \tag{1.4}$$

where $f_0, \ldots, f_n \in F[x]$, with $f_0 \neq 0$. We may assume that this equation is chosen so that $\deg f_0$ is as small as possible. *We claim that* $\deg f_0$ *must be zero.* If so, we will have obtained a representation of $p(x)$ by γ over $F[x]$, as desired.

To establish our claim, assume that $\deg f_0 > 0$. Over the field $E = F(x)$, consider the diagonal form $\langle -p(x), a_1, \ldots, a_n\rangle$ and its associated symmetric bilinear form B. The set Q of isotropic vectors in (E^{n+1}, B) is a quadric surface in the n-dimensional projective space $\mathbb{P}^n E$; the equation 1.4 says precisely that $f = (f_0, \ldots, f_n)$ lies on this quadric surface Q.

We shall derive a contradiction by producing a vector $h = (h_0, \ldots, h_n) \in Q$, where $h_i \in F[x]$ and $h_0 \neq 0$ with $\deg h_0 < \deg f_0$. To construct h, we first divide all the f_i $(0 \le i \le n)$ by f_0:

$$f_i = f_0 g_i + r_i, \quad \text{where } r_i = 0 \text{ or } \deg r_i < \deg f_0.$$

Here, of course, $g_0 = 1$, $r_0 = 0$, and $g_i, r_i \in F[x]$. We have the following vectors in E^{n+1}:

$$f = (f_0, \ldots, f_n), \quad g = (g_0, \ldots, g_n), \quad \text{and} \quad r = (r_0, \ldots, r_n),$$

related by a single vector equation $f = f_0 g + r$. We may assume that $r \neq 0$ (for, if otherwise, f_0 divides all f_i, and we would have a new solution for 1.4 with $f_0 = 1$).

To construct h, we simply join $f \in Q$ to g by a line, and take h to be the second intersection of this line with the quadric surface Q. To proceed more formally, we set

$$h = \alpha f + \beta g = (h_0, \ldots, h_n) \qquad (\alpha, \beta \in F[x]),$$

subject to the condition

$$0 = B(h, h) = \beta\,[\,2\alpha B(f, g) + \beta B(g, g)\,].$$

This is (obviously) satisfied by $\alpha := B(g, g)$ and $\beta = -2B(f, g)$. With these choices, we then have

$$h_0 = B(g, g) f_0 - 2B(f, g) = B(f_0 g - 2f,\ g) = -B(f + r,\ g),$$

and hence

$$f_0 h_0 = -B(f + r,\ f - r) = B(r,\ r) = \sum_{i=1}^{n} a_i r_i(x)^2 \neq 0,$$

since $\gamma = \langle a_1, \ldots, a_n \rangle$ is anisotropic over F. From this, we have clearly $h_0 \neq 0$ and $\deg h_0 < \deg f_0$, as desired. □

Remark 1.5. It is worth noting that the case where $\dim \gamma = 1$ in the above theorem can be proved quickly and independently, as follows. Suppose $p(x) = a_1(f_1(x)/f_0(x))^2$. Then,

$$q(x) := f_1(x)/f_0(x) \in F(x)$$

is integral over $F[x]$. Since $F[x]$ is integrally closed, we must have $q(x) \in F[x]$, and $p(x) = a_1 q(x)^2$ gives the desired conclusion. Because of this, we might think of the more general result in 1.3 as a kind of "integrality" theorem.

The special case of 1.3 discussed in the remark above holds, of course, even for higher powers (in place of squares), and also for the case of many variables, since the argument depends only on the normality property of polynomial algebras over a field. However, the general statement of the Cassels-Pfister theorem fails to hold for either case. Later in XIII.5, we shall give examples of multivariate polynomials (over the real numbers) that are sums of squares of rational functions, but are *not* sums of squares of real polynomials. On the other hand, it is easy to see that the quartic polynomial

$$p(x) = 1 + (1 - x^2)^2 \in \mathbb{R}[x]$$

is *not* a sum of 4-th powers in $\mathbb{R}[x]$, but E. Becker has shown that $p(x)$ is a sum of 4-th powers in the rational function field $\mathbb{R}(x)$.

The counterexamples alluded to above point to the relatively subtle nature of 1.3. Nevertheless, the second conclusion in 1.3 can be easily extended to the case of more than one variable, as follows.

Substitution Principle 1.6. *Let γ be a quadratic form over F, and let $X = (x_1, \ldots, x_s)$ be a set of (commuting) independent indeterminates over F. Let $p(X) \in F(X)$ and $e = (e_1, \ldots, e_s) \in F^s$ be such that $p(e)$ is defined and not equal to zero. Then*

$$p(X) \in D_{F(X)}(\gamma) \implies p(e) \in D_F(\gamma).$$

Proof. The assumption on e means that we can write $p(X)$ in the form $f(X)/g(X)$, where $f, g \in F[X]$ with $g(e) \neq 0$ and $f(e) \neq 0$. By hypothesis, γ represents $p(X)$ and hence also $f(X)g(X)$ over $F(X)$. If we can show that γ represents $f(e)g(e)$ over F, then γ will also represent $f(e)/g(e) = p(e)$ over F. We may thus assume that $p(X)$ is a *polynomial* in $x_1, \ldots, x_s$, and proceed by induction on s. The case $s = 1$ is just 1.3(2). In general, for $s > 1$, view γ as a form over $F' = F(x_1, \ldots, x_{s-1})$. Since γ represents $p(x_1, \ldots, x_s)$ over $F'(x_s)$, γ represents $p(x_1, \ldots, x_{s-1}, e_s)$ over

F', by 1.3(2). By the inductive hypothesis, it follows immediately that γ represents $p(e_1, \dots, e_{s-1}, e_s) = p(e)$ over F. □

2. Second and Third Representation Theorems

The First Representation Theorem in I.3.5 states that a quadratic form γ represents a value $d \in \dot{F}$ iff the form $\gamma \perp \langle -d \rangle$ is isotropic. In this section, we shall obtain two more representation theorems. Both of these involve the notion of rational function fields, and will be proved by means of the Cassels-Pfister Theorem (and its consequence 1.6). The reader should first look at Corollaries 2.3, 2.4 and 2.9, 2.10 below for motivation.

Second Representation Theorem 2.1. *Let $\gamma = \langle a_1, \dots, a_n \rangle$ be an anisotropic form over F, where $n \geq 1$. Let $\varphi = \langle a_2, \dots, a_n \rangle$, and $d \in \dot{F}$. Then, for a single indeterminate x,*

$$d \in D_F(\varphi) \iff a_1x^2 + d \in D_{F(x)}(\gamma).$$

Remarks. (1) If γ was isotropic over F, then $D_{F(x)}(\gamma) = F(x) \setminus \{0\}$. But $D_F(\varphi)$ need not be $\dot{F}$. Thus, the result above is definitely false without the hypothesis that γ be *anisotropic* over F.

(2) In the case $n = 1$, φ is the 0-dimensional form. In this case, the theorem simply says that a 1-dimensional form $\langle a_1 \rangle$ can never represent $a_1x^2 + d$ over $F(x)$ if $d \neq 0$. This case is, in fact, covered by the proof below, and is *not* an exception.

Proof of 2.1. Assume first that $d \in D_F(\varphi)$. View $\gamma = \langle a_1 \rangle \perp \varphi$ as an orthogonal decomposition over $F(x)$. The first summand represents a_1x^2 over $F(x)$, and the second summand represents d (already over F). Thus, γ represents $a_1x^2 + d$. This is the "easy" direction.

Conversely, assume that $a_1x^2 + d \in D_{F(x)}(\gamma)$. By the Cassels-Pfister Theorem, there exists an equation

$$a_1x^2 + d = a_1f_1(x)^2 + \cdots + a_nf_n(x)^2, \quad \text{where } f_i \in F[x]. \tag{2.2}$$

Now, γ is anisotropic over F, and the LHS above is of degree 2. By considering the leading coefficients of the f_i's, we see that $\deg f_i \leq 1$ for all i. Write $f_1(x) = a + bx$, where $a, b \in F$. Let $c \in F$ be a solution for one of the equations $a + bx = \pm x$ (such a c certainly exists!). Plugging c into the equation (2.2), we get

$$a_1c^2 + d = a_1(\pm c)^2 + a_2f_2(c)^2 + \cdots + a_nf_n(c)^2.$$

Cancelling the term a_1c^2, we conclude that $d \in D_F(\varphi)$. □

Remark. In spite of its apparently elementary nature, the Second Representation Theorem does not admit an easy proof. The argument given above relied heavily on the Cassels-Pfister Theorem. Besides this, there is now a second proof, pointed out by David Leep in [Lp2]. However, this proof makes use of another theorem, due to Amer and Brumer, on simultaneous zeros of pairs of quadratic forms, which is also a highly nontrivial result. Thus, until an easier proof is found, 2.1 remains a subtle theorem.

Applying 2.1 to the form $\gamma = n\langle 1\rangle$, we deduce the following result, which is actually a main motivation for 2.1.

Corollary 2.3. *Let $n \geq 1$, and let F be a field in which -1 is not a sum of $n-1$ squares (e.g., F can be a formally real field). If $d \in \dot{F}$ and x^2+d is a sum of n squares in $F(x)$, then d is a sum of $n-1$ squares in F.*

By induction on n, we further obtain

Corollary 2.4. *For F as in 2.3, $1 + x_1^2 + \cdots + x_n^2$ cannot be a sum of n squares in $F(x_1, \ldots, x_n)$. Similarly, $x_0^2 + x_1^2 + \cdots + x_n^2$ cannot be a sum of n squares in $F(x_0, x_1, \ldots, x_n)$.*

This corollary is a good instance of the kind of results in mathematics that are easy to believe, but not so easy to prove! The role of this corollary in quadratic form theory can perhaps be compared to that of the Invariance of Dimension theorem (that $\mathbb{R}^{n+1}$ cannot be embedded in $\mathbb{R}^n$) in topology.

To formulate the Third Representation Theorem, it is convenient to introduce first a formal definition.

Definition 2.5. Let φ and γ be (nonsingular) quadratic forms over F, of dimensions n and m. We say that γ *dominates* φ if, for indeterminates $X = (x_1, \ldots, x_n)$ over F, we have $\varphi(X) \in D_{F(X)}(\gamma)$. (This means that there exist rational functions $f_i(X) \in F(X)$ $(1 \leq i \leq m)$ such that $\gamma(f_1(X), \ldots, f_m(X)) = \varphi(X)$.)[(1)]

Example 2.6. If γ is isotropic over F, then $D_{F(X)}(\gamma) = F(X) \setminus \{0\}$, and so γ dominates *any* form φ. Thus, the notion of dominance is significant only in the case where γ is anisotropic.

Example 2.7. What is meant by saying that γ dominates a 1-dimensional form $\langle a\rangle$ $(a \in \dot{F})$? By definition, this means that $ax^2 \in D_{F(x)}(\gamma)$, or equivalently, $a \in D_{F(x)}(\gamma)$. By 1.2, this boils down to $a \in D_F(\gamma)$, or that γ contains a subform isometric to $\langle a\rangle$ (over F).

[(1)]Instead of "γ dominates φ", the expression "γ represents φ" is sometimes also used in the literature.

In general, if the form φ is isometric to a subform of the form γ (over F), let us write $\varphi \subseteq \gamma$. If this is the case, then $\varphi(X)$ is certainly a value of γ over $F(X)$, and so γ dominates φ. The example above shows that the converse is also true in the case where dim $\varphi = 1$. It turns out that the converse is true in general, as long as γ is anisotropic. This is a part of the Third Representation Theorem below. Inasmuch as this theorem gives two criteria for $\varphi \subseteq \gamma$, it is often called the *Subform Theorem* in the literature.

Third Representation Theorem 2.8. *For any form φ and any anisotropic form γ over F, the following are equivalent:*

(1) *γ dominates φ.*

(2) *$\varphi \subseteq \gamma$ (over F).*

(3) *$D_K(\varphi) \subseteq D_k(\gamma)$ for any field $K \supseteq F$.*

In particular, γ dominates $\varphi \Rightarrow$ dim $\gamma \geq$ dim φ.

Proof. (2) $\Rightarrow$ (3) is clear, and (3) $\Rightarrow$ (1) follows by applying (3) to $K = F(X)$, where $X = (x_1, \ldots, x_n)$ $(n = \dim \varphi)$.

For (1) $\Rightarrow$ (2), we induct on $m =$ dim γ. For $m = 0$, there is nothing to prove. (It may be more proper to start the induction from $m = 1$. But the argument needed for $m = 1$ is truly subsumed in the inductive step below.) Assuming the implication (1) $\Rightarrow$ (2) for $m - 1$, write $\varphi = \langle b_1, \ldots, b_n \rangle$. By hypothesis,

$$b_1 x_1^2 + \cdots + b_n x_n^2 \in D_{F(X)}(\gamma).$$

By the Substitution Principle 1.6 (applied to $x_1 \mapsto 1$ and $x_i \mapsto 0$ for $i \geq 2$), we get $b_1 \in D_F(\gamma)$, so we can write $\gamma \cong \langle b_1 \rangle \perp \gamma'$, where γ' is a (necessarily anisotropic) F-form of dimension $m - 1$. View γ' as a form over $F' = F(x_2, \ldots, x_n)$ (still anisotropic, by 1.1), and set $d = b_2 x_2^2 + \cdots + b_n x_n^2 \in F'$. By the Second Representation Theorem,

$$b_1 x_1^2 + d \in D_{F(X)}(\gamma) = D_{F'(x_1)}(\langle b \rangle \perp \gamma') \implies d \in D_{F'}(\gamma').$$

This says that γ' dominates $\langle b_2, \ldots, b_n \rangle$. By the inductive hypothesis, we have $\langle b_2, \ldots, b_n \rangle \subseteq \gamma'$, and so $\varphi \cong \langle b_1, b_2, \ldots, b_n \rangle \subseteq \gamma$. □

From 2.8, we get the following slight generalization of 2.4.

Corollary 2.9. *Let φ be an n-dimensional anisotropic form over F, and let $b_0, b_1, \ldots, b_n \in \dot{F}$. Then*

(1) $b_0 x_0^2 + \cdots + b_n x_n^2 \notin D_{F(x_0, \ldots, x_n)}(\varphi)$,

(2) $b_0 + b_1 x_1^2 + \cdots + b_n x_n^2 \notin D_{F(x_1, \ldots, x_n)}(\varphi)$.

Proof. (1) follows since dim $\varphi = n <$ dim $\langle b_0, \ldots, b_n \rangle$. (2) follows from (1) by using the substitution $x_i \mapsto x_i / x_0$ for $i \geq 1$. □

Corollary 2.10. *Let γ be an anisotropic form over F (with* dim $\gamma >$ *0), and φ a form over F of dimension n. Let $X = (x_1, \ldots, x_n)$. If $\varphi(X) \in F(X)$ is a similarity factor for $\gamma_{F(X)}$ (i.e., $\varphi(X) \cdot \gamma \cong \gamma$ over $F(X)$), then for any $a \in D_F(\gamma)$, we have $a \cdot \varphi \subseteq \gamma$ over F. (In particular,* dim $\gamma \geq$ dim φ.)

Proof. Since γ represents a over F, it follows that $\varphi(X) \cdot \gamma \cong \gamma$ implies that γ represents $a \cdot \varphi(X)$ over $F(X)$. Thus, by 2.8, $a \cdot \varphi \subseteq \gamma$ over F. $\square$

3. Milnor's Exact Sequence for $W(F(x))$

Throughout this section, F denotes a fixed field, and $E = F(x)$, the rational function field over F in the indeterminate x. Milnor's exact sequence computes the Witt group of E in terms of the Witt group of F and of the simple algebraic extensions of F. The construction of the exact sequence and the procedure used to establish its exactness are directly inspired by the computation of $W(\mathbb{Q})$ presented in VI.4.

In VI.4, the set of rational primes $\{p\}$ played a key role. These are to be replaced, in this section, by $\{\pi\}$, the set of monic irreducible polynomials in $F[x]$. For each such π, let E_π denote the π-adic completion of E. The residue class field of this complete field is naturally isomorphic to $F[x]/(\pi(x))$, a finite (in fact, simple) extension of F. Following the ideas used in VI.4, we define ∂_π to be the composition

$$W(E) \longrightarrow W(E_\pi) \longrightarrow W(\overline{E}_\pi),$$

where the second map is the *second* residue homomorphism (with π as the uniformizer). Thus, if $u \in F[x]$ is prime to π (that is, a π-adic unit), we have $\partial_\pi\langle u\rangle = 0$ and $\partial_\pi\langle \pi u\rangle = \langle \overline{u}\rangle$, where the bar denotes the projection of the valuation ring of E_π onto $\overline{E}_\pi$. Note that in VI.4, where we dealt with $W(\mathbb{Q})$, the definition of ∂_2 was somewhat exceptional, due to the usual trouble with the even prime. Here, we don't have that difficulty, and all π 's are treated alike since char $F \neq 2$.

Milnor's Theorem 3.1. *Let $E = F(x)$, and let i be the functorial map $W(F) \to W(E)$. Then the following sequence of abelian groups is split exact:*

$$0 \longrightarrow W(F) \stackrel{i}{\longrightarrow} W(E) \stackrel{\oplus \partial_\pi}{\longrightarrow} \bigoplus_\pi W(\overline{E}_\pi) \longrightarrow 0, \tag{3.2}$$

where the direct sum extends over all monic irreducible polynomials $\pi \in F[x]$.

Proof. The first step is to split i. Let π be a fixed monic linear polynomial (e.g., $\pi(x) = x$). The completion E_π has residue class field $\overline{E}_\pi =$

$F[x]/(\pi(x)) \cong F$. Let $j\ (= j_\pi)$ denote the composition of $W(E) \to W(E_\pi)$ with the first residue homomorphism $W(E_\pi) \to W(\overline{E}_\pi) = W(F)$. Since $j\langle u\rangle = \langle \overline{u}\rangle$ for any π-adic unit u, it is clear that $ji\langle a\rangle = \langle a\rangle$ for any $a \in \dot{F}$. This shows that i is a split monomorphism.

The rest of the proof will follow the basic pattern of VI.4.1, the idea being to filter the Witt group $W(E)$ in order to show that $W(E)/\operatorname{im}(i)$ looks exactly like the direct sum $\bigoplus_\pi W(\overline{E}_\pi)$. For any nonnegative integer d, let L_d denote the subring of $W(E)$ generated by $\langle f\rangle$, where $f \in F[x]$ is of degree $\leq d$. For instance, if the prime divisors of a polynomial h are all of degree $\leq d$, then $\langle h\rangle \in L_d$. We have an ascending chain of subrings

$$L_0 \subseteq L_1 \subseteq L_2 \subseteq \cdots \subseteq W(E),$$

where L_0 is clearly $i(W(F))$. Our main task is that of determining the filtration quotients L_d/L_{d-1}. Note that ∂_π vanishes on L_{d-1} if $\deg \pi \geq d$. *We claim that*

$$\text{(3.3)}\qquad \bigoplus_{\deg \pi = d} \partial_\pi\colon\ L_d/L_{d-1} \longrightarrow \bigoplus_{\deg \pi = d} W(\overline{E}_\pi) \text{ is an isomorphism.}$$

Once we prove this claim, a five-lemma argument together with induction will imply the exactness of 3.2, just as in VI.4.3.

We shall now prove our Claim 3.3. Consider a fixed monic irreducible polynomial π of degree d. For each residue class $\overline{g} \in \overline{E}_\pi$, let g denote the unique polynomial of degree $< d$ that lifts $\overline{g}$. The rule $\langle \overline{g}\rangle \mapsto \langle \pi g\rangle + L_{d-1}$ gives rise to a well-defined group homomorphism $v_\pi\colon W(\overline{E}_\pi) \to L_d/L_{d-1}$. This is easily verified as before by checking the additive relations among the generators $\langle \overline{g}\rangle \in W(\overline{E}_\pi)$. If π, π' are both monic irreducible of degree d, consider the composition

$$W(\overline{E}_\pi) \xrightarrow{v_\pi} L_d/L_{d-1} \xrightarrow{\partial_{\pi'}} W(\overline{E}_{\pi'}).$$

If $\pi \neq \pi'$, then $\partial_{\pi'} v_\pi(\overline{g}) = \partial_{\pi'}\langle \pi g\rangle = 0$, since πg is a π'-adic unit. On the other hand,

$$\partial_\pi v_\pi(\overline{g}) = \partial_\pi(\langle \pi g\rangle) = \langle \overline{g}\rangle.$$

Thus, to show that we get an isomorphism in 3.3, it suffices to show that L_d/L_{d-1} is spanned by the sum of the images of v_π ($\deg \pi = d$). By definition, L_d is additively spanned by the expressions $\langle f_1 \cdots f_r g_1 \cdots g_s\rangle$, where $f_1, \ldots, f_r$ are different *monic* polynomials of degree d, and $\deg(g_i) < d$. We shall present a procedure that systematically "shrinks" r. Indeed, suppose $r \geq 2$. Setting $h = f_1 - f_2$, we have $\deg(h) < d$. By the isometry $\langle f_2, h\rangle \cong \langle f_1, f_1 f_2 h\rangle$, we may express $\langle f_1\rangle$ as $\langle f_2\rangle + \langle h\rangle - \langle f_1 f_2 h\rangle$ in $W(E)$.

Multiplying both sides by $\langle f_2 \cdots f_r g_1 \cdots g_s \rangle$, we obtain

$$\langle f_1 \cdots f_r g_1 \cdots g_s \rangle = \langle f_3 \cdots f_r g_1 \cdots g_s \rangle + \langle f_2 \cdots f_r g_1 \cdots g_s h \rangle - \langle f_1 f_3 \cdots f_r g_1 \cdots g_s h \rangle,$$

where, on the RHS, each product has at most $r-1$ factors of degree d. By induction, it follows that L_d is generated by $\alpha = \langle f_1 g_1 \cdots g_s \rangle$, where $\deg\, g_i < d$. If f_1 is reducible, this generator α belongs to L_{d-1}. If f_1 is (monic) irreducible (say $f_1 = \pi$), let g be the unique polynomial of degree $< d$ defined by the congruence

$$g \equiv g_1 \cdots g_s \pmod{\pi}.$$

As in VI.4.2, we have

$$\alpha = \langle \pi g_1 \cdots g_s \rangle \equiv \langle \pi g \rangle = v_\pi(\overline{g}) \pmod{L_{d-1}}.$$

This shows that $L_d/L_{d-1} = \sum \{\, \mathrm{im}(v_\pi) \colon \deg \pi = d \,\}$, as claimed. □

For illustration, let us give a typical application of Milnor's exact sequence. Using this exact sequence, we'll answer the question: *if* q *is a form over* F, *and* $p(x) \in F[x] \setminus \{0\}$, *when is* $p(x) \in G_E(q_E)$? (Recall that "G" denotes the group of similarity factors of quadratic forms: see VII.4.1.)

Let $p = a\pi_1 \cdots \pi_r$, where $a \in \dot{F}$ (we shall write $a = \mathrm{lead.\,coef.}(p)$) and the π_i's are monic irreducible polynomials. We may assume (after knocking out pairs of repeated prime factors) that the π_i's are distinct.

Theorem 3.4. *In the above notations,* $p(x) \in G_E(q_E)$ *iff* lead. coef.$(p) \in G_F(q)$ *and* $\overline{E}_{\pi_i} \otimes q$ *is hyperbolic over* $\overline{E}_{\pi_i}$ *for each* i.

Proof. To get the leading coefficient into play, we have to choose a suitable splitting j_∞ for the map i in 3.2. Indeed, let E_∞ denote the completion of E at the $(\frac{1}{x})$-adic valuation. Its residue class field $\overline{E}_\infty$ can be naturally identified with F. We pick j_∞ to be the composition of $W(E) \to W(E_\infty)$ with the sum of the first and the second residue homomorphisms $W(E_\infty) \to W(\overline{E}_\infty) = W(F)$. As before, j_∞ obviously splits i, and it is a *ring homomorphism.* Now let

$$q = \langle a_1, \ldots, a_n \rangle \quad \text{and} \quad f_i = a\pi_1 \cdots \hat{\pi}_i \cdots \pi_r,$$

where the caret means the omission of a factor. If $\pi \neq \pi_i$ for each i, clearly ∂_π vanishes on both q_E and $p(x) \cdot q_E$. On the other hand,

$$\partial_{\pi_i}(p(x) \cdot q_E) = \langle\, a_1\overline{f_i(x)},\ \ldots\ ,\ a_n\overline{f_i(x)}\,\rangle,$$

and this is zero in $W(\overline{E}_{\pi_i})$ iff $\overline{E}_{\pi_i} \otimes q$ is hyperbolic over $\overline{E}_{\pi_i}$. Thus, using 3.1, $p(x) \in G_E(q_E)$ iff each $\overline{E}_{\pi_i} \otimes q$ is hyperbolic and

$$j_\infty(q_E) = j_\infty(p(x) \cdot q_E).$$

Since $j_\infty(q_E) = q$ and $j_\infty(p(x) \cdot q_E) = a \cdot q$ (where $a =$ lead. coef.(p)), the last condition boils down to $a \in G_E(q)$. □

The result above will have interesting applications in the next two chapters: see X.2.13, XI.2.4(10), and XI.5.9(3). In closing, we observe that, if we use the splitting j_{x-e} in place of j_∞ above, we'll get the following analogue of 1.3(2).

Corollary 3.5. *Let q be a quadratic form over F, and let $p(x) \in F[x] \cap G_E(q_E)$. If $e \in F$ is such that $p(e) \neq 0$, then $p(e) \in G_F(q)$.*

4. Scharlau's Reciprocity Formula for $F(x)$

In the last term of Milnor's exact sequence 3.1, we have accounted for all valuations of $E = F(x)$ over F, except for the $(\frac{1}{x})$-adic valuation. For a fixed choice of x, the $(\frac{1}{x})$-adic valuation is sometimes called the ∞-*spot* of E, since it corresponds to the "north pole" of the Riemann sphere of E over F. The residue class field $\overline{E}_\infty$ of the corresponding completion E_∞ is identical with F itself. To extend the notation of Section 3, we shall write ∂_∞ to denote the composition of $W(E) \to W(E_\infty)$ with the second residue homomorphism $W(E_\infty) \to W(\overline{E}_\infty) = W(F)$. We have therefore a group homomorphism

$$\partial\colon\ W(E) \longrightarrow \bigoplus_v W(\overline{E}_v)$$

defined by the direct sum $\bigoplus_v \partial_v$, where v ranges over *all* valuations of E over F. (Of course, $\partial_v = \partial_\pi$ if v is π-adic.)

Our principal goal in this section is to determine coker(∂): in Theorem 4.2 below, we'll show that coker$(\partial) \cong W(F)$. To check this, our strategy is to define a suitable epimorphism

$$s\colon\ \bigoplus_v W(\overline{E}_v) \longrightarrow W(F)$$

whose kernel will be precisely im(∂). We proceed as follows.

The desired map s is to be specified by coordinate maps $s_*^v\colon W(\overline{E}_v) \to W(F)$, one for each valuation v. Let us first consider the case where v is a π-adic valuation for a monic irreducible polynomial $\pi \in F[x]$. Write $\overline{E}_v$ as a simple extension $F[\theta_\pi]/F$, where θ_π has the minimal polynomial $\pi(x)$. We shall take $s^{(v)}$ to be the F-linear functional on $\overline{E}_v$ defined by

$$(4.1)\qquad s^{(v)}(\theta_\pi^{n-1}) = 1, \quad \text{and} \quad s^{(v)}(\theta_\pi^i) = 0 \quad \text{for } i < n-1,$$

where $n = \deg\,\pi$. We may then define

$$s_*^v\colon\ W(\overline{E}_v) \longrightarrow W(F)$$

to be the transfer homomorphism associated with $s^{(v)}$. Lastly, if v is the $(\frac{1}{x})$-adic valuation, we take s_*^v $(= s_*^\infty)$ to be the *minus* of the identity map $W(\overline{E}_\infty) = W(F) \to W(F)$.

Theorem 4.2 (Scharlau [Sc3]). *The following sequence is split exact:*

$$0 \longrightarrow W(F) \xrightarrow{\ i\ } W(E) \xrightarrow{\ \partial\ } \bigoplus_v W(\overline{E}_v) \xrightarrow{\ s\ } W(F) \longrightarrow 0.$$

Proof. Since $s_*^\infty = -\operatorname{Id}$, the map s is certainly a split surjection. In view of Milnor's exact sequence 3.1, we see easily that 4.2 follows as soon as we show that $s \circ \partial$ is the zero map. In other words, our task is now reduced to proving the following reciprocity formula:

$$\sum_v s_*^v \partial_v(q) = 0 \in W(F), \tag{4.3}$$

for any quadratic form q over E.

Since both ∂_v and s_*^v are additive maps, we may assume that q is a 1-dimensional form over E. After knocking out denominators and pairs of repeated factors, we may express q as $\langle a\pi_1 \cdots \pi_n\rangle$, where $a \in \dot{F}$, and the π_i are distinct monic irreducible polynomials. In proving 4.3, we may clearly assume that $a = 1$. If π is not one of the π_i's above, ∂_π clearly vanishes on q. Hence, upon transposition, 4.3 amounts to

$$\sum_{i=1}^n s_*^i \partial_i(q) = \partial_\infty(q), \tag{4.4}$$

where $\partial_i = \partial_{\pi_i}$, etc. We can think of $\partial_i(q)$ as being supported by $\overline{E}_i$ $(= \overline{E}_{\pi_i})$ itself, given by

$$\partial_i(q) = \langle \pi_1(\theta_i) \cdots \widehat{\pi_i(\theta_i)} \cdots \pi_n(\theta_i)\rangle \in W(\overline{E}_i).$$

Let A be the F-algebra $F[x]/(\pi_1(x) \cdots \pi_n(x))$. Using the rule

$$\theta_i^r \mapsto x^r \pi_1(x) \cdots \widehat{\pi_i(x)} \cdots \pi_n(x),$$

we may identify $\overline{E}_i$ with an F-subspace A_i in A. This is by no means a ring embedding. However, since there exists an equation

$$1 = \sum_i f_i(x)\pi_1(x) \cdots \widehat{\pi_i(x)} \cdots \pi_n(x),$$

it follows that A is additively spanned by the A_i's. By dimension count, we see that $A = A_1 \oplus \cdots \oplus A_n$. Letting

$$d_i = [\overline{E}_i \colon F] = \deg \pi_i \quad \text{and} \quad d = \sum d_i = [A \colon F],$$

we define an F-linear functional $t \colon A \to F$ by the rule

$$t(x^{d-1}) = 1, \quad \text{and} \quad t(x^i) = 0 \quad \text{for } i < d-1.$$

Then A becomes an F-quadratic space with the quadratic form ρ given by $\rho(g(x)) = t(g(x)^2)$. If $i \neq j$, A_i and A_j are clearly orthogonal in (A, ρ). We have thus an orthogonal decomposition $\rho = \perp_i (A_i, \rho)$. We claim that *the embedding map* $\overline{E}_i \to A_i$ *is actually an isometry from* $(\overline{E}_i, s_*^i\partial_i(q))$ *to* (A_i, ρ).

To see this, let $f(x)$ be any polynomial of degree $< d_i$. By definition,

$$\big(s_*^i\partial_i(q)\big)(f(\theta_i)) = s^{(i)}\big(f(\theta_i)^2\pi_1(\theta_i)\cdots\widehat{\pi_i(\theta_i)}\cdots\pi_n(\theta_i)\big).$$

Writing

$$f(x)^2\pi_1(x)\cdots\widehat{\pi_i(x)}\cdots\pi_n(x) = c_0 + \cdots + c_{d_i-1}x^{d_i-1} + \pi_i(x)g(x),$$

we have

$$\big(s_*^i\partial_i(q)\big)(f(\theta_i)) = s^{(i)}\big(c_0 + c_1\theta_i + \cdots + c_{d_i-1}\theta_i^{d_i-1}\big) = c_{d_i-1}.$$

On the other hand, at the polynomial $f(x)\pi_1(x)\cdots\widehat{\pi_i(x)}\cdots\pi_n(x,$, the value of ρ is

$$t\big(f(x)^2\pi_1(x)^2\cdots\widehat{\pi_i(x)^2}\cdots\pi_n(x)^2\big).$$

This is also equal to c_{d_i-1}, since the parenthetical expression is equal to

$$\begin{aligned}\pi_1(x)\cdots\widehat{\pi_i(x)}\cdots\pi_n(x)\,[\,f(x)^2\pi_1(x)\cdots\widehat{\pi_i(x)}\cdots\pi_n(x)\,]\\ \equiv \pi_1(x)\cdots\widehat{\pi_i(x)}\cdots\pi_n(x)\,[c_0 + c_1x + \cdots + c_{d_i-1}x^{d_i-1}],\end{aligned}$$

modulo $\pi_1(x)\cdots\pi_n(x)$.

Having established the claim (that $\overline{E}_i \to A_i$ is an isometry), we see that (4.4) now amounts to

$$\partial_\infty(q) = (A, \rho) \in W(F).$$

To calculate (A, ρ), we separate the case $d =$ odd from the case $d =$ even. If $d = 2e$, then the "vectors" $1, x, \ldots, x^{e-1}$ span a totally isotropic subspace of half the dimension of (A, ρ). In this case, (A, ρ) is hyperbolic, i.e., equal to zero in $W(F)$. Next, assume $d = 2e+1$. Here, the span of $1, x, \ldots, x^{e-1}$ is totally isotropic, so $(A, \rho) \cong e\mathbb{H} \perp \langle c\rangle$ for some $c \in \dot{F}$. Equating the determinants, we see that $c = 1$, so in this case, $(A, \rho) = \langle 1\rangle \in W(F)$. To complete the proof of 4.2, it suffices, therefore, to prove the following fact.

Lemma 4.5. *Let* $q = \langle h(x)\rangle$ *be a* 1*-dimensional form over* $E = F(x)$, *where* h *has degree* d *and leading coefficient* b_0. *Then* $\partial_\infty(q) = 0 \in W(F)$ *for* d *even, and* $\partial_\infty(q) = \langle b_0\rangle \in W(F)$ *for* d *odd.*

Proof. Say $h(x) = b_0x^d + \cdots + b_d$. Writing $y = 1/x$, we have

$$h = (b_0 + b_1y + \cdots + b_dy^d)/y^d,$$

where the numerator is a (y)-adic unit (since $b_0 \neq 0$). Thus, if $d =$ even, $\partial_\infty \langle h \rangle$ is zero, and if $d =$ odd, $\partial_\infty \langle h \rangle = \langle b_0 \rangle \in W(F)$. $\square$

To illustrate the use of the reciprocity formula 4.3, let us apply it in the special case where the field of constants F is a finite field (of characteristic not 2). In this case, the complete fields E_v have *finite* residue class fields $\overline{E}_v$, and the analysis of VI.1 shows that E_v has a unique anisotropic 4-dimensional form, which we shall denote by φ_{E_v}. The first and second residue forms of φ_{E_v} are both equal to $\varphi_{\overline{E}_v}$, the unique anisotropic binary form over the finite field $\overline{E}_v$.

Let $A = \left(\frac{a,b}{E}\right)$ be an arbitrary quaternion algebra over E, and $q = \langle 1, -a, -b, ab \rangle$ be its norm form. We wish to study the set of valuations v that do not split the algebra A. More formally, we define

$$\Lambda(q) = \{v\colon\ E_v \otimes q \neq 0 \in W(E_v)\}.$$

If $v \notin \Lambda(q)$, clearly $\partial_v(q) = 0 \in W(\overline{E}_v)$. On the other hand, if $v \in \Lambda(q)$, then $E_v \otimes q$ being nonhyperbolic implies $E_v \otimes q = \varphi_{E_v}$, and hence $\partial_v(q) = \varphi_{\overline{E}_v} \in W(\overline{E}_v)$. We may, therefore, characterize the set $\Lambda(q)$ as

$$\{v\colon\ \partial_v(q) \neq 0\} = \{v\colon\ \partial_v(q) = \varphi_{\overline{E}_v}\}.$$

It follows at once that $\Lambda(q)$ is finite. If we rewrite 4.3 but restrict the summation range to $v \in \Lambda(q)$, we obtain an equation

$$\sum_{v \in \Lambda(q)} s_*^v(\varphi_{\overline{E}_v}) = 0 \in W(F).$$

By Chapter VII, Exercise 1, $s_*^v(\varphi_{\overline{E}_v}) = \varphi_F \in W(F)$, where φ_F denotes the unique anisotropic binary form over F. Thus, our equation reduces to $|\Lambda(q)| \cdot \varphi_F = 0 \in W(F)$. Since $W(F)$ is either $\mathbb{Z}_4$ or $\mathbb{Z}_2 \oplus \mathbb{Z}_2$, we conclude that $|\Lambda(q)|$ is an even integer. We have thus proved the following:

Hilbert Reciprocity Theorem 4.6. *Let $E = F(x)$, where F is a finite field. Given any quaternion algebra $A = \left(\frac{a,b}{E}\right)$, A splits over E_v except for a finite, even number of valuations v of E over F.*

A more conceptual way to formulate Hilbert's Reciprocity is to say that the product of the local Hilbert symbols of the quaternion algebra A is equal to 1. In this form, the Reciprocity Law holds indeed over all global fields E. (The case handled in 4.6 is where E is a rational function field in one variable over a finite field.) More generally, in class field theory, there is a corresponding reciprocity formula for the Hasse invariants of a finite-dimensional central simple algebra over a global field; see, e.g., [CF].

Exercises for Chapter IX

1. Let q_1, q_2 be anisotropic forms over a field F. If x is an indeterminate, show that $q_1 \perp \langle x \rangle q_2$ is an anisotropic form over $K = F(x)$.
2. (Davenport–Cassels) Let $\gamma = \sum_{i,j=1}^{n} a_{ij} x_i x_j$, where $a_{ij} = a_{ji} \in \mathbb{Z}$. Assume that

   (1) γ has no nontrivial zero in $\mathbb{Z}^n$, and
   (2) for any $x \in \mathbb{Q}^n$, there exists $y \in \mathbb{Z}^n$ such that $|\gamma(x - y)| < 1$.

   Let p be any nonzero integer. Show that γ represents p over $\mathbb{Z}$ iff γ represents p over $\mathbb{Q}$. (**Hint.** Follow the proof of 1.3, but replace the use of the euclidean algorithm by an application of the condition (2).)
3. Show that the above applies to the quadratic forms $X_1^2 + X_2^2 + X_3^2$ and $X_1^2 - 2X_2^2$. (Thus, if $n \in \mathbb{Z}$ is a sum of squares of three rationals, n is already a sum of squares of three integers. Compare VI.3.12.)
4. Let K be the quotient field of a discrete valuation ring A (which is not necessarily complete). Assume that A contains a subfield k that maps isomorphically onto the residue class field of A. Prove the analogues of 1.1, 1.2 and 1.3 for quadratic forms γ over k (with A and K replacing $F[x]$ and $F(x)$).
5. Keep the notations of Exercise 4. Does the analogue of the Second Representation Theorem hold for k and K? Namely, if $\gamma = \langle a_1, \ldots, a_n \rangle$ is anisotropic over k, and $a_1 t^2 + d \in D_K(\gamma)$ ($d \in \dot{k}$, and t is a uniformizer for A), does it follow that $d \in D_k \langle a_2, \ldots, a_n \rangle$?
6. Keep the above notations, and suppose k has finite level s (see XI.2.1). Show that t is a sum of $s+1$ squares in K, but not a sum of s squares in K.
7. Let q be an n-dimensional quadratic form over F, and $f_i(x) \in F[x]$ ($1 \le i \le n$). Let
$$f(x) = q(f_1(x), \ldots, f_n(x)) \in F[x].$$
   (1) Show that if q is anisotropic and the f_i are not all zero, then $\deg f = 2 \cdot \max\{\deg f_i\}$.
   (2) If the f_i are not all zero, then q represents lead. coef.(f) over F.
   (In fact, (1) and (2) remain true if q is replaced by any homogeneous polynomial of degree d. In (2), the number 2 is to be replaced by d.)
8. Let φ, γ be quadratic forms over F, and let $K = F(x_1, \ldots, x_n)$. Show that $\varphi \subseteq \gamma$ over F iff $\varphi \subseteq \gamma$ over K.

Chapter X

Pfister Forms and Function Fields

In Chapter III, we studied a class of 4-dimensional forms that arise as norm forms of quaternion algebras. For a quaternion algebra $\left(\frac{a_1, a_2}{F}\right)$ over a field F, the associated norm form is

$$q = \langle 1, -a_1, -a_2, a_1a_2\rangle = \langle 1, -a_1\rangle \otimes \langle 1, -a_2\rangle.$$

This form has several very remarkable properties. First, the set of values represented by q (denoted by $D_F(q)$) coincides with the group $G_F(q)$ of similarity factors of q, so in particular $D_F(q)$ is a group under multiplication (III.2.4). Second, if q is isotropic, then it is in fact hyperbolic over F (III.2.7). In his important paper [Pf$_2$], Pfister showed that all of the above properties also hold true for any quadratic form of the type $\bigotimes_{i=1}^n \langle 1, a_i\rangle$, for any set of elements $a_1, \ldots, a_n \in \dot{F}$. Such a form is now called an *n-fold Pfister form* over F, after [EL$_1$].

Pfister's fundamental discovery has since revealed many new facets of the theory of quadratic forms, and become a powerful new tool in the research in this area. In this chapter, we give a systematic account of Pfister's discovery and its principal ramifications. In the middle sections of this chapter, we'll also offer a self-contained introduction to the basic theory of function fields of quadratic forms — a theory that is, in many ways, connected to the theory of Pfister forms. Some of the main sources for this chapter are: [EL$_1$], [L$_1$] for §1, [Pf$_2$], [L$_1$] for §2, [AP$_1$] and [EL$_1$] for §5, [Mi] for §6, etc. On the other hand, the material in §§3-4 draws heavily from the current literature on function fields. Applications of the theory of Pfister forms and function fields will be taken up in the three subsequent chapters.

1. Chain P-Equivalence

We begin by formally defining Pfister forms.

Definition 1.1. For an n-tuple of elements $a_1, \ldots, a_n \in \dot{F}$, we write $\langle\langle a_1, \ldots, a_n \rangle\rangle$ to denote the 2^n-dimensional quadratic form $\bigotimes_{i=1}^n \langle 1, a_i \rangle$, and will refer to this as an *n-fold Pfister form* (over F).

A 0-fold Pfister form is, by convention, taken to be the form $\langle 1 \rangle$. A 1-fold Pfister form $\langle\langle a \rangle\rangle = \langle 1, a \rangle$ is the norm form of a quadratic algebra $F[x]/(x^2 + a)$, by II.3.7. A 2-fold Pfister form $\langle\langle -a, -b \rangle\rangle$ is the norm form of the quaternion algebra $\left(\frac{a, b}{F}\right)$, by III.2.2. This can be pushed one step further: a 3-fold Pfister form $\langle\langle a, b, c \rangle\rangle$ can be realized as the norm form of a certain nonassociative Cayley-Dickson algebra associated with the triple $\{a, b, c\}$. Thus, the general notion of Pfister forms is very well-motivated by these "low-fold" cases.

On the other hand, in commutative algebra, Pfister forms also arise naturally in the context of trace forms. For instance, if $a_1, \ldots, a_n$ represent $\mathbb{Z}_2$-independent classes in $\dot{F}/\dot{F}^2$, then the trace form on the multiquadratic extension $K = F(\sqrt{a_1}, \ldots, \sqrt{a_n})$ is isometric to $\langle 2^n \rangle \langle\langle a_1, \ldots, a_n \rangle\rangle$. (This is true for $n = 1$ by VII.6.17, and therefore true for all n by Chapter I, Exercise 29(2).)

In working with Pfister forms, it is useful to note that, if some $a_i = -1$, then $\langle\langle a_1, \ldots, a_n \rangle\rangle$ becomes hyperbolic (by I.6.1). On the other hand, we have

$$\langle\langle 1, a_2, \ldots, a_n \rangle\rangle \cong 2 \, \langle\langle a_2, \ldots, a_n \rangle\rangle,$$

where $2q$ means $q \perp q$. In particular, $\langle\langle 1, \ldots, 1 \rangle\rangle \cong 2^n \langle 1 \rangle$.

Another important motivation for studying Pfister forms is, of course, the following.

Proposition 1.2. *Let IF denote (as usual) the ideal of all even-dimensional forms in $W(F)$. Then $I^n F$ is generated as an abelian group by all the n-fold Pfister forms over F.*

Proof. We have shown before (II.1.2, applied to $W(F)$) that IF is additively generated by $\langle 1, a \rangle = \langle\langle a \rangle\rangle$ ($a \in \dot{F}$). Thus, $I^n F$ is additively generated by the n-fold products

$$\langle\langle a_1 \rangle\rangle \cdots \langle\langle a_n \rangle\rangle = \langle\langle a_1, \ldots, a_n \rangle\rangle \quad (a_i \in \dot{F}). \qquad \square$$

We'll begin our study by assembling some basic formulas for 2-fold Pfister forms. Recall that $D(q) = D_F(q)$ denotes the set of values in $\dot{F}$ represented by q.

Proposition 1.3. (1) *For any* $x \in D\langle\langle a\rangle\rangle$, $\langle\langle a, b\rangle\rangle \cong \langle\langle a, bx\rangle\rangle$.

(2) *For any* $y \in D\langle a, b\rangle$, $\langle\langle a, b\rangle\rangle \cong \langle\langle y, ab\rangle\rangle$.

Proof. These follow from the following easy isometries:

$$\langle\langle a, b\rangle\rangle \cong \langle 1, a\rangle \perp \langle b\rangle\langle x, xa\rangle \cong \langle\langle a, bx\rangle\rangle,$$
$$\langle\langle a, b\rangle\rangle \cong \langle 1, ab, a, b\rangle \cong \langle 1, ab, y, aby\rangle \cong \langle\langle y, ab\rangle\rangle. \qquad \square$$

The goal of this section is to build up the properties of n-fold Pfister forms from those of 1-fold and 2-fold Pfister forms. To this end, and in analogy with Witt's notion of chain equivalence (I.5), we introduce the following technical definition.

Definition 1.4. Let $\langle\langle a_1, \dots, a_n\rangle\rangle$ *and* $\langle\langle b_1, \dots, b_n\rangle\rangle$ be two n-fold Pfister forms. We say that they are *simply P-equivalent*[1] if there exist two indices i and j such that

(1) $\langle\langle a_i, a_j\rangle\rangle \cong \langle\langle b_i, b_j\rangle\rangle$, and

(2) $a_k = b_k$ for any $k \neq i, j$.

(Note that, in condition (1) above, if i is equal to j, the expression $\langle\langle a_i, a_j\rangle\rangle$ is understood to be just $\langle\langle a_i\rangle\rangle$.) More generally, we say that two n-fold Pfister forms φ and γ are *chain P-equivalent* if there exists a sequence of n-fold Pfister forms $\varphi_0, \varphi_1, \dots, \varphi_m$ such that $\varphi_0 = \varphi$, $\varphi_m = \gamma$, and that each φ_i is simply P-equivalent to φ_{i+1} $(0 \le i \le m-1)$.

Chain P-equivalence is clearly an equivalence relation on all n-fold Pfister forms; it will be denoted by the symbol $\approx$. Of course, $\varphi \approx \gamma$ implies that $\varphi \cong \gamma$. It is by no means obvious, at this point, that the converse also holds. Nevertheless, this turns out to be the case, and will be one of the theorems we prove in this section. To this end, let us first observe that, if π is any permutation of $\{1, 2, \dots, n\}$, then

$$\langle\langle a_1, \dots, a_n\rangle\rangle \approx \langle\langle a_{\pi(1)}, \dots, a_{\pi(n)}\rangle\rangle.$$

This follows immediately from the fact that, for $n \ge 2$, the symmetric group on n letters is generated by the transpositions.

Since any n-fold Pfister form φ represents 1, we may write $\varphi \cong \langle 1\rangle \perp \varphi'$. We shall call φ' the *pure subform* of φ (in analogy with the "pure quaternions"). This terminology is justified, since the isometry type of φ' is uniquely determined by that of φ, according to Witt's Cancellation Theorem (I.4.2). *In the balance of this chapter, we shall always write φ' for the pure subform of a Pfister form φ.*

---

[1] The letter "P" (in honor of Pfister) is used to distinguish the simple equivalence between Pfister forms from the earlier notion of simple equivalence for diagonal forms in I.5.

Pure Subform Theorem 1.5. *Let $\varphi = \langle\langle a_1, \dots, a_n\rangle\rangle$ be an n-fold Pfister form ($n \geq 1$), and let $b \in D_F(\varphi')$. Then there exist $b_2, \dots, b_n \in \dot{F}$ such that*

$$\varphi \approx \langle\langle b, b_2, \dots, b_n\rangle\rangle.$$

Proof. We induct on n. If $n = 1$, then $\varphi = \langle 1, a_1\rangle$. Since $b \in D_F(\varphi') = D_F\langle a_1\rangle$, we have $\langle b\rangle \cong \langle a_1\rangle$, and the result follows. Now assume the result for $(n-1)$-fold Pfister forms. Let

$$\tau = \langle\langle a_1, \dots, a_{n-1}\rangle\rangle \cong \langle 1\rangle \perp \tau'.$$

Then $\varphi \cong \tau\langle 1, a_n\rangle \cong \tau \perp \langle a_n\rangle\tau$, so $\varphi' \cong \tau' \perp \langle a_n\rangle\tau$. Since by hypothesis $b \in D_F(\varphi')$, there exist

$$x \in D_F(\tau') \cup \{0\} \quad \text{and} \quad y \in D_F(\tau) \cup \{0\}$$

such that $b = x + a_n y$. We may further write $y = t^2 + y_0$, where $y_0 \in D_F(\tau') \cup \{0\}$.

Case 1. If $y = 0$, then $0 \neq b = x \in D_F(\tau')$. By the inductive hypothesis, there exist $d_2, \dots, d_{n-1} \in \dot{F}$ such that $\tau \approx \langle\langle x, d_2, \dots, d_{n-1}\rangle\rangle$. Thus,

$$\varphi \approx \langle\langle x, d_2, \dots, d_{n-1}, a_n\rangle\rangle = \langle\langle b, d_2, \dots, d_{n-1}, a_n\rangle\rangle,$$

and we are done.

Case 2. Suppose $y \neq 0$. *We claim that*

$$\varphi \approx \langle\langle a_1, \dots, a_{n-1}, a_n y\rangle\rangle.$$

There is nothing to prove if $y_0 = 0$, for then $y = t^2$. So we may assume that $y_0 \in D_F(\tau')$. By the inductive hypothesis again, $\tau \approx \langle\langle y_0, c_2, \dots, c_{n-1}\rangle\rangle$ for some $c_i \in \dot{F}$. Thus,

$$\begin{aligned}\varphi &\approx \langle\langle y_0, c_2, \dots, c_{n-1}, a_n\rangle\rangle \\ &\approx \langle\langle y_0, c_2, \dots, c_{n-1}, a_n(t^2 + y_0)\rangle\rangle \quad \text{(by 1.3(1))} \\ &\approx \langle\langle a_1, \dots, a_{n-1}, a_n y\rangle\rangle,\end{aligned}$$

proving our claim. If $x = 0$, then the last entry $a_n y$ above is just b, and we are done. So we may assume that $x \in D_F(\tau')$. Again, our inductive hypothesis implies that $\tau \approx \langle\langle x, d_2, \dots, d_{n-1}\rangle\rangle$ for some $d_i \in \dot{F}$, and so

$$\begin{aligned}\varphi &\approx \langle\langle x, d_2, \dots, d_{n-1}, a_n y\rangle\rangle \\ &\approx \langle\langle x + a_n y, d_2, \dots, d_{n-1}, a_n x y\rangle\rangle \quad \text{(by 1.3(2))} \\ &\approx \langle\langle b, d_2, \dots, d_{n-1}, a_n x y\rangle\rangle.\end{aligned}$$

□

For later reference, we record here one of the key steps used in the proof of 1.5 in the form of a proposition. (Note that this proposition is a generalization of 1.3(1).)

Proposition 1.6. *Let* $\tau = \langle\langle a_1, \ldots, a_{n-1}\rangle\rangle$ *and* $y \in D_F(\tau)$. *Then for any* $a_n \in \dot{F}$:

$$\langle\langle a_1, \ldots, a_{n-1}, a_n\rangle\rangle \approx \langle\langle a_1, \ldots, a_{n-1}, a_n y\rangle\rangle.$$

In particular, $\langle\langle a_1, \ldots, a_{n-1}, y\rangle\rangle$ *is isometric to* 2τ, *and* $\langle\langle a_1, \ldots, a_{n-1}, -y\rangle\rangle$ *is hyperbolic.*

Proof. This is just the "Claim" in Case 2 in the proof of 1.5. Since 1.5 is now fully proved, this "Claim" is valid for all n. The last statement of the proposition follows immediately from this, by setting $a_n = \pm 1$. □

Using the Pure Subform Theorem 1.5, we shall now derive two of the principal properties of Pfister forms. The first one is

Theorem 1.7. *If a Pfister form* φ *is isotropic, then it is hyperbolic.*

Proof. Since φ contains a hyperbolic plane, we have $-1 \in D_F(\varphi')$ by Witt's Cancellation Theorem. By 1.5, $\varphi \approx \langle\langle -1, \ldots\rangle\rangle$, which is hyperbolic. □

The next property has to do with the similarity factors of a Pfister form. Recall that, for any quadratic form q over F, $G(q) = G_F(q)$ denotes the *group of similarity factors of* q (that is, the group of elements $c \in \dot{F}$ with $\langle c\rangle\, q \cong q$; see VII.4.1).

Theorem 1.8. *For any Pfister form* φ *over* F, $D_F(\varphi) = G_F(\varphi)$. *In particular,* φ *is a group form over* F.

Proof. It is clear that $G_F(\varphi) \subseteq D_F(\varphi)$, since φ represents 1. To prove that $c \in D_F(\varphi) \implies \langle c\rangle\, \varphi \cong \varphi$, we can proceed in either one of the following ways.

Plane Method. Write $c = t^2 + b$, where $b \in D_F(\varphi') \cup \{0\}$. We may clearly assume that $b \neq 0$. Therefore, by 1.5, $\varphi \approx \langle\langle b, b_2, \ldots, b_n\rangle\rangle$ for suitable $b_i \in \dot{F}$. Since $c \in D_F\langle 1, b\rangle$, we have $\langle 1, b\rangle \cong \langle c, cb\rangle$, so $c \in G_F\langle\langle b\rangle\rangle$. From this, it follows that $c \in G_F(\varphi)$. (The idea here is that we used the Pure Subform Theorem to reduce the proof of 1.8 from the case of n-fold to the obvious case of 1-fold.)

Slick Method. The Pfister form $\varphi\,\langle\langle -c\rangle\rangle \cong \varphi \perp \langle -c\rangle\varphi$ (of one higher fold) contains a subform $\langle c, -c\rangle \cong \mathbb{H}$, so by 1.6, it is hyperbolic. Since $\dim(\langle c\rangle\varphi) = \dim(\varphi)$, it follows that $\langle c\rangle\, \varphi \cong \varphi$! □

The special case of Theorem 1.8 for the Pfister form $\langle\langle 1, \ldots, 1\rangle\rangle$ is already worthy of some celebration:

Corollary 1.9. *For any* $n \geq 0$, *the nonzero sums of* 2^n *squares in* F *form a subgroup of* $\dot{F}$.

Incidentally, when the corollary is stated in this fashion, it also holds for fields of characteristic 2. (By the Frobenius Law, such fields are pythagorean; the group in question in 1.9 is simply $\dot{F}^2$!)

As another "aside", we note that it is also possible to prove the Pfister form property 1.8 *without* using the Pure Subform Theorem 1.5 or the notion of chain P-equivalence. Such an alternative approach will be given in an Appendix to this section.

Next, we shall further generalize the Pure Subform Theorem 1.5. This generalization will be the key step in our subsequent proof of the theorem that isometry of Pfister forms implies their chain P-equivalence.

Theorem 1.10. *If* $\tau = \langle\langle b_1, \ldots, b_r\rangle\rangle$ $(r \geq 0)$, $\gamma = \langle\langle d_1, \ldots, d_s\rangle\rangle$ $(s \geq 1)$, *and* $e_1 \in D_F(\tau\gamma')$, *then there exist* $e_2, \ldots, e_s \in \dot{F}$ *such that*

$$\langle\langle b_1, \ldots, b_r, d_1, \ldots, d_s\rangle\rangle \approx \langle\langle b_1, \ldots, b_r, e_1, \ldots, e_s\rangle\rangle.$$

Proof. We prove this theorem by induction on s. If $s = 1$, then $e_1 \in D_F(\langle d_1\rangle\tau)$, so $e_1 = d_1 x$, where $x \in D_F(\tau)$. Proposition 1.6 implies that

$$\langle\langle b_1, \ldots, b_r, d_1\rangle\rangle \approx \langle\langle b_1, \ldots, b_r, d_1 x\rangle\rangle \approx \langle\langle b_1, \ldots, b_r, c_1\rangle\rangle.$$

By induction, we may assume the result for $\langle\langle b_1, \ldots, b_r, d_1, \ldots, d_{s-1}\rangle\rangle$. Let $\sigma = \langle\langle d_1, \ldots, d_{s-1}\rangle\rangle$, so

$$\gamma = \sigma\langle d_s, 1\rangle \cong \langle d_s\rangle\sigma \perp \sigma \quad \text{and} \quad \gamma' \cong \langle d_s\rangle\sigma \perp \sigma'.$$

Therefore, $\tau\gamma' \cong \langle d_s\rangle\tau\sigma \perp \tau\sigma'$. Since $c_1 \in D_F(\tau\gamma')$, there exist $x \in D_F(\tau\sigma) \cup \{0\}$ and $y \in D_F(\tau\sigma') \cup \{0\}$ such that $c_1 = d_s x + y$. If $x \neq 0$ and $y \neq 0$, we get the desired result in the following two steps.

Step 1. $\langle\langle b_1, \ldots, b_r, d_1, \ldots, d_s\rangle\rangle \approx \langle\langle b_1, \ldots, b_r, d_1, \ldots, d_s x\rangle\rangle$ by 1.9.

Step 2. By induction, there exist $e_2, \ldots, e_{s-1} \in \dot{F}$ such that

$$(*) \qquad \langle\langle b_1, \ldots, b_r, d_1, \ldots, d_{s-1}\rangle\rangle \approx \langle\langle b_1, \ldots, b_r, y, e_2, \ldots, e_{s-1}\rangle\rangle.$$

Therefore, by Step 1,

$$\begin{aligned}\langle\langle b_1, \ldots, b_r, d_1, \ldots, d_{s-1}, d_s\rangle\rangle &\approx \langle\langle b_1, \ldots, b_r, d_1, \ldots, d_{s-1}, d_s x\rangle\rangle \\ &\approx \langle\langle b_1, \ldots, b_r, y, e_2, \ldots, e_{s-1}, d_s x\rangle\rangle \\ &\approx \langle\langle b_1, \ldots, b_r, c_1, e_2, \ldots, e_{s-1}, d_s xy\rangle\rangle,\end{aligned}$$

where the last "$\approx$" follows from 1.3(2).

We are now left with the case where one of x, y is zero. If $y = 0$, then $0 \neq e_1 = d_s x$, and Step 1 provides the needed proof. If $x = 0$, then $e_1 = y$, and from $(*)$, we get

$$\langle\langle b_1, \ldots, b_r, d_1, \ldots, d_s\rangle\rangle \approx \langle\langle b_1, \ldots, b_r, e_1, \ldots, e_{s-1} d_s\rangle\rangle,$$

which completes the proof. □

The following special case of 1.10 (with $r = 1$) is already noteworthy.

Corollary 1.11. *Let q be a Pfister form. If $q \cong \langle 1, b, e, \ldots \rangle$ with $b, e \in \dot{F}$, then $q \cong \langle\langle b, e, e_2, \ldots, e_s \rangle\rangle$ for suitable $e_i \in \dot{F}$.*

Proof. By the Pure Subform Theorem, $q \cong \langle\langle b \rangle\rangle \gamma$ for a suitable Pfister form $\gamma = \langle\langle b_2, \ldots, b_{s+1} \rangle\rangle$. Comparing $\langle\langle b \rangle\rangle \gamma \cong \langle\langle b \rangle\rangle \perp \langle\langle b \rangle\rangle \gamma'$ with $q \cong \langle\langle b \rangle\rangle \perp \langle e, \ldots \rangle$, we see that $e \in \langle\langle b \rangle\rangle \gamma'$. We are now done by applying 1.10 with $\tau = \langle\langle b \rangle\rangle$. □

This Corollary will be generalized in another direction later in this chapter; see Remark 4.29.

We are now in a position to prove the following main result on chain P-equivalence (from [EL₁]).

Chain P-Equivalence Theorem 1.12. *Let φ, ψ be n-fold Pfister forms. Then $\varphi \cong \psi$ iff $\varphi \approx \psi$.*

Proof. It suffices to prove the "only if" part. Write

$$\varphi = \langle\langle a_1, \ldots, a_n \rangle\rangle \quad \text{and} \quad \psi = \langle\langle b_1, \ldots, b_n \rangle\rangle.$$

Assuming that $\varphi \cong \psi$, we claim that, for any integer $r \in [0, n]$:

(A_r) *There exist $c_{r+1}, \ldots, c_n \in \dot{F}$ such that*

$$\varphi \approx \langle\langle b_1, \ldots, b_r, c_{r+1}, \ldots, c_n \rangle\rangle.$$

If this is established, then, for $r = n$, the statement (A_n) implies the desired conclusion that $\varphi \approx \psi$. Now we prove (A_r) by induction on r. There is nothing to prove in case $r = 0$. Assume, inductively, that (A_r) is true, where $r < n$. We must proceed to prove (A_{r+1}). Set

$$\tau = \langle\langle b_1, \ldots, b_r \rangle\rangle, \quad \beta = \langle\langle b_{r+1}, \ldots, b_n \rangle\rangle, \quad \text{and} \quad \gamma = \langle\langle c_{r+1}, \ldots, c_n \rangle\rangle.$$

Then γ is an s-fold Pfister form, where $s = n - r$. We have, from the various hypotheses, $\tau \cdot \beta = \psi \cong \varphi \cong \tau\gamma$; that is,

$$\tau \perp \tau\beta' \cong \tau \perp \tau\gamma'.$$

By the cancellation theorem, it follows that $\tau\beta' \cong \tau\gamma'$. But then

$$b_{r+1} \in D_F(\beta') \subseteq D_F(\tau\beta') = D_F(\tau\gamma').$$

By 1.9, we get

$$\langle\langle b_1, \ldots, b_r, c_{r+1}, \ldots, c_n \rangle\rangle \approx \langle\langle b_1, \ldots, b_r, b_{r+1}, c'_{r+2}, \ldots, c'_n \rangle\rangle$$

for suitable $c'_j \in \dot{F}$. From this and the inductive hypothesis (A_r), we deduce

$$\varphi \approx \langle\langle b_1, \ldots, b_r, b_{r+1}, c'_{r+2}, \ldots, c'_n \rangle\rangle,$$

which establishes the truth of (A_{r+1}). □

Appendix: Round Forms

As we have mentioned earlier in §1, it is possible to give a *direct* proof for the equation $D_F(\varphi) = G_F(\varphi)$ for a Pfister form φ without using the Pure Subform Theorem 1.5 or the notion of chain P-equivalence. This proof is self-contained, and actually leads to a somewhat more general result, so it is well worth a short coverage in this Appendix. The basic idea of this proof has already emerged in our work in Example II.5.4 (see the argument for $\langle c\rangle\langle 1,1,1,1\rangle \cong \langle 1,1,1,1\rangle$ there, for any nonzero sum of four squares c). To generalize this idea, we introduce the notion of round forms.

Definition 1.13. An F-quadratic form σ is said to be a *round form*[(2)] if $D_F(\sigma) = G_F(\sigma)$. Clearly, any round form σ is a group form, since $G_F(\sigma)$ is always a subgroup of $\dot{F}$.

Round Form Theorem 1.14. *If σ is a round form over F, then so is $\varphi = \langle\langle b\rangle\rangle\sigma$ for any $b \in \dot{F}$.*

Proof. Clearly, $1 \in D_F(\sigma) \subseteq D_F(\varphi)$, so $G_F(\varphi) \subseteq D_F(\varphi)$. Conversely, let $c \in D_F(\varphi)$; say $c = s + bt$, where $s,\, t \in D_F(\sigma) \cup \{0\}$. If s or t is zero, it is easy to see that $\langle c\rangle\,\varphi \cong \varphi$. Otherwise, $s,\, t \in G_F(\sigma)$, and we have

$$\begin{aligned}\varphi &\cong \sigma \perp \langle b\rangle\,\sigma \cong \langle s\rangle\,\sigma \perp \langle bt\rangle\,\sigma \\ &\cong \langle s,\, bt\rangle\,\sigma \cong \langle c,\, cbst\rangle\,\sigma \\ &\cong \langle c\rangle\,\sigma \perp \langle cb\rangle\,\sigma \cong \langle c\rangle\,\varphi.\end{aligned}$$

□

From 1.14, it follows immediately, by induction on $n \geq 0$, that *any n-fold Pfister form is a round form, and a fortiori, a group form.* We note also that, if we were only interested in proving that $D_F(2^n\langle 1\rangle)$ is a group, the proof above would have worked completely within the framework of the forms $2^n\langle 1\rangle$ (for varying n's), and we would not even need to introduce the notion of Pfister forms for such an elementary proof.

We conclude this Appendix by giving some examples of round forms. This list of examples shows, in particular, that round forms need not be Pfister forms, even if they are anisotropic.

Examples 1.15. (1) If σ is a hyperbolic form over F, we clearly have $D_F(\sigma) = \dot{F} = G_F(\sigma)$. Hence σ is round (but σ is a Pfister form only if $\dim(\sigma) = 2^n$).

(2) Over a pythagorean field F, $\sigma = n\langle 1\rangle$ is a round form (for any n), since $D_F(\sigma) = \dot{F}^2 = G_F(\sigma)$.

(3) Over the rational field $\mathbb{Q}$, $\sigma = 4n\langle 1\rangle$ is a round form (for any n), since $D_{\mathbb{Q}}(\sigma) = \dot{\mathbb{Q}}^+ = G_{\mathbb{Q}}(\sigma)$ by the theorem of Lagrange.

---

(2) There are other slightly different definitions for round forms in the literature. We adopt here the definition that is most convenient for the formulation of the Round Form Theorem 1.14.

(4) To give an example of a round form σ with $\dim(\sigma) > 2$ and $\det(\sigma) \neq 1 \in \dot{F}/\dot{F}^2$, consider a formally real field F with four square classes $\{\pm 1, \pm 2\}$. For the form $\sigma = \langle 1, 1, 1, 2 \rangle$, it is easy to verify that

$$D_F(\sigma) = \{1, 2\} \cdot \dot{F}^2 = G_F(\sigma).$$

Thus, σ is a round form over F. However, $d(\sigma) = 2 \neq 1 \in \dot{F}/\dot{F}^2$.

(5) The analogue of 1.7 *does not* work for round forms; that is, if a round form σ is isotropic, it need not be hyperbolic. An easy example for this is the form $\sigma = \langle 1, 1, 1, -1 \rangle$ over the field of three elements.

The variety of the examples of round forms in (1) to (5) above suggests that the class of round forms over a field may be worthy of an independent study. However, due to the limitation of space, we shall not take up this matter here.

2. Multiplicative Forms

The principal goal of this section is to prove a characterization theorem for Pfister forms. The fact that the family of Pfister forms over a field F can be *characterized* by certain properties points to the intrinsic nature of such forms. After proving this characterization theorem, we will be in a much better position to appreciate the significance (as well as the beauty) of the family of Pfister forms.

In the last section, we gave two independent proofs for the fact that any m-fold Pfister form φ over a field F is a *group form*. This fact has a very important consequence. Let

$$X = (x_1, \ldots, x_{2^m}), \qquad Y = (y_1, \ldots, y_{2^m})$$

be sets of independent (commuting) indeterminates over F, and let $L = F(X, Y)$ be the rational function field in the x_i's and y_i's. Since $\varphi_L = L \otimes_F \varphi$ is still an m-fold Pfister form over L, we see that $D_L(\varphi)$ $(= D_L(\varphi_L))$ is a group under multiplication. In particular, since φ represents $\varphi(X)$ and $\varphi(Y)$ (as elements of $\dot{L}$), it also represents $\varphi(X) \cdot \varphi(Y)$ over L. This means that there exist rational functions $z_i \in F(X, Y)$ $(1 \leq i \leq 2^m)$ such that

$$\varphi(X) \cdot \varphi(Y) = \varphi(z_1, \ldots, z_{2^m}). \tag{2.1}$$

The situation here is reminiscent of the problem of "composition of quadratic forms" studied in V.5.1, except that, in V.5.1, the z_i's were required to be *bilinear forms* in the x_i's and y_j's, instead of being just rational functions. Thus, we may envisage the formula (2.1) (where $z_i \in F(X, Y)$) as a kind of "generalized composition" for φ. To draw special attention to this property, we make the following formal definition.

Definition 2.2. Let q be a quadratic form of dimension n, and let $L = F(X,Y)$, where $X = (x_1, \dots, x_n)$ and $Y = (y_1, \dots, y_n)$ are sets of independent indeterminates over F. We say that q is *multiplicative* if $q(X) \cdot q(Y) \in D_L(q)$; that is, if

$$q(X) \cdot q(Y) = q(z_1, \dots, z_n) \tag{2.3}$$

for suitable $z_1, \dots, z_n \in L = F(X,Y)$.

Example 2.4. (1) If q is a Pfister form, then by our motivating remarks preceding Definition 2.2, q is multiplicative.

(2) If q is isotropic over F, then q is also multiplicative, since $D_L(q) = \dot{L}$ for the field $L = F(X,Y)$ above.

The following result shows that multiplicative forms are closely related to group forms. (It also clarifies the two examples given above in 2.4.)

Proposition 2.5. *A form q over F is multiplicative iff q is a "hereditary group form", in the sense that $D_K(q)$ is a group for any field $K \supseteq F$. (In particular, in this case, q represents* 1 *over F.)*

Proof. The "if" part follows by taking K to be $F(X,Y)$. For the converse, assume q is multiplicative, so we have the formula (2.3). Let K be *any* extension of F, and let $q(d)$, $q(e)$ be two nonzero values of q over K ($d, e \in K^n$). By (2.3), q represents the polynomial $q(X) \cdot q(Y)$ over $F(X,Y)$, and hence also over $K(X,Y)$. [Here, of course, we treat $X = (x_1, \dots, x_n)$ and $Y = (y_1, \dots, y_n)$ as "universal indeterminates", so they remain algebraically independent over K.] By the Substitution Principle IX.1.6, we infer that q represents $q(d) \cdot q(e)$ in K. This implies that $D_K(q)$ is a group under multiplication, as desired. □

Before we state the main result in this section, we need to introduce one more definition.

Definition 2.6. An n-dimensional form q over F is said to be *strongly multiplicative* if $q(X) \in G_{F(X)}(q)$; that is, if $\langle q(X) \rangle \cdot q \cong q$ over $F(X)$, where $X = (x_1, \dots, x_n)$.

Example 2.7. (1) If q is a Pfister form over F, then $q(X) \in D_{F(X)}(q) = G_{F(X)}(q)$ by 1.7, so q is strongly multiplicative.

(2) If q is hyperbolic over F, then q is also strongly multiplicative, since in this case $G_K(q) = \dot{K}$ for any extension field $K \supseteq F$.

We now come to the characterization theorem for anisotropic Pfister forms. The same result can also be thought of as giving a complete classification of the anisotropic multiplicative or strongly multiplicative forms.

Theorem 2.8 (Pfister). *For any anisotropic form q over F, the following are equivalent:*

(1) *q is a Pfister form over F.*

(2) *q is multiplicative.*

(3) *q is strongly multiplicative.*

(*Note.* The anisotropicity of q will be needed only for the implication (2) $\Longrightarrow$ (1) below.)

Proof. (1) $\Longrightarrow$ (3) was already pointed out in 2.7(1).

(3) $\Longrightarrow$ (2). Since $\langle q(X)\rangle \cdot q \cong q$ over $F(X)$, the form q over $F(X,Y)$ clearly represents $q(X)\cdot q(Y)$.

(2) $\Longrightarrow$ (1). Since q is multiplicative, $1 \in D_F(q)$, so q contains the 0-fold Pfister form $\langle 1\rangle$. Choose r maximal so that q contains (over F) an r-fold Pfister form φ, say $q \cong \varphi \perp q_0$. If $\dim q_0 = 0$, we are done, so assume instead $q_0 \cong \langle c, \dots\rangle$. Consider the "generic value" $\varphi(X) + c\varphi(Y)$ of the form $\varphi \perp \langle c\rangle\,\varphi$ over $K = F(X,Y)$, where

$$X = (x_1, \dots, x_{2^r}), \qquad Y = (y_1, \dots, y_{2^r})$$

are independent indeterminates. The multiplicativity of q implies that

$$\varphi(X) + c\varphi(Y) = \varphi(Y)\left(\frac{\varphi(X)}{\varphi(Y)} + c\right) \in D_K(q),$$

since $\varphi(Y) \in D_K(\varphi) \subseteq D_K(q)$ and $\varphi(X)/\varphi(Y) + c \in D_K(q)$. (The fact that φ is a Pfister form ensures that $\varphi(X)/\varphi(Y) \in D_K(\varphi)$.) Thus, the anisotropic form q "dominates" $\varphi \perp \langle c\rangle\,\varphi$ in the sense of IX.2.5, and the Third Representation Theorem IX.2.8 implies that $\varphi \perp \langle c\rangle\,\varphi \subseteq q$. This is a contradiction, since $\varphi \perp \langle c\rangle\,\varphi$ is an $(r+1)$-Pfister form. Thus, we must have $q \cong \varphi$, as desired! $\square$

We see now that *a multiplicative form is either an isotropic form or else an anisotropic Pfister form.* It only remains to determine the isotropic strongly multiplicative forms. This is accounted for by the following result.

Theorem 2.9. *A form q is isotropic and strongly multiplicative over F iff q is a hyperbolic form.*

Proof. The "if" part is in 2.7(2). For the converse, let q be isotropic and strongly multiplicative. Take a Witt decomposition $q \cong h \perp \gamma$, where h is hyperbolic and γ is anisotropic. Since q is isotropic, $\dim \gamma < \dim q$. Over $F(X)$, $\langle q(X)\rangle \cdot q \cong q$ leads to

$$\langle q(X)\rangle \cdot h \perp \langle q(X)\rangle \cdot \gamma \cong h \perp \gamma.$$

Cancelling $\langle q(X)\rangle \cdot h$, we get $\langle q(X)\rangle \cdot \gamma \cong \gamma$. If $\dim \gamma > 0$, IX.2.10 would imply that $\dim \gamma \geq \dim q$, which is not the case. Therefore, we must have $\dim \gamma = 0$, and so $q \cong h$ is hyperbolic, as desired. □

From 2.8 and 2.9, we see that *a strongly multiplicative form over F is either a hyperbolic form or else an anisotropic Pfister form.*

The Venn diagram below gives a quick schematic summary of all of the results proved so far in this section.

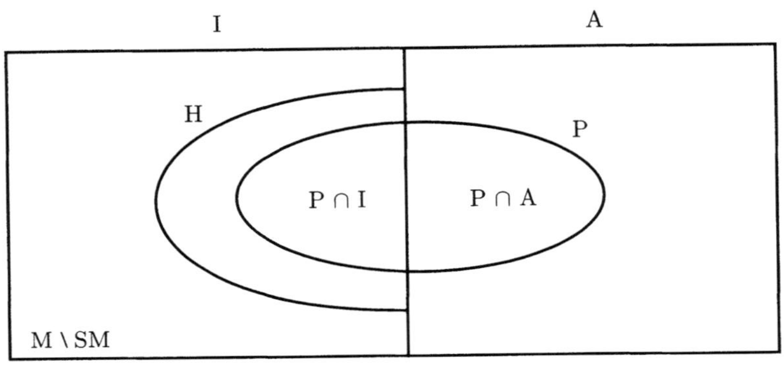

$$\mathrm{H} = \mathrm{SM} \cap \mathrm{I}, \quad \mathrm{M} = \mathrm{P} \cup \mathrm{I}, \quad \mathrm{SM} = \mathrm{P} \cup \mathrm{H}.$$

I = isotropic forms	P = Pfister forms
A = anisotropic forms	M = multiplicative forms
H = hyperbolic forms	SM = strongly multip. forms

Next we shall give some applications for the notions of multiplicative and strongly multiplicative forms. By definition, an n-dimensional multiplicative form q (over F) admits a composition formula

$$q(X) \cdot q(Y) = q(z_1, \ldots, z_n) \tag{2.10}$$

for suitable $z_1, \ldots, z_n \in F(X, Y)$. *What can be said about these rational functions $z_1, \ldots, z_n$?* Can they be chosen in some nice form, whatever "nice forms" should mean?

In the case where q is isotropic, we may take q to be a diagonal form $\langle 1, -1\rangle \perp \langle a_3, \ldots, a_n\rangle$, and the explicit formula

$$q(X) \cdot q(Y) = \left(\frac{q(X)q(Y)+1}{2}\right)^2 - \left(\frac{q(X)q(Y)-1}{2}\right)^2$$

expresses $q(X)q(Y)$ in the form $q(z_1, z_2, 0, \ldots, 0)$ with

$$z_1 = (q(X)q(Y)+1)/2, \qquad z_2 = (q(X)q(Y)-1)/2,$$

which are *polynomials* in X and Y. So the isotropic case is settled. If q is anisotropic instead, then q is strongly multiplicative by 2.7. In this case, we can invoke the following easy characterization result, due to Pfister, for strongly multiplicative forms in terms of composition formulas.

Theorem 2.11. *An n-dimensional form q over F (not assumed to be anisotropic) is strongly multiplicative iff there exist $z_1, \dots, z_n$ that are linear forms in $Y = (y_1, \dots, y_n)$ with coefficients in $F(X) = F(x_1, \dots, x_n)$ such that*

$$q(x_1, \dots, x_n) \cdot q(y_1, \dots, y_n) = q(z_1, \dots, z_n). \tag{2.12}$$

(In particular, such a formula exists for any Pfister form q over F.)

Proof. First assume q is strongly multiplicative, so we have $\langle q(X)\rangle \cdot q \cong q$ over $F(X)$. Here, $q(X)$ is being viewed as a "scalar", since we are working over the field $F(X)$. If we view Y as a vector variable over $F(X)$, The isometry of quadratic forms above means that we can make an invertible linear change of variables (over the field $F(X)$) to get the form $\langle q(X)\rangle \cdot q$ from q; that is, there exists a matrix $U \in \mathrm{GL}_n(F(X))$ such that

$$q(X) \cdot q(Y) = q(YU).$$

Therefore, (2.12) follows by choosing $(z_1, \dots, z_n)$ to be YU, since then each z_i is a linear form in Y with coefficients from $F(X)$. The converse follows similarly, by reversing the steps in the argument above. $\square$

For r-fold Pfister forms q where $r \leq 3$, it is classically well-known that there exist composition formulas (2.12) where the z_i's are *bilinear* forms in X and Y. These formulas are easily written down by using the multiplicative property of the norm maps on quadratic, quarternion, and Cayley-Dickson algebras over F. For instance, viewing $\langle\langle a, b\rangle\rangle$ as the norm form on the quaternion algebra $\left(\frac{-a,\,-b}{F}\right)$, we quickly derive the explicit composition formula(3):

$$\begin{aligned}
(x_1^2 + ax_2^2 + bx_3^2 + abx_4^2)&(y_1^2 + ay_2^2 + by_3^2 + aby_4^2) \\
&= (x_1y_1 + ax_2y_2 + bx_3y_3 + abx_4y_4)^2 \\
&\quad + a(-x_1y_2 + x_2y_1 - bx_3y_4 + bx_4y_3)^2 \\
&\quad + b(-x_1y_3 + x_3y_1 + ax_2y_4 - ax_4y_2)^2 \\
&\quad + ab(-x_1y_4 + x_4y_1 - x_2y_3 + x_3y_2)^2.
\end{aligned}$$

Note that, among the four forms appearing on the RHS, the first is precisely the bilinear form associated with $\langle\langle a, b\rangle\rangle$, and the three others are *alternating*

(3)Here we multiply the two quaternions $x_1 + x_2i + x_3j + x_4k$ and $y_1 - y_2i - y_3j - y_4k$, using the rules $i^2 = -a$, $j^2 = -b$, $k = ij = -ji$.

bilinear forms in the two sets of variables $x_1, \ldots, x_4$ and $y_1, \ldots, y_4$. This observation can be very useful in specific applications. By setting $a = b = 1$ in the above, we get back, of course, the famous 4-square identity of Euler.

For other derivations of multiplicative formulas for 8 (and 16) squares, see, for instance, the references [Ta] and [ZE]. Taussky's work [Ta] was apparently written without knowledge of Pfister's paper [Pf$_2$].

For r-fold Pfister forms with $r > 3$, the theorem of Hurwitz (V.5.10) implies that there is *no* composition formula (2.12) where the z_i's are *bilinear* forms in X and Y. Therefore, being able to take the z_i's as linear forms in Y with coefficients in $F(X)$ is perhaps as strong a conclusion as one would hope to obtain.

To close this section, let us give an interesting application of the property $D_F(q) = G_F(q)$ for a Pfister form q, proved in 1.7. This property, coupled with the earlier result IX.3.4, leads to the following representation theorem observed independently by Knebusch [Kn$_1$] and the author.

Theorem 2.13. *Let q be a Pfister form over F, and let $E = F(x)$ (the rational function field in one variable x). Let $p(x)$ be any nonzero polynomial. Then q_E represents $p(x)$ over E iff q represents the leading coefficient of $p(x)$ over F, and, for every monic irreducible factor $\pi \mid p$ of odd multiplicity, q is hyperbolic over the extension field $F[x]/(\pi(x))$ of F.*

Taking q to be $\langle\langle 1, \ldots, 1\rangle\rangle$, we deduce the following observation of Kaplansky as a special case of 2.13.

Corollary 2.14. *Let $p(x)$ be a nonzero polynomial in $F[x]$. If $p(x)$ is a sum of 2^n squares in $F[x]$, then so is every monic irreducible factor $\pi \mid p$ of odd multiplicity.*

For some other applications of 2.13 to field invariants (such as "level" and "height"), see XI.2.4(10) and XI.5.9(3).

3. Introduction to Function Fields

In algebraic geometry, a function field is associated with every irreducible algebraic variety. In the case of an irreducible quadratic from φ, we have therefore a function field associated with the quadric hypersurface defined by the quadratic equation $\varphi = 0$. Not surprisingly, the study of such function fields holds the key to many basic issues in the algebraic theory of quadratic forms over fields.

Already in the 1930s, E. Witt realized the importance of function fields of quadratic forms, and obtained the first results in this area. But the full potential of function field methods was not realized until the appearance of several works in 1971–75, most notably the Arason-Pfister paper [AP$_1$], Wadsworth's Chicago thesis (partially published in [Wad$_1$]), and to some extent, also [EL$_1$]. In the meantime, Knebusch's penetrating research on function fields of quadratic forms in the period 1972–76 appeared in his two important papers [Kn$_4$, Kn$_5$]. These papers established solid foundations for the study of function fields of quadratic forms, and paved the road to an exciting new chapter in the algebraic theory of quadratic forms.

In this preliminary section, we give a self-contained introduction to the idea of the function field of a quadratic form. A preamble for the construction of such a function field is the following.

Lemma 3.1. *Let $\varphi(x_0, \dots, x_n)$ be a regular $(n+1)$-dimensional quadratic form over F, where $n \geq 1$. Then φ is reducible as a polynomial in $F[x_0, \dots, x_n]$ iff ($n = 2$ and) $\varphi \cong \mathbb{H}$.*

Proof. If $\varphi(x_0, \dots, x_n)$ factors nontrivially, it must factor into a product of two linear forms. Since φ is regular and $n \geq 1$, this happens iff φ is isometric to the quadratic form $x_0 x_1$; that is, iff ($n = 2$ and) $\varphi \cong \mathbb{H}$. □

Now let $\varphi(X) = \varphi(x_0, \dots, x_n)$ $(n \geq 1)$ be a (regular) $(n+1)$-dimensional quadratic form over F, with $\varphi \not\cong \mathbb{H}$. Then, by 3.1, $\varphi(X)$ is an *irreducible* polynomial in $F[X]$, so the principal ideal $(\varphi(X)) \subseteq F[X]$ is a prime ideal.

Definition 3.2. The ("big") *function field* of φ is defined to be the quotient field of the integral domain $F[X]/(\varphi(X))$. This is a field of transcendence degree n over F; we shall denote it by $F[\varphi]$.

As we mentioned at the beginning of this section, $F[\varphi]$ is the usual function field, in the sense of algebraic geometry, of the affine quadric hypersurface $\varphi(X) = 0$ in F^{n+1}. It is easy to see that $F[\varphi]$ depends (up to an F-isomorphism) only on the isometry class of φ.

An easy computation shows that $F[\varphi]$ can be expressed as a quadratic extension of a *rational* function field in n variables over F. Indeed, if we write $F[\varphi]$ as $F(x_0, \dots, x_n)$ (where the x_i's should have been written as $\overline{x}_i$'s), the relation $a_0 x_0^2 + \cdots + a_n x_n^2 = 0$ shows that

$$F[\varphi] = F(x_1, \dots, x_n)\left(\sqrt{-(a_1 x_1^2 + \cdots + a_n x_n^2)/a_0}\right), \tag{3.3}$$

as claimed.

The reason we called $F[\varphi]$ the "big" function field is that we could have formed a smaller one, defined by

$$F(\varphi) := F(x_1/x_0,\ x_2/x_0,\ \dots\ ,\ x_n/x_0) \subseteq F[\varphi]. \tag{3.4}$$

Note that this subfield of $F[\varphi]$ is uniquely determined; i.e. it does not depend on the choice of x_0 as the denominators in (3.4). Indeed, since

$$\frac{x_i}{x_j} = \frac{x_i}{x_0} \Big/ \frac{x_j}{x_0} \in F[\varphi], \tag{3.5}$$

$F(\varphi)$ could have been expressed as $F(\{x_i/x_j\}) \subseteq F[\varphi]$, which exhibits no dependence on any particular subscript. The field $F(\varphi)$ may be called the *homogeneous function field* of φ, since, in algebraic geometry, it is just the function field of the *projective variety* defined by the homogeneous equation $\varphi(X) = 0$ in $\mathbb{P}^n(F)$. Of course, $F(\varphi)$ has transcendence degree $n-1$ over F (two less than $\dim(\varphi)$).

The two function fields $F[\varphi]$ and $F(\varphi)$ are related by the obvious relation

$$F[\varphi] = F(\varphi)(x_0), \tag{3.6}$$

and they have pretty much the same behavior. In practice, it is sufficient to work with just one of them. Our slight preference is to work with the ("big") function field $F[\varphi]$, although, whenever it is more convenient, we'll have no compunction in switching over to the ("small") function field $F(\varphi)$.[(4)]

Note that $F(\varphi)$ is also a quadratic extension of a rational function field (this time in $n-1$ variables). Indeed, if we write $t_i = x_i/x_0$ $(1 \le i \le n)$, the equation

$$a_0 + a_1 t_1^2 + \cdots + a_n t_n^2 = 0$$

shows that

$$F(\varphi) = F(t_1, \ldots, t_{n-1})\left(\sqrt{-(a_0 + a_1 t_1^2 + \cdots + a_{n-1}t_{n-1}^2)/a_n}\,\right), \tag{3.7}$$

which is to be compared with (3.3).

The following remark is often useful in simplifying the notations in working with function fields. We omit its trivial proof.

Remark 3.8. For any $c \in \dot{F}$, we have the canonical identifications:

$$F[c\cdot\varphi] = F[\varphi] \quad \text{and} \quad F(c\cdot\varphi) = F(\varphi).$$

In other words, similar forms have the "same" function fields. However, the converse is not true. In an Appendix to XII.2, we shall construct examples of *nonsimilar* anisotropic 5-dimensional forms (over the field $\mathbb{R}(x,y,z)$) whose function fields are isomorphic over F.

---

(4)In the sequel, whenever we use the notations $F[\varphi]$ and $F(\varphi)$, it will be assumed that these are *defined*; i.e. $\dim\varphi \ge 2$ and $\varphi \not\cong \mathbb{H}$.

Example 3.9. Let φ be a binary form $\not\cong \mathbb{H}$. In studying the function fields of φ, there is no loss of generality in assuming that $\varphi \cong \langle 1, a\rangle$ ($a \notin -\dot{F}^2$). Under this assumption, (3.7) and (3.3) show that

$$F(\varphi) = F\big(\sqrt{-a}\,\big) \quad \text{and} \quad F[\varphi] = F\big(\sqrt{-a}\,\big)(x). \tag{3.10}$$

Example 3.11. Here, we consider the important case where $\dim(\varphi) = 3$. The fields $F[\varphi]$ and $F(\varphi)$ in this case are called *function fields of a conic* (since $\varphi = 0$ defines a projective "conic section"). Using Remark 3.8, we may assume that $\varphi \cong \langle 1, -a, -b\rangle \cong \langle -b, -a, 1\rangle$. With these diagonalizations, we have

$$F[\varphi] = F(y,z)\big(\sqrt{ay^2+bz^2}\,\big), \quad \text{and} \quad F(\varphi) = F(x)\big(\sqrt{ax^2+b}\,\big). \tag{3.12}$$

On the other hand, if we consider the similar form

$$\psi := \langle -b\rangle\varphi \cong \langle 1, -b, ab\rangle \cong \langle ab, -b, 1\rangle,$$

we get, respectively, the following isomorphic models of the two function fields:

(3.13)

$$F[\psi] = F(y,z)\big(\sqrt{b(y^2-az^2)}\,\big), \quad \text{and} \quad F(\psi) = F(x)\big(\sqrt{b(x^2-a)}\,\big).$$

Readers who were diligent in doing exercises will recall that the two ("small") function fields $F(\varphi)$ and $F(\psi)$ above have made a cameo appearance in Ch. III, Exer. 26, as splitting fields for the quaternion algebra $A = \left(\frac{a,\,b}{F}\right)$. The fact that they play this splitting role for A (or for the 2-fold Pfister form $\langle\langle -a, -b\rangle\rangle$) will be a recurrent theme for much of the discussions in the rest of this section.

Before we go on, let us point out an interesting basic property of function fields. To proceed more generally, we recall the following result from classical field theory.

Theorem 3.14. *For a (not necessarily algebraic) field extension K/F, the following are equivalent:*

(1) *F is algebraically closed in K, and K is separably generated over F.*

(2) *For any field extension F'/F, the tensor product algebra $F' \otimes_F K$ is an integral domain.*

(3) *In the algebraic closure $\overline{K}$ of K, $\overline{F}$ and K are linearly disjoint over F (that is, any set of F-linearly independent elements of K remains linearly independent over $\overline{F}$).*

If any of these properties holds for K/F, we say that K is a regular extension of F.

For a proof of this, see Jacobson's text *Basic Algebra* II, §8.18, W. H. Freeman and Co., New York, 1989, or Weil's classical treatise "Foundations of Algebraic Geometry", Colloq. Publications, A.M.S., 1946.

If $\varphi = \langle 1, a\rangle$ is an anisotropic binary form over F, the function fields $F[\varphi]$ and $F(\varphi)$ are *not* regular over F, since they contain the quadratic extension $F(\sqrt{-a}\,)$ over F. However, as soon as we go beyond the binary form case, the following result prevails.

Theorem 3.15. *Let $\varphi = \varphi(x_0, \ldots, x_n)$ be a quadratic form of dimension $n+1 \geq 3$. Then the function fields $F[\varphi]$ and $F(\varphi)$ are both regular extensions of F.*

Proof. The easiest property in 3.14 to verify here is (2). We'll just check the case $K = F[\varphi]$, and leave the other case to the reader. The main point here is that, as long as $\dim \varphi \geq 3$, the polynomial $\varphi(X)$ is *absolutely irreducible* by 3.1 (that is, φ remains irreducible over any extension field $F' \supseteq F$). Given this,

$$F' \otimes_F \frac{F[X]}{(\varphi(X))} \cong \frac{F'[X]}{(\varphi(X))} \tag{3.16}$$

is an integral domain, from which we easily deduce that $F' \otimes_F F[\varphi]$ is an integral domain. This checks property (2) in 3.14, so $F[\varphi]/F$ is a regular field extension of F. □

Remark 3.17. Of course, some of the other properties of a regular extension listed in 3.14 could have been directly checked for $K = F[\varphi]$. For instance, our standing assumption that $\mathrm{char}(F) \neq 2$ certainly guarantees that $F[\varphi]$ is separably generated over F. The fact that F is algebraically closed in $F[\varphi]$ (in case $\dim \varphi \geq 3$) would also follow from the proof of 3.15. Indeed, if $F \subsetneq E$ is an algebraic extension of F in $F[\varphi]$, then $E \otimes_F F[\varphi] \supseteq E \otimes_F E$ would imply that $E \otimes_F F[\varphi]$ is *not* an integral domain, which would contradict what we have already proved.

An immediate consequence of 3.15 (and its proof) is the following.

Corollary 3.18. *If $\dim \varphi \geq 3$, and F'/F is any field extension, then $F'[\varphi]$ and $F'(\varphi)$ are isomorphic to the quotient fields of $F' \otimes_F F[\varphi]$ and $F' \otimes_F F(\varphi)$ respectively. If $\dim \varphi = 2$, the same conclusion holds as long as the binary F-form φ is anisotropic over F'.*

We now conclude this introductory section by giving a discussion on the function field of a *family* of quadratic forms, say $\{\varphi_i\}$. For simplicity, let us just focus on the case of "small" function fields. How should the function field $F(\{\varphi_i\})$ be defined?

We first consider the case where the φ_i are all binary anisotropic forms, say $\varphi_i = \langle 1, -a_i \rangle$, where $a_i \in \dot{F} \setminus \dot{F}^2$. In this case, we may define $F(\{\varphi_i\})$ to be the multiquadratic extension $F(\{\sqrt{a_i}\})$, obtained by simultaneously adjoining the square roots of all a_i's. This is then just the field compositum of all the quadratic extensions $F(\sqrt{a_i}\,)$ in the algebraic closure of F.

Next we consider the case where $\dim \varphi_i \geq 3$ for all i. First suppose the family $\{\varphi_i\}$ consists of just two forms, say φ and ψ. *We would like to define the function field $F(\psi, \varphi)$ to be $F(\psi)(\varphi)$.* In other words, if F' is the function field $F(\psi)$ of ψ, we take $F(\psi, \varphi)$ to be $F'(\varphi_{F'})$—the function field of $\varphi_{F'}$ over F'. By 3.18, we see that $F(\psi, \varphi)$ is the quotient field of the integral domain

$$F' \otimes_F F(\varphi) = F(\psi) \otimes_F F(\varphi).$$

Since this tensor algebra is an integral domain, it follows that, up to an F-isomorphism, $F(\psi, \varphi)$ is just the unique "free compositum" of $F(\psi)$ and $F(\varphi)$ in the sense of field theory.[(5)] This may, therefore, be taken to be the definition of $F(\psi, \varphi)$ too. In particular, we see that the formation of this function field is *symmetrical* in ψ and φ (up to F-isomorphisms, of course).

The case of two forms can be easily extended to the case of n forms for $n < \infty$. The case of an arbitrary family of forms $\{\varphi_i\}$ (of dimension ≥ 3) can thus be defined by a direct limit process, by using the finite case. Or, one can more boldly define $F(\{\varphi_i\})$ as the free compositum of the family of function fields $\{F(\varphi_i)\}$.

In the most general case, let $\{\varphi_i\}$ be a family of F-forms with $\dim \varphi_i \geq 2$ such that the subfamily $\{\sigma_j\}$ of the binary forms among $\{\varphi_i\}$ are all anisotropic. Let $\{\varphi_k\}$ be the family of the remaining forms. Then we can define the function field $F(\{\varphi_i\})$ to be $F(\{\sigma_j\})(\{\varphi_k\})$; that is, we first go to the function field $F' = F(\{\sigma_j\})$, which is a multiquadratic extension of F, and then form the function field $F'(\{\varphi_k\})$ over F'.

The point of the above discussions is to show that the notion of the function field of a *family* of quadratic forms $\{\varphi_i\}$ can indeed be given a precise meaning. Although these discussions are somewhat laborious, they cannot be totally dispensed with, since later in this chapter (and in Ch. XIII), we will be using the notion of the function field $F(\{\varphi_i\})$ for the construction of some fields with specific quadratic form theoretic properties. Knowing that the formation of $F(\{\varphi_i\})$ can be put on a firm footing will greatly facilitate these later transcendental field constructions.

---

(5) The *uniqueness* of the free compositum should not be a surprise here. In field theory, it is known quite generally that, if F is separably algebraically closed in a field E, then for any F'/F, there is (up to isomorphism) only one free compositum of F' and E over F: see Jacobson's Basic Algebra II, p. 554. Here, if we take E to be $F(\varphi)$, F is in fact algebraically closed in E (since $\dim \varphi \geq 3$). In particular, the uniqueness result in Jacobson's book applies.

4. Basic Theorems on Function Fields

In this section, we shall develop the principal properties of function fields of quadratic forms. The main focus of our presentation will be on the isomorphic types of such function fields, and on the nature of the quadratic forms that become isotropic or hyperbolic over these function fields. The exposition here is a considerable expansion of the earlier one I gave in my Queen's University Lecture Notes [L$_1$].

We should warn the reader that this section is quite long, since toward the end of the section, we have included a brief survey on some of the more recent results in the research on function fields. These results are, however, not essential for the rest of the book. Thus, for a first reading, the second part of this section can be skipped. In fact, after working half way through this section, the reader might find it profitable to go ahead to read the beginning part of §5 to see first some concrete applications of function field methods. Some more applications of function field results will be given later in XII.2 and XIII.2.

For much of this section, it will be more convenient to work with the "big" function fields $F[\varphi]$, although we could have equally well used the small function fields $F(\varphi)$. We remind the reader again that, whenever the notation $F[\varphi]$ is used, it will be assumed that $\dim \varphi \geq 2$ and $\varphi \not\cong \mathbb{H}$, for otherwise $F[\varphi]$ is undefined.

The main idea of forming the function field $F[\varphi]$ is that we get a field generated by the coordinates of a "generic point" $(\overline{x}_0, \ldots, \overline{x}_n)$ of the affine variety "$\varphi \equiv 0$", where $\overline{x}_i$ denotes the image of x_i in $F[\varphi]$. Knebusch called $F[\varphi]$ a "generic zero field", and gave a precise definition for this term using valuations and places. For our elementary exposition, however, we will have enough material to chew on without entering into a full discussion of the notion of generic zero fields, for which we'll simply refer our reader to Knebusch's seminal paper [Kn$_4$].

The equation $\varphi(\overline{x}_0, \ldots, \overline{x}_n) = 0$ implies that the quadratic form φ becomes isotropic over its function field $F[\varphi]$. (A similar remark applies to $F(\varphi)$.) This observation leads to the first basic result on function fields, in the form of a criterion for a function field to be purely transcendental over F. (To avoid duplications, we shall state most of our results in terms of $F[\varphi]$, with the understanding that they will also hold for the "small' function fields $F(\varphi)$.)

Theorem 4.1. *A function field $F[\varphi]$ is purely transcendental iff the form φ is isotropic over F. (In particular, any two isotropic quadratic forms of the same dimension have isomorphic function fields.)*

Proof. First assume $F[\varphi]$ is purely transcendental. Since φ becomes isotropic over $F[\varphi]$, IX.1.1 implies that φ must already be isotropic over F. Conversely, assume that φ is isotropic over F. After changing variables, we may express φ in the form $x_0x_1 + \psi(x_2, \ldots, x_n)$, where ψ is a regular quadratic form in $x_2, \ldots, x_n$. Using this expression of φ to calculate $F[\varphi]$, we see quickly that $F[\varphi]$ is isomorphic to the rational function field $F(x_1, \ldots, x_n)$. □

Since φ always becomes isotropic over $F[\varphi]$, it is of interest to ask *what other forms over F might also become isotropic, or even hyperbolic, over $F[\varphi]$*. This natural question turned out to be much deeper than it might have first appeared. Indeed, one main direction for the research on function fields of quadratic forms is precisely to try to answer this question in as many cases as possible (depending on the nature of the form φ). Although various results have been obtained in the last three decades, a full answer to the above question has remained unknown up to this date. In this section, we shall present some of the principal known results in this area.

To facilitate our discussions on the question at hand, we introduce the following notational device.

Notation 4.2. For any quadratic form q, we write $q > \varphi$ (resp. $q \gg \varphi$)[(6)] to express the fact that q becomes isotropic (resp. hyperbolic) over the function field $F[\varphi]$ of the quadratic form φ.

For any field extension K/F, we have introduced earlier the notation $W(K/F)$ for the kernel of the functorial map $W(F) \to W(K)$. This ideal of $W(F)$ is called the *Witt kernel* of the extension K/F. In terms of this Witt kernel notation, the relation $q \gg \varphi$ in 4.2 simply amounts to $q \in W(F[\varphi]/F)$.

Examples 4.3. (1) Of course, $\varphi > \varphi$; and $q \gg \varphi \Longleftrightarrow q\,W(F) \gg \varphi$.

(2) Suppose $q_1 = q_2 + q_3 \in W(F)$. If $q_i \gg \varphi$ for two values of i, then it holds for all three.

(3) Let $c \in \dot{F}$. Then $q \supseteq c \cdot \varphi \Rightarrow q > \varphi$, since $\varphi_{F[\varphi]}$ is isotropic.

(4) If $\dim q > 0$, clearly $q \gg \varphi \Rightarrow q > \varphi$. The converse fails in general, but does hold when q is a Pfister form (by 1.7).

(5) Suppose $q \supseteq q'$, with $\dim q < 2 \dim q'$. Then $q \gg \varphi \Rightarrow q' > \varphi$. This follows from Ch. I, Exer. 14.

(6) Both "$\gg$" and "$>$" turn out to be *transitive* relations on forms. We shall prove this later in 4.22.

---

(6)It will be convenient sometimes to write also $\varphi < q$ instead of $q > \varphi$, and $\varphi \ll q$ instead of $q \gg \varphi$.

In the notation of 4.2, we can now formulate our main problem in a more formal way.

Question 4.4A. *If φ is given, what exactly are the forms $q \gg \varphi$?* (In other words, what exactly is the Witt kernel $W(F(\varphi)/F)$?) The same question may be asked on the relation $q > \varphi$.

We can also reverse the roles of the forms φ and q, and ask instead:

Question 4.4B. *If q is given, what exactly are the forms $\varphi \ll q$?* (And the same for $\varphi < q$.)

One important case we have dealt with earlier is when $\varphi = \langle\langle -a \rangle\rangle$. In this case, VII.3.1 and VII.3.2 gave us complete descriptions for forms q such that $q > \varphi$ or $q \gg \varphi$, respectively.

Most of the results in this section will be concerned with 4.4A, so we'll start there. If φ happens to be an isotropic form ($\not\cong \mathbb{H}$), then $F[\varphi]$ is purely transcendental over F by 4.1, so $W(F) \to W(F[\varphi])$ is an injection by IX.1.1. In this case, $q \gg \varphi$ iff q is hyperbolic over F (and by the same reference, $q > \varphi$ iff q is isotropic over F). If φ is an anisotropic form, the characterization of the forms $q \gg \varphi$ is a much more subtle problem. The following result gives the first significant necessary condition on the forms $q \gg \varphi$ (for a given φ). As we shall see shortly, this result has quite a few important consequences!

Theorem 4.5. *Suppose $q \gg \varphi$, where q, φ are quadratic forms over F, with $1 \in D_F(\varphi)$. Then $\varphi(X) \in G_{F(X)}(q)$, where $X = (x_0, \ldots, x_n)$, and $\dim \varphi = n+1$. (In other words, we have $\varphi(X) \cdot q \cong q$ over the rational function field $F(X)$.) If q is anisotropic, then $a \cdot \varphi \subseteq q$ (over F) for any $a \in D_F(q)$; in particular, we must have $\dim q \geq \dim \varphi$ (if $\dim q \neq 0$).*

Proof. Write $\varphi \cong \langle 1 \rangle \perp \varphi'$, so

$$L := F[\varphi] = K(\sqrt{-\varphi'(X')}\,), \tag{4.6}$$

where $X' = (x_1, \ldots, x_n)$ and $K = F(X')$. After removing the hyperbolic part of q (which affects neither the hypothesis nor the conclusion of the theorem), we may assume that q is *anisotropic* over F. By IX.1.1, q_K is anisotropic, but by hypothesis, q_L is hyperbolic. Using VII.3.2, we have an isometry

$$q_K \cong \sigma \cdot \langle 1, \varphi'(X') \rangle \quad \text{over } K, \tag{4.7}$$

where σ is some form over K. Now work over the rational function field $K(x_0) = F(X)$. Over this field, the binary form $\langle 1, \varphi'(X')\rangle$ represents $x_0^2 + \varphi'(X') = \varphi(X)$, and therefore has $\varphi(X)$ as a similarity factor. Reading (4.7) in $F(X)$, we clearly get $\varphi(X) \in G_{F(X)}(q)$, as desired. Since q is anisotropic, it follows further from IX.2.10 that, for any $a \in D_F(q)$, $a \cdot \varphi \subseteq q$ over F. □

Remark 4.8. Needless to say, the hypothesis that $1 \in D_F(\varphi)$ in 4.5 does not really limit the applicability of this result. If $1 \notin D_F(\varphi)$, we may replace φ by $b \cdot \varphi$ for any $b \in D_F(\varphi)$. Since $F[\varphi] = F[b \cdot \varphi]$, the assumption that $q \gg \varphi$ will simply lead to $b \cdot \varphi(X) \in G_{F(X)}(q)$, and in case q is anisotropic, to $ab \cdot \varphi \subseteq q$ (for any $a \in D_F(q)$), and we'll have $\dim q \leq \dim \varphi$ just as before.

We shall now deduce a number of interesting consequences of 4.5. The first of these is complete answer to Question 4.4B (for the $q \gg \varphi$ part) in case q is a Pfister form.

Corollary 4.9. *Let q be an anisotropic Pfister form. Then $q \gg \varphi$ iff $c \cdot \varphi \subseteq q$ for some $c \in \dot{F}$.*

Proof. The "only if" part follows from 4.5 and 4.8. (Take c to be as in 4.8.)(7) For the "if" part, assume that $c \cdot \varphi \subseteq q$ for some $c \in \dot{F}$. Then $q > \varphi$ by 4.3(1), and hence $q \gg \varphi$ by 4.3(4). □

We record the following immediate application of 4.9 to the study of the isomorphism type of function fields of quadratic forms. From here on, we write $q \sim \varphi$ if q and φ are *similar*; that is, $q \cong c \cdot \varphi$ for some $c \in \dot{F}$.

Corollary 4.10. *Let φ be a form of dimension 2^n, and let q be an anisotropic n-fold Pfister form over F, where $n \geq 1$. Then the following are equivalent:*

(1) $q \sim \varphi$.

(2) $F[q] \cong F[\varphi]$ *(of course, over F).*

(3) $q > \varphi$ *(that is, q becomes isotropic over $F[\varphi]$).*

Proof. (1) $\Longrightarrow$ (2) $\Longrightarrow$ (3) are clear. For (3) $\Longrightarrow$ (1), assume $q > \varphi$. Since q is a Pfister form, $q \gg \varphi$. By 4.9, $c \cdot \varphi \subseteq q$ for some $c \in \dot{F}$. Since $\dim \varphi = \dim q$, we have $q \cong c \cdot \varphi \sim \varphi$. □

In the language of classical algebraic geometry, the function field of an irreducible variety determines only the "birational type" of the variety itself. Corollary 4.10 is an interesting result in that, in some cases, the function field of a quadratic form q not only determines the isomorphic type of the quadric it defines, but also the quadratic form q itself up to a similarity.

In parallel to 4.9, in the case where φ is a Pfister form, we can give a complete answer (to the $q \gg \varphi$ part) of Question 4.4A. The following result

(7)For this part, of course, we do not need q to be a Pfister form.

was first formulated and proved in my paper [EL$_1$] with Elman, but was already implicit in the work of Arason and Pfister [AP$_1$]; see also [Ara$_1$].

Theorem 4.11. *Let φ be a Pfister form, and let q be an anisotropic form (over F). The following statements are equivalent:*

(1) $q \gg \varphi$.

(2) $q \cong \varphi \cdot \tau$ *for some F-form τ.*

(3) $q = \varphi \cdot \tau \in W(F)$ *for some F-form τ.*

In particular, $W(F[\varphi]/F) = \varphi \cdot W(F)$.

Proof. (2) $\Longrightarrow$ (3) is a tautology, and (3) $\Longrightarrow$ (1) follows since $\varphi \gg \varphi$ by 1.7. For (1) $\Longrightarrow$ (2), assume that $q \gg \varphi$, and take $a_1 \in D_F(q)$. By 4.5, $q \cong a_1 \cdot \varphi \perp q_1$ for some F-form q_1, which is necessarily $\gg \varphi$, by 4.3(2). Invoking an inductive hypothesis at this point, we have $q_1 \cong \varphi \cdot \langle a_2, \ldots, a_n \rangle$ (for some $a_2, \ldots, a_n \in \dot{F}$), and so $q \cong \varphi \cdot \langle a_1, \ldots, a_n \rangle$, proving (3). $\square$

Note that the corollary above does not require the Pfister form φ to be anisotropic. If φ is isotropic, then it is hyperbolic, and $F[\varphi]$ is purely transcendental. In this case, we have

$$W(F[\varphi]/F) = 0 = \varphi \cdot W(F),$$

so the corollary does hold indeed.

Remark 4.12. In the special case where $\varphi = \langle 1, a \rangle \not\cong \mathbb{H}$, recall that $F(\varphi) \cong F(\sqrt{-a}\,)$ and $F[\varphi] = F(\sqrt{-a}, x)$. By IX.1.1,

$$W(F[\varphi]/F) = W(F(\varphi)/F) = W(F(\sqrt{-a}\,)/F),$$

so the last statement in 4.10 retrieves the fact (VII.3.2) that

$$W(F(\sqrt{-a}\,)/F) = \langle 1, a \rangle W(F).$$

But of course, the proof of 4.5 already assumed the validity of this equation (in step (4.7)), and 4.10 depends critically on 4.5. Indeed, after reviewing the proofs of 4.5 and 4.11, we see that the computation of $W(F[\varphi]/F)$ for a Pfister form φ was carried out essentially by making a reduction to the case of 1-fold Pfister forms—over a rational function field $F(x_1, \ldots, x_n)$, where $n = (\dim \varphi) - 1$.

Corollary 4.13. *In 4.11, if q is (in addition) a Pfister form, then the conditions (1), (2), (3) there are also equivalent to:*

(4) $q \supseteq \varphi$.

(5) $q \cong \varphi \cdot \tau$ *for some Pfister form τ.*

Proof. Clearly, (5) $\Longrightarrow$ (4) $\Longrightarrow$ (1). Thus, it suffices to prove (2) $\Longrightarrow$ (5). Say $q \cong \varphi \cdot \langle a, b, \dots \rangle$. After a scaling, we may assume that $a = 1$, so we have $q \gg \varphi \langle\langle b \rangle\rangle$. By 4.11, $q \cong \varphi \langle\langle b \rangle\rangle \sigma$ for some σ. Continuing in this way, we get (5). $\square$

The next result builds on the fact, used in the proof of 4.11, that a Pfister form φ "dies off" over the function field $F[\varphi]$. This "suicidal property" of φ in relation to its own function field turns out to be essentially a new characterization for Pfister forms (along with, of course, the hyperbolic forms).

Theorem 4.14. *For any form $q \not\cong \mathbb{H}$ of dimension ≥ 2, we have $q \gg q$ iff q is either a hyperbolic form or a scalar multiple of a Pfister form.*

Proof. It suffices to prove the "only if" part, so start with $q \gg q$. After a scaling, we may assume that $1 \in D_F(q)$. By 4.5, we have $q(X) \cdot q \cong q$ over $F(X)$, so q is *strongly multiplicative* in the sense of 2.6. By 2.8 and 2.9, such a form is either a hyperbolic form or a scalar multiple of a Pfister form, so we are done.

We note in passing that 4.14 could have been proved by using 2.8 alone (and not 2.9). We just go into the following two cases.

Case 1. *q is isotropic over F.* Here, $F[\varphi]$ is purely transcendental over F by 4.1. Thus, $q \gg q$ implies that q is already hyperbolic over F.

Case 2. *q is anisotropic over F.* As before, we may assume $1 \in D_F(q)$ to get $q(X) \cdot q \cong q$ over $F(X)$. Then (3) $\Longrightarrow$ (1) in 2.8 shows that q is a Pfister form over F. $\square$

Corollary 4.15. *$q \gg \varphi \gg q$ iff q, φ are scalar multiples of a single Pfister form, or q, φ are both hyperbolic.*

Proof. The "if" part is (by now) obvious. Conversely, assume that $q \gg \varphi \gg q$. If q and φ are both anisotropic, 4.5 implies that $q \cong a \cdot \varphi$ for some $a \in \dot{F}$. Then $q \gg q$, and 4.14 shows readily that q and φ are scalar multiples of a single (anisotropic) Pfister form. Now assume q is isotropic. Then $F[q]/F$ is purely transcendental, so $\varphi \gg q$ implies that φ is hyperbolic. The same argument also shows that q is hyperbolic. $\square$

Our next goal is to generalize the isomorphism result 4.10 to some other classes of forms (besides Pfister forms). To develop some results in this direction, let us first introduce the important notion of a "Pfister neighbor", due to M. Knebusch.

Definition 4.16. A form σ over F is called a *Pfister neighbor* if, for some $a \in \dot{F}$, $a \cdot \sigma \subseteq \varphi$ for some Pfister form φ (over F) of dimension $< 2 \dim \sigma$.

(In other words, σ is a Pfister neighbor iff it is similar to a subform σ_1 of a Pfister form φ with $\dim \sigma_1 > \frac{1}{2} \dim \varphi$.)

Note that, in the above, if φ is an n-fold Pfister form, then $2^{n-1} < \dim \sigma \leq 2^n$. Thus, if σ is indeed a Pfister neighbor, the integer n is uniquely determined by $\dim \sigma$. It is a pleasant surprise, perhaps, that even the Pfister form φ is uniquely determined (up to an isometry) by σ, as the following result shows.

Proposition 4.17. *Let σ, φ, and n be as above, and suppose we also have $b \cdot \sigma \subseteq \psi$ for some $b \in \dot{F}$ and some Pfister form ψ of dimension $< 2 \dim q$. Then $\varphi \cong \psi$.*

Proof. *Case* 1. *φ is isotropic.* By 1.7, φ is hyperbolic, so by Ch. I, Exer. 14, σ is isotropic. Then ψ is also isotropic, and hence hyperbolic. In this case, $\varphi \cong \psi \cong 2^{n-1}\mathbb{H}$.

Case 2. *φ is anisotropic.* Over $F[\psi]$, the argument above shows that σ becomes isotropic. By 1.7, it follows that $\varphi \gg \psi$. Since $\dim \varphi = \dim \psi = 2^n$, 4.5 yields $\varphi \sim \psi$, and hence $\varphi \cong \psi$. □

In view of the uniqueness of φ, we call φ the *associated Pfister form* of the Pfister neighbor σ. Actually, the following slightly reformulated version of 4.17 may make the result look a little more impressive: *two n-fold Pfister forms φ and ψ are isometric if they contain, respectively, r-dimensional subforms that are similar, for some $r \in (2^{n-1}, 2^n]$.*

Reportedly, the increasingly popular use of the term "Pfister neighbor" has greatly enticed mathematicians to sit next to Albrecht Pfister in the recent conferences in the theory of quadratic forms. For reasons of mathematical exposition, however, we prefer a list of examples of non-human Pfister neighbors below.

Examples 4.18. (1) A 2^n-dimensional form q is a Pfister neighbor iff $q \cong a \cdot \varphi$ for some $a \in \dot{F}$ and some n-fold Pfister form φ. (In this case, φ is the associated Pfister form of q.) These are rather trivial examples of Pfister neighbors, but are examples nevertheless. For instance, any form of dimension 1 or 2 is a Pfister neighbor of this kind.

(2) $\sigma = r\langle a \rangle$ $(a \in \dot{F},\ r \geq 1)$ *is always a Pfister neighbor.* In fact, if $2^{n-1} < r \leq 2^n$, then the subform relation $\langle a \rangle \cdot \sigma \subseteq \varphi := 2^n \langle 1 \rangle$ shows that σ is a Pfister neighbor, with an associated Pfister form φ.

(3) Let φ be an n-fold Pfister form $(n \geq 1)$, with pure subform φ'. Then $d \cdot \varphi'$ (for any $d \in \dot{F}$) is a Pfister neighbor, with associated Pfister form φ. In fact, *any Pfister neighbor σ of dimension $2^n - 1$ arises in this*

manner. For, suppose $\varphi = a \cdot \sigma \perp \langle b \rangle$ is its associated Pfister form. By 1.8,

$$\langle 1 \rangle \perp \varphi' \cong \varphi \cong b \cdot \varphi \cong ab \cdot \sigma \perp \langle 1 \rangle,$$

so Witt cancellation yields $\sigma \cong ab \cdot \varphi'$. We may refer to this property by saying that *all Pfister neighbors of codimension* 1 *in* φ *are similar.*

(4) Let τ_1 be a subform of positive dimension in an $(n-1)$-fold Pfister form τ. For any scalar $a \in \dot{F}$, $\tau \perp a\tau_1$ is a subform of the n-fold Pfister form $\varphi := \tau \langle\langle a \rangle\rangle$. Any form σ similar to $\tau \perp a\tau_1$ is then a Pfister neighbor (with associated Pfister form φ). Such a form σ is called a *special Pfister neighbor*, following Ahmad and Ohm [AO]. For instance, any form similar to $\tau \perp \langle a \rangle$ (for some $(n-1)$-fold Pfister form τ) is a special Pfister neighbor of dimension $2^{n-1}+1$, and conversely. An easy check shows that the examples of Pfister neighbors given above in (1), (2), and (3) are all special Pfister neighbors. (For some alternative characterizations of special Pfister neighbors, see Exer. 23, 24.)

(5) *Any ternary form* σ *is a special Pfister neighbor.* For, taking $d = d(\sigma)$, we have $d \cdot \sigma \cong \langle a, b, ab \rangle$ for some $a, b \in \dot{F}$. Thus, $\sigma \cong d \cdot \varphi'$ for $\varphi = \langle\langle a, b \rangle\rangle$. This is the special case $n = 2$ in (3) (and therefore also a special case of (4)).

(6) *A* 4*-dimensional form* σ *is a Pfister neighbor iff* $d(\sigma) = 1 \in \dot{F}/\dot{F}^2$. The "only if" part is clear from (1). Conversely, if $d(\sigma) = 1$, then

$$\sigma \cong \langle a, b, c, abc \rangle \cong a \langle\langle ab, ac \rangle\rangle.$$

The case of 5-dimensional forms is handled in the following basic result of Knebusch [Kn_5].

Proposition 4.19. *For a* 5*-dimensional form* σ *with determinant* $d \in \dot{F}/\dot{F}^2$*, the following are equivalent:*

(1) σ *is a Pfister neighbor.*

(2) σ *is a special Pfister neighbor.*

(3) $d \in D_F(\sigma)$.

(4) σ *contains a subform* $c \langle\langle a, b \rangle\rangle$ *for some* $a, b, c \in \dot{F}$.

Proof. (3) $\iff$ (4) is clear from 4.18(6).

(4) $\Longrightarrow$ (2). Given (4), we have $\sigma \cong c \langle\langle a, b \rangle\rangle \perp \langle d \rangle$. Then $c \cdot \sigma \cong \langle\langle a, b \rangle\rangle \perp \langle cd \rangle$ shows that σ is a special Pfister neighbor.

(2) $\Longrightarrow$ (1) is, of course, trivial.

(1) $\Longrightarrow$ (4). Say $r \cdot \sigma \perp \langle s, a, b \rangle$ is a 3-fold Pfister form q. After scaling this by $s \in D_F(q) = G_F(q)$, we may assume that $s = 1$. By 1.11, there exists $t \in \dot{F}$ such that

$$q \cong \langle\langle a, b, t \rangle\rangle \cong \langle 1, a, b, ab \rangle \perp t \langle\langle a, b \rangle\rangle.$$

Cancellation of $\langle 1, a, b\rangle$ then yields $r \cdot \sigma \cong \langle ab\rangle \perp t\,\langle\langle a, b\rangle\rangle$, so (4) holds with $c = rt \in \dot{F}$. □

It may well be said that the proposition above is a result of a combinatorial nature. The implication (1) $\Longrightarrow$ (3), stated for *diagonal forms* only, amounts to the following curious combinatorial fact: *If G is an elementary 2-group with a $\mathbb{Z}_2$-basis $\{a, b, c\}$, then any 5-element subset of G contains a translation of a Klein 4-(sub)group of G.*

The idea of 5-dimensional Pfister neighbors is closely related to the notion of linkage of quaternion algebras. To express this relationship more formally, we state the following consequence of 4.19.

Theorem 4.20. *Over any field F, the following are equivalent:*

(1) *Any pair of quaternion algebras over F are linked.*

(2) *There are no biquaternion division algebras over F.*

(3) *Any Albert form (6–dimensional form of determinant -1) over F is isotropic.*

(4) *The classes of quaternion algebras form a subgroup of the Brauer group $B(F)$.*

(5) *Any 5-dimensional form over F is a Pfister neighbor.*

(6) *Any 5-dimensional form σ over F represents $d(\sigma)$.*

(7) *Any 5-dimensional form over F contains a subform $c\,\langle\langle a, b\rangle\rangle$ for some a, b, $c \in \dot{F}$.*

A field F satisfying any (and hence all) of the above conditions is called a linked field (as in III.4*).*

Proof. The equivalence of (1), (2), (3), and (4) follows from III.4.8 and III.2.11. The equivalence of (5), (6), and (7) follows from 4.19. Finally, (3) $\Longleftrightarrow$ (6) is clear. □

In fact, over a linked field, any form of dimension $2^n + 1$ (for some n) is a Pfister neighbor. The proof of this is left as an exercise (see Exer. 25).

Example 4.21. Recall that any local field or global field is a linked field (by VI.3.6). Thus, over such a field, *any form of dimension $2^n + 1$ (for any n) is a Pfister neighbor.*

At this time, we return to the two relations "$>$" and "$\gg$" defined in 4.2. These relations turn out to be especially useful in understanding the basic properties of Pfister neighbors. First, let us verify the transitive nature of these relations, which we have already alluded to in 4.3(6). In fact, we'll be proving a little more.

Theorem 4.22. *Let φ, σ be F-forms whose function fields are defined. For any form q, we have:*

$$q > \varphi, \quad \varphi > \sigma \implies q > \sigma, \qquad \text{and}$$
$$q \gg \varphi, \quad \varphi > \sigma \implies q \gg \sigma.$$

In particular, both "$>$" and "$\gg$" are transitive relations.

Proof. Let us first deal with the special case where φ becomes the hyperbolic plane over $L = F[\sigma]$. In this case, $\varphi = \langle\langle -a \rangle\rangle$ for some $a \notin \dot{F}^2$, and by 4.5, σ must also be 2-dimensional, say $\sigma = \langle\langle -b \rangle\rangle$ $(b \notin \dot{F}^2)$. Then $\varphi_L \cong \mathbb{H}_L$ implies $a = b \in \dot{F}/\dot{F}^2$, so we have $\varphi \cong \sigma$. The conclusions of 4.22 are tautologies in this case.

We may now assume $\varphi_L \ncong \mathbb{H}_L$, so that the function field $L[\varphi]$ is defined, and we can think of $L[\varphi]$ as a free compositum of $L = F[\sigma]$ and $F[\varphi]$. From the hypothesis $\varphi > \sigma$, $L[\varphi]/L$ is purely transcendental by 4.1.

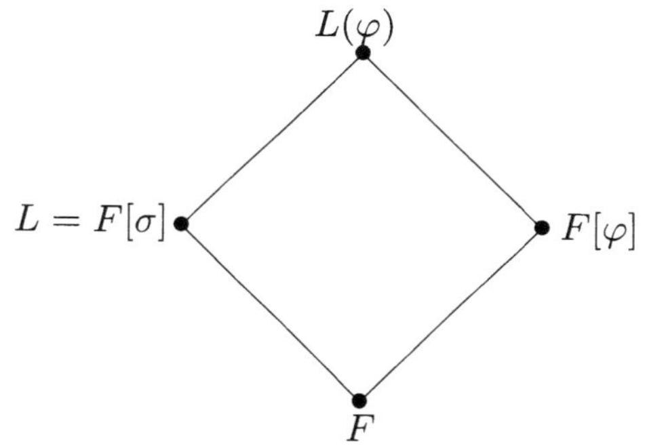

If q is isotropic over $F[\varphi]$, then it is isotropic over $L[\varphi]$. By IX.1.1, q is isotropic over L, and hence $q > \sigma$. This proves the first implication in 4.22, and the second follows by the same argument, with the words "isotropic" replaced by "hyperbolic". $\square$

Remark 4.23. In the case where $\varphi \supseteq \sigma$, 4.22 implies that, *if q is isotropic (resp. hyperbolic) over $F[\varphi]$, then in fact q is isotropic (resp. hyperbolic) over $F[\sigma]$.* This might seem a little surprising at first, but that is just the way these function fields work. (I first learned about this kind of special behavior of function fields from a conversation with A. Wadsworth.) In particular, if σ is taken to be binary, the following special conclusions can be drawn.

Corollary 4.24. *Let $\varphi = \langle 1, a, \dots \rangle$, where $a \notin -\dot{F}^2$. If $q \gg \varphi$, then $q \cong \langle 1, a \rangle \sigma$ for some F-form σ. If $q > \varphi$, then $q \supseteq b \langle 1, a \rangle$ for some $b \in \dot{F}$.*

Next, we try to say something about the special relation $\varphi > \sigma > \varphi$ between two quadratic forms σ and φ whose function fields are defined.

In view of 4.22, this is an equivalence relation between such forms. It turns out that this equivalence relation has another very natural interpretation, which we shall now explain.

In the language of classical algebraic geometry, two extension fields K and L of a field F are said to be *stably isomorphic* (over F), written $K \cong_{\text{st}} L$, if there exist an F-isomorphism

$$K(x_1, \dots, x_k) \cong L(y_1, \dots, y_\ell) \quad \text{for some } k, \ell > 0.$$

(Here, the x_i 's and y_j 's are commuting indeterminates.) The following result gives an interesting link between the relation $\varphi > \sigma > \varphi$ and the stable isomorphism of the function fields of σ and φ. It also gives a remarkable application of 4.22 to the situation where the relation $\varphi > \sigma > \varphi$ holds.

Theorem 4.25. *Let $\varphi(X_0, \dots, X_m)$ and $\sigma(Y_0, \dots, Y_n)$ be regular quadratic forms (over F) such that the function fields $L = F[\sigma]$ and $K = F[\varphi]$ are defined. Then the following statements are equivalent:*

(1) $\varphi > \sigma > \varphi$.

(2) $K(x_1, \dots, x_n) \cong L(y_1, \dots, y_m)$ *over* F.

(3) $K \cong_{\text{st}} L$*; that is,* K *and* L *are stably isomorphic over* F.

If these conditions hold, then for any quadratic form q, we have

$$(4.26) \qquad q \gg \varphi \iff q \gg \sigma, \quad \text{and} \quad q > \varphi \iff q > \sigma.$$

Note that the first equivalence here amounts to a Witt kernel equation

$$W(F[\varphi]/F) = W(F[\sigma]/F).$$

Proof. We shall prove (2) $\Rightarrow$ (3) $\Rightarrow$ (1) $\Rightarrow$ (2), the first implication being trivial. Now assume (3), so that $K' \cong L'$ (over F) for suitable rational function field extensions K'/K and L'/L. Since φ is isotropic over K, it is also isotropic over K', and hence over L'. By 4.1, φ is isotropic over L; that is, $\varphi > \sigma$, and (1) follows by symmetry.

Next, assume (1), and refer to the diagram in the proof of 4.22. In this diagram, the "top" field is $K(\sigma) = L(\varphi)$. Since σ is isotropic over K, and φ is isotropic over L (of dimensions $n+1$ and $m+1$ respectively), 4.1 implies that the "top" field is F-isomorphic to $K(x_1, \dots, x_n)$ as well as to $L(y_1, \dots, y_m)$, which proves (2).

Finally, under the assumption $\varphi > \sigma > \varphi$, the conclusions in (4.26) (for any quadratic form q) follow directly from 4.22. □

In connection to 4.25, we should mention another important interpretation of the relation $\varphi > \sigma > \varphi$ that we will not develop in our text. Quite generally, Knebusch has proved in [Kn$_4$] that the F-form φ becomes isotropic over an extension field F'/F iff there exists an F-place from the function field $F[\varphi]$ to F'. (An F-*place* means a place that is the identity on F.) Using this result of Knebusch, we see that the relation $\varphi > \sigma > \varphi$ may also be interpreted as to mean that the two function fields $F[\varphi]$ and $F[\sigma]$ admit F-places into each other. In view of 4.25, this leads to a place-theoretic interpretation of the stable isomorphism of function fields of quadratic forms. However, this interpretation will not be needed (or used) in the sequel.

Returning to notion of Pfister neighbors, we can now explain why such a notion is interesting and useful, in the light of 4.25.

Lemma 4.27. *Let φ and σ be Pfister neighbors with the same associated Pfister form τ. If $F[\varphi]$ and $F[\sigma]$ are both defined, then $\varphi > \sigma > \varphi$; in particular, the equivalences in* (4.26) *hold for all forms q.*

Proof. Since τ is a Pfister form, $\tau > \sigma \implies \tau \gg \sigma$ by 1.7. Therefore, by 4.3(5), $\varphi > \sigma$, and $\sigma > \varphi$ follows by symmetry. □

In the special case of 4.27 where φ is already a Pfister form, we have in fact the stronger statement: $\varphi \gg \sigma > \varphi$. The conclusions in this case are already worth recording.

Corollary 4.28. *Let σ be a Pfister neighbor with an associated Pfister form φ, and let q be any Pfister form. Then,*

(1) $W(F[\sigma]/F) = \varphi \cdot W(F)$.

(2) *An F-form becomes isotropic over $F[\sigma]$ iff it becomes isotropic over $F[\varphi]$.*

(3) *σ is similar to a subform of q iff $q \cong \varphi \cdot \tau$ for some Pfister form τ.*

Proof. (1) and (2) follow from 4.25 and 4.11. For (3), the "if" part is clear. For the "only if" part, assume that σ is similar to a subform of q. Then we have $q \gg \sigma$ (since q is Pfister), and $\sigma > \varphi$ (since σ is a Pfister neighbor associated with φ). By 4.22, $q \gg \varphi$, so 4.13 applies. (Note that part (3) here means that we can add one more condition to those characterizing the divisibility of Pfister forms by one another obtained in 4.11 and 4.13.) □

Remark 4.29. In the case where σ is a 3-dimensional form (necessarily a Pfister neighbor by 4.18(5)), the last part (3) above says: *if $q \supseteq \langle a, b, c, \dots \rangle$, then we can express q in the form $\langle\langle ab, ac, \dots \rangle\rangle$.* Thus, we may view 4.28(3) as a generalization of 1.11.

Using 4.27, we can extend some of our earlier results on Pfister forms to Pfister neighbors. We begin with the following extension of 4.10, which is due to Knebusch [Kn_5] and Wadsworth [Wad_1].

Theorem 4.30. *Suppose σ is an anisotropic Pfister neighbor of dimension $2^n - 1 \geq 3$. For any form τ of the same dimension, the following are equivalent:*

(1) $\sigma \sim \tau$.

(2) $F[\sigma] \cong F[\tau]$ *over* F.

(3) $\sigma > \tau$.

Proof. (1) $\Longrightarrow$ (2) $\Longrightarrow$ (3) are clear. For (3) $\Longrightarrow$ (1), assume $\sigma > \tau$. In view of 4.18(3), we may assume that σ is the pure subform of its associated Pfister form φ. Then φ must also be anisotropic, and $\dim \varphi = 1 + \dim \tau$. Since $\sigma > \tau$, we have $\varphi \gg \tau$. After scaling τ, we may thus assume that $\varphi \cong \tau \perp \langle b \rangle$ (for some $b \in \dot{F}$) by 4.5. Then

$$\varphi \cong b \cdot \varphi \cong b \cdot \tau \perp \langle 1 \rangle \implies \sigma \cong b \cdot \tau. \qquad \square$$

As a special case of 4.30, we retrieve the following well-known classical result of Witt on function fields of ternary forms. (Of course, the corresponding result on binary forms is also true, and is already well covered by VIII.3.1.)

Corollary 4.31. *For ternary forms σ and τ, the following are equivalent:*

(1) $\sigma \sim \tau$.

(2) $F[\sigma] \cong F[\tau]$ *over* F.

(3) $\sigma > \tau > \sigma$ (*that is,* $F[\sigma] \cong_{\text{st}} F[\tau]$).

Proof. (1) $\Longrightarrow$ (2) $\Longrightarrow$ (3) are again clear, so it suffices to prove (3) $\Longrightarrow$ (1). Thus, assume $\sigma > \tau > \sigma$. If σ is anisotropic, then $\sigma > \tau$ alone implies (1) by 4.30 (since σ is a Pfister neighbor by 4.18(5)). We may now assume σ and τ are *both* isotropic. But then $\dim \sigma = \dim \tau = 3$ clearly forces σ and τ to be similar. $\square$

It turns out that *the corollary is also true for* 4-*dimensional forms σ and τ, as long as σ is anisotropic*. This result is due to A. Wadsworth [Wad_1]; we shall return to prove it later in XII.2.2. Wadsworth's result depends on a certain similarity theorem on forms under quadratic extensions; see XII.2.1 for a full statement of this theorem. While the proof of this theorem will be postponed to Chapter XII, we propose to use it here to solve some special cases of Question 4.4A: *given an anisotropic form σ, what are the forms q such that $q > \sigma$?*

We would like to focus our attention on the case where $\sigma = \langle 1, -a, -b \rangle$. (Recall that it is harmless to assume $1 \in D_F(\sigma)$.) We opt for the "small" function field of σ, which is

$$F(\sigma) = F(x)\big(\sqrt{ax^2+b}\big),$$

according to (3.12). This is called the "function field of a conic", since the equation $y^2 = ax^2 + b$ defines a "conic section" over F.

We do know all the F-forms q that become *hyperbolic* over $F(\sigma)$ (or $F[\sigma]$). Indeed, since σ is a Pfister neighbor, with an associated Pfister form $\varphi = \langle\langle -a, -b\rangle\rangle$, 4.28 gives

$$W(F(\sigma)/F) = W(F(\varphi)/F) = \varphi \cdot W(F).$$

But what are the (anisotropic) forms q that become *isotropic* over $F(\sigma)$ (or $F[\sigma]$)? The following result provides the answers in case $\dim q \leq 4$.

Theorem 4.32. *Let $E = F(x)\big(\sqrt{ax^2+b}\big)$ be the function field of an anisotropic ternary form $\sigma = \langle 1, -a, -b\rangle$, and let q be an anisotropic F-form.*

(1) *If* $\dim q = 2$, *q cannot become isotropic over E.*

(2) *If* $\dim q = 3$, *q_E is isotropic iff $q \sim \sigma$.*

(3) *If* $\dim q = 4$, *q_E is isotropic iff $e \cdot \sigma \subseteq q$ for some $e \in \dot{F}$.*

Proof. For (1), we may assume $q = \langle 1, -c\rangle$ ($c \notin \dot{F}^2$). Since $\dim \sigma = 3$, the function field extension $F(\sigma)/F$ is *regular* by 3.15. It follows that F is algebraically closed in $E = F(\sigma)$ (see 3.14(1)). In particular, $\sqrt{c} \notin E$, and this means that q remains anisotropic over E.

For (2) and (3), we need only prove the "only if" parts. First assume $\dim q = 3$. Since q is an anisotropic Pfister neighbor, $q > \sigma$ implies $q \sim \sigma$ by 4.30. Finally, assume $\dim q = 4$, with $q > \sigma$, and say $1 \in D_F(q)$. Let $d = d(q)$, and let $\varphi = \langle\langle -a, -b\rangle\rangle$. If $d \in \dot{F}^2$, then q is a 2-fold Pfister form. Then $q > \sigma > \varphi$ implies $q \sim \varphi$ by 4.10. In particular, $q \supseteq e \cdot \sigma$ for some $e \in \dot{F}$. (Of course, $q \cong \varphi$ here since both are Pfister forms.) Now assume $d \notin \dot{F}^2$, and let $K = F(\sqrt{d})$. Over K, q *becomes* a 2-fold Pfister form, and it remains anisotropic (by an easy application of VII.3.3). We still have $q_K > \sigma_K$, so the case we have settled gives $q_K \cong \varphi_K$. Let $\tau = \sigma \perp \langle dab\rangle$. Then $d(\tau) = d = d(q)$, and we have

$$q_K \cong \varphi_K = \langle 1, -a, -b, ab\rangle_K \cong \sigma_K \perp \langle dab\rangle_K = \tau_K.$$

Thus, by Wadsworth's Similarity Theorem (XII.2.1), there exists $e \in \dot{F}$ such that $q \cong e \cdot \tau \supseteq e \cdot \sigma$, as desired. $\square$

It might be tempting to surmise that an anisotropic form q becomes isotropic over $E = F(\sigma)$ (the function field of a conic) iff $e \cdot \sigma \subseteq q$ for some $e \in \dot{F}$. The theorem affirmed the truth of this in case $\dim q \leq 4$. However, the following gives a counterexample to the foregoing statement, with q a Pfister neighbor of dimension 5.

Example (Wadsworth). Over $F = \mathbb{Q}((x))$, let $\sigma = \langle 1, 2, x\rangle$, and let

$$q = \langle 1\rangle \perp \langle\langle 1, 7x\rangle\rangle \cong \langle 1, 1, 1, 7x, 7x\rangle,$$

which is an anisotropic Pfister neighbor contained in the Pfister form $\langle\langle 1, 1, 7x\rangle\rangle$. Now

$$\langle\langle 1, 1, 7x\rangle\rangle \cong \langle\langle 1, 1, x\rangle\rangle \cong \langle\langle 1, 2, x\rangle\rangle \gg \sigma.$$

Therefore, $q > \sigma$. However, it can be shown that q has no subform $\cong e \cdot \sigma$ (for any $e \in \dot{F}$). The details of the proof of this are left to the reader as a (somewhat challenging) exercise.

While Wadsworth's example showed that $q \supseteq e \cdot \sigma$ (for some $e \in \dot{F}$) is, in general, not a necessary condition for an anisotropic 5-dimensional form q to become isotropic over $F(\sigma)$, it has been shown subsequently that the Pfister neighbor case utilized in the above example was truly an "exception". Indeed, Hoffmann has shown that, if q is an anisotropic 5-dimensional form that is *not* a Pfister neighbor, then q is isotropic over $F[\sigma]$ in 4.32 iff $e \cdot \sigma \subseteq q$ for some $e \in \dot{F}$; see [Ho$_2$].

Prior to Hoffmann's work, another "positive" case in the context of 4.32 has been established by Merkurjev [Me$_2$]: this is the case of *Albert forms* (that is, 6-dimensional forms q with $d(q) = -1$).

Theorem 4.33. *Let $E = F(\sigma)$ be as in 4.32. If q is an anisotropic Albert form over F, then q_E is isotropic iff $e \cdot \sigma \subseteq q$ for some $e \in \dot{F}$.*

This is, however, not an easy result! We shall return to prove it later (see XIII.2.13), where the result will be utilized in Merkurjev's construction of fields of u-invariant 6. For much more information on the isotropy of 6-dimensional forms over various function fields $F[\sigma]$ (not necessarily of conics), see the papers [Ho$_3$], [Lag], [IP$_1$], [IP$_2$], etc.

The theory of function fields is a deep subject for study in the recent research in quadratic forms. Limitation of space has prevented us from exploring the many further topics in function field theory, such as generic splitting fields, excellent forms, Witt index patterns of quadratic forms under field extensions, and the characterizations of I^nF via the Knebusch degree, etc. Of course, for a first introduction to function fields, it would not necessarily be wise anyway to go in depth into any of these more specialized topics. To partly compensate for this, however, we shall close this section by mentioning (without proof) three of the strongest and most interesting results obtained on function fields in the last decade. These three results, all centering around Questions 4.4A and 4.4B, are of a very general nature, and yet their statements are completely accessible to (and can be fully appreciated by) anyone who knows the definition of the function field of a quadratic form.

The first result we want to mention is due to Hoffmann [Ho4]. This amounts to a dimension-theoretic necessary condition for an anisotropic form q to become isotropic over a function field $F[\varphi]$, but it is equally interesting to state it as a theorem on the preservation of anisotropicity.

Theorem 4.34 (Hoffmann). *Let φ, q be F-forms, where q is anisotropic and $F[\varphi]$ is defined. If there exists a positive integer n such that*

$$\dim\, q \leq 2^n < \dim\, \varphi,$$

then q remains anisotropic over $F[\varphi]$.

This result has come to be known as a "Separation Theorem", in that, if $q > \varphi$, 4.34 implies that $\dim\, q$ and $\dim\, \varphi$ must not be "separated" by a power of 2 as in the statement of 4.34. For instance, if q is an anisotropic form of dimension 7, then $q > \varphi$ can happen only if $\dim\, \varphi \leq 8$. Of course, what 4.34 gives is just a *general* dimension bound (valid in all cases) on $\dim\, \varphi$ if $q > \varphi$. In specific situations, one can sometimes do better. For instance, if q is an anisotropic Albert form, we'll show later (see XIII.2.6) that $q > \varphi$ can happen only if $\dim\, \varphi \leq 6$.

In case q is an anisotropic n-fold Pfister form, we have $q > \varphi$ iff $q \gg \varphi$. Here, the conclusion of 4.34 (that $q > \varphi \implies \dim\, \varphi \leq 2^n$) is already contained in that of 4.5. However, the case where φ is an $(n+1)$-fold Pfister form in 4.34 gives something remarkable. In this case, the theorem implies that:

$$\textit{if } q \textit{ is anisotropic and } q > \varphi, \textit{ then } \dim\, q \geq 2^n + 1.$$

Note that, on general ground, this is indeed the best lower bound, since (in case φ is anisotropic) we can take q to be a (2^n+1)-dimensional Pfister neighbor of φ.

The second result we want to mention concerns the "first Witt index" of a quadratic form. By definition, the *first Witt index* $i_1(\varphi)$ of an *anisotropic* form φ over F of dimension ≥ 2 is defined to be the ordinary Witt index of the form $\varphi_{F[\varphi]}$ (in the sense of I.4.3). Note that $i_1(\varphi) \geq 1$, since φ becomes isotropic over $F[\varphi]$. In general, $i_1(\varphi)$ may be much larger than 1. For instance, if φ is an anisotropic $(n+1)$-fold Pfister form, we have $i_1(\varphi) = 2^n$, since φ becomes hyperbolic over $F[\varphi]$.

The "higher" Witt indices of φ can be defined by induction: the second Witt index $i_2(\varphi)$ is taken to be the first Witt index of the anisotropic part of $\varphi_{F[\varphi]}$, etc.

The exact nature of the positive integer $i_1(\varphi)$ for an anisotropic form φ of a given dimension has recently been determined by N. Karpenko [Kar$_1$]. The following result of his verified a conjecture made earlier by D. Hoffmann.

Theorem 4.35 (Karpenko). *Let φ be an anisotropic form, and write*

$$\text{(A)} \qquad \dim\varphi - 1 = 2^{n_1} + 2^{n_2} + \cdots + 2^{n_r},$$

where $n_1 < n_2 < \cdots < n_r$. Then there exists an integer $s \in [0, r)$ such that

$$\text{(B)} \qquad i_1(\varphi) - 1 = 2^{n_1} + 2^{n_2} + \cdots + 2^{n_s}.$$

In other words, $i_1(\varphi) - 1$ is given by the increasing dyadic expansion of $\dim\varphi - 1$, *truncated at some point. Furthermore, all possible values of $i_1(\varphi) - 1$ given in* (B) *can be realized by anisotropic forms φ over suitable fields.*

Some explicit examples should help clarify the meaning of Karpenko's quantitative result:

- $\dim\varphi = 17$. Here, the RHS of (A) is 2^4, so (B) predicts that $i_1(\varphi) = 1$. (Cf. Exercise 26.)
- $\dim\varphi = 16$. Here, the RHS of (A) is $2^0 + 2^1 + 2^2 + 2^3$, so (B) predicts that $i_1(\varphi) \in \{1, 2, 4, 8\}$.
- $\dim\varphi = 15$. Here, the RHS of (A) is $2^1 + 2^2 + 2^3$, so (B) predicts that $i_1(\varphi) \in \{1, 3, 7\}$.

Karpenko's proof of 4.35 is based on algebro-geometric methods involving cycles, Chow groups, and Steenrod operations. Such a proof is certainly beyond the scope of this book.

The last result we want to state is a recent one on the "essential dimensions" of quadratic forms, due to Karpenko and Merkurjev [KM]. If φ is an anisotropic form over F with $\dim\varphi \geq 2$, Izhboldin, Karpenko, and Merkurjev defined the *essential dimension* of φ to be the integer

$$\text{(4.36)} \qquad \dim_{\text{es}}\varphi = \dim\varphi - i_1(\varphi).$$

For instance, if $\dim\varphi = 2^n + 1$, then $i_1(\varphi) = 1$ by 4.35 (as in the concrete example given above where $\dim\varphi = 17$), and hence $\dim_{\text{es}}\varphi = 2^n$.

A useful general observation on $\dim_{\text{es}}\varphi$ is the following.

Proposition 4.37. *Let φ_1 be the anisotropic part of the form $\varphi_{F[\varphi]}$ over the function field $F[\varphi]$. Then*

(1) $\dim_{\text{es}}\varphi = (\dim\varphi + \dim\varphi_1)/2$.

(2) $\dim_{\text{es}}\varphi \geq (\dim\varphi)/2$, *with equality iff φ is similar to a Pfister form over F.*

Proof. (1) Since $\varphi \cong \varphi_1 \perp i_1(\varphi)\,\mathbb{H}$, we have

$$2\dim_{\rm es}\varphi = \dim\varphi + (\dim\varphi - 2\,i_1(\varphi)) = \dim\varphi + \dim\varphi_1.$$

(2) The inequality follows from (1) since $\dim\varphi_1 \geq 0$. Clearly, equality holds iff $\dim\varphi_1 = 0$, iff $\varphi_{F[\varphi]}$ is hyperbolic. Since φ is assumed anisotropic, this holds iff φ is similar to a Pfister form, by 4.14. □

We can derive from 4.35 a general lower bound on the essential dimension of forms.

Corollary 4.38. *If φ is an anisotropic form of dimension $> 2^n$, then* $\dim_{\rm es}\varphi \geq 2^n$.

Proof. Using the notations (and conclusions) of 4.35, we have

$$\dim_{\rm es}\varphi = (\dim\varphi - 1) - (i_1(\varphi) - 1) \geq 2^{n_r}.$$

Since $\dim\varphi > 2^n$ implies that $n_r \geq n$, this yields the bound $\dim_{\rm es}\varphi \geq 2^n$. □

The main result on essential dimensions in [KM] is geometric in nature. Stated in the special case of quadratic forms, it gives the following.

Theorem 4.39 (Karpenko-Merkurjev). *Let q, φ be anisotropic F-forms of dimension ≥ 2. If $q > \varphi$, then $\dim_{\rm es} q \geq \dim_{\rm es}\varphi$. In this case, equality holds iff also $\varphi > q$.*

The statement of this theorem was originally a conjecture of O. T. Izhboldin. To see the power of 4.39, let us record the following special case of it.

Corollary 4.40. *Let φ be an anisotropic $(n+1)$-fold Pfister form. If an anisotropic form q becomes isotropic over $F[\varphi]$, then $\dim q \geq 2^n + i_1(q)$.*

Proof. We have $\dim q - i_1(q) = \dim_{\rm es} q \geq \dim_{\rm es}\varphi$ by 4.39, and $\dim_{\rm es}\varphi$ is just 2^n. □

Note that, in the situation of 4.40, if we apply Hoffmann's result 4.34 instead, we'll only get the weaker estimate $\dim q \geq 2^n + 1$. (Recall that $i_1(q)$ is always ≥ 1.) This suggests that the result 4.39 may be even more powerful than Hoffmann's result 4.34. As a matter of fact, there does exist a logical dependence relation among the three major results we have stated so far. To make this explicit, we'll prove the following.

Proposition 4.41. *Theorems 4.35 and 4.39, taken together, imply Hoffmann's theorem 4.34.*

Proof. Suppose $\dim q \leq 2^n < \dim \varphi$, where q is anisotropic (as in 4.34). If $q > \varphi$, φ is clearly anisotropic (by 4.1). By 4.38 (a consequence of 4.35) and 4.39,

$$\dim q \geq \dim_{\mathrm{es}} \varphi + i_1(q) \geq 2^n + 1,$$

a contradiction. Thus, q must remain anisotropic over $F[\varphi]$. □

The fact that 4.35 and 4.39 imply 4.34, however, should not detract from Hoffmann's result 4.34. The proofs of *both* of the results 4.35 and 4.39 (in [Kar] and [KM]) require the more advanced algebro-geometric techniques of cycles and Chow groups, while Hoffmann's proof of 4.34 (in [Ho$_4$]) is completely within the framework of "classical" quadratic form theory.

We hope that the three theorems 4.34, 4.35, and 4.39 quoted above served to convey the flavor of some of the exciting research that is ongoing in the study of function fields of quadratic forms. Readers interested in following the developments in function field theory should start with Knebusch's foundational papers [Kn$_4$], [Kn$_5$], and then proceed to the work of Elman-Lam, Elman-Lam-Wadsworth, Arason-Knebusch, Fitzgerald, Hoffmann, Van Geel, Lewis, Ahmad, Ohm, Rost, Merkurjev, Karpenko, Izhboldin, Vishik, Hurrelbrink, Rehmann, Laghribi, Kahn, and others.

5. Hauptsatz, Linkage, and Forms in I^nF

This section offers the beginnings of an in-depth study of the quadratic forms in I^nF, the n-th power of the "fundamental ideal" IF. The first word in the section title above refers to the following beautiful result of Arason and Pfister proved in 1971 in their joint paper [AP$_1$].

Hauptsatz 5.1. *Let q, be a positive-dimensional anisotropic form over F. If $q \in I^nF$, then* $\dim q \geq 2^n$.

An equivalent way to state this result is the following: *if a form q belongs to I^nF and $\dim q < 2^n$, then q must be a hyperbolic form.*

The significance of the Hauptsatz lies in the fact that it offers an important dimension-theoretic sufficient condition for a form to belong to I^nF. This Hauptsatz may be regarded as the first step towards finding a set of necessary and sufficient conditions for the quadratic forms in I^nF (for each given n).

Let us record a few immediate consequences of the Hauptsatz. The first one is the "Krull Intersection Property" in part (1) below.

Corollary 5.2. (1) *In the Witt ring $W(F)$, $\bigcap_{i=0}^{\infty} I^iF = 0$.*

(2) *More generally, if K/F is any field extension, and J is the kernel of the functorial map $r^*\colon W(F) \to W(K)$, then $\bigcap_{i=0}^{\infty}(J + I^iF) = J$.*

(3) *If* $J = \varphi \cdot W(F)$, *where* φ *is a Pfister form over* F, *then* $\bigcap_{i=0}^{\infty}(J + I^iF) = J$.

Proof. (1) Let q be a form belonging to $\bigcap_{i=0}^{\infty} I^iF$. Pick a large integer n such that $\dim q < 2^n$. Since $q \in I^nF$, the Hauptsatz implies that $q = 0 \in W(F)$.

(2) This is a self-strengthening of (1). If $q \in \bigcap_{i=0}^{\infty}(J + I^iF)$, then $r^*(q) = q_K \in I^iK$ for all i (since $r^*(I^iF) \subseteq I^iK$). By (1), we have $r^*(q) = 0 \in W(K)$, so $q \in J$.

(3) If $\varphi \cong \langle 1\rangle$, both sides of the equation are $W(F)$. If $\varphi \cong \mathbb{H}$, both sides are zero, by (1). If $\varphi \not\cong \langle 1\rangle$ or $\mathbb{H}$, then the function field $F[\varphi]$ is defined. In this case, (3) follows from (2), since the kernel of $W(F) \to W(F[\varphi])$ is $\varphi \cdot W(F)$ by 4.11. □

Corollary 5.3. *Let* φ, γ *be a pair of* 2^n*-dimensional forms which represent a common value* $a \in \dot{F}$. *Then*

$$\varphi \equiv \gamma \pmod{I^{n+1}F} \implies \varphi \cong \gamma.$$

Proof. Since $a \in D_F(\varphi) \cap D_F(\gamma)$, there exist forms φ_0 and γ_0 such that $\varphi \cong \langle a\rangle \perp \varphi_0$ and $\gamma \cong \langle a\rangle \perp \gamma_0$. Consider $\sigma := \varphi_0 \perp \langle -1\rangle\gamma_0$. Since

$$\varphi \perp \langle -1\rangle\gamma \cong \langle a, -a\rangle \perp \varphi_0 \perp \langle -1\rangle\gamma_0 \cong \mathbb{H} \perp \sigma,$$

the hypothesis $\varphi \equiv \gamma \pmod{I^{n+1}F}$ leads to $\sigma \in I^{n+1}F$. Since $\dim \sigma < 2^n + 2^n = 2^{n+1}$, the Hauptsatz implies that σ is hyperbolic, and hence $\varphi \cong \gamma$. □

Corollary 5.4. *Let* $r, s \in \dot{F}$, *and let* φ, γ *be* n*-fold Pfister forms over* F. *Then*

$$\varphi \cong \gamma \iff \langle r\rangle\, \varphi \equiv \langle s\rangle\, \gamma \pmod{I^{n+1}F}.$$

Proof. Clearly, $\varphi \equiv \langle r\rangle\,\varphi$ and $\gamma \equiv \langle s\rangle\,\gamma$ modulo $I^{n+1}F$. This proves the forward implication, and reduces the backward implication to the case $r = s = 1$. This case follows from 5.3 since φ and γ both represent 1. □

Note that 5.4 is a known result for $n = 1, 2$. For $n = 1$, this follows from II.2.3, and for $n = 2$, this is V.3.5. In fact, from these earlier results, we can prove the Hauptsatz 5.1 directly for $n = 1, 2, 3$ as follows.

For $n = 1$, 5.1 is trivial since any form in IF is even-dimensional. For $n = 2$, consider a form $q \in I^2F$ with $0 < \dim q < 4$. Then $\dim q = 2$, and $q \in I^2F$ implies that $\det(q) \in -\dot{F}^2$, so $q \cong \mathbb{H}$. For $n = 3$, consider a form $q \in I^2F$ with $0 < \dim q < 8$. Adding some hyperbolic planes to q if necessary, we may assume that $\dim q = 6$ and that (after a scaling)

$$q \cong \langle w, x, wx, -y, -z, -yz\rangle.$$

Then $q \in I^3F$ implies that $\langle\langle w, x\rangle\rangle \equiv \langle\langle y, z\rangle\rangle \pmod{I^3F}$, and V.3.5 yields $\langle\langle w, x\rangle\rangle \cong \langle\langle y, z\rangle\rangle$, so q is hyperbolic, as desired.

Note that we also know the truth of the Hauptsatz 5.1 if the form $q \in I^nF$ does not lie in $W_t(F)$ (the torsion subgroup of $W(F)$). In fact, in this case, Pfister's Local-Global Principle implies that there exists an ordering α on F such that $\text{sgn}_\alpha(q) \neq 0$. Since $q \in I^nF$, $\text{sgn}_\alpha(q) = 2^nk$ for some integer k. Therefore, $\dim q \geq 2^n|k| \geq 2^n$. *This argument is sufficient to prove the Hauptsatz* 5.1 *for all formally real pythagorean fields.*

Finally, if F is a field with only a finite number of square classes, then $W(F)$ is a noetherian ring by II.2.4. In this case, a direct use of Krull's Intersection Theorem in commutative algebra (along with the fact that odd-dimensional forms are not 0-divisors in $W(F)$) implies the truth of 5.2: see Ch. VIII, Exer. 19.

Although we can prove 5.1 or 5.2 in several special cases, the methods used above unfortunately do not generalize to the case of arbitrary n and arbitrary field F. In order to prove 5.1 in general, we'll need the method of function fields. As it turns out, with the function field results of §4 at our disposal, the proof of 5.1 boils down to a simple induction, as follows.

Proof of 5.1. Let $q \in I^nF$ be as in 5.1. Since the n-fold Pfister forms additively generate I^nF, there exists an expression

$$q = \varepsilon_1\varphi_1 + \cdots + \varepsilon_r\varphi_r \in I^nF,$$

where $\varepsilon_i = \pm 1$ and φ_i are anisotropic n-fold Pfister forms. To show that $\dim q \geq 2^n$, we induct on r. If $r = 1$, we have $q \cong \langle \pm 1\rangle\varphi_1$, so $\dim q = 2^n$. For the general case, we go up to the function field $L = F[\varphi_1]$. Over this field, we have a shorter expression

$$q_L = \varepsilon_2(\varphi_2)_L + \cdots + \varepsilon_r(\varphi_r)_L \in I^nL. \tag{5.5}$$

If q_L is hyperbolic, 4.5 yields directly $\dim q \geq \dim \varphi_1 = 2^n$. Thus, we may assume that $(q_L)_{\text{an}}$ (the anisotropic part of q_L) is a positive-dimensional form in I^nL. Thus, the inductive hypothesis (*invoked over the field* L) implies that $\dim_L (q_L)_{\text{an}} \geq 2^n$. But then clearly,

$$\dim_F q = \dim_L q_L \geq \dim_L (q_L)_{\text{an}} \geq 2^n. \qquad \square$$

The very short proof of the Hauptsatz above perhaps belies its true depth. Of course, this proof made crucial use of 4.5, which is a centerpiece in the function field theory of quadratic forms. However, function field theory was not yet developed when the paper [AP$_1$] was written. In the original proof of the Hauptsatz given by Arason and Pfister in 1971, the function field of a Pfister form was implicitly used without a name, but the idea of exploiting 4.5 to consummate the argument was unmistakable. Thus, from a historical perspective, the Arason-Pfister proof of the Hauptsatz represented the first significant application of the function field techniques. With such a spectacular introduction, function fields entered the arena of quadratic form theory in 1971—and they have remained on the center stage of this theory ever since. All of the function field results reported in §4, with the exception of 4.1 and 4.5 only, were developed after the appearance of the proof of the Hauptsatz in [AP_1].

As a natural supplement to the Hauptsatz, let us prove the following characterization theorem for scalar multiples of n-fold Pfister forms. For this application, all we need is the function field result 4.14.

Theorem 5.6. *A 2^n-dimensional form q lies in I^nF iff it is a scalar multiple of an n-fold Pfister form.*

Proof. The "if" part is clear. For the converse, assume that $q \in I^nF$. We may assume that $n \geq 1$ and $q \not\cong \mathbb{H}$, so that the function field $K = F[q]$ is defined. Since q_K is isotropic, $\dim_K (q_K)_{\mathrm{an}} < 2^n$. But $(q_K)_{\mathrm{an}} \in I^nK$, so the Hauptsatz implies that $\dim (q_K)_{\mathrm{an}} = 0$. This means that $q \gg q$, so by 4.14, q must be a scalar multiple of an n-fold Pfister form (over F). (The case where q is hyperbolic is no exception since $\dim q = 2^n$.) $\square$

After proving the Hauptsatz 5.1 and the last theorem 5.6, a very natural question to ask would be

Question 5.7. *What are the possible dimensions of the anisotropic quadratic forms $q \in I^nF \setminus \{0\}$ (where F ranges over all fields of characteristic $\neq 2$)?*

This question is quite easy to answer for $n \leq 2$. For $n = 1$, the answer is clearly "all integers $2d$ ($d \geq 1$)". For $n = 2$, the answer is easily seen to be "all integers $2d$ ($d \geq 2$)". Indeed, the anisotropic forms $4m\langle 1\rangle \in I^2\mathbb{Q}$ account for all dimensions $2d$ with d even, and the forms

$$q = (4m-1)\langle 1\rangle \perp \langle 2\rangle \perp \langle -x,\, 2x\rangle \qquad (m \geq 1) \tag{5.8}$$

in I^2F for $F = \mathbb{Q}((x))$ account for all dimensions $2d$ with $d \geq 3$ odd. Note that the form q in (5.8) is anisotropic since its first and second residue forms $(4m-1)\langle 1\rangle \perp \langle 2\rangle$ and $\langle -1,\, 2\rangle$ are both anisotropic over the field of the rationals. In terms of 2-fold Pfister forms over F, we have simply

$$q = m\,\langle\langle 1, 1\rangle\rangle - \langle\langle -2,\, x\rangle\rangle \in I^2F \qquad (m \geq 1). \tag{5.9}$$

Going on to the case $n = 3$, we can also provide the following complete answer to 5.7.

Proposition 5.10. *The possible dimensions of anisotropic forms in $I^3F \setminus \{0\}$ (for "varying" F's) are $2d$, where $d = 4$ or $d \geq 6$.*

Proof. By the Hauptsatz (or just by a direct argument in this case), we must have $d \geq 4$, and of course, $d = 4$ is possible. The fact that $d \neq 5$ is a celebrated observation of Pfister: in [Pf$_3$], Pfister showed that *10-dimensional forms in I^3F (for any F) are always isotropic.* Later, we shall have an occasion to study this fact and other related facts in the context of low-dimensional forms. Therefore, we shall postpone the proof of Pfister's observation to a later chapter (see XII.2.8). Here, we go on to show that any dimension $2d$ is possible if $d \geq 6$.

It will be sufficient to handle the cases $d = 6, 7, 8, 9$. For, once we have an anisotropic form $q_0 \in I^3F_0$ of dimension $2d_0$, we can construct $q = q_0 \perp \langle x\rangle \varphi \in I^3F$ over the field $F = F_0((x))$ by taking *any* anisotropic 3-fold Pfister form φ over F_0. This new form is anisotropic over F (by VI.1.9), and has dimension $2d_0 + 8$. Repeated use of this construction will therefore cover *all* cases $d \geq 6$, if we can take care of the four initial cases $d = 6, 7, 8, 9$.

The case $d = 8$ is trivial, as we can take $q = 16\langle 1\rangle \in I^3\mathbb{Q}$. For $d = 6$, we can take the anisotropic form

$$\langle\langle y\rangle\rangle\,\langle 1, 1, 1, 2, -x, 2x\rangle \in I^3F \quad \text{over } F = \mathbb{Q}((x))((y)).$$

For $d = 7$, we can take the anisotropic form

$$\langle\langle 1, 1, 1\rangle\rangle' \perp \langle -1\rangle\langle\langle x, y, z\rangle\rangle' \in I^3F \quad \text{over } F = \mathbb{Q}((x))((y))((z)),$$

where "prime" means taking the pure subform. Finally, for $d = 9$, we can take the 18-dimensional anisotropic part of

$$\begin{aligned} &8\langle 1\rangle \perp \langle x, y\rangle\,\langle 1, 1, 1, 2, -x, 2x\rangle \\ &\quad \cong \mathbb{H} \perp 7\langle 1\rangle \perp \langle 2\rangle \perp x\,\langle 1, 1, 1, 2\rangle \perp y\,\langle 1, 1, 1, 2, -x, 2x\rangle \end{aligned}$$

in I^3F over $F = \mathbb{Q}((x))((y))$. □

In order to understand the structure of certain anisotropic forms of dimension $\leq 2^{n+1}$ in I^nF, we introduce at this point the linkage theory of Pfister forms developed in Elman-Lam [EL$_1$]. Retrospectively, this linkage theory was the first step toward a complete understanding of the anisotropic dimensions of forms in I^nF, and of the "gap phenomenon" for such dimensions (which we'll discuss shortly).

Definition 5.11. A family of n-fold Pfister forms $\{\varphi_1, \dots, \varphi_m\}$ is said to be *r-linked* ($r \geq 0$) if there exists an r-fold Pfister form σ and $(n-r)$-fold Pfister forms $\tau_1, \dots, \tau_m$ such that $\varphi_i \cong \sigma\tau_i$ for all i. (Such σ is called an *r-linkage* for $\varphi_1, \dots, \varphi_m$.) If, further, $\varphi_1, \dots, \varphi_m$ are r-linked but not $(r+1)$-linked, we'll say that $\{\varphi_1, \dots, \varphi_m\}$ has *linkage number r*. If this linkage number is $\geq n-1$, we'll simply say that $\varphi_1, \dots, \varphi_m$ are *linked.*

Note that, by 4.11, the "r-linked" condition amounts to $\varphi_1, \ldots, \varphi_m \in \sigma \cdot WF$ for some r-fold Pfister form σ. Of course, the linkage notion above was directly inspired by the case of the linkage of quaternion algebras discussed in III.4. Indeed, two quaternion algebras are linked iff their norm forms (as 2-fold Pfister forms) are linked. A nice example of linkage for higher folds is given by the following.

Example 5.12A. Let F be any number field. Then for $n \geq 3$, any family $\{\varphi_1, \ldots, \varphi_m\}$ of n-fold Pfister forms over F are linked. In fact, the Pfister form $2^{n-1}\langle 1\rangle$ always provides an $(n-1)$-linkage for $\{\varphi_1, \ldots, \varphi_m\}$; see VI.3.9.

Example 5.12B. For any field F, the n-fold Pfister forms

$$\varphi_1 = \langle\langle a_1, \ldots, a_n\rangle\rangle \quad \text{and} \quad \varphi_2 = \langle\langle -a_1, \ldots, -a_n\rangle\rangle$$

are always linked (for any $a_i \in \dot{F}$). To see this, note that $\langle\langle a, b\rangle\rangle \cong \langle\langle a, ab\rangle\rangle$ (which is a special case of 1.3(2)). Repeated use of this isometry yields

$$\varphi_1 \cong \langle\langle a_1, a_1a_2,\ a_2a_3,\ \ldots\ ,\ a_{n-1}a_n\rangle\rangle.$$

Applying the same process to φ_2, we see that a linkage for φ_1, φ_2 is provided by the $(n-1)$-fold Pfister form $\langle\langle a_1a_2,\ a_2a_3,\ \ldots\ ,\ a_{n-1}a_n\rangle\rangle$.

We now go on to the main results on the linkage of pairs of Pfister forms from [EL$_1$]. In the following, $i(q)$ denotes the Witt index of a form q.

Theorem 5.13. *Let $q = \varphi \perp \langle -1\rangle\gamma$, where φ, γ are n-fold Pfister forms over F. Then φ, γ are r-linked iff $i(q) \geq 2^r$. Further, if r is precisely the linkage number of φ and γ, then $i(q) = 2^r$.*

Proof. *Step* 1. Suppose $\varphi \cong \sigma\varphi_1$ and $\gamma \cong \sigma\gamma_1$, where σ, φ_1, γ_1 are Pfister forms, with $\dim \sigma = 2^r$. Then[(8)]

$$\begin{aligned} q &\cong \sigma(\langle 1\rangle \perp \varphi_1') \perp \langle -1\rangle\sigma(\langle 1\rangle \perp \gamma_1') \\ &\cong 2^r\mathbb{H} \perp \sigma(\varphi_1' \perp \langle -1\rangle\gamma_1') \end{aligned}$$

shows that $i(q) \geq 2^r$.

Step 2. Conversely, assume $i(q) \geq 2^r$. To show that φ, γ are r-linked, we induct on r (the case $r = 0$ being trivial). Invoking an inductive hypothesis, we may write $\varphi \cong \sigma_0\varphi_1$ and $\gamma \cong \sigma_0\gamma_1$, where σ_0, φ_1, γ_1 are Pfister forms, with $\dim \sigma_0 = 2^{r-1}$. As in Step 1, we have

$$\text{(5.14)} \qquad q \cong 2^{r-1}\mathbb{H} \perp (\sigma_0\varphi_1' \perp \langle -1\rangle\, \sigma_0\gamma_1').$$

Since $i(q) = 2^r$, $\sigma_0\varphi_1' \perp \langle -1\rangle\sigma_0\gamma_1'$ must be isotropic. By I.3.6, there exists $c \in D_F(\sigma_0\varphi_1') \cap D_F(\sigma_0\gamma_1')$. Thus, 1.10 yields $(n-r)$-fold Pfister forms φ_2 and γ_2 such that

$$\text{(5.15)} \qquad \varphi \cong \sigma_0\, \langle\langle c\rangle\rangle\, \varphi_2 \quad \text{and} \quad \gamma \cong \sigma_0\, \langle\langle c\rangle\rangle\, \gamma_2,$$

---

(8) As usual, φ_1' and γ_1' denote the pure subforms of φ_1 and γ_1.

which shows that φ and γ are r-linked.

Step 3. Suppose the linkage number of φ and γ is exactly r. Then, by the argument in Step 2, the form $\sigma\varphi_2' \perp \langle -1\rangle\sigma\gamma_2'$ must be anisotropic. Therefore, the analogue of (5.14) (using φ_2' and γ_2') shows that $i(q) = 2^r$. □

Thanks to 5.13, we can now compute the anisotropic dimensions of forms in I^nF of the shape $q = \langle x\rangle\,\varphi \perp \langle y\rangle\,\gamma$, where φ, γ are n-fold Pfister forms, and $x, y \in \dot{F}$.

Theorem 5.16. *For the form q above, $i(q)$ is either zero or equal to 2^r, where r is the linkage number of φ and γ. Thus, q_{an} (the anisotropic part of q) has dimension 2^{n+1} or $2^{n+1} - 2^{r+1}$, respectively.*

Proof. Assuming that q is isotropic, we must show that $i(q) = 2^r$.

Case 1. *φ is isotropic.* If γ is also isotropic, then $\{\varphi, \gamma\}$ has linkage number n, and indeed $i(q) = 2^n$. If γ is anisotropic, then $\{\varphi, \gamma\}$ has linkage number $n-1$, and $q \cong 2^{n-1}\mathbb{H} \perp \langle y\rangle\,\gamma$ also yields $i(q) = 2^{n-1}$.

Case 2. *We may now assume that φ, γ are both anisotropic.* Since q is isotropic, there exists an equation $x\varphi(u) + y\gamma(v) = 0$, where $(u, v) \neq (0, 0)$, and hence *both* $a = \varphi(u)$, $b = \gamma(v)$ are nonzero. By 1.8, we have $\varphi \cong \langle a\rangle\varphi$ and $\gamma \cong \langle b\rangle\gamma$, so

$$\langle xa\rangle(\varphi \perp \langle -1\rangle\gamma) \cong \langle xa\rangle\varphi \perp \langle yb\rangle\gamma \cong \langle x\rangle\varphi \perp \langle y\rangle\gamma = q.$$

Therefore, by 5.13, we have $i(q) = 2^r$. □

Given any n and any r such that $0 \le r \le n$, it is not difficult to produce two n-fold Pfister forms φ, γ over some field F that have the exact linkage number r. For instance, for $r = 2$, we can take the Pfister forms

$$\varphi = \langle\langle x, y, z_1, \ldots, z_{n-2}\rangle\rangle \quad \text{and} \quad \gamma = \langle\langle x, y, z_1', \ldots, z_{n-2}'\rangle\rangle$$

over an iterated Laurent series field in the variables $\{x, y, z_i, z_i'\}$. Therefore, 5.13 yields the following.

Corollary 5.17. *The numbers $\{2^{n+1} - 2^{r+1} \colon 0 \le r \le n\}$ can all be realized as the dimensions of anisotropic forms in I^nF (for a suitable field F).*

This leaves various "gaps" for dimensions $< 2^{n+1}$, in the form of open intervals, that are *not yet* accounted for by the constructions in 5.13. To be precise, these open intervals are

(5.18) $(2^{n+1} - 2^{r+1}, 2^{n+1} - 2^r) = (2^n + 2^{n-1} + \cdots + 2^{r+1}, 2^n + 2^{n-1} + \cdots + 2^r)$, where $1 \le r \le n$, or, from left to right,

$$(5.19)\qquad \begin{aligned} &(0,\, 2^n),\ (2^n,\, 2^n+2^{n-1}),\ (2^n+2^{n-1},\, 2^n+2^{n-1}+2^{n-2}),\\ &\quad \dots,\ (2^{n+1}-8,\, 2^{n+1}-4),\ \text{and}\ (2^{n+1}-4,\, 2^{n+1}-2).\end{aligned}$$

Here, the last interval is a trivial gap, since it contains only the odd integer $2^{n+1}-3$, so effectively the potential last gap $< 2^{n+1}$ is the single integer $2^{n+1}-6 \in (2^{n+1}-8,\, 2^{n+1}-4)$.

The interval $(0,\, 2^n)$ is the "first gap": by the Hauptsatz, we know that this is a true gap, in that no value in it can be the dimension of an anisotropic form in I^nF (for any field F). *How about the "second gap"* $(2^n,\, 2^n+2^{n-1})$, *etc., up to the last gap* $2^{n+1}-6$? For $n = 3$, the only gap in (5.19) is $2^4 - 6 = 10$, and as we have observed in the proof of 5.10, Pfister has shown that this is indeed a gap value for I^3F. This was, then, the very first sighting of the "gap phenomenon". Going one step further, for $n = 4$, Hoffmann [Ho_5] showed that the "second gap" $(16, 24)$ is also a true gap for I^4F (that is, no anisotropic forms in I^4F can have dimensions 18, 20, and 22). The third and last gap, 26 for I^4F, was not treated in [Ho_5].

Of course, the above discussion was focused only on dimensions $< 2^{n+1}$. For even dimensions $2d > 2^{n+1}$, quick experiments will show that it seems generally "easier" to produce (anisotropic) $2d$-dimensional forms in $I^{n+1}F$. For instance, we can take an anisotropic $q_0 \in I^nF_0$ of dimension $2^{n+1}-2^{r+1}$ (obtained by the method of 5.16), and take

$$q_1 = q_0 \perp \langle x\rangle\,\varphi \in I^nF \quad\text{or}\quad q_2 = \langle\langle x\rangle\rangle\,\varphi \in I^{n+1}F \subseteq I^nF$$

over $F = F_0((x))$, where φ is any anisotropic n-fold Pfister form over F_0. This will realize the dimensions

$$2^{n+1} + (2^n - 2^{r+1}) \quad\text{and}\quad 2^{n+1} + 2(2^n - 2^{r+1}).$$

Similar constructions will produce various other anisotropic forms of (even) dimension $> 2^{n+1}$ in I^nF.

The ultimate answer to Question 5.7 has now been obtained in the following beautiful result announced by A. Vishik in 2002, shortly after the publication of his papers [Vi_1, Vi_2].

Vishik's Gap Theorem 5.20. *The possible dimensions of anisotropic forms in I^nF (for varying F's) are precisely*:

$$(5.21)\qquad 2^{n+1} - 2^{r+1}\quad (0 \le r \le n), \qquad \textit{and} \qquad 2^{n+1} + 2k \quad (k \ge 0).$$

In other words, the gaps listed in (5.19) *are all true gaps, and there are no more gaps in even dimensions* $\ge 2^{n+1}$.

For $n = 2$ and $n = 3$, this is, of course, consistent with our earlier conclusions in those cases. To be explicit, the true (even) "gap values" for

I^nF beyond 2^n are, according to 5.20:

$$10 \text{ (for } n=3); \quad 18, 20, 22, 26 \text{ (for } n=4); \text{ and}$$
$$34, 36, 38, 40, 42, 44, 46, 50, 52, 54, 58 \text{ (for } n=5); \text{ etc.}$$

Of course, the number of gap values grows very quickly with n, starting from Pfister's first sighting of the unique second gap of "10" for $n = 3$.

Vishik's detailed proof for 5.20 is not yet published. However, a proof for this theorem has become available in Karpenko's paper [Kar3] entitled "Holes in I^n". Karpenko's proof involves (again) computations in the Chow groups of direct products of projective quadrics.

Recently, another proof for the "second gap" $(2^n,\ 2^n + 2^{n-1})$, based on the Karpenko-Merkurjev Theorem on essential dimensions, also appeared in [KM]. This proof is due to D. Hoffmann. Since we have given a complete statement of the Karpenko-Merkurjev Theorem in 4.39, Hoffmann's proof is completely within our reach, without further recourse to algebro-geometric methods. We shall give this proof below, largely following [KM].

Proof for the "Second Gap". Assuming that there exists an anisotropic form $\beta \in I^nF$ with $\dim \beta \in (2^n, 2^n + 2^{n-1})$, we may choose β to have the smallest possible dimension[(9)]; say $\beta = \varphi_1 + \cdots + \varphi_m \in I^nF$, where each φ_i is similar to an n-fold Pfister form. We may further assume β is chosen such that m is minimal, as well as the quantity

$$\sum_{i<j} \{ \dim\, (\varphi_i \perp \varphi_j)_{\text{an}} \}, \tag{5.22}$$

where τ_{an} (as usual) denotes the anisotropic part of a form τ.

If the φ_i's are mutually similar, say to some n-fold Pfister form φ, then $\beta \in \varphi \cdot W(F)$, and hence by 4.11, $\beta \cong \varphi \cdot \tau$ for some form τ. Since $\dim \tau \neq 1$, we must have $\dim \beta \geq 2^{n+1}$, which is not the case. We may thus assume, say, φ_1 is *not* similar to φ_2.

Let $q := \varphi_1 \perp \varphi_2 \cong \psi \perp i\,\mathbb{H}$ be a Witt decomposition (so ψ is anisotropic and $i = i(q)$). *We claim that* $\dim \psi \geq 2^n + 2^{n-1}$. This is clear if $i = 0$, so let us assume $i > 0$. Let r be the linkage number for the Pfister forms similar to φ_1 and φ_2, so that, by 5.16, $i(q) = 2^r$. If $r = n-1$, then by 5.6 (or directly by the proof of 5.16), ψ is similar to an n-fold Pfister form, so ψ may be used to replace $\varphi_1 + \varphi_2$, in contradiction to the minimality of m. Therefore, $r \leq n-2$ and so

$$\dim\, \psi = 2^{n+1} - 2^{r+1} \geq 2^{n+1} - 2^{n-1} = 2^n + 2^{n-1},$$

as claimed. On the other hand, ψ must not be similar to an $(n+1)$-fold Pfister form (for otherwise 5.4 would imply that φ_1 and φ_2 are similar).

(9)The minimality of $\dim \beta$ here is, of course, with respect to all possible fields F.

Therefore, by 4.14, the anisotropic part ψ_0 of ψ over its own function field $F[\psi]$ is nonzero. Computing the essential dimension of ψ by 4.37(1), we have then by the Hauptsatz:

$$(5.23)\qquad 2\ \dim_{\text{es}} \psi = \dim \psi + \dim \psi_0 \geq 2^n + 2^{n-1} + 2^n.$$

Next, let us estimate $\dim_{\text{es}} \beta$. Again, let β_0 be the anisotropic part of β over $F[\beta]$. Since $\dim \beta_0$ is nonzero and $< \dim \beta$, the minimal choice of β and the Hauptsatz imply that $\dim \beta_0 = 2^n$, so that (by 4.37(1) again):

$$(5.24)\qquad 2\ \dim_{\text{es}} \beta = \dim \beta + \dim \beta_0 < 2^n + 2^{n-1} + 2^n.$$

Comparing (5.23) with (5.24), we see from the Karpenko-Merkurjev Theorem that β *must remain anisotropic over* $F[\psi]$. Of course, the dimension of β has not changed by going from F to $F[\psi]$, nor has the length m in the equation $\beta = \varphi_1 + \cdots + \varphi_m$. Also, clearly

$$\dim \big((\varphi_i \perp \varphi_j)_{F[\psi]} \big)_{\text{an}} \leq \dim \big(\varphi_i \perp \varphi_j \big)_{\text{an}}$$

for $i < j$, with *strict* inequality holding for $i = 1$ and $j = 2$, since $(\varphi_1 \perp \varphi_2)_{\text{an}} \cong \psi$. This now contradicts the minimality of the sum in (5.22)! ☐

For yet another proof for the "Second Gap" theorem, see [Kar_2].

6. Milnor's Higher K-Groups

In this section, we offer a quick introduction to Milnor's algebraic K-theory of fields, following Milnor's paper [Mi]. We have already discussed the groups K_nF for $n \leq 2$ in V.6. The material in this section builds on our earlier discussions on K_2F, which the reader should quickly review before reading the present section.

The inclusion of a discussion on K_nF in this chapter is prompted by several factors. First, of course, Milnor's K-theory of fields places the group K_2F in its proper context. Second, a part of the material in §1 and §5 of this chapter was actually motivated by Milnor's introduction of the groups K_nF. With these groups in place, results such as 1.11 (Chain P-Equivalence Theorem), 5.4, and 5.13 proved earlier in this chapter will find natural interpretations and applications in K-theory. Last but not least, the introduction of the groups K_nF will enable us to state, if not prove, some of the most significant results obtained in quadratic forms and K-theory in recent years, namely, the theorems of Voevodsky et al. on the "Milnor Conjectures" in the algebraic K-theory of fields.

To introduce the groups K_nF, we follow a notational device of Milnor in "converting" multiplicative structures into additive structures. Starting with the multiplicative group $\dot{F}$, we define K_1F to be an additive version of

$\dot{F}$, with a canonical isomorphism $\ell\colon \dot{F} \to K_1F$, where $\ell(ab) = \ell(a) + \ell(b)$ for all $a, b \in \dot{F}$ (and hence $\ell(1) = 0$). The letter "ℓ" here comes from the word "logarithm".

Viewing K_1F as a $\mathbb{Z}$-module, we can then consider its graded tensor algebra

$$(6.1) \qquad (\mathbb{Z},\ K_1F,\ K_1F \otimes K_1F,\ K_1F \otimes K_1F \otimes K_1F,\ \dots).$$

By definition, K_*F is the graded algebra obtained from (6.1) by factoring out the homogeneous ideal generated by elements of the form

$$(6.2) \qquad \ell(a) \otimes \ell(1-a), \quad \text{where } a \in F \setminus \{0,1\}.$$

Thus, $K_*F = (K_0F, K_1F, K_2F, \dots)$, where $K_0F = \mathbb{Z}$, and for any $n \geq 2$, K_nF is the quotient of the n-fold tensor product group $K_1F \otimes \cdots \otimes K_1F$ by the subgroup generated by $\ell(a_1) \otimes \cdots \otimes \ell(a_n)$ such that $a_i + a_{i+1} = 1$ for *some* $i < n$, where all $a_i \in \dot{F}$. In particular, the group K_2F obtained in this way is "the same" as that defined earlier in V.6. The only difference is that we are using here the *additive* notation for K_2F (since it is a homogeneous component of a graded ring), whereas we have previously used the multiplicative notation for K_2F (in conformance with the multiplicative notation for the Brauer group $B(F)$). To reconcile this difference, all we need is to identify the generator $\ell(a) \otimes \ell(b)$ here[10] with the generator $[a, b]$ in our earlier notations for K_2F. In particular, the properties of the symbols $[a, b]$ proved in V.6.3 can be transcribed as follows in the new additive setting (with the same numbering!).

Lemma 6.3. *For $a, b \in \dot{F}$, the following holds in K_2F:*

(1) $\ell(a)\ell(b) = 0$ *whenever* $a + b = 0$.

(2) $\ell(a)\ell(b) = -\ell(b)\ell(a)$.

(3) $\ell(a)\ell(a) = \ell(a)\ell(-1) = \ell(-1)\ell(a)$.

(4) $\ell(a)\ell(b) = \ell(a+b)\ell(-b/a)$ *whenever* $a + b \neq 0$.

Two immediate consequences of 6.3 are as follows.

Corollary 6.4. *The ring K_*F is "graded commutative", in the sense that, for any $\alpha \in K_mF$ and $\beta \in K_nF$, the identity $\beta\alpha = (-1)^{mn}\alpha\beta$ holds in $K_{m+n}F$.*

Proof. This follows by repeated applications of 6.3(2). □

Corollary 6.5. *If $a_i \in \dot{F}$ are such that $a_1 + \cdots + a_n \in \{0, 1\}$, then $\ell(a_1) \cdots \ell(a_n) = 0$ in K_nF.*

---

[10] From here on, any $\ell(a_1) \otimes \cdots \otimes \ell(a_n)$ will be interpreted as an element in K_nF, rather than as an element in the tensor algebra (6.1).

Proof. We proceed by induction on n. The cases $n = 1$, 2 being clear (thanks to 6.3(1)), we may assume that $n \geq 3$. If $a_1 + a_2 = 0$, then the case $n = 2$ gives the desired conclusion, so we may assume that $a_1 + a_2 \neq 0$. In this case, the equation

$$a_1/(a_1 + a_2) + a_2/(a_1 + a_2) = 1$$

implies that $(\ell(a_1) - \ell(a_1 + a_2))(\ell(a_2) - \ell(a_1 + a_2)) = 0$ in K_2F. Right multiplying this by $\ell(a_3) \cdots \ell(a_n)$ and using 6.3(2) and the inductive hypothesis

$$\ell(a_1 + a_2)\ell(a_3) \cdots \ell(a_n) = 0 \in K_{n-1}F,$$

we get clearly $\ell(a_1)\ell(a_2) \cdots \ell(a_n) = 0$ in K_nF. $\square$

Just as in the case of $n = 2$, K_nF may be thought of as the recipient group of a universal "n-fold Steinberg symbol"

$$\dot{F} \times \cdots \times \dot{F} \longrightarrow K_nF, \quad \text{with } (a_1, \ldots, a_n) \longmapsto \ell(a_1) \cdots \ell(a_n).$$

Here, by an n-fold Steinberg symbol, we mean a map f from $\dot{F} \times \cdots \times \dot{F}$ (n factors) to an abelian group that is multiplicative in each variable, with the property that $f(a_1, \ldots, a_n) = 0$ whenever $a_i + a_{i+1} = 1$ for some $i < n$. This is, of course, in direct generalization of the Steinberg symbols (for $n = 2$) first introduced in V.6.1.

The above viewpoint on the group K_nF enables us to give some information on this group in the case where F is a real-closed field.

Proposition 6.6 (cf. V.6.16). *Let F be a real-closed field. For $n \geq 1$, $K_nF = A \oplus B$, where $B \cong \mathbb{Z}_2$ is generated by $\ell(-1)^n$, and A is a divisible group, generated by $\ell(a_1) \cdots \ell(a_n)$, where each $a_i \in \dot{F}^2$.*

Proof. To begin with, we note that there is a natural n-fold Steinberg symbol $f \colon \dot{F} \times \cdots \times \dot{F} \to \mathbb{Z}_2$ that takes $(a_1, \ldots, a_n)$ to 0 if any $a_i \in \dot{F}$ is positive, and to 1 if all $a_i \in \dot{F}$ are negative (with respect to the unique ordering on the real-closed field F). More formally,

$$(6.7) \qquad f(a_1, \ldots, a_n) = \frac{1 - \operatorname{sgn}(a_1)}{2} \cdot \ \cdots \ \cdot \frac{1 - \operatorname{sgn}(a_n)}{2} \in \mathbb{Z}_2,$$

where $\operatorname{sgn}(a_i)$ is 1 or -1 if a_i is positive or negative. (The fact that f is an n-fold Steinberg symbol is easy to verify directly.) Thus, the universal property of K_nF implies that f factors through a unique group homomorphism $g \colon K_nF \to \mathbb{Z}_2$. This g is a split epimorphism since $g(\ell(-1)^n) = 1$, and $\ell(-1)^n$ is an element of order ≤ 2. Thus, $K_nF = C \oplus B$, where $C = \ker(g)$, and B is the 2-element subgroup of K_nF generated by $\ell(-1)^n$. Clearly, C contains the subgroup $A \subseteq K_nF$ generated by $\ell(a_1) \cdots \ell(a_n)$, where all $a_i > 0$. Thus, to show that $C = A$ amounts to showing that $A + B = K_nF$. For this, we induct on n (the case $n = 1$ being clear). Invoking an inductive hypothesis, we may assume that $K_{n-1}F = A_0 + B_0$, where A_0, B_0 have the obvious meanings. In view of 6.3(2), it suffices to check that

$$\text{(6.8)} \qquad \ell(a)A_0 \subseteq A + B \quad \text{and} \quad \ell(a)B_0 \subseteq A + B, \quad \text{for any } a \in \dot{F}.$$

Case 1. $a > 0$. In this case, $\ell(a)A_0 \subseteq A$ already, and $\ell(a)\ell(-1)^{n-1} = \ell(a)^n \in A$ by 6.3(3).

Case 2. $a < 0$. Let $b = -a > 0$. For $a_i > 0$, we have

$$\begin{aligned}\ell(a)\ell(a_2)\cdots\ell(a_n) &= \ell(-1)\ell(a_2)\cdots\ell(a_n) + \ell(b)\ell(a_2)\cdots\ell(a_n)\\ &\in \ell(a_2)\ell(a_2)\cdots\ell(a_n) + A = A,\end{aligned}$$

so $\ell(a)A_0 \subseteq A$. Finally,

$$\ell(a)\ell(-1)^{n-1} = \ell(-1)^n + \ell(b)\ell(-1)^{n-1} = \ell(-1)^n + \ell(b)^n$$

shows that $\ell(a)B_0 \subseteq A + B$.

To complete the proof, it only remains to show that A is a *divisible* group. This is now clear since any positive element $a_1 \in F$ has a positive r-th root for any $r \geq 1$. If $a_1 = b^r$, where $b > 0$, then

$$\ell(a_1)\ell(a_2)\cdots\ell(a_n) = r\cdot\ell(b)\ell(a_2)\cdots\ell(a_n) \in r\cdot A$$

for any $a_2, \ldots, a_n > 0$ in F. □

In the following, we shall write k_nF for the quotient group $K_nF/2K_nF$ (for any field F). From 6.6, we immediately deduce the following.

Corollary 6.9. *If F is a real-closed field, then $k_nF \cong \mathbb{Z}_2$, and it is generated by (the image of) $\ell(-1)^n$ for any $n \geq 1$.*

Another consequence of 6.6 is the following K-theoretic characterization of nonreal fields, due to Milnor.

Theorem 6.10. *A field F is nonreal iff $\ell(-1)$ is nilpotent in K_*F, iff $\bigoplus_{i\geq 1} K_iF$ is a nil ideal in the graded ring K_*F.*

Proof. First assume F is formally real. Fixing an ordering on F, we can embed F in a real-closed field F_0. By functoriality, we have a ring homomorphism $K_*F \to K_*F_0$. Since $\ell(-1)$ is not nilpotent in K_*F_0 (by 6.6), it follows that $\ell(-1)$ is *also* not nilpotent in K_*F.

Conversely, let F be nonreal, say $-1 = a_1^2 + \cdots + a_r^2$, where $a_i \in \dot{F}$. By 6.5, we have $\ell(-a_1^2)\cdots\ell(-a_r^2) = 0 \in K_rF$, and hence $\ell(-1)^r \in 2K_rF$. Since $2\,\ell(-1) = 0$, it follows that $\ell(-1)^{r+1} = 0 \in K_{r+1}F$. Next, consider any generator $\alpha = \ell(a_1)\cdots\ell(a_n) \in K_nF$. From

$$\ell(a_1)^{r+2} = \ell(a_1)\ell(-1)^{r+1} = 0$$

and 6.3(2), it follows that $\alpha^{r+2} = 0$. Finally, consider any sum of generators $\beta = \alpha_1 + \cdots + \alpha_s$. Using the "graded commutative" property in 6.4, we see easily that $\beta^k = 0$ if $k > s(r+1)$. This shows that $\bigoplus_{i \geq 1} K_i F$ is a nil ideal in $K_* F$. $\square$

In general, computing the higher K-groups $K_n F$ for a field F is a difficult task. We should thus especially appreciate the following computation in the case of finite fields.

Example 6.11. *For any finite field F, $K_n F = 0$ for any $n \geq 2$.*

Indeed, we have proved earlier (in V.6.14) that $K_2 F = 0$. Since $K_n F$ is generated by $K_2 F \cdot K_{n-2} F$, it follows that $K_n F = 0$ for $n \geq 2$. This computation shows also that the Milnor K-groups $K_n F$ are in general different from the Quillen K-groups $K_n^{\mathrm{Q}} F$, since for finite fields F, $K_n^{\mathrm{Q}} F$ is nonzero for *odd* n, as we have pointed out earlier in the paragraph following V.6.14.

For a field F equipped with a rank 1 discrete valuation $v\colon \dot{F} \to \mathbb{Z}$ with a residue field $\overline{F}$, Milnor has constructed a group homomorphism $\partial_v\colon K_n F \to K_{n-1}\overline{F}$ that carries a generator of the form $\ell(\pi)\ell(u_2)\cdots\ell(u_n) \in K_n F$ to $\ell(\overline{u}_2)\cdots\ell(\overline{u}_n) \in K_{n-1}\overline{F}$, for a uniformizer π of the valuation v, and for any units $u_2, \ldots, u_n$ in the valuation ring of v. This construction is, of course, inspired by the construction of the second residue homomorphism in Springer's Theorem (VI.1.4 and VI.1.5) on quadratic forms over a field F with a complete discrete valuation v.

Using the maps ∂_v for v ranging over the π-adic valuations on a rational function field $E = F(x)$ (where π denotes a typical monic irreducible polynomial in $F[x]$), Milnor obtained a short exact sequence

$$(6.12) \qquad 0 \longrightarrow K_n F \longrightarrow K_n F(x) \xrightarrow{(\partial_\pi)} \bigoplus_{\pi} K_{n-1}(F[x]/(\pi)) \longrightarrow 0$$

that is analogous to the Witt group exact sequence for $E = F(x)$ established in IX.3.1. Since the work needed here is rather similar to that given earlier for the proof of IX.3.1, we will not present it here. Let us just point out that this circle of ideas can also be applied to the computation of $K_n\mathbb{Q}$ (cf. the computation of the Witt ring $W(\mathbb{Q})$ in VI.4). The upshot of this computation is that $K_n\mathbb{Q} \cong \mathbb{Z}_2$ for all $n \geq 3$, as was pointed out by Milnor in §1 of [Mi].

We have not gone into the details in proving the exactness of the sequence (6.12) since our main interest in the groups $K_n F$ in this section is *not* in (6.12), but rather in the direction of relating these groups to the filtration factors $I^n F / I^{n+1} F$ of the Witt ring $W(F)$. Let us now embark upon this interesting theme.

Using the same calculations in the proof of V.6.5, we check easily that, for any field F, the rule

$$(6.13) \qquad (a_1, a_2, \ldots, a_n) \longmapsto \langle\langle -a_1, -a_2, \ldots, -a_n \rangle\rangle + I^{n+1}F$$

defines an n-fold Steinberg symbol from $\dot{F} \times \cdots \times \dot{F}$ (n factors) into the additive group $I^nF/I^{n+1}F$. By the universal property of K_nF, it follows that there is a unique group homomorphism from K_nF to $I^nF/I^{n+1}F$ taking $\ell(a_1) \cdots \ell(a_n)$ to the element on the RHS of (6.13). Since $I^nF/I^{n+1}F$ is an elementary 2-group, we obtain then a well-defined group homomorphism

$$(6.14) \qquad \alpha_n\colon\ k_nF \longrightarrow I^nF/I^{n+1}F$$

with $\alpha_n(\overline{\ell(a_1) \cdots \ell(a_n)}) = \langle\langle -a_1, \ldots, -a_n \rangle\rangle + I^nF$. Since from here on we'll be working mostly with the group $k_nF = K_nF/2K_nF$, we shall drop the overbar in the equation above, and interpret $\ell(a_1) \cdots \ell(a_n)$ as an element in k_nF (rather than in K_nF) in the following, unless it is explicitly stated otherwise. Elements of this form will be called the *generators* of k_nF: they span k_nF as a vector space over $\mathbb{Z}_2$.

Since I^nF is additively generated by the n-fold Pfister forms, α_n is clearly an epimorphism. Milnor raised in [Mi] the following fundamental question.

Question 6.15. *Is α_n an isomorphism for all fields F?*

As $k_0F \cong \mathbb{Z}_2$ and $k_1F \cong \dot{F}/\dot{F}^2$, our work in Chapter II already showed that α_0 and α_1 are isomorphisms. For $n = 2$, α_2 is just the homomorphism denoted by α in V.6.5(2), and we have also shown that this is an isomorphism in V.6.7. Finally, if F is a real-closed field, then (by 6.9) $k_nF \cong \mathbb{Z}_2$ is generated by $\ell(-1)^n$, and

$$\alpha_n(\ell(-1)^n) = 2^n\langle 1 \rangle + I^{n+1}F$$

is nonzero in $I^nF/I^{n+1}F$, so α_n is an isomorphism as well. All of the positive evidence above would seem to suggest that the answer to 6.15 might be "yes" in general.

Milnor raised 6.15 as an open question, but a positive answer to 6.15 seemed to be so compelling to some authors that they began to refer to it as (a part of) the "Milnor Conjectures". (Milnor raised a similar question about the homomorphism β_n from k_nF to the n-th Galois cohomology group of F. The "other" Milnor Conjecture is the parallel statement that β_n is also an isomorphism. We'll say more about this a little bit later in this section.)

Of course, the idea of trying to prove that α_n is an isomorphism is that one can then understand the group $I^nF/I^{n+1}F$ as generated by the classes of the n-fold Pfister forms with a predictable set of relations. An obvious thought for showing the bijectivity of α_n is to construct an inverse map from $I^nF/I^{n+1}F$ by using suitable quadratic form invariants taking values in k_*F. For the case $n = 2$, this was indeed how α_2 was proved to be an isomorphism in V.6.7: the inverse map for α_2 was provided by $w_\pm$, a signed version of the second Stiefel-Whitney invariant. For a general n, however, finding a map from $I^nF/I^{n+1}F$ to k_nF does not seem to be an easy task. By using higher Stiefel-Whitney classes (building on the earlier work of Delzant), Milnor [Mi] constructed a homomorphism

$$w_t\colon\ I^nF/I^{n+1}F \longrightarrow k_tF, \quad \text{where } t = 2^{n-1}, \tag{6.16}$$

which takes $\langle\langle -a_1, \ldots, -a_n\rangle\rangle + I^{n+1}F$ to $\ell(a_1)\cdots\ell(a_n)\ell(-1)^{t-n}$ in k_tF. Thus, if multiplication by $\ell(-1)^{t-n}$ happens to carry k_nF *injectively* into k_tF, then the map α_n would indeed be injective (and hence an isomorphism). However, such an assumption on the field F is too restrictive. The "trouble" with this approach, of course, lies in the fact that $t = 2^{n-1}$ is much bigger than n. For $n = 2$, t happens to be 2, so the map in (6.16) goes into the desired group k_2F. But if $n > 2$, the map w_t goes into $k_{2^{n-1}}F$, instead of k_nF.

Without constructing further quadratic form invariants, it is possible to prove some partial results toward an affirmative answer on Milnor's question. We'll present some such results from [EL$_1$], which were among the first attempts at solving the "Milnor Conjecture" 6.15. The proofs of these results are based on the material in §1 and §5.

Theorem 6.17. *For any n, the homomorphism $\alpha_n\colon k_nF \to I^nF/I^{n+1}F$ is injective on generators, in the sense that*

$$\alpha_n(\ell(a_1)\cdots\ell(a_n)) = \alpha_n(\ell(b_1)\cdots\ell(b_n)) \tag{6.18}$$

implies $\ell(a_1)\cdots\ell(a_n) = \ell(b_1)\cdots\ell(b_n) \in k_nF$.

Proof. Let $\varphi = \langle\langle -a_1, \ldots, -a_n\rangle\rangle$ and $\psi = \langle\langle -b_1, \ldots, -b_n\rangle\rangle$. By the definition of α_n, (6.18) means that $\varphi \equiv \psi \pmod{I^{n+1}F}$. Therefore, by 5.4, $\varphi \cong \psi$. Then, by 1.12, φ and ψ are chain P-equivalent. It is thus sufficient to handle the case where φ and ψ are simply P-equivalent. This means that there exist two distinct indices[11] i and j such that $\langle\langle a_i, a_j\rangle\rangle \cong \langle\langle b_i, b_j\rangle\rangle$, and $a_k = b_k$ for any $k \neq i, j$. Now the former implies that $\ell(a_i)\ell(a_j) = \ell(b_i)\ell(b_j)$ (since α_2 is already known to be an isomorphism). Multiplying this by $\ell(a_k) = \ell(b_k)$ for $k \neq i, j$, we see that $\ell(a_1)\cdots\ell(a_n) = \ell(b_1)\cdots\ell(b_n)$ in k_nF. □

Let us now record some consequences of 6.17.

---

[11]We may assume that $n \geq 2$ here, since the case $n = 1$ is clear (by II.2.3).

Corollary 6.19. *For any $a_i \in \dot{F}$, $\ell(a_1)\cdots\ell(a_n) = 0 \in k_nF$ iff $\langle\langle -a_1, \ldots, -a_n\rangle\rangle$ is hyperbolic. In particular,*

(1) $k_nF = 0$ iff $I^nF = 0$, iff all n-fold Pfister forms over F are hyperbolic;

(2) k_nF is finite iff I^nF is a finitely generated abelian group.

Proof. The main statement follows from 6.17 by taking b_1 to be 1. This clearly implies (1), and (2) follows likewise from the fact that k_nF is a $\mathbb{Z}_2$-vector space generated by $\ell(a_1)\cdots\ell(a_n)$, and I^nF is a $\mathbb{Z}$-module generated by $\langle\langle -a_1, \ldots, -a_n\rangle\rangle$ $(a_i \in \dot{F})$. □

Theorem 6.17 also enables us to obtain a refinement of Milnor's K-theoretic characterization of nonreal fields in 6.10. This refinement requires the fact that the level of a nonreal field F (the smallest integer s such that -1 is a sum of s squares in F) is always a power of 2; this result will be proved later in XI.2.

Corollary 6.20. *A field F is nonreal iff $\ell(-1)$ is nilpotent in k_*F, iff $\bigoplus_{i\geq 1} k_iF$ is a nil ideal in the graded ring k_*F. Furthermore, if F has level 2^m, then m is precisely the smallest integer such that $\ell(-1)^{m+1} = 0$ in k_*F. In this case, $\ell(-1)^{m+2} = 0$ in K_*F, and for any $a \in \dot{F}$, $\ell(a)^{m+3} = 0$ in K_*F.*

Proof. If F is formally real, then the n-fold Pfister form $\langle\langle 1, \ldots, 1\rangle\rangle$ is anisotropic, and hence $\ell(-1)^n \neq 0$ in k_nF for any n (by 6.19). Now assume F is nonreal, with level 2^m. Then the m-fold Pfister form $\varphi = \langle\langle 1, \ldots, 1\rangle\rangle$ is anisotropic and the $(m+1)$-fold Pfister form $\varphi\langle\langle 1\rangle\rangle$ is hyperbolic. By 6.19, we see that $m+1$ is exactly the index of nilpotency of $\ell(-1)$ in k_*F. Lifting to $K_{m+1}F$, we may write $\ell(-1)^{m+1} = 2\alpha$, where $\alpha = K_{m+1}F$. Since $2\,\ell(-1) = 0 \in K_1F$, it follows that $\ell(-1)^{m+2} = 0$ in $K_{m+2}F$. Therefore, for any $a \in \dot{F}$, 6.3(3) implies that

$$\ell(a)^{m+3} = \ell(a)\,\ell(-1)^{m+2} = 0 \in K_{m+3}F. \tag{6.21}$$

This shows once more that $\bigoplus_{i\geq 1} K_iF$ is a nil ideal in K_*F, and *a fortiori*, $\bigoplus_{i\geq 1} k_iF$ is a nil ideal in k_*F. □

We shall now try to prove a stronger version of 6.17 (from [EL₁]). The proof of this new version depends heavily on 6.17, so it is best viewed as a self-strengthening of that result.

Linkage Theorem 6.22. *Let $\varphi_i = \langle\langle a_{i1}, \ldots, a_{in}\rangle\rangle$ $(i = 1, 2, 3)$ be three n-fold Pfister forms over F such that $\varphi_1 + \varphi_2 + \varphi_3 \in I^{n+1}F$. Then φ_1, φ_2,*

φ_3 are linked (in the sense of 5.11). There exist an $(n-1)$-linkage σ, and $x, y \in \dot{F}$ such that

$$\varphi_1 \cong \sigma\langle 1, -xy\rangle, \quad \varphi_2 \cong \sigma\langle 1, x\rangle, \quad \textit{and} \quad \varphi_3 \cong \sigma\langle 1, y\rangle. \tag{6.23}$$

In particular, there exists an isometry

$$\langle -y\rangle\, \varphi_1 \perp \varphi_3 \cong \varphi_2 \perp 2^{n-1}\mathbb{H}. \tag{6.24}$$

Proof. The proof of this theorem is a very typical application of the function field techniques. Let L be the function field $F[\varphi_1]$. (We may, of course, assume that $n \geq 2$, so that $F[\varphi_1]$ is defined.) Since $(\varphi_1)_L$ is hyperbolic, 5.4 implies that $(\varphi_2)_L \cong (\varphi_3)_L$. Fix a Witt decomposition of $q := \varphi_2 \perp \langle -1\rangle\varphi_3$ over F, say $q \cong q_a \perp q_h$, where q_a is anisotropic and q_h is hyperbolic.

Case 1. *q_a is the zero form.* This means that $\varphi_2 \cong \varphi_3$ over F. Then $\varphi_1 + \varphi_2 + \varphi_3 \in I^{n+1}F$ implies that $\varphi_1 \in I^{n+1}F$, so φ_1 is hyperbolic by the Hauptsatz 5.1. The conclusions of the theorem are trivial in this case.

Case 2. *q_a is not the zero form.* Since $(q_a)_L$ is hyperbolic, 4.5 implies that there exist $r \in D_F(q_a)$ and an F-form τ_0 such that $q_a \cong \langle r\rangle\, \varphi_1 \perp \tau_0$. Thus,

$$q = \varphi_2 \perp \langle -1\rangle\, \varphi_3 \cong \langle r\rangle\, \varphi_1 \perp \tau, \tag{6.25}$$

where $\tau := \tau_0 \perp q_h$ has dimension 2^n. From the hypothesis $\varphi_1 + \varphi_2 + \varphi_3 \in I^{n+1}F$, we have

$$\varphi_2 \perp \langle -1\rangle\, \varphi_3 \equiv \langle r\rangle\, \varphi_3 \pmod{I^{n+1}F},$$

so (6.25) yields $\tau \in I^{n+1}F$, and hence $\tau \cong 2^{n-1}\mathbb{H}$ by the Hauptsatz. This means that the Witt index of $q = \varphi_2 \perp \langle -1\rangle\varphi_3$ is at least 2^{n-1}, so 5.13 implies that φ_2, φ_3 are linked. We may thus write $\varphi_2 \cong \sigma\langle 1, x\rangle$ and $\varphi_3 \cong \sigma\langle 1, y\rangle$, where σ is a suitable $(n-1)$-fold Pfister form, and $x, y \in \dot{F}$. Now, modulo $I^{n+1}F$, we have

$$\varphi_1 \equiv \varphi_2 - \varphi_3 \equiv \sigma\,(\langle x\rangle - \langle y\rangle) \equiv \langle -y\rangle\, \sigma\,\langle 1, -xy\rangle.$$

Thus, 5.4 implies that $\varphi_1 \cong \sigma\,\langle 1, -xy\rangle$, proving (6.23). Finally, a quick calculation from (6.23) yields (6.24). □

Corollary 6.26. *For a given integer $n > 1$, the following statements are equivalent:*

(1) *Every element of k_nF is a generator.*

(2) *Every element of k_mF (for $m \geq n$) is a generator.*

(3) *Every pair of n-fold Pfister forms over F is linked.*

(4) *Every pair of m-fold Pfister forms (for $m \geq n$) is linked.*

If (3) holds, the field F is said to be n-linked. In this case, α_m is an isomorphism for all $m \geq n$.

Proof. First, (3) $\Leftrightarrow$ (4) is easily verified, and (2) $\Rightarrow$ (1) is a tautology. So it suffices for us to check that (1) $\Rightarrow$ (3) $\Rightarrow$ (2).

(1) $\Rightarrow$ (3). Let φ_1, φ_2 be n-fold Pfister forms over F, and let γ_1, γ_2 be the "corresponding" generators in k_nF. By (1), $\gamma_1 + \gamma_2 = \gamma_3$ for some generator $\gamma_3 \in k_nF$. Applying α_n, we get $\varphi_1 + \varphi_2 \equiv \varphi_3 \pmod{I^{n+1}F}$, where φ_3 is the n-fold Pfister form corresponding to γ_3. Now 6.22 implies that φ_1, φ_2 are linked.

(3) $\Rightarrow$ (2). It suffices to show that if γ_1, γ_2 are generators of k_nF, then $\gamma_1 + \gamma_2$ is also a generator. But the n-fold Pfister forms $\alpha_n(\gamma_1)$, $\alpha_n(\gamma_2)$ are linked (by (3)), so we can write

$$\gamma_1 = \ell(a_1)\cdots\ell(a_{n-1})\ell(x) \quad \text{and} \quad \gamma_2 = \ell(a_1)\cdots\ell(a_{n-1})\ell(y)$$

for suitable a_i, x, $y \in \dot{F}$. It follows that

$$\gamma_1 + \gamma_2 = \ell(a_1)\cdots\ell(a_{n-1})\ell(xy),$$

which is a generator in k_nF, as desired. $\square$

Recall that a field F is called a *linked field* if every two quaternion algebras over F are linked. This amounts to the fact that F is 2-linked, in the sense of 6.26. Thus, 6.26 (along with VI.3.5) yields the following.

Corollary 6.27. *If F is a linked field, then α_m is an isomorphism for all m. In particular, this conclusion holds for any local or global field F.*

Example 6.28. Let F be a global field with real completions $F_1, \ldots, F_r$, and let $e_i \in \dot{F}$ $(1 \le i \le r)$ be as in VI.3.9. For $n \ge 3$, it follows easily from VI.3.9 that the group $I^nF/I^{n+1}F$ has a $\mathbb{Z}_2$-basis given by the cosets

$$\{2^{n-1}\langle 1, -e_i\rangle + I^{n+1}F\colon\ 1 \le i \le r\}.$$

Using the isomorphism α_n, we see that a $\mathbb{Z}_2$-basis for k_nF is given by the elements $\{\ell(-1)^{n-1}\ell(e_i)\}$. From this, we can check easily that the natural map

$$k_nF \longrightarrow \bigoplus_i k_nF_i$$

is an isomorphism (for $n \ge 3$). This result is due to Tate, whose proof (as presented in the Appendix to Milnor's paper [Mi]) used the method of idèles. The simplified proof as given above (using 6.27 and VI.3.9) comes essentially from [EL$_0$].

From the linkage theorem, we can also prove the following partial result on the injectivity of α_n; this is the "self-strengthening" of Theorem 6.17 that we have alluded to earlier.

Theorem 6.29. (1) *Let* $\gamma = \gamma_1 + \gamma_2 + \gamma_3$, *where the* γ_i*'s are generators of* $k_n F$. *If* $\alpha_n(\gamma) = 0 \in I^n F / I^{n+1} F$, *then* $\gamma = 0 \in k_n F$.

(2) *For a given integer* n, *suppose every element of* $k_n F$ *can be expressed as a sum of three generators. Then* α_n *is an isomorphism.*

Proof. For $i \leq 3$, let $\varphi_i = \alpha_n(\gamma_i)$ be the n-fold Pfister form corresponding to the generator γ_i. Then $\alpha_n(\gamma) = 0$ implies that $\varphi_1 + \varphi_2 + \varphi_3 \in I^{n+1} F$, so we can write the φ_i's in the form (6.23), with some $\sigma = \langle\langle -a_1, \ldots, -a_{n-1} \rangle\rangle$. By 6.17, we have

$$\gamma_2 = \ell(a_1) \cdots \ell(a_{n-1}) \ell(-x), \quad \gamma_3 = \ell(a_1) \cdots \ell(a_{n-1}) \ell(-y),$$

and

$$\gamma_1 = \ell(a_1) \cdots \ell(a_{n-1}) \ell(xy),$$

from which we clearly get $\gamma = \gamma_1 + \gamma_2 + \gamma_3 = 0 \in k_n F$. The last conclusion (2) follows immediately from (1). □

In [EL$_1$], it was shown that, if $|k_n F| \leq 2^6$, then indeed every element in $k_m F$ (for $m \geq n$) is a sum of three generators. Thus, in this case, (2) above implies that α_m is an isomorphism for any $m \geq n$. But of course, this is not to be viewed as a main application of (1). The value of (1) and 6.22 lies in the fact that they give an accurate picture of what is really happening in the case where $\gamma_1 = \gamma_2 + \gamma_3$ in $k_n F$, or $\varphi_1 = \varphi_2 + \varphi_3$ in $I^n F / I^{n+1} F$.

We close this section by making a short report on the recent developments on some of the central themes discussed in this section. For convenience, let us refer to an affirmative answer to Question 6.15 as "Milnor's (First) Conjecture". This conjecture has been proved in the late 1990s by D. Orlov, V. Vishik, and V. Voevodsky; see [OVV$_1$] and [OVV$_2$]. Milnor's original "conjectures" in [Mi] also involved the Galois cohomology groups of the field F with $\mathbb{Z}_2$-coefficients, denoted by $H^n(F, \mathbb{Z}_2)$. (These are the cohomology groups of the profinite Galois group of the separable closure of the field F, with $\mathbb{Z}_2$-coefficients.) By identifying $H^1(F, \mathbb{Z}_2)$ with $\dot{F}/\dot{F}^2$, one can construct the cup product cohomology classes

(6.29) $$(a_1) \cup \cdots \cup (a_n) \in H^n(F, \mathbb{Z}_2) \quad \text{for } a_1, \ldots, a_n \in \dot{F}/\dot{F}^2.$$

Using the universal property of $k_n F$ (again), Bass and Tate defined a group homomorphism

(6.30) $$\beta_n \colon k_n F \longrightarrow H^n(F, \mathbb{Z}_2)$$

by sending a generator $\ell(a_1) \cdots \ell(a_n) \in k_n F$ to the cup product $(a_1) \cup \cdots \cup (a_n)$ (for all $a_i \in \dot{F}$). In parallel to Question 6.15, Milnor also asked if β_n is an isomorphism. The claim that β_n is an isomorphism for all n may be called "Milnor's Second Conjecture". Note that β_2 is indeed an isomorphism, on account of Merkurjev's work [Me$_1$] in 1981; see V.6.11.

Milnor's Second Conjecture was also proved (along with the First) by Voevodsky [Vo] in 1996, for fields of characteristic $\neq 2$. In fact, Voevodsky's results on β_n apply more generally to link the group $K_nF/2^m(K_nF)$ to the n-th étale cohomology of F with coefficients in $\mu_{2^m}^{\otimes n}$, the n-fold tensor product of the group of 2^m-th roots of unity. Voevodsky's proofs used many high power tools, including motivic cohomology, stable homotopy theory for algebraic varieties, and algebraic cobordisms. For this deeply significant work in cohomology theories, Voevodsky was awarded one of the two Fields Medals at the International Congress of Mathematicians held in Beijing, China, in August of 2002. As of today, however, the proofs of Voevodsky's results on the Milnor Conjectures have not been completely published.

For a presentation of Voevodsky's work on Milnor's Conjectures in the *Séminaire Bourbaki* style, see Kahn's report [Kah]. For an expository account on the same, see Morel's article [Mor]. A historical survey on Milnor's Conjectures and their influence can also be found in Pfister's paper [Pf6]. For some more applications of Milnor's Conjectures to quadratic form theory, see the work of Arason and Elman in [AE].

Exercises for Chapter X

1. For any field F, show that $I^3F = 0$ iff, for any a, b, c, $d \in \dot{F}$, the form $\langle a, b, c, d\rangle$ over F represents $-abc$.
2. Determine the group of similarity factors for the form $\langle a, a, a, a\rangle$.
3. Let φ be a 4-dimensional form over F of determinant 1. Show that the number of cosets of $\dot{F}^2$ contained in $D(\varphi)$ is either infinite, or a power of 2.
4. For any form $\varphi \in I^2F$ of dimension $2n$, show that $\varphi = \sum_{i=1}^{n-1}\langle a_i\rangle\varphi_i \in W(F)$, where $a_i \in \dot{F}$ and the φ_i's are 2-fold Pfister forms.
5. For any Pfister form φ, show that $\langle\langle a\rangle\rangle\,\varphi \cong \langle\langle b\rangle\rangle\,\varphi$ iff $ab \in D(\varphi)$.
6. Let σ, τ be Pfister forms, and let $a \in \dot{F}$. Show that $\sigma \perp \langle a\rangle\tau$ is a Pfister form iff $\sigma \cong \tau$. (**Hint.** Use 5.4.)
7. (Pfister) Show that, if n is a 2-power, then
$$D_F(n) \cdot D_F(n+1) = D_F(2n).$$
8. (Pfister) For any $a, b, \ldots, f \in \dot{F}$, show that
$$\begin{aligned}&\langle\langle -ac,\ -b\rangle\rangle + \langle\langle -ce,\ -d\rangle\rangle + \langle\langle -ea,\ -f\rangle\rangle\\ &\quad\equiv \langle\langle -bd,\ -c\rangle\rangle + \langle\langle -df,\ -e\rangle\rangle + \langle\langle -fb,\ -a\rangle\rangle \pmod{I^3F}.\end{aligned}$$
9. For $n \geq 3$, show that any anisotropic n-fold Pfister form over $\mathbb{Q}$ is isometric to $2^n\langle 1\rangle$.

10. For any n-fold Pfister form φ, and any integer $r \geq 1$, show that $\varphi^r \cong 2^{(r-1)n}\varphi$.

11. Show that the four statements in I.Exercise 10 are all equivalent to $I^2F = 2IF$.

12. (Cf. III.Exercise 6) Show that, if a 5-dimensional form is universal, then it is isotropic.

13. Prove the following weaker version of 1.7 by using Pfister's Local-Global Principle: if a Pfister form φ is isotropic, then $2^k\varphi$ is hyperbolic for some integer k.

14. In the composition formula for a quaternionic norm form given after the proof of 2.11, show that the four linear forms appearing on the RHS are linearly independent (when viewed as linear forms in $(y_1, \ldots, y_4)$ with coefficients in $F(x_1, \ldots, x_4)$).

15. For $n \geq 3$, classify the isometry classes of n-fold Pfister forms over a global field. (**Hint.** Use 6.28.)

16. Let φ be an anisotropic form of positive dimension over a global field. If $\varphi \in I^2F \cap W_t(F)$, show that φ must be a 2-fold Pfister form, and is in fact universal.

17. If $\langle a_1, \ldots, a_n \rangle \cong \langle b_1, \ldots, b_n \rangle$, show that $\langle\langle a_1, \ldots, a_n \rangle\rangle \cong \langle\langle b_1, \ldots, b_n \rangle\rangle$.

18. For any Pfister forms φ, ψ over a field, show that $D(\varphi') = D(\psi')$ implies $D(\varphi) = D(\psi)$, where φ' and ψ' denote the pure subforms of φ and ψ. Does the converse also hold?

19. If φ is a Pfister form over a pythagorean field, show that $1 \in D(\varphi')$ implies that $D(\varphi) = D(\varphi')$.

20. Let S be a subset of $\dot{F}$, and let $y \in \dot{F}$. Using VIII.9.7, show that

    (1) F admits an ordering containing S iff any Pfister form $\langle\langle a_1, \ldots, a_n \rangle\rangle$ ($a_i \in S$) is anisotropic; and
    (2) $y \in P$ for all orderings P on F containing S iff $y \in D_F(\varphi)$ for some Pfister form $\varphi = \langle\langle a_1, \ldots, a_n \rangle\rangle$ with $a_i \in S$.
    (3) Using Pfister forms again, derive an analogous version of VIII.9.8 (with S replacing G) for extending orderings from one field to another.

21. For any Pfister form $\varphi = \langle\langle a_1, \ldots, a_n \rangle\rangle$ over a pythagorean field F, show that $D_F(\varphi) = \bigcap_P \dot{P}$, where P ranges over all orderings of F containing $a_1, \ldots, a_n$. From this fact, deduce that, if $\psi = \langle\langle b_1, \ldots, b_n \rangle\rangle$, then $\varphi \cong \psi$ iff $D_F(\varphi) = D_F(\psi)$.

22. For any $a_1, \ldots, a_n \in \dot{F}$, show that the following are equivalent:

    (1) $\langle\langle a_1, \ldots, a_n \rangle\rangle \cong \langle\langle -a_1, \ldots, -a_n \rangle\rangle$.
    (2) $\langle\langle a_1a_2,\ a_2a_3,\ \ldots,\ a_{n-1}a_n \rangle\rangle$ represents -1.

If, moreover, F is a formally real pythagorean field, show that (1), (2) are further equivalent to:

(3) For any ordering P on F, $a_i a_{i+1} \in -P$ for some i.
(4) For any ordering P on F, $a_i a_j \in -P$ for some $i < j$.
(5) There exists no ordering P such that all $a_i \in P$ or all $a_i \in -P$.

23. Let σ be a Pfister neighbor such that $2^{n-1} < \dim \sigma \leq 2^n$. Show that σ is a special Pfister neighbor iff σ has a subform similar to an $(n-1)$-fold Pfister form.

24. (This exercise may be viewed as a generalization of the implication (1) $\Rightarrow$ (4) in 4.19.) Let $\varphi = \sigma \perp \rho$ be an n-fold Pfister form, where $\dim \sigma = 2^{n-1} + 1$. Show that ρ is a Pfister neighbor iff σ is a special Pfister neighbor. [**Hint.** For the (harder) "only if" part, use 4.18(3) and 4.28(3). Of course, Witt cancellation is indispensable.]

25. (Hoffmann) Show that a field F is linked (in the sense of 4.20) iff, for any n, any $(2^n + 1)$-dimensional form over F is a Pfister neighbor.

26. (Hoffmann) Let σ be an anisotropic form of dimension $2^n + m$, where $1 \leq m \leq 2^n$ (so that $2^n < \dim \sigma \leq 2^{n+1}$).

(1) Using 4.34, show that $i_1(\sigma) \leq m$ (where $i_1(\sigma)$ denotes the first Witt index of σ, that is, the Witt index of $\sigma_{F[\sigma]}$). [**Hint.** Use Ch. I, Exercise 16(3).]

(2) The anisotropic form σ is said to have *maximal splitting* if $i_1(\sigma)$ attains its biggest possible value, namely m. Show that if either $\dim \sigma = 2^n + 1$ or if σ is a Pfister neighbor, then σ has maximal splitting.

27. (Kahn) Let φ be an anisotropic Pfister form over F, and let σ be an anisotropic form of dimension ≥ 2. Using 4.34 again, show that $\sigma > \varphi > \sigma$ iff σ is a Pfister neighbor with associated Pfister form φ.

28. (Hoffmann) Among the quadratic forms of dimension 2^n over F, define a relation $\sim_{\text{hn}}$ by: $\sigma \sim_{\text{hn}} \tau$ iff there exists $a \in \dot{F}$ such that $\sigma \perp \langle a \rangle \tau$ is a scalar multiple of a Pfister form (or, equivalently by 5.6, $\sigma \equiv b \cdot \tau \pmod{I^{n+1}F}$ for some $b \in \dot{F}$). Show that:

(1) $\sigma \sim \tau \Longrightarrow \sigma \sim_{\text{hn}} \tau$; and
(2) $\sim_{\text{hn}}$ is an equivalence relation (called the "half-neighbor" relation) on the set of 2^n-dimensional forms over F.
(3) If $\sigma \sim_{\text{hn}} \tau$, show that $\sigma > \tau > \sigma$.

[**Hint** (for (3)). Since $\sigma \perp \langle a \rangle \tau$ is hyperbolic over $F(\tau)$, $\sigma_{F(\tau)} \cong \langle -a \rangle \tau_{F(\tau)}$ is given by a form of dimension $< 2^n$.] *Comment.* If $n \leq 2$ in (3), one can in fact show that $\sigma \sim \tau$: see XII.2.4.

Chapter XI

Field Invariants

In this chapter, we study four quadratic invariants of a field F. These are $s(F)$ (the level), $h(F)$ (the height), $P(F)$ (the Pythagoras number), and $u(F)$ (the u-invariant, for lack of a better name). The computation of these basic invariants makes very interesting uses of the theory of quadratic forms over the field F, and provides in turn a very substantial enrichment of this theory.

The first interesting result is the fact that the level $s(F)$ must be a power of 2 (or infinite). This was first proved by Pfister [Pf$_1$], but it had remained an open problem for more than thirty years. We also present Pfister's construction of fields of a given (finite) level 2^n, making full use of the theory of function fields in X.4. Assuming the Hasse-Minkowski Principle (VI.3.1) and the theorem of Tsen-Lang (to be stated), we calculate the level of various number fields and function fields. The material in this chapter is selected mainly from the "early" literature in the algebraic theory of quadratic forms. Some more recent results on field invariants (and commutative ring invariants) will be given later in Chapter XIII.

The principal tool for this chapter is the machinery of Pfister forms developed in Chapter X. Thus, the results in this chapter will provide important illustrations of the many striking ways in which Pfister forms can be utilized. Indeed, to a considerable extent, Pfister's use of the theory of multiplicative forms in the mid-1960s to determine the nature of the level invariant $s(F)$ has signaled the revival of the algebraic theory of quadratic forms in modern times.

We state here without proof (a special case of) the theorem of Tsen and Lang, which will be needed in this chapter in order to draw examples from function fields over algebraically closed fields and real-closed fields. A proof of this result can be found in Scharlau's book [Sc4].

Theorem of Tsen-Lang (Special Case). *Let K be a field of transcendence degree n over an algebraically closed field k. Then any quadratic form over K of dimension $> 2^n$ is isotropic.*

In the terminology to be introduced in §6, this result says that the field K has u-invariant $\leq 2^n$. An easy direct proof of this result in the case where $K = k(t)$ (with $n = 1$) can be found in II.3.8.

1. Sums of Squares

We first recall a familiar notation used in our earlier chapters.

Definition. For any integer $m \geq 1$, we write $D(m) = D_F(m)$ for the set $D_F(m\langle 1\rangle)$. This consists of all nonzero elements of F that are sums of m squares in F. If necessary, we take $D(0)$ to be the empty set. Thus, we have an ascending chain $D(0) \subseteq \dot{F}^2 = D(1) \subseteq D(2) \subseteq \cdots$.

In this section, we put together a few interesting properties of these sets $D_F(m)$. First, if $m = 2^n$, then $m\langle 1\rangle$ is just the n-fold Pfister form $\langle\langle 1, \ldots, 1\rangle\rangle$. By X.1.9, we have the following.

Theorem 1.1. *For any n, $2^n\langle 1\rangle$ is a group form over F; that is, $D_F(2^n)$ is a subgroup of the multiplicative group $\dot{F}$.*

Note that we have also given a direct and completely elementary proof for this in the Appendix to X.1, using the notion of round forms. In the following, we shall give yet another elementary proof. This proof is matrix-theoretic, and it has the advantage of yielding a slightly more general result (1.1′ below). We start with a lemma.

Lemma 1.2. *Let $m = 2^n$ and $c = c_1^2 + \cdots + c_m^2$, where $c_i \in F$. Then there exists an $m \times m$ matrix $S \in \mathbb{M}_m(F)$ with first row $c_1, \ldots, c_m$ such that $S \cdot S^t = S^t \cdot S = c \cdot I_m$ (where "t" denotes the transpose).*

Proof. We first deal with the case $c = 0$. If all $c_i = 0$, we may pick S to be the zero matrix. So assume, say, $c_1 \neq 0$. Let R denote the row matrix $(c_1, \ldots, c_m)$. We set $S = c_1^{-1} \cdot R^t \cdot R$ which has indeed a first row equal to R. Further,

$$S \cdot S^t = c_1^{-2} \cdot R^t \cdot RR^t \cdot R = 0,$$

since $R \cdot R^t = c = 0$. Similarly, $S^t \cdot S = 0$, and the proof is complete.

We may now assume that $c \neq 0$, and proceed inductively on n. Split up the set $c_1, \ldots, c_m$ into two equal parts, and re-label the elements in these as $a_1, \ldots, a_{2^{n-1}}$ and $b_1, \ldots, b_{2^{n-1}}$, respectively. Write $a = \sum a_i^2$ and $b = \sum b_i^2$, so $c = a + b$. Since a, b cannot both be zero, we may as well assume that $a \neq 0$. (The case $b \neq 0$ is similar.) By the inductive hypothesis, there exist square matrices A, B of size 2^{n-1}, such that

$$A \cdot A^t = A^t \cdot A = a \cdot I_{2^{n-1}} \quad \text{and} \quad B \cdot B^t = B^t \cdot B = b \cdot I_{2^{n-1}}.$$

Further, the first row of A is $(a_1, \ldots, a_{2^{n-1}})$ and the first row of B is $(b_1, \ldots, b_{2^{n-1}})$. Now, we "put together" A and B to form a matrix

$$S = \begin{pmatrix} A & B \\ -a^{-1}A^t \cdot B^t \cdot A & A^t \end{pmatrix} \in \mathbb{M}_m(F).$$

This has first row equal to $(c_1, \ldots, c_m)$ as desired, and easy matrix computation confirms that $S \cdot S^t = S^t \cdot S = c \cdot I_m$. □

Theorem 1.1′. *Let $m = 2^n$ and $u_1, \ldots, u_m, v_1, \ldots, v_m \in F$. Then there exist $w_2, \ldots, w_m \in F$ such that*

$$(u_1^2 + \cdots + u_m^2) \cdot (v_1^2 + \cdots + v_m^2) = (u_1v_1 + \cdots + u_mv_m)^2 + w_2^2 + \cdots + w_m^2.$$

In particular, if $\sum u_iv_i = 0$, then $(\sum u_i^2) \cdot (\sum v_i^2)$ is a sum of $m - 1$ squares.

Proof. Write $u = \sum u_i^2$ and $v = \sum v_i^2$. By the lemma, there exist U, V in $\mathbb{M}_m(F)$ such that

$$U \cdot U^t = U^t \cdot U = u \cdot I_m, \qquad V \cdot V^t = V^t \cdot V = v \cdot I_m.$$

Further, the first row of U is $(u_1, \ldots, u_m)$, and the first row of V is $(v_1, \ldots, v_m)$. We now have

$$(uv) \cdot I_m = u \cdot V \cdot V^t = V \cdot (U^t \cdot U) \cdot V^t = W \cdot W^t, \quad \text{where } W = V \cdot U^t.$$

This equation implies that if $(w_1, w_2, \ldots, w_m)$ is the first row of W, then $uv = w_1^2 + w_2^2 + \cdots + w_m^2$. But, since $W = V \cdot U^t$, we clearly have $w_1 = u_1v_1 + \cdots + u_mv_m$, which completes the proof. □

Remark. For $F = \mathbb{R}$ (the field of real numbers), the above gives a hilarious proof of the Cauchy-Schwarz inequality! For a further generalization of 1.1′ to Pfister forms, see Exercise 25.

The groups $D_F(2^n)$ play a very special role in the study of the quadratic properties of F; this theme will become increasingly clear as we go along. At this point, let us prove the following relationship between the groups $D_F(2^n)$ and the additive structure of $W(F)$. Note that the arguments here are independent of our previous results (in Ch. VIII) on the torsion structure of Witt rings.

Proposition 1.3. *Let $w \in \dot{F}$, and let φ be the 1-fold Pfister form $\langle 1, -w \rangle$.*

(1) *If w is not a sum of squares in F, then φ has infinite additive order in $W(F)$.*

(2) *Suppose* w *is a sum of squares in* F. *Let* 2^n *be the smallest power of* 2 *for which* $w \in D_F(2^n)$. *Then the additive order of* φ *in* $W(F)$ *is precisely* 2^n.

Proof. (1) Assume instead that $r \cdot \varphi = 0 \in W(F)$, where r is a positive integer. Then, $\langle 1, \dots, 1, -w, \dots, -w\rangle$ (r terms of each kind) is hyperbolic. But then $\langle 1, \dots, 1\rangle \cong \langle w, \dots, w\rangle$, and this implies $w \in D_F(r)$, a contradiction.

(2) Here, we have $2^n \cdot \varphi = \langle 1, \dots, 1, -w, \dots, -w\rangle$, and by hypothesis, this is isotropic. But $2^n \cdot \varphi$ is an ($(n+1)$-fold) Pfister form, so, by X.1.7, we conclude that $2^n \cdot \varphi = 0 \in W(F)$. By the argument in (1), we see that no smaller power of 2 can annihilate φ. Hence the additive order of φ in $W(F)$ is precisely 2^n. □

In the sequel, we shall have many occasions to refer to the union $\bigcup_{i=1}^{\infty} D_F(i)$, which is a multiplicative subgroup of $\dot{F}$ (by VIII.1.1). Recall that, in VIII.1, we have written $\dot{\sigma}(F)$ to denote this subgroup. In the present context, however, I find it much more suggestive to adopt the notation $D_F(\infty) := \bigcup_{i=1}^{\infty} D_F(i)$. For formally real F, it is precisely the group of all totally positive elements of F (VIII.1.12). On the other hand, if F is not formally real (with $\operatorname{char} F \neq 2$), $D_F(\infty)$ coincides with $\dot{F}$.

As an illustration, let us calculate the group $D_F(\infty)$ in the important case where F is a global field. Here, not surprisingly, the computation banks heavily on the omnipotent Hasse-Minkowski Principle. The result below is usually attributed to Siegel since Siegel was the one who published its first proof. However, the result was already known to Hilbert, and of course, in the case where $F = \mathbb{Q}$, the result was due to Lagrange. Thus, the most appropriate name for this sums of squares result over a global field is probably the "Lagrange-Hilbert-Siegel Theorem".

Theorem 1.4. *Let* F *be a global field, and* q *a four-dimensional form over* F. *If* q *represents a certain element* $b \in D_F(\infty)$, *then* q *represents all elements of* $D_F(\infty)$. *In particular (taking* $q = \langle 1,1,1,1\rangle$*), we have* $D_F(\infty) = D_F(4)$.

Proof. For $a \in D_F(\infty)$, we wish to show that $a \in D_F(q)$. By VI.3.2, it suffices to check that $a \in D_{F_\mathfrak{p}}(q)$, where $F_\mathfrak{p}$ is any completion of F. If $F_\mathfrak{p}$ is $\mathbb{C}$, or a completion of F at some finite prime, then any four-dimensional form is universal over $F_\mathfrak{p}$ (VI.2.12), and we have nothing to prove. So assume $F_\mathfrak{p} \cong \mathbb{R}$. Since a and b represent the same square class in $F_\mathfrak{p} \cong \mathbb{R}$ (both being positive), we have $b \in D_{F_\mathfrak{p}}(q) \Rightarrow a \in D_{F_\mathfrak{p}}(q)$. □

If we apply the same argument above to a nonreal (= totally imaginary) global field, the conclusion will be the following stronger statement.

Corollary 1.5. *If F is a nonreal global field, then any four-dimensional form over F is universal, and every element of F is a sum of four squares.*

2. The Level of a Field

In studying the structure of the sets $D_F(n)$ (consisting of nonzero sums of n squares in F, with $D_F(0) = \varnothing$), it is convenient to introduce the following notions and notations.

Definition 2.1. We say that $a \in \dot{F}$ has length n (written $\text{len}(a) = \text{len}_F(a) = n$) if $a \in D_F(n) \setminus D_F(n-1)$. If $a \in \dot{F} \setminus D_F(\infty)$ (that is, a is not a sum of squares), we write $\text{len}(a) = \infty$. If necessary, we may take $\text{len}(0)$ to be 0. The *level* of the field F, denoted by $s(F)$, is defined to be $\text{len}(-1)$.

According to this definition, $s(F) = \infty$ iff F is a formally real field, and, in case F is nonreal, $s(F)$ is the smallest natural number n such that -1 is a sum of n squares in F.

The use of the notation $s(F)$ stems from the German word "Stufe". In this book, I would like to use the term "level", which seems to be a more fascinating word. For instance, it is one of the few 5-letter palindromes[(1)] in English. Exercise for the curious reader: find all others!

The principal result on the level of a field F is the following delightful fact from [Pf$_1$], c. 1965.

Pfister's Level Theorem 2.2. *$s(F)$ is either ∞ or a power of 2.*

Proof. If $\text{char}(F) = 2$, we have $D(\infty) = \dot{F}^2$, so $s(F) = 1$. Now assume (as usual) that $\text{char}(F) \neq 2$. If $s = s(F)$ is finite, choose k such that $2^k \leq s < 2^{k+1}$. Then the Pfister form $\varphi = 2^{k+1}\langle 1\rangle$ is isotropic, and hence hyperbolic, by X.1.7. Thus, $2^k\langle 1\rangle \cong 2^k\langle -1\rangle$. This shows that $-1 \in D_F(2^k)$, and so $s = \text{len}(-1) = 2^k$. □

In the above proof, we could have avoided the use of X.1.7 if we want. In fact, 2.1 can be seen directly from the fact that $D_F(2^k)$ is a subgroup of $\dot{F}$ (for which we have given at least two elementary proofs). Let $2^k \leq s < 2^{k+1}$ as before, and write

$$-1 = a_1^2 + \cdots + a_{2^k}^2 + a_{2^k+1}^2 + \cdots + a_s^2.$$

By transposition, we get $-(1 + a_{2^k+1}^2 + \cdots + a_s^2) = a_1^2 + \cdots + a_{2^k}^2 \neq 0$. Since $a_1^2 + \cdots + a_{2^k}^2$ and $1 + a_{2^k+1}^2 + \cdots + a_s^2$ are both in $D(2^k)$ and $D(2^k)$ is a group, it follows that $-1 \in D(2^k)$, and hence $s = 2^k$ as before.

(1) A *palindrome* is a word, verse, or sentence that is the same when read backward or forward, letter by letter.

The problem of determining the nature of the number $s = s(F)$ for a nonreal field originated from an "exercise" proposed by van der Waerden in the early 1930s. Using the 2-square, 4-square and 8-square identities and arguments as in the above proof, van der Waerden showed that $s = s(F)$ is 1, 2, 4 or else a multiple of 8. If van der Waerden had known that $D_F(2^k)$ was a group for every k, he would have most probably proved that s is always of the form 2^k. This crucial step, however, was to wait for Pfister's breakthrough some 30 years later.

Let us now record the following important consequence of 2.2 for the Witt group.

Corollary 2.3. *If F has finite level $s = 2^k$, then, as an additive group, $W(F)$ has exponent $2s = 2^{k+1}$, and is a direct sum of groups of the form $\mathbb{Z}_{2^r}$ where $r \leq k+1$.*

Proof. Apply Proposition 1.3 with $w = -1$ and $\varphi = \langle 1, -w\rangle = \langle 1, 1\rangle$. The conclusion is that the additive order of the element $2 \in W(F)$ is s. This means that the ring $W(F)$ has characteristic $2s$, so $W(F)$ has exponent[(2)] $2s = 2^{k+1}$. By the theorem of Prüfer and Baer on abelian groups with finite exponents, it follows that $W(F)$ has the structure claimed in this corollary. □

It behooves us to offer some illustrative examples below.

Examples 2.4. (1) If F is quadratically closed ($\dot{F} = \dot{F}^2$), then $s(F) = 1$ (but not conversely, of course).

(2) Let F be a finite field with q elements. If $q \equiv 1 \pmod 4$, then $-1 \in \dot{F}^2$, so $s(F) = 1$. If $q \equiv 3 \pmod 4$, then $-1 \notin \dot{F}^2$; but $-1 \in D_F(2)$ by II.3.4, so $s(F) = 2$. Consequently, if $s(K) > 2$, K must have characteristic zero.

(3) Recall, from III.2.7, that the quaternion algebra $\left(\frac{-1,-1}{F}\right)$ splits iff $s(F) \leq 2$.

(4) If $4 < |W(F)| < \infty$, then F is nonreal with level s dividing $|W(F)|/4$. To see this, we assume the result in Exercise 10. According to this exercise, our assumptions on $W(F)$ imply that $W(F)$ cannot be cyclic. Since $W(F)$ is an abelian 2-group with exponent $2s$, its cardinality must be divisible by $4s$, and hence s divides $|W(F)|/4$. (For a much stronger result in the same spirit, see 7.4 below.)

(5) By IX.1.2, F and $F(x)$ always have the same level. Similarly, F and $F((x))$ have the same level.

---

(2) The fact that $W(F)$ has a 2-power exponent (for F nonreal) was also proved earlier by prime ideal methods at the end of VIII.3. Here, this exponent is explicitly computed via the level $s(F)$.

(6) (See VI.2.6) Suppose F is a local field whose residue class field has an odd cardinality q. If $q \equiv 1 \pmod 4$, we have $s(F) = 1$. If $q \equiv 3 \pmod 4$, then $s(F) = 2$.

(7) The field $\mathbb{Q}_2$ of 2-adic numbers has level 4, since the quaternion algebra $\left(\frac{-1,-1}{\mathbb{Q}_2}\right)$ does not split (see VI.2.24(4)). If $F \supseteq \mathbb{Q}_2$ is an extension of degree n, it can be easily shown (using mainly VII.2.7 and VI.2.14) that $\left(\frac{-1,-1}{F}\right)$ splits over F iff n is even. Thus, $s(F) = 4$ iff $[F\colon \mathbb{Q}_2] = \text{odd}$.

(8) If F is a global field, then $s(F) = 1, 2, 4$, or ∞. This fact, which follows from 1.4, is known as *Siegel's Theorem*. It will be amplified below in 2.9, 2.10, and 2.11.

(9) Let k_0 be a real-closed field, and K a function field of transcendence degree n over k_0. If K is nonreal, then $s(K)$ divides 2^n. This can be deduced from a forthcoming result (Theorem 4.10 below).

(10) (Pfister) Let $K = F(\alpha)$ be a simple algebraic extension of F, and let $f(x) \in F[x]$ be the minimal polynomial of α over F. Then $s(K) \le 2^{n-1}$ iff f is a sum of 2^n squares in $F[x]$. This is an immediate consequence of X.2.13. (For a nice application of this, see Example 5.9(3) below.)

As a small application for the notion of the level, we offer the following worked example, which does not seem to be easily available in the field theory literature.

Example 2.5. *For an arbitrary field F, what is the maximal dimension d of a linear subspace contained in the "quadric" in $V = F^n$ defined by the equation $x_1^2 + \cdots + x_n^2 = 0$?*

If $\operatorname{char}(F) = 2$, the "quadric" boils down to the hyperplane $x_1 + \cdots + x_n = 0$, so $d = n - 1$. In the following, we assume that $\operatorname{char}(F) \ne 2$. The number d is the maximal dimension of a totally isotropic subspace in the regular quadratic space $q = n\langle 1\rangle$, so by I.4.4, $d = i(q)$, the Witt index of q. We may assume $s = s(F) < \infty$ (for otherwise $d = 0$). Write (uniquely)

$$n = t(2s) + r, \quad \text{where } -s < r \le s.$$

Then $n\langle 1\rangle = r\langle 1\rangle \in W(F)$ (by 2.3). Since $|r| \cdot \langle \pm 1\rangle$ is anisotropic, it follows that the anisotropic part of $n\langle 1\rangle$ is $r\langle 1\rangle$ if $r \ge 0$, and $|r| \cdot \langle -1\rangle$ if $r < 0$. Therefore,

$$d = i(q) = (n - |r|)/2.$$

For instance, if $n = 45$ and $s = 8$, then $r = -3$ and $d = (45 - 3)/2 = 21$.

Returning to Pfister's Theorem 2.2, we shall now address the following existence question: *Given any integer $k \ge 0$, does there exist a field K with $s(K) = 2^k$?* This question has also been answered affirmatively by Pfister.

We shall present Pfister's solution below, but for the most efficient formulation of this solution, we try to make full use of the general function field results developed in X.4, instead of using a more cumbersome elementary approach.

Theorem 2.6 (Pfister). *Let F_0 be a field with $s(F_0) \geq 2^k$, and let $\varphi = (n+1)\langle 1\rangle$ where $2^k \leq n < 2^{k+1}$. Then the big function field $F_0[\varphi]$ and the small function field $F_0(\varphi)$ both have level 2^k.*

Proof. Since $F_0[\varphi] \cong F_0(\varphi)(t)$, it suffices to handle the field $K = F_0[\varphi]$. As $\varphi_K \cong (n+1)\langle 1\rangle_K$ is isotropic, $s(K) \leq n$, so 2.2 implies that $s(K) \leq 2^k$. If $s(K) < 2^k$, then $q := 2^k\langle 1\rangle_{F_0}$ becomes isotropic and hence hyperbolic over K. But then X.4.5 implies that q must be isotropic over F_0 to begin with; that is, $s(F_0) < 2^k$, a contradiction. Therefore, we must have $s(K) = 2^n$. $\square$

Just to take an explicit look at the fields of level 2^k obtained in 2.6, recall that the big function field of φ is

$$F_0[\varphi] = F_0(x_1, \ldots, x_n)\left(\sqrt{-(x_1^2 + \cdots + x_n^2)}\right), \tag{2.7}$$

and similarly, the small function field of φ is

$$F_0(\varphi) = F_0(t_1, \ldots, t_{n-1})\left(\sqrt{-(1 + t_1^2 + \cdots + t_{n-1}^2)}\right). \tag{2.8}$$

If we take F_0 to be any formally real field, for instance, then the fields in (2.7) and (2.8) both have level 2^k. For $F_0 = \mathbb{R}$, these are the fields originally constructed by Pfister, who based his proofs on Cassels' result IX.2.4 over the coefficient field $\mathbb{R}$.

Remark. Note that $F_0(\varphi)$ in (2.8) is isomorphic to the quotient field of the integral domain

$$F_0[t_1, \ldots, t_n]/(1 + t_1^2 + \cdots + t_n^2).$$

The level of this commutative ring can be computed too: it turns out to be n if (say) $F_0 = \mathbb{R}$, as we'll see later in XIII.4.

We have now answered the existence question on fields K with $s(K) = 2^k$; that is, as long as we do not require K to satisfy other extra conditions! A natural extension of the same existence question is to ask for fields K with $|\dot{K}/\dot{K}^2| < \infty$ that have a prescribed level 2^k. As of this date of writing, the answer to this question seems to be still unknown (except, of course, when $k \leq 2$). For more details on this, see Question 6.2 in Chapter XIII.

We shall now conclude this section by giving some explicit information on the computation of the level for algebraic number fields. To begin with, we consider the case of imaginary (= nonreal) quadratic number fields.

Theorem 2.9. *Let $K = \mathbb{Q}(\sqrt{-d}\,)$, where $d \in \mathbb{Q}$ is positive. Then $s(K) \leq 4$, and we have $s(K) \leq 2$ iff $d \in D_{\mathbb{Q}}(3)$.*

Proof. By the classical theorem of Lagrange, $d \in D_{\mathbb{Q}}(4)$. Therefore, the equation $(\sqrt{-d}\,)^2 + d = 0$ over K shows that $s(K) \leq 4$. If $d \in D_{\mathbb{Q}}(3)$, the same equation shows that $s(K) \leq 3$, and therefore $s(K) \leq 2$. Conversely, if $s(K) \leq 2$, there exists an equation

$$-1 = \left(w + x\sqrt{-d}\right)^2 + \left(y + z\sqrt{-d}\right)^2 \qquad (w, x, y, z \in \mathbb{Q}).$$

Then $wx + yz = 0$, and $-1 = (w^2 + y^2) - d(x^2 + z^2)$. Clearly, $x^2 + z^2 \neq 0$. Thus,

$$d(x^2 + z^2)^2 = (x^2 + z^2) + (w^2 + y^2)(x^2 + z^2) = x^2 + z^2 + (wz - yx)^2$$

shows that $d \in D_{\mathbb{Q}}(3)$. □

Remark 2.10. Let us assume, without loss of generality, that $d \in \mathbb{Z}$. We know that $d \in D_{\mathbb{Q}}(3)$ iff $d \in D_{\mathbb{Z}}(3)$ (see IX.Exercise 3), iff d (positive) is not of the form $4^a(8b - 1)$ ($a \geq 0$, $b \geq 1$) (see VI.3.12). Thus, the imaginary quadratic fields K with level 4 are precisely of the following types: $\mathbb{Q}(\sqrt{-d}\,)$, $d > 0$, $d \equiv -1 \pmod 8$.

Now, how about an arbitrary nonreal (= totally imaginary) global field K? As we have already observed in 2.4(8), $s(K) = 1, 2$, or 4 ("Siegel's Theorem"). How to distinguish the case $s(K) \leq 2$ from the case $s(K) = 4$ is an interesting arithmetic question. This has been investigated separately by Chowla, Fein-Gordon-Smith, Connell, and Pourchet. Let us just include a very brief discussion of their results. For this discussion, we assume the reader has a nodding acquaintance with number theory.

Let F be a nonreal number field. Then, by Example 2.4(6) and the Hasse-Minkowski Principle, $s(F) \leq 2$ iff $s(F_{\mathfrak{p}}) \leq 2$ for all completions $F_{\mathfrak{p}}$ of F at even primes $\mathfrak{p}$ (primes lying over 2). By 2.4(7), this happens iff (at such primes) $[F_{\mathfrak{p}} \colon \mathbb{Q}_2]$ is even. Now $[F_{\mathfrak{p}} \colon \mathbb{Q}_2]$ equals $e_{\mathfrak{p}} f_{\mathfrak{p}}$, where $e_{\mathfrak{p}}$ is the ramification index of $\mathfrak{p}$ over 2, and $f_{\mathfrak{p}}$ is the residue class field extension degree. Thus, we have

Proposition 2.11. *For a nonreal number field F, we have $s(F) \leq 2$ iff for any even prime $\mathfrak{p}$, at least one of $e_{\mathfrak{p}}$, $f_{\mathfrak{p}}$ is even.*

This result is especially expedient in calculating the level of a cyclotomic field, as the following example shows.

Example 2.12. Let $F = \mathbb{Q}(\zeta)$, where ζ is a primitive p-th root of unity (p an odd prime). Any even prime $\mathfrak{p}$ is unramified over 2. We have thus $e_{\mathfrak{p}} = 1$, and $f_{\mathfrak{p}}$ is the order of 2 mod p. Thus, $s(F) = 2$ iff 2 has even order (mod p). If $p \equiv 7 \pmod 8$, it can be easily verified that 2 has odd order mod p, and hence $s(F) = 4$. If $p \equiv 3, 5 \pmod 8$, 2 is a quadratic nonresidue mod p; consequently, 2 must have even order mod p, and $s(F) = 2$. In the remaining case: $p \equiv 1 \pmod 8$, the order of $2 \pmod p$ may be odd or even. Thus, $s(F)$ could be 2 or 4. For example, if $p = 17$, 2 has order $8 \pmod p$ and $s(F) = 2$; if $p = 73$, 2 has order $9 \pmod p$ and $s(F) = 4$.

For more illustration, we include here an ad hoc calculation of the level of another number field, independently of 2.11.

Example 2.13. Let $F = \mathbb{Q}(\theta)$, where θ is a root of $f(X) = X^4 + X + 1$. Since f has no real roots, F is totally imaginary, so by Siegel's theorem, $s(F) \le 4$. A simple completion of squares shows that

$$0 = \theta^4 + \theta + 1 = \left(\theta^2 - \frac{1}{2}\right)^2 + \left(\theta + \frac{1}{2}\right)^2 + \frac{1}{2}.$$

Thus, $-1 \in 2 \cdot D_F(2) = D_F(2)$, so $s(F) \le 2$. By using a little Galois theory, it can be shown that F contains no quadratic extension of $\mathbb{Q}$, so we have $s(F) = 2$. Without using Galois theory, however, we can also show that $s(F) = 2$ by he following direct argument. Assume that F contains an element $i = \sqrt{-1}$. Then θ satisfies a quadratic equation over $\mathbb{Q}(i)$, say

$$\theta^2 + (a + bi)\theta + (c + di) = 0, \quad \text{where } a, b, c, d \in \mathbb{Q}.$$

Then $(\theta^2 + a\theta + c)^2 = -(b\theta + d)^2$. Comparing this with the minimal equation for θ, we see that

$$a = 0, \quad 2c + b^2 = 0, \quad 2bd = 1 \quad \text{and} \quad c^2 + d^2 = 1.$$

Eliminating c and d, we get $b^6 - 4b^2 + 1 = 0$. This is clearly impossible for any $b \in \mathbb{Q}$. Thus, $i \notin F$, which proves that $s(F) = 2$.

3. Pfister-Witt Annihilator Theorem

In Ch. VIII, we studied the torsion subgroup $W_t(F)$ of the Witt ring $W(F)$, and showed that this is always a 2-primary group. More quantitatively, it is of interest to understand more precisely the subgroup of elements of $W(F)$ that are killed by 2^n, where n is a fixed integer. Since $2^n\langle 1\rangle$ is a Pfister form, we are led more generally to the study of the annihilator of a Pfister form in $W(F)$.

Recall (from X.1.13) that a quadratic form φ is *round* if the equation $D_F(\varphi) = G_F(\varphi)$ holds, where $G_F(\varphi)$ denotes the group of similarity factors of φ. Pfister forms are principal examples of round forms. Working more generally with round forms, we now prove the following result (where "ann" is short for "annihilator").

Pfister-Witt Annihilator Theorem 3.1. *Let φ be a round form over a field F. Then*

(1) $\operatorname{ann}(\varphi) \cap IF$ *is generated as an ideal by* $\{\langle\langle -c\rangle\rangle : c \in D_F(\varphi)\}$.

(2) *If* φ *is not hyperbolic, the same holds for* $\operatorname{ann}(\varphi)$.

Proof. First, note that for any $c \in \dot{F}$:

$$\text{(3.2)} \qquad \langle\langle -c\rangle\rangle \in \operatorname{ann}(\varphi) \iff \varphi \cong \langle c\rangle\varphi \iff c \in D_F(\varphi).$$

To prove (1), consider any $q = \langle a_1, \ldots, a_n\rangle \in IF$ with $q \cdot \varphi = 0 \in W(F)$. Our job is to show that q can be "expressed" in terms of the generators listed in (1). For this, we induct on the even integer $n = \dim q$, the case $n = 0$ being trivial. For $n \geq 2$, the form

$$q \cdot \varphi \cong \langle a_1\rangle\varphi \perp \cdots \perp \langle a_n\rangle\varphi$$

is isotropic (in fact hyperbolic), so there exists an equation

$$0 = a_1\varphi(v_1) + \cdots + a_n\varphi(v_n),$$

where the vectors v_i are not all zero. We may assume that $c_i = \varphi(v_i)$ are not all zero. (This is automatic if φ is anisotropic. If φ is isotropic, we have, in fact, $D(\varphi) = \dot{F}$, so we can choose $\varphi(u_1) = a_2$, $\varphi(u_2) = -a_1$, and $u_3 = \cdots = 0$ to replace the v_j's.) Say $c_1, \ldots, c_r$ are nonzero, while the remaining c_j's are zero. By 3.2, we have $\langle\langle -c_i\rangle\rangle \in \operatorname{ann}(\varphi)$ for $1 \leq i \leq r$. We can "modify" our form q as follows:

$$\begin{aligned} q &\perp \langle -a_1\rangle\langle\langle -c_1\rangle\rangle \perp \cdots \perp \langle -a_r\rangle\langle\langle -c_r\rangle\rangle \\ &\cong \langle a_1, \ldots, a_n\rangle \perp \langle -a_1, \ldots, -a_r\rangle \perp \langle a_1c_1, \ldots, a_rc_r\rangle \\ &\cong r \cdot \mathbb{H} \perp \langle a_1c_1, \ldots, a_rc_r\rangle \perp \langle a_{r+1}, \ldots, a_n\rangle \\ &\cong (r+1) \cdot \mathbb{H} \perp q', \end{aligned}$$

where $\dim q' = n - 2$. Working in $W(F)$, it is clear that $q' \in \operatorname{ann}(\varphi)$, so the induction proceeds.

For (2), assume φ is *not* hyperbolic. If $q = \langle a_1, \ldots, a_n\rangle \in \operatorname{ann}(\varphi)$ with $n > 0$, then automatically $n \geq 2$. The same inductive argument shows that n is *even*, so $q \in IF$. Thus, $\operatorname{ann}(\varphi) \subseteq IF$, and (2) follows from (1). $\square$

Remark 3.3. Of course, the main conclusion of the theorem is in part (2). But it is convenient to have part (1) stated explicitly for later use since it holds without any condition on the round form φ.

Theorem 3.1 has many important applications to the study of the torsion subgroup of the Witt group $W(F)$. For expositional reasons, we'll postpone the formulation of these applications to §5. To complete this section, we shall give instead some applications of 3.1 to the study of Pfister forms and the units in $W(F)$.

If q is a nonhyperbolic round form, 3.1 guarantees that any form $q \in \operatorname{ann}(\varphi)$ has an expression

$$q = \langle b_1\rangle\langle\langle -c_1\rangle\rangle \perp \cdots \perp \langle b_m\rangle\langle\langle -c_m\rangle\rangle \in W(F) \tag{3.4}$$

for some m, where $b_i \in \dot{F}$ and $c_i \in D(\varphi)$. This amounts to the existence of some isometry

$$q \perp i\mathbb{H} \cong \langle b_1\rangle\langle\langle -c_1\rangle\rangle \perp \cdots \perp \langle b_m\rangle\langle\langle -c_m\rangle\rangle \tag{3.5}$$

for some m, where $b_i \in \dot{F}$ and $c_i \in D(\varphi)$. (Of course, the m's in (3.4) to (3.5) are not the same. To go from (3.4) to (3.5), we simply take a bigger m, and choose the "new" c_i's to be 1.) It is of interest to ask if an isometry of the form (3.5) can be found with $i = 0$. Unfortunately, the following example, taken from [EL3], shows that this cannot be done in general.

Example 3.6. Let F be a formally real field with an element w with $\operatorname{len}_F(w) = 16$. (For instance, we can take $w = x_1^2 + \cdots + x_{16}^2$ in $F = \mathbb{R}(x_1, \ldots, x_{16})$, according to IX.2.4.) Then, $q := 4\langle\langle -w\rangle\rangle$ annihilates the anisotropic Pfister form $\varphi := 4\langle 1\rangle$ (by 1.3). *We claim that no isometry* (3.5) *can exist with* $i = 0$. Indeed, if $i = 0$ in (3.5), then q contains a subform $\langle b\rangle\langle\langle -c\rangle\rangle$ with $c \in D(4)$. After a scaling, we may assume that $b = 1$. Then we may express $-c \in D(q)$ in the form $x - yw$, where $x, y \in D(4) \cup \{0\}$. From $yw = x + c \in D(8) \cup \{0\}$, we get $w \in D(8)$ (since clearly $y \neq 0$), a contradiction.

In the example above, φ is an anisotropic 2-fold Pfister form (and hence a nonhyperbolic round form). If φ is a 1-fold Pfister form instead, it is well-known that the situation is quite different, as we have the following stronger form of 3.1 (thanks to an earlier exercise).

Theorem 3.7. *Let* $\varphi = \langle\langle -a\rangle\rangle$*, where* $a \in \dot{F} \setminus \dot{F}^2$*. Then* $q \in \operatorname{ann}(\varphi)$ *iff* $q \cong \perp_{i=1}^m \langle b_i\rangle\langle\langle -c_i\rangle\rangle$ *for some* m*, where* $b_i \in \dot{F}$ *and* $c_i \in D(\varphi)$*.*

Proof. ("Only if" part) From $q \cdot \varphi = 0 \in W(F)$, we have $a \cdot q \cong q$. Then $d(a \cdot q) = d(q)$ shows that $\dim q$ is even. By Exercise 19 in Chapter II, we have a decomposition $q \cong \perp_{i=1}^m \langle b_i\rangle\langle\langle -c_i\rangle\rangle$ with $a \cdot \langle\langle -c_i\rangle\rangle \cong \langle\langle -c_i\rangle\rangle$ for all i. This means that $c_i \in D(\varphi)$, as desired. □

Corollary 3.8. *Let* $\varphi = \langle\langle -a\rangle\rangle$*, where* $a \in \dot{F}$*, and let* q *be an* n*-fold Pfister form, with* $n \geq 1$*. Then* $q \cdot \varphi = 0$ *iff* $q \cong \langle\langle -c_1, \ldots, -c_n\rangle\rangle$ *for some* $c_i \in \dot{F}$ *such that* $\langle\langle -c_1\rangle\rangle\varphi = 0$*.*

Proof. The "if" part is clear. For the "only if" part, assume that $q \cdot \varphi = 0$. We may assume $a \notin \dot{F}^2$ (for otherwise the conclusion is obvious). Applying 3.7, we have, after a scaling, $q \cong \langle 1, -c_1\rangle \perp \cdots$ for some $c_1 \in D(\varphi)$. Thus, the desired conclusion follows from the Pure Subform Theorem X.1.5. □

To conclude this section, we shall give a sample application of 3.1 to the study of the relationship between the additive and the multiplicative structures of the Witt ring.

Theorem 3.9. *Let $q \in IF$. If $2^k q = 0 \in W(F)$ (for a fixed $k \geq 1$), then $2^{k-1}q^2 = 0$, and $(1+q)^{2^k} = 1$ (in $W(F)$).*

Proof. We'll follow here an argument of Leep and Marshall (from [LeM]). Applying 3.1(1) to the Pfister form $\varphi = 2^k\langle 1\rangle$, we can write $q = \sum_i \langle a_i\rangle q_i$, where $q_i = \langle\langle -c_i\rangle\rangle \in \operatorname{ann}(\varphi)$. Noting that $q_i^2 = 2q_i$, we have

$$q^2 = 2\sum_i q_i + 2\sum_{i<j} \langle a_i a_j\rangle\, q_i q_j.$$

Multiplying this by 2^{k-1} yields $2^{k-1}q^2 = 0$.

To show that $(1+q)^{2^k} = 1$, we induct on $k \geq 1$. If $k = 1$, then $2q = 0 \Longrightarrow q^2 = 0$ as above, so $(1+q)^2 = 1 + 2q + q^2 = 0$. For $k \geq 2$, let $q' = 2q + q^2 \in IF$. Then

$$2^{k-1}q' = 2^k q + 2^{k-1}q^2 = 0,$$

so by the inductive hypothesis, we have $(1+q')^{2^{k-1}} = 1$. But then

$$(1+q)^{2^k} = (1 + 2q + q^2)^{2^{k-1}} = (1+q')^{2^{k-1}} = 1. \qquad \square$$

With the help of 3.9, we can quickly rederive and sharpen an earlier result (viz. VIII.8.9) about the unit group of the Witt ring $W(F)$.

Corollary 3.10. *The group $U := \mathrm{U}(W(F))$ is 2-primary torsion. If k is an integer such that 2^k annihilates the group $I_t^2F := I^2F \cap W_t(F)$, then $U^{2^k} = 1$.*

Proof. Since $I_t^2F \subseteq W_t(F)$ are 2-primary torsion groups, the corollary follows by applying 3.9 in conjunction with the equation

$$(*) \qquad \mathrm{U}\big(W(F)\big) = (\dot{F}/\dot{F}^2) \times (1 + I_t^2F)$$

in VIII.8.8. $\square$

Remark 3.11. While 3.10 is already a nice result, an even more precise statement is possible. According to M. Marshall [Ma$_2$], the additive group I_t^2F and the multiplicative group $1+I_t^2F$ are actually isomorphic. Thus, $(*)$ implies that I_t^2F *and* $\mathrm{U}\big(W(F)\big)$ *have in fact the same exponents*! However, Marshall's proof for $I_t^2F \cong 1 + I_t^2F$ is rather sophisticated, relying heavily on the use of divided powers and exp-log constructions that are beyond the scope of our elementary exposition.

4. The Property (A_n)

In this section, we study the following Pfister form theoretic property of a field F that was proposed by Elman and Lam in [EL$_6$]:

(A_n) *Any torsion n-fold Pfister form over F is hyperbolic.*

The word "torsion" here, of course, refers to additive torsion in the Witt group $W(F)$. In the case where F is a nonreal field, all forms are torsion, so the property (A_n) above simply amounts to $I^nF = 0$ (or equivalently, that all $(n-1)$-fold Pfister forms over F are universal). The point of the property (A_n) is, therefore, to generalize to all fields the $I^nF = 0$ property on nonreal fields. One reason (A_n) is being singled out for study is that this property seems to have a rather tractable behavior, as we shall see below.

This exposition in this section follows largely the original source [EL$_6$]. We begin by giving characterizations of fields satisfying (A_1) and (A_2).

Proposition 4.1. (1) *F satisfies (A_1) iff F is pythagorean (iff IF is torsionfree).*

(2) *F satisfies (A_2) iff I^2F is torsionfree.*

To prove (1), suppose F satisfies (A_1). If $a^2+b^2 \in \dot{F}$, then $\langle 1, -(a^2+b^2)\rangle$ is torsion, and hence hyperbolic. This implies that $a^2 + b^2 \in \dot{F}^2$, so F is pythagorean. Conversely, assume F is pythagorean, and consider a torsion 1-fold Pfister form $\sigma = \langle 1, -w\rangle$. By 1.3, $w \in D(\infty) = \dot{F}^2$, so $\sigma \cong \mathbb{H}$. Thus, F satisfies (A_1).

To prove (2), we need the following lemma (from [EL$_2$]) on the torsion elements in I^2F.

Lemma 4.2. *An element $q \in I^2F$ is torsion iff it can be expressed in the form $q = \sum_i \langle\langle a_i, -b_i\rangle\rangle$, where $a_i \in \dot{F}$ and $b_i \in D(\infty)$. (In particular, the group I_t^2F is additively generated by the torsion 2-fold Pfister forms.)*

Proof. The "if" part is clear, since $\langle\langle -b_i\rangle\rangle \in W_t(F)$. For the "only if" part, assume that $q \in W_t(F)$. Since $W_t(F)$ is 2-primary, q is annihilated by some Pfister form $2^n\langle 1\rangle$. By 3.1 and 1.3, we can write $q = \sum_i \langle y_i\rangle\langle\langle -w_i\rangle\rangle$, where $y_i \in \dot{F}$ and $w_i \in D(\infty)$. Adding and subtracting $q' := \sum_i \langle\langle -w_i\rangle\rangle \in W_t(F)$, we get

$$q = \textstyle\sum_i \langle\langle y_i, -w_i\rangle\rangle - q' \in W(F).$$

Thus, $q' \in I^2F$, which implies that $w_1 \cdots w_r \in \dot{F}^2$. Using equations of the type

$$\langle\langle -w_1\rangle\rangle + \langle\langle -w_2\rangle\rangle = \langle\langle -w_1, -w_2\rangle\rangle + \langle\langle -w_1w_2\rangle\rangle$$

repeatedly, we may write q' in the form $\sum_j \langle\langle -s_j, -t_j\rangle\rangle$, where $s_j, t_j \in D(\infty)$. On the other hand, there exists a positive integer m such that $-q' = mq'$.

We have thus

$$q = \sum_i \langle\langle y_i, -w_i\rangle\rangle + m \sum_j \langle\langle -s_j, -t_j\rangle\rangle, \tag{4.3}$$

as desired. □

It is now very easy to prove the second part of 4.1. The "if" part is clear; in fact, for any $n \geq 1$, the torsion-freeness of I^nF (trivially) implies (A_n). For the "only if" part, assume (A_2). In the equation (4.3) above, all summands on the RHS are zero (since they are torsion 2-fold Pfister forms). Therefore, $q = 0$, which shows that I^2F is torsionfree. □

Remark 4.4. As the reader saw above, the proof of the "only if" part of 4.1(2) took a rather nontrivial argument. It turned out to be also true that F satisfies (A_3) iff I^3F is torsionfree. This is yet a little harder: we shall return to prove it in XII.3. At the time when [EL$_6$] was written, it was already suspected (or at least hoped) that, *for all* n, the property (A_n) is equivalent to I^nF being torsionfree. This is difficult to prove since there is no easy characterization for the elements in I^nF, let alone its torsion elements. Recently, assuming the solution of the Milnor Conjectures due to Voevodsky and Orlov-Vishik-Voevodsky, Arason and Elman [AE] have proved that, in general, *the ideal* I_t^nF *of torsion elements in* I^nF *is generated by the torsion* n*-fold Pfister forms.* This implies, in particular, that (A_n) is equivalent to I^nF being torsionfree. However, the results of Orlov, Vishik and Voevodsky are quite deep, and also, as of this date of writing, they are not yet completely published. Thus, in the present section, it would not be reasonable to assume the Arason-Elman result on the equivalence of (A_n) with the torsion-freeness of I^nF. Rather, we'll be working with the condition (A_n) just as it was defined at the beginning of this section, without identifying it with the torsion-freeness of the ideal I^nF.

For a general $n \geq 1$, let us give below some characterizations of the property (A_n) from [EL$_6$]. Here, (4) $\iff$ (5) was first shown by Pfister [Pf$_4$].

Theorem 4.5. *For any field* F *and any integer* $n \geq 1$, *the following are equivalent*:

(1) (A_n) *holds.*

(2) (A_m) *holds for all* $m \geq n$.

(3) *No anisotropic* n*-fold Pfister form* q *over* F *satisfies* $2q = 0$.

(4) *Every* $(n-1)$*-fold Pfister form represents all of* $D_F(2)$.

(5) *Every* $(n-1)$*-fold Pfister form represents all of* $D_F(\infty)$.

In particular, we have $(A_n) \Longrightarrow (A_m)$ *for any* $m \geq n$.

Proof. The implications (2) $\Longrightarrow$ (1) $\Longrightarrow$ (3) are clear. Next, let us prove (3) $\Longrightarrow$ (2). Assume (3) holds, and that for some $m \geq n$ there exists a torsion anisotropic m-fold Pfister form σ, say with $2^{t+1}\sigma = 0 \neq 2^t\sigma$ (for some $t \geq 0$). Since $2^t\sigma \in \text{ann}(2)$, 2.8 yields $c_i \in \dot{F}$ such that

$$2^t\sigma \cong \langle\langle -c_1, \ldots, -c_{m+t}\rangle\rangle \quad \text{with} \quad 2\langle\langle -c_1\rangle\rangle = 0.$$

But then $\langle\langle -c_1, \ldots, -c_n\rangle\rangle \in \text{ann}(2)$ is hyperbolic by (3). In particular, $2^t\sigma = 0$, a contradiction.

Since (5) $\Longrightarrow$ (4) is trivial, we are done if we can show that (4) $\Longrightarrow$ (3) and (1) $\Longrightarrow$ (5). Assume (4) holds, and let q be an n-fold Pfister form with $2q = 0$. By 3.8, $q \cong \langle\langle -c_1, \ldots, -c_n\rangle\rangle$ for some c_i, with $2\langle\langle -c_1\rangle\rangle = 0$, or equivalently, $c_1 \in D_F(2)$. But by (4), $\langle\langle -c_2, \ldots, -c_n\rangle\rangle$ represents c_1, so q is hyperbolic, proving (3). Finally, assume (1), and let φ be any $(n-1)$-fold Pfister form. For any $c \in D_F(\infty)$, $\langle\langle -c\rangle\rangle q$ is torsion by 1.3, so (1) implies that $\langle\langle -c\rangle\rangle q = 0$, or equivalently, $c \in D_F(q)$. This proves (5). □

Next, we prove a lemma that gives a sufficient condition for a field F to satisfy (A_n).

Lemma 4.6. *Let $K = F(\sqrt{a})$ be a quadratic extension of F. If every $(n-1)$-fold Pfister form over F represents all of $\mathrm{N}_{K/F}(\dot{K})$, then F satisfies (A_n).*

Proof. As a first step, we claim that, for any m-fold Pfister form τ over F with $m \geq n$, $\langle\langle -a\rangle\rangle\tau = 0 \Rightarrow \tau = 0$ (in $W(F)$). Indeed, by 3.8, $\langle\langle -a\rangle\rangle\tau = 0$ implies that $\tau = \langle\langle -b\rangle\rangle\tau_1$, where $\langle\langle -a, -b\rangle\rangle = 0$, and τ_1 is an $(m-1)$-fold Pfister form over F. Then $b \in \mathrm{N}_{K/F}(\dot{K})$ is represented by τ_1 (since $m - 1 \geq n - 1$), and hence $\tau = 0$, as claimed.

To check that F satisfies (A_n), we appeal to the criterion (3) in Theorem 4.5. Let q be any n-fold Pfister form over F such that $2q = 0$. Since $\langle\langle -a, -a\rangle\rangle \cong 2\,\langle\langle -a\rangle\rangle$, we have $\langle\langle -a, -a\rangle\rangle q = 0$. Applying (twice) the claim in the last paragraph, we get $q = 0$, as desired. □

Going-Down Theorem 4.7. *Suppose $[K\colon F] = 2$ and K is nonreal. If K satisfies (A_n), so does F.*

Proof. According to 4.6, it suffices for us to check that every $(n-1)$-fold Pfister form σ over F represents $\mathrm{N}_{K/F}(x)$ for every $x \in \dot{K}$. The hypotheses on K amount to $I^nK = 0$, so we certainly have $x \in D_K(\sigma_K)$. From Scharlau's Norm Principle (VII.4.3) (applied to the Pfister form σ), it follows that $\mathrm{N}_{K/F}(x) \in D_F(\sigma)$, as desired. □

Remark 4.8. Theorem 4.7 above is a special case of a much more general Going-Down Theorem in [EL$_6$]. This more general result states that, *for any* *finite normal extension K/F, if K satisfies (A_n), then so does F.* (Here, there is no nonreal assumption on K.)

Corollary 4.9. *Suppose $[K: F] = 2$ and every n-fold Pfister form over K is universal. Then every n-fold Pfister form over F represents all of $D_F(\infty)$; in particular, $D_F(\infty) = D_F(2^n)$. (If F is nonreal, then every n-fold Pfister form over F is universal, and $\dot{F} = D_F(2^n)$.)*

Proof. Clearly K is nonreal, and it satisfies (A_{n+1}) (by 4.5). Thus, by 4.7, F also satisfies (A_{n+1}), and the desired conclusions follow from 4.5. □

We can now deduce the following nice result from [Pf4], which has in fact been the main motivation for much of the material in this section.

Theorem 4.10 (Pfister). *Let F be a field of transcendence degree n over a real-closed field k_0. Then every n-fold Pfister form over F represents all of $D_F(\infty)$; in particular, $D_F(\infty) = D_F(2^n)$.*

Proof. If $i = \sqrt{-1} \notin F$, then $K := F(i)$ is a field of transcendence degree n over the algebraically closed field $k_0(i)$, with $[K: F] = 2$. By the Tsen-Lang Theorem, every 2^n-dimensional quadratic form (and, in particular, any n-fold Pfister form) over K is universal. Thus, the desired conclusions follow from 4.9. If $i \in F$, then $F = K$, and we are done without even invoking 4.9. □

Corollary 4.11 (Pfister). *For any real-closed field k_0, every totally positive element of the field $k_0(x_1, \ldots, x_n)$ is a sum of 2^n squares.*

A rational function $f \in k_0(x_1, \ldots, x_n)$ is called *positive semidefinite* if f is nonnegative wherever it is defined. (Recall that k_0 has a unique ordering, whose positive cone is given by $\dot{k}_0^2$.) In his famous solution of Hilbert's 17th Problem, Artin [Ar] showed in 1927 that *f is positive semidefinite iff f is a sum of squares in $k_0(x_1, \ldots, x_n)$*. Corollary 4.11 above shows that 2^n squares will suffice. An easier and more direct proof can be given in the special case $n = 1$. Up to a square in $k_0(x)$, f has the form $f_1 \cdots f_r$, where each f_i is a monic irreducible quadratic polynomial. From

$$f_i = x^2 + 2b_i x + c_i = (x + b_i)^2 + \left(\sqrt{c_i - b_i^2}\right)^2 \in D_{k_0(x)}(2),$$

it follows that $f \in D_{k_0(x)}(2)$ (since $D_{k_0(x)}(2)$ is a multiplicative group).

Returning now to the property (A_n), it is natural to ask, after proving the Going-Down Theorem in 4.7, whether there is also the possibility of proving a Going-Up Theorem, at least for quadratic extensions. The answer turned out to be "yes", but some additional arguments will be required.

For a quadratic extension $K = F(\sqrt{a})$ as before, we let $s\colon K \to F$ be the F-linear functional defined by $s(1) = 0$ and $s(\sqrt{a}) = 1$, and let s_* denote the transfer of quadratic forms from K to F (in the sense of VII.1). To deal with Pfister forms over F and over K, we'll need the following lemma.

Lemma 4.12. *Let τ, γ be Pfister forms over F and K respectively, with $\dim\gamma > 1$, and let $\sigma = \tau_K \cdot \gamma$. If $s_*(\sigma) = 0$, then there exist $b \in \dot{F}$ and a Pfister form γ_1 over K, such that $\sigma \cong \tau_K\langle\langle b\rangle\rangle\gamma_1$.*

Proof. Transferring $\sigma \cong \tau_K \perp \tau_K\gamma'$ (where γ' denotes the pure subform of γ), we get

$$0 = s_*(\tau_K) \perp s_*(\tau_K\gamma') = s_*(\tau_K\gamma').$$

In particular, the transfer $s_*(\tau_K\gamma')$ is isotropic. This means that there exists $b \in \dot{F} \cap D_K(\tau_K\gamma')$. By X.1.10, $\tau_K\gamma \cong \tau_K\langle\langle b\rangle\rangle\gamma_1$ for a suitable Pfister form γ_1 over K. $\square$

We shall now deduce the following transfer principle for Pfister forms that will be crucial for proving a Going-Up Theorem for (A_n) with respect to quadratic extensions.

Transfer Principle 4.13. *Let σ be an n-fold Pfister form over $K = F(\sqrt{a})$. Then $s_*(\sigma) \in I^{n+1}F$ iff $s_*(\sigma) = 0$, iff there exists a Pfister form φ over F such that $\sigma \cong \varphi_K$.*

Proof. If $\sigma \cong \varphi_K$ as stated, then surely $s_*(\sigma) = 0$. Conversely, assume that $s_*(\sigma) \in I^{n+1}F$. Since $\sigma \cong \langle 1\rangle \perp \sigma'$ and $s_*(\langle 1\rangle_K) = 0$, we have $s_*(\sigma') \in I^{n+1}F$. But

$$\dim_F s_*(\sigma') = 2 \cdot \dim_K \sigma' < 2^{n+1},$$

so by the Hauptsatz, $s_*(\sigma) = s_*(\sigma') = 0$. Applying 4.12 repeatedly, we see that $\sigma \cong \varphi_K$ for a suitable Pfister form φ over F. $\square$

We are now in a position to prove the following.

Going-Up Theorem 4.14. *Let $K = F(\sqrt{a})$ and $n \geq 2$. Suppose either (1) $a \in D_F(\infty)$, or (2) every n-fold Pfister form over F becomes hyperbolic over K. If F satisfies (A_n), then so does K.*

Proof. Let $y \in D_K(2)$, and let σ_1 be an $(n-1)$-fold Pfister form over K. Our job is to show that $\sigma = \langle\langle -y\rangle\rangle\sigma_1$ is hyperbolic over K (see (4) in 4.5). Let us show first that $s_*(\sigma) = 0$, where s_* is the transfer used in 4.13. Appealing to VII.3.13 (see especially the proof of VII.3.14), we may assume that $\sigma_1 \cong \langle\langle z\rangle\rangle\sigma_2$, where $z \in \dot{K}$ and σ_2 is an $(n-2)$-fold Pfister form over F. By Frobenius Reciprocity (VII.1.3),

$$s_*(\sigma) \cong s_*(\langle\langle -y, z\rangle\rangle\sigma_2) \cong \sigma_2 \cdot s_*(\langle\langle -y, z\rangle\rangle).$$

Now $s_*\langle\langle -y, z\rangle\rangle \in W_t(F) \cap I^2F$ (by VII.3.14), so we may write (by 4.2):

$$s_*\langle\langle -y, z\rangle\rangle = \sum \langle\langle a_i, -b_i\rangle\rangle,$$

where $a_i \in \dot{F}$ and $b_i \in D_F(\infty)$. Consequently,

$$s_*(\sigma) = \sum \sigma_2 \langle\langle a_i, -b_i\rangle\rangle = 0,$$

since each summand is a *torsion* n-fold Pfister form over F. By the transfer principle 4.13, σ is K-isometric to φ_K, for some n-fold Pfister form φ over F. Under the hypothesis (2) of the theorem, $\varphi_K = 0$, so we are done. Under (1) (that is, $a \in D_F(\infty)$),

$$\ker(W(F) \longrightarrow W(K)) = \langle\langle -a\rangle\rangle W(F) \subseteq W_t(F)$$

by VII.3.2 and 1.3. Since $\varphi_K \in W_t(K)$, we must have $\varphi \in W_t(F)$, and therefore $\varphi = 0 \in W(F)$ by (A_n). This shows that σ is hyperbolic over K, as desired. □

Remark 4.15. Note that the "Going-Up" conclusion in 4.14 need not hold for *arbitrary* quadratic extensions. For instance, the pythagorean field $F = \mathbb{R}((t_1))((t_2))\cdots$ satisfies (A_n) for all $n \geq 1$. However, the quadratic extension $K = F(\sqrt{-1}\,)$ does not satisfy any (A_n).

Corollary 4.16. *Let $K = F(\sqrt{a}\,)$ and $n \geq 1$. If $I^nF = 0$, then $I^nK = 0$.*

Proof. We may assume that $a \notin \dot{F}^2$ (for otherwise $K = F$). If $I^nF = 0$, then $n \geq 2$ and F satisfies (A_n). Now apply 4.14. □

Corollary 4.17. *Suppose $F(\sqrt{a}\,)$ is nonreal and satisfies (A_n). Then any quadratic extension L/F also satisfies (A_n).*

Proof. Again we may assume that $a \notin \dot{F}^2$ (for otherwise we can apply 4.16). In this case, we can go-up from $F(\sqrt{a}\,)$ to $L(\sqrt{a}\,)$ by 4.16, and then go-down from $L(\sqrt{a}\,)$ to L. □

Corollary 4.18. (1) *Suppose a certain nonreal quadratic extension of F satisfies (A_n). Then, for any $w \in D_F(\infty)$, we have $I^nF = \langle\langle w\rangle\rangle I^{n-1}F$.*

(2) *Suppose, for some $w \in \dot{F}$, we have $I^nF = \langle\langle w\rangle\rangle I^{n-1}F$. Then $w \in D_F(\infty)$; moreover, F and all of its quadratic extensions satisfy (A_n).*

Proof. (1) By 4.17, $F(\sqrt{-w}\,)$ satisfies (A_n). Since $F(\sqrt{-w}\,)$ is nonreal, every n-fold Pfister form over F becomes hyperbolic over $F(\sqrt{-w}\,)$, which (easily) implies that $I^nF = \langle\langle w\rangle\rangle I^{n-1}F$.

(2) The hypothesis implies that $2^n\langle 1\rangle \cong \langle\langle w, w_2, \ldots, w_n\rangle\rangle$ for suitable $w_i \in \dot{F}$ (see, e.g., X.4.13). Thus, $w \in D_F(2^n - 1)$. To complete the proof, it suffices to show that $K = F(\sqrt{-w}\,)$ satisfies (A_{n+1}), for the rest follows from 4.7 and 4.17. Now by VII.3.13, $I^{n+1}K = I^nF \cdot IK$ (upon viewing

$W(K)$ as a $W(F)$-module). Therefore, $I^{n+1}K = \langle\langle w \rangle\rangle I^n K = 0$; that is, K satisfies (A_{n+1}). □

Remark 4.19. The conclusion in part (2) above is indeed the best possible. For example, if F is the field $\mathbb{Q}_2$ of 2-adic numbers, then $I^2F = 2IF$. It is true that F and all of its quadratic extensions satisfy (A_3), but they certainly do not satisfy (A_2).

5. Height and Pythagoras Number

We study in this section two invariants of a field F, namely, its height $h(F)$, and its Pythagoras number $P(F)$. As before, $W_t(F)$ denotes the torsion subgroup of the Witt ring $W(F)$. We have shown previously that $W_t(F)$ is a 2-primary torsion group (see VIII.4.10), by using Pfister's Local-Global Principle and the notion of the pythagorean hull of a field. In this section, we shall refine this result by giving more information on the structure of the torsion subgroup $W_t(F)$. This is just a direct application of the Pfister-Witt Annihilator Theorem 3.1 that we could have included right after the proof of 3.1. However, we have postponed this application to the present section so that we can introduce and study the height $h(F)$ (and the Pythagoras number $P(F)$) of a field F more systematically.

Notation/Terminology 5.1. For any abelian group A and any positive integer m, let $A_m = \{a \in A \colon ma = 0\}$. The *exponent* of A, denoted by $\exp(A)$, is the smallest positive integer m such that $mA = 0$ (i.e. such that $A = A_m$). (If no such integer m exists, $\exp(A)$ is taken to be ∞.)

In the following, we'll apply the above notation and terminology to the abelian group $W_t(F) \cap IF$, where, as usual, IF denotes the ideal of even-dimensional forms in $W(F)$. We use the group $W_t(F) \cap IF$ solely for the purpose of unifying the cases of formally real fields and nonreal fields. Recall that, if F is formally real, then $W_t(F) \subseteq IF$, so $W_t(F) \cap IF$ is just $W_t(F)$; on the other hand, if F is nonreal, then $W(F)$ is torsion, so $W_tF \cap IF$ is just IF in that case.

Since the integer $2^k \in W(F)$ can be interpreted as given by the k-fold Pfister form $\varphi = \langle\langle 1, \ldots, 1 \rangle\rangle$, the Pfister-Witt Annihilator Theorem 3.1 gives directly the following.

Theorem 5.2. *For any field F (real or nonreal) and any $k \geq 0$, $(IF)_{2^k}$ is generated as an ideal by the 1-fold Pfister forms $\langle\langle -w \rangle\rangle$, where $\mathrm{len}_F(w) \leq 2^k$ (that is, $w \in D_F(2^k)$).*

Since $W_t(F) \cap IF$ is 2-primary torsion, it is the union of its ascending subgroups $(IF)_{2^k}$ for $k = 1, 2, \ldots$. Therefore, we reach the following conclusion.

Corollary 5.3. *$W_t(F) \cap IF$ is generated as an ideal by the 1-fold Pfister forms $\langle\langle -w \rangle\rangle$, where $\operatorname{len}_F(w) < \infty$ (that is, $w \in D_F(\infty)$).*

Remark. Of course, this corollary is of interest only for formally real fields. In the spirit of Example 3.6, we should point out that, in the formally real case, 5.3 *does not* mean that a torsion form $q \in W_t(F)$ has a decomposition $q \cong \perp \langle a_i \rangle \langle\langle -w_i \rangle\rangle$ where $w_i \in D_F(\infty)$. Rather, q can only be expressed as $\sum_i \langle a_i \rangle \langle\langle -w_i \rangle\rangle$ in the Witt ring $W(F)$. In fact, Arason and Pfister [AP$_2$] have shown that, over the (formally real) field $F = \mathbb{Q}(x, y)$, the quaternionic form

$$q := \langle\langle x,\ -(1 + y^2 + 3x) \rangle\rangle \in I^2 F$$

provides the needed example. Since $\langle 1, 1, x, x, x \rangle$ represents $1 + y^2 + 3x$, one has $q \in (I^2F)_4$. However, Arason and Pfister showed that q has no subform of the shape $\langle a_1 \rangle \langle\langle -w_1 \rangle\rangle$ where $w_1 \in D_F(\infty)$. Of course, such examples can only exist in $(IF)_{2^k}$ for $k \geq 2$, but not for $k = 1$ according to 3.7 (or, essentially, II.Exercise 19). A lot of additional useful information on decompositions of the type $q \cong \perp \langle a_i \rangle \langle\langle -w_i \rangle\rangle$ (where $w_i \in D_F(\infty)$) can be found in [AP$_2$].

We shall now introduce the two key definitions in this section.

Definition 5.4. The *height* of F, denoted by $h(F)$, is defined to be $\exp(W_t(F))$, which is either a power of 2, or the symbol ∞.

For example, if F is nonreal, then $h(F) = 2 \cdot s(F)$, by 2.3. In general, $h(F) = 1$ iff $W(F)$ is torsionfree, iff F is formally real pythagorean, by VIII.4.1.

Definition 5.5. The *Pythagoras number* of F, denote by $P(F)$, is the smallest positive integer n such that $D_F(\infty) = D_F(n)$. (If no such integer n exists, $P(F)$ is taken to be ∞.)

The invariants $s(F)$, $h(F)$, and $P(F)$ (for a fixed field F) are closely related to one another, as the following result shows.

Theorem 5.6. (1) *If F is formally real, then $h(F)$ is the smallest 2-power $2^k \geq P(F)$.*

(2) *If F is nonreal with level $s = s(F)$, then*

$$s \leq P(F) \leq s + 1 \leq 2s = h(F).$$

Proof. Suppose $h = h(F) = 2^k < \infty$. For any $w \in D_F(\infty)$, we know that $\langle\langle -w \rangle\rangle \in W_t(F)$ by 1.3. Hence $h \cdot \langle\langle -w \rangle\rangle = 0$, which gives $w \in D_F(h)$. This shows that $P(F) \leq h$. Now let F be formally real. Let 2^d be the smallest 2-power $\geq P(F)$. The above shows that $d \leq k$. For any $w \in D_F(\infty)$,

$2^d \langle\langle -w \rangle\rangle = 0$ by 1.3. Since $W_t(F) \subseteq IF$, 5.2 implies that $2^d W_t(F) = 0$, and so $k \leq d$.

For (2), write $-1 = a_1^2 + \cdots + a_s^2$. For $x \in \dot{F}$, we may write $x = y^2 - z^2$ since $\mathbb{H}$ is universal. But then

$$x = y^2 + (a_1 z)^2 + \cdots + (a_s z)^2 \in D_F(s+1).$$

This shows that $P(F) \leq s+1$, and all other inequalities in (2) are clear (or already proved). □

The following result shows further that, in (2) above, the two values s and $s+1$ for $P(F)$ are indeed both possible.

Theorem 5.7. *For any integer $k \geq 0$, there exist (nonreal) fields F and K such that $s(F) = s(K) = 2^k$, with $P(F) = 2^k + 1$ and $P(K) = 2^k$.*

Proof. Start with *any* field E with level $s = 2^k$. If we take $F = E(x)$, then, by IX.2.3, $-1 + x^2$ has length $s+1$ in F. Now $s(F) = s(E) = s$ (by IX.1.1), and $P(F) = s+1$.

To construct the field K, work in the algebraic closure $\overline{E}$ of E. By Zorn's Lemma, E admits an extension K inside $\overline{E}$ that is maximal with respect to the property that the form $\varphi = s\langle 1 \rangle$ is anisotropic over K. Clearly, $s(K) = s$. We finish by proving that $P(K) = s$, that is, $a \in \dot{K} \Rightarrow a \in D_K(\varphi)$. We may assume that $a \notin -\dot{K}^2$ (for otherwise $a \in D_K(\varphi)$ is clear). Thus, φ is isotropic over $K(\sqrt{-a}\,) \supsetneq K$, and VII.3.1 implies that $\varphi \cong \langle b \rangle \langle 1, a \rangle \perp \cdots$ for some $b \in \dot{K}$. Since φ is a Pfister form, we have $\varphi \cong b \cdot \varphi \cong \langle 1, a \rangle \perp \cdots$ so $a \in D_K(\varphi)$, as desired. □

Remark 5.8. In the last step above, we could have used Witt cancellation to arrive at the stronger conclusion that $a \in D_K(s-1)$. This shows more than we had bargained for; namely, *up to squares, -1 is the only element in K that has the maximal length s.*

We conclude this section with an assortment of other examples.

Examples 5.9. (1) Let F be any formally real global field. By 1.4, we have $P(F) \leq 4$, and so 5.6(1) implies that $h(F) \leq 4$. *We claim that* $h(F) = 4$. In fact, let $\mathfrak{p}$ be a nonarchimedean, nondyadic prime, such that -1 is not a square in the completion $F_{\mathfrak{p}}$. By the approximation theorem, choose $a \in \dot{F}$ that is positive at all real completions of F, and sufficiently close to a uniformizer for $F_{\mathfrak{p}}$. Then $\langle\langle -a \rangle\rangle \in W_t(F)$ has additive order 4, since $2\langle\langle -a \rangle\rangle = \langle\langle 1, -a \rangle\rangle$ is the norm form of $\left(\frac{-1, a}{F}\right)$, and $\left(\frac{-1, a}{F_{\mathfrak{p}}}\right)$ does not split (by VI.2.2(4)). If F is *nonreal*, then, of course, $h(F) = 2, 4, 8$ according to $s(F) = 1, 2, 4$.

(2) Let F be any *formally real* number field. *Then,* $P(F) \leq h(F) = 4$, *with equality iff there exists a dyadic prime* $\mathfrak{p} \,|\, 2$ *such that the local degree* $[F_\mathfrak{p} \colon \mathbb{Q}_2]$ *is odd.* In fact, if such a prime $\mathfrak{p}$ exists, then $s(F_\mathfrak{p}) = 4$ according to 2.4(7). If $a \in \dot{F}$ is totally positive and sufficiently close to -1 in $F_\mathfrak{p}$, then clearly $a \notin D_F(3)$, showing $P(F) = 4$. Conversely, suppose $[F_\mathfrak{p} \colon \mathbb{Q}_2] =$ even (so $s(F_\mathfrak{p}) \leq 2$) for every $\mathfrak{p} \,|\, 2$. If $a \in D_F(\infty)$, then $\langle 1, 1, 1\rangle$ represents a in *every* completion of F. The Hasse-Minkowski Principle implies that $a \in D_F(3)$. We have thus $P(F) = 3$. In case $[F \colon \mathbb{Q}]$ is odd, there clearly exists a dyadic prime $\mathfrak{p}$ such that $[F_\mathfrak{p} \colon \mathbb{Q}_2] =$ odd, so we have automatically $P(F) = 4$. If $[F \colon \mathbb{Q}]$ is even, the situation is slightly more involved. For concrete examples, take $F = \mathbb{Q}(\sqrt{d})$ ($d > 0$ a square free integer). If $d \equiv 1 \pmod 8$, (2) splits into two distinct primes in F, so the local degrees at the two dyadic primes are both 1. In this case $P(F) = 4$. In all other cases, (2) either remains prime, or ramifies, so the local degree at the (unique) dyadic prime is 2. In this case, $P(F) = 3$.

(3) Let $E = F(x)$ where F is *formally real.* By Milnor's exact sequence (IX.3.1), we know that

$$W(E) \cong \bigoplus W\big(F[x]/(\pi(x))\big) \oplus W(F),$$

where the first summation extends over all monic irreducible polynomials $\pi(x)$. Since F has characteristic zero, $F[x]/\big(\pi(x)\big)$ runs over *all* finite extensions F_1 of F. It follows that $h(E) = \sup h(F_1)$, where the supremum is taken over all finite extensions $F_1 \supseteq F$. Combining 2.4(10), 5.6, and Exercise 4, we see that

> $P(E) \leq 2^n$ *if and only if* $s(K) \leq 2^{n-1}$ *for every nonreal algebraic extension* $K \supseteq F$.

(This statement was first proved by Pfister.) On the other hand, recall that $P(E) \geq 1 + P(F)$ by IX.2.3. As an explicit application, let F be a formally real number field in the following. Suppose, first, at all dyadic primes $\mathfrak{p} \,|\, 2$, the local degrees $[F_\mathfrak{p} \colon \mathbb{Q}_2]$ are even. Then, $s(K) \leq 2$ for every nonreal $K \supseteq F$ (see the paragraph preceding 2.11). Using Pfister's statement above (plus Example (2)), we see that $P(E) = 4$ $(= h(E))$. Next, suppose there exists at least one dyadic prime $\mathfrak{p} \,|\, 2$ with $[F_\mathfrak{p} \colon \mathbb{Q}_2]$ odd. Then, $P(F) = 4$ by Example (2), and our general discussion above yields $5 \leq P(E) \leq 8$. The exact value of $P(E)$ in this case had remained unknown for some time; but eventually, Y. Pourchet [Po] succeeded in showing that $P(E) = 5$.

(4) Let k_0 be a real-closed field, and let F be a function field of transcendence degree n over k_0. Theorem 4.10 implies that $P(F) \leq 2^n$. If F is formally real, then Theorem 5.6(1) gives $h(F) \,|\, 2^n$. Assume further that $F = k_0(x_1, \ldots, x_n)$. We have shown before that $1 + x_1^2 + \cdots + x_n^2$ is not a sum of n squares (IX.2.4). Thus, $h(F) \geq P(F) \geq n + 1$, i.e., $P(F)$ is

sandwiched between $n+1$ and 2^n. Consider the special case $n=2$, where we change the notation to $F=k_0(x,y)$, and assume that k_0 is the real field $\mathbb{R}$. It has been shown by Cassels, Ellison, and Pfister, that $P(F)=4$ in this case. In fact, consider the Motzkin polynomial

$$(5.10) \qquad M(x,y)=1+x^2(x^2-3)y^2+x^2y^4$$

in F. Since the arithmetic mean is always no smaller than the geometric mean, we have

$$1+x^4y^2+x^2y^4 \geq 3\cdot\sqrt[3]{1\cdot x^4y^2\cdot x^2y^4}=3x^2y^2.$$

Transposition shows that $M(x,y)$ is a positive semidefinite function. By the paragraph following 4.11, we see that M is a sum of four squares in $F=\mathbb{R}(x,y)$. (To derive an explicit representation, see XIII.Exercise 12.) Using the theory of elliptic curves, Cassels, Ellison, and Pfister have shown in [CEP] that M *cannot be a sum of three squares in* F. This implies that $P(F)=4$, and an easy argument based on this shows that

$$(5.11) \qquad P\big(\mathbb{R}(x_1,\ldots,x_n)\big)\geq n+2 \text{ whenever } n\geq 2.$$

In this connection, we should note that another proof for $P(\mathbb{R}(x,y))=4$ using algebraic geometry (but not elliptic curves) has been given by Colliot-Thélène in [CT]. For some results on the Pythagoras number of $\mathbb{Q}(x_1,\ldots,x_n)$, see the discussions on Open Question 6.9 in Chapter XIII.

(5) Take $F=\mathbb{R}(x_1,x_2,\ldots)$, with a countably infinite set of independent indeterminates. Then $D_F(1)\subsetneq D_F(2)\subsetneq\cdots\subsetneq D_F(i)\subsetneq\cdots$ is a strictly increasing chain. The height of F is therefore infinite, and so is its Pythagoras number.

(6) Let $F=k((t))$. Since $W(F)\cong W(k)\oplus W(k)$ as additive groups (Springer's theorem), we have clearly $h(F)=h(k)$. Suppose k is formally real. If $f\neq 0$ is a sum of squares in F, it is easy to show that f is congruent modulo $\dot{F}^2$ to an element $a\in D_k(\infty)$ (see VI.Exercise 3). This observation enables us to show that $P(F)=P(k)$. Suppose, next, that k is nonreal with level s. Then, in any case, $P(F)=s+1$ (see IX.Exercise 6), but $P(k)$ may be s or $s+1$.

It has been shown by Hoffmann [Ho9] that there exist formally real fields with any prescribed positive integer as Pythagoras number. For more discussions on this, see XIII.3.

6. The u-Invariant of a Field

Definition 6.1. We define the *u-invariant* of F by $u=u(F)=\max\{\dim\varphi\}$, where φ ranges over all anisotropic (quadratic) forms over F. If no such maximum exists, we define $u(F)$ to be ∞. It is easy to verify that $u(F)$ can be characterized also in the two following alternative ways:

$$u(F) = \min\{n|\text{ forms of dimension } > n \text{ over } F \text{ are isotropic}\};$$

$$u(F) = \min\{n|\text{ forms of dimension } \geq n \text{ over } F \text{ are universal}\},$$

with the understanding that the "minimum" of an empty set of integers is the symbol ∞.

The letter "u" in the name "u-invariant" apparently comes from the word "universal"; see the second characterization of $u(F)$ above.

If F is formally real, then $m \cdot \langle 1 \rangle$ is anisotropic for all integers m, and so $u(F)$ is ∞. For this reason, we are interested in the u-invariant mainly in the nonreal case in this section. Of course, $u(F)$ may still be ∞ for some nonreal fields (see Example (8) below). At any rate, by the second equation above, we see that the level $s(F)$ is a lower bound for $u(F)$.

Examples 6.2. (1) If F is quadratically closed, then $u(F) = 1$, and conversely.

(2) $u(F) \leq 2$ iff all binary forms are universal, iff all 2-fold Pfister forms are hyperbolic, iff $I^2F = 0$ (see II.Exercise 5). In this case, the Witt ring is described completely in II.3.5. Examples of fields with $u(F) \leq 2$ are:

(A) finite fields (II.3.4);

(B) any nonreal field of transcendence degree 1 over a real-closed field (Theorem 4.10).

(3) If F is a field of transcendence degree n over an algebraically closed field, then $u(F) \leq 2^n$, by the theorem of Tsen-Lang stated in the introduction to this chapter.

(4) If F is a local field, then $u(F) = 4$ (by VI.2.12).

(5) If F is a nonreal global field, then $u(F) = 4$ by Corollary 1.5. This applies to all totally imaginary number fields, and all finite extensions of $\mathbb{F}_q(x)$ ($\mathbb{F}_q$ any finite field with q odd).

(6) If K is an algebraic extension over F of odd degree, then $u(F) \leq u(K)$, by Springer's Theorem (VII.2.7).

(7) Let F be a complete discretely valuated field with residue class field k of characteristic not 2. Then $u(F) = 2 \cdot u(k)$ (see VI.1.10). In particular, *if k is algebraically closed, and $F_n = k((t_1))((t_2)) \cdots ((t_n))$, then $u(F_n) = 2^n$.* This observation, due to I. Kaplansky [Ka₁], shows the existence of nonreal fields of any 2-power u-invariant.

(8) Let $F = \mathbb{C}(x_1, x_2, \ldots)$ (x_i independent indeterminates). It is not hard to see that, for any n, the n-dimensional form $\langle x_1, x_2, \ldots, x_n \rangle$ cannot be universal over F. Hence $u(F)$ is infinite.

Theorem 6.3. *Suppose $u(F) < \infty$. Then IF is a nilpotent ideal. In fact, if $u(F) < 2^n$, then $I^nF = 0$. If $u(F) \leq 2^n$ instead, then any nonzero anisotropic form in I^nF is a universal n-fold Pfister form, and any two n-fold Pfister forms over F are linked.*

Proof. First suppose $u(F) < 2^n$. Let φ be any n-fold Pfister form over F. Since $\dim \varphi = 2^n > u(F)$, φ is isotropic, and hence hyperbolic by X.1.7. It follows from X.1.2 that $I^nF = 0$.

Now assume only $u(F) \leq 2^n$, and consider any nonzero anisotropic form q in I^nF. Then $\dim q \leq 2^n$, and the Hauptsatz X.5.1 shows that $\dim q = 2^n$. It follows that q is universal, and X.5.6 implies that $q \cong a \cdot \varphi$, where $a \in \dot{F}$ and φ is an n-fold Pfister form. Since $\dim \varphi = 2^n$, $a \in D_F(\varphi)$, and hence $q \cong a \cdot \varphi \cong \varphi$. Finally, let φ_1, φ_2 be n-fold Pfister forms over F. By what we have proved so far, $\varphi_1 + \varphi_2 = \varphi_3$ for some n-fold Pfister form φ_3. But now the Linkage Theorem X.6.22 implies that φ_1 and φ_2 are linked. □

Good illustrations for the second part of Theorem 6.3 are given by finite fields (for $n = 1$), local fields (for $n = 2$), and the field $\mathbb{C}((x_1))\cdots((x_n))$ (for general n). In each of these cases, the anisotropic n-fold Pfister form is unique. (For the last field, for instance, it is $\langle\langle x_1, \ldots, x_n \rangle\rangle$: see VI.Exercise 2.)

Let $q(F)$ denote the cardinality of $\dot{F}/\dot{F}^2$, the group of square classes of F. If F is not formally real, it seems reasonable to expect that $u(F)$ should not exceed $q(F)$. This was first verified in special cases by I. Kaplansky, and, later, proved in full by M. Kneser. (Further generalizations can be found in [EL3].)

Theorem 6.4. *Suppose F is not formally real. Then $u(F) \leq q(F)$.*

To give Kneser's proof of 6.4, the key fact is:

Kneser's Lemma 6.5. *Let F be a nonreal field, and let φ be a nonuniversal form over F, of dimension d. Then, for any $a \in \dot{F}$, $D_F(\varphi) \subsetneq D_F(\varphi \perp \langle a \rangle)$, and the latter has more than d square classes.*

Proof. For a fixed a, say $D_F(\varphi) = D_F(\varphi \perp \langle a \rangle)$. In particular, $a \in D_F(\varphi)$. Write $-1 = e_1^2 + \cdots + e_s^2$, where $s = s(F)$. We claim that $a(e_1^2 + \cdots + e_i^2) \in D_F(\varphi)$, and prove it by induction on i. If $i = 1$, this follows from $a \in D_F(\varphi)$. In general, if we have shown $a(e_1^2 + \cdots + e_{i-1}^2) \in D_F(\varphi)$, then

$$a(e_1^2 + \cdots + e_{i-1}^2) + ae_i^2 \in D_F(\varphi \perp \langle a \rangle) = D_F(\varphi).$$

So the claim is established. In particular, we have

$$-a = a(e_1^2 + \cdots + e_s^2) \in D_F(\varphi).$$

This implies that $\varphi \perp \langle a \rangle$ is isotropic, so $D_F(\varphi) = D_F(\varphi \perp \langle a \rangle) = \dot{F}$, a contradiction. The rest of the lemma follows easily by induction on d. □

Corollary 6.6. *If ψ is an anisotropic form of dimension n over a nonreal field F, then ψ represents at least n distinct elements of $\dot{F}/\dot{F}^2$.*

This corollary clearly implies Theorem 6.4. In particular, we have:

Corollary 6.7. *$W(F)$ is finite if and only if $s(F)$ and $q(F)$ are both finite.*

Proof. The "only if" part is clear. Conversely, if $s(F)$ and $q(F)$ are both finite, we have $u(F) < \infty$ also, by 6.4. To enumerate the elements of $W(F)$, we need only look at forms of dimension $\leq u(F) \leq q(F)$. At each dimension, there are only finitely many inequivalent forms. This clearly implies that $W(F)$ is finite. □

The use of Kneser's Theorem 6.4 in the proof of 6.7 is only for convenience. In fact, the "if" part in 6.7 could have been proven by a judicious application of Dirichlet's Pigeon Hole Principle, as we have indicated in the hint to the same result in II.Exercise 10.

In Example 6.2(7), we have recalled Kaplansky's observation that $u(F) = 2^n$ for the iterated Laurent series field $F = \mathbb{C}((x_1))\cdots((x_n))$. In dire need of examples of fields with other u-invariant values, Kaplansky ventured the conjecture (in [Ka$_1$], c. 1953) that the u-invariant of a field is either ∞, or a power of 2. This conjecture stood open many years, and was only settled, negatively, in 1989 when Merkurjev constructed a field of u-invariant 6. Merkurjev's remarkable construction will be presented later in XIII.2. At this time, let us just explain why the number "6" is the first viable choice, by proving the following folklore result (which was already well-known before Merkurjev's work).

Proposition 6.8. *$u(F)$ cannot be 3, 5, or 7.*

This result is to be compared with those in III.Exercise 6 and X.Exercise 12. However, for the proof of 6.8, we do not need to eliminate the three cases one by one. If $u(F)$ is 3, 5, or 7, Theorem 6.3 shows that $I^3F = 0$. Therefore, it is enough for us to prove the following more general result (from [EL$_3$]).

Theorem 6.9. *If $I^3F = 0$ and $1 < u(F) < \infty$, then $u(F)$ is even.*

Proof. If σ is any form in I^2F and $a \in \dot{F}$, then

$$\langle 1, -a \rangle \cdot \sigma \in I^3F = 0 \implies \langle a \rangle \sigma \cong \sigma.$$

This clearly implies that σ is universal. Assume that $u = u(F)$ is an odd integer, and let φ be a u-dimensional anisotropic form. Let $a = d(\varphi)$.

Since φ is universal, we may write $\varphi \cong \langle a \rangle \perp \sigma_1 \cong \langle -a \rangle \perp \sigma_2$, for suitable σ_1 and σ_2 of (even) dimension $u-1$. By II.2.2, we have either $\sigma_1 \in I^2F$ or $\sigma_2 \in I^2F$. By the first part of the proof, either σ_1 or else σ_2 is universal, a contradiction. □

Quite remarkably, Merkurjev has shown that, in the situation of the theorem above, $u(F)$ can in fact be *any* prescribed even integer. For more information on this, see 2.23(4) in Chapter XIII..

A major problem in the study of the u-invariant of fields is to find good estimates on such invariants under finite algebraic field extensions K/F. Of course, one should not hope for $u(K) \leq u(F)$. For instance, a quadratically closed field F may have a finite extension K/F that is *not* quadratically closed; in that case, $u(F) = 1$ but $u(K) > 1$. So far, it seems to be unknown if there is a universal constant M such that $u(K) \leq M \cdot u(F)$ for all finite field extensions K/F. As a compromise, one tries to find "good" functions f on the extension degree $[K\colon F]$ such that

$$u(K) \leq f\left([K\colon F]\right) \cdot u(F) \tag{6.10}$$

for any finite extension K/F. So far, the best known result in this direction is the following (from [Lp$_1$]).

Leep's Theorem 6.11. *For any field extension K/F with $n = [K : F] < \infty$, we have $u(K) \leq \frac{1}{2}(n+1)u(F)$.*

In the case where F is a quadratically closed field, for instance, Leep's theorem implies that any degree n extension of F has u-invariant bounded by $(n+1)/2$. This is the best result at least for $n = 3$, since some quadratically closed fields F (for instance the quadratic closure of $\mathbb{Q}$) do admit cubic extensions that are not quadratically closed; see VII.7.

In the case where $n = 2$, the bound $u(K) \leq \frac{3}{2} \cdot u(F)$ for quadratic extensions in 6.11 was first proved by Elman and Lam in [EL$_3$]. The transfer method used for that proof (summarized in Exercise 22 below) is still well worth a close scrutiny. However, this method seems to be special to the quadratic case, and does not extend to a proof for the general result 6.11.

Interestingly, Leep's proof of 6.11 turned out to depend on a certain generalization of the u-invariant notion to *systems* of quadratic forms. The introduction of u-invariants for systems of forms creates more room for arguments on such invariants, and leads reasonably easily to a proof of 6.11. In retrospect, this seems to be yet another example of the dictum in mathematics that working with more general concepts and definitions often makes it easier to prove theorems!

With just a little thought, it is actually not surprising that we need to deal with *systems* of quadratic forms over F if we are to analyze a given quadratic form φ over K (where K/F is a finite extension). Indeed, let $\{e_1, \ldots, e_n\}$ be an F-basis on K, and let $B\colon V \times V \to K$ be the symmetric bilinear form (on a K-space V) associated with φ. Then

$$(6.12) \qquad B(u,v) = B_1(u,v)e_1 + \cdots + B_n(u,v)e_n \qquad (B_i(u,v) \in F)$$

for $u, v \in V$, where the B_i 's are easily seen to be symmetric bilinear forms on V, viewed as a vector space over F. By depolarization, we get

$$(6.13) \qquad \varphi(v) = \varphi_1(v)e_1 + \cdots + \varphi_n(v)e_n,$$

where each φ_i is a quadratic form on V_F (with associated symmetric bilinear form B_i). From this equation, it is clear that finding an isotropic vector for the K-quadratic form φ amounts to finding a *common* isotropic vector v for the system of quadratic forms $\{\varphi_1, \ldots, \varphi_n\}$ on the F-space V_F. Note that, although φ may be a *regular* quadratic form over K to begin with, the F-quadratic forms φ_i may no longer be regular. Thus, for the considerations to follow, the name "quadratic form" should be taken to mean a *possibly nonregular* quadratic form.

The remarks in the paragraph above led us to the following formulation of the notion of a system of quadratic forms.

Definition 6.14. By a *quadratic n-system* over F, we mean a (finite-dimensional) F-vector space V equipped with a system of n quadratic forms $\{\varphi_1, \ldots, \varphi_n\}$. Such an n-system $(V; \varphi_1, \ldots, \varphi_n)$ is said to be *isotropic* if there exists a nonzero vector $v \in V$ such that $\varphi_i(v) = 0$ for all i, and *anisotropic* otherwise.

Thus, a quadratic 1-system is just an ordinary quadratic space $(V; \varphi_1)$, and a quadratic 2-system is a V equipped with a pair of quadratic forms φ_1, φ_2, etc. With the notion of n-systems in place, we can then define quite naturally the notion of "system u-invariants", as follows.

Definition 6.15. The *system u-invariant* $u_n(F)$ is defined to be the supremum of dim V, where V ranges over all anisotropic quadratic n-systems over F. (If this supremum does not exist, $u_n(F)$ is taken to be ∞.)

The system u-invariant $u_n(F)$ is a generalization of the former u-invariant in 6.1, since $u_1(F) = u(F)$. Again, $u_n(F)$ (as defined in 6.15) is of interest only in case F is nonreal, for otherwise the n-system $\{d\langle 1\rangle, \ldots, d\langle 1\rangle\}$ is anisotropic for any dimension d, which makes $u_n(F) = \infty$ for all $n \geq 1$.

The key for proving Leep's Theorem 6.11 is provided by the following result from [Lp$_1$] relating $u_n(F)$ to $u(F)$.

Proposition 6.16. *For any field* F, $u_n(F) \leq \frac{1}{2}n(n+1)u(F)$ *for any* $n \geq 1$.

Assuming this result, we can easily verify 6.11 as follows.

Proof of Theorem 6.11. Let $n = [K\colon F] < \infty$. We may clearly assume $u(F) < \infty$. Let (V, φ) be a K-quadratic space of dimension $> \frac{1}{2}(n+1)u(F)$. Then V_F has F-dimension $> \frac{1}{2}n(n+1)u(F)$, and it may be viewed as a quadratic n-system over F equipped with the quadratic F-forms $\varphi_1, \ldots, \varphi_n$ defined in (6.13). This n-system must then be isotropic by 6.16, so there exists $v \in V \setminus \{0\}$ such that $\varphi_i(v) = 0$ for all i. This implies that $\varphi(v) = 0$, so the K-form φ is isotropic, as desired. □

Our focus is now shifted to the inequality claimed in 6.16. Prior to the work done in [Lp₁], it has been observed in [EL₃] that $u_n(F) \leq (2^n - 1)u(F)$. This bound is very easy to prove (see Exercise 21), and it leads to the estimate $u(K) \leq u(F) \cdot (2^n - 1)/n$ by the proof of 6.11. However, the scalar factor for growth here is exponential in the extension degree $n = [K\colon F]$, which is, of course, quite undesirable. Leep's quadratic growth factor for 6.16 is based on another result of his, as follows.

Proposition 6.17. *For $n \geq 2$, $u_n(F) \leq u_{n-1}(F) + n\, u_1(F)$.*

With this result, we can prove 6.16 quickly by induction on n. The case $n = 1$ being trivial, let us assume $n \geq 2$ and that $u_{n-1}(F) \geq u(F) \cdot (n-1)n/2$. By 6.17, we then have

$$u_n(F) \leq u(F) \cdot \big((n-1)n/2 + n\big) = u(F) \cdot n(n+1)/2.$$

We are now reduced to proving 6.17, following the arguments in [Lp₁].

Proof of 6.17. For any anisotropic quadratic n-system $(V;\, \varphi_1, \ldots, \varphi_n)$ over F, we must show that $d := \dim V$ is bounded by $u_{n-1}(F) + n\, u_1(F)$. Let $W \subseteq V$ be a subspace maximal with respect to the property that $\varphi_1 = \cdots = \varphi_n$ on W, and let $m = \dim W$. Clearly, $(W;\, \varphi_1)$ is anisotropic, so $m \leq u_1(F)$. Now let

$$U = \{v \in V :\ v \perp W \text{ with respect to each of } \varphi_1, \ldots, \varphi_n\}.$$

Fixing bases on V and on W, we see that U is the solution space of a homogeneous system of nm linear equations in d unknowns. Thus,

$$\dim U \geq d - nm; \quad \text{that is, } d \leq \dim U + nm.$$

Since $m \leq u_1(F)$, we need only show that $\dim U \leq u_{n-1}(F)$. We check this by proving that *the quadratic $(n-1)$-system*

$$(*) \qquad (U;\, \varphi_1 - \varphi_2,\, \varphi_1 - \varphi_3,\, \ldots,\, \varphi_1 - \varphi_n)$$

is anisotropic. Let $u \in U$ be a common zero, so that $\varphi_1(u) = \ldots = \varphi_n(u)$. Since $u \in U$ is orthogonal to W in all φ_i, it follows that the forms φ_i will all agree on $W + Fu$. By the maximal choice of W, we must have $u \in W$. But then $\varphi_i(u) = 0$ for all i, and hence $u = 0$, as desired. □

Now that we have completed the proof for Leep's Theorem 6.11, some additional remarks are in order. First, although we have primarily used the system u-invariant estimate $u_n(F) \leq \frac{1}{2}n(n+1)u(F)$ as a tool for proving Leep's Theorem, this estimate on $u_n(F)$ has a significance on its own in that it gives quantitative information on the problem of finding common zeros for *systems* of quadratic forms over F, based solely on the knowledge of $u(F)$ (the "classical" u-invariant). To make this point explicit, let us simply state the following special cases of 6.16 for fields of u-invariant ≤ 4, without using the notation $u_n(F)$.

Corollary 6.18. *Let F be any (nonreal) field.*

(1) *If $u(F) = 1$, any system of n quadratic forms over F in more than $n(n+1)/2$ variables has a nontrivial common zero.*

(2) *If $u(F) = 2$, any system of n quadratic forms over F in more than $n(n+1)$ variables has a nontrivial common zero.*

(3) *If $u(F) = 4$, any system of n quadratic forms over F in more than $2n(n+1)$ variables has a nontrivial common zero.*

Of course, since the proof of 6.16 was based on very general principles, its corollary 6.18 cannot be expected to be of a very sharp nature. However, the following example shows that the estimate (1) above is sharp at least for $n = 2$ (that is, for pairs of quadratic forms).

Example. Let $F = \widetilde{\mathbb{Q}}$, the quadratic closure of $\mathbb{Q}$. To show that 6.18(1) is sharp for $n = 2$, it suffices to check that the *ternary* quadratic forms

$$f(x, y, z) = x^2 - yz \quad \text{and} \quad g(x, y, z) = 4y^2 + 12xz + 9z^2$$

have no nontrivial common zero over F. Indeed, let $(a, b, c) \in F^3$ be a common zero. If $c \neq 0$, we may assume that $c = 1$. Then $b = a^2$, and $4a^2 + 12a + 9 = 0$, in contradiction to VII.2.11, where we have shown that this quartic equation has no roots in $\widetilde{\mathbb{Q}}$. Thus, $c = 0$, which clearly implies that $a = 0$ and $b = 0$ (in that order).

For specific fields and/or specific n, better results than 6.18(2) and 6.18(3) are often available by the application of ad hoc methods. For comparison, let us state some classical results on common zeros for systems of quadratic forms over finite fields and $\mathfrak{p}$-adic fields. If F is a finite field (for which $u(F) = 2$), the following special case of a theorem of Chevalley [Ch] (c. 1936) is certainly much sharper than 6.18(2).

Theorem 6.19. *Over a finite field F, any n quadratic forms in more than $2n$ variables have a nontrivial common zero.*

To conclude our discussions on $u_n(F)$, let us now focus on 6.18(3) for $n = 2$ (that is, the case of pairs of quadratic forms). If $u(F) = 4$, Corollary 6.18(3) guarantees that any two quadratic forms in more than 12 variables over F have a nontrivial common zero. However, in the case of $\mathfrak{p}$-adic fields (which have u-invariant 4), the classical work of Dem'yanov [Dem] (and Birch-Lewis-Murphy [BLM]) already gave the following sharper result.

Theorem 6.20. *Over a $\mathfrak{p}$-adic field F, any two quadratic forms in more than 8 variables have a nontrivial common zero.*

Even more remarkably, the same result holds if F is a nonreal number field, according to the more recent (but considerably harder) results on Hasse principles for pairs of quadratic forms over number fields proved by Colliot-Thélène, Sansuc, and Swinnerton-Dyer.

The last result we want to present in this section is the computation of the u-invariant of a nonreal linked field, first obtained in my joint paper [EL$_4$] with Elman. Recall that a field F is said to be *linked* if every pair of quaternion algebras over F can be expressed "with a common slot". Other characterizations for linked fields have been given in X.4.20. Local fields and global fields are among the principal examples of linked fields, according to VI.3.6. In fact, it is precisely the special behavior of the quadratic forms over local fields and global fields that has led to the formulation (if not the proof) of the following result.

Linked Field Theorem 6.21. *Let F be any nonreal linked field. Then $u(F) = 1, 2, 4,$ or 8. We have $u(F) \leq 4$ iff $I^3F = 0$.*

Proof. We begin by noting that, since F is linked, any pair of n-fold Pfister forms can be expressed with $n-1$ common slots. We shall prove the theorem at hand in a sequence of six steps.

Step 1. We first show that *any element $q \in I^nF$ is equal to an n-fold Pfister form modulo $I^{n+1}F$*. By X.1.2, q is a $\mathbb{Z}$-combination of n-fold Pfister forms. It clearly suffices to handle the case where $q = \alpha - \beta$, where α and β are n-fold Pfister forms. Upon writing

$$\alpha = \langle\langle a, c_2, \ldots, c_n\rangle\rangle \quad \text{and} \quad \sigma = \langle\langle b, c_2, \ldots, c_n\rangle\rangle,$$

we have $\alpha - \beta = \langle a, -b\rangle\langle\langle c_2, \ldots, c_n\rangle\rangle \equiv \langle\langle -ab, c_2, \ldots, c_n\rangle\rangle$ modulo $I^{n+1}F$, as desired.

Step 2. *We claim that $I^4F = 0$* (which is a necessary condition for $u \leq 8$, by 6.3). To establish this, it suffices to show that, if φ is any 4-fold Pfister form, then the pure subform φ' (with $\langle 1\rangle \perp \varphi' = \varphi$) represents -1. (If so, X.1.5 implies $\varphi \cong \langle\langle -1, \ldots\rangle\rangle$ is hyperbolic.) Since -1 is a sum of squares by hypothesis, it is sufficient to show that φ' represents any nonzero $z = \sum_{i=1}^{m} a_i^2$. We induct on m. To begin the induction, we first show that φ' represents a_1^2. If $\varphi = \langle\langle a, b, c, d\rangle\rangle$, after putting a common slot into $\langle\langle a, b\rangle\rangle$ and $\langle\langle c, d\rangle\rangle$, we may assume that $a = c$. Then, $\varphi \cong \langle\langle 1, b, c, d\rangle\rangle$, so $\varphi' \cong \langle 1, \dots\rangle$ clearly represents a_1^2. If $m > 1$, write $z = a_1^2 + u$, with $u = \sum_{i=1}^{m} a_i^2 \neq 0$. (If $u = 0$, we are back to the case $m = 1$.) By the inductive hypothesis and by X.1.5, we may write $\varphi \cong \langle\langle u, p, r, s\rangle\rangle$. Express

$$\langle\langle u,\, p\rangle\rangle \cong \langle\langle t,\, p'\rangle\rangle, \qquad \langle\langle r,\, s\rangle\rangle \cong \langle\langle t,\, s'\rangle\rangle,$$

with a common slot t. Then

$$\varphi \cong \langle\langle t, p', t, s'\rangle\rangle \cong \langle\langle t, p', 1, s'\rangle\rangle \cong \langle\langle u, p, 1, s'\rangle\rangle.$$

From this, we see that $\varphi' \cong \langle 1, u, \dots\rangle$, so it clearly represents $a_1^2 + u = z$.

Step 3. *Let q be any form, and let φ, τ be 2-fold and 3-fold Pfister forms, respectively. If $q = \varphi - \tau \in W(F)$, then there exist 2-fold Pfister forms μ_1, μ_2 and an $x \in \dot{F}$, such that $q = \langle x\rangle\mu_1 - \mu_2 \in W(F)$.* To see this, write $\varphi = \langle\langle a, b\rangle\rangle$ and $\tau = \langle\langle a, d, e\rangle\rangle$, with a common slot a. Further rewriting $\langle\langle d, e\rangle\rangle$, if necessary, we may suppose that $e \in D(\varphi')$ (thus $\langle e\rangle\varphi \cong \varphi$ by X.1.8). In $W(F)$, we have

$$\begin{aligned} q &= \langle e\rangle\varphi - \langle\langle a, e\rangle\rangle - \langle d\rangle\langle\langle a, e\rangle\rangle \\ &= \langle e\rangle \cdot (\langle\langle a, b\rangle\rangle - \langle\langle a, e\rangle\rangle) - \langle d\rangle\langle\langle a, e\rangle\rangle \\ &= \langle e\rangle \cdot \langle -e\rangle \cdot \langle\langle a, -be\rangle\rangle - \langle d\rangle\langle\langle a, e\rangle\rangle \\ &= \langle -d\rangle\langle\langle a, e\rangle\rangle - \langle\langle a, -be\rangle\rangle. \end{aligned}$$

Step 4. *If q is a form belonging to I^2F with* $\dim q \geq 8$, *then q represents* -1. In fact, by Step 1 and Step 2, we may write $q = \varphi - \tau \in W(F)$, where φ, τ are, respectively, 2-fold and 3-fold Pfister forms. Using Step 3, we have an equation $q = \langle x\rangle\mu_1 \perp \langle -1\rangle\mu_2 \in W(F)$, where μ_1, μ_2 are 2-fold Pfister forms. If $\dim q > 8$, this equation implies that q is isotropic, so certainly q represents -1. If $\dim q = 8$, the equation implies an isometry $q \cong \langle x\rangle\mu_1 \perp \langle -1\rangle\mu_2$, and $1 \in D(\mu_2) \Rightarrow -1 \in D(q)$.

Step 5. *We claim that $u(F) \leq 8$, i.e., any 9-dimensional form f over F is isotropic.* We may clearly assume that $d(f) = 1$. Thus, $f \perp \langle -1\rangle$ has determinant -1, and belongs to I^2F (by II.2.2). Since $f \perp \langle -1\rangle$ has dimension 10, the argument in Step 2 shows that it must be isotropic. Hence, f represents 1, and we may write $f \cong \langle 1\rangle \perp q$, $\dim q = 8$. But then $d(q) = 1$, so $q \in I^2F$ (again, by II.2.2). Step 4 implies that q represents -1, so f is indeed isotropic.

Step 6. If $I^3F \neq 0$, then $u(F) \geq 8$ by 6.3. Combining this with Step 5, we get $u(F) = 8$. Finally, assume $I^3F = 0$. In this case, every element in I^2F is given by a 2-fold Pfister form. We want to show that $u(F) \leq 4$, i.e., any 5-dimensional form f is isotropic. We may clearly assume that $d(f) = 1$.

By II.2.2, $f \perp \langle -1 \rangle \in I^2F$. Since $f \perp \langle -1 \rangle$ has dimension 6, it must be isotropic. Thus, $f \cong \langle 1 \rangle \perp q$ for some 4-dimensional $q \in I^2F$. But then q is a 2-fold Pfister form, which must be universal (since $I^3F = 0$). This shows that f is isotropic, as desired. □

We shall now conclude with a result closely related to the Linked Field Theorem 6.21.

Theorem 6.22. *If F is a nonreal field and there are at most three nonsplit quaternion algebras over F, then $u(F) = 1, 2,$ or 4.*

Remark 6.23. If F has a unique nonsplit quaternion algebra, then the above gives $u(F) = 4$. This special case was first proved by Kaplansky (see VI.Exercise 1). A case in point is where F is a local field, which (according to VI.2.10) has a unique nonsplit quaternion algebra.

Proof of 6.22. We shall work in the case where F has exactly three distinct nonsplit quaternion algebras, say, A, B, C. (The other cases are easier, and require only a trivial subset of the following arguments.)

Step 1. *We show first that F is a linked field* (so that we can apply Theorem 6.21). It suffices to show that A, B can be written with a common slot. Let $A = \left(\frac{a,x}{F}\right)$, $B = \left(\frac{b,y}{F}\right)$. Assume first that $D = \left(\frac{x,b}{F}\right)$ is nonsplit. If $D \cong A$, then A, B have a common slot b. If $D \cong B$, then A, B have a common slot x. If $D \cong C$, then A, C have a common slot x, and B must be $A \cdot C = \left(\frac{x,ab}{F}\right)$ in the Brauer group $B(F)$. We may therefore assume that $D = 1 \in B(F)$. Similarly, we may assume that $E = \left(\frac{a,y}{F}\right) = 1 \in B(F)$. But then

$$A = A \cdot D = \left(\frac{ab,\, x}{F}\right), \qquad B = B \cdot E = \left(\frac{ab,\, y}{F}\right),$$

with the common slot ab, as desired.

Step 2. Let the level of F be $s = 2^m$. *We claim that* $m \le 2$. The quickest proof is by using in advance a result in Section 7, namely, 7.3(2). This provides an upper bound $m(m-1)/2 \le 2$, which forces m to be ≤ 2.

Step 3. In view of 6.21, the proof is complete once we show that $I^3F = 0$. Note that any 3-fold Pfister form can be expressed as $\langle\langle 1, r, s \rangle\rangle$, since A, B, C can be expressed with some common slot.

Case 1. Any 2-fold Pfister form $\langle\langle 1, p \rangle\rangle$ is hyperbolic. Then the remark just made implies that $I^3F = 0$.

Case 2. There exist *two* anisotropic 2-fold Pfister forms with 1 as a slot. In this case, *any* 2-fold Pfister form (there are only four) looks like $\langle\langle 1, p \rangle\rangle = \langle 1, 1, p, p \rangle$, and hence represents any nonzero sum of 2 squares. By

4.5 ((4) $\iff$ (5)), and the fact that F is nonreal, it follows that any 2-fold Pfister form over F is universal. Hence $I^3F = 0$.

Case 3. There exists exactly *one* anisotropic 2-fold Pfister form of the shape $\langle\langle 1, p\rangle\rangle$. Assume, by way of contradiction, that $I^3F \neq 0$. Consider any *anisotropic* 3-fold Pfister form φ, and write it as $\varphi = \langle\langle 1, r, s\rangle\rangle$. By the case hypothesis, we must have $\langle\langle 1, r\rangle\rangle \cong \langle\langle 1, p\rangle\rangle \cong \langle\langle 1, s\rangle\rangle$, so

$$\varphi \cong \langle\langle 1, p, s\rangle\rangle \cong \langle\langle 1, p, p\rangle\rangle \cong \langle\langle 1, 1, p\rangle\rangle.$$

By the same argument, we must have $\langle\langle 1, 1\rangle\rangle \cong \langle\langle 1, p\rangle\rangle$ and thus $\varphi \cong \langle\langle 1, 1, 1\rangle\rangle$
$= 8\langle 1\rangle$. But, since $s(F) \leq 4$ by Step 2, $8\langle 1\rangle$ is supposed to be isotropic, a contradiction. □

We remark, in closing, that the value $u(F) = 8$ can indeed be achieved under the hypothesis of Theorem 6.21. For an example, let $F = F_0((t))$, where F_0 is any finite extension of the p-adic rationals, $p \neq 2$. Then $u(F) = 8$ (see 6.2(7)), and it can be verified without difficulty that F is a linked field.(3) There are eight F-quaternion algebras here, which, incidentally, shows that Theorem 6.22 is also the best possible result.

Appendix: The General u-Invariant

In this Appendix, we give a very brief introduction to the notion of the "general u-invariant" of a field that was introduced by Elman and the author in [EL3]. The main purpose of our exposition is just to present the basic definitions in this more general approach to the u-invariant; limitation of space will prevent us from giving a more detailed coverage of the theory here. Readers interested in following a fuller development of the relevant ideas discussed in this section are encouraged to consult the paper [EL3], as well as the treatment of the u-invariant theory given in Chapter 8 of Pfister's book [Pf5].

The need to reconsider the definition 6.1 for Kaplansky's "classical" u-invariant $u(F)$ stemmed from the fact that $u(F)$ is always ∞ for formally real fields F. This has the undesirable effect of limiting the consideration of the u-invariant to the class of nonreal fields alone. To remedy this situation, the following new definition of the u-invariant was proposed in [EL3].

Definition 6.24. For any field F, the (*general*) *u-invariant* $u(F)$ is defined to be $\sup\{\dim \varphi\}$, where φ ranges over all anisotropic forms in $W_t(F)$ (the torsion subgroup of $W(F)$). If this supremum does not exist, $u(F)$ is defined to be ∞.

(3)Since $u(F) = |\dot{F}/\dot{F}^2| = 8$, F is, in fact, a $\overline{C}$-field in the sense of 7.16 below. Thus, the conclusion of Exercise 17 applies to show that F is a linked field.

The "inconsistency" between 6.24 and the original definition 6.1 for the (classical) u-invariant is only minor. In the case where F is a nonreal field, we have $W(F) = W_t(F)$ (by 2.3), so the definitions of $u(F)$ in 6.1 and in 6.24 do agree. In the case of a formally real field F, 6.1 dictates that $u(F) = \infty$ (which is basically a "useless" definition); however, 6.24 renders a more meaningful definition for $u(F)$ that reflects the structure of the torsion quadratic forms over F.

The remarks in the last paragraph suggest that we should now replace the classical definition 6.1 for the u-invariant by the new definition 6.24. In particular, this will be done consistently in this Appendix. Confusion is not likely since, in working with formally real fields F, it will always be clear that $u(F)$ should mean the "largest" dimension of the anisotropic torsion quadratic forms over F (rather than just the symbol ∞). Since such torsion forms φ must have zero signature with respect to any ordering on F, dim φ must be even. It follows immediately that

Proposition 6.25. *If F is formally real, $u(F)$ is an even integer, or ∞.*

This may seem to be a trivial fact, but it nevertheless merits an explicit mention, since its analogue is now known to be false for nonreal fields. In [Iz], Izhboldin produced the first example of a nonreal field with u-invariant 9. For more discussions on this, see XIII.2.23(7).

Coming back to Definition 6.24, we can often formulate results on the (general) u-invariant independently of whether the ground field is real or nonreal. For instance, the interpretation for lower values of $u(F)$ can be given for all fields as follows.

Proposition 6.26. (1) *$u(F) \leq 1$ iff F is pythagorean.*

(2) *$u(F) \leq 2$ iff I^2F is torsionfree, iff any 1-fold Pfister form represents all of $D_F(\infty)$.*

(3) *$u(F) \leq 4$ iff I^3F is torsionfree and every nonzero anisotropic form $q \in I^2F \cap W_t(F)$ is a 2-fold Pfister form.*

Proof. (1) First assume F is pythagorean. If F is nonreal, then it is quadratically closed, so $u(F) = 1$. If F is formally real, then $W_t(F) = 0$, and so $u(F) = 0$. Conversely, assume that $u(F) \leq 1$, and consider any $a \in D_F(2)$. Then $\langle 1, -a\rangle \in W_t(F)$ must be isotropic, and so $a \in \dot{F}^2$. This shows that F is pythagorean.

Before we come to (2) and (3), let us first make the following general observation (in generalization of the first part of 6.3):

(6.27) *If $u(F) < 2^n$, then I^nF is torsionfree.*

Indeed, if $0 \neq q \in I^n F$ is anisotropic, then the Hauptsatz X.5.1 implies that $\dim q \geq 2^n > u(F)$. Therefore, $q \notin W_t(F)$.

(2) If $u(F) \leq 2$, (6.27) applies with $n = 2$ to show that I^2F is torsionfree. Conversely, suppose I^2F is torsionfree. If F is nonreal, this means that $I^2F = 0$, in which case binary forms are universal, and so $u(F) \leq 2$. Next, assume F is formally real. Let q be an anisotropic form in $W_t(F)$. Then $q \equiv \langle 1, -d\rangle \pmod{I^2F}$ for some $d \in \dot{F}$. For any real closure $K \supseteq F$, we have $q_K = 0 \in W(K)$, so $\langle 1, -d\rangle \in I^2K$ implies that $d \in \dot{K}^2$. Therefore, $\langle 1, -d\rangle \in W_t(F)$ too, from which we get

$$q - \langle 1, -d\rangle \in I^2F \cap W_t(F) = 0.$$

This shows that $\dim q \leq 2$, and hence $u(F) \leq 2$. The last condition in (2) means that F satisfies the condition (A_2) in 4.5, and according to 4.1(2), (A_2) amounts to I^2F being torsionfree.

(3) Assume first $u(F) \leq 4$. Then (6.27) applies with $n = 3$ to show that I^3F is torsionfree. Let q be any nonzero anisotropic form in $I^2F \cap W_t(F)$. Then $\dim q \leq u(F) \leq 4$ implies that $q \cong a \cdot \varphi$, where $a \in \dot{F}$ and φ is a 2-fold Pfister form. For any $b \in \dot{F}$,

$$\varphi\langle 1, -b\rangle \in I^3F \cap W_t(F) = 0,$$

so φ is universal. In particular, $q \cong a \cdot \varphi \cong \varphi$, as desired. Conversely, assume that I^3F is torsionfree and that any nonzero anisotropic form in $I^2F \cap W_t(F)$ is 2-fold Pfister. Consider any nonzero anisotropic form $q \in IF \cap W_t(F)$. By the arguments we have used so far in (2) and (3), there exists $d \in \dot{F}$ such that $q - \langle 1, -d\rangle$ is a universal 2-fold Pfister form φ. Thus, $q = \varphi \perp \langle 1, -d\rangle \in W(F)$ implies that $\dim q \leq 4$. If F is formally real, this shows $u(F) \leq 4$. If F is nonreal instead, we certainly have $u(F) < \infty$, and 6.9 shows that $u(F)$ is even. But then the arguments above imply that $u(F) \leq 4$, as desired. □

Various remarks we have made on the classical u-invariant also hold true for the general u-invariant. Let us collect some of these below.

Remarks 6.28. (1) If K/F is an odd degree extension, then Springer's Theorem VII.2.7 implies easily that $u(F) \leq u(K)$.

(2) Let $K = F((t))$. Using the decomposition $W(K) = W(F) \oplus (\langle t\rangle - 1)W(F)$, it is easy to see that $u(K) = 2u(F)$.

(3) If $u(F) < 2^{n+1}$, then $P(F) \leq 2^n$. To see this, let $w \in D_F(\infty)$. Since $2^n\langle 1, -w\rangle \in W_t(F)$, it must be isotropic and hence hyperbolic. This implies that $w \in D_F(2^n)$, whence $P(F) \leq 2^n$.

Let us now record a couple of easy examples of u-invariants for formally real fields.

Examples 6.29. (1) Let F be a formally real field with four square classes represented by $\{\pm 1, \pm 2\}$. Then $W_t(F) = \mathbb{Z}_2 \cdot \langle 1, -2 \rangle$, so $u(F) = 2$.

(2) Let $K = F((t))$, where F is as in (1). Then $u(K) = 2u(F) = 4$ by (1) and 6.28(2), with $I^2K \cap W_t(K) = \mathbb{Z}_2 \langle\langle -2, t \rangle\rangle$.

(3) If F is a global field, then $u(F) = 4$. The proof of this (in generalization of 6.2(5)) is left to the reader.

Many of the known results for the u-invariant of nonreal fields can be (suitably) extended to the case of the general u-invariant. Let us give some examples of this below.

(6.30) *If $u = u(F) < \infty$, then any anisotropic torsion form of dimension u represents at least u square classes.* (For a proof of this, see [EL₃].)

(6.31) *For any field F, $u(F) \le |\dot{F}/\dot{F}^2|$.* (We may assume that $|\dot{F}/\dot{F}^2| < \infty$, in which case $W(F)$ is a finitely generated group. Then $|W_t(F)| < \infty$, so $u(F) < \infty$. Now (6.30) shows that $u(F) \le |\dot{F}/\dot{F}^2|$.)

(6.32) *For any linked field F, $u(F) \in \{0, 1, 2, 4, 8\}$, with $u(F) \le 4$ iff I^3F is torsionfree.* (For a proof of this generalization of 6.21 from nonreal fields to arbitrary fields, see [EL₄].)

(6.33) *If there are only finitely many distinct torsion 2-fold Pfister forms, then $u(F) < \infty$.* In particular, if there are only finitely many distinct F-quaternion algebras, then $u(F) < \infty$. (This follows easily from 4.2.)

To facilitate the consideration of the (general) u-invariant, Elman and the author introduced in [EL₃] a sequence of modified u-invariants as follows:

$$u^{(i)}(F) = \sup\{\dim \varphi \mid \varphi \text{ anisotropic } F\text{-form with } 2^i\varphi = 0 \in W(F)\}$$

(where $i \ge 1$). Since $W_t(F)$ is 2-primary torsion, one has

$$u^{(1)}(F) \le u^{(2)}(F) \le u^{(3)}(F) \le \cdots \le u(F) = \sup_i \{u^{(i)}(F)\}. \tag{6.34}$$

In [Pf₅: Chapter 8], Pfister called (6.34) the "Elman-Lam filtration" (of u-invariants of F). The interest in this filtration stems largely from the following observation in [EL₃].

Proposition 6.35. *Let F be any field for which $0 < u^{(1)}(F) < \infty$. Then*

$$u^{(i)}(F) \le \left(\textstyle\sum_{j=0}^{i} 1/2^j \right) \cdot u^{(1)}(F) < 2u^{(1)}(F) \quad (\textit{for any } i \ge 1).$$

In particular, $u(F) < 2u^{(1)}(F) < \infty$.

The proof of this can be found in [EL$_3$: (4.1)] or [Pf$_5$: p. 117]. The point of 6.35 is that the single invariant $u^{(1)}(F)$ seems to exercise some control over the other invariants $u^{(i)}(F)$ (and thus on $u(F)$). Technically, $u^{(1)}(F)$ should be much easier to work with, since, in case $s(F) > 1$, any form φ satisfying $2\varphi = 0 \in W(F)$ has (by 3.7) the shape $\perp \langle a_j \rangle \langle\langle -w_j \rangle\rangle$, where $w_j \in D_F(2)$; in particular, $u^{(1)}(F)$ is even or ∞.

The initial hope in introducing the $u^{(i)}(F)$'s was that these invariants *might* all be equal to the u-invariant $u(F)$. However, we know now that this is not the case, thanks to the work of Hoffmann [Ho$_8$] in 1998. Indeed, given any integer $k \geq 1$, Hoffmann has constructed, for many values $n \geq 2^{k+1}$, a field F with the property that

$$u^{(1)}(F) = \cdots = u^{(k)}(F) = 2n, \quad \text{and} \quad u^{(k+1)}(F) = \cdots = u(F) = 2n + 2.$$

Thus, the $u^{(i)}(F)$'s need not be all equal. Hoffmann's construction of F in [Ho$_8$] was very much modeled upon Merkurjev's construction of fields of u-invariant 6 using iterated function fields (to be presented in XIII.2).

As of 2001, we know a little bit more. In the paragraph following 6.25, we have mentioned that there exist nonreal fields of u-invariant 9 according to Izhboldin [Iz]; further, such fields can have level > 1. For such a field F, we must have $u^{(1)}(F) \leq 8$, since $u^{(1)}(F)$ is even. In fact, $u^{(1)}(F)$ must be exactly 8, since it is a known result that, for any field K with $u^{(1)}(K) < 8$, one has $u^{(1)}(K) = u(K)$. (For a proof of this, see [Pf$_5$: p. 118].) Thus, for the field F above, $u^{(1)}(F) = 8 < 9 = u(F)$.

Given what is said above, the exact relationship between $u(F)$ and the $u^{(k)}(F)$'s has remained somewhat mysterious.

7. The Size of $W(F)$, and $\overline{\text{C}}$-Fields

In this section, we shall provide upper and lower bounds for the cardinality of $W(F)$. For convenience, write $s = s(F)$, $q = q(F)$, $u = u(F)$, $h = h(F)$, and $d = P(F)$ throughout. We begin by presenting a famous counting argument, due to Kaplansky.

Kaplansky's Lemma 7.1. *Suppose $d \geq t = 2^n$. Then $q \geq |D(\infty)/\dot{F}^2| \geq 2^{n(n+1)/2}$. In other words, viewing $D(\infty)/\dot{F}^2$ as a $\mathbb{Z}_2$-vector space, we have $\dim_{\mathbb{Z}_2} D(\infty)/\dot{F}^2 \geq n(n+1)/2$.*

Proof. To simplify notations, we write $D_j := D_F(2^j)$, which is a group under multiplication. For the chain $\dot{F}^2 = D_0 \subseteq D_1 \subseteq \cdots$ and any integer $j \geq 0$, we claim that

$$x, y \in D_{j+1}, \quad x \equiv y \pmod{D_j} \Longrightarrow x + y \in D_{j+1} \cup \{0\}.$$

In fact, let $y = cx$, $c \in D_j$. Then, $x + y = (1+c)x$. If $1 + c = 0$, we have $x + y = 0$. If $1 + c \neq 0$, then both $1 + c$ and x belong to D_{j+1}, which is a group. Hence, $x + y \in D_{j+1}$, as claimed. Using the hypothesis $d \geq t = 2^n$, there exists an element $z = e_1^2 + \cdots + e_t^2$ of length t. Break up the t terms in this summation into 2^{n-j-1} subsums (j held fixed), each of length 2^{j+1}. Each subsum belongs to D_{j+1}, and two separate subsums must belong to different cosets of D_{j+1} modulo D_j, lest their sum belongs to $D_{j+1} \cup \{0\}$, in which case the summation $z = e_1^2 + \cdots + e_t^2$ "shrinks". Thus, our subsums yield 2^{n-j-1} different cosets of D_{j+1} modulo D_j. Yet, none of these is the identity coset, so $[D_{j+1} \colon D_j] \geq 2^{n-j-1} + 1$. However, this index is a 2-power, so actually $[D_{j+1} \colon D_j] \geq 2^{n-j}$. Consequently,

$$\begin{aligned} q &\geq [D(\infty) \colon \dot{F}^2] \geq [D_n \colon \dot{F}^2] = [D_n \colon D_{n-1}] \cdots [D_1 \colon D_0] \\ &\geq 2^1 \cdot 2^2 \cdots 2^n = 2^{n(n+1)/2}. \end{aligned}$$ ◻

Corollary 7.2. *Let $h = h(F) = 2^{m+1}$ and $d = P(F)$. Then,*

(1) *the form $\langle 1, 1\rangle$ represents at least $h/2$ square classes, and $q \geq |D(\infty)/\dot{F}^2| \geq 2^{m(m+1)/2}$;*

(2) *in case $h = d$, $\langle 1,1\rangle$ represents at least h square classes, and $q \geq |D(\infty)/\dot{F}^2| \geq 2^{(m+1)(m+2)/2}$.*

Proof. Since $d \geq h/2 = 2^m$, the lemma applies with n substituted by m. The inequality $[D_1 \colon D_0] \geq 2^m$ derived above shows that $\langle 1, 1\rangle$ represents at least $h/2$ square classes. For statement (2), since $d = h = 2^{m+1}$, we may apply the lemma with $n = m + 1$ instead. ◻

For the balance of this section, we shall be interested mainly in nonreal fields. Thus, we write $s = s(F) = 2^m$ in the following. Note that $d \geq 2^m$ (and $h = 2s = 2^{m+1}$). Therefore, by the above, we know that $\langle 1, 1\rangle$ represents at least s square classes, and $q \geq 2^{m(m+1)/2}$.

Corollary 7.3. (1) $\dim_{\mathbb{Z}_2} I^{m-r+1}F/I^{m-r+2}F \geq r(r+1)/2$, *where* $1 \leq r \leq m$.

(2) *Let $Q = \mathrm{Quat}(F)$ be the subgroup of the Brauer group generated by the quaternion algebras. Then,* $\dim_{\mathbb{Z}_2} Q \geq m(m-1)/2$.

Proof. (1) Define a map $f \colon \dot{F} \to I^{m-r+1}F/I^{m-r+2}F$ by the rule

$$f(a) = 2^{m-r}\langle\langle -a\rangle\rangle \in I^{m-r+1}F/I^{m-r+2}F.$$

This is easily seen to be a group homomorphism, with

$$\begin{aligned} \ker f &= \{a \in \dot{F} \mid 2^{m-r}\langle\langle -a\rangle\rangle \in I^{m-r+2}F\} \\ &= \{a \in \dot{F} \mid 2^{m-r}\langle 1, -a\rangle = 0 \in W(F)\} && \text{(by X.5.1)} \\ &= D_{m-r} && \text{(by 1.3).} \end{aligned}$$

We obtain, therefore, a monomorphism from $\dot{F}/D_{m-r}$ into the quotient $I^{m-r+1}F/I^{m-r+2}F$. This yields:

$$\begin{aligned}\dim_{\mathbb{Z}_2} I^{m-r+1}F/I^{m-r+2}F &\geq \dim_{\mathbb{Z}_2} \dot{F}/D_{m-r} \\ &\geq \dim_{\mathbb{Z}_2} D_m/D_{m-1} + \cdots + \dim_{\mathbb{Z}_2} D_{m-r+1}/D_{m-r} \\ &\geq 1 + 2 + \cdots + r = r(r+1)/2.\end{aligned}$$

(2) Define $g\colon \dot{F} \to Q$ by $g(a) = \left(\frac{-1,a}{F}\right) \in Q$. This is a homomorphism with kernel D_1, and we finish as in (1) above. □

The above result enables us to derive a lower bound on $|W(F)|$ in terms of the level $s(F)$.

Proposition 7.4. *If F has finite level $s = 2^m$, then*

$$|W(F)| \geq 2 \cdot 2^{m(m+1)(m+2)/6}.$$

Proof. Using the filtration $W(F) \supseteq IF \supseteq \cdots \supseteq I^{m+1}F$, we get

$$\begin{aligned}|W(F)| &\geq |W(F)/IF| \cdot |IF/I^2F| \cdots |I^mF/I^{m+1}F| \\ &\geq 2 \cdot \prod_{r=1}^{m} 2^{r(r+1)/2} = 2 \cdot 2^e,\end{aligned}$$

where $e = \sum_{r=1}^{m} r(r+1)/2 = m(m+1)(m+2)/6$. □

Remark. If we retain the term $|IF/I^2F| = q$ in the above estimate, we get a better lower bound $|W(F)| \geq 2q \cdot 2^t$, where $t = (m-1)m(m+1)/6$.

Example. Let F be the field of 2-adic numbers. Then, $s(F) = 4$, $q(F) = 8$, and $|W(F)| = 32$ (see VI.2.31). In this case, the lower bound in the proposition becomes an actual equality.

Next, we shall provide a lower bound on $|W(F)|$, in terms of $q = q(F)$ and $u = u(F)$. This result is due to C. Cordes [Co$_1$].

Proposition 7.5. $|W(F)| \geq u \cdot q$.

Proof. If $q = \infty$ or F is formally real, we have $|W(F)| = \infty$, and there is nothing to prove. In the following, we may, therefore, assume that F is nonreal, with $q < \infty$. By Theorem 6.4, u is also finite. Take a u-dimensional anisotropic form φ over F. If σ is any odd-dimensional subform of φ, the q forms $\langle a \rangle \cdot \sigma$ ($a \in \dot{F}/\dot{F}^2$) are mutually nonisometric (having different determinants) and all anisotropic. Thus, the total number of different odd-dimensional anisotropic forms is at least $q \cdot m$, where m is the number of positive odd integers $\leq u$. Since $|W(F)/IF| = 2$, we get $|W(F)| \geq 2 \cdot qm$. If u is even, we have $m = u/2$, so $|W(F)| \geq u \cdot q$. If u is odd, we have $m = (u+1)/2$, and we get a slightly better bound $|W(F)| \geq (u+1) \cdot q$. □

Corollary 7.6. *Let F be a nonreal field such that $q < \infty$. Then $|W(F)| = u \cdot q$ iff F has at most two quaternion algebras and $q > 1$.*

Proof. *Sufficiency.* Suppose F has no nonsplit quaternion algebra. Then, $u \leq 2$. Since we assume that $q > 1$, F is not quadratically closed, and we must have $u = 2$. Using $I^2F = 0$, clearly, $|W(F)| = 2 \cdot q$. Next, suppose F has exactly one nonsplit quaternion algebra. By 6.12, $u = 4$ and in particular $I^3F = 0$ (by 6.3). Consequently,

$$|W(F)| = |W(F)/IF| \cdot |IF/I^2F| \cdot |I^2F| = 2 \cdot q \cdot 2 = 4q.$$

Necessity. Suppose $|W(F)| = u \cdot q$. From the counting argument in the proof of 7.5, we see that all anisotropic forms of a given odd dimension must be scalar multiples of one another. We may assume that $u > 2$ (if otherwise, the desired conclusion is completely trivial). Therefore, there is exactly one anisotropic 3-dimensional form of determinant 1. From III.2.5 and III.2.7, it follows that F has a unique nonsplit quaternion algebra. Clearly, q cannot be 1. □

Remark. If F is a local field, then the lower bound in 7.5 becomes an actual equality, since F has a unique nonsplit quaternion algebra. On the other hand, it can be shown that, given any integer $n \geq 0$, there exist nonreal fields F with $q(F) = 2^n$, which have no nonsplit quaternion algebras. For such fields (with $n \geq 1$), we have $|W(F)| = u \cdot q$ by 7.6. Thus, in general, 7.5 is the best possible result.

Finally, we would like to provide an upper bound for $|W(F)|$ in terms of q that is also due to Cordes. For any given integer $i \geq 0$, let us write N_i for the number of different anisotropic forms of dimension i over F. Also, for any form φ, let $V(\varphi)$ denote the number of distinct square classes represented by φ.

Proposition 7.7. *Let F be a nonreal field, with $q = |\dot{F}/\dot{F}^2| < \infty$. Let i be any nonnegative integer $< q$. Then $(i+1)N_{i+1} \leq (q-i)N_i$. Equality holds iff $V(\varphi) = \dim \varphi$ for every anisotropic form φ with $\dim \varphi \in \{i, i+1\}$.*

Proof. Let A be the set of pairs

$$\{(\sigma, a) \mid \ \dim \sigma = i, \ \ a \in \dot{F}, \ \ \sigma \perp \langle a \rangle \text{ is anisotropic}\}.$$

In A, two pairs (σ, a), (σ', a') are viewed as the same element iff $\sigma \cong \sigma'$ and $a\dot{F}^2 = a'\dot{F}^2$. What is the cardinality of A? For σ, there are N_i different choices. Given a specific (anisotropic) σ, we have at most $q - i$ choices for $a \in \dot{F}/\dot{F}^2$, since $V(\langle -1 \rangle \sigma) \geq i$, by 6.6. Consequently,

$$|A| \leq (q-i) \cdot N_i \tag{7.8}$$

with equality iff $V(\sigma) = i$ for every anisotropic i-dimensional form σ. Next, let B denote the set of isometry classes of $(i+1)$-dimensional anisotropic forms: $|B| = N_{i+1}$. The rule $f(\sigma, a) = \sigma \perp \langle a \rangle$ clearly defines a surjective mapping $f\colon A \to B$. Given a specific anisotropic form $\tau \in B$, what is $|f^{-1}(\tau)|$? Since $V(\tau) \geq i+1$ (again, by 6.6), we clearly have

$$|f^{-1}(\tau)| \geq i+1, \tag{7.9}$$

with equality iff $V(\tau) = i+1$. Consequently,

$$(q-i) \cdot N_i \geq |A| \geq |B| \cdot (i+1) = (i+1) \cdot N_{i+1}.$$

The two ends are equal iff (7.8) and (7.9) (for any anisotropic $\tau \in B$) are equalities. This completes the proof. □

Theorem 7.10 (Cordes). *Let F be a nonreal field with $q < \infty$. Then, for any nonnegative integer i, $N_i \leq \binom{q}{i}$ (binomial coefficients). This yields $|W(F)| \leq 2^q$.*

Proof. We induct on i. Induction is started by observing that N_0 and $\binom{q}{0}$ are both equal to 1. Also, for $i > q$, the binomial coefficient $\binom{q}{i}$ is, by definition, zero, so the desired conclusion in this case is just a restatement of 6.4. Inductively, if $N_i \leq \binom{q}{i}$, where $i < q$, then, by 7.7,

$$N_{i+1} \leq \frac{q-i}{i+1} \cdot N_i \leq \frac{q-i}{i+1} \cdot \binom{q}{i} = \binom{q}{i+1}. \tag{7.11}$$

Finally, since elements of $W(F)$ are in 1-1 correspondence with the anisotropic forms over F, we have

$$\begin{aligned} |W(F)| &\leq N_0 + N_1 + \cdots + N_q \\ &\leq \binom{q}{0} + \binom{q}{1} + \cdots + \binom{q}{q} = 2^q. \end{aligned} \tag{7.12}$$

□

Remark 7.13. For any given i $(0 \leq i < q)$, if $N_i < \binom{q}{i}$, then $N_{i+1} < \binom{q}{i+1}$. This follows by replacing "$\leq$" in the middle of (7.11) by a strict inequality "$<$".

We shall now characterize the fields F for which the upper bound $|W(F)| \leq 2^q$ becomes an equality. Such fields have many special properties; indeed, they can be fully characterized as follows.

Theorem 7.14 (Cordes). *For a nonreal field F with $q < \infty$, the following statements are equivalent:*

(1) $|W(F)| = 2^q$.

(2) $N_i = \binom{q}{i}$ *for all i.*

(3) $u = q$.

(4) *Either* $q = 1$ *or* $N_j = \binom{q}{j}$ *for some* j, $2 \le j \le q$.

(5) $N_2 = \binom{q}{2}$.

(6) $V(\varphi) = 2$ *for any anisotropic binary form* φ.

(7) $V(\varphi) = \dim \varphi$ *for any anisotropic form* φ.

(8) F *has exactly* $(q-1)(q-2)/6$ *quaternion division algebras.*

Proof. We allow this plethora of statements only to facilitate the proof. Here is the cycle of implications:

(1) $\Longleftrightarrow$ (2) is clear by the calculation in (7.12).

(2) $\Longleftrightarrow$ (3) is clear by 7.10 and 7.13.

(2) $\Longrightarrow$ (4) is trivial.

(4) $\Longrightarrow$ (5) follows from 7.13.

(5) $\Longrightarrow$ (6). By (5), we have

$$2 \cdot N_2 = q(q-1) = (q-1) \cdot N_1. \tag{7.15}$$

Hence, (6) follows from the last statement of 7.7 (with $i = 1$).

(6) $\Longrightarrow$ (7). We prove (7) by induction on $\dim \varphi$. If $\dim \varphi = 1, 2$, there is nothing to prove. Suppose φ (anisotropic) has dimension $k+1$ ($k \ge 2$). Write $\varphi \cong \langle a \rangle \perp \sigma$, $\dim \sigma = k$. Clearly, $D(\varphi)$ is the union of $D(\langle a, d \rangle)$, where d ranges over $D(\sigma)$. By the inductive hypothesis, there are k square classes in $D(\sigma)$. For $d \in D(\sigma)$, there are 2 square classes in $D(\langle a, d \rangle)$, by (6). Since each $D(\langle a, d \rangle)$ contains the square class $a\dot{F}^2$, we have $V(\varphi) \le k+1$. This must be an equality in view of 6.6.

(7) $\Longrightarrow$ (2) by the last statement of 7.7, and induction on i.

So far, we have proved the equivalence of (1) through (7). We also have (2) $\Longrightarrow$ (8), since the number of quaternion division algebras is the number of distinct anisotropic forms of the type $\langle a, b, ab \rangle$, which is N_3/q. From this observation, we also get (8) $\Longrightarrow$ (4), since under (8), (4) holds for $j = 3$. $\square$

Definition 7.16. After Cordes [Co_1], we define a nonreal field F with $q = q(F) < \infty$ to be a $\overline{\text{C}}$-*field* if $|W(F)| = 2^q$, or equivalently, if any of the conditions in Theorem 7.14 is satisfied.

The point about such fields is that, for a prescribed square class number $q < \infty$, these $\overline{\text{C}}$-fields are the ones for which we have the largest possible number of anisotropic forms (namely 2^q). For some quick examples, we have the following.

Examples 7.17. (1) Any finite field F (of characteristic not 2) is a $\overline{\text{C}}$-field, with $u = q = 2$ and $|W(F)| = 4$.

(2) Any nondyadic local field F is a $\overline{\mathrm{C}}$-field, with $u = q = 4$ and $|W(F)| = 16$. A dyadic local field K is *not* a $\overline{\mathrm{C}}$-field, since $u = 4 < q$.

(3) If F is a $\overline{\mathrm{C}}$-field, then $K = F((t))$ is also one, since both of the invariants u and q "double" when we go from F to K. Thus, more examples of $\overline{\mathrm{C}}$-fields can be obtained from (1), (2) above by iterating the Laurent series field constructions.

The quadratic form behavior over $\overline{\mathrm{C}}$-fields is quite remarkable (which is why Cordes singled them out for study). Let us now record some additional special quadratic form properties for $\overline{\mathrm{C}}$-fields. The first one below is well supported by the examples given in 7.17.

Proposition 7.18. *If F is a $\overline{\mathrm{C}}$-field, then $s(F) \leq 2$.*

Proof. We have $|W(F)| = 2^q$ as in 7.14. If $s(F) \neq 1$, then by (6) in 7.14, the binary anisotropic form $\langle 1, 1\rangle$ represents exactly two square classes. From 7.2, we deduce that $s(F) = 2$. □

Theorem 7.19. *Let F be a $\overline{\mathrm{C}}$-field with $q = 2^n$.*

(1) There is a unique anisotropic universal form φ over F.

(2) Let $\{t_1, \ldots, t_n\}$ be a $\mathbb{Z}_2$-basis for $\dot{F}/\dot{F}^2$, chosen in such a way that $t_n = -1$ in case $s(F) = 2$. Then the form φ in (1) is given by $\langle\langle t_1, \ldots, t_n\rangle\rangle$ when $s(F) = 1$, and by $\langle\langle t_1, \ldots, t_{n-1}, 1\rangle\rangle$ when $s(F) = 2$.

(3) If σ and τ are anisotropic forms over F, then $D_F(\sigma) = D_F(\tau) \Rightarrow \sigma \cong \tau$.

(4) Every subset of $\dot{F}/\dot{F}^2$ is the value set of a (unique) anisotropic form over F.

Proof. (1) Let φ be anisotropic and universal. By 7.14, we have $\dim \varphi = V(\varphi) = |\dot{F}/\dot{F}^2| = q$. If ψ is another anisotropic universal form, then $\dim \psi = q$ too. From $N_q = \binom{q}{q} = 1$, it follows that $\psi \cong \varphi$.

(2) Note that, if an r-fold Pfister form ρ does not represent a square class $a\dot{F}^2$, then $\rho\langle\langle -a\rangle\rangle$ is an anisotropic $(r+1)$-fold Pfister form. First assume $s(F) = 1$. The observation made above implies that $\langle\langle t_1, \ldots, t_n\rangle\rangle$ is anisotropic, and so $\varphi \cong \langle\langle t_1, \ldots, t_n\rangle\rangle$. If $s(F) = 2$, it follows similarly that $\langle\langle t_1, \ldots, t_{n-1}, 1\rangle\rangle$ is anisotropic, and so $\varphi \cong \langle\langle t_1, \ldots, t_{n-1}, 1\rangle\rangle$.

(3) If $-a \notin D_F(\sigma)$, then $\sigma \perp \langle a\rangle$ is also anisotropic, and $V(\sigma \perp \langle a\rangle) > V(\sigma)$ by 7.14(7). Proceeding like this, we can find $a_1, \ldots, a_r \in \dot{F}$ such that $\sigma \perp \langle a_1, \ldots, a_r\rangle$ is anisotropic and universal, and hence $\cong \varphi$. If $D_F(\sigma) = D_F(\tau)$, then $\dim \sigma = V(\sigma) = V(\tau) = \dim \tau$, and $\tau \perp \langle a_1, \ldots, a_r\rangle$ is also anisotropic, of dimension q, and hence $\cong \varphi$. Now Witt cancellation gives $\tau \cong \sigma$.

(4) There are exactly 2^q anisotropic forms (counting the form "0") and 2^q subsets of $\dot{F}/\dot{F}^2$ (counting the empty subset $\varnothing$). From (3), it follows that $\sigma \mapsto D(\varphi) \subseteq \dot{F}/\dot{F}^2$ is a 1-1 correspondence between the set of isometry classes of anisotropic forms and the set of all subsets of $\dot{F}/\dot{F}^2$. □

Remark 7.20. If a field F has the property (3) for its anisotropic forms, Cordes called F a C-field. Such a field may have infinitely many square classes, and also it need not be nonreal.[4] Thus, C-fields are considerably more general than $\overline{\text{C}}$-fields. For instance, Cordes pointed out (without proof) in [Co$_1$] that $F = \mathbb{R}(x_1, x_2, \dots)$ is a C-field. For lack of space, we will not discuss C-fields further in this text.

Corollary 7.21. *Let F be a $\overline{\text{C}}$-field, and let $t_1, \dots, t_n$ be a square class basis chosen as in* 7.19(2).

(1) *If $s(F) = 1$, then $W(F)$ is isomorphic to the group ring $\mathbb{Z}_2G$, where $G = \dot{F}/\dot{F}^2$.*

(2) *If $s(F) = 2$, then $W(F)$ is isomorphic to the group ring $\mathbb{Z}_4H$, where $H \subseteq \dot{F}/\dot{F}^2$ is the subgroup generated by $t_1, \dots, t_{n-1}$.*

Proof. (1) Here, $\exp(W(F)) = 2$. Thus, the natural ring homomorphism $\pi\colon \mathbb{Z}_2G \to W(F)$ (induced by the inclusion map $G = \dot{F}/\dot{F}^2 \hookrightarrow W(F)$) is surjective. It follows that π is an isomorphism, since $|\mathbb{Z}_2G| = 2^q = |W(F)|$. (*Note.* The anisotropic forms over F in this case are given by all the diagonal subforms of $\langle\langle t_1, \dots, t_n\rangle\rangle$.)

(2) Here, $\exp(W(F)) = 4$. Thus, for the group H defined in (2), the natural ring homomorphism $\pi\colon \mathbb{Z}_4H \to W(F)$ is surjective, so we can finish as in (1) above. □

Remark 7.22. It is worthwhile to point out that the following strong converse of 7.21 is also true, for any field F with $q = |\dot{F}/\dot{F}^2| < \infty$. *If $W(F) \cong \mathbb{Z}_2G$ as groups, where G is a group of order q, then F is a $\overline{\text{C}}$-field of level* 1, *and if $W(F) \cong \mathbb{Z}_4H$ as groups, where H is a group of order $q/2$, then F is a $\overline{\text{C}}$-field of level* 2. These statements follow easily from 7.4 and 2.3, since in either case we clearly have $|W(F)| = 2^q$. More generally, the question of when Witt rings are isomorphic to group rings of suitable groups has been studied in detail by Roger Ware in his two paper [War$_1$] and [War$_3$].

Examples 7.23. Of course, from 7.17, it is already clear that there do exist $\overline{\text{C}}$-fields F with $s = 1$ or 2, with any prescribed $q = |\dot{F}/\dot{F}^2| < \infty$. For $s = 1$, we may take

(7.24) $$F = \mathbb{C}((t_1)) \cdots ((t_n)) \quad \text{with } q = 2^n,$$

---

[4] See, however, Exercise 19.

and for $s = 2$, we may take

(7.25) $$F = \mathbb{F}_3((t_1))\cdots((t_{n-1})) \quad \text{with } q = 2^n.$$

By 7.19(2), the unique anisotropic universal form is $\langle\langle t_1, \ldots, t_n\rangle\rangle$ for F as in (7.24), and is $\langle\langle t_1, \ldots, t_{n-1}, 1\rangle\rangle$ for F as in (7.25).

We finish by noting that, in the terminology to be introduced in XII.1, the result obtained in 7.21 means that a $\overline{\mathrm{C}}$-field K *is in fact uniquely determined "up to quadratic equivalence" by its square class number q and its level s*. For a given q, K is quadratically equivalent to the field F in (7.24) if $s(K) = 1$, and to the field F in (7.25) if $s(K) = 2$.

Exercises for Chapter XI

1. Show that the conclusion of Lemma 1.2 no longer holds if m is not a power of 2. (**Hint.** Note that if $S \cdot S^t = c \cdot I_m$ and c is nonzero, then $m\langle 1\rangle \cong m\langle c\rangle$.)
2. Let $F = \mathbb{R}(x_1, x_2, \ldots)$. If m is not a power of 2, show that $D_F(r)$ is not closed under multiplication.
3. (Pfister) For $x, y \in \dot{F}$, show that $\mathrm{len}_F(xy) \leq \mathrm{len}_F(x) + \mathrm{len}_F(y) - 1$.
4. Let F be a field, and let $K = F(\sqrt{-d}\,)$ $(d \in \dot{F})$ be a quadratic extension of F. If $s = s(K) < \infty$, use Theorem 1.1′ to show that
   (1) either $s(F) = s$, or $\mathrm{len}_F(d) \leq 2s - 1$;
   (2) $\mathrm{len}_F(d) \leq 2s - 1$ always holds, unless $s = 1$.
5. Let $2^k \leq n < 2^{k+1}$, and let $d \in \dot{F}$ be such that $\mathrm{len}_F(d) = n$ in a field F with $s(F) \geq 2^k$. If $K = F(\sqrt{-d}\,)$, show that $s(K) = 2^k$.
6. If there exists a field F with $|\dot{F}/\dot{F}^2| < \infty$ and $s(F) > 8$, show that there exists a field K with $|\dot{K}/\dot{K}^2| < \infty$ and $s(K) = 8$.
7. Suppose φ is a Pfister form, and $\beta = \varphi \otimes \langle 1, a\rangle$ is anisotropic. If x, y are in the same coset of $D(\beta)$ modulo $D(\varphi)$, show that $x + ay \in D(\beta)$. (**Hint**. Generalize the argument used in the proof of Lemma 7.1.)
8. If $D(2^m) = \dot{F}$, show that $2^m \cdot IF = 0$. Using this, show that, for a field F with level s, $P(F) = s$ iff $\exp(IF) = s$.
9. If a field F has level $s < \infty$, show that:
   (1) any odd-dimensional form has additive order $2s$ in $W(F)$;
   (2) if q is any form of additive order $2s$, $\mathbb{Z} \cdot q$ must be an additive direct summand of $W(F)$.
10. Show that $W(F)$ is a finite cyclic group iff either F is quadratically closed, or $|\dot{F}/\dot{F}^2| = 2$ and $s(F) = 2$ (in which case $W(F) \cong \mathbb{Z}_4$).
11. (From [EL3]) Let φ be a binary form, and let q be a form of dimension ≥ 2. If $q \cdot \varphi$ is isotropic, show that q contains a binary subform β

such that $\beta \cdot \varphi = 0 \in W(F)$. (This exercise may be viewed as a generalization of 3.7.)

12. (From [EL_3]) Let F be a formally real field with at most two orderings. If a form $\varphi = \langle 1 \rangle \perp \varphi'$ lies in $W_t(F)$, show that $D_F(\varphi')$ contains a totally negative element. From this, show that any form $\sigma \in W_t(F)$ can be written as an orthogonal sum $\beta_1 \perp \cdots \perp \beta_r$, where $\dim \beta_i = 2$ and $\beta_i \in W_t(F)$ for every i.
13. If F is nonreal and there are only a finite number of mutually nonisomorphic quaternion algebras over F, show that $u(F) < \infty$.
14. (Pfister) If F is nonreal with $P(F) = 1 + s(F)$, show that $u(F) \geq 2 \cdot s(F)$.
15. Let F be a nonreal field with $u(F) < \infty$. Suppose F has exactly one anisotropic universal form. Show that $u(F) = \dim \varphi$, and that $u(F)$ is a power of 2.
16. Let F be a nonreal field of transcendence degree 2 over $\mathbb{R}$. Using the Tsen-Lang Theorem, and VII.3.1, show that any 7-dimensional form over F contains a subform $\langle a, a, b, b \rangle$. Then use 4.10 to show that $u(F) \leq 6$. (It is "conjectured" that $u(F) \leq 4$; see Open Question 6.5 in Chapter XIII.)
17. Let F be a $\overline{\mathrm{C}}$-field with q square classes. Show that F is a linked field iff $q \leq 8$.
18. (Cordes) Let F be a field with $q = q(F) < \infty$. Let N denote the number of distinct quaternion division algebras over F. Show that $N \leq (q^2 + 2)/6$. If F is nonreal, show that $N \leq (q-1)(q-2)/6$ (cf. the characterization 7.14(8) for a $\overline{\mathrm{C}}$-field).
19. (Cordes) Let F be a C-field, in the sense of 7.20. If $|\dot{F}/\dot{F}^2| < \infty$, show that F must be nonreal.
20. Let F be a formally real field for which -1 and 2 form a $\mathbb{Z}_2$-basis of $\dot{F}/\dot{F}^2$. Let

    $$K_1 = F(\sqrt{-2}\,), \quad K_2 = F(\sqrt{2}\,), \quad \text{and} \quad K_3 = F\big([-(2+\sqrt{2}\,)]^{1/2}\big).$$

    Calculate the Witt groups $W(F)$, $W(K_1)$, $W(K_2)$, and $W(K_3)$. Determine the level and the u-invariant for K_1 and K_3.
21. Show that the system u-invariant $u_n(F)$ of a field F satisfies the estimate $u_n(F) \leq (2^n - 1)u(F)$ by considering the totally isotropic subspaces of n quadratic forms $\varphi_1, \ldots, \varphi_n$ in an F-vector space of dimension $> (2^n - 1)u(F)$. (This is, of course, a much weaker statement than Leep's proposition 6.16.)
22. ([EL_3]) For any quadratic extension $K = F(\sqrt{a}\,)$ over a nonreal field F, show that $u(K) \leq \frac{3}{2}u(F)$ by considering a K-quadratic space V and

its transfer $s_*(V)$ with respect to the F-linear functional $s\colon K \to F$ defined by $s(1) = 0$ and $s(\sqrt{a}) = 1$. (**Hint**. Suppose $\dim_K V > \frac{3}{2}u(F)$, where we may assume $u(F) < \infty$. Then $s_*(V)$ is isotropic, so $\varphi(v_1) \in F$ for some $v_1 \neq 0$. Repeating this construction in the K-orthogonal complement to $K \cdot v_1$, etc., one gets mutually orthogonal vectors $v_1, \ldots, v_r$ in V with $r = u(F) + 1$ with each $\varphi(v_i) \in F$. Since $\langle \varphi(v_1), \ldots, \varphi(v_r) \rangle_F$ is isotropic, so is the given K-quadratic space V.)

23. (Pourchet) Let q be a Pfister form over a number field, and let $f \neq 0$ be a polynomial in $F[x]$. Show that q represents f in $F[x]$ iff q represents $F_\mathfrak{p}[x]$ for every completion $F_\mathfrak{p}$ of F. (**Hint.** Use IX.Exercise 7, X.2.13, and the Hasse-Minkowski Principle.)

24. (Pourchet) For a nonzero polynomial $f \in \mathbb{Q}[x]$, show that the following statements are equivalent:
    (1) $f \in D_{\mathbb{Q}[x]}(4)$.
    (2) lead. coef.$(f) > 0$, and, for any irreducible factor $\pi \mid f$ of odd multiplicity, the level of $\mathbb{Q}[x]/(\pi(x))$ is ≤ 2.
    (3) $f \in D_{\mathbb{Q}[x]}(\infty)$, and in $\mathbb{Q}_2[x]$ the irreducible factors $\pi \mid f$ of odd multiplicity all have even degree (where $\mathbb{Q}_2$ is the field of 2-adic numbers).

25. Verify the lower bound for $P\big(\mathbb{R}(x_1, \ldots, x_n)\big)$ in 5.11.

26. Generalize Theorem 1.1′ by proving the following. Let φ be a Pfister form over F, and let B_φ be its associated symmetric bilinear form. Given $u = \varphi(\alpha)$, $v = \varphi(\beta)$, $w = B_\varphi(\alpha, \beta)$, where α, β are arbitrary vectors, then $uv = w^2 + \varphi'(\gamma)$ for a suitable vector γ (where φ' denotes the pure subform of φ).

Chapter XII

Special Topics in Quadratic Forms

The subject of the algebraic theory of quadratic forms has grown tremendously in the last few decades. By now there is a rather huge literature on the subject, and new progress continues to be made on it from day to day. In order to make this book a suitable introductory text to the subject, we have stressed the coverage of the foundational material in the theory at hand, but have refrained from a fuller development of our subject in all of its possible research directions. As a result, many of the more specialized new findings in quadratic form theory have not been reported upon so far in this text. To remedy this (if only to a limited extent), we offer in this chapter (and the next) a collection of short expositions on a potpourri of topics that can be read (more or less) independently of one another. These topics are chosen to supplement the coverage of the main text in the previous eleven chapters, but also to give an idea of some of the other aspects of quadratic form theory that we have not had an occasion to discuss so far.

The topics treated in this chapter concern largely quadratic forms and Witt rings, ranging from the theory of low-dimensional forms, biquadratic extensions, classification problems, Hilbert and pre-Hilbert fields, to the study of the axiomatic foundations of quadratic forms. In the next chapter, we will shift our attention to other special topics — mainly on quadratic form-theoretic field invariants. For instructors using this book as a text in a course on quadratic forms, the topics in this and the next chapter should offer plenty of material for individual seminar presentations by the participating students.

1. Isomorphisms of Witt Rings

A very natural question to ask in the study of the Witt rings of quadratic forms is: *when are the Witt rings of two different fields isomorphic* (*as abstract rings*)?

Certainly, the Witt rings of two fields can be isomorphic *without* the fields themselves being isomorphic. For instance, the finite fields $\mathbb{F}_3$ and $\mathbb{F}_7$ have different cardinalities, but have isomorphic Witt rings. (In fact, all finite fields $\mathbb{F}_q$ with $q \equiv 3 \pmod 4$ have Witt rings isomorphic to the ring $\mathbb{Z}_4$, according to II.3.6(B).) All euclidean fields, big or small, have Witt rings isomorphic to $\mathbb{Z}$. It is also known that many nonisomorphic pairs of algebraic number fields can have isomorphic Witt rings. For instance, Czogala, Szymiczek and Perlis have shown that the countably many quadratic number fields $\mathbb{Q}(\sqrt{d})$ produce only seven mutually nonisomorphic Witt rings.(1)

In view of the various examples above, it is of interest to understand how isomorphisms can arise between the Witt rings of two fields F and K. The case $F = K$ should by all means be included in this investigation, since the isomorphism problem in this case becomes the problem of studying the *automorphisms* of $W(F)$ for a given field F.

In this section, we shall give a quick exposition on the principal results on the isomorphism problem of Witt rings. These results can be traced back to the Witt ring lectures given by David Harrison [Ha] in the early 1970s at the University of Kentucky. They were also covered in part in Cunningham's seminar notes on quadratic forms [Cu], and later expanded upon in Cordes's paper [Co$_1$].

As it turned out, isomorphisms between Witt rings are accounted for by a certain type of equivalence between two fields, which is known as a *quadratic equivalence* (or more precisely, an *equivalence with respect to quadratic forms*). To keep our terminology easy and simple, we shall refer to this below as "q-equivalence".

Definition 1.1. We say that two fields F and K (of characteristic $\neq 2$) are *q-equivalent* if there exists a group isomorphism $t\colon \dot{F}/\dot{F}^2 \to \dot{K}/\dot{K}^2$ such that

(1) $t(-1) = -1$, and

(2) for any $a, b \in \dot{F}$, $b \in D_F\langle 1, a\rangle$ iff $t(b) \in D\langle 1, t(a)\rangle$.

---

(1)The seven isomorphism types of Witt rings are realized, respectively, by $d = -1$ and $d = \pm 2, \pm 7, \pm 17$; see [Cz] and [Sz$_2$].

Here, $t(a)$ really stands for $t(a\dot{F}^2)$, for any $a \in \dot{F}$. This notation is, of course, not all that precise; however, it is not likely to cause any serious problems. For instance, in (2) above, the quadratic form $\langle 1, t(a)\rangle$ is determined up to an isometry by the square class $t(a\dot{F}^2)$, and $D_F\langle 1, t(a)\rangle$ is always a union of square classes in $\dot{F}$. Thus, the fact that $t(a)$ has only a meaning as a square class does not really create any ambiguity in the statement of the condition (2) above.

It is worth noting that the axiom (2) can also be formulated in two other equivalent forms. Since t is a group isomorphism, it is easy to see (by scaling) that (2) is equivalent to either one of the following:

(2a) for any $a,\ b,\ c \in \dot{F}$, $b \in D_F\langle c,\ a\rangle$ iff $t(b) \in D_K\langle t(c),\ t(a)\rangle$,

(2b) for any $a,\ c \in \dot{F}$, $1 \in D_F\langle c, a\rangle$ iff $1 \in D_K\langle t(c),\ t(a)\rangle$.

In the balance of this section, we shall use rather freely these equivalent formulations of (2) in our discussions on the q-equivalence of fields.

The following proposition gives a reasonable explanation as to why the concept introduced in 1.1 above is called a "quadratic form equivalence."

Proposition 1.2. *Let $t\colon \dot{F}/\dot{F}^2 \to \dot{K}/\dot{K}^2$ be given as in Definition* 1.1. *Then we can define a one-one correspondence*

(1.3) $\quad T\colon$ *(isom. classes of F-forms)* $\longrightarrow$ *(isom. classes of K-forms)*

such that, for any F-form q and any $b \in \dot{F}$,

$$b \in D_F(q) \iff t(b) \in D_K(T(q)). \tag{1.4}$$

Moreover, this one-one correspondence preserves the orthogonal sums and tensor products of forms, takes $\mathbb{H}_F$ to $\mathbb{H}_K$, and preserves the isotropy and anisotropicity of forms, as well as their Witt indices.

Proof. Given an F-quadratic form q, take a diagonalization, $\langle a_1, \ldots, a_n\rangle$, say, where $a_i \in \dot{F}$. We try to define $T(q)$ to be the isometry class of the K-form $\langle t(a_1), \ldots, t(a_n)\rangle$. To check that we get a well-defined T as in 1.3, we appeal to Witt's Chain Equivalence Theorem I.5.2. According to this result, the well-definition is assured if we can check that, whenever $\langle a_1, \ldots, a_n\rangle$ is simply equivalent to $\langle b_1, \ldots, b_n\rangle$ over F, we have

$$\langle t(a_1),\ \ldots,\ t(a_n)\rangle \cong \langle t(b_1),\ \ldots,\ t(b_n)\rangle \quad \text{over } K.$$

Without loss of generality, we may assume that $a_i = b_i$ for $i \geq 3$, and that $\langle a_1,\ a_2\rangle \cong \langle b_1,\ b_2\rangle$ over F. In this case, we are done if we can show that

$$\langle t(a_1),\ t(a_2)\rangle \cong \langle t(b_1),\ t(b_2)\rangle \quad \text{over } K.$$

Now $\langle a_1,\ a_2\rangle \cong \langle b_1,\ b_2\rangle$ implies that $a_1a_2 = b_1b_2 \in \dot{F}/\dot{F}^2$. Therefore,

$$t(a_1)\,t(a_2) = t(b_1)\,t(b_2) \in \dot{K}/\dot{K}^2. \tag{1.5}$$

On the other hand, by the property (2a), $b_1 \in D_F\langle a_1, a_2\rangle$ yields $t(b_1) \in D_K\langle t(a_1), t(a_2)\rangle$. From (1.5) and a determinant argument, it follows that

$$\langle t(a_1), t(a_2)\rangle \cong \langle t(b_1), t(a_1)t(a_2)t(b_1)\rangle \cong \langle t(b_1), t(b_2)\rangle$$

over K, as desired.

We see easily that T in 1.3 is a one-one correspondence, since we can use t^{-1} to define an inverse map T^{-1} for T. Now let us check the asserted properties of T.

That T preserves $\perp$ and $\otimes$ is clear, and the axiom 1.1(1) guarantees that

$$T(\mathbb{H}_F) = T\langle 1, -1\rangle_F = \langle 1, -1\rangle_K = \mathbb{H}_K.$$

From this, it follows easily that T preserves isotropicity, anisotropicity, and consequently also Witt indices.

It only remains to prove 1.4, for which, of course, the forward implication would suffice. We do this by induction on $n = \dim(q)$. The case $n = 1$ is trivial, and the case $n = 2$ follows from the property (2a). For $n \geq 3$, we can use the following "Inductive Description of Value Sets" presented in Ch. I, Exer. 20; namely,

$$D_K\langle k_1, \ldots, k_n\rangle = \bigcup \{\, D_K\langle k_1, k\rangle : \;\; k \in D_K\langle k_2, \ldots, k_n\rangle\}.$$

A simple calculation of $D_K\langle t(a_1), \ldots, t(a_n)\rangle$ via this formula and (2a) shows quickly that

$$D_K\langle t(a_1), \ldots, t(a_n)\rangle = t\,(D_F\langle a_1, \ldots, a_n\rangle),$$

which amounts to 1.4. □

Corollary 1.6. *If* $t\colon \dot F/\dot F^2 \to \dot K/\dot K^2$ *defines a q-equivalence as in* 1.1, *then* t *induces a ring isomorphism* $\tau\colon W(F) \to W(K)$ *with the property that*

$$\tau(\dot F/\dot F^2) = \dot K/\dot K^2. \tag{1.7}$$

Here, $\dot F/\dot F^2$ *and* $\dot K/\dot K^2$ *are identified with the respective groups of unary forms in the Witt rings* $W(F)$ *and* $W(K)$.

Proof. We define $\tau[q] = [T(q)]$, where T is as constructed in 1.3, and $[q]$ denotes the Witt class of an F-quadratic form q. The conclusions of 1.2 show easily that τ is a well-defined ring isomorphism from $W(F)$ to $W(K)$. Finally, since $\tau[\langle a\rangle] = [T\langle a\rangle] = [\langle t(a)\rangle]$, 1.7 holds. □

We now come to the main result on Witt ring isomorphisms. A somewhat surprising conclusion from this result is that the isomorphism type of a Witt ring $W(F)$ is already determined by that of the factor ring $W(F)/I^3F$. (As usual, IF denotes the ideal of even-dimensional forms in $W(F)$.)

Harrison-Cordes Theorem 1.8. *For a pair of fields F and K, the following statements are equivalent:*

(1) *F and K are q-equivalent.*

(2) *$W(F) \cong W(K)$ as rings.*

(3) *$W(F)/I^3F \cong W(K)/I^3K$ as rings.*

Proof. (1) $\Rightarrow$ (2) is Cor. 1.6. For (2) $\Rightarrow$ (3), take any ring isomorphism $\tau\colon W(F) \to W(K)$. Since IF (resp. IK) is the only ideal of index 2 in $W(F)$ (resp. $W(K)$), we must have $\tau(IF) = IK$, and thus also $\tau(I^3F) = I^3K$. Therefore, τ induces a factor ring isomorphism $\overline{\tau}$ from $W(F)/I^3F$ to $W(K)/I^3K$.

Finally, for (3) $\Rightarrow$ (1), assume that there exists a ring isomorphism

$$\overline{\tau}\colon\ W(F)/I^3F \longrightarrow W(K)/I^3K.$$

The same reasoning as before shows that $\overline{\tau}$ induces an isomorphism from I^iF/I^3F to I^iK/I^3K ($i = 1, 2$), and therefore also an isomorphism from IF/I^2F to IK/I^2K. Via the canonical identifications of these groups with the respective square class groups, we see that $\overline{\tau}$ induces a group isomorphism $t\colon\ \dot{F}/\dot{F}^2 \to \dot{K}/\dot{K}^2$. We are done if we can show that this is a *q-equivalence* from F to K. First, for any $a \in \dot{F}$, we have (by definition)

$$\overline{\tau}(\langle 1, -a\rangle + I^2F/I^3F) = \langle 1, -t(a)\rangle + I^2K/I^3K. \tag{1.9}$$

On the other hand, $\overline{\tau}(\langle 1, 1\rangle + I^2F/I^3F) = \langle 1, 1\rangle + I^2K/I^3K$. Comparing this with 1.9 for $a = -1$, we see that

$$\langle 1,\ -t(-1)\rangle \equiv \langle 1,\ 1\rangle \pmod{I^2K},$$

which implies that $t(-1) = -1$, as in 1.1(1). To verify 1.1(2), let us work with its equivalent form (2b). Suppose $1 \in D_F\langle c,\ a\rangle$, where $a, c \in \dot{F}$. Then the quaternion algebra $\left(\frac{c,\,a}{F}\right)$ splits, so $\langle 1,\ -c\rangle\,\langle 1,\ -a\rangle = 0 \in I^3F$. Applying the ring isomorphism $\overline{\tau}$, we have

$$\overline{\tau}(\langle 1,\ -c\rangle + I^3F)\cdot\overline{\tau}(\langle 1,\ -a\rangle + I^3F) = 0.$$

Now $\overline{\tau}(\langle 1,\ -c\rangle + I^3F)$ has the form $\langle 1,\ -t(c)\rangle + \alpha + I^3K$, where $\alpha \in I^2K$, and $\overline{\tau}(\langle 1,\ -a\rangle + I^3K)$ has the form $\langle 1,\ -t(a)\rangle + \beta + I^3K$, where $\beta \in I^2K$. Therefore, we have

$$(\langle 1,\ -t(c)\rangle + \alpha)\,(\langle 1,\ -t(a)\rangle + \beta) \in I^3K,$$

and hence $\langle 1,\ -t(c)\rangle\,\langle 1,\ -t(a)\rangle \in I^3K$. This implies that $\left(\frac{t(c),\,t(a)}{K}\right)$ splits, so we have $1 \in D_K\langle t(c),\ t(a)\rangle$. This proves the "only if" part of (2b), and the "if" part can be proved similarly. $\square$

While the proof of 1.8 seems to consist of a sequence of largely routine steps, there are actually some subtleties behind the theorem. The implication (2) $\Longrightarrow$ (1) shows that the existence of a ring isomorphism $\tau\colon W(F) \to W(K)$ leads to a q-equivalence between F and K, which in turn leads to a ring isomorphism $W(F) \to W(K)$ taking the group (of unary forms) $\dot{F}/\dot{F}^2$ to the group $\dot{K}/\dot{K}^2$. However, all of this *does not imply* that the original isomorphism τ satisfies $\tau(\dot{F}/\dot{F}^2) = \dot{K}/\dot{K}^2$! To illustrate this point, we shall work in the special case $K = F$, and provide an example of a Witt ring automorphism $\tau\colon W(F) \to W(F)$ which need not take all unary forms to unary forms.

Example 1.10. Let $k = \mathbb{F}_q$, where $q \equiv 1 \pmod 4$, and work with the Laurent series field $F = k((x))$. By the results of VI.1, $W(F)$ can be identified with the group ring $W(k)[G]$, where G is the group $\{\langle 1\rangle, \langle x\rangle\}$. Fix a nonsquare $a \in k$, and consider the ternary form $\varphi = \langle 1, a, x\rangle \in W(F)$. Recalling that $W(k) \cong \mathbb{Z}_2 \oplus \mathbb{Z}_2$, we have

$$\varphi^2 = \langle 1,1,1\rangle + 2\,\langle a,\, x,\, ax\rangle = \langle 1\rangle \in W(F).$$

Now, $W(k)\cdot\varphi$ is the additive group

$$\{0,\,\langle 1\rangle,\,\langle a\rangle,\,\langle 1,\,a\rangle\}\cdot\varphi = \{0,\,\langle 1,a,x\rangle,\,\langle 1,a,ax\rangle,\,\langle x,\,ax\rangle\},$$

which has zero intersection with $W(k)$. Since $|W(F)| = 16$, it follows that

$$W(F) = W(k) \oplus W(k)\,\varphi = W(k)\,[H], \quad \text{where } H = \{\langle 1\rangle,\,\varphi\}.$$

Thus, we get a ring automorphism τ of $W(F)$ that is the identity on $W(k)$, and takes $\langle x\rangle$ to φ. Here, φ is an *anisotropic* ternary form, so it cannot be represented by a unary form over F.

The automorphism τ turns out to have order 2, since

$$\tau(\varphi) = \tau\langle 1,\,a,\,x\rangle = \langle 1,\,a\rangle + \langle 1,\,a,\,x\rangle = \langle x\rangle.$$

An easy computation shows that τ has the fixed ring

$$\{0,\,\langle 1\rangle,\,\langle a\rangle,\,\langle 1,a\rangle,\,\langle x,ax\rangle,\,\langle 1,x,ax\rangle,\,\langle a,x,ax\rangle,\,\langle 1,a,x,ax\rangle\},$$

and that τ permutes the following elements in paris:

$$\{\langle 1,x\rangle,\,\langle a,x\rangle\},\quad \{\langle 1,ax\rangle,\,\langle a,ax\rangle\},\quad \{\langle x\rangle,\,\langle 1,a,x\rangle\},\quad \{\langle ax\rangle,\,\langle 1,a,ax\rangle\}.$$

Finally, we note that, in this example, we could also have replaced φ by any of the other three anisotropic ternary forms

$$\langle a\rangle\varphi = \langle 1,a,ax\rangle,\quad \langle x\rangle\varphi = \langle 1,x,ax\rangle,\quad \text{or}\quad \langle ax\rangle\,\varphi = \langle a,x,ax\rangle$$

over the field $k((x))$.

As a concluding thought for this section, we should point out that, for two fields F and K, $W(F) \cong W(K)$ as groups need not imply that $W(F) \cong W(K)$ as rings. For instance, for the field F of 5-adic numbers, $W(F) \cong \mathbb{Z}_2^4$. A nonreal field K of level 1 with u-invariant 2 and $|\dot{K}/\dot{K}^2| = 8$ has the same Witt group, but F and K are *not* q-equivalent since $I^2F \neq 0 = I^2K$.

2. Quadratic Forms of Low Dimension

Quite often, low dimensional quadratic forms have special properties that can be proved by ad hoc computations not generalizable to higher dimensions. We shall give some illustrations of this here by working with forms of low even dimension. We first prove a theorem on the similarity of 4-dimensional forms due to A. Wadsworth [Wad$_1$], and then give some applications of this result to other forms. Recall that two forms σ and τ are said to be *similar* (written $\sigma \sim \tau$) if $\tau \cong t \cdot \sigma$ for some scalar $t \in \dot{F}$.

Wadsworth's Similarity Theorem 2.1. *Let σ, τ be 4-dimensional forms of determinant $d \neq 0$ over F, and let $K = F(\sqrt{d})$. Then $\sigma \sim \tau$ over F iff $\sigma \sim \tau$ over K.*

Proof. We need only prove the "if" part. Assume that $\sigma \sim \tau$ over K, and $d \notin \dot{F}^2$. (If $d \in \dot{F}^2$, there is nothing to prove.) After a scaling, we may also assume that σ and τ both represent 1; say

$$\sigma \cong \langle 1 \rangle \perp \sigma', \quad \tau \cong \langle 1 \rangle \perp \tau'.$$

Over K, σ and τ become (2-fold) Pfister forms, so $\sigma_K \sim \tau_K$ amounts to $\sigma_K \cong \tau_K$. Therefore, $q := \sigma' \perp \langle -1 \rangle \tau'$ becomes hyperbolic over K. Since

$$d(q) = -1 \neq -d \in \dot{F}/\dot{F}^2,$$

VII.3.3 implies that q is isotropic over F. By I.3.6, there exist $a, b, c \in \dot{F}$ such that

$$\sigma' \cong \langle a,\ b,\ abd \rangle \quad \text{and} \quad \tau' \cong \langle a,\ c,\ acd \rangle.$$

Consider the F-form $\varphi = \langle 1, a \rangle \perp \langle -1 \rangle\, bc\, \langle 1, ad \rangle$. In $W(K)$, we have $\langle a \rangle = \langle ad \rangle$, and so

$$b \cdot \varphi_K = b\,\langle 1,\ ad \rangle - c\,\langle 1,\ ad \rangle = \sigma' - \tau' = 0 \implies \varphi_K = 0.$$

Since $d(\varphi) = d \notin \dot{F}^2$, VII.3.3 (applied one more time) implies that φ_F is isotropic. Thus, there exists $t \in \dot{F}$ such that

$$\langle 1,\ a \rangle \cong \langle t,\ at \rangle \quad \text{and} \quad bc\,\langle 1,\ ad \rangle \cong \langle t,\ adt \rangle.$$

From the latter, we have $tb\,\langle 1,\ ad \rangle \cong c\,\langle 1,\ ad \rangle$, and so

$$t \cdot \sigma \cong t\,\langle 1,\ a \rangle \perp tb\,\langle 1,\ ad \rangle \cong \langle 1,\ a \rangle \perp c\,\langle 1,\ ad \rangle = \tau. \qquad \square$$

The first application of 2.1 is to the study of function fields of quadratic forms (in the sense of X.3). Recall that, for a quadratic form σ over F ($\dim\sigma \geq 2$, $\sigma \not\cong \mathbb{H}$), $F[\sigma]$ denotes the ("big") function field of σ; moreover, $F[\sigma]$ is "unchanged" upon scaling σ. In X.4.31, we have proved Witt's classical theorem, to the effect that, for 3-dimensional forms σ, the F-isomorphism type of $F[\sigma]$ determines σ "up to similarity". Wadsworth's Theorem 2.1 enabled him to prove the same result ((1) $\Leftrightarrow$ (2)) for *anisotropic* 4-dimensional forms σ [Wad_1]. Here, we'll prove a little more, as follows.

Theorem 2.2. *Let σ, τ be 4-dimensional forms over F, with σ anisotropic. The following statements are equivalent:*

(1) $\sigma \sim \tau$.

(2) $F[\sigma] \cong F[\tau]$ *over* F.

(3) $\sigma > \tau > \sigma$.[(2)]

Proof. (1) $\Rightarrow$ (2) $\Rightarrow$ (3) are clear, so we need only verify (3) $\Rightarrow$ (1). The proof here is a slight refinement of that given earlier for the third case of X.4.31. Assume (3). Then τ must be anisotropic over F, for otherwise X.4.1 would imply that $F[\tau]$ is purely transcendental, in which case σ would be isotropic over F by IX.1.1. After a scaling, we may assume that $1 \in D_F(\sigma) \cap D_F(\tau)$. Let $d = d(\sigma)$, and $e = d(\tau)$. If, say, $d \in \dot{F}^2$, then σ is a (2-fold anisotropic) Pfister form over F, and X.4.10 implies that $\sigma \sim \tau$. Thus, we may assume that $d, e \notin \dot{F}^2$. Let $K = F(\sqrt{d})$. *Then σ_K must remain anisotropic.* [For, if otherwise, then σ_K (being a Pfister form) is hyperbolic, and VII.3.3 yields $d(\sigma) \in \dot{F}^2$, a contradiction.] Therefore, applying X.4.10 over K, we have $\sigma_K \sim \tau_K$. This implies that

$$e = d(\tau_K) = d(\sigma_K) \in \dot{K}^2.$$

Since $e \notin \dot{F}^2$, we see from VII.3.8 that $e \in d\dot{F}^2$. Thus, up to square factors, σ and τ *both* have determinant d, so now $\sigma_K \sim \tau_K \implies \sigma \sim \tau$ by Theorem 2.1. □

Remark 2.3. (A) Note that, in contrast to the case of three-dimensional forms in X.4.31, it is essential to assume here that the 4-dimensional form σ is *anisotropic*. Without this assumption, 2.2 need not hold. For instance, if $a\dot{F}^2 \neq b\dot{F}^2$, then $\sigma = \langle 1, 1, -1, a\rangle$ is *not* similar to $\tau = \langle 1, 1, -1, b\rangle$, but $F[\sigma] \cong F[\tau]$ since both fields are purely transcendental over F.

(B) Theorem 2.2 does admit a generalization to the case of *certain* 2^n-dimensional forms that is due to D. Hoffmann. In this generalization, the 2^n-dimensional anisotropic form σ is assumed to be of the shape $\sigma_1 \cdot \sigma_2$,

---

(2) Recall that this means σ is isotropic over $F[\tau]$ and τ is isotropic over $F[\sigma]$. It also means that the function fields $F[\varphi]$ and $F[\tau]$ are stably isomorphic, according to X.4.25.

where $\dim \sigma_1 = 4$, and σ_2 is an $(n-2)$-fold Pfister form. Of course, the isomorphism theorem in this case subsumes *both* 2.2 and X.4.10; for the details, we refer the reader to [Ho$_5$].

(C) Th. 2.2, as stated, *does not* extend to general anisotropic forms of higher dimensions. In fact, in the Appendix to this section, we shall construct examples of 5-dimensional anisotropic forms σ and τ such that $F[\sigma] \cong F[\tau]$ (over F), but $\sigma \not\sim \tau$. This will show that (2) $\Rightarrow$ (1) and (3) $\Rightarrow$ (1) do not hold for forms of dimension > 4.

(D) In view of (C), it may be said that, for higher dimensional forms, the conditions (2) and (3) are substantially weaker than (1). Of course, (2) $\Rightarrow$ (3) still holds, but it does not seem to be known whether (3) $\Rightarrow$ (2), for forms σ and τ of dimension > 4. This is a significant problem in the theory of function fields of quadratic forms; see the Open Question 6.10 in XIII.6.

For a quadratic form σ over F, the *Witt invariant* $c(\sigma)$ (in the Brauer group $B(F)$) is defined in V.3. Recall that "c" is well-defined on $W(F)$, and that, on I^2F, c is the unique homomorphism to $B(F)$ sending $\langle\langle -a, -b\rangle\rangle$ to the Brauer class of $\left(\frac{a,b}{F}\right)$. The following result is another consequence of Theorem 2.1 on the similarity of 4-dimensional forms.

Proposition 2.4. *Let σ, τ be 4-dimensional forms over F. If either*

(1) *$d(\sigma) = d(\tau)$ and $c(\sigma) = c(\tau)$, or*

(2) *$\sigma \equiv \tau \pmod{I^3F}$,*

then $\sigma \sim \tau$ over F.

Proof. First assume (1), and let $d = d(\sigma) = d(\tau)$. Applying 2.1, we may replace F by $F(\sqrt{d})$ to assume that $d = 1$. Thus, $\sigma \cong u\,\langle\langle w, x\rangle\rangle$ and $\tau \cong v\,\langle\langle y, z\rangle\rangle$ (for suitable $u, v, \ldots \in \dot{F}$). By V.3.3 and V.3.16,

$$c(\sigma) = \left(\frac{-w, -x}{F}\right) \quad \text{and} \quad c(\tau) = \left(\frac{-y, -z}{F}\right).$$

If these are equal (in $B(F)$), we have

$$\sigma \sim \langle\langle w,\, x\rangle\rangle \cong \langle\langle y,\, z\rangle\rangle \sim \tau.$$

Next, we assume (2). This congruence clearly implies that $d(\sigma) = d(\tau)$. Thus, as in (1), we may assume that $d(\sigma) = d(\tau) = 1$. Then the congruence (2) implies that $c(\sigma) = c(\tau)$, so we are back to the case (1). (Alternatively, we could also have gotten the desired conclusion $\sigma \sim \tau$ from X.5.4.) $\square$

Not surprisingly, Wadsworth's Theorem has applications also to forms of dimension 6. Before coming to these applications, however, we'll need some basic information on low-dimensional forms in I^2F that is independent of Theorem 2.1. The next few preparatory results are due to Pfister.

Proposition 2.5. *Let q be a 6-dimensional form in I^2F.*

(1) *If $c(q) = 1$, then q is hyperbolic.*

(2) *If $c(q) = \left(\frac{u,v}{F}\right)$ for some $u, v \in \dot{F}$, then q is isotropic.*

Proof. (1) Since $d(q) = -1$, we may assume, after a scaling, that

$$q \cong \langle w,\, x,\, wx\rangle \perp \langle -y,\, -z,\, -yz\rangle \quad (w, x, y, z \in \dot{F}).$$

(Recall, from V.3.16, that $c(a \cdot q) = c(q)$.) Therefore,

$$q = \langle\langle w,\, x\rangle\rangle - \langle\langle y,\, z\rangle\rangle \in I^2F. \tag{2.6}$$

Taking Witt invariants, we have

$$1 = c(q) = \left(\frac{-w,\, -x}{F}\right)\left(\frac{-y,\, -z}{F}\right) \in B(F).$$

Thus,

$$\left(\frac{-w,\, -x}{F}\right) \cong \left(\frac{-y,\, -z}{F}\right),$$

and 2.6 shows that $q = 0 \in W(F)$.

(2) We may assume that $-uv \notin \dot{F}^2$ (for otherwise $c(q) = 1$ already, and (1) applies). Let $K = F(\sqrt{-uv}\,)$. Assume q is anisotropic. Since $c_K(q_K) = \left(\frac{u,v}{K}\right) = 1$, q_K is hyperbolic by (1). It follows from VII.3.3 that $-1 = d(q) = uv$ in $\dot{F}/\dot{F}^2$, which is a contradiction. Thus, q must be isotropic. □

It is worth noting that 2.5 can be used to give a new proof of Albert's Theorem III.4.8 on 6-dimensional forms associated with biquaternion algebras. There are four equivalent statements in III.4.8, but the hardest part of the proof was to show that (4) ⇒ (1) in the notation there. In the following, we will restate this implication, and supply a second proof for it, making full use of the techniques of quadratic forms.

Corollary 2.7. *Let B, C be quaternion division algebras over F that do not have a common quadratic splitting field. Then $A = B \otimes_F C$ is a division algebra.*

Proof. If A is not a division algebra, then Wedderburn's Theorem implies that $A \cong \mathbb{M}_2(D)$, where D is a central simple F-algebra, necessarily of dimension 4. By III.5.1, $D \cong \left(\frac{u,v}{F}\right)$ for some $u, v \in \dot{F}$. Since the Albert form $q = q_B \perp \langle -1\rangle q_C$ in III.4.7 is a difference of the norm forms of B and C in $W(F)$, we have

$$c(q) = [B \otimes_F C] = [D] = \left(\frac{u, v}{F}\right) \in B(F).$$

Therefore, by 2.5(2), q is isotropic. This means that q_B and q_C represent a common value $a \in \dot{F}$. But then $F(\sqrt{a}\,)$ splits both B and C, which is a contradiction. □

We shall need one more isotropicity result, this time on 10-dimensional forms!

Proposition 2.8. *Let φ be a 10-dimensional form in I^2F. If $c(\varphi) = 1$, then φ is isotropic.*

Proof. Let $\varphi \cong a\langle 1, b, c, d\rangle \perp \psi$, where $\dim(\psi) = 6$ and $d(\psi) = -bcd$. If $bcd \in \dot{F}^2$, then $\psi \in I^2F$, and

$$c(\varphi) = 1 \Longrightarrow c(\psi) = \left(\frac{-b, -c}{F}\right),$$

so ψ is already isotropic by 2.7(2). Assume now $bcd \notin \dot{F}^2$, and let $K = F(\sqrt{bcd}\,)$. The above argument then implies that ψ_K is isotropic. If ψ is anisotropic over F (which of course we may assume), then ψ contains a subform $e\,\langle 1, -bcd\rangle$ (by VII.3.1), with an orthogonal complement σ with $\dim(\sigma) = 4$ and $d(\sigma) = 1$. Up to a scalar multiple, σ is a 2-fold Pfister form, so its orthogonal complement σ' in φ has $\dim(\sigma') = 6$, $\sigma' \in I^2F$, and $c(\sigma')$ is a quaternion algebra (the one having norm form σ). By 2.5(2) again, σ' is isotropic, so we are done. □

The proposition above is a very well-known result proved by Pfister in [Pf3]. It implies, in particular, that *no anisotropic 10-dimensional quadratic forms can lie in I^3F.* As we have already pointed out in X.5, this fact amounts to the first sighting (for $n = 3$) of the "second gap" phenomenon in the study of the dimensions of anisotropic forms in I^nF, for a general field F of characteristic not 2.

With the above preparatory results, we now return to prove a partial analogue of 2.4 for 6-dimensional forms.

Theorem 2.9. *For any two 6-dimensional forms φ and ψ belonging to I^2F, the following are equivalent:*

(1) $\varphi \sim \psi$.

(2) $\varphi \equiv \psi \pmod{I^3F}$.

(3) $c(\varphi) = c(\psi)$.

Proof. We follow here here the proof given by Mammone and Shapiro in [MS] (with a slight simplification).

(1) $\Rightarrow$ (2). If, say, $\psi \cong a \cdot \varphi$ $(a \in \dot{F})$, then $\varphi \equiv a \cdot \varphi \equiv \psi \pmod{I^3F}$.

(2) $\Rightarrow$ (3). This follows by computing the Witt invariants.

(3) $\Rightarrow$ (1). Assume (3). After scaling φ and ψ (independently), we may write $\varphi \cong \langle 1\rangle \perp \varphi_1$ and $\psi \cong \langle 1\rangle \perp \psi_1$. Here, $\varphi_1 \perp -\psi_1 \in I^2F$ has dimension 10, with

$$c(\varphi_1 \perp -\psi_1) = c(\varphi \perp -\psi) = c(\varphi)\, c(\psi) = 1, \tag{2.10}$$

so $\varphi_1 \perp -\psi_1$ is isotropic by 2.8. Thus, we can write $\varphi_1 \cong \langle a\rangle \perp \sigma$ and $\psi_1 \cong \langle a\rangle \perp \tau$ for some $a \in \dot{F}$. Now

$$\varphi \cong \langle 1, a\rangle \perp \sigma \quad \text{and} \quad \psi \cong \langle 1, a\rangle \perp \tau, \tag{2.11}$$

with σ, τ 4-dimensional with a common determinant $-a$, and

$$c(\sigma \perp -\tau) = c(\varphi \perp -\psi) = 1$$

as in 2.10. Therefore, $c(\sigma) = c(\tau)$, and 2.4 yields $\tau \cong b \cdot \sigma$ for some $b \in \dot{F}$. Since $d(\sigma) = -a$, V.3.16 gives

$$c(\tau) = c(b \cdot \sigma) = c(\sigma)\left(\frac{b,\ -a}{F}\right),$$

and hence $\left(\frac{b, -a}{F}\right) = 1$. This means that $b\langle 1,\, a\rangle \cong \langle 1,\, a\rangle$, so 2.11 shows immediately that $b \cdot \varphi \cong \psi$, as desired. □

The point about 2.9 is that it is basically a quadratic form theoretic version of a well-known theorem of Jacobson on biquaternion algebras. Jacobson's original proof of this theorem (in [Ja]) used the theory of Jordan norms on simple algebras with involution. The proof to be given below (from [MS]) is completely within the context of quadratic forms.

Jacobson's Theorem 2.12. *Let $A = B \otimes C$ and $A' = B' \otimes C'$, where B, B', C, C' are F-quaternion algebras (and $\otimes$ means $\otimes_F$). Let q (resp. q') be the Albert form for the pair $\{B,\, C\}$ (resp. $\{B',\, C'\}$), as defined in III.4.7. Then $A \cong A'$ iff $q \sim q'$.*

Proof. If $A \cong A'$, then

$$c(q) = [B] \cdot [C] = [A] = [A'] = [B'] \cdot [C'] = c(q'),$$

so 2.9 gives $q \sim q'$. The converse is similar, since $[A] = [A'] \in B(F)$ is no different from $A \cong A'$. □

One consequence of Jacobson's Theorem is that the Albert form q for the pair $\{B,\, C\}$ depends (up to a similarity, of course) only on the isomorphism type of the biquaternion algebra $A = B \otimes C$, and not on the choice of such a tensor product representation of A as a biquaternion algebra. To drive this point home, we make the following explicit statement.

Corollary 2.13. *Every biquaternion algebra A gives rise to a 6-dimensional Albert form $q \in I^2F$ that is defined up to a similarity. Moreover, the F-isomorphism type of A is uniquely determined by the similarity class of its Albert form q, which can be any 6-dimensional form in I^2F.*

We conclude this section with a few relevant observations.

Remarks 2.14. (1) According to a theorem of Albert, biquaternion algebras are precisely the 16-dimensional central simple algebras that have exponent ≤ 2 in the Brauer group.

(2) For readers familiar with the definition of the Schur index in the theory of algebras, it is easy to verify (from the results of III.4) the following correlation between Schur indices and Witt indices. For a biquaternion algebra A with Albert form q, the Schur index of A is 4, 2, or 1 according as the Witt index of q is 0, 1, or 3.

(3) Jacobson's Theorem has a complete analogue for biquaternion algebras in characteristic 2. This case was not completely covered in Jacobson's original paper [Ja], as the presentation there contained a small error. A corrected formulation and proof of 2.12 in characteristic 2, partly based on results of C. H. Sah, can be found in [MS].

Appendix: Forms with Isomorphic Function Fields

In this Appendix, we will show that, for anisotropic F-forms σ and τ of the same dimension, it is possible for the function fields $F(\sigma)$ and $F(\tau)$ to be isomorphic over F (in particular $F[\sigma] \cong F[\tau]$) without the forms σ and τ being similar to each other. As we have pointed out before in Remark 2.3(C), such examples will show that, in Theorem 2.2, the conditions (2) and (3) are not equivalent to condition (1) in general. From 2.2, we knew that the least dimension for which we can hope to construct such examples would be 5.

The construction to be presented below is based on a general theorem of H. Ahmad and J. Ohm in their paper [AO], which states that *any pair of "special" Pfister neighbors* (in the sense of X.4.18(4)) *of the same dimension and having the same associated Pfister form have isomorphic function fields.* Luckily, since all we wanted is just an example of two nonsimilar anisotropic forms with the same function fields, a special case of the Ahmad-Ohm theorem for 5-dimensional Pfister neighbors will be sufficient. (In this connection, we should recall Knebusch's result X.4.19 that 5-dimensional Pfister neighbors are always special.)

One of the main ideas for the results of Ahmad and Ohm in [AO] is the following "transposition theorem", which shows that certain "symmetrical pairs" of Pfister neighbors of dimension $2^{n+1}+1$ constructed from a single n-fold Pfister form φ will have isomorphic function fields. Note that the two forms σ and τ constructed below are both "special" Pfister neighbors (in the sense of X.4.18(4)), associated with the same $(n+2)$-fold Pfister form $\varphi\,\langle\langle b,\,c\rangle\rangle$.

Theorem 2.15. *Let φ be an n-fold Pfister form over F, and let*

$$\sigma = \varphi\,\langle\langle b\rangle\rangle \perp \langle c\rangle \quad \text{and} \quad \tau = \varphi\,\langle\langle c\rangle\rangle \perp \langle b\rangle,$$

where $b,\,c \in \dot{F}$. Then $F(\sigma) \cong F(\tau)$ (over F). If $n \geq 1$, then $\sigma \sim \tau$ iff $b\,\varphi \cong c\,\varphi$ (iff $bc \in D_F(\varphi)$).

Proof. The function field $F(\tau)$ has the form $F(x,y)$, where x, y are vector variables (of length 2^n) satisfying the relation

$$\varphi(x) + c\,\varphi(y) + b = 0. \tag{2.16}$$

Consider the field $F(y)$, over which we have an isometry $\varphi \cong \varphi(y)\cdot\varphi$ (since φ remains a Pfister form over $F(y)$). As in the proof of X.2.11, this means that there exists an invertible matrix U over $F(y)$ such that $\varphi(x) = \varphi(y)\cdot\varphi(x\,U)$. For the vector

$$x' := x\,U \in F(y,x)^{2^n} = F(\tau)^{2^n},$$

we have then $F(\tau) = F(y,x) = F(y,x')$ and $\varphi(x') = \varphi(x)/\varphi(y)$. Dividing (2.16) by $\varphi(y)$ yields

$$\varphi(x') + c + b/\varphi(y) = 0. \tag{2.17}$$

Let $y' := y/\varphi(y) \in F(y)$. Then

$$\varphi(y') = \varphi(y)/\varphi(y)^2 = 1/\varphi(y),$$

so (2.17) may be rewritten as

$$\varphi(x') + b\,\varphi(y') + c = 0. \tag{2.18}$$

Now from $y = y'\varphi(y) = y'/\varphi(y')$, we see that $F(y) = F(y')$, so we have

$$F(\tau) = F(y,\,x) = F(y,\,x') = F(y',\,x'). \tag{2.19}$$

In view of (2.18), there exists an F-algebra epimorphism

$$f\colon\ F[X',\,Y']/(\varphi(X') + b\,\varphi(Y') + c) \to F[y',\,x'],$$

with $f(X') = x'$ and $f(Y') = y'$. The LHS is an affine domain of transcendence degree $2^{n+1}-1$ (over F), and by (2.19), this is also the transcendence degree of the affine domain on the RHS. Thus, f must be a monomorphism, and hence an isomorphism. Therefore, f extends to a field isomorphism $F(\sigma) \cong F(\tau)$.

For the last statement in 2.15, note that, if $b\varphi \cong c\varphi$, then

$$b\sigma \cong b(\varphi \perp b\varphi) \perp \langle bc\rangle \cong c(\varphi \perp c\varphi) \perp \langle bc\rangle \cong c\tau,$$

and hence $\sigma \sim \tau$. Conversely, if $\sigma \sim \tau$, then $a\sigma \cong \tau$ for some $a \in \dot{F}$. Taking determinants (assuming that $n \geq 1$), we see that $ab = c \in \dot{F}/\dot{F}^2$. Thus, we have $b\sigma \cong c\tau$, and a reversal of the previous argument (together with the cancellation of $\varphi \perp \langle bc\rangle$) shows that $b\varphi \cong c\varphi$. □

Upon choosing φ above to be the 1-fold Pfister form $\langle\langle a\rangle\rangle$, we see that the 5-dimensional Pfister neighbors

$$(2.20) \qquad \sigma = \langle\langle a,\, b\rangle\rangle \perp \langle c\rangle \quad \text{and} \quad \tau = \langle\langle a,\, c\rangle\rangle \perp \langle b\rangle$$

always have isomorphic function fields. (These Pfister neighbors are both associated with the Pfister form $\langle\langle a, b, c\rangle\rangle$.) To accomplish our goal, it suffices, therefore, to choose the elements a, b, c (in a suitable field F) such that σ and τ are both anisotropic and $bc \notin D_F\, \langle\langle a\rangle\rangle$, for then $\sigma \nsim \tau$ over F (by the last part of Th. 2.15). This is easy. For instance, we can take the elements a, b, c to be independent indeterminates, and let $F = \mathbb{R}(a, b, c)$. With these choices, $\langle\langle a, b, c\rangle\rangle$ is an anisotropic 3-fold Pfister form, and the desired conditions on σ, τ and $\langle\langle a\rangle\rangle$ are clearly all satisfied.

In conclusion, we should perhaps mention one more thing. As was pointed out by Ahmad and Ohm in [AO], the use of 5-dimensional Pfister neighbors in the above construction was, in retrospect, quite fortuitous. In fact, Hoffmann has proved that, if σ and τ are anisotropic 5-dimensional forms one of which is *not* a Pfister neighbor, then $F(\sigma) \cong F(\tau)$ implies that $\sigma \sim \tau$; see [Ho_2].

3. Some Classification Theorems

A major problem in the algebraic theory of quadratic forms over fields is to find enough invariants with which to determine the isometry class of a form. This "classification problem" has remained largely untractable, since only a few types of invariants for quadratic forms have been found so far. The best known classical invariants are: dimension, determinant, the Hasse (or Witt) invariant, and the total signature. There are some other invariants too, such as the higher Stiefel-Whitney classes, which take values in certain cohomology groups. However, since we did not develop any cohomology theory in this text, it will not be possible for us to discuss such quadratic form invariants of a cohomological nature.

Leaving the (higher) cohomological invariants aside, it is still of interest to find characterizations of those fields F for which the *classical invariants* enumerated in the last paragraph would suffice to classify quadratic forms. Some such characterizations in terms of the Witt ring of F have been found in joint work [EL$_4$] by Elman and the author. In this section, we shall give an exposition on the main classification theorem obtained in that paper, which is the following simple statement.

Theorem 3.1. *Quadratic forms over a field F are classified by their dimension, determinant, Hasse invariant, and total signature iff I^3F is torsionfree.*

Note that in the case where F is *nonreal*, the Witt group $W(F)$ is torsion and there are no orderings with which to define signatures. Accordingly, 3.1 simplifies to the following statement: *quadratic forms over a nonreal field F are classified by their dimension, determinant, and Hasse invariant iff $I^3F = 0$.* We also note, of course, that in the case of (real or nonreal) local and global fields, the "if" part of Theorem 3.1 recaptures the classical classification theorem for quadratic forms over such fields, as obtained in Chapter VI; see VI.2.12 and VI.3.3.

There are some "more elementary" versions of Theorem 3.1 too, which are perhaps best thought of as motivations for that theorem. First of all, *quadratic forms over F are classified by their dimension and total signature iff IF is torsionfree (iff F is pythagorean).* This is essentially Pfister's Local-Global Principle VIII.3.2 in the pythagorean case. Beyond this, the next case is: *quadratic forms are classified by their dimension, determinant, and total signature iff I^2F is torsionfree.* (The proof of this would be a small subset of the proof we shall give for 3.1.) In the context of *nonreal* fields, the above two cases correspond to the cases of quadratically closed fields and fields of u-invariant ≤ 2, respectively.

Before we embark upon the proof of Theorem 3.1, some general remarks are in order. If one assumes (the injectivity part of) Merkurjev's Theorem stated in V.6.11, the proof of 3.1 would become considerably easier. However, Merkurjev's Theorem is a deep result, and Theorem 3.1 was first discovered and proved independently of it. Thus, it would still be appropriate (as well as instructive) to give a proof of 3.1 *without* assuming Merkurjev's Theorem. Such a proof would have to be constructed carefully, since we would not have the benefit of knowing that the Clifford invariant homomorphism $\gamma\colon I^2F \to B(F)$ in V.3 has kernel exactly equal to I^3F.

The proof of Theorem 3.1 to be presented below is a streamlined version of that given in the original source [EL$_4$] (and in [Sc$_4$]). This proof should constitute a good review of earlier material, since many of the results and techniques developed in the previous chapters (for instance, on Hasse invariants, quadratic extensions, transfers, Pfister forms, etc.) will be used freely in carrying out this proof. Before plunging into the details, let us first quickly mention some of the main facts and key ideas that are to be used.

(A) The structure of the torsion subgroup of I^2F, as given in XI.4.2.

(B) The Transfer Principle for Pfister forms under a quadratic field extension K/F (proved in XI.4.13), along with an I^n-exact sequence ($n \leq 3$) for K/F (obtained in VII.3.15).

(C) The property (A_n): "torsion n-fold Pfister forms are hyperbolic" (introduced in XI.4), and the Going-Up Theorem XI.4.14 for this property with respect to quadratic extensions of the type $K = F(\sqrt{w})$ where $w \in \dot{\sigma}(F)$ (nonzero sums of squares in F). This will be needed specifically for $n = 3$ — and for this case only.

(D) The key idea is then to apply (C) together with an induction *over all fields* to prove the "iff" statement in Theorem 3.1.

Let us now begin the proof of Theorem 3.1. We shall actually prove a little more than is stated in 3.1, in that we will show the equivalence of the following three conditions:

(1) *F has the property* (A_3).

(2) *I^3F is torsionfree.*

(3) *Quadratic forms over F are classified by their dimension, determinant, Hasse invariant, and total signature.*

(3) $\Rightarrow$ (1). Compare a given torsion 3-fold Pfister form φ with the form $\langle\langle 1, 1, -1\rangle\rangle$. Both forms have dimension 8, determinant 1, trivial Hasse invariant, and zero total signature. Therefore, (3) implies that $\varphi \cong \langle\langle 1, 1, -1\rangle\rangle$ is hyperbolic.

(1) $\Rightarrow$ (2). Suppose F satisfies (A_3), but there exists a nonzero anisotropic form $\sigma \in I^3F$ with $2\sigma = 0$. We may assume F to have been chosen such that $\dim \sigma = 2n$ is as small as possible. By Ch. II, Exercise 19 (or equivalently, XI.3.7),

$$\sigma \cong \perp_{i=1}^{n} \langle a_i\rangle \langle\langle -w_i\rangle\rangle,$$

where $a_i \in \dot{F}$, and each w_i is a nonzero sum of (two) squares in F. Let $K = F(\sqrt{w_1})$. This field also satisfies (A_3), by XI.4.14. Since the anisotropic part of σ over K has dimension $< 2n$, we have $\sigma_K = 0$ by the choice of n. Now by VII.3.15(2) and (A_3), we have $\sigma \in \langle\langle -w_1\rangle\rangle I^2F = 0$, a contradiction.[(3)]

The final implication (2) $\Rightarrow$ (3) needed to complete the proof of 3.1 will be deduced from an alternative result concerning the Clifford invariant Γ on the Witt ring, as follows.

[(3)] As was pointed out in XI.4.4, Arason and Elman have proved that, *for any n*, (A_n) is equivalent to the torsion-freeness of I^nF. We do not want to invoke this fact here since it depends on the very deep results of Orlov, Vishik and Voevodsky on the truth of the Milnor Conjectures.

Theorem 3.2. *Suppose I^3F is torsionfree, and q is a torsion quadratic form (i.e., $q \in W_t(F)$). Then $\Gamma(q) = 1 \Rightarrow q$ is hyperbolic.*

Note that, if we assume (the injectivity part of) Merkurjev's Theorem V.6.11, then $\Gamma(q) = 1$ implies that $q \in I^3F$, and hence $q \in I^3F \cap W_t(F) = 0$. However, 3.2 *will be proved below without invoking Merkurjev's Theorem.* In some sense, 3.2 itself may be thought of as giving "a little piece of the action" for Merkurjev's deeper result.

Before proving 3.2, let us first complete the proof of Theorem 3.1 by checking the pending implication (2) $\Rightarrow$ (3) above. Let q_1, q_2 be quadratic forms of dimension n, with the same determinant, Hasse invariant, and total signature. Then $q := q_1 \perp \langle -1 \rangle q_2$ has total signature zero, so by Pfister's Local-Global Principle VIII.3.2, $q \in W_t(F)$. Also, since the Clifford invariant of a form is completely determined by its dimension, determinant, and Hasse invariant (see V.3), we have $\Gamma(q_1) = \Gamma(q_2)$. Since Γ is a homomorphism on $W(F)$ (to the Brauer-Wall group $BW(F)$), we see that $\Gamma(q) = \Gamma(q_1)\Gamma(q_2)^{-1} = 1$. If I^3F is torsionfree, it follows from 3.2 that $q = 0 \in W(F)$, and hence $q_1 \cong q_2$, as desired!

The last piece of the puzzle is now:

Proof of 3.2. Under the hypothesis of that theorem, assume that $\Gamma(q) = 1$, but q is non-hyperbolic. Since $\ker(\Gamma) \subseteq I^2F$, we have $q \in I^2F \cap W_t(F)$, so by XI.4.2, we can write

$$q = \sum_{i=1}^{m} \langle\langle a_i, -w_i \rangle\rangle \in W(F), \quad \text{where } a_i \in \dot{F} \text{ and } w_i \in \dot{\sigma}(F).$$

We may assume F and q to have been chosen such that m is as small as possible. Let $K = F(\sqrt{w_1})$. Since I^3F is torsionfree, F clearly satisfies (A_3), and hence K also does, by XI.4.14. Therefore, I^3K is torsionfree, by the (already proved) implication (1) $\Rightarrow$ (2). But

$$q = \sum_{i=1}^{m-1} \langle\langle a_i, -w_i \rangle\rangle \in W(K),$$

so q must become hyperbolic over K (by the minimal choice of m). Applying VII.3.15(1), we can write

$$q = \langle\langle -w_1 \rangle\rangle \langle b_1, \ldots, b_{2r} \rangle \in W(F),$$

for suitable $b_i \in \dot{F}$. Thus,

$$\begin{aligned} q &= \langle b_1 \rangle \langle\langle -w_1, b_1 b_{r+1} \rangle\rangle + \cdots + \langle b_r \rangle \langle\langle -w_1, b_r b_{2r} \rangle\rangle \\ &\equiv \langle\langle -w_1, b_1 b_{r+1} \rangle\rangle + \cdots + \langle\langle -w_1, b_r b_{2r} \rangle\rangle \\ &\equiv \langle\langle -w_1, (-1)^{r+1} b_1 \cdots b_{2r} \rangle\rangle \pmod{I^3F}. \end{aligned} \tag{3.3}$$

Since $\Gamma(q) = 1$ and $\Gamma(I^3F) = 1$ (see V.3.4), it follows that

$$\Gamma(\langle\!\langle -w_1, (-1)^{r+1}b_1 \cdots b_{2r} \rangle\!\rangle) = 1,$$

so $\langle\!\langle -w_1, (-1)^{r+1}b_1 \cdots b_{2r} \rangle\!\rangle$ is hyperbolic (by V.3.5). Thus, 3.3 gives $q \in I^3F \cap W_t(F) = 0$, a contradiction. □

Corollary 3.4. *If quadratic forms over F are classified by their dimension, determinant, Hasse invariant, and total signature, then the same holds for quadratic forms over $K = F(\sqrt{w})$ for every $w \in \dot{\sigma}(F)$.*

Proof. This follow now from Theorem 3.1 and the Going-Up Theorem XI.4.14 for (A_3). □

We now conclude this section with two further observations on the (equivalent) conditions (1), (2) and (3).

Remark 3.5. Note that the assumption $w \in \dot{\sigma}(F)$ is essential for the truth of 3.4. Here is a counterexample for the case of a general quadratic extension. Let F_0 be a pythagorean field which admits a quadratic extension K_0 that is *not* pythagorean. Then $F = F_0((x))((y))$ is a pythagorean field, so it has the classification property in 3.4. However, its quadratic extension $K = K_0((x))((y))$ does not, since it has an anisotropic torsion Pfister form $\langle\!\langle -b, x, y \rangle\!\rangle$, where $b \in \dot{K}_0$ is any sum of two squares in K_0 that is not a square.

Remark 3.6. According to XI.4.5, the condition (A_3) amounts to the fact that every 2-fold Pfister form $\langle\!\langle a, b \rangle\!\rangle$ over F represents all elements in $\dot{\sigma}(F)$. In particular, this would imply that F has Pythagoras number $P(F) \le 4$. However, $P(F) \le 4$ itself would not imply the property (A_3). For instance, for $F = \mathbb{Q}((t))$, we have $P(F) = 4$ by XI.5.9(6), but F *does not* satisfy (A_3), since $\langle\!\langle 1, -3, t \rangle\!\rangle$ is an anisotropic torsion 3-fold Pfister form over F.

4. Witt Rings under Biquadratic Extensions

The fact that, for a quadratic extension $K = F(\sqrt{a})$ over F, we have the equation

(4.1) $$W(F/K) := \ker(W(F) \longrightarrow W(K)) = \langle 1, -a \rangle W(F)$$

has been used extensively in our text. It is a natural question to ask if this equation can be generalized to a *multiquadratic* extension

(4.2) $$K = F(\sqrt{a_1}, \dots, \sqrt{a_n}), \quad \text{where } a_i \in \dot{F}.$$

There is an obvious candidate for the kernel $W(K/F)$, namely the ideal $J \subseteq W(F)$ generated by the 1-fold Pfister forms $\langle\!\langle -a_i \rangle\!\rangle = \langle 1, -a_i \rangle$ ($1 \le i \le n$). Of course, we have $J \subseteq W(K/F)$, since each a_i becomes a square in K and so $\langle\!\langle -a_i \rangle\!\rangle_K \cong \mathbb{H}_K$. However, it is not clear if $W(K/F) \subseteq J$.

Of course, even in the case $n = 1$, the equation $W(K/F) = J$ required a rather nontrivial proof: see VII.3.2.

It turns out that $W(K/F) = J$ *does* hold when $n = 2$ (the case of *biquadratic* extensions). This was proved by Elman, Wadsworth, and the author in [ELW$_1$]. Changing notation slightly, we have

Theorem 4.3. $W(F(\sqrt{a}, \sqrt{b})/F) = \langle\langle -a\rangle\rangle W(F) + \langle\langle -b\rangle\rangle W(F)$.

However, it can be shown that this *does not* generalize further to triquadratic extensions. For instance, if F is the rational function field $\mathbb{Q}(t)$ and $K = F(\sqrt{a_1}, \sqrt{a_2}, \sqrt{a_3})$, where

$$a_1 = -1, \quad a_2 = t, \quad a_3 = 1 + t^2,$$

then the ideal J above is *properly* contained in $W(K/F)$, according to the work in a later joint paper [ELTW] with Tignol. This is a bit unfortunate in some sense, but phenomena such as this also bear witness to some of the innate subtleties in the algebraic theory of quadratic forms.

There are some interesting ideas in the proof of 4.3 given in [ELW$_1$]. In this section, we shall offer an exposition of this proof. The main ingredients of the argument are partly contained in Proposition 4.4 below on *quadratic* extensions. To prepare ourselves for this proposition, we need an important definition. For any field extension K/F and any K-form σ, we say that σ is *defined over* F if there exists an F-form τ such that $\sigma \cong \tau_K$. Of course, τ must have the same dimension as σ, although τ may not be uniquely determined (as an F-form).

Proposition 4.4. *Let* $K = F(\sqrt{a})$ *be a quadratic extension of* F, *and let* $R = \operatorname{im}(W(F) \to W(K))$, *which is a subring of* $W(K)$. *Then a* K*-form* σ *is defined over* F *iff* $\sigma \in R$.

Proof. It suffices to prove the "if" part, so assume that $\sigma \in R$. After replacing σ by its anisotropic part, we may assume that σ is anisotropic. (Note that $\mathbb{H}_K$ is defined over F!) Say $\sigma = \alpha_K \in W(K)$, where α is an F-form, which may be chosen to be anisotropic. Write α in the form $\beta\langle\langle -a\rangle\rangle \perp \tau$ (over F), where $\dim(\beta)$ is as large as possible. Then τ must remain anisotropic over K (for otherwise τ would contain $b\langle\langle -a\rangle\rangle$ for some $b \in \dot{F}$). Since $\sigma = \alpha_K = \tau_K \in W(K)$, we have $\sigma \cong \tau_K$, as desired. □

Corollary 4.5. *Let* K/F *be as in* 4.4, *and let* σ, σ' *be* K*-forms. If* σ' *and* $\sigma \perp \sigma'$ *are defined over* F, *then so is* σ.

Proof. For R as in 4.4, $\sigma = (\sigma \perp \sigma') - \sigma' \in R$. Now apply 4.4. □

Instead of proving 4.3, we shall prove a more general result with essentially no extra work. *We fix a quadratic extension* $K = F(\sqrt{a})$ *below, and let* φ *be an* n*-fold Pfister form over* F. Interpreting K as the (small) function field of $\langle\langle -a\rangle\rangle$, we can then form the iterated (small) function field $L := F(\sqrt{a}, \varphi)$, in the sense of X.3. There are two cases here. If $\dim\varphi \geq 3$, then $L = K(\varphi)$, the (small) function field over K of the K-form φ_K. (This is just the field compositum of $K = F(\sqrt{a})$ and $F(\varphi)$ over F.) If $\dim\varphi = 2$, say (after a scaling) $\varphi = \langle\langle -b\rangle\rangle$, $L = F(\sqrt{a}, \varphi)$ is interpreted to mean $F(\sqrt{a}, \sqrt{b})$. (This reinterpretation is necessary since $K(\varphi)$ would no longer make sense if φ becomes the hyperbolic over K. On the other hand, $F(\sqrt{a}, \sqrt{b})$ always makes sense.)

We propose to prove the following result (from [ELW$_1$]).

Theorem 4.6. $W(F(\sqrt{a}, \varphi)/F) = \langle\langle -a\rangle\rangle\, W(F) + \varphi \cdot W(F)$.

In the case where $\varphi = \langle\langle -b\rangle\rangle$ $(b \in \dot{F})$, since $F(\sqrt{a}, \varphi)/F)$ is taken to mean $F(\sqrt{a}, \sqrt{b})$, 4.6 recaptures 4.3.

Fixing the notation in the paragraph following the proof of 4.5, we let $R = \operatorname{im}(W(F) \to W(K))$ as in 4.4, and claim the following.

Proposition 4.7. (1) *If* β *is a* K*-form and* $\varphi \cdot \beta$ *is defined over* F*, then there is a* K*-form* γ *defined over* F *such that* $\varphi \cdot \beta \cong \varphi \cdot \gamma$.

(2) $\varphi W(K) \cap R = \varphi R$.

Proof. (1) We induct on n, where $\beta \cong \langle b_1, \ldots, b_n\rangle$. From the given hypothesis, there exists an element $c \in \dot{F} \cap D_K(\varphi \cdot \beta)$. Write $c = a_1b_1 + \cdots + a_nb_n$, with $a_i \in D_K(\varphi) \cup \{0\}$. Then $\delta := \langle a_1b_1, \ldots, a_nb_n\rangle$ represents c, so $\delta \cong \langle c\rangle \perp \delta_0$ for some form δ_0. (If any a_i is zero, simply replace it by $a_i' = 1$.) Since φ is a Pfister form, $b_i\varphi \cong b_ia_i\varphi$. Therefore,

$$\varphi \cdot \beta \cong \varphi \cdot \delta \cong \varphi \cdot \langle c\rangle \perp \varphi \cdot \delta_0.$$

Since $\varphi \cdot \langle c\rangle$ (as well as $\varphi \cdot \beta$) is defined over F, so is $\varphi \cdot \delta_0$ (by 4.5). Then, by the inductive hypothesis, $\varphi \cdot \delta_0 \cong \varphi \cdot \gamma_0$ for some form γ_0 defined over F. Thus, $\gamma := \langle c\rangle \perp \gamma_0$ is defined over F, and $\varphi \cdot \beta \cong \varphi \cdot \gamma$, as desired.

Finally, (2) follows easily from (1) and 4.4. □

We are now ready to tackle 4.6.

Proof of 4.6. Since φ is a Pfister form, it becomes hyperbolic over $F(\varphi)$, and hence over $F(\sqrt{a}, \varphi)$. Thus, "$\supseteq$" holds in 4.6, and we need only prove the inclusion "$\subseteq$". If $\varphi_K \cong \mathbb{H}$, the field $F(\sqrt{a}, \varphi)$ is just $F(\sqrt{a})$. Here, 4.6 boils down to the old result VII.3.2. Now assume $\varphi_K \not\cong \mathbb{H}$. Let α be an F-form in $W(F(\sqrt{a}, \varphi)/F)$. Going up to K, we have

$$\alpha_K \in R \cap W(K(\varphi)/K) = R \cap \varphi_K W(K)$$

by X.4.11. Using 4.7(2), we can then write $\alpha_K = \varphi_K\gamma_K$ for some F-form γ. Thus,

$$\alpha - \varphi\gamma \in W(K/F) = \langle\!\langle -a\rangle\!\rangle\, W(F),$$

which gives $\alpha \in \langle\!\langle -a\rangle\!\rangle\, W(F) + \varphi\, W(F)$, as desired. □

In retrospect, we see that the crux of the proof of 4.6 is the special property of the quadratic extension K/F proved in Proposition 4.4. Following [ELW$_1$], let us say that a field extension K/F is *excellent* if it satisfies the conclusion of 4.4. (An equivalent requirement is that, if a K-form σ is defined over F, then its anisotropic part is also defined over F: see Exercise 2.) For such an extension K/F, 4.5 and 4.7 can be proved just as before, and the argument used in the proof of 4.6 then yields the following.

Theorem 4.8. *If K/F is an excellent extension, then for any Pfister form φ over F, we have*

$$W(K(\varphi)/F) = W(K/F) + \varphi \cdot W(F).$$

(If $\varphi_K \cong \mathbb{H}$, interpret $K(\varphi)$ as K.)

Consider the special case $K = F(\psi)$, where ψ is another, say m-fold, Pfister form over F. If $m = 1$, then $\psi \cong \langle\!\langle -a\rangle\!\rangle$, so by 4.4, $F(\psi) = F(\sqrt{a})$ is excellent (over F). Quite remarkably, Arason has proved that, for $m = 2$, $F(\psi)/F$ is *also* excellent: this result was published in an Appendix to [ELW$_1$].(4) Assuming this, we have the following consequence of 4.8.

Corollary 4.9. *Let φ, ψ be Pfister forms over F, where* $\dim(\psi) = 4$. *Then*

$$W(F(\psi, \varphi)/F) = \psi \cdot W(F) + \varphi \cdot W(F).$$

Examples of excellent extensions are easy to find. For instance, each of the following is readily checked to be a sufficient condition for K/F to be an excellent extension:

(1) *The map $\dot F/\dot F^2 \to \dot K/\dot K^2$ is surjective.* (This is the case, for instance, if the group $\dot K/\dot K^2$ is generated by a set of integers.)

(2) *Any anisotropic F-form remains anisotropic over K.* (This is the case, for instance, if $[K : F]$ is an odd integer, or if $K = F(x)$ for an indeterminate x; see, respectively, VII.2.3 and IX.1.1.)

As for global fields F, not every finite extension K/F is excellent. However, in [ELW$_1$], it is proved that K/F is excellent whenever K contains a Galois extension of F of even degree. In particular, every multiquadratic

(4)Arason proved this result in the equivalent form that the function field of a conic is an excellent extension of the base field F. It has been pointed out that Arason's algebro-geometric proof of this result contained a small gap, but this gap can be fixed without too much difficulty. For another (completely elementary) proof of the same fact, see Rost's paper [Ro].

extension K/F as in 4.2 is excellent. Coupling this with 4.8, and applying an induction to the number of square roots being adjoined, we obtain the following result.

Theorem 4.10. *For any global field F, and $a_1, \ldots, a_n \in \dot{F}$,*

$$W(F(\sqrt{a_1}, \ldots, \sqrt{a_n})/F) = \langle\langle -a_1 \rangle\rangle\, W(F) + \cdots + \langle\langle -a_n \rangle\rangle\, W(F).$$

5. Nonreal Fields with Eight Square Classes

In the Appendix to VI.2, we have completed the classification of Witt rings for nonreal fields with four square classes. In that case there are four possible Witt groups, which are tabulated at the end of that Appendix. The next case is that of nonreal fields F with $|\dot{F}/\dot{F}^2| = 8$. The Witt ring classification in this case has also been successfully accomplished, in the work of Cordes [Co$_1$] and Szymiczek [Sz$_1$]. This classification, however, requires much more work than the $|\dot{F}/\dot{F}^2| = 4$ case, so we do not propose to present it here. Instead, we would like to list the ten possible Witt groups which result from that classification, and give an account on the existence of the ten fields that give rise to those Witt groups. We shall list the Witt groups in the same style as in the chart in the Appendix to VI.2. Here, F is nonreal with $|\dot{F}/\dot{F}^2| = 8$, and s, u, m denote, respectively, the level, the u-invariant, and the number of distinct quaternion algebras (including the split one) over the field F. The ten possible Witt groups are given in the following table, where, again, $\mathbb{Z}_n^k$ denotes the direct sum of k copies of $\mathbb{Z}_n = \mathbb{Z}/n\mathbb{Z}$.

	Witt Group $W(F)$	$\lvert W(F)\rvert$	s	u	m
(1)	$\mathbb{Z}_2^6$	2^6	1	4	4
(2)	$\mathbb{Z}_4^2 \oplus \mathbb{Z}_2^2$	2^6	2	4	4
(3)	$\mathbb{Z}_2^8$	2^8	1	8	8
(4)	$\mathbb{Z}_4^4$	2^8	2	8	8
(5)	$\mathbb{Z}_2^4$	2^4	1	2	1
(6)	$\mathbb{Z}_4 \oplus \mathbb{Z}_2^2$	2^4	2	2	1
(7)	$\mathbb{Z}_8 \oplus \mathbb{Z}_2^2$	2^5	4	4	2
(8)	$\mathbb{Z}_2^5$	2^5	1	4	2
(9)	$\mathbb{Z}_4^2 \oplus \mathbb{Z}_2$	2^5	2	4	2
(10)	$\mathbb{Z}_4 \oplus \mathbb{Z}_2^3$	2^5	2	4	2

While the above chart is supposed to be a complete listing of the 10 possible Witt rings, it turns out that the ten *Witt groups* are already all different. What this means is that, in the case under consideration (nonreal fields with $|\dot{F}/\dot{F}^2| = 8$), *the Witt group determines the Witt ring uniquely.*

This, in part, justifies our omission of a description of the Witt ring structure in each of the ten cases.

Let us now give an account of how these ten Witt groups arise. The first four come from the Witt groups of nonreal fields F_0 with four square classes. In fact, if F_0 is any one of the four fields listed in the Appendix to VI.2, then the Laurent series field $F = F_0((x))$ is nonreal with $|\dot{F}/\dot{F}^2| = 8$ and $W(F) \cong W(F_0) \oplus W(F_0)$. This accounts for the first four Witt groups in the above chart, where the numbering of the four groups is consistent with that in the earlier chart. Note that, for fields of the type (3) or (4), we have $|W(F)| = 2^8$, where $8 = |\dot{F}/\dot{F}^2|$, so these are C-fields in the sense of Cordes (see XI.7.16). The existence of fields of the types (5) and (6) is the special case $n = 2$ of the result below.

Theorem 5.1. *For any $n \geq 1$, there exists a field F with $|\dot{F}/\dot{F}^2| = 2^n$, $u(F) = 2$, and $s(F) = 1$ or 2, whichever is desired. If $s(F) = 1$, then $W(F) \cong \mathbb{Z}_2^{n+1}$; if $s(F) = 2$, then $W(F) \cong \mathbb{Z}_4 \oplus \mathbb{Z}_2^{n-1}$.*

Proof. We first deduce the second statement from the first. Since F is nonreal with $u(F) = 2$, we have $I^2F = 0$. Therefore, $|W(F)| = 2|IF| = 2^{n+1}$. If $s(F) = 1$, then $2W(F) = 0$, so $W(F) \cong \mathbb{Z}_2^{n+1}$. If $s(F) = 2$, then $\langle 1 \rangle$ has exponent 4, so $W(F) = \mathbb{Z}_4\langle 1 \rangle \oplus G$ for some subgroup $G \subseteq IF$. Now $2G \subseteq 2IF = 0$ implies that $G \cong \mathbb{Z}_2^{n-1}$.

To prove the first statement, we apply the Gross-Fischer construction in VII.3.17, and proceed by induction on n. If $n = 1$, we can take $F = \mathbb{F}_5$ or $\mathbb{F}_3$. For the inductive step, let F be given as in the first statement, and let $\{a_1, \ldots, a_n\}$ be a $\mathbb{Z}_2$-basis for $\dot{F}/\dot{F}^2$. For the rational function field $K_0 = F(x)$, a $\mathbb{Z}_2$-basis for $\dot{K}_0/\dot{K}_0^2$ is given by $\{a_i\}$ together with all monic irreducible polynomials $\{f_j\}$ in $F[x]$, including, of course, x and $x + a_i$ for all i. Using the Gross-Fischer construction, we can thus come up with an algebraic extension K/F such that $\dot{K}/\dot{K}^2$ has a $\mathbb{Z}_2$-basis $\{a_1, \ldots, a_n, x\}$. In particular, $|\dot{K}/\dot{K}^2| = 2^{n+1}$. *We claim that K is the field we want.* To check that $u(K) = 2$, it suffices to show that all quaternion algebras $\left(\frac{u,v}{K}\right)$ split over K. The linearity property III.2.11 reduces this to the following three cases.

Case 1. $u = v = x$. Here $\left(\frac{u,v}{K}\right) = 1$, since $s(K) \leq s(F) \leq 2$.

Case 2. $u = a_i$, $v = a_j$. Here, $\left(\frac{u,v}{F}\right) = 1$ already, since $u(F) = 2$.

Case 3. $u = a_i$, $v = x$. Here $\left(\frac{u,v}{K}\right) = 1$, since $\langle 1, -x \rangle_K$ represents $(\sqrt{x + a_i}\,)^2 - x = a_i$.

If $s(F) = 1$, then $s(K) = 1$ too, and we are done. If $s(F) = 2$, we can choose a_1 to be -1. Since $a_1 \notin \dot{K}^2$, we have $s(K) = 2$. □

Next, we come to the type (7) in our chart. This type is rather "special" in the list, since it is the only one where the level of the field reaches 4 (so that $W(F)$ has a direct summand $\cong \mathbb{Z}_8$). Luckily, the existence of this type is not in question, since the field $\mathbb{Q}_2$ (of the 2-adic numbers) has 8 square classes and Witt group $\mathbb{Z}_8 \oplus \mathbb{Z}_2^2$, by VI.2.29. Just to give a taste of the classification work in general, let us prove a strong uniqueness result in this case. This uniqueness result, incidentally, serves to show in part the very special role played by the field $\mathbb{Q}_2$ in the classification theory of finite Witt rings.

Theorem 5.2. *For any field F, the following are equivalent:*

(1) $W(F) \cong W(\mathbb{Q}_2)$ *as rings.*

(2) $W(F) \cong W(\mathbb{Q}_2)$ *as groups.*

(3) $W(F)$ *has order* 32 *and exponent* 8.

(4) $|\dot{F}/\dot{F}^2| = 8$ *and* $s(F) = 4$.

Proof. (1) $\Rightarrow$ (2) is trivial, and (2) $\Rightarrow$ (3) is part of VI.2.29.

(3) $\Rightarrow$ (4). The fact that $W(F)$ has exponent 8 implies that $s(F) = 4$, so $4\langle 1\rangle \neq 0 \in W(F)$. Therefore, $|I^2F| \geq 2$, so

$$|\dot{F}/\dot{F}^2| = |IF/I^2F| \leq |IF|/2 = 8.$$

Here, equality must hold, for otherwise the Appendix to VI.2 would show that $|W(F)| \leq 16$.

(4) $\Rightarrow$ (1). Write $-1 = x^2 + y^2 + z^2 + w^2$, and let $a = x^2 + y^2$ and $b = z^2 + w^2$. Then $a, b \notin F^2$, and they represent different square classes. (If $b = ar^2$ for some $r \in F$, then $-1 = a(1 + r^2)$ is a sum of two squares.) Also, $a, b \in D(2\langle 1\rangle)$ and $-1 \notin D(2\langle 1\rangle)$ imply that $-1, a, b$ form a basis for $\dot{F}/\dot{F}^2$. Let $\alpha = \langle 1, -a\rangle$ and $\beta = \langle 1, -b\rangle$, with $\alpha^2 = 2\alpha = 0$ and $\beta^2 = 2\beta = 0$ in $W(F)$. Since

$$\langle a, b\rangle \cong \langle a + b, ab(a + b)\rangle = \langle -1, -ab\rangle, \tag{5.3}$$

$W(F)$ is additively generated by $\langle 1\rangle, \langle a\rangle, \langle b\rangle$, and so

$$W(F) = \mathbb{Z}\langle 1\rangle + \mathbb{Z}\alpha + \mathbb{Z}\beta.$$

Now $s(F) = 4$ implies that $\langle 1\rangle$ has order 8. The elements α, β and $\alpha + \beta$ have order 2 and are not equal to $4\langle 1\rangle$ (since $4\langle 1\rangle \in I^2F$). Therefore, we must have

$$W(F) = \mathbb{Z}_8\langle 1\rangle \oplus \mathbb{Z}_2\alpha \oplus \mathbb{Z}_2\beta.$$

Moreover, by 5.3 and the fact that $ab \in D(2\langle 1\rangle)$,

$$\alpha\beta = \langle 1,\ -a,\ -b,\ ab\rangle \cong \langle 1,\ 1,\ ab,\ ab\rangle \cong 4\langle 1\rangle.$$

This completely determines the multiplicative structure of $W(F)$; namely, $W(F)$ is isomorphic to the ring

$$\mathbb{Z}_8[s,\, t]/(2s,\, 2t,\, s^2,\, t^2, st-4).$$

Recalling VI.2.31, we then have $W(F) \cong W(\mathbb{Q}_2)$ as rings. □

We have now accounted for *seven* of the ten types of Witt groups listed in our chart. The proof for the existence of fields of the three last types is nontrivial. In fact, when Cordes first arrived at the classification of nonreal fields with 8 square classes into the ten types (see [Co$_1$]), the existence problem for the types (8), (9), and (10) was left open. Cordes noticed that the fields giving rise to any of these types must have a nontrivial "Kaplansky radical"; that is, there must exist some $a \notin F^2$ such that $\langle 1, -a\rangle$ is a universal form. At the time when [Co$_1$] was written, no examples of such elements were known in nonreal fields F with $|\dot F/\dot F^2| < \infty$ and $u(F) > 2$.

To complete our existence proof for the fields of types (8), (9) and (10), we must, therefore, have a better understanding of the Kaplansky radical and come up with examples of nonreal fields with a nontrivial radical. The next section is devoted to the former goal; after that, we shall return in §7 to construct the three "missing" fields, of the types (8), (9) and (10).

6. Kaplansky Radical and Hilbert Fields

In this section, we give an exposition on the notion of the Kaplansky radical of a field and its applications to the study of pre-Hilbert fields and Hilbert fields. There does not seem to be any textbook treatment of this part of quadratic form theory, so we hope that the exposition here will provide a convenient reference for these useful topics. The main material in this section is drawn from the work of Fröhlich [Fr], Kaplansky [Ka$_1$, Ka$_2$], Cordes [Co$_1$], and Szymiczek [Sz$_1$].

The beginning point of the investigation is the consideration of the symmetric bimultiplicative pairing

$$\text{(6.0)} \qquad \dot F/\dot F^2 \times \dot F/\dot F^2 \longrightarrow B(F) \quad \text{given by} \quad (a,b) \longmapsto \left(\frac{a,\, b}{F}\right),$$

where $B(F)$ is the Brauer group of F. The "radical" of this pairing is essentially the *Kaplansky radical* of F. In more detail, we have the following.

Proposition 6.1. *For any $a \in \dot F$, the three statements below are equivalent:*

(1) $\left(\frac{a,b}{F}\right) = 1$ *for every* $b \in \dot F$.

(2) $\langle\langle -a\rangle\rangle = \langle 1, -a\rangle$ *is universal over* F.

(3) $a \in D_F\langle 1, -b\rangle$ *for every* $b \in \dot F$.

The set $R(F)$ of elements $a \in \dot{F}$ satisfying these conditions is a subgroup of $\dot{F}$ containing $\dot{F}^2$, called the Kaplansky radical of F.

Proof. By Hilbert's Criterion, $\left(\frac{a,\,b}{F}\right) = 1$ is equivalent to $1 \in D_F\langle a,\, b\rangle$, which translates into $b \in D_F\langle 1,\, -a\rangle$, or symmetrically, into $a \in D_F\langle 1,\, -b\rangle$. Quantifying this over all $b \in \dot{F}$, we get (1) $\Leftrightarrow$ (2) $\Leftrightarrow$ (3). The fact that $R(F)$ is a subgroup of $\dot{F}$ (obviously containing $\dot{F}^2$) follows from the bi-multiplicativity of the pairing in 6.0. $\square$

Notice that, since $R(F) \supseteq \dot{F}^2$, it is often convenient to think of $R(F)$ as a "group of square classes"; that is, to work instead with $R(F)/\dot{F}^2$. (For instance, the true radical of the pairing in 6.0 is $R(F)/\dot{F}^2$, and not quite $R(F)$.) The ambiguity here is basically harmless. After all, we have been using the same notational convention throughout for $D_F(q)$—the set of values represented by a form q.

Recall that $D(n)$ denotes the set $D_F(n\langle 1\rangle)$, and $D(\infty)$ denotes the set $\bigcup_{n\geq 1} D(n)$ (the group of nonzero sums of squares).

Corollary 6.2. *We have $\dot{F}^2 \subseteq R(F) \subseteq D(2) \subseteq D(\infty) \subseteq \dot{F}$, and $R(F) = \dot{F}$ iff F is a nonreal field with u-invariant $u(F) \leq 2$.*

Proof. The inclusion $R(F) \subseteq D(2)$ is the special case of 6.1(3) for $b = -1$. As for the condition $R(F) = \dot{F}$, it amounts to all binary forms being universal over F, which means exactly that F is nonreal with $u(F) \leq 2$. $\square$

The next proposition reveals some very special properties of elements in the Kaplansky radical.

Proposition 6.3. (1) *Let $a_i \in \dot{F}$ and $r_i \in R(F)$. Then, for $n \geq 2$,*

$$(*) \qquad D\langle a_1, \ldots, a_n\rangle = D\langle a_1 r_1, \ldots, a_n r_n\rangle.$$

In particular, for $n \geq 2$, the LHS consists of cosets of $\dot{F}$ modulo $R(F)$.

(2) *For $n \geq 3$, $\langle a_1, \ldots, a_n\rangle$ is isotropic iff $\langle a_1 r_1, \ldots, a_n r_n\rangle$ is.*

Proof. (1) We first treat the case $n = 2$. In this case, it suffices to show that $D\langle a, b\rangle = D\langle a, br\rangle$ for any $r \in R(F)$. After a scaling, we may assume that $a = 1$. Now, for any $c \in \dot{F}$, $c \in D\langle 1, b\rangle$ means $\left(\frac{c,\,-b}{F}\right) = 1$. The latter is equivalent to $\left(\frac{c,\,-br}{F}\right) = 1$, which means $c \in D\langle 1,\, br\rangle$. This proves $(*)$ for $n = 2$, and the general case $n \geq 2$ follows easily from the "Inductive description of Value Sets" in Exercise 20 of Chapter I.

(2) The isotropicity of $\langle a_1, \ldots, a_n\rangle$ means that $-a_1 \in D\langle a_2, \ldots, a_n\rangle$. By (1), we may change the latter to $-a_1 r_1 \in D\langle a_2 r_2, \ldots, a_n r_n\rangle$ (since $n-1 \geq 2$ here). Now we get the isotropicity of $\langle a_1 r_1, \ldots, a_n r_n\rangle$ (and conversely). $\square$

Corollary 6.4. *Let $n \geq 2$, and let $r \in R(F)$. Then:*

(1) $-1 \in D(n)$ iff $-r \in D(n)$.

(2) If F is formally real, then $-r \notin D(\infty)$.

(3) If F has finite level $2^k \geq 2$, then $-r$ is a sum of 2^k squares, but no fewer (that is, the "length" of $-r$ is exactly 2^k).

Proof. (1) follows from the last statement in 6.3(1), and (2), (3) both follow from (1). □

Corollary 6.5. (1) *Let $n \geq 2$. Then $R(F) + \cdots + R(F)$ (n summands) is contained in $D(n) \cup \{0\}$.*

(2) *$R(F) = D(2)$ iff $R(F) = D(\infty)$.*

Proof. (1) Let $a = r_1 + \cdots + r_n \neq 0$, where $r_i \in R(F)$. Then by 6.3(1),

$$a \in D\langle r_1, \ldots, r_n\rangle = D\langle 1, \ldots, 1\rangle = D(n).$$

(2) The "if" part is clear, since $R(F) \subseteq D(2) \subseteq D(\infty)$. Conversely, assume that $R(F) = D(2)$. Given a sum of n nonzero squares $b_1^2 + \cdots + b_n^2$, we may assume by induction that

$$b_2^2 + \cdots + b_n^2 \in R(F) \cup \{0\}.$$

Adding b_1^2, we get a sum in $R(F) + (R(F) \cup \{0\})$, which must lie in $D(2) \cup \{0\} = R(F) \cup \{0\}$ by (1). □

For nonreal fields of u-invariant > 2, we have the following extension of XI.6.6.

Proposition 6.6. *Let F be a nonreal field with $u(F) > 2$. Then:*

(1) $R(F) \subsetneq D\langle 1, a\rangle$ for every $a \in \dot{F}$.

(2) Any anisotropic form φ of dimension $n \geq 2$ represents at least n cosets of $R(F)$.

(3) $u(F) \leq [\dot{F} : R(F)]$.

Proof. We first prove (1) (which is essentially the special case of (2) for $n = 2$). Suppose $R(F) = D\langle 1, a\rangle$ for some $a \in \dot{F}$. Then $a \in R(F)$, and 6.3(1) gives $R(F) = D\langle 1,\, a \cdot a\rangle = D(2)$. But then 6.5(2) implies that $R(F) = D(\infty) = \dot{F}$, which means that $u(F) \leq 2$, a contradiction.

(2) We induct on $n \geq 2$ (the case $n = 2$ being covered by (1)). For $n \geq 3$, write $\varphi \cong \langle a\rangle \perp \psi$. By Kneser's Lemma XI.6.5, $D(\varphi) \supsetneq D(\psi)$. Thus, $D(\varphi)$ contains at least one more coset of $R(F)$ than $D(\psi)$, so we are done by the inductive hypothesis.

(3) We may assume that $[\dot{F} : R(F)] = n < \infty$ (for otherwise there is nothing to prove). If $u(F) > n$, then there exists an $(n+1)$-dimensional

anisotropic form σ. But by (2), $D(\sigma)$ contains at least $n+1$ cosets of $R(F)$, which is impossible. $\square$

We come now to the following definition, which was prompted by the work of A. Fröhlich on the axiomatic foundations of quadratic form theory over local fields.

Definition 6.7. We call F a *pre-Hilbert field* if there exists a unique quaternion division algebra over F.

Kaplansky called such fields *generalized Hilbert fields.*(5) The name pre-Hilbert field is in the same spirit, but seems a little easier. (Hilbert fields themselves will be defined later.) We start with a characterization of pre-Hilbert fields in terms of the binary forms over F.

Theorem 6.8. *A field F is pre-Hilbert iff, for any $a \in \dot{F}$, $[\dot{F} : D\langle 1, -a\rangle] \leq 2$, with equality holding for at least one $a \in \dot{F}$.*

Proof. "Necessity": The group homomorphism $\varepsilon : \dot{F}/\dot{F}^2 \rightarrow B(F)$ given by $b\dot{F}^2 \mapsto \left(\frac{a,b}{F}\right)$ has kernel $D\langle 1, -a\rangle$. Since the image of ε has at most 2 elements, the index inequality follows. If the (unique) quaternion division algebra is given by $\left(\frac{a,b}{F}\right)$, then, for this choice of a, we have $[\dot{F} : D\langle 1, -a\rangle] = 2$.

"Sufficiency": Fix an element $a \in \dot{F}$ such that $[\dot{F} : D\langle 1, -a\rangle] = 2$, and take any $b \notin D\langle 1, -a\rangle$. Then $\left(\frac{a,b}{F}\right) \neq 1$. Now consider any $\left(\frac{c,d}{F}\right) \neq 1$.

Case 1. $\left(\frac{a,d}{F}\right) \neq 1$. Then $\left(\frac{a,b}{F}\right) = \left(\frac{a,d}{F}\right) = \left(\frac{c,d}{F}\right)$ since each quaternion algebra here is nonsplit.

Case 2. We may now assume $\left(\frac{a,d}{F}\right) = 1$, and similarly, also $\left(\frac{c,b}{F}\right) = 1$. But then

$$\left(\frac{a,\, b}{F}\right) = \left(\frac{ac,\, b}{F}\right) \neq 1, \qquad \left(\frac{c,\, d}{F}\right) = \left(\frac{ac,\, d}{F}\right) \neq 1.$$

This forces $\left(\frac{ac,b}{F}\right) = \left(\frac{ac,d}{F}\right)$, and hence, again, $\left(\frac{a,b}{F}\right) = \left(\frac{c,d}{F}\right)$. $\square$

In the nonreal case, two rather special properties of the radical are given in the following result.

Theorem 6.9. *If F is nonreal, then $[\dot{F} : R(F)] \neq 2$, and $[\dot{F} : R(F)] = 4$ only if F is pre-Hilbert.*

(5)Actually, Kaplansky also allowed the case where F has *no* quaternion division algebras. The discrepancy between this and 6.7 is not serious. Kaplansky's original definition was couched in terms of binary forms instead of quaternion algebras: see Exercise 10.

Proof. First assume $[\dot{F} : R(F)] = 2$. Certainly, $u(F) > 2$. Consider any $a \in \dot{F} \setminus \dot{F}^2$. Then $\varphi = \langle 1, -a \rangle$ is anisotropic, so by 6.6, $D(\varphi)$ contains at least two cosets of $R(F)$. Since $[\dot{F} : R(F)] = 2$, we have $D(\varphi) = \dot{F}$. This shows that $a \in R(F)$, and hence $\dot{F} = R(F)$, a contradiction.

Next, assume that $[\dot{F} : R(F)] = 4$. For any $a \in \dot{F}$, $\varphi = \langle 1, -a \rangle$ represents again at least two cosets of $R(F)$ (by 6.6), and hence $[\dot{F} : D(\varphi)] \leq 2$. If we pick a to be outside of $R(F)$, then $D(\varphi) \neq \dot{F}$, and hence $[\dot{F} : D(\varphi)] = 2$. By 6.8, F must then be pre-Hilbert. □

In the formally real case, the pre-Hilbert notion can be easily reduced to notions that we are already familiar with, as follows.

Theorem 6.10. *For any field F, the following are equivalent:*

(1) *F is formally real and pre-Hilbert.*

(2) $[\dot{F} : R(F)] = 2$.

(3) *F is uniquely ordered (that is, $|X_F| = 1$), and I^2F is torsionfree.*

Any field F satisfying (1) *has Pythagoras number $P(F) \leq 2$.*

Proof. (3) $\Rightarrow$ (1). Let α be the unique ordering. Then F is formally real, and Pfister's Local-Global Principle gives a signature isomorphism $\mathrm{sgn}_\alpha \colon I^2F \to 4\mathbb{Z}$ (since I^2F is torsionfree). This implies right away that $\langle\langle 1, 1 \rangle\rangle$ is the unique anisotropic 2-fold Pfister form, so (1) follows.

(1) $\Rightarrow$ (2). Given (1), the unique anisotropic 2-fold Pfister form is $\langle\langle 1, 1 \rangle\rangle$. *We claim that* $[\dot{F} : D(2)] = 2$. In fact, if $a \notin D(2)$, then $\langle\langle -a, -a \rangle\rangle$ is anisotropic, and hence isometric to $\langle\langle 1, 1 \rangle\rangle$. This gives $\langle -a, -a \rangle \cong \langle 1, 1 \rangle$, and so $-a \in D(2)$. This proves our claim, which clearly forces $D(2) = D(\infty)$ (and hence $P(F) \leq 2$, as asserted in the last conclusion of the theorem). We finish by showing that $R(F) \subseteq D(2)$ is an equality. Suppose, for the moment, that there exists $w \in D(2) \setminus R(F)$. Then $\left(\frac{w, x}{F}\right) \neq 1$ for some $x \in \dot{F}$, and hence $\left(\frac{w, x}{F}\right) = \left(\frac{-1, -1}{F}\right)$. Now passing to a real-closure of F gives the desired contradiction.

(2) $\Rightarrow$ (3). Given (2), 6.6 forces F to be formally real. Thus, $R(F) \subseteq D(\infty) \subsetneq \dot{F}$ implies that $D(\infty) = R(F)$ has index 2 in $\dot{F}$, and hence $|X_F| = 1$. Finally, by XI.4.2, $I^2F \cap W_t(F)$ is additively generated by the forms $\varphi = \langle\langle a, -w \rangle\rangle$, where $w \in D(\infty)$. But the latter gives $w \in R(F)$, so $\varphi = 0 \in W(F)$. This shows that I^2F is torsionfree. □

Corollary 6.11. *Let F be a pre-Hilbert field.*

(1) *If F is formally real, then quadratic forms over F are classified by dimension, determinant, and (unique) signature. Moreover, F has general u-invariant $u(F) \leq 2$, $P(F) \leq 2$, and $I^2F = 4\mathbb{Z} \cdot \langle 1 \rangle$.*

(2) *If F is nonreal, then quadratic forms over F are classified by dimension, determinant, and the Hasse invariant. Moreover, $u(F) = 4$, $P(F) \leq 4$, $|I^2F| = 2$, and $I^3F = 0$.*

Proof. (1) Here, I^2F is torsionfree, so $u(F) \leq 2$ by XI.6.26(2). If φ, ψ are forms with the same dimension, determinant, and signature, then $\varphi - \psi \in I^2F \cap W_t(F) = 0$, and hence $\varphi \cong \psi$. The other conclusions are already clear from 6.10 and its proof.

(2) Most of the conclusions here have been obtained before already: see V.3.25, VI.2.13, and Ch. VI, Exer. 1.[6] We also get $P(F) \leq 4$, since $4\langle 1\rangle$ is universal. From $u(F) = 4$, we have $I^3F = 0$, so if σ is the unique anisotropic 2-fold Pfister form, we have $2\sigma = 0 \in W(F)$. Since I^2F is additively generated by σ, it follows that $|I^2F| = 2$. □

Corollary 6.12. *If F is a nonreal pre-Hilbert field, then $|W(F)| = 4|\dot{F}/\dot{F}^2|$.*

Proof. This follows from (2) above, since $|W(F)/IF| = 2$ and $|IF/I^2F| = |\dot{F}/\dot{F}^2|$. □

A common conclusion for the two cases in 6.11 is the following.

Corollary 6.13. *If φ is an anisotropic 4-dimensional form over a pre-Hilbert field F, then $d(\varphi) \in R(F)$.*

Proof. After a scaling, we may assume that $\varphi \cong \langle -a, -b, ab, d\,\rangle$, where $d := d(\varphi)$. Then $\left(\frac{a,b}{F}\right) \neq 1$. If $d \notin R(F)$, then $\left(\frac{c,d}{F}\right) \neq 1$ for some $c \in \dot{F}$, and hence $\left(\frac{c,d}{F}\right) \cong \left(\frac{a,b}{F}\right)$ (by uniqueness). This gives an isometry

$$\langle -c,\ -d,\ cd\rangle \cong \langle -a,\ -b,\ ab\rangle,$$

so $\varphi \cong \langle -a,\ -b,\ ab,\ d\rangle \cong \langle -c,\ -d,\ cd,\ d\rangle$ is isotropic, a contradiction. □

Next, we shall introduce the Hilbert fields.

Definition 6.14. A pre-Hilbert field F is said to be a *Hilbert field* if $R(F) = \dot{F}^2$; that is, if F has "trivial" radical. (In this case, we get a "perfect pairing" $\dot{F}/\dot{F}^2 \times \dot{F}^2/\dot{F}^2 \to \{\pm 1\}$.)

The local fields studied in Chapter VI are the motivating examples for the above definition (given first by Fröhlich). Paraphrasing 6.14, *F is a Hilbert field iff $\dot{F} \neq \dot{F}^2$ and, for every nonsquare $a \in \dot{F}$, $[\dot{F} : D\langle 1, -a\rangle] = 2$* (or equivalently, the norm group of every quadratic extension $F(\sqrt{a}\,)/F$ has index 2 in $\dot{F}$).

[6]Of course, the conclusion that $u(F) = 4$ is just a special case of XI.6.22, and the classification information in (2) is, in turn, a special case of the main theorem on classification in §3. However, the full force of these deeper results is not needed here.

We can now derive the following characterization of Hilbert fields, due to A. Fröhlich [Fr].

Theorem 6.15. (1) *A formally real field F is Hilbert iff F is euclidean.*

(2) *A nonreal field F is Hilbert iff F has a unique anisotropic 4-dimensional form (which is necessarily a 2-fold Pfister form).*

Proof. (1) The "if" part is clear. For the converse, assume F is formally real and Hilbert. Then by 6.10, $[\dot{F} : \dot{F}^2] = [\dot{F} : R(F)] = 2$, so F is a euclidean field.

(2) Here, F is nonreal. Assume F is Hilbert. Then $u(F) = 4$ by 6.11. Let φ be *any* anisotropic 4-dimensional isomorphism form; then by 6.13, $d = d(\varphi) \in R(F) = \dot{F}^2$, so we may write $\varphi \cong c\langle\langle a, b\rangle\rangle$. Since $I^3F = 0$ (by 6.11(2)), we have $\varphi \cong \langle\langle a, b\rangle\rangle$, so φ is necessarily the unique anisotropic 2-fold Pfister form.

Conversely, assume that there exists a unique anisotropic 4-dimensional form. Then $u(F) \geq 4$, so there is at least one nonsplit quaternion algebra $\left(\frac{-a,\,-b}{F}\right)$. Such an algebra must be unique, so F is pre-Hilbert. If F is *not* Hilbert, take an element $d \in R(F) \setminus \dot{F}^2$. Since $\varphi = \langle 1, a, b, ab\rangle$ is anisotropic, 6.3(2) implies that $\psi = \langle d, a, b, ab\rangle$ is *also* anisotropic. But φ, ψ are non-isometric (since $d(\varphi) \neq d(\psi)$), which contradicts the hypothesis. □

According to 6.15(1) above, the only formally real Hilbert fields are the "obvious" ones (namely the euclidean fields). On the other hand, by using local fields, we can easily come up with nonreal Hilbert fields F with 2^n square classes for any $n \geq 2$.

How about pre-Hilbert fields that are not Hilbert? The corresponding existence question will be taken up in the next section. However, due to the lack of space, the more advanced theory of Hilbert and pre-Hilbert fields (e.g., questions about the index $[\dot{F} : R(F)]$, the behavior of $R(F)$ under field extensions, etc.) will not be presented here.

7. Construction of Some Pre-Hilbert Fields

Instead of pursuing further the theory of pre-Hilbert fields, we opt for the explicit construction of some such fields. In particular, the material in this section will serve as an illustration of some construction techniques, and will complete the existence proof for the ten nonreal fields with 8 square classes started in §5.

Let us say that a field F has a *nontrivial radical* if $\dot{F}^2 \subsetneq R(F) \subsetneq \dot{F}$. We shall only be interested in pre-Hilbert fields of this type, since the ones with $R(F) = \dot{F}$ are the nonreal fields with $u(F) \leq 2$, and the ones with $R(F) = \dot{F}^2$ are the Hilbert fields (for which local fields already give a lot of examples).

We shall start with formally real fields. In II.5.3, we have constructed such a field F with a square class basis $\{-1, 2\}$. Since $\langle 1, -2\rangle$ represents both -1 and -2, it is *universal*, and hence $2 \in R(F)$. This gives $[\dot{F} : R(F)] = 2$, so F is a pre-Hilbert field with 4 square classes, and with a nontrivial radical $R(F) = \{1, 2\}$ (up to squares).

For bigger square class groups, we have the following result of K. Szymiczek [Sz$_1$].

Theorem 7.1. *For any $n \geq 1$, there exists a formally real pre-Hilbert field F with $|\dot{F}/\dot{F}^2| = 2^{n+1}$ (which necessarily has a nontrivial radical, in view of* 6.15(1)).

The construction of F depends on the number-theoretic lemma below.

Lemma 7.2. *For any $n \geq 1$, there exist prime numbers*

$$p_1 < p_2 < \cdots < p_n, \qquad p_i \equiv 1 \pmod 4 \quad \textit{for each } i,$$

such that $\gcd(p_j + p_i, p_1 \cdots p_n) = 1$ *whenever $j < i$.*

Proof. For $n = 1$, we can take $p_1 = 5$. For $n \geq 2$, we shall construct the prime sequence *with the additional property that* $p_i \equiv 1 \pmod{p_j}$ whenever $i > j$. By induction, suppose $p_1, \ldots, p_n$ are given as above. By Dirichlet's theorem on primes in an arithmetic progression, there exists a prime

$$p_{n+1} \equiv 1 \pmod{4p_1 \cdots p_n}.$$

If $j < i < n+1$, $p_j + p_i$ is certainly relatively prime to p_{n+1} (since the latter is "much larger"), as well as to $p_1 \cdots p_n$. Now consider $p_j + p_{n+1}$, where $j \leq n$. This is certainly relatively prime to p_j and p_{n+1}. *How about p_k where $j \neq k \leq n$?* If $k < j$, then

$$p_j + p_{n+1} \equiv 1 + 1 \equiv 2 \not\equiv 0 \pmod{p_k}.$$

On the other hand, if $k > j$, then

$$p_j + p_{n+1} \equiv p_j + 1 \not\equiv 0 \pmod{p_k}.$$

This checks all the required properties for $p_1, \ldots, p_{n+1}$. $\square$

Proof of 7.1. Given $n \geq 1$, let $\{p_1, \ldots, p_n\}$ be as in 7.2. Then $-1, p_1, \ldots, p_n$ are $\mathbb{Z}_2$-independent square classes in $\dot{\mathbb{Q}}/\dot{\mathbb{Q}}^2$, which can be completed to a basis by adding a set of positive elements G, which we can take to be the set of all primes $\notin \{p_1, \ldots, p_n\}$. Following the Gross-Fischer construction, we form $F_1 = \mathbb{Q}(\sqrt{G}) \subseteq \mathbb{R}$, in which $-1, p_1, \ldots, p_n$ continue to represent $\mathbb{Z}_2$-independent square classes. Repeat the construction to get $F_2 \subseteq F_3 \subseteq \cdots \subseteq$

$\mathbb{R}$, and form their union F. Then $\dot{F}/\dot{F}^2$ has square class basis $-1, p_1, \ldots, p_n$, so $|\dot{F}/\dot{F}^2| = 2^{n+1}$, and F is formally real. *We claim that* $p_i \in R(F)$ *for each* i. This would imply that $[\dot{F} : R(F)] = 2$, and F would then be pre-Hilbert (by 6.10), as desired. To prove our claim, we must check that $\varphi_i = \langle 1, -p_i \rangle$ *is universal over* F. Now p_i is a sum of two squares in $\mathbb{Z}$, and hence in F, so certainly $D_F(\varphi_i)$ contains -1 (and $-p_i$). Since $D_F(\varphi_i)$ is a group, we are done if $p_j \in D_F(\varphi_i)$ for each $j \neq i$. But by 7.2, $p_j + p_i$ is prime to $p_1 \cdots p_n$, so it factors into $q_1 \cdots q_m$, where each q_i is a prime $\notin \{p_1, \ldots, p_n\}$. By construction, $q_i \in \dot{F}_1^2 \subseteq \dot{F}^2$. Therefore $D_F\langle p_i, p_j \rangle$ contains 1, which gives $p_j \in D_F\langle 1, -p_i \rangle$, as desired. □

Having taken care of the formally real case, we now move on to the construction of *nonreal* pre-Hilbert fields with nontrivial radicals. For quite some time after the appearance of Kaplansky's paper [Ka$_1$], no such fields were known with $|\dot{F}/\dot{F}^2| < \infty$. As we have mentioned in §5, the three types of nonreal fields with eight square classes not yet constructed in that section were known to have nontrivial radicals. For the reader's convenience, let us recall the Witt groups of these three types of fields F (with $|\dot{F}/\dot{F}^2| = 8$):

(7.3) $$W(F) \cong \mathbb{Z}_2^5,$$

(7.4) $$W(F) \cong \mathbb{Z}_4^2 \oplus \mathbb{Z}_2,$$

(7.5) $$W(F) \cong \mathbb{Z}_4 \oplus \mathbb{Z}_2^3.$$

The last column of the chart in §5 (on the number of quaternion algebras) indicates that these three kinds of fields must be pre-Hilbert fields.

In the late 1970s, these "missing" types of nonreal fields with 8 square classes were found, respectively, by L. Berman, M. Kula, and T. L. Lee. As an illustration of their construction techniques, we shall give an account of this part of the story, even though our exposition here is not going to be entirely self-contained.

For Berman's construction of the type 7.3 (in [Bm]), we start with a formally real pythagorean field E with 2^4 square classes and 5 orderings. (Such a field E exists by the work of Bröcker [Br]; we won't go into the details here.) For $F = E(\sqrt{-1}\,)$ (of level 1), the square class exact sequence

(7.6) $$1 \longrightarrow \{\pm \dot{E}^2\} \longrightarrow \dot{E}/\dot{E}^2 \longrightarrow \dot{F}/\dot{F}^2 \xrightarrow{\ \mathrm{N}\ } \dot{E}/\dot{E}^2$$

shows that $|\dot{F}/\dot{F}^2| = 8$ (since the norm map "N" here is trivial). Let $\chi_i \colon \dot{E}/\dot{E}^2 \to \{\pm 1\}$ $(1 \leq i \leq 5)$ be the characters of the five orderings $\{P_i\}$ on E. By part (1) of Exercise 16 in Chapter VIII, we may assume that $\chi_1, \ldots, \chi_4$ form a $\mathbb{Z}_2$-basis of $\mathrm{Hom}(\dot{E}/\dot{E}^2, \{\pm 1\})$. After relabelling $\chi_1, \ldots, \chi_4$ if necessary, we may also assume that $\chi_5 = \chi_1\chi_2\chi_3$. Pick an element

$$a \in (P_1 \cap P_2 \cap P_3) \setminus P_4,$$

which must also lie in P_5. *We claim now that* $D_E\langle 1, -a\rangle = \dot{P}_4$, where the inclusion "$\subseteq$" is clear. To prove "$\supseteq$", let $b \in \dot{P}_4$. Then $\langle\langle -a, -b\rangle\rangle$ has total signature 0. Since E is formally real pythagorean, the Local-Global Principle implies that $\langle\langle -a, -b\rangle\rangle$ is hyperbolic, so $b \in D_E\langle 1, -a\rangle$, which proves the equation $D_E\langle 1, -a\rangle = \dot{P}_4$. Since $\dot{E}/\dot{E}^2 \to \dot{F}/\dot{F}^2$ is *onto* (from the exact sequence 7.6) and $-1 \in \dot{F}^2$, this equation implies that $\langle 1, -a\rangle$ is *universal over* F; that is, $a \in R(F)$. On the other hand, $a \notin \pm\dot{E}^2 \implies a \notin \dot{F}^2$, so $R(F) \supsetneq \dot{F}^2$. Can $R(F) = \dot{F}$? If so, then $I^2F = 0$ and $|W(F)| = 2|IF/I^2F| = 16$. But from the Witt ring exact sequence

$$0 \longrightarrow \langle 1, 1\rangle W(E) \longrightarrow W(E) \xrightarrow{\alpha} W(F) \longrightarrow 0$$

(where α is onto since $\dot{E}/\dot{E}^2 \to \dot{F}/\dot{F}^2$ is onto), we have

$$W(F) \cong \mathbb{Z}^5/2(\mathbb{Z}^5) \cong \mathbb{Z}_2^5. \tag{7.7}$$

Therefore, we *cannot* have $R(F) = \dot{F}$, so now 6.6 yields $[\dot{F} : R(F)] = 4$; that is, $|R(F)/\dot{F}^2| = 2$. From 7.7, we have $|I^2F| = 2$, so F is indeed a pre-Hilbert field, of the type 7.3.

Using a more general version of the above method, Berman [Bm] has in fact constructed, for each integer $n \geq 3$, a pre-Hilbert field F of level 1 with nontrivial radical, such that $|\dot{F}/\dot{F}^2| = 2^n$.

Next, we present Kula's construction (in [Ku2]) of a nonreal pre-Hilbert field F (with $|\dot{F}/\dot{F}^2| = 8$ and $|R(F)/\dot{F}^2| = 2$) of the type 7.4. For this, we assume the reader has some familiarity with valuation theory and Henselizations. Let v_3, v_5 be the 3-adic and 5-adic valuations on $\mathbb{Q}$. Working in $\overline{\mathbb{Q}}$ (the algebraic closure of $\mathbb{Q}$), let H_3 be a Henselization of $(\mathbb{Q}, v_3)$, and let H_5 be an unramified Henselian extension of $(\mathbb{Q}, v_5)$ with a quadratically closed residue field. By Springer's Theorem (which also works for Henselian fields, since all we need is Hensel's Lemma), H_3 has a square class basis $\{-1, 3\}$, and H_5 has a square class basis $\{5\}$. Now let $F := H_3 \cap H_5$. By Chapter I, Exercise 8,

$$\pi\colon \dot{F}/\dot{F}^2 \longrightarrow \dot{H}_3/\dot{H}_3^2 \times \dot{H}_5/\dot{H}_5^2$$

is injective. But

$$\pi(-1) = (-1, 1), \quad \pi(3) = (3, 1), \quad \text{and} \quad \pi(5) = (-1, 5),$$

so π is also surjective. Thus, a $\mathbb{Z}_2$-basis for $\dot{F}/\dot{F}^2$ is $\{-1, 3, 5\}$. Next, note that $-2 \in \dot{F}^2$ (since -2 is a square in both H_3 and H_5), so F *has level* $s(F) = 2$. Also $\left(\frac{-1, 3}{H_3}\right) \neq 1$ by Springer's Theorem, so $-1 \notin R(F)$. On the other hand, $\langle 1,\, 5\rangle_F$ represents 5 as well as

$$(\sqrt{-2}\,)^2 + 5 = 3, \quad \text{and} \quad 3^2 + 5(\sqrt{-2}\,)^2 = 9 - 10 = -1;$$

thus $-5 \in R(F)$. Since $|\dot F/\dot F^2| = 8$, it follows from 6.6 that $|\dot F/R(F)| = 4$ (so $R(F)/\dot F^2$ is generated by -5), and thus F must be pre-Hilbert by 6.9. To compute $W(F)$, first note that, by 6.12, $|W(F)| = 32$. From $s(F) = 2$, $W(F)$ has exponent 4, so it must be of the type 7.4 or 7.5.[7] But from $\left(\frac{-1,3}{F}\right) \neq 1$, we have $\langle 1,\, 1\rangle \not\cong \langle 3,\, 3\rangle$, so $|2W(F)| \geq 4$. Thus,

$$W(F) \cong \mathbb{Z}_4^2 \oplus \mathbb{Z}_2.$$

Also, since $\langle 1,\, 1\rangle$ is not universal, we have $P(F) = 3$. The unique quaternion division algebra over F is $\left(\frac{-1,3}{F}\right)$.

Our final task is that of constructing a nonreal pre-Hilbert field F (with $|\dot F/\dot F^2| = 8$ and a nontrivial radical) of the type 7.5. To produce such an F, we use the method of "digging holes" in fields, the main ideas of which go back already to Emil Artin. To present a broader view of this construction, we start a bit more generally, as follows.

Let K/F_0 be a field extension, and let G be a subgroup of $\dot F_0$ containing $\dot F_0^2$ such that $G \subseteq \dot K^2$. The set

$$S = \{\sqrt{x}\colon\ x \in G \setminus \dot F_0^2\} \subseteq K$$

is disjoint from F_0, so by Zorn's Lemma, there exists a subfield F of K that is maximal with respect to the properties $F \supseteq F_0$ and $F \cap S = \varnothing$. (Informally, we say that F is obtained from F_0 by "digging holes" at S in the field K.) Quite remarkably, the following general fact always holds in this situation.

Digging Holes Lemma 7.8. *We have a natural exact sequence*

$$(*) \qquad 1 \longrightarrow G/\dot F_0^2 \xrightarrow{\ \varepsilon\ } \dot F/\dot F^2 \xrightarrow{\ \delta\ } \dot K/\dot K^2.$$

Proof. Here, ε and δ are both induced by inclusions. Since $G \subseteq \dot K^2$, $(*)$ is a 0-sequence. The fact that no element $x \in G \setminus \dot F_0^2$ has a square root in F gives the injectivity of ε. Finally, suppose $a \in \dot F \setminus \dot F^2$ becomes a square in K. Then $F(\sqrt{a}\,) \supsetneq F$ must intersect S; that is, $\sqrt{x} \in F(\sqrt{a}\,)$ for some $x \in G \setminus \dot F_0^2$. As above, $x \notin \dot F^2$, so VII.3.8 implies that $x \in a\dot F^2$. Therefore, $a\dot F^2 = \varepsilon(x\dot F_0^2)$. $\square$

Remark 7.9. In applications, we often start with a field extension K/F_0 such that each square class of K is defined over F_0; that is, $\dot K = \dot F_0 \cdot \dot K^2$. In this case, δ is obviously surjective, so $(*)$ "becomes" a short exact sequence. For instance, this is always the case if each square class of K is

[7] This deduction uses nothing more than the Fundamental Theorem of Abelian Groups.

defined over its prime field. In an even more special situation, K may be a quadratically closed field, in which case the square class group of F will simply be isomorphic to $G/\dot{F}_0^2$.

We can now present the construction of a field of the type 7.5 due to T. L. Lee. Start with the field K of 5-adic numbers, whose group of square classes has basis $\{2, 5\}$. Noting that $-26 \in \dot{K}^2$, we have

$$F_0 = \mathbb{Q}(\sqrt{-26}\,) \subseteq K.$$

Since $26 = 1^2 + 5^2$, $s(F_0) = 2$. Let G be the subgroup of $\dot{F}_0$ generated by $\dot{F}_0^2$ and -1. "Digging a hole" at $i = \sqrt{-1}$ in K, let $F \subseteq K$ be a field maximal with respect to $F \supseteq F_0$ and $i \notin F$. Applying 7.8 and 7.9, we see that $\dot{F}/\dot{F}^2$ has order 8, with a basis $\{-1, 2, 5\}$. Now the form $\langle 1, 1\rangle_F$ represents -1, 2, and 5, and so $-1 \in R(F)$. This shows that F has nontrivial radical, so, as in Kula's example, $|\dot{F}/R(F)| = 4$, and F is pre-Hilbert. It remains to compute $W(F)$, which has order 32. Since $-1 \in R(F)$, we have $\langle\langle 1, a\rangle\rangle = 0 \in W(F)$ for every $a \in \dot{F}$. Therefore, $2 \cdot IF = 0$, and so $IF \cong \mathbb{Z}_2^4$. From $s(F) = 2$, it follows that

$$W(F) \cong \mathbb{Z}_4 \oplus \mathbb{Z}_2^3\,,$$

as desired. Note that here $P(F) = 2$ (since $-1 \in R(F)$), and the unique quaternion division algebra over F is $\left(\frac{2,5}{F}\right)$ (since $\left(\frac{2,5}{\mathbb{Q}_5}\right) \neq 1$.)

If one prefers a field that is algebraic over $\mathbb{Q}$, one can simply replace F by the algebraic closure F_1 of $\mathbb{Q}$ in F. It is easy to see that $\dot{F}_1/\dot{F}_1^2$ still has $\{-1,\ 2,\ 5\}$ as a basis. In fact, F_1 has all the quadratic form theoretic properties of F stated before, so F_1 is a nonreal pre-Hilbert infinite number field (with 8 square classes) of the type 7.5.

To conclude this section, let us make some remarks about the classification of Witt rings in general. From the modern viewpoint of Witt ring theory, there are two fundamental constructions which we can perform when some Witt rings are given. The first is called the "group ring construction" (or "group extension"). Given a Witt ring $R = W(F)$, we can form the group ring $R[C]$, where C is the group of two elements. The new ring $R[C]$ is a Witt ring too, since $W(F((x))) \cong R[C]$ by VI.1.7. Of course, the process $R \mapsto R[C]$ can be repeated, so we can get more generally a group ring $R[G]$, where G is any elementary 2-group.

The second construction is a bit more subtle. To begin with, note that we cannot form the *ordinary* direct products within the category of Witt rings. Indeed, if $W(F_1)$ and $W(F_2)$ are Witt rings, then the ordinary direct product $W(F_1) \times W(F_2)$ has nontrivial idempotents, and so cannot be isomorphic to the Witt ring of *any* field, according to VIII.8.6. However, a slight modification of the usual direct product construction will give a more viable substitute for $W(F_1) \times W(F_2)$.

Recall that any Witt ring $W(F)$ comes with a unique ring epimorphism $\varepsilon\colon W(F) \to \mathbb{Z}_2$, which takes the even-dimensional forms to 0, and the odd-dimensional forms to 1. Given two Witt rings $W(F_1)$ and $W(F_2)$, it is therefore natural to form the following pullback diagram:

$$\begin{array}{ccc} R & \overset{\pi_1}{\dashrightarrow} & W(F_1) \\ {\scriptstyle \pi_2}\,\vdots & & \downarrow{\scriptstyle \varepsilon_1} \\ W(F_2) & \underset{\varepsilon_2}{\longrightarrow} & \mathbb{Z}_2 \end{array} \tag{7.10}$$

where R is the subring of $W(F_1) \times W(F_2)$ consisting of pairs (q_1, q_2) with $\dim q_1$, $\dim q_2$ *of the same parity*. (Such a ring R is usually called the "fiber-product" of the fibrations ε_1 and ε_2.) *Note that the subring R has index 2 in $W(F_1) \times W(F_2)$* (so it has "half the size" of the usual direct product). In fact, the pair $(\langle 1 \rangle, 0)$ does not lie in R, but if $(q_1, q_2) \notin R$, then

$$(q_1, q_2) - (\langle 1 \rangle, 0) = (q_1 \perp \langle -1 \rangle, q_2) \in R,$$

since now $\dim(q_1 \perp \langle -1 \rangle)$ and $\dim q_2$ have the same parity. Also, R comes with a natural surjection on $\mathbb{Z}_2$, namely $\varepsilon_1\pi_1 = \varepsilon_2\pi_2$. Finally, each π_i here is *surjective*, for, given (say) q_1 over F_1 of dimension n, we have obviously

$$(q_1, n\langle 1 \rangle) \in R, \quad \text{with } \pi_1((q_1, n\langle 1 \rangle)) = q_1 \in W(F_1).$$

In the context of Witt rings, it is perhaps more appropriate to call R the *Witt product* of $W(F_1)$ and $W(F_2)$. Witt products of more than two Witt rings can be formed in a similar manner.

The challenging question here is, of course, whether the Witt product R in (7.10) can itself be realized as a Witt ring of some field. At least in the case where F_1 and F_2 have finitely many square classes, the answer is known to be "yes", although we will not be able to prove this in our text.[(8)] Various special cases of this, however, have appeared implicitly in earlier parts of this book. For instance, in II.5.4, the Witt ring $W(F)$ of type (A) is essentially the Witt product of $W(\mathbb{R})$ and $W(\mathbb{Q}_3)$, and the Witt ring of type (C) is essentially the Witt product of $W(\mathbb{R})$ and $W(\mathbb{Q}_5)$. At the end of VIII.4, the construction of a pythagorean SAP field with 2^n square classes (and n orderings) gives a Witt ring that is the Witt product of n copies of $W(\mathbb{R}) \cong \mathbb{Z}$. More relevant to this section, the field F constructed by Kula

[(8)] For more information on this in an axiomatic setting for quadratic form theory, see §8 (especially 8.6 and 8.7).

as the intersection of the two Henselian fields H_3 and H_5 yields a Witt ring that can be checked to be the Witt product of

$$W(H_3) \cong \mathbb{Z}_4[C] \quad \text{and} \quad W(H_5) \cong \mathbb{Z}_2[C],$$

where $|C| = 2$. In fact, since $|W(H_3)| = 2^4$ and $|W(H_5)| = 2^2$, the Witt product has cardinality $2^4 \cdot 2^2/2 = 2^5$. Since we have deduced independently that $W(F) \cong \mathbb{Z}_4^2 \oplus \mathbb{Z}_2$, it is easy to see that the natural map from $W(F)$ to the Witt product (guaranteed by the universal property of such products) is a ring isomorphism.

The above discussion leads us to a natural research problem in quadratic form theory that is, by now, very well-known. Let us say that a Witt ring $W(F)$ has *finite type* if $|\dot{F}/\dot{F}^2| < \infty$; by II.2.4, this simply means that $W(F)$ is a noetherian ring. Granted the fact that the Witt product of two Witt rings of finite type is another Witt ring (necessarily of finite type), one can start from the Witt rings of $\mathbb{R}$, finite fields and local fields, and form the smallest class $\mathcal{C}$ of Witt rings that is closed with respect to the formation of group extensions $R \mapsto R[C]$ ($|C| = 2$) and finite Witt products. The Witt rings in the class $\mathcal{C}$ may be called *Witt rings of elementary type.* The formation of this class of Witt rings quickly led researchers to speculate the following.

Elementary Type Conjecture 7.11. *Every Witt ring of finite type is of elementary type.*

As of today, this tantalizing conjecture has remained unsolved, even in the case of finite Witt rings (that is, the case of nonreal fields with finitely many square classes). It is worth pointing out, for instance, that the problem of whether a nonreal field F with $|\dot{F}/\dot{F}^2| < \infty$ must have level ≤ 8 (mentioned earlier in Chapter XI) may be viewed as a special case of 7.11. In fact, if Conjecture 7.11 is true, then $W(F)$ can be obtained from the Witt rings of finite fields and local fields by Witt products and group extensions. Since these Witt rings have exponent ≤ 16, the same would hold for $W(F)$, and hence the level of F would indeed have to be ≤ 8. In the same spirit, other problems on Witt rings of finite type can be efficiently tackled as soon as the Elementary Type Conjecture 7.11 is known to be true.

8. Axiomatic Schemes for Quadratic Forms

"Mathematicians love to axiomatize" is perhaps pretty much of a foregone conclusion. In a beautiful subject such as the algebraic theory of quadratic forms over fields, it is thus only natural that the practitioners of this theory tried to come up with axiomatic schemes for the subject that would place it on a pedestal that is as primitive as possible. Indeed, over the years, several such axiomatization projects have been carried out, at different levels of generality, and with varying degrees of success.

The first such effort was perhaps that of Knebusch, Rosenberg, and Ware, who developed (in [KRW]) a theory of "abstract" Witt rings. In this theory, a field is not needed, but rather, an abstract Witt ring is just a suitable quotient of an integral group ring over a torsion abelian group. This theory was designed so that it would cover the case of Hermitian (as well as quadratic) forms over *semilocal* rings. Some years later, M. Marshall invented a theory of spaces of orderings, which provided an axiomatic basis for the "reduced" theory of quadratic forms over formally real fields. Due to the lack of space, we will not cover the [KRW] theory or Marshall's theory of spaces of orderings here. Instead, we shall briefly report below on an early axiomatic scheme of C. Cordes, and a related scheme introduced later, again by M. Marshall. Our exposition will, however, be limited to a rudimentary discussion of the basic set-up of these two axiomatic theories.

Cordes's "quadratic form scheme" in his paper [Co$_2$] was directly inspired by his work on the q-equivalence of field, which we presented in §1. In Cordes's own words, the idea of defining such a scheme is to "abstract value sets of binary quadratic forms in order to eliminate the immediate necessity of a field." In its most primitive form, a quadratic form scheme can be defined as follows.

Definition 8.1. Let G be a multiplicative elementary 2-group with a distinguished element denoted by -1 (which may be equal to 1). For any $a \in G$, we denote the product $(-1)a$ by $-a$. A *quadratic form scheme* is a triple $(G, -1, V)$, where V is a mapping from G into the set of all subgroups of G satisfying the following two conditions:

(C1) $a \in V(a)$ *for every* $a \in G$.

(C2) *If* $b \in V(a)$, *then* $-a \in V(-b)$.

Given any field F (of characteristic not 2), we get a quadratic form scheme by taking $G = \dot{F}/\dot{F}^2$, and -1 to be the coset $-\dot{F}^2$, with $V(a) := D_F\langle 1, a\rangle$ viewed as a subgroup of G. Here, the axiom (C1) is trivial, and the axiom (C2) merely states the fundamental property:

$$(8.2) \qquad b \in D_F\langle 1,\, a\rangle \implies -a \in D_F\langle 1,\, -b\rangle.$$

The pair $(\dot{F}/\dot{F}^2, -\dot{F}^2)$ together with the map $a \mapsto D_F\langle 1, a\rangle \subseteq G$ is called the *quadratic form scheme of the field* F.

Notice that, given any elementary 2-group G with a distinguished element $-1 \in G$, we always get a quadratic form scheme by taking $V(a) = G$ for every $a \in G$. This case "corresponds" to form schemes over nonreal fields of u-invariant ≤ 2. Such fields have been constructed in Theorem 5.1, at least in the case where $|G| < \infty$.

From a categorical point of view, a *morphism* from one quadratic form scheme $(G, -1, V)$ to another $(G', -1', V')$ can be defined to be a group homomorphism $f\colon G \to G'$ such that $f(-1) = -1'$ and $f(V(a)) \subseteq V'(f(a))$ for every $a \in G$. With this notion of morphisms in place, we arrive at a category $\mathcal{C}$ of quadratic form schemes. The isomorphisms in this category are the morphisms

$$\varphi\colon\ (G, -1, V) \longrightarrow (G', -1', V')$$

such that $\varphi\colon G \to G'$ is a group isomorphism and $\varphi(V(a)) = V'(\varphi(a))$ for every $a \in G$. The construction of the form scheme of a field defines a functor from the category of fields to this new category $\mathcal{C}$, since every morphism of fields $\varphi\colon F \to K$ induces a group homomorphism $\varphi_*\colon \dot{F}/\dot{F}^2 \to \dot{K}/\dot{K}^2$ taking $-\dot{F}^2$ to $-\dot{K}^2$, and obviously

$$b \in D_F\langle 1, a\rangle \implies \varphi_*(b) \in D_K\langle 1, \varphi_*(a)\rangle.$$

Upon reviewing Definition 1.1, we see that *two fields F and K are q-equivalent iff their quadratic form schemes are isomorphic in the category $\mathcal{C}$.* According to Theorem 1.8, this is the case iff F and K have isomorphic Witt rings.

In view of these nice facts, it would be of interest to investigate the category $\mathcal{C}$ in its own right and find out to what extent a theory of quadratic forms can be developed on the basis of (C1) and (C2) alone. To begin with, given a general quadratic form scheme $(G, -1, V)$, we can define an ("abstract") n-ary form to be just a formal n-tuple $\langle a_1, \ldots, a_n\rangle$, with all $a_i \in G$. Such forms (for all $n \geq 0$) can be "added" and "multiplied" as usual, but, to get something resembling a quadratic form theory, we would need to define the notion of *value sets* of forms, and also the notion of *isometries.*

As it turned out, these notions can all be defined, but it does not appear that we can prove a large number of desirable properties about them. To illustrate this point, let us first consider the notion of value sets of forms.

For a unary form $\langle a\rangle$, we can define its value set to be $D\langle a\rangle = \{a\} \subseteq G$. For a binary form $\langle b, c\rangle$, we can take $D\langle b, c\rangle$ to be $bV(bc) \subseteq G$. Since $bc \in V(bc)$ by (C1), we have $bV(bc) = cV(bc)$, so indeed $D\langle b, c\rangle = D\langle c, b\rangle$ (for all $b, c \in G$). With this notion of $D\langle b, c\rangle$, we can then define a *Kaplansky radical* for the scheme, namely, the set $R(G, -1, V)$ consisting of all $a \in G$ such that $\langle 1, -a\rangle$ is "universal" (i.e., with $D\langle 1, -a\rangle = V(-a) = G$). By (C2), $R(G, -1, V)$ can be identified as $\bigcap_{b\in G} V(b)$, so, as expected, it is a subgroup of G. (In particular, we have $1 \in R(G,-1,V)$; that is, $\langle 1,-1\rangle$ is always universal.)

For forms $\langle a_1,\dots,a_n\rangle$ (with $n \geq 3$), it is reasonable to define their value sets by induction on n:

$$D\langle a_1,\dots,a_n\rangle = \bigcup \{D\langle a_1,\, t\rangle \colon\ t \in D\langle a_2,\dots,a_n\rangle\}, \tag{8.3}$$

following the idea of Exercise 20 in Chapter I. But now the limitation of the axioms (C1) and (C2) becomes apparent: it does not seem clear, e.g., if $\langle a_1,\dots,a_n\rangle$ and $\langle a_{\sigma(1)},\dots,a_{\sigma(n)}\rangle$ have the same value sets for any permutation σ. It is not difficult to prove (by induction) that this would hold for all n if it holds for $n = 3$. In the latter situation, the crucial case to check is $D\langle a,b,1\rangle = D\langle b,a,1\rangle$. This would require that

(C3) *If* $x \in D\langle a,\, t\rangle$, *where* $t \in D\langle b,\, 1\rangle$, *then* $x \in D\langle b,\, s\rangle$ *for some* $s \in D\langle a,\, 1\rangle$.

Such a property does not seem easily provable from (C1) and (C2). It may thus be another axiom that one might want to add to the system $(G,-1,V)$ (see, e.g. [Ku3]). An easy transformation (using (C2)) produces the following equivalent form of this new axiom:

$$D\langle x,-a\rangle \cap V(b) \neq \varnothing \implies D\langle x,-b\rangle \cap V(a) \neq \varnothing. \tag{C3$'$}$$

This may be thought of as a "binary version" of (C2), since (C2) could have been expressed in the following fashion:

$$D\langle -a\rangle \cap V(b) \neq \varnothing \implies D\langle -b\rangle \cap V(a) \neq \varnothing. \tag{C2}$$

The notion of *isometry* between n-ary forms presents similar problems. For unary forms, of course, $\langle a\rangle \cong \langle a'\rangle$ should mean just $a = a' \in G$. For binary forms, we may define $\langle b,b'\rangle \cong \langle c,c'\rangle$ to mean that $bb' = cc' \in G$ and $D\langle b,b'\rangle \cap D\langle c,c'\rangle \neq \varnothing$ (cf. I.5.1).[9] Since $D\langle b,b'\rangle = bV(bb')$ and $D\langle c,c'\rangle = cV(cc')$, these two conditions actually imply that $D\langle b,b'\rangle = D\langle c,c'\rangle$. Thus, two binary forms are isometric iff they have the same "determinant" and the same value sets. From this, it follows that isometry between binary forms is an equivalence relation.

For n-ary forms with $n \geq 3$, we may define isometry by induction on n, in the style of Chapter I, Exercise 19. Thus, we define $\langle a_1,\dots,a_n\rangle \cong \langle b_1,\dots,b_n\rangle$ to mean that there exist $a,b,c_3,\dots,c_n \in G$ such that

$$\langle a_2,\dots,a_n\rangle \cong \langle a,c_3,\dots,c_n\rangle, \qquad \langle b_2,\dots,b_n\rangle \cong \langle b,c_3,\dots,c_n\rangle, \tag{8.4}$$

and $\langle a_1,a\rangle \cong \langle b_1,b\rangle$ (assuming that 8.4 is already meaningful for $(n-1)$-ary forms). The problem here is that it is not clear if isometry is an equivalence relation. (Reflexity and symmetry are easy, but transitivity is not.) Again,

[9] For instance, $\langle 1,-1\rangle \cong \langle a,-a\rangle$ (for every $a \in G$), since $D\langle 1,-1\rangle = V(-1) = G$.

the case $n = 3$ is crucial, but even in this case, transitivity looks murky, and does not seem to follow from (C1) and (C2).

On the positive side, however, the simple axioms for the objects and morphisms in the category $\mathcal{C}$ of quadratic form schemes make it possible, and in fact quite easy, to define *Witt products* and *group extensions* in $\mathcal{C}$. We proceed as follows.

If $(G_1, -1, V_1)$ and $(G_2, -1, V_2)$ are objects in $\mathcal{C}$, their Witt product may be defined to be the triple $(G, -1, V)$, where

(8.5) $\quad G = G_1 \times G_2, \quad -1 = (-1,-1), \quad \text{and} \quad V(a_1, a_2) = V_1(a_1) \times V_2(a_2)$

for any $a_i \in G_i$. It is a completely routine matter to check that $(G, -1, V)$ is a quadratic form scheme, and that it satisfies the usual universal property of a direct product in $\mathcal{C}$.

The construction of a *group extension* of a scheme $(G, -1, V)$ in $\mathcal{C}$ follows closely the pattern of the quadratic form theory over a Laurent series field $F((x))$. Letting $C = \{1, x\}$ be a group of order 2, we define the group extension $(G, -1, V)[C]$ to consist of the elementary 2-group $G \times C$ with the distinguished element $(-1, 1)$, and with the map V^C sending $a \in G$ to $V(a) \subseteq G \subseteq G \times C$ if $a \neq -1$, and to $G \times C$ if $a = -1$, and sending (a, x) to $\{1, ax\} \subseteq G \times C$. Again, it if not difficult to check directly that $(G \times C, -1, V^C)$ is an object of $\mathcal{C}$. This construction is set up in such a way that, if $(G, -1, V)$ is the quadratic form scheme of a field F, then $(G, -1, V)[C]$ is isomorphic to the quadratic form scheme of the Laurent series field $F((x))$.

The advantage of working with more general concepts in mathematics resides in the fact that such concepts sometimes lead us to interesting new discoveries. In the case of the formulation of Cordes's abstract notion of quadratic form schemes, a particularly fruitful result is the following realization theorem of Kula.

Theorem 8.6. *Given two fields F_1 and F_2 with finitely many square classes, there exists a field F whose quadratic form scheme is isomorphic to the Witt product of those of F_1 and F_2.*

The proof of this theorem involves a rather heavy dosage of valuation theory, which we are not in a position to present here. The interested reader is referred to Kula's original papers ([Ku$_1$], [Ku$_3$]). More pertinent for us is the following powerful consequence of 8.6.

Corollary 8.7. *Let F_1, F_2 and F be as in 8.6. Then $W(F)$ is isomorphic to the Witt product of the two Witt rings $W(F_1)$ and $W(F_2)$. In particular, the Witt product of two Witt rings $W(F_1)$, $W(F_2)$ of finite type is again a Witt ring of finite type.*

Proof. Let $(\varphi_1, \varphi_2)\colon \dot{F}/\dot{F}^2 \to \dot{F}_1/\dot{F}_1^2 \times \dot{F}_2/\dot{F}_2^2$ be a group isomorphism giving rise to an isomorphism from the form scheme over F to the Witt product of the form schemes over F_1 and F_2. Then each φ_i induces a ring homomorphism $(\varphi_i)_*\colon W(F) \to W(F_i)$, so $((\varphi_1)_*, (\varphi_2)_*)$ defines a ring homomorphism φ_* from $W(F)$ to the Witt product R of $W(F_1)$ and $W(F_2)$. Since (φ_1, φ_2) is onto, so is φ_*. Finally, note that, for any F-form q,

$$D_F(q) = D_{F_1}(\varphi_{1*}(q)) \times D_{F_2}(\varphi_{2*}(q)).$$

(This follows by an easy induction on $\dim(q)$.) Therefore, an F-form $\langle a\rangle \perp q$ is isotropic iff each $\varphi_{i*}(\langle a\rangle \perp q)$ is isotropic over F_i. In particular, φ_* is injective (as well as surjective), so it is an isomorphism from $W(F)$ to R. □

For an illustration of 8.6 and 8.7, let F_1 be $\mathbb{Q}_5$ (the field of 5-adic numbers) and let $F_2 = \mathbb{R}$. The form scheme of F_1 has group $G_1 = [a, b]$, the elementary 2-group with basis a, b, where $a = 2$, $b = 5$. Here, $-1 = 1 \in G_1$, and the map V_1 (from G_1 to subgroups of G_1) is given by

$$V_1(1) = [a, b], \quad V_1(a) = [a], \quad V_1(b) = [b], \quad \text{and} \quad V_1(ab) = [ab].$$

On the other hand, the form scheme of F_2 has group $G_2 = [-1]$, the elementary 2-group with basis -1, and the map V_2 is simply given by $V_2(1) = \{1\}$, $V_2(-1) = [-1]$. Upon taking the Witt product of $(G_1, -1, V_1)$ and $(G_2, -1, V_2)$, we get a scheme with group

$$G = G_1 \times G_2 = [-1,\, a,\, b],$$

and a map V (from G to subgroups of G) given by the following chart:

(8.8)

x	$V(x)$	x	$V(x)$
1	$[a,\, b]$	-1	$[-1,\, a,\, b]$
a	$[a]$	$-a$	$[-1,\, a]$
b	$[b]$	$-b$	$[-1,\, b]$
ab	$[ab]$	$-ab$	$[-1,\, ab]$

Kula's Theorem 8.6 predicts that this is the form scheme of some field. In the constructive part of the proof of II.5.7, Case (2), we have indeed come up with such a field, namely, by taking the intersection of $\mathbb{Q}_5$ with a real-closed subfield of the algebraic closure of $\mathbb{Q}_5$.

By 8.7, *any* field F realizing the form scheme 8.8 has a Witt ring of characteristic zero (since $W(F)$ is the Witt product of $W(F_1)$ and $W(F_2) \cong \mathbb{Z}$). Thus, F must be formally real. This can also be seen directly as follows (after Cordes). Since F realizes the form scheme 8.8,

$$\begin{aligned} D_F\langle 1,1,1\rangle &= \bigcup_{x\in D\langle 1,1\rangle} D\langle 1, x\rangle \\ &= D\langle 1,1\rangle \cup D\langle 1,a\rangle \cup D\langle 1,b\rangle \cup D\langle 1,ab\rangle. \end{aligned}$$

Since the last three groups here are all in $[a, b] = D_F\langle 1,1\rangle$, the above computation gives $D_F\langle 1,1,1\rangle = D_F\langle 1,1\rangle = [a,b]$. By induction, we see that $D_F(\infty) = [a,b]$. In particular, $-1 \notin D_F(\infty)$, so F must be formally real.

On the other hand, Kula's construction of a field F with 8 square classes and $W(F) \cong \mathbb{Z}_4 \oplus \mathbb{Z}_4 \oplus \mathbb{Z}_2$ in XII.7 was a concrete case of Theorem 8.6 where all three fields F_1, F_2, and F are nonreal.

Since there are only two axioms ((C1) and (C2)) in Cordes's definition of a quadratic form scheme, such a scheme is perceived to be a somewhat "crude" object. Of course, more axioms may be added to (C1) and (C2) to increase the chance that the scheme be realizable by some field. Or, one could introduce a finer axiomatic system that would focus our attention only on quadratic form schemes arising in a certain way. Murray Marshall's theory of "quaternionic structures" is precisely an effort in this direction. In the balance of this section, we shall give a quick introduction to Marshall's theory, following largely his Queen's University notes [Ma$_1$].

The distinctive feature of Marshall's theory is the (hypothetical) presence of a set of "quaternion algebras". However, these "quaternion algebras" will only form a pointed set, and they need not be elements of an abelian group (like the Brauer group). To be precise, the following is Marshall's definition.

Definition 8.9. A *quaternionic structure* (or a Q-structure, for short) is a triple (G, Q, q), where G is (as in 8.1) an elementary 2-group with a distinguished element -1, Q is a pointed set with distinguished element 1 (not to be confused with the identity of G), and $q\colon G \times G \to Q$ is a *surjective* mapping satisfying the following axioms:

(Q1) $q(a, b) = q(b, a)$.

(Q2) $q(a, -a) = 1$ (where, as before, $-a = (-1)a \in G$).

(Q3) $q(a, b) = q(a, c)$ iff $q(a, bc) = 1$.

(Q4) $q(b, b') = q(c, c') \implies q(b, b') = q(b, x),\ q(c, c') = q(c, x)$ for some $x \in G$.

The first relevant observation is, of course, that any field F gives rise to a Q-structure. We simply take $G = \dot{F}/\dot{F}^2$, $-1 = -\dot{F}^2$, and take Q to be the set of quaternion algebras in the Brauer group of F (with the split quaternion algebra as the distinguished element). Then, we take Q to be the pairing

$$(a\dot{F}^2,\, b\dot{F}^2) \mapsto \left(\frac{a,\, b}{F}\right).$$

Here, (Q1) and (Q2) are well-known properties of quaternion algebras, and (Q3) is a weak version of the linearity property in II.2.11. Finally, (Q4) is the common slot property in II.4.13! Because of this, we may refer to (Q4) as the *Common Slot Axiom* for the general quaternionic structure defined in 8.9.

Note that the axioms above capture essentially all of the formal properties we know about the quaternion algebras $\left(\frac{a,b}{F}\right)$ — or at least those which do not involve the tensor products of such algebras. For instance, the following additional properties of $(G,\, Q,\, q)$ can be easily deduced from the axioms.

(Q5) $\quad q(a,\, 1) = 1$ (since $q(a,\, 1) = q(a,\, 1) \Longrightarrow q(a,\, 1^2) = 1$).

(Q6) $\quad q(a,\, -1) = q(a,\, a)$ (since $q(a,\, (-1)a) = q(a,\, -a) = 1$).

(Q7) $\quad q(a,\, -ab) = q(a,\, b)$ (since $q(a,\, -ab^2) = q(a,\, -a) = 1$).

Next, we note that *any Q-structure $(G,\, Q,\, q)$ as in 8.9 gives rise to a quadratic form scheme*: we simply use the same group $(G,\, -1)$, and define V by

(8.10) $$V(a) = \{b \in G\colon\ q(-a,\, b) = 1\} \quad (\forall\, a \in G).$$

Since $q(-a,\, a) = 1$, we have $a \in V(a)$. And, if $b \in V(a)$, then

$$q(-a,\, b) = 1 \implies q(b,\, -a) = 1 \implies -a \in V(-b).$$

This verifies the axioms (C1) and (C2) in 8.1, so $(G,\, -1,\, V)$ is indeed a quadratic form scheme (in the sense of Cordes). We note further that, as soon as the map V is given, then we know exactly which pairs of "abstract" quaternion algebras are equal, via the following.

Lemma 8.11. *$q(b, b') = q(c, c')$ iff there exists $x \in V(-bc)$ such that $b \in V(-b'x)$ and $c \in V(-c'x)$.*

Proof. Suppose such an x exists. Then $q(bc, x) = 1$, $q(b, b'x) = 1$, and $q(c, c'x) = 1$. By (Q3), these give

$$q(b, b') = q(b, x) = q(c, x) = q(c, c').$$

Conversely, suppose $q(b, b') = q(c, c')$. By (Q4), we have $q(b, b') = q(b, x)$ and $q(c, c') = q(c, x)$ for some $x \in G$. Then, by (Q3),

$$q(b,\, b'x) = 1, \quad q(c,\, c'x) = 1, \quad q(bc,\, x) = 1.$$

Now 8.10 yields $b \in V(-b'x)$, $c \in V(-c'x)$, and $x \in V(-bc)$. ☐

The work above suggests also how one might try to construct a Q-structure starting from any quadratic form scheme. Let $(G, -1, V)$ be such a form scheme. Motivated by 8.11, we introduce a relation "$\sim$" on ordered pairs $(b, b') \in G^2$ as follows:

$$(8.12)\qquad (b, b') \sim (c, c') \quad \text{if} \quad b \in V(-b'x) \text{ and } c \in V(-c'x) \quad \text{for some } x \in V(-bc).$$

It is easy to see that this relation is reflexive and symmetric. As for transitivity, we have the result below.

Theorem 8.13. *A quadratic form scheme $(G, -1, V)$ "comes from" a quaternionic structure iff the relation on G^2 defined in 8.12 above is a transitive relation.*

Proof. If $(G, -1, V)$ comes from some Q-structure (G, Q, q), then, by 8.11, $(b, b') \sim (c, c')$ amounts to $q(b, b') = q(c, c') \in Q$. Clearly, "$\sim$" is then transitive.

Conversely, if "$\sim$" is transitive, then it is an equivalence relation. Let $q(b, b')$ denote the equivalence class of $(b, b') \in G^2$, and let Q be the set of all equivalence classes, with the class of $(1, 1)$ as the distinguished element. Then q is a surjective mapping from G^2 to Q, and we are done if we can check that each of the four properties (Q1), ..., (Q4) holds. (In the following work, keep in mind that $1 \in V(a)$ for any $a \in G$, or quivalently, $V(-1) = G$.)

(Q1) If $c = b'$ and $c' = b$, then 8.12 holds with $x = -bb'$. Therefore, we have $q(b, b') = q(b', b)$ for all $b, b' \in G$.

(Q2) If $b' = -b$ and $c = c' = 1$, then 8.12 holds with $x = 1$. This means that $q(b, -b) = 1$ for all $b \in G$.

To prove (Q3) and (Q4), we need the following observation:

$$(8.14)\qquad b \in V(-b') \iff (b, b') \sim (1, 1).$$

By 8.12, the RHS amounts to the existence of an element $x \in V(-b)$ such that $b \in V(-b'x)$. The latter implies that $b'x \in V(-b)$, so $b' \in V(-b)$, which gives $b \in V(-b')$. Conversely, if $b \in V(-b')$, we get the RHS of 8.14 by taking $x = b'$.

(Q3) If $q(b, b') = q(b, c')$, 8.12 yields an $x \in G$ such that $b \in V(-b'x) \cap V(-c'x)$. Thus, $b'x, c'x \in V(-b)$. This implies that $b'c' \in V(-b)$, and hence $q(b, b'c') = 1$ by 8.14. Conversely, assume that $q(b, b'c') = 1$. By 8.14, $b \in V(-b'c')$. But then $b \in V(-b'x) \cap V(-c'x)$ holds for $x = c'$, so we have $q(b, b') = q(b, c')$.

(Q4) Suppose $q(b, b') = q(c, c')$. Then $b \in V(-b'x)$ and $c \in V(-c'x)$ for some x. By 8.14, $q(b, b'x) = q(c, c'x) = 1$, so by (Q3) we have $q(b, b') = q(b, x)$ and $q(c, c') = q(c, x)$.

The above work shows that (G, Q, q) is a Q-structure, and 8.14 shows that $(G, -1, V)$ is the quadratic form scheme associated to (G, Q, q). □

In view of Theorem 8.13, we may think of a quaternionic structure (G, Q, q) as a kind of "strengthened" quadratic form scheme $(G, -1, V)$. The "strengthening" lies in the assumption of the transitive property of the relation "$\sim$" on G^2. However, while this transitive property is quite natural, its actual formulation in terms of the mapping V would make a rather messy axiom. On the other hand, all of Marshall's axioms (Q1)–(Q4) for a quaternionic structure are easy and neat. This would seem to suggest that Marshall's theory of Q-structures is "the way to go" in formulating the axiomatic foundations of quadratic form theory over fields.

To provide some further evidence for the effectiveness of Marshall's system, let us present here the beginning parts of the axiomatic development of quadratic form theory with respect to a given quaternionic structure.

Given a Q-structure (G, Q, q), we have already constructed its associated quadratic form scheme $(G, -1, V)$, so it will be convenient to use the V-notation in conjunction with (G, Q, q). As before, an n-ary form will simply mean a formal expression $\langle a_1, \ldots, a_n \rangle$, where $a_i \in G$. The isometry of unary and binary forms has already been discussed before in the context of a form scheme. For binary forms, we have the following alternative interpretation of isometry in terms of the pairing q.

Proposition 8.15. (Cf. III.2.10.) *$\langle b, b' \rangle \cong \langle c, c' \rangle$ iff $bb' = cc' \in G$ and $q(b, b') = q(c, c') \in Q$.*

Proof. First assume $\langle b, b' \rangle \cong \langle c, c' \rangle$. By our earlier definition of binary form isometry in a form scheme, this means that $bb' = cc'$ and $bc \in V(bb')$. Thus, $-bb' \in V(-bc)$, so the condition in 8.11 is satisfied with $x = -bb'$. Hence $q(b, b') = q(c, c')$.

Conversely, assume that $e := bb' = cc'$ and that $q(b, b') = q(c, c')$. By (Q7),

$$q(b, -e) = q(b, b') = q(c, c') = q(c, -e).$$

Therefore, by (Q3), $q(bc, -e) = 1$, and this implies that $bc \in V(e)$ by (Q1) and 8.10. Hence $\langle b, b' \rangle \cong \langle c, c' \rangle$. □

For n-ary forms with $n \geq 3$, isometry is defined inductively, using the idea of Exercise 19 in Chapter I. For a precise formulation, see the discussion around 8.4. This inductive definition of isometry might at first appear a little unwieldy. But it is actually reasonably efficient for doing proofs. Let us mention some properties of isometries ($\cong$) which can be quickly verified from the inductive definition. The proofs of these are left to the reader.

(8.16) $g \cong g' \implies c \cdot g \cong c \cdot g'$, where $c \cdot \langle a_1, \ldots, a_n \rangle = \langle ca_1, \ldots, ca_n \rangle$.

(8.17) $g \cong g' \implies d(g) = d(g')$, where $d(\langle a_1, \ldots, a_n \rangle) := a_1 \cdots a_n$.

(8.18) $g \cong g' \implies f \perp g \cong f \perp g'$ and $g \perp f \cong g' \perp f$.[10]

(8.19) $g \cong g$, and $f \cong g \implies g \cong f$.

While the isometry relation (on n-ary forms) is reflexive and symmetric (as claimed in 8.19), the inductive definition does not lend itself readily to a proof of transitivity. Fortuitously, in a quaternionic structure, transitivity of isometry *does* hold; let us now present the (rather tricky) proof of this (due to Marshall).

We have already dealt with the case of forms of dimension ≤ 2 in the more general context of quadratic form schemes. In the present case of a Q-structure, an alternative proof for the transitivity of binary form isometries follows from 8.15. For ternary forms ("scaled" to have determinant -1), we can similarly appeal to the following extension of 8.15.

Proposition 8.20. $\langle b, b', -bb' \rangle \cong \langle c, c', -cc' \rangle$ *iff* $q(b, b') = q(c, c')$. (*Thus, isometry is an equivalence relation for ternary forms.*)

Proof. First assume $q(b,b') = q(c,c')$. By (Q4), $q(b,b') = q(b,x)$ and $q(c,c') = q(c,x)$ for some $x \in G$. Applying (Q7) and 8.15, we can check that

$$\langle b', -bb' \rangle \cong \langle -bx, x \rangle \quad \text{and} \quad \langle c', -cc' \rangle \cong \langle -cx, x \rangle,$$

and likewise, $\langle b, -bx \rangle \cong \langle c, -cx \rangle$. Therefore, by the inductive definition of ternary form isometry, we have

$$\langle b, b', -bb' \rangle \cong \langle c, c', -cc' \rangle.$$

The converse is proved by (essentially) reversing these steps. $\square$

Having settled the case of dimensions ≤ 3, we can now tackle the general case.

Theorem 8.21 (Marshall). *Isometry on n-ary forms in any Q-structure is a transitive relation (and hence an equivalence relation).*

---

[10] If $f = \langle a_1, \ldots, a_n \rangle$ and $g = \langle b_1, \ldots, b_m \rangle$, the orthogonal sum $f \perp g$ means, of course, the form $\langle a_1, \ldots, a_n, b_1, \ldots, b_m \rangle$.

Proof. We may assume that $n \geq 4$, and that the statement is true for $(n-1)$-ary forms. Suppose $f \cong g \cong h$ are n-ary forms, with

$$f = \langle a \rangle \perp f', \quad g = \langle b \rangle \perp b', \quad h = \langle c \rangle \perp h'.$$

Then we can write $f' \cong \langle a' \rangle \perp k$ and $g' \cong \langle b' \rangle \perp k$, with $\langle a, a' \rangle \cong \langle b, b' \rangle$. Similarly, we can write $g' \cong \langle b'' \rangle \perp k'$ and $h' \cong \langle c' \rangle \perp k'$, with $\langle b, b'' \rangle \cong \langle c, c' \rangle$. By the inductive hypothesis, $\langle b' \rangle \perp k \cong \langle b'' \rangle \perp k'$. Therefore, we can write $k \cong \langle b_1 \rangle \perp \ell$ and $k' \cong \langle b_2 \rangle \perp \ell$, with $\langle b', b_1 \rangle \cong \langle b'', b_2 \rangle$. Applying 8.18, we have

$$\langle a,\, a',\, b_1 \rangle \cong \langle b,\, b',\, b_1 \rangle \cong \langle b,\, b'',\, b_2 \rangle \cong \langle c,\, c',\, b_2 \rangle,$$

so by 8.20, $\langle a,\, a',\, b_1 \rangle \cong \langle c,\, c',\, b_2 \rangle$. Thus, we can further write

$$\langle a',\, b_1 \rangle \cong \langle a_1,\, d \rangle, \qquad \langle c',\, b_2 \rangle \cong \langle c_1,\, d \rangle,$$
$$\text{with } \langle a,\, a_1 \rangle \cong \langle c,\, c_1 \rangle.$$

Now set $p = \langle d \rangle \perp \ell$. Applying 8.18 again, we get

$$f' \cong \langle a' \rangle \perp k \cong \langle a',\, b_1 \rangle \perp \ell \cong \langle a_1,\, d \rangle \perp \ell = \langle a_1 \rangle \perp p,$$
$$h' \cong \langle c' \rangle \perp k' \cong \langle c',\, b_2 \rangle \perp \ell \cong \langle c_1,\, d \rangle \perp \ell = \langle c_1 \rangle \perp p.$$

By the inductive hypothesis again, we have $f' \cong \langle a_1 \rangle \perp p$ and $h' \cong \langle c_1 \rangle \perp p$. Since $\langle a,\, a_1 \rangle \cong \langle c,\, c_1 \rangle$, the inductive definition for n-ary form isometry yields $f \cong h$. □

Corollary 8.22 (Cancellation Property). *Suppose $f \cong f'$. Then*

$$f \perp g \cong f' \perp g' \quad \textit{iff} \quad g \cong g'.$$

Proof. If $g \cong g'$, then $f \perp g \cong f \perp g' \cong f' \perp g'$, and we are done by invoking transitivity. Conversely, assume that $f \perp g \cong f' \perp g'$. Since $f' \perp g' \cong f \perp g'$, we have $f \perp g \cong f \perp g'$ (again by transitivity). To cancel f, it suffices to handle the case where $f \cong \langle a \rangle$. By the inductive definition of isometry, we can write $g \cong \langle b \rangle \perp k$, $g' \cong \langle c \rangle \perp k$, with $\langle a, b \rangle \cong \langle a, c \rangle$. By 8.15, $ab = ac$, so $b = c$. Thus, $g \cong \langle b \rangle \perp k \cong g'$. □

We can next define the *value set* $D(f)$ for any form f. In the context of a quadratic form scheme, we have pointed out earlier that $D(f)$ can be defined by induction on $n = \dim(f)$, as in 8.3. In working with a Q-structure (G, Q, q), since the notion of isometries works so well, we can give a more convenient definition (without using induction) by taking

(8.23) $$D(f) = \{x \in G\colon\ f \cong \langle x, x_2, \ldots, x_n \rangle \text{ for some } x_i \in G\}.$$

One obvious advantage of this definition is that $D(f)$ clearly depends only on the isometry class of f. The following proposition reconciles this definition with that given earlier in 8.3.

Proposition 8.24. *Let* $f = \langle a_1, \ldots, a_n\rangle$. *Then* $D(f) = \bigcup D\langle a_1,\, t\rangle$, *where* t *ranges over* $D\langle a_2, \ldots, a_n\rangle$.

Proof. First, let $x \in D\langle a_1,\, t\rangle$, where $t \in D\langle a_2, \ldots, a_n\rangle$. The latter means that $\langle a_2, \ldots, a_n\rangle \cong \langle t, x_3, \ldots, x_n\rangle$ for suitable $x_i \in G$. Similarly, $\langle a_1,\, t\rangle \cong \langle x,\, x_2\rangle$ for some $x_2 \in G$. Therefore,

$$f \cong \langle a_1, a_2, \ldots, a_n\rangle \cong \langle a_1, t, x_3, \ldots, x_n\rangle \cong \langle x, x_2, \ldots, x_n\rangle,$$

and so $x \in D(f)$. Conversely, let $x \in D(f)$, so $f \cong \langle x,\, x_2, \ldots, x_n\rangle$ for suitable x_i's. By the inductive definition of isometry, we can write

$$\langle a_2, \ldots, a_n\rangle \cong \langle t\rangle \perp g, \qquad \langle x_2, \ldots, x_n\rangle \cong \langle s\rangle \perp g,$$

with $\langle a_1,\, t\rangle \cong \langle x,\, x_2\rangle$. We have then $x \in D\langle a_1,\, t\rangle$, where $t \in D\langle a_2, \ldots, a_n\rangle$, as desired. □

Again exploiting the fact that isometries are well-behaved, we can define a form f to be *isotropic* if $f \cong \mathbb{H} \perp g$ for some form g (where $\mathbb{H}$ is the "abstract" hyperbolic plane $\langle 1,\, -1\rangle \cong \langle a,\, -a\rangle$), and *anisotropic* otherwise. With these definitions, we can prove a Witt decomposition theorem as in I.4.1, and we can define two forms to be *Witt-similar* if their anisotropic parts are isometric. All of this work leads to a Witt ring $W(G, Q, q)$, whose elements are "Witt-similarity classes" of forms, in one-one correspondence with the isometry classes of the anisotropic forms.

We hope that the above more or less self-contained exposition has provided a quick introduction to the beginning parts of the abstract theory of quadratic forms with respect to a quaternionic structure. Readers desiring to learn more about this theory are encouraged to consult Marshall's excellent monograph [Ma$_1$], which is by far the best and most complete reference on this subject.

In conclusion, we should point out that, while a large part of the usual quadratic form theory can be carried over to the setting of a Q-structure, it is by no means true that *every* theorem in quadratic form theory can be proved in the axiomatic setting. For instance, the Arason-Pfister Hauptsatz (whose proof depends on the use of transcendental methods) has so far defied a generalization to the abstract setting. To be precise, it is not known, for $n \geq 3$, if every form of dimension $< 2^n$ lying in the nth power of the fundamental ideal of $W(G, Q, q)$ must be hyperbolic. Also, much of the theory of quadratic forms under (algebraic or transcendental) field extensions does not seem easily generalizable to the axiomatic setting. Finally, the problem of realizing a quaternionic structure (or a quadratic form scheme) by a field has remained difficult, and essentially unsolved.

Exercises for Chapter XII

1. (Pfister) Let q be a form over F with $\dim q = 2m \leq 12$. If $d(q) = (-1)^m$ and $c(q) = 1$, show that $q \in I^3F$. (A considerable part of this theorem of Pfister is already implicit in §2 of this chapter. As we mentioned in V.6, in 1981, Merkurjev has extended Pfister's theorem to *all* even-dimensional forms.)

2. ([ELW$_1$]) Show that each of the following conditions is equivalent to K/F being an excellent field extension:

   (1) If a K-form σ is defined over F, so is the anisotropic part of σ.
   (2) If an F-form τ becomes isotropic over K, then $\tau_K \cong \tau'_K$ for some isotropic F-form τ'.
   (3) (In case $[K : F] = n < \infty$.) Condition (2) for $\dim \tau \leq n$.

3. ([ELW$_1$]) Let F be a nonreal field with u-invariant ≤ 4 such that every anisotropic 4-dimensional form has determinant 1 (in $\dot{F}/\dot{F}^2$). Show that every field extension K/F is excellent. (This applies, for instance, to all nonreal Hilbert fields.)

4. (Ware) Show that a field extension K/F is *not* excellent if $W(K/F) = 0$ but some anisotropic F-form becomes isotropic over K.

5. Let F be a pre-Hilbert field. Show that:

   (1) $D\langle 1, a\rangle = D\langle 1, b\rangle$ iff $ab \in R(F)$.
   (2) If $[\dot{F} : R(F)] < \infty$, then any index 2 subgroup of $\dot{F}$ containing $R(F)$ has the form $D\langle 1, a\rangle$ for some $a \in \dot{F}$.

6. Let $K = F(\sqrt{a}\,)$ be a quadratic extension of F. For any $b \in \dot{F}$, show that
$$\dot{F} \cap D_K\langle 1, b\rangle = D_F\langle 1, b\rangle \cdot D_F\langle 1, ab\rangle.$$

7. Let $K = F(\sqrt{-1}\,)$, where F is a formally real pythagorean field. Use Exercise 6 to show that, for any $b \in \dot{F}$,
$$b \in R(K) \iff D_F\langle 1, b\rangle \cdot D_F\langle 1, -b\rangle = \dot{F}.$$

8. For any quadratic extension K/F, show that $R(F) \subseteq R(K)$, and that $\mathrm{N}_{K/F}(R(K)) \subseteq R(F)$.

9. If $K = F(\sqrt{a}\,)$, where $a \in R(F) \setminus \dot{F}^2$, show that $R(K) \cap \dot{F} = R(F)$.

10. (Cf. 6.7 and 6.8.) Show that F is a pre-Hilbert field iff, for any $a \in \dot{F}$, there are at most two isometry classes of binary forms over F with determinant a, and there exists *some* $a \in \dot{F}$ for which there are exactly two such isometry classes. Write down (and prove!) a similar statement about Hilbert fields.

11. (Cf. 6.9.) Let F be a formally real field with $[\dot{F} : R(F)] = 4$. Show that F has exactly four quaternion algebras (up to isomorphisms), $|X_F| \leq 2$, and $P(F) \leq 2$.
12. (Cf. 6.10.) Show that F is a formally real pre-Hilbert field iff there exists an ordering P on F such that $\langle a,\, b\rangle \cong \langle 1,\, ab\rangle$ for all $a,\, b \in \dot{P}$.
13. Recall that $W(\mathbb{F}_3) \cong \mathbb{Z}_4$ and $W(\mathbb{F}_5) \cong \mathbb{Z}_2[G]$, where G is a group of order 2. Show that, for any commutative ring R of characteristic not 2 with a given ring epimorphism onto $\mathbb{Z}_2$ (in particular, for any Witt ring $R = W(F)$ where $-1 \notin \dot{F}^2$), the Witt product of $W(\mathbb{F}_3)$ and R is isomorphic to the Witt product of $W(\mathbb{F}_5)$ and R. (This exercise is of importance in studying the uniqueness question in the decomposition of a Witt ring into a Witt product of finitely many other Witt rings.)
14. (Cf. Ch. I, Exercise 24(2).) In any quadratic form scheme $(G, -1, V)$, show that $V(a) \cap V(b) \subseteq V(-ab)$ for any $a,\, b \in G$.
15. In any Q-structure (G, Q, q), verify the properties of isometries claimed in (8.16)–(8.19).
16. In any Q-structure, show that $\langle a_1, \ldots, a_n\rangle \cong \langle a_{\pi(1)}, \ldots, a_{\pi(n)}\rangle$ for any permutation π of $\{1, 2, \ldots, n\}$.
17. Show that, if a quadratic form scheme $(G, -1, V)$ "comes from" a quaternionic structure (G, Q, q), then $(G, -1, V)$ has the property (C3).

Chapter XIII

Special Topics on Invariants

This chapter is very much in the spirit of the last one, in that it continues to offer a selection of topics in the further study of quadratic form theory. The four main topics expounded in this chapter are grouped together here under a common theme, namely, that of the computation of the quadratic invariants of fields (and, by extension, of commutative rings).

While the various quadratic invariants for fields are mostly easy and natural to define, the determination of their *exact values* often presents daunting challenges. The determination of the possible values of $s(F)$ (the level of a nonreal field F) is a case in point. As we have noted in XI.2, this determination problem was raised by B. L. van der Waerden in the early 1930s, and was in fact known in some quarters as "van der Waerden's Problem". It had remained unsolved for more than 30 years, until Pfister [Pf$_1$] showed in 1965 that $s(F)$ must be a power of 2, using the beginnings of his newfound theory of multiplicative forms. Two other important quadratic invariants of fields are the u-invariant $u(F)$, and the Pythagoras number $P(F)$. *What kind of integers can these be?* Kaplansky [Ka$_1$] constructed fields of u-invariant 2^n for any prescribed integer $n \geq 0$, and speculated that, for nonreal fields F, $u(F)$ is always a power of 2: this has come to be known as "Kaplansky's Conjecture". As for the Pythagoras number $P(F)$, for a very long time, no fields F were known whose Pythagoras numbers have values other than 2^n or $2^n + 1$.

The determination problem for $u(F)$ and $P(F)$ has a very natural appeal, and indeed, it was a sure sign of progress in quadratic form theory that we now know the complete answer to the Pythagoras number problem, and perhaps a good portion of the answer to the u-invariant problem. In 1988, Merkurjev constructed a nonreal field F with $u(F) = 6$. This was followed, a few years later, by his construction (in [Me$_3$]) of nonreal fields F with $u(F) = 2n$ *for any* $n \geq 1$. This completely laid to rest Kaplansky's Conjecture, but brought to the forefront again the problem of whether $u(F)$ could still be always *even* (if it is not 1 or ∞).[(1)] In 2001, this problem was settled as well, again negatively, by Izhboldin [Iz], who showed that *there exist nonreal fields of* u*-invariant* 9. In view of XI.6.8, this is the first possible odd value for the u-invariant. From the results of (Merkurjev and) Izhboldin, it would now seem conceivable that *each* odd integer ≥ 11 can also be realized as the u-invariant of some nonreal field. However, at this moment, such a bold statement remains conjectural.

The situation of the Pythagoras number $P(F)$ is more clear-cut. In 1999, Hoffmann [Ho$_9$] constructed a field F_n with *any* prescribed Pythagoras number $n \geq 1$. This laid to rest any conjecture to the effect that $P(F)$ might be any specific kind of natural numbers.

In this chapter, we shall try to convey the flavor of some of these more recent advances in quadratic form theory by presenting a couple of special cases of them. Specifically, we shall explain in §2 Merkurjev's first method for constructing a field of u-invariant 6, and in §3, we shall present a special case of Hoffmann's construction in [Ho$_9$] to produce fields with Pythagoras numbers 6 and 7. While, of course, the case $u(F) = 2n$ and the case $P(F) = n$ both required significantly more work, the expositions in §2 and §3 should give the reader some concrete ideas as to how the general problem can be approached, and eventually solved.

As a kind of "warm-up exercise", we include in §1 the computation of the u-invariant of a specific field, namely the quotient field of the bivariate power series ring $\mathbb{C}[[x, y]]$. After the work in §§2–3 on $u(F)$ and $P(F)$, we make a foray (§§4–5) into the area of quadratic invariants of *commutative rings*. Here, the situation is markedly different, but the problem of determining the nature of the level (§4) and the Pythagoras number (§5) turn out to be quickly and pleasantly solvable—if one applies the appropriate tools!

As was the case in the earlier incarnations of this book, the chapter concludes with a short list of open problems in quadratic form theory.

1. The u-Invariant of $\mathbb{C}((x, y))$

In this section, we shall illustrate the notion of the u-invariant of a field by a concrete computation. The field chosen for our considerations is $\mathbb{C}((x, y))$,

---

(1) Recall that this was shown to be true in XI.6.9 in the special case where $I^3F = 0$.

which is the quotient field of the power series ring $\mathbb{C}[[x,y]]$ in two independent variables x, y. This field is a subfield of the iterated power series field $\mathbb{C}((x))((y))$, whose u-invariant is already known to be 4, by XI.6.2(7). In the following, we will show that the u-invariant of $\mathbb{C}((x,y))$ is also equal to 4. In fact, we'll prove a more general result on diagonal forms of *any* degree d over $k((x,y))$, for any algebraically closed field k, as follows.

Theorem 1.1. *For k as above, any diagonal form of degree d in more than d^2 variables over the field $k((x,y))$ has a nontrivial zero.*

Recall that a field K is said to be a *C_i-field* if any form of degree d in more than d^i variables over K has a nontrivial zero. Thus, we may refer to the theorem above by the informal statement that $k((x,y))$ is a "C_2-field for diagonal forms." Through the rest of this section, k *shall denote an algebraically closed field.*

Corollary 1.2. $u(k((x,y))) = 4$.

Proof. Since quadratic forms over $k((x,y))$ can be diagonalized (of course our blanket assumption char$(k) \neq 2$ is still in force here), the theorem applied in the case $d = 2$ gives $u(k((x,y))) \leq 4$. The form $\langle 1, x\rangle \perp y\langle 1, x\rangle$ is anisotropic over $k((x))((y))$, and hence over $k((x,y))$. This shows that $u(k((x,y))) = 4$. □

The proof for 1.1 to be given below is not entirely self-contained. However, the facts that we shall assume for this proof are fairly standard. They are the following:

Fact 1.3. $k((x))$ *is a* C_1*-field.*

Fact 1.4. $k((x))(y)$ *is a* C_2*-field.*

Actually, what we need precisely is Fact 1.4. We stated 1.3 also mainly because there is a logical dependence between 1.3 and 1.4. In general, if K is a C_i-field, then $K(y)$ is a C_{i+1}-field. Using this for $i = 1$, we see that 1.3 implies 1.4. These facts, needed for the proof below, can be found in Greenberg's book "Lectures on Forms in Many Variables", Benjamin, 1969.

Assuming 1.4, the idea of the proof of 1.1 is to use standard facts on formal power series to make a reduction from the field $k((x,y))$ to its subfield $k((x))(y)$. The proof below is from [CDLR], with contributions also from A. Wadsworth.

We first review some basic facts about formal power series. These facts apply to power series in any (finite) number of variables; however, we shall state them only in the case of two variables, since this is the case we are interested in here.

Let ord_y denote the standard "lowest degree" valuation on the field $k((y))$. A power series $F(x,y) \in k[[x,y]]$ is said to be *regular in* y *of degree* s if $\operatorname{ord}_y F(0,y) = s$; that is, F contains a term ay^s $(a \in k \setminus \{0\})$, but no term by^i with $b \in k \setminus \{0\}$ and $i < s$. The following classical theorem of Weierstrass is one of the most important tools in the power series calculus.

Weierstrass Division Theorem 1.5. *Let $F(x,y) \in R = k[[x,y]]$ be regular of degree s in y. For any $G \in R$, there exist $U \in R$ and $V \in \sum_{i=0}^{s-1} k[[x]] \cdot y^i$ such that $G = UF + V$. Moreover, such U, V are unique (for given F and G).*

Proof. Note that the conclusion of the theorem means precisely that $R/R \cdot F$ is a free module over $k[[x]]$ with basis $\{1, y, \ldots, y^{s-1}\}$.

First, we treat the case where F and G do not involve the variable x. Say $f_0 \in k[[y]]$ is regular of degree s in y, and $g_0 \in k[[y]]$. Then,

$$\begin{aligned} g_0 &\in a_0 + \cdots + a_{s-1}y^{s-1} + k[[y]] \cdot y^s \\ &= a_0 + \cdots + a_{s-1}y^{s-1} + k[[y]] \cdot F, \end{aligned} \tag{1.6}$$

so we are done.

In the general case, write F, G in the form

$$F = \sum_{i=0}^{\infty} f_i x^i, \quad G = \sum_{i=0}^{\infty} g_i x^i \qquad (f_i,\, g_i \in k[[y]]).$$

We'll try to find $U = \sum_{i=0}^{\infty} u_i x^i$ and $V = \sum_{i=0}^{\infty} v_i x^i$ with $u_i \in k[[y]]$ and $v_i \in \sum_{j=0}^{s-1} ky^j$ such that $G = UF + V$. Comparing the x^i-coefficients in the latter equation, we have

$$\begin{aligned} g_0 &= u_0 f_0 + v_0, \\ g_1 - u_0 f_1 &= u_1 f_0 + v_1, \\ &\ldots, \\ g_i - u_0 f_i - \cdots - u_{i-1} f_1 &= u_i f_0 + v_i\,. \\ &\ldots. \end{aligned}$$

Since $f_0 = F(0,y)$ is regular of degree s in y, we can solve these equations uniquely for u_i, v_i, starting with $i = 0$, using the procedure in 1.6. ☐

Definition 1.7. A power series $W \in R = k[[x,y]]$ is called a *Weierstrass polynomial of degree* s *in* y if it has the form

$$W = w_0 + w_1 y + \cdots + w_{s-1} y^{s-1} + y^s, \tag{1.8}$$

where $w_i \in k[[x]]$, with $w_i(0) = 0$ for all i.

Since $W(0,y) = y^s$, such a W is always regular in y of degree s. The importance of the notion of Weierstrass polynomials lies in the following basic fact.

Weierstrass Preparation Theorem 1.9. *If $F \in R$ is regular of degree s in y, then there is a unique Weierstrass polynomial W of degree s in y such that W is an associate of F in R.*

Proof. We apply 1.5 to $G = y^s$, to get

$$(1.10) \qquad y^s = UF - (w_0 + w_1 y + \cdots + w_{s-1}y^{s-1}),$$

where $U \in R$, and $w_i \in k[[x]]$. Here, U must have a nonzero constant term (and hence U is a unit of R), for otherwise we can't get a term y^s from the RHS of 1.10. Moreover, setting $x = 0$, we have

$$w_0(0) + \cdots + w_{s-1}(0)y^{s-1} + y^s \in U(0,y) \cdot y^s\, k[[y]],$$

so $w_i(0) = 0$ for all i. Thus, $W = w_0 + \cdots + w_{s-1}y^{s-1} + y^s$ is the Weierstrass polynomial we sought. The uniqueness of W follows from the uniqueness of the formula 1.10. □

In order to apply the above results to more general power series in R, we note the following easy fact.

Lemma 1.11. *Given a finite number of nonzero power series $F_1, \ldots, F_n \in R$, there exists a k-algebra automorphism σ of R, leaving y fixed, such that each $\sigma(F_i)$ is regular in y (of some degree s_i).*

Proof. We first handle the case of a single $0 \neq F \in R$. Here, we define σ by $\sigma(x) = x + y^r$, $\sigma(y) = y$, where $r \in \mathbb{N}$ is to be chosen. Since $\sigma(F) = F(x+y^r, y)$, we have $\sigma(F)(0,y) = F(y^r, y)$. Thus, it suffices to check that, if r is large enough, $F(y^r, y) \neq 0$ in $k[[y]]$. To this end, we order the monomials $\{x^i y^j\}$ lexicographically by their exponents (i,j). Let $ax^{i_0}y^{j_0}$ $(a \neq 0)$ be the term in F that is lexicographically smallest. Then any $r > j_0$ works. In fact for any *other* nonzero monomial bx^iy^j in F, the substitution $x \mapsto y^r$ will take it to by^{ri+j} with exponent $> ri_0 + j_0$. Thus, $y^{ri_0+j_0}$ is alone of its kind in $F(y^r, y)$ with a nonzero coefficient a.

For the general case in the lemma, let $F = F_1 \cdots F_n$. Pick a σ such that $\sigma(F)$ is regular (of some degree) in y. Since $\sigma(F)(0,y) = \prod_i F_i(0,y) \neq 0$, it follows that each F_i is regular (of some degree s_i) in y. □

As the reader can see, the results (1.5) through (1.11) are valid for any ground field k. Bringing back the assumption that k is algebraically closed, let us now present

Proof of Theorem 1.1. Consider any diagonal form

$$F_1 t_1^d + \cdots + F_n t_n^d \quad \text{over } K = k((x,y)). \tag{1.12}$$

Assuming that $n > d^2$, we would like to prove that this form has a nontrivial zero in K^n. We may assume that each $F_i \in R \setminus \{0\}$, where $R = k[[x,y]]$. After applying a k-algebra automorphism to R, we may assume that *each* F_i is regular (say of degree s_i) in y. By the Weierstrass Preparation Theorem 1.9, $U_i F_i = W_i$ for some unit $U_i \in R$ and some Weierstrass polynomial $W_i \in R$ of degree s_i in y. Since $U_i(0,0) \in k \setminus \{0\}$ has a d-th root in the (algebraically closed) field k, we can write $U_i = V_i^d$ for suitable $V_i \in R$. In the new variables $T_i = t_i/V_i$, the form in 1.12 becomes

$$F_1 V_1^d T_1^d + \cdots + F_n V_n^d T_n^d = W_1 T_1^d + \cdots + W_n T_n^d.$$

Since $W_i \in K_0 := k((x))(y)$, the RHS is a form defined over K_0. By 1.4, this form has a nontrivial zero over K_0. Therefore, the form 1.12 has a nontrivial zero over K. □

For more general results on quadratic forms over quotient fields of two-dimensional Henselian rings, see the recent work of Colliot-Thélène, Ojanguren and Parimala [CTOP]. Some computations of the Witt rings of quotient fields of tw0-dimensional regular local rings can be found in the earlier papers of Jaworski ([Jaw_1, Jaw_2]).

2. Fields of u-Invariant 6

Throughout this section, unless it is stated to the contrary, the u-invariant $u(F)$ of a field F is taken in the classical sense in XI.6; namely, $u(F)$ is the supremum of the dimensions of the anisotropic quadratic forms over F. In particular, our main interests in this section will be focused on nonreal fields F, since formally real fields have infinite u-invariants in the classical sense.

We have stated in the text of XI.6 that, although the u-invariant of a field F cannot be 3, 5, or 7, it *need not* be a power of 2 as was once conjectured by Kaplansky. The first critical case is, of course, the value 6. In a spectacular work in 1988, Merkurjev [Me_2] surprised everyone by coming up with the construction of a nonreal field F of u-invariant 6. The discovery of such a field proved to be an important landmark in the study of the quadratic invariants of fields.

In this section, we shall give a self-contained exposition on Merkurjev's construction of a field of u-invariant 6. This exposition is based on an earlier paper [L_6] I wrote on the same topic, except that, in the present exposition, we make an effort to include *all* proofs. For more detailed historical information on the subject, and for the many ramifications of Merkurjev's construction (especially the connections to the theory of finite-dimensional central simple algebras), we refer the reader to [L6] (along with, of course, the original source [Me2]).

Following the footsteps of Albert and Brauer, Merkurjev exploited the class of biquaternion algebras for his construction of fields of u-invariant 6. Recall that a biquaternion algebra A over a field F is just a tensor product $B \otimes_F C$, where B and C are quaternion algebras over F. Such an algebra A is 16-dimensional over F, and the *Albert form* q_A is defined to be the "formal difference" of the pure subforms of the norm forms of the quaternion algebras B and C, as in III.4.7. It is true that q_A is determined up to a scalar multiple by the isomorphism class of A (independently of the chosen decomposition $A \cong B \otimes_F C$), according to Jacobson's Theorem XII.2.12. However, this fact is not essential for Merkurjev's construction. Thus, we can just take the Albert form q_A to be defined via *some* decomposition $A \cong B \otimes_F C$ as above. Much more relevant are the following facts:

(2.1) $\dim q_A = 6$ *and* $d_\pm(q) = 1$, *so that* $q \in I^2F$.

(2.2) *If* q *is a* 6*-dimensional form in* I^2F, *then* q *is similar to* q_A *for some biquaternion algebra* $A = B \otimes_F C$.

(2.3) $A \cong B \otimes_F C$ *is a division algebra iff* q_A *is an anisotropic form, according to Albert's Theorem* III.4.8.

The beginning point of Merkurjev's construction is the following observation on the relationship between biquaternion division algebras and fields of u-invariant 6.

Proposition 2.4. *If there is a biquaternion division algebra over a nonreal field* F, *then* $u(F) \geq 6$. *Conversely, if* $u(F) = 6$, *then there is a biquaternion division algebra over* F.

Proof. The first statement is clear from 2.3. For the second statement, assume that $u(F) = 6$. By 2.2 and 2.3, it will be sufficient to produce an anisotropic 6-dimensional form q of determinant -1. Fix *any* anisotropic 6-dimensional form φ and let $d = d(\varphi)$. Since $u(F) = 6$, the form $\psi :=$ $\langle 1, d\rangle \perp \varphi$ is isotropic. Write $\psi \cong \mathbb{H} \perp q$, where $\dim q = 6$. Then $d(q) =$ $-d(\psi) = -1$, so we are done if we can show that q is anisotropic. *Assume q is isotropic*, say $q \cong \mathbb{H} \perp q_0$. Then

$$\langle 1, d\rangle \perp \varphi \cong \psi \cong 2\mathbb{H} \perp q_0 \cong \langle 1, d, -1, -d\rangle \perp q_0,$$

and Witt cancellation gives $\varphi \cong \langle -1, -d\rangle \perp q_0$. Taking determinants, we see that $d(q_0) = 1$, so we can write $q_0 \cong \langle a\rangle\langle\langle b, c\rangle\rangle$ for suitable a, b, $c \in \dot{F}$. But $\langle\langle -a, b, c\rangle\rangle$ is isotropic, and hence hyperbolic (by X.1.7). Therefore, $\langle a\rangle\langle\langle b, c\rangle\rangle \cong \langle\langle b, c\rangle\rangle$, and

$$\varphi \cong \langle -1, -d\rangle \perp q_0 \cong \langle -1, -d\rangle \perp \langle 1, b, c, bc\rangle$$

is isotropic, a contradiction. Hence q must be anisotropic, as desired. □

Remark 2.5. If we assume a more high-power result, namely, XI.6.21, the second statement in 2.4 can be improved into the following: *if* $u(F) = 6$ *or* $u(F) > 8$, *then there is a biquaternion division algebra over* F. In fact, by XI.6.21, the new hypothesis would imply that F is *not* a linked field. Then III.4.8 directly gives the desired conclusion. However, if $u(F) = 8$, there may not exist a biquaternion division algebra over F. An example for this is given by the field $F = \mathbb{C}((x))((y))((z))$, which is a $\overline{C}$-field (in the sense of XI.7.16) with 8 square classes, and hence a linked field (of u-invariant 8) by XI.Exercise 17.

The remark above is not needed for Merkurjev's construction, since the second statement in 2.4 already proves to be sufficient for our purposes. The main new ingredients for Merkurjev's construction are certain results on the behavior of Albert forms over function fields of low-dimensional quadratic forms. The ultimate function field fact that makes Merkurjev's construction possible is the following result of his [Me$_2$].

Theorem 2.6. *Let A be a biquaternion division algebra over k, with an Albert form q. Then q remains anisotropic over the function field $k(\varphi)$ of any k-form φ with* $\dim \varphi \geq 7$. *(Or, equivalently, the scalar extension $A^{k(\varphi)} := A \otimes_k k(\varphi)$ remains a division algebra over $k(\varphi)$.)*

The key step in Merkurjev's construction is the verification of the function field theorem 2.6, which is by no means easy. Assuming 2.6, however, the construction of fields of u-invariant 6 is at hand, as we shall now explain.

Start with any field k over which there is a biquaternion division algebra A, and let $q = q_A$. We construct inductively an ascending chain of fields

$$k = F_0 \subseteq F_1 \subseteq F_2 \subseteq \cdots \tag{2.7}$$

as follows. If F_i is already given, we let F_{i+1} be the function field $F_i(\{\psi_\alpha\})$, where ψ_α ranges over (the isometry classes of) all 7-dimensional forms over F_i. Here, $F_i(\{\psi_\alpha\})$ is just the free compositum of the various function fields $F_i(\psi_\alpha)$; see X.3. Now we let $F := \bigcup_{i=0}^{\infty} F_i$. Clearly, any 7-dimensional form over F is isotropic (since such a form is defined over some F_i). On the other hand, by 2.6 (applied repeatedly), q remains anisotropic over F (and $A^F := A \otimes_k F$ remains a division algebra). Therefore, $u(F) = 6$, as desired! (It should be of no surprise that F has a biquaternion division algebra, since, by 2.4, this is true for any field with u-invariant 6.)

Summing up, we have the following result.

Merkurjev's Theorem 2.8. *Any field k_0 (of characteristic not 2) is contained in a (nonreal) field F with $u(F) = 6$.*

Proof. Let $k = k_0(x, y, z, w)$. Proceeding as in VI.1.11, we can check that

$$A := \left(\frac{x, y}{k}\right) \otimes_k \left(\frac{z, w}{k}\right)$$

is a division k-algebra. With this observation, the paragraph preceding the statement of the theorem provides the construction for F. □

Since the level $s(F)$ is always $\leq u(F)$, and is always a power of 2, the level of a field of u-invariant 6 can only be 1, 2, or 4. An easy consequence of Merkurjev's construction is the following.

Corollary 2.9. *There exist fields F with $u(F) = 6$ such that $s(F)$ has any prescribed value in $\{1, 2, 4\}$.*

Proof. Keeping the notations in the proof of 2.8, let $s(k_0) = 1$. Then clearly $s(F) = 1$. Next, let $s(k_0) = 2$. We can check easily that

$$\sqrt{-1} \notin k_0 \implies \sqrt{-1} \notin k_0(x, y, z, w) \implies \sqrt{-1} \notin F.$$

Thus, $s(F) = 2$. Finally, let $s(k_0) = 4$, and assume $s(F) \leq 2$. Then $\langle\langle 1, 1\rangle\rangle$ becomes isotropic, and hence hyperbolic over F. However, $\langle\langle 1, 1\rangle\rangle$ is anisotropic over k as well as over k_0. Since F is obtained from k by iterated constructions of function fields of forms of dimension 7, this contradicts the last conclusion of X.4.5. Therefore, we must have $s(F) = 4$ as well. □

Of course, by using repeated Laurent series field constructions, we get for free the following self-strengthenings of 2.8 and 2.9.

Corollary 2.10. *For any field k_0 and any integer $n \geq 1$, there exists an extension field K_n/k_0 such that $u(K_n) = 3 \cdot 2^n$, and if $s(k_0) = 1, 2,$ or 4, then $s(K_n) = 1, 2,$ or 4 respectively.*

The completion of Merkurjev's constructions used for 2.8 is now reduced to the verification of the function field result 2.6. The main step for proving this result is to first understand in exactly what way an anisotropic Albert form q can become isotropic over the function field $k(\psi)$ of a form ψ of dimension ≤ 3. The case when $\dim \psi = 2$ is quite easy; the relevant result here was already proved by Albert, as follows.

Theorem 2.11. *Let q be an anisotropic Albert form over a field k (that is, $q \in I^2k$, with $\dim q = 6$). Let $A = B \otimes_k C$ be a biquaternion algebra whose Albert form (defined via the decomposition $A = B \otimes_k C$) is similar to q. For any binary form $\psi = \langle 1, -a\rangle$, where $a \in \dot{k} \setminus \dot{k}^2$, the following are equivalent:*

(1) *$A^{k(\psi)}$ is not a division algebra over $k(\psi)$.*

(2) *$A \cong \left(\frac{a, b}{k}\right) \otimes_k \left(\frac{c, d}{k}\right)$ for some b, c, $d \in \dot{k}$.*

(3) $q > \psi$ *(that is, q becomes isotropic over $k(\psi)$).*

(4) q *has a subform isometric to* $\langle r\rangle\psi$ *for some* $r \in \dot{k}$.

Proof. (2) $\Rightarrow$ (1) is trivial (since $k(\psi) = k(\sqrt{a}\,)$), and (1) $\Leftrightarrow$ (3) follows from III.4.8. Next, VII.3.1 gives (3) $\Rightarrow$ (4), so it only remains to prove (4) $\Rightarrow$ (2).

Assuming (4), write $q \cong r\langle 1, -a, -b\rangle \perp \cdots$ for some $b \in \dot{k}$. Then, by determinant considerations,

$$\langle abr\rangle q \cong \langle -a, -b, ab\rangle \perp \langle -1\rangle\langle -c, -d, cd\rangle \tag{2.12}$$

for some c, $d \in \dot{k}$. Applying the Witt invariant map from I^2k/I^3k to the Brauer group $B(k)$, we arrive at the isomorphism in (2). □

The result 2.11 will be used later in the eventual proof of 2.6. At this juncture, the crux of the matter is to prove the following analogue of 2.11 for *ternary* forms $\psi = \langle 1, -a, -b\rangle$. For such a form ψ, $k(\psi)$ is the *function field of a conic.*

Theorem 2.13 (Merkurjev). *Let q and A be as in 2.11, and let $\psi = \langle 1, -a, -b\rangle$, where a, $b \in \dot{k}$. Then the following are equivalent:*

(1) $A^{k(\psi)}$ *is not a division algebra over* $k(\psi)$.

(2) $A \cong \left(\frac{a,b}{k}\right) \otimes_k \left(\frac{c,d}{k}\right)$ *for some* b, c, $d \in \dot{k}$.

(3) $q > \psi$ *(that is, q becomes isotropic over $k(\psi)$).*

(4) q *has a subform isometric to* $\langle r\rangle\psi$ *for some* $r \in \dot{k}$.

Proof. The proof here follows the same outline as in the proof of 2.11. Here again, (2) $\Rightarrow$ (1) is trivial (since $\left(\frac{a,b}{k}\right)$ splits over $k(\psi)$), and (1) $\Leftrightarrow$ (3) follows from III.4.8. The implication (4) $\Rightarrow$ (2) can be proved by exactly the same determinant calculation as in the proof of 2.12 (letting b in (2.12) be the b in this theorem). Given all of these, the preponderance of the proof falls upon (3) $\Rightarrow$ (4)! In the case of 2.11, this implication is the basic result VII.3.1 for a quadratic extension $k(\sqrt{a}\,)/k$. In the present case, we must derive the analogous implication for the function field of a conic, $k(\psi)/k$.

The following argument for (3) $\Rightarrow$ (4) using the "excellence" property of $k(\psi)$ is essentially equivalent to Merkurjev's original argument, as was observed by Rost and Wadsworth.[2] Assume that q is isotropic over $K := k(\psi)$. By Arason's theorem (see the paragraph before XII.4.9), K/k is an excellent extension, so $q_K \cong \mathbb{H} \perp \tau_K$ for a suitable 4-dimensional k-form τ. Thus, by X.4.28, the anisotropic part of $q \perp \langle -1\rangle\tau$ can be written in

---

[2] Actually, Merkurjev showed (1) $\Rightarrow$ (2). One can then get (2) $\Rightarrow$ (4) from Jacobson's Theorem. The argument for (3) $\Rightarrow$ (4) here is more direct.

the form $\sigma \cdot \varphi$, where $\varphi = \langle\langle -a, -b \rangle\rangle$ and σ is some k-form, necessarily of dimension ≤ 2. This shows, in particular, that $\tau \in I^2k$, so we can write $\tau \cong \langle t \rangle \langle\langle -c, -d \rangle\rangle$ for some t, c, $d \in \dot{k}$. If $\dim \sigma = \{0, 2\}$, then $\sigma \cdot \varphi \in I^3k$, and we have

$$q \equiv \langle t \rangle \langle\langle -c, -d \rangle\rangle \pmod{I^3k}.$$

Applying the Witt invariant map from I^2k/I^3k to the Brauer group $B(k)$, we get $A = \left(\frac{c, d}{k}\right) \in B(k)$, which contradicts the fact that A is a division algebra. Thus, we may assume that $\sigma \cong \langle s \rangle$ $(s \in \dot{k})$, so now

$$q = \langle s \rangle \langle\langle -a, -b \rangle\rangle + \langle t \rangle \langle\langle -c, -d \rangle\rangle \in W(F).$$

Arguing as in X.5.16 (Case 2), we may "rechoose" s and t so that $s + t = 0$. After this, we have

$$q \cong \langle s \rangle \langle -a, -b, ab, c, d, -cd \rangle,$$

and so $q \supseteq \langle r \rangle \psi$ for $r = sab \in \dot{k}$. □

The proof of (3) $\Rightarrow$ (4) above depends on the excellence property of $k(\psi)/k$, which, however, is not proved in our text.(3) To make our arguments self-contained, we present below another way to deduce (4) from (3) that is due to David Leep.

As in X.3.13, we write $K = k(\psi)$ in the form $E\left(\sqrt{ax^2 + b}\right)$, where $E = k(x)$. If q_K is isotropic, VII.3.1 and IX.1.1 imply that

$$q_E \cong f \cdot \langle 1, -(ax^2 + b) \rangle \perp \langle g_1, g_2, g_3, g_4 \rangle \tag{2.14}$$

for suitable square-free polynomials f, $g_i \in k[x] \backslash \{0\}$. We may assume (2.14) is chosen such that $\deg f$ is as small as possible. *We claim that* $\deg f = 0$. If this is true, then $f = r \in \dot{k}$, and hence the k-form $r \cdot q \perp \langle -1 \rangle$ is isotropic over E. Applying IX.1.1 again, we can write

$$r \cdot q \perp \langle -1 \rangle \cong \mathbb{H} \perp \langle s_1, \ldots, s_5 \rangle, \quad \text{where } s_i \in \dot{k}. \tag{2.15}$$

Comparing (2.14) and (2.15), we see that $\langle s_1, \ldots, s_5 \rangle$ represents $-(ax^2 + b)$ (and hence also $-(a + bx^2)$) over E. By IX.1.3(2), $-a \in D_k\langle s_1, \ldots, s_5 \rangle$. Thus, we may assume that $s_1 = -a$, and IX.2.1 further enables us to assume that $s_2 = -b$. It then follows that $r \cdot q \supseteq \langle 1, -a, -b \rangle = \psi$, as desired.

To prove our claim that $\deg f = 0$, let us assume the contrary, in which case there exists a monic irreducible polynomial $p \in k[x]$ such that $p \mid f$. Note that $ax^2 + b$ itself is irreducible. (If otherwise, $-b \in a\dot{k}^2$, and ψ would be isotropic. This is not the case (by X.4.1) as q is anisotropic, but becomes isotropic over $K = k(\psi)$.) If $p \mid (ax^2 + b)$, then we must have $ap = ax^2 + b$,

---

(3)We remind the reader here that an elementary proof of this excellence property is available from Rost's article [Ro].

and hence $-ap$ is a similarity factor for $\langle 1, -(ax^2+b)\rangle$. Thus, putting $f_1 := -f/ap$, we have $\deg f_1 < \deg f$ and

$$f\langle 1, -(ax^2+b)\rangle \cong f_1\langle 1, -(ax^2+b)\rangle, \tag{2.16}$$

which contradicts the choice of f. Thus, *we must have* $p \nmid (ax^2+b)$. Since $d(q) = -1$, (2.14) leads to the following three possibilities:

Case 1. $p \nmid g_i$ for $1 \le i \le 4$.

Case 2. $p \mid g_i$ for $1 \le i \le 4$.

Case 3. $p \mid g_1, g_2$, but $p \nmid g_3, g_4$ (say).

In the following, we shall write $\partial_p^1(q)$ and $\partial_p^2(q)$ for the first and the second residue forms of q, with respect to the p-adic valuation on $k(x)$.

Case 1. Here, (2.14) shows that

$$0 = \partial_p^2(q) = \partial_p^2(f\langle 1, -(ax^2+b)\rangle)$$

in the Witt ring of $\overline{E}_p := k[x]/(p)$. This means that ax^2+b is a square in $\overline{E}_p$; say $ax^2+b = h^2 - p\ell$, where $h, \ell \in k[x]$, with $\deg h < \deg p$. If $\deg p = 1$, then $p(u) = 0$ for some $u \in k$. This would give $au^2+b = h(u)^2$, which would again contradict the anisotropicity of ψ. Thus, $\deg p \ge 2$, which implies that

$$\deg \ell = \deg(h^2 - ax^2 - b) - \deg p \le (\deg p) - 2.$$

Noting that $p\ell = h^2 - (ax^2+b)$ is a similarity factor of $\langle 1, -(ax^2+b)\rangle$, we can now take $f_1 = \ell f/p$ (with $\deg f_1 < \deg f$) to get a contradictory equation (2.16).

Cases 2, 3. *Here, we claim that* $\partial_p^1(q) = 0$. In Case 2, this is clear. In Case 3, write $q_E \cong p \cdot \sigma \perp \langle g_3, g_4\rangle$, where σ is a form over E such that $\partial_p^1(\sigma) = \sigma$. From

$$0 = \partial_p^2(q) = \overline{\sigma} \in W(\overline{E}_p),$$

we have $d(\overline{\sigma}) = 1$ over $\overline{E}_p$, and hence $d(\overline{\langle g_3, g_4\rangle}) = -1$ (since $d(q) = -1$). This means that

$$\partial_p^1(q) = \overline{\langle g_3, g_4\rangle} = 0 \in W(\overline{E}_p),$$

as claimed. Therefore, in either case, IX.3.4 implies that $p \cdot q \cong q$ over E. Taking $f_1 = f/p$ with $\deg f_1 < \deg f$, we get again the contradictory equation (2.16). *The alternative proof for* (3) $\Rightarrow$ (4) *in* 2.13 *is now complete.* $\square$

Experts in the theory of central simple algebras and Brauer groups have pointed out that the implication (1) $\Rightarrow$ (2) in 2.13 is a special case of a much more general result on the Schur index of a Brauer class on a Brauer-Severi variety. We refer the reader to [L6] for a more detailed discussion on this issue.

Our last formal task in this section is to prove Theorem 2.6 *from* 2.13. (This will complete the work on the construction of fields of u-invariant 6.) This task is accomplished by relating the function field of a higher-dimensional form to the function field of a conic. The fact that this can be done is somewhat surprising, although the argument needed to establish this connection turns out to be quite elementary.

Lemma 2.17. *Let ψ be the k-form $\langle 1, -a\rangle \perp -\varphi$, where $n = \dim \varphi \geq 1$, and let $L = k(x_1, \ldots, x_n)$. Then the "big" function field $k[\psi]$ is isomorphic to the function field of a conic over L; namely, $k[\psi] \cong L(\langle 1, -a, -\varphi\rangle)$, where $-\varphi$ here is viewed as a (nonzero) element of the rational function field L.*

Proof. The "big" function field $k[\psi]$ is generated by elements y, z, $x_1, \ldots, x_n$ with the relation

$$y^2 - az^2 - \varphi(x_1, \ldots, x_n) = 0.$$

It is, therefore, obtained from $L(z)$ by adjoining a square root of the element $az^2 + \varphi \in L(z)$; this adjunction gives, of course, the function field of the conic defined by the ternary form $\langle 1, -a, -\varphi\rangle$ over L (see X.3.13). □

We have delineated the function field result above in a complete and independent statement, as it divulges the real significance of the case of the function field of a conic. Hopefully, some function field results can be proved by making a reduction to the case of conics, via 2.17. As we will see shortly, the forthcoming proof of 2.6 (from 2.13) will run precisely on this pattern. Note that in 2.17 as well as in the following, $\dim \psi = n + 2 \geq 3$.

Proposition 2.18. *Keeping the notations in* 2.17, *let A be any biquaternion division algebra over k. Then the following are equivalent:*

(1) *$A^{k(\psi)}$ is not a division algebra.*

(2) *$A^{k[\psi]}$ is not a division algebra.*

(3) *There exist b, c, $d \in \dot{k}$ such that $A \cong \left(\frac{a,b}{k}\right) \otimes_k \left(\frac{c,d}{k}\right)$, and $\left(\frac{a,b\varphi}{L}\right) \otimes_L \left(\frac{c,d}{L}\right)$ is not a division algebra.*

Proof. (1) $\Leftrightarrow$ (2) is clear since $E := k[\psi]$ is a rational function field (in one variable) over $k(\psi)$. Since $E \cong L(\langle 1, -a, -\varphi\rangle)$ by 2.17, $\left(\frac{a,\varphi}{E}\right)$ splits. Thus, if (3) holds,

$$A^E \cong \left(\frac{a,\, b}{E}\right) \otimes_E \left(\frac{c,\, d}{E}\right) \cong \left(\frac{a,\, b\varphi}{E}\right) \otimes_E \left(\frac{c,\, d}{E}\right)$$

is not a division algebra, proving (1). Conversely, assume (1). Then 2.13 (applied to $L(\langle 1, -a, -\varphi\rangle)/L$) implies that

$$A^L \cong \left(\frac{a,\ \varphi}{L}\right) \otimes_L \left(\frac{\gamma,\ \delta}{L}\right) \quad \text{for some } \gamma, \delta \in \dot{L}. \tag{2.19}$$

In particular, $A^{L(\sqrt{a})}$ is not a division algebra. Since $L(\sqrt{a})/k(\sqrt{a})$ is purely transcendental, this implies that $A^{k(\sqrt{a})}$ is not a division algebra. By 2.11, we have

$$A \cong \left(\frac{a,\ b}{k}\right) \otimes_k \left(\frac{c,\ d}{k}\right) \quad \text{for some } b, c, d \in \dot{k}. \tag{2.20}$$

Putting this together with (2.19), we have

$$\left(\frac{a,\ b\varphi}{L}\right) \otimes_L \left(\frac{c,\ d}{L}\right) = \left(\frac{\gamma,\ \delta}{L}\right) \in B(L),$$

so the LHS is not a division algebra. This proves (3). □

Note that in the proposition above, we have managed to translate the statement (1) about the function field $k(\psi)$ into a statement (3) about the field L (and the element $\varphi \in \dot{L}$). The point is, of course, that it is a lot easier to work with L, than with $k(\psi)$ since L is a rational function field over k. To prove 2.6, our goal is to show that condition (1) in 2.18 can hold *only when* $\dim \psi = n + 2 \leq 6$. The following proposition will, therefore, complete our work!

Proposition 2.21. *Keeping the notations above, if the k-biquaternion algebra A satisfies* (3) *in* 2.18, *then we must have* $n = \dim \varphi \leq 4$.

Proof. We may clearly assume that $n \geq 3$; in particular, the function field $k(\varphi)$ is defined. By III.4.8, hypothesis (3) in 2.18 implies that the form

$$\langle -c,\ -d,\ cd,\ a,\ b\,\varphi, -ab\,\varphi\rangle$$

is isotropic over L. Thus, there exist $f_i \in k[x_1, \ldots, x_n]$, not all zero, such that

$$-cf_1^2 - df_2^2 + cdf_3^2 + af_4^2 + b\,\varphi f_5^2 - ab\,\varphi f_6^2 = 0. \tag{2.22}$$

We may assume that $\{f_i \colon 1 \leq i \leq 6\}$ have no (nonconstant) common divisors. *We claim that $f_1, \ldots, f_4$ are not all divisible by φ.* For, if they are, then $\varphi^2 \mid (b\,\varphi f_5^2 - ab\,\varphi f_6^2)$, and so $\varphi \mid (f_5^2 - af_6^2)$. But φ does not divide both f_5, f_6, so $\langle 1, -a\rangle$ becomes isotropic (and hence hyperbolic) over the function field $k(\varphi)$. Since $\dim \varphi \geq 3$, we know this is *not* the case.[4] Thus, φ does not divide all of $f_1, \ldots, f_4$, and (2.22) implies that $\langle a, -c, -d, cd\rangle$ is isotropic over $k(\varphi)$. Now let $\sigma = \langle\langle -c, -d\rangle\rangle$ and $k' = k(\sqrt{a})$. Clearly, σ

[4] In fact, $\dim \varphi \geq 3$ implies that k is algebraically closed in $k(\varphi)$: see X.3.

remains anisotropic over k', for otherwise $\sigma \cong \langle\langle -a, -e \rangle\rangle$ for some $e \in \dot{k}$, and

$$A \cong \left(\frac{a,b}{k}\right) \otimes_k \left(\frac{a,e}{k}\right)_k \cong \mathbb{M}_2\left(\left(\frac{a,be}{k}\right)\right)$$

is not a division algebra. Now over $k'(\varphi)$, $\sigma \cong \langle a, -c, -d, cd\rangle$ is isotropic and hence hyperbolic. By X.4.5 (applied to $k'(\varphi)/k'$), this implies that $n = \dim \varphi \leq \dim \sigma = 4$. $\square$

Some further comments on the work done in this section and on subsequent developments are collected together in the following.

Remarks 2.23. (1) After the appearance of Merkurjev's work in 1988/89, the isotropy question of 6-dimensional forms over function fields of quadrics has been extensively investigated. In this direction, we shall only mention a few key papers, such as [Ho_3], [Lag], and [IP_1, IP_2]. The 5-dimensional case was treated in [Ho_2]. The cases where $\dim q \in [7, 12]$ have also received considerable attention. For a detailed report on the work on isotropy questions over function fields of quadrics, up to the year 2000, we refer the reader to Hoffmann's survey [Ho_{10}].

(2) After Merkurjev's result was known, Mammone and Moresi succeeded in extending Merkurjev's method to the case of characteristic 2, and constructing fields of u-invariant 6 in that case.

(3) As we have explained in the Appendix to XI.6, the classical definition of the u-invariant of a field can be modified to give the definition of a *general u-invariant* for fields that need not be ∞ when the field is formally real. In view of Merkurjev's work, it is natural to ask if there also exist formally real fields with (general) u-invariant 6. This question has also been answered positively, by a suitable modification of Merkurjev's function field theoretic construction techniques reported in this section. For the details, see §5 of my exposition [L_6].

(4) Shortly after he constructed nonreal fields of u-invariant 6, Merkurjev proved in [Me_3] the existence of fields F with $I^3F = 0$ and $u(F) = 2n$ for any positive integer n (or $n = \infty$).[5] His method of the construction of these fields is, however, different from that used in the proof of 2.8. Going beyond the study of the isotropy of Albert forms over function fields of quadrics, Merkurjev introduced and proved a general index reduction formula that determines exactly which central division algebras A over a field k remain division algebras over such function fields $k(\psi)$. The proof of this index reduction formula in [Me_3] used high-power tools from Quillen's algebraic K-theory, depending ultimately on Swan's calculation of the Quillen

[5] In connection with this existence result, it is relevant to recall that, for a field F with $I^3F = 0$, $u(F)$ must be even if it is finite and $\neq 1$, according to XI.6.9.

K-groups of quadric hypersurfaces. Later, an alternative proof for Merkurjev's criterion for $A^{k(\psi)}$ to be a division algebra was obtained by Tignol [Ti]. Tignol's proof used only standard results from quadratic form theory and the theory of finite-dimensional central simple algebras, and is completely independent of Quillen's algebraic K-theory. Not surprisingly, Tignol also exploited the viewpoint that the function field of a quadric can be viewed as a function field of a lower-dimensional quadric over a rational function field of the ground field.

(5) Extensions of Merkurjev's results to the general u-invariant (and the "Hasse number") of formally real fields can be found in [L_7] and [Hor]. The case of characteristic 2 was treated by Mammone, Moresi, Tignol, and Wadsworth.

(6) Merkurjev's construction of nonreal fields with prescribed even u-invariants led naturally to a quantitative study of the behavior of the u-invariant under finite field extensions. For work in this direction, see, for instance, the papers of Leep-Merkurjev [LM] and Mináč-Wadsworth [MW].

(7) While Merkurjev's methods were sufficient only to produce fields of u-invariant $2n$ ($n \geq 1$), the question of *odd* u-invariant remains. Since $u(F) \neq 3, 5, 7$, the first open case was the integer 9. In 2000, this case was settled positively by O. T. Izhboldin. In his paper [Iz], published posthumously in 2001, Izhboldin constructed the first examples of nonreal fields F with $u(F) = 9$. His method of construction was similar to Merkurjev's first construction of fields of u-invariant 6, although various new proofs are required. The Albert forms in Merkurjev's original construction are replaced by what Izhboldin called *essential* 9-dimensional forms: these are anisotropic 9-dimensional forms q over a field k that are not Pfister neighbors, but have the property that the even Clifford algebra $C_0(q)$ of q has (Schur) index ≥ 4. A main step in [Iz] was to show that *such a form q remains essential (in particular anisotropic) over the function field $k[\varphi]$ of any 10-dimensional k-form φ*. Given any field k_0, one can check that the "generic" 9-dimensional form

$$q := \langle x_1, x_2, \ldots, x_9 \rangle$$

is essential over the rational function field $k = k_0(x_1, \ldots, x_9)$. Thus, starting with the pair (k, q), and forming repeatedly function fields of 10-dimensional forms (as in Merkurjev's construction for 2.8), one arrives at a direct limit field $F \supseteq k$ with $u(F) = 9$. In his paper, Izhboldin also obtained various other (an)isotropy results on forms of dimension 10 and 12 over function fields of other quadrics. The proofs of the main results in [Iz] are algebro-geometric, relying heavily on the theory of Chow groups and unramified cohomology of quadric hypersurfaces.

Of course, part of the success of Izhboldin's methods lies in the choice of the notion of essential 9-dimensional forms. It is not clear what would be a viable generalization of this notion to higher odd-dimensional forms. In particular, no fields of odd u-invariant > 9 have been constructed, although it seems now rather likely that such fields do exist.

Some other problems on the u-invariant remain open also. For instance, it is unknown whether there exist nonreal fields F with $|\dot{F}/\dot{F}^2| < \infty$ and $u(F) = 6$, and the relationship between $u(F)$ and $u(F(x))$ is still rather far from being fully understood. Some of these open problems will be collected together in the last section of this chapter.

3. Fields of Pythagoras Number 6 and 7

This section is devoted to the constructive study of the nature of the Pythagoras number $P(F)$ of a field F. In XI.5.7, we have proved the existence of fields F for which $P(F) = 2^n$ or $P(F) = 2^n+1$, for any $n \geq 0$. However, no fields with Pythagoras number other than 2^n and 2^n+1 have been produced in the earlier chapters. In fact, the problem of determining precisely the set of Pythagoras numbers of fields had remained open for quite some time.

In 1999, this problem was solved by D. Hoffmann. In his paper [Ho$_9$], Hoffmann showed that there exist formally real fields F with $P(F) = n$, for *any* given natural number n. Thus, for formally real fields F, there is really nothing interesting one can say about the nature of the number $P(F)$. Here, of course, it is only the case of formally real fields that merits our attention. Recall that, if F is a *nonreal* field, say of level 2^n, then $P(F) \in \{2^n, 2^n+1\}$, and both values are possible, as we have seen in Theorem XI.5.7.

This section is written in the same spirit as §2, in that we'll work with a special case of a more general theorem, and hope to convey the flavor of the general result through the more concrete work done in the special case. As far as the Pythagoras numbers of formally real fields are concerned, it had been well agreed, before Hoffmann's work [Ho$_9$], that "Pythagoras numbers 6 and 7" would be excellent test-cases:

Does there exist a (necessarily formally real) field F with $P(F) = 6$, or respectively, $P(F) = 7$?

We shall answer this question positively below, by specializing Hoffmann's construction to the case of $P(F) = 6$ or 7. This specialization makes it possible for us to explain the basic ideas in Hoffmann's construction, without getting bogged down by the technicalities of the most general case. As a matter of fact, once the case $P(F) = 6$ (or 7) is understood, the general case $P(F) = n$ is only a few steps away.

The construction of a field of Pythagoras number 6 (or any given integer n) is based on the same techniques first used by Merkurjev in his construction of a field of u-invariant 6 (see §2). In essence, it is just another application of the powerful function field techniques. In order to ensure that the function fields constructed are formally real, we shall use the following result of Elman, Lam, and Wadsworth on the extension of orderings to function fields.

Theorem 3.1 ([ELW$_2$]). *An ordering P on a field F can be extended to the function field $F(\varphi)$ of a quadratic form φ iff φ is indefinite (i.e., neither positive definite nor negative definite) at P.*

Proof. The "only if" part is clear: if P extends to an ordering P' on $F(\varphi)$, then, since φ is isotropic over $F(\varphi)$, it must be indefinite at P', and hence also at P.

For the converse, assume φ is indefinite at P. Let R be a real-closure of (F, P), and consider the function field $R(\varphi)$. Adjoining the generic zero of φ in $R(\varphi)$ to F, we get (a copy of) the function field $F(\varphi)$ within $R(\varphi)$. Since φ is indefinite and hence isotropic over R, $R(\varphi)/R$ is a purely transcendental extension by X.4.1. Thus, the unique ordering on R extends to an ordering P_1 on $R(\varphi)$ by VIII.1.13(C). Now P_1 restricts to an ordering P' on $F(\varphi)$ extending P, as desired. □

From the theorem above, we obtain immediately the following criterion for the formal reality of a function field.

Corollary 3.2. *A function field $F(\varphi)$ is formally real iff φ is not a totally definite form over F (that is, φ is indefinite with respect to at least one ordering of F).*

We can now begin the construction of a field of Pythagoras number 6 after Hoffmann. It is based on the following crucial result on the function field of a 7-dimensional form of the shape $6\langle 1\rangle \perp \langle -y\rangle$.

Theorem 3.3. *For any field F, let $\varphi = 6\langle 1\rangle \perp \langle -y\rangle$, where $-y \in \dot{F}$ is not a sum of two squares. Then, any anisotropic form $\sigma = 5\langle 1\rangle \perp \langle -x\rangle$ over F remains anisotropic over the function field $F(\varphi)$.*

Before proving this theorem, let us first explain how it can be used to produce the results we want.

Theorem 3.4. *There exists a (formally real) field K with $P(K) = 6$.*

Proof. Start with any formally real field F that has an element x of length 6 (i.e., x is a sum of six, but no fewer, squares in F). [For instance, we can take x to be $x_1^2+\cdots+x_6^2$ in the rational function field $F=\mathbb{R}(x_1,\ldots,x_6)$.] For the family of 7-dimensional forms

(3.5) $$\mathcal{C}=\{6\langle 1\rangle \perp \langle -y\rangle :\ y \text{ is a nonzero sum of squares in } F\},$$

let F_1 be the function field of the family $\mathcal{C}$ (see X.3). Since any y in 3.5 is a sum of squares in F, the form $6\langle 1\rangle \perp \langle -y\rangle$ is indefinite with respect to any ordering on F. By iterated use of 3.2, it follows that F_1 remains a formally real field. In particular, for any y in 3.5, $-y$ cannot be a sum of two squares in any subfield of F_1. Therefore, by iterated use of 3.3, it follows that $5\langle 1\rangle \perp \langle -x\rangle$ remains anisotropic over F_1; in other words, x still has length 6 over F_1. Repeating the construction now with F_1 as the base field, we can construct another formally real field $F_2 \supseteq F_1$, in which x again has length 6. Continuing in this fashion, we arrive at a (formally real) field $K=\bigcup_{i\geq 1} F_i$, in which x has length 6. Moreover, for any sum of squares y' in K, y' is a sum of squares in *some* F_i, and so y' becomes a sum of *six* squares in F_{i+1}—and hence in K. Thus, $P(K)=6$, as desired! □

The nice conclusions above certainly provide us ample incentives to scrutinize the following

Proof of 3.3. Let σ and φ be as in 3.3. For $\tau := 3\langle 1\rangle \perp \langle x\rangle$, we have $\sigma \perp \tau = 8\langle 1\rangle \in W(F)$. Since $\tau \cong \mathbb{H} \perp \beta$ for some binary form β over $F(\tau)$, it follows that

(3.6) $$\sigma \perp \beta \cong 8\langle 1\rangle \quad \text{over } F(\tau).$$

Next, note that $8\langle 1\rangle$ is anisotropic over F. [If otherwise, $8\langle 1\rangle \cong 4\mathbb{H}$, and so $5\langle 1\rangle$ is isotropic over F (by Chapter I, Exercise 14), which contradicts the anisotropicity of σ.] It follows that

(3.7) $$8\langle 1\rangle \text{ is anisotropic over } F(\tau).$$

For, if $8\langle 1\rangle$ is isotropic (and hence hyperbolic) over $F(\tau)$, then, by X.4.5, $\tau \subseteq 8\langle 1\rangle$ over F (meaning that τ is isometric to a subform of $8\langle 1\rangle$). Now cancellation of $3\langle 1\rangle$ leads to $\langle x\rangle \subseteq 5\langle 1\rangle$ over F, a contradiction.

Now assume σ becomes isotropic over $F(\varphi)$. Then 3.6 shows that $8\langle 1\rangle$ is isotropic, and hence hyperbolic, over $F(\varphi)(\tau)=F(\tau)(\varphi)$. Thus, over $F(\tau)$, 3.7 and X.4.5 show that $\varphi = 6\langle 1\rangle \perp \langle -y\rangle \subseteq 8\langle 1\rangle$. Cancelling $6\langle 1\rangle$, we see that $\langle -y\rangle \subseteq \langle 1,1\rangle$ over $F(\tau)$. Thus, the Pfister form $\rho=\langle 1,1,y,y\rangle$ becomes hyperbolic over $F(\tau)$. But the hypothesis that $-y \notin D_F\langle 1,1\rangle$ implies that ρ is anisotropic over F. Therefore, by X.4.5 again, we have $\tau \subseteq \rho$ over F. Since $\dim \tau = \dim \rho = 4$, this implies that $\tau \cong \rho$ over F. Computing determinants, we get $x \in \dot{F}^2$. This is impossible since φ is anisotropic. □

In the argument above, we saw the importance of the assumption $-y \notin D_F\langle 1,1\rangle$. Indeed, if this hypothesis is removed from 3.3, the result may no longer be true, as the following explicit example shows.

Example 3.8. Let F be a field of level 8, say with $-1 = x_1^2 + \cdots + x_8^2$. The element $x = x_1^2 + \cdots + x_6^2 \in \dot{F}$ certainly has length 6. Since $-x = 1 + x_7^2 + x_8^2$, we have $\langle -x\rangle \subseteq 3\langle 1\rangle$, so $\sigma := 5\langle 1\rangle \perp \langle -x\rangle \subseteq 8\langle 1\rangle$. Now consider any element $-y \in D_F\langle 1,1\rangle$. Then

$$\varphi := 6\langle 1\rangle \perp \langle -y\rangle \subseteq 6\langle 1\rangle \perp 2\langle 1\rangle = 8\langle 1\rangle. \tag{3.9}$$

Over the function field $F(\varphi)$, $8\langle 1\rangle$ is isotropic (and hence hyperbolic) by 3.9. But then by Chapter I, Exercise 14 again, σ is isotropic over $F(\varphi)$. However, σ is anisotropic over F (since x has length 6). This shows that the conclusion of Theorem 3.3 fails for the pair of forms σ and φ.

Of course, we could have also verified the importance of $-y \notin D_F\langle 1,1\rangle$ for Theorem 3.3 in the following "abstract" way. Suppose Theorem 3.3 remains true *without* the assumption that $-y \notin D_F\langle 1,1\rangle$. Then we could have performed the construction of $F_1 \subseteq F_2 \subseteq \cdots$ in the proof of 3.4, starting from a nonreal field F as in 3.8, with the element $x = x_1^2 + \cdots + x_6^2$ of length 6. Applying the version of 3.3 without the assumption $-y \notin D_F\langle 1,1\rangle$, we would get a field $K = \bigcup_{i\geq 1} F_i$ with $P(K) = 6$. But this is impossible since $K \supseteq F$ is a nonreal field!

Returning to 3.4, it is easy to see that a slight modification of the construction method will also yield a field of Pythagoras number 7. We simply replace 3.3 by

Theorem 3.10. *For any field F, let $\varphi = 7\langle 1\rangle \perp \langle -y\rangle$, where $-y \notin \dot{F}^2$. Then any anisotropic form $\sigma = 6\langle 1\rangle \perp \langle -x\rangle$ over F remains anisotropic over $F(\varphi)$.*

With this result, we can construct a field K with $P(K) = 7$ as before by starting with an element x of length 7 over some formally real field F. To prove 3.10, we can repeat the earlier steps in the proof of 3.3, working with the new pair σ, φ, and taking now $\tau := 2\langle 1\rangle \perp \langle x\rangle$. Here, 3.7 holds again, and, if σ becomes isotropic over $F(\varphi)$, the same considerations as before lead to the hyperbolicity of $\rho = \langle 1, y\rangle$ over $F(\tau)$. This is impossible if $-y \notin \dot{F}^2$. □

The two special cases (of Pythagoras numbers 6 and 7) covered in detail above should have given a good idea of how one can go about constructing a field F with *any* prescribed Pythagoras number n. One needs to have a general result (for the given integer n) in the nature of 3.3 and 3.10; after this, one starts with an element x of length n in some field, and carries out the (by now standard) iterated function field construction to get the desired field. To be more precise, let us make a clear statement of a viable substitute for 3.3 and 3.10 that is to be used for the general construction. To avoid complications, we work over a formally real base field F.

Theorem 3.11. *Let x, y be nonzero sums of squares in a formally real field F, and let*

$$\varphi = n\langle 1\rangle \perp \langle -y\rangle, \quad \sigma = (n-1)\langle 1\rangle \perp \langle -x\rangle.$$

If σ is anisotropic over F, then it remains anisotropic over $F(\varphi)$.

The proof of 3.11 is a generalization of those for 3.3 and 3.10, carried out in the more convenient setting of a formally real base field F. For the details, we refer the reader to Hoffmann's original paper [Ho9]. With 3.11 in place, one can then easily "produce" a formally real field of Pythagoras number n by starting with the element $x = x_1^2 + \cdots + x_n^2$ in $F := \mathbb{R}(x_1, \ldots, x_n)$, and carrying out the function field construction on F to get a formally real field K with $P(K) \le n$. Theorem 3.11 guarantees that x still has length n in K; consequently, $P(K) = n$, as desired.

By suitable modifications of the above method, Hoffmann has also been able to construct fields of Pythagoras number n with other additional properties. For instance, K may be chosen to have a unique ordering, and the behavior of the (general) u-invariant $u(K)$ can also be somewhat controlled.

4. Levels of Commutative Rings

The *level* $s(A)$ for a commutative ring A is defined in the same way as for fields: it is the smallest positive integer n such that -1 is a sum of n squares in A, and it is ∞ if no such n exists. In order to discuss the nature of this invariant in a proper setting, we begin this section by first developing the ingredients of "real algebra" for commutative rings. Throughout this (and the next) section, $\sigma(A)$ will denote the set of sums of squares in A.

Definition 4.1. A commutative ring A is called *semireal* if $-1 \notin \sigma(A)$ (that is, $s(A) = \infty$), and *real* if $A \neq 0$ and $a_1^2 + \cdots + a_n^2 = 0$ implies $a_1 = \cdots = a_n = 0$ (for all n).

If A is a field, these two notions are the same. However, if A is a commutative ring, "real" $\Rightarrow$ "semireal" but not vice versa. For instance, the commutative ring

$$A = \mathbb{R}[x_1, \ldots, x_n]/(x_1^2 + \cdots + x_n^2)$$

is *not* real, but it has a homomorphism into $\mathbb{R}$, so it is semireal.

The two generalizations of the formal reality of fields to commutative rings A can be extended to ideals in such rings as follows.

Definition 4.2. A (proper) ideal $\mathfrak{A} \subseteq A$ is called *semireal* (resp. *real*) if $A/\mathfrak{A}$ is semireal (resp. real).

If $\mathfrak{A}$ is a maximal ideal, the two notions in 4.2 are the same. If $\mathfrak{A}$ is a prime ideal, they are not; in this case, $\mathfrak{A}$ being real simply means that the quotient field of $A/\mathfrak{A}$ is formally real.

A truly fundamental result in real commutative algebra, due to Coste-Roy [CR] and Bröcker, Dress and Scharlau [BDS], is the following:

Theorem 4.3. *A ring*[(6)] *A is real (resp. semireal) iff it has a real (resp. semireal) prime ideal.*

To prove this, we use a familiar trick in commutative algebra—that of conjuring up prime ideals by the maximal avoidance of multiplicative sets. The "real" version of this goes as follows.

Lemma 4.4. *Let $S \subseteq A$ be a multiplicative set such that $0 \notin S$, $1 \in S$, and $S + \sigma(A) \subseteq S$. Let $\mathfrak{p}$ be an ideal of A maximal with respect to the property that $\mathfrak{p}$ is disjoint from S. Then $\mathfrak{p}$ is a real prime ideal.*

Proof. The fact that $\mathfrak{p}$ is prime is well-known from commutative algebra. (For this, we do not need $S + \sigma(A) \subseteq S$.) If $\mathfrak{p}$ is not real, there would exist a relation $b_1^2 + \cdots + b_n^2 \in \mathfrak{p}$ where, say, $b_1 \notin \mathfrak{p}$. By the choice of $\mathfrak{p}$, $\mathfrak{p} + (b_1)$ must intersect S, so we have $s \equiv rb_1 \pmod{\mathfrak{p}}$ for some $s \in S$. But then $s^2 \equiv r^2 b_1^2 \pmod{\mathfrak{p}}$, and so

$$s^2 + r^2 b_2^2 + \cdots + r^2 b_n^2 \equiv r^2(b_1^2 + \cdots + b_n^2) \equiv 0 \pmod{\mathfrak{p}}.$$

This is a contradiction, since the LHS $\in S + \sigma(A) \subseteq S$. □

Proof of 4.3. If $\mathfrak{p}$ is a semireal prime ideal, then the ring homomorphism $A \to A/\mathfrak{p}$ shows that A is semireal. Conversely, suppose A is semireal. Then $S = 1 + \sigma(A)$ satisfies the hypotheses in 4.4, so 4.4 (along with Zorn's Lemma) yields a real prime ideal $\mathfrak{p}$ of A. □

Local-Global Criterion for Finite Level 4.5. *For any ring A, $s(A) < \infty$ iff $s(A_{\mathfrak{m}}) < \infty$ for all maximal ideals $\mathfrak{m} \subseteq A$.*

Proof. We need only prove the "if" part. Assume that $s(A) = \infty$. Then A is semireal, so there exists a real prime ideal $\mathfrak{p}$. The field $A_{\mathfrak{p}}/\mathfrak{p}A_{\mathfrak{p}}$ is formally real since it is isomorphic to the quotient field of $A/\mathfrak{p}$. This shows that $s(A_{\mathfrak{p}}) = \infty$. If $\mathfrak{m}$ is any maximal ideal containing $\mathfrak{p}$, then we have a homomorphism $A_{\mathfrak{m}} \to A_{\mathfrak{p}}$, which shows that $s(A_{\mathfrak{m}}) = \infty$ as well. □

(6) From here on in this section, the word "ring" means a *commutative* ring.

To relate the above results to real algebraic geometry, let us now consider rings of the type $A = k[x_1, \ldots, x_n]/\mathfrak{A}$, where $\mathfrak{A}$ is an ideal in the polynomial ring $k[x_1, \ldots, x_n]$, and k is a real-closed field. These rings arise as the coordinate rings of affine k-varieties, and are known as *affine algebras over* k. For such algebras that are domains, we have the following important classical result of Artin and Lang.

Lang's Homomorphism Theorem 4.6. *Any real affine domain A over a real-closed field k admits a k-algebra homomorphism into k.*

A proof of this can be found in Lang's "Algebra", Springer-Verlag, 2002. For an alternative proof, see §5 of my introductory article [L_5], on which the current exposition is largely based. An immediate consequence of 4.6 is the following geometrical interpretation of the semireality of a polynomial ideal.

Weak Real Nullstellensatz 4.7. *An ideal $\mathfrak{A} \subseteq k[x_1, \ldots, x_n]$ is semireal iff the affine variety defined by $\mathfrak{A}$ has a k-point (that is, a point in k^n).*

Proof. If $a = (a_1, \ldots, a_n)$ is a k-point, evaluation at a gives a k-algebra homomorphism from $A = k[x_1, \ldots, x_n]/\mathfrak{A}$ to k, so A is clearly semireal. Conversely, if A is semireal, 4.3 implies that it has a real prime $\mathfrak{p}/\mathfrak{A}$. Then $k[x_1, \ldots, x_n]/\mathfrak{p}$ is a real affine domain, so it admits a k-algebra homomorphism φ into k, by 4.6. Letting $a_i = \varphi(\overline{x}_i)$, we see that $(a_1, \ldots, a_n) \in k^n$ is a k-point for the variety defined by $\mathfrak{A}$. ☐

As in the case of classical algebraic geometry over $\overline{k}$ (the algebraic closure of k), there is a strong version of 4.7 too, called the *Real Nullstellensatz*, to the effect that a polynomial f vanishes on all the k-points of an ideal $\mathfrak{A}$ iff $f^{2r} + g \in \mathfrak{A}$ for some $r \geq 0$ and some $g \in \sigma(k[x_1, \ldots, x_n])$. For a proof of this, see [L_5: (6.7)]. The Strong Real Nullstellensatz, however, will not be needed in the following.

Next, we shall briefly discuss the extension of the classical Artin-Schreier theory of orderings (in VIII.1) to the case of a commutative ring A.

Just as in the case of fields, a *preordering* on A is a set $T \subseteq A$ such that

$$T + T \subseteq T, \quad T \cdot T \subseteq T, \quad A^2 \subseteq T, \quad -1 \notin T. \tag{4.8}$$

Such a T is called an *ordering* on A if, in addition,

$$T \cup -T = A, \quad \text{and} \quad \text{supp}(T) := T \cap -T \text{ is a prime ideal in } A. \tag{4.9}$$

Recall that, in the case where A is a field, the second condition in 4.9 is automatic (since $T \cap -T = 0$ by VIII.9.2). If A is only a commutative ring, however, this may not be the case, and it must be *assumed* (along with $T \cup -T = A$) in order that T be an ordering of A.

If T is an ordering on A, then T can be used (in an obvious way) to define an ordering $\widetilde{T}$ on the quotient field of the domain $A/\mathfrak{p}$, where $\mathfrak{p}$ is the prime ideal $\mathrm{supp}(T)$; in particular, $\mathfrak{p}$ is a real prime. Conversely, if $\mathfrak{p}$ is any real prime ideal of A and $\widetilde{T}$ is any ordering on the quotient field of $A/\mathfrak{p}$, then $T := \{a \in A\colon\ \overline{a} \in \widetilde{T}\}$ is an ordering on A. Thus, *an ordering on A may be thought of as a real prime ideal $\mathfrak{p}$ of A together with an ordering on the quotient field of $A/\mathfrak{p}$.*

We have now the following generalization of the Artin-Schreier Criterion VIII.1.10 for the formal reality of fields.

Theorem 4.10. *For any ring A, the following are equivalent:*

(1) *A is semireal.*

(2) *A has a homomorphism into a formally real field.*

(3) *A has an ordering.*

(4) *A has a preordering.*

Proof. (3) $\Rightarrow$ (4) is a tautology.

(4) $\Rightarrow$ (1). If T is a preordering on A, then $-1 \notin T \supseteq \sigma(A)$ gives (1).

(1) $\Rightarrow$ (2). If A is semireal, it has a real prime $\mathfrak{p}$ by 4.3. Then A has a natural homomorphism into the formally real quotient field of $A/\mathfrak{p}$.

(2) $\Rightarrow$ (3). Let $\varphi\colon A \to F$ be a homomorphism, where F is a formally real field. Then any ordering $\widetilde{T}$ on F induces $T := \{a \in A \mid \varphi(a) \in \widetilde{T}\}$, which is an ordering on A. □

As in the case of fields, any maximal preordering in a ring A can be shown to be an ordering.[7] However, *an ordering on A may no longer be maximal as a preordering.* This represents an interesting departure from the classical case of fields. Nevertheless, the orderings of A may be characterized among the preorderings as follows: *a preordering $T \subseteq A$ is an ordering iff*

$$(4.9)' \qquad \forall\, a, b \in A\colon\ \ ab \in -T \implies a \in T \text{ or } b \in T.$$

For a proof of this interesting characterization of orderings, see [L_5: (3.2)].

Having provided the general framework in which to study real commutative algebra, we can now more meaningfully tackle the numbers that arise as quadratic invariants of commutative rings. For the balance of this section, our attention will be focused on the level invariant, $s(A)$. The study of the Pythagoras number $P(A)$ will be postponed to the next section.

After Pfister showed (in [Pf_1], c. 1965) that the level of a (nonreal) field must be a 2-power, it had remained an open question for some time

[7] Of course, as soon as we know this, Serre's idea of using Zorn's Lemma (in VIII.9) would give a quick independent proof of (4) $\Rightarrow$ (3) in 4.10.

whether the level $s(A)$ of a (commutative) ring A is also some special kind of integer. This level problem has been explicitly mentioned in the writings of Baeza and Knebusch. More specifically, in his list of open problems for quadratic forms (in [Kn3], c. 1976), Knebusch proposed as "Problem 13" the computation of the level of the following affine $\mathbb{R}$-domain:

$$A_n = \mathbb{R}[x_1, \dots, x_n]/(1 + x_1^2 + \cdots + x_n^2). \tag{4.11}$$

Of course, $s(A_n) \leq n$. Knebusch asked: is $s(A_n) = n$?

The choice of the domain A_n is reasonable for this problem, since A_n is a "generic" commutative $\mathbb{R}$-algebra in which -1 is a sum of n squares (for a given n). Also, recall that the quotient field F_n of A_n was exactly the kind of field whose level was brilliantly computed by Pfister: $s(F_n) = 2^k$ if $2^k \leq n < 2^{k+1}$ (see XI.2.6 and the ensuing Remark).

Baeza and Pfister both observed that Knebusch's question can be answered affirmatively in two special cases, as follows.

Proposition 4.12. *If* $n = 2^k$ *or* $n = 2^k + 1$, *then* $s(A_n) = n$.

Although this observation will be superseded by the later results in this section, we deem it still of interest to see how the two particular cases in 4.12 were solved, for two reasons. First, the method of solution is essentially field-theoretic, so this solution may be regarded as a natural application of our earlier results on fields. Second, the case $n = 2^k + 1$ in 4.12 shows that the level of a commutative ring can be of the form $2^k + 1$ for any prescribed integer k, which represents already an interesting departure from the case of the level of a field.

Proof of 4.12. If $n = 2^k$, Pfister's result (recalled above) shows that $s(F_n) = n$. Then, of course, $s(A_n) = n$ too. Now let $n = m + 1$, where $m = 2^k$. If $s(A_n) < n$, we can write

$$-1 = \sum_{j=1}^{m} (g_j + x_{m+1}h_j)^2 \in A_n, \quad \text{where } g_j,\, h_j \in \mathbb{R}[x_1, \dots, x_m],$$

since x_{m+1}^2 can be expressed in terms of $x_1, \dots, x_m$. Lifting this equation to $\mathbb{R}[x_1, \dots, x_{m+1}]$, we have

$$-1 = \sum_{j=1}^{m} (g_j + x_{m+1}h_j)^2 + p \cdot (1 + x_1^2 + \cdots + x_{m+1}^2) \tag{4.13}$$

for some polynomial $p = p(x_1, \dots, x_{m+1})$. By a degree inspection, we see that p cannot involve the variable x_{m+1}. Comparing the "coefficients" of powers of x_{m+1} in 4.13, we get the following three equations:

$$\sum h_j^2 + p = 0, \qquad \sum g_j h_j = 0, \quad \text{and}$$
$$\sum g_j^2 + p \cdot (1 + x_1^2 + \cdots + x_m^2) = -1.$$

Thus, $1+\sum g_j^2 = (\sum h_j^2)(1+x_1^2+\cdots+x_m^2) \neq 0$. Letting $r = \max\{\deg(h_j)\}$ (were "deg" means total degree), we can write

$$h_j = H_j + \cdots, \qquad g_j = G_j + \cdots,$$

where $\deg H_j = r$ and $\deg G_j = r+1$. Comparing leading terms yields

$$\sum H_j G_j = 0, \quad \text{and} \quad \sum G_j^2 = \left(\sum H_j^2\right)(x_1^2 + \cdots + x_m^2).$$

Multiplying the second equation by $\sum H_j^2 \neq 0$, we get

$$\left(\sum H_j^2\right)^2 (x_1^2 + \cdots + x_m^2) = \left(\sum H_j^2\right)\left(\sum G_j^2\right).$$

Here, all sums are from $j = 1$ to $j = m = 2^k$. Since $\sum_j H_j G_j = 0$, the RHS above is a sum of $m-1$ squares in $\mathbb{R}(x_1, \ldots, x_m)$, by XI.1.1′. But then so is $x_1^2 + \cdots + x_m^2$, which contradicts IX.2.4. □

The argument above does not generalize to other values of $n = m+1$, since it depended critically on XI.1.1′, which applies only when $m = 2^k$. However, it turns out that the result in 4.12 *does* hold for all n. This was proved by Dai, Lam, and Peng in 1980, using a topological argument. The topological ingredient of their proof is the following well-known fact.(8)

Borsuk-Ulam Theorem 4.14. *Any continuous mapping $Q\colon S^{n-1} \to \mathbb{R}^{n-1}$ must collapse some pair of antipodal points on the unit sphere $S^{n-1} \subseteq \mathbb{R}^n$.*

We shall now state and prove the Dai-Lam-Peng result from [DLP], which provides an affirmative answer to Knebusch's question, and shows that the level of a commutative ring can be any positive integer.

Theorem 4.15. *For all n, the affine domain A_n has level n.*

Proof. If $s(A_n) < n$, there would exist a polynomial equation

$$(4.16) \qquad -1 = f_1(x)^2 + \cdots + f_{n-1}(x)^2 + f_0(x)(1 + x_1^2 + \cdots + x_n^2),$$

where $f_j(x) \in \mathbb{R}[x] = \mathbb{R}[x_1, \ldots, x_n]$. For $i = \sqrt{-1}$, one has $f_j(ix) = p_j(x) + iq_j(x)$, where the p_j are *even* real polynomials and the q_j are *odd*

(8)See, e.g., Spanier's "Algebraic Topology", p. 266, McGraw-Hill, 1966.

real polynomials. Thus, replacing x by ix in 4.16 and comparing the real parts, we get

$$(4.17)\qquad -1=\sum_{j=1}^{n-1}\left(p_j(x)^2-q_j(x)^2\right)+p_0(x)(1-x_1^2-\cdots-x_n^2).$$

Consider the continuous mapping $Q\colon \mathbb{R}^n\to\mathbb{R}^{n-1}$ defined by the $(n-1)$-tuple $(q_1,\ldots,q_{n-1})$. By 4.14 we must have $Q(a)=Q(-a)$ for some point $a\in S^{n-1}$. But $q_j(-x)=-q_j(x)$, so $q_j(a)=0$ for all j. Plugging a into 4.17 now yields $-1=\sum_{j=1}^{n-1}p_j(a)^2\in\mathbb{R}$, a contradiction! □

It is worth noting that a variant of this method of proof will apply to some other affine domains as well. For instance, let us consider the domain

$$(4.18)\qquad A_{n,t}=\mathbb{R}[x_1,\ldots,x_n]/(1+x_1^{2t}+\cdots+x_n^{2t}),$$

where t is an odd integer. *Essentially the same argument will show that* $s(A_{n,t})=n$. For if otherwise, we can use the same substitution $x_j\mapsto ix_j$ on an analogue of 4.16, which changes $1+x_1^{2t}+\cdots+x_n^{2t}$ into $1-x_1^{2t}-\cdots-x_n^{2t}$. Since the real hypersurface $x_1^{2t}+\cdots+x_n^{2t}=1$ is homeomorphic to S^{n-1} by a radial map, the Borsuk-Ulam Theorem still applies, to give a contradiction.

Yet another variant of 4.15 provable by the same method is the following.

Theorem 4.19. *Let $q(x_1,\ldots,x_n)$ be a real form with no nontrivial zeros. Then $s(A_n[q^{-1}])=n$.*

Proof. If otherwise, we would have an equation $-q^{2r}=\sum_{j=1}^{n-1}f_j^2\in A_n$, where we may assume r is chosen to be *even*. We can lift this equation to $\mathbb{R}[x_1,\ldots,x_n]$, and make the substitution $x_j\mapsto ix_j$ as before. If $d:=\deg(q)$, we have

$$q(ix)^{2r}=(i^dq(x))^{2r}=i^{2rd}q(x)^{2r}=q(x)^{2r}.$$

Thus, the same contradiction as in the proof of 4.15 would result upon evaluating an analogue of the equation 4.17 at the point a given by the Borsuk-Ulam Theorem. □

The considerations in the proofs of the results 4.15–4.19 above were further developed by Dai and the author into a more systematic study between levels in algebra and topology. To give a flavor of this study, we shall give a quick exposition on the "Level Theorem" in the paper [DL].

To better understand the role of the Borsuk-Ulam Theorem, we first introduce a topological category $\mathcal{C}$. The objects of $\mathcal{C}$ are of the form $(X,-)$, where X is a topological space and "bar" is an involutive homeomorphism from X onto itself. (If "bar" is given and fixed, we shall often write X for the object $(X,-)$. Whenever confusion is unlikely, involutions in different spaces will all be denoted by "bars".) A morphism from $(X,-)$ to $(Y,-)$

is taken to be an *equivariant map*; that is, a continuous map $\varphi\colon X \to Y$ such that $\varphi(\overline{x}) = \overline{\varphi(x)}$ for all $x \in X$. As a notational device, we'll write $\varphi\colon X \multimap\!\!\to Y$ for such a morphism.

Some interesting and important objects in the category $\mathcal{C}$ are as follows.

(4.20) The topological spheres S^n: these will be understood to carry the *antipodal involution*: $\overline{x} = -x$ for every $x \in S^n$.

(4.21) The space $\mathbb{C}^n$ with the involution $(z_1, \ldots, z_n) \mapsto (\overline{z}_1, \ldots, \overline{z}_n)$ given by complex conjugation of the coordinates.

(4.22) An affine variety $X = V_{\mathbb{C}}(\mathfrak{A}) \subseteq \mathbb{C}^n$ defined over $\mathbb{R}$ by an ideal $\mathfrak{A} \subseteq \mathbb{R}[x_1, \ldots, x_n]$, equipped with the Euclidean (not the Zariski) topology. This variety is stable under complex conjugation, and is thus a subobject of the object $\mathbb{C}^n$ in 4.21.

(4.23) The odd-dimensional real projective space $\mathbb{R}P^{2m-1}$, equipped with the involution given in homogeneous coordinates by

$$[x_1, y_1, \ldots, x_m, y_m] \mapsto [-y_1, x_1, \ldots, -y_m, x_m].$$

This is easily checked to be a fixed-point-free involution. If you think of a point $[x_1, y_1, \ldots, x_m, y_m]$ as arising from a complex point

$$z = (z_1, \ldots, z_m) \in \mathbb{C}^m \setminus \{0\},$$

where $z_j = x_j + iy_j$, then the above involution is the one induced by the map $z \mapsto iz$ on $\mathbb{C}^m \setminus \{0\}$. (On the other hand, it is known in topology that the even-dimensional projective spaces $\mathbb{R}P^{2m}$ do not have fixed-point-free involutions.)

(4.24) The Stiefel manifold $V_{n,m}$ of orthonormal m-frames in $\mathbb{R}^n$, equipped with the involution

$$(v_1, \ldots, v_r, v_{r+1}, \ldots, v_m) \mapsto (v_1, \ldots, v_r, -v_{r+1}, \ldots, -v_m),$$

where r is a fixed integer $< m$. These objects subsume those in 4.20, since $V_{n,1} = S^{n-1}$ if we choose $r = 0$.

For any object $X \in \operatorname{Obj} \mathcal{C}$, the (*topological*) *level* of X is defined as follows:

$$s(X) = \inf\{n\colon\ \exists\, X \multimap\!\!\to S^{n-1}\}. \tag{4.25}$$

If no morphism $X \multimap\!\!\to S^{n-1}$ exists, $s(X)$ is taken to be ∞. The invariant $s(X)$ was first studied for the topological category $\mathcal{C}$ by Yang, Conner and Floyd, under the name of "coindex".[9] To be precise, our $s(X)$ here is $1 + \text{coindex}\,(X)$ in the notation of these authors. The shift of "1" turns

[9]The coindex is also referred to in the literature as "B-index", the "B" presumably coming from the name of Borsuk.

out to be convenient (and more natural) for the applications of the coindex theory to algebra.

One obvious consequence of the definition 4.15 is the following

Observation 4.26. *If there exists a morphism* $X \multimap\!\!\to Y$, *then* $s(X) \leq s(Y)$. *In particular, if there exist* $X \multimap\!\!\to Y \multimap\!\!\to X$, *then* $s(X) = s(Y)$.

Note that the study of the level is of interest mainly for objects $(X, -)$ of $\mathcal{C}$ whose involution "bar' is fixed-point-free. For, if the involution on X has a fixed point, then clearly there are no morphisms $X \multimap\!\!\to S^{n-1}$, so $s(X) = \infty$. In general, even if an involution "bar" is fixed-point-free, $s(X)$ may still not be finite. However, for spaces of interest to us, the situation is much better.

Proposition 4.27. *Let* $(X, -)$ *be a space with a fixed-point-free involution.*

(1) *If* X *is a topological subspace of* $\mathbb{R}^n$, *then* $s(X) \leq n$.

(2) *If* X *is a topological subspace of* $\mathbb{C}^n$ *and "bar" is induced by complex conjugation on* $\mathbb{C}^n$, *then* $s(X) \leq n$.

Proof. (1) follows by defining $\varphi\colon X \multimap\!\!\to S^{n-1}$ by

$$\varphi(x) = (x - \overline{x})/\|x - \overline{x}\| \quad \text{for any } x \in X.$$

On the other hand, (2) follows by defining $\psi\colon X \multimap\!\!\to S^{n-1}$ by

$$\psi(z_1, \ldots, z_n) = (y_1/d, \ldots, y_n/d),$$

where $z_j = x_j + iy_j$ and $d = (y_1^2 + \cdots + y_n^2)^{1/2}$. □

The most vital result in the study of the topological level is the following alternative statement of the

Borsuk-Ulam Theorem 4.28. $s(S^{n-1}) = n$.

This implies, in particular, the existence of objects of any prescribed level in the topological category $\mathcal{C}$. To see that 4.28 is just slight variation of the earlier statement of the Borsuk-Ulam Theorem in 4.14, we proceed as follows.

If we assume 4.14, there cannot exist $Q\colon S^{n-1} \multimap\!\!\to S^{m-1}$ for $m < n$, since $S^{m-1} \subseteq \mathbb{R}^m \subseteq \mathbb{R}^{n-1}$ and such a map Q does not collapse any pair of antipodal points on S^{n-1}. This shows that $s(S^{n-1}) = n$. Conversely, if some $Q\colon S^{n-1} \to \mathbb{R}^{n-1}$ fails to collapse some pair of antipodal points on S^{n-1}, then $\sigma\colon S^{n-1} \to S^{n-2}$ given by

$$\sigma(x) = \big(Q(x) - Q(-x)\big)/\|Q(x) - Q(-x)\| \qquad (x \in S^{n-1})$$

is a morphism in $\mathcal{C}$, in contradiction to 4.28.

Corollary 4.29. *Let $X_n = V_{\mathbb{C}}(f)$ be the affine variety defined by the real polynomial $f = 1 + x_1^2 + \cdots + x_n^2 \in \mathbb{R}[x_1, \ldots, x_n]$, with involution given by complex conjugation. Then $s(X_n) = n$.*

Proof. Since X_n has no real points, we have a morphism $X_n \multimap S^{n-1}$ by the construction in the proof of 4.27(2). On the other hand,

$$(x_1, \ldots, x_n) \mapsto (ix_1, \ldots, ix_n)$$

gives a morphism $S^{n-1} \multimap X_n$. By 4.26 and 4.28, it follows that $s(X_n) = s(S^{n-1}) = n$. $\square$

(Of course, the above argument would have worked if $f = 1 + x_1^{2t_1} + \cdots + x_n^{2t_n}$, where the t_i's are odd integers.)

In order to apply the topological level theory to algebra, we shall introduce an important functor A from $\mathcal{C}$ to the category $\mathcal{A}$ of commutative $\mathbb{R}$-algebras. For $X = (X, -) \in \text{Obj}\,\mathcal{C}$, let $A_X := \text{Mor}(X, \mathbb{C})$, the set of morphisms from X to $\mathbb{C}$ (where $\mathbb{C}$ is viewed as an object in $\mathcal{C}$ as in 4.21). In other words, A_X is the set of continuous functions $f\colon X \to \mathbb{C}$ with the property that $f(\overline{x}) = \overline{f(x)}$ for every $x \in X$. Note that A_X is a ring under the usual addition and multiplication of functions. Constant maps of X into $\mathbb{R}$ are in A_X, so A_X is a (commutative) $\mathbb{R}$-algebra. Clearly, any $\varphi\colon X \multimap Y$ induces (naturally) an $\mathbb{R}$-algebra homomorphism $\varphi^*\colon A_Y \to A_X$. Thus, $X \mapsto A_X$ (and $\varphi \mapsto \varphi^*$) gives a *contravariant* functor from $\mathcal{C}$ to $\mathcal{A}$: this is just the representable functor $\text{Mor}(-, \mathbb{C})$ given by the commutative $\mathbb{R}$-algebra $\mathbb{C}$ in the category $\mathcal{A}$.

For any function $f \in A_X$, write $f(x) = p(x) + iq(x)$, where $i = \sqrt{-1}$ and p, q are real-valued continuous functions on X. The equation $f(\overline{x}) = \overline{f(x)}$ gives

$$p(\overline{x}) + i\,q(\overline{x}) = p(x) - i\,q(x),$$

so we have $p(\overline{x}) = p(x)$ and $q(\overline{x}) = -q(x)$ for every $x \in X$. Thus, each $f \in A_X$ may be thought of as a pair of real-valued continuous functions (p, q), where p is "even" and q is "odd" (with respect to the involution on X).

With the functor $A\colon \mathcal{C} \to \mathcal{A}$ in place, we are now in a position to state the following main result from [DL].

Level Theorem 4.30. *For any $X \in \text{Obj}\,\mathcal{C}$, $s(X) = s(A_X)$. In other words, the following diagram commutes:*

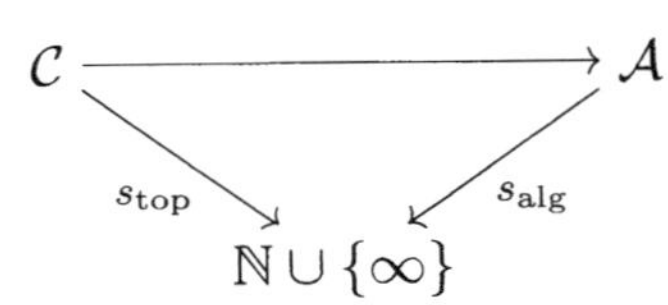

This theorem exhibits a nice synergy between algebra and topology in that it "computes" the topological level $s(X)$ in terms of the algebraic level of the function algebra A_X, and vice versa. By taking $X = S^{n-1}$, we then obtain a commutative $\mathbb{R}$-algebra $A_{S^{n-1}}$ whose algebraic level is $s(S^{n-1}) = n$ according to the Borsuk-Ulam Theorem 4.28. Of course, once such an $\mathbb{R}$-algebra is exhibited, it follows immediately from the generic nature of the algebra A_n in 4.11 that $s(A_n) = n$ for any n. This way of proving $s(A_n) = n$ may be viewed as a higher form of the ad hoc argument given in the earlier proof for 4.15.

Proof of Theorem 4.30. The following proof, taken verbatim from [DL], is completely elementary.

Step 1. First, we show that $s(A_{S^{n-1}}) \leq n$. To see this, define $f_j\colon S^{n-1} \to \mathbb{C}$ by $f_j(x) = ix_j$, where $x = (x_1, \ldots, x_n) \in S^{n-1}$ and $i = \sqrt{-1}$. Clearly, $f_j \in A_{S^{n-1}}$, and

$$(f_1^2 + \cdots + f_n^2)(x) = (ix_1)^2 + \cdots + (ix_n)^2 = -1 \qquad (\forall x \in S^{n-1}).$$

Thus, $-1 = f_1^2 + \cdots + f_n^2 \in A_{S^{n-1}}$, showing that $s(A_{S^{n1}}) \leq n$.

Step 2. Let m be any integer $< s(X)$, and let $h(x_1, \ldots, x_m)$ be any real polynomial that does not represent -1 over $\mathbb{R}$. Then h does not represent -1 over A_X. In fact, assume there exist $f_1, \ldots, f_m \in A_X$ such that $-1 = h(f_1, \ldots, f_m) \in A_X$. Write $f_j = p_j + iq_j$ $(1 \leq j \leq m)$. Then the $\{q_j\}$ do not have a common zero on X. For, if $x \in X$ is such a common zero, then evaluation of $h(f_1, \ldots, f_m)$ at x gives $-1 = h(p_1(x), \ldots, p_m(x))$, a contradiction. Thus, we can define a continuous map $q\colon X \to S^{n-1}$ by

$$q(x) = \big(q_1(x)/\delta(x), \ldots, q_m(x)/\delta(x)\big),$$

where $\delta(x) = (q_1(x)^2 + \cdots + q_m(x)^2)^{1/2} \neq 0$. This is a morphism in $\mathcal{C}$, since the q_i 's are *odd* functions. We have thus $s(X) \leq m$, a contradiction.

Step 3. Applying Step 2 to $h = x_1^2 + \cdots + x_m^2$ where $m < s(X)$, we see that -1 is not a sum of fewer than $s(X)$ squares in A_X. Thus, $s(A_X) \geq s(X)$. To show the reversed inequality, we may assume that $n := s(X) < \infty$. Take a morphism $X \multimap S^{n-1}$. This induces a ring homomorphism $A_{S^{n-1}} \to A_X$. Therefore, $s(A_X) \leq s(A_{S^{n-1}}) \leq n$ (by Step 1), as desired. □

While the levels of the real spheres S^{n-1} were computed by the Borsuk-Ulam Theorem, the next group of interesting spaces with fixed-point-free involutions are the real projective space $\mathbb{R}P^{2m-1}$ in 4.23, and the Stiefel manifolds $V_{n,m}$ in 4.24. For some information (and literature) on the computation of the (topological) levels of these spaces, see respectively [DL] and [Pf5]. Pfister's monograph [Pf5] contains especially a wealth of information on the applications of quadratic form theory to algebraic geometry and topology.

The very fact that $s(X)-1$ is called the "coindex" of $X \in \operatorname{Obj} \mathcal{C}$ suggests that there is a dual notion of the "index" of X. Indeed, turning the morphism arrow around in 4.25, we can define the *colevel* of X to be

$$s'(X) = \sup\{m:\ \exists\, S^{m-1} \multimap X\}. \tag{4.31}$$

This invariant is just $1+\operatorname{index}(X)$ in the terminology of Yang, Conner and Floyd.

Of course, the Borsuk-Ulam Theorem gives again $s'(S^{n-1}) = n$. We can also quickly deduce the following fact.

Proposition 4.32. *For any $X \in \operatorname{Obj} \mathcal{C}$, $s'(X) \le s(X)$. If there is a morphism $X \multimap Y$, then $s'(X) \le s'(Y)$.*

Proof. The second statement is obvious, and the first statement follows from the observation that

$$S^{m-1} \multimap X \multimap S^{n-1} \Longrightarrow m \le n,$$

as a result of the Borsuk-Ulam Theorem. ☐

In view of 4.31 and 4.32, it would seem reasonable to seek also a notion of colevel in the algebraic category. For commutative $\mathbb{R}$-algebras, a possible definition has been given in [DL]. The idea here is that one replaces the spheres S^{n-1} by their function algebras $A_{S^{n-1}}$, and considers homomorphisms of the given algebra into $A_{S^{n-1}}$. More precisely,

Definition 4.33. For any commutative $\mathbb{R}$-algebra A, the *colevel* $s'(A)$ is defined to be

$$s'(A) = \sup\{m:\ \exists\ \mathbb{R}\text{-algebra homomorphism } A \longrightarrow A_{S^{m-1}}\}.$$

(Again, $s'(A) = \infty$ if no such algebra homomorphism exists.)

It is easy to check that this new invariant has the formal properties of the topological colevel, as given in 4.32.

Proposition 4.34. (1) *For any commutative $\mathbb{R}$-algebra A, $s'(A) \le s(A)$.*
(2) *If there is an algebra homomorphism $B \to A$, then $s'(A) \le s'(B)$.*
(3) *If A is $A_{S^{n-1}}$ or the ring A_n in 4.11, then $s'(A) = n$.*

Proof. (1) If there is some algebra homomorphism from A to $A_{S^{m-1}}$, then $s(A) \ge s(A_{S^{m-1}}) = m$. Taking the supremum of all such m, we get (1).

(2) is clear, since any algebra homomorphism $A \to A_{S^{m-1}}$ can be composed with $B \to A$ to get an algebra homomorphism $B \to A_{S^{m-1}}$.

(3) Since $s(A_{S^{n-1}}) = n$, $A_{S^{n-1}}$ has no homomorphism into A_{S^n}. Therefore, $s'(A_{S^{n-1}}) = n$. Similarly, the fact that A_n has a homomorphism into $A_{S^{n-1}}$ but none into A_{S^n} shows that $s'(A_n) = n$. $\square$

The existence of the Level Theorem 4.30 suggests that there should be an analogous Colevel Theorem. We shall offer such a theorem below, although this theorem (from [DL]) is not as strong as 4.30.

Colevel Theorem 4.35. *For any space with involution $(X, -)$, we have $s'(X) \leq s'(A_X)$. Equality holds if X is an affine variety defined over $\mathbb{R}$, with involution given by complex conjugation.*

Proof. To prove that $s'(X) \leq s'(A_X)$, we may assume that $n = s'(A_X) < \infty$. If $s'(X) \geq n+1$, then by definition there is a morphism $S^n \multimap X$. This induces an algebra homomorphism $A_X \to A_{S^n}$. From Definition 4.33, it follows that $s'(A_X) \geq n+1$, a contradiction.

For the rest, we assume that X is an affine variety in $\mathbb{C}^n$ defined by an ideal $\mathfrak{A} \subseteq \mathbb{R}[x_1, \ldots, x_n]$, with involution "bar" given by complex conjugation. We shall denote the real coordinate ring $\mathbb{R}[x_1, \ldots, x_n]/\mathfrak{A}$ by $\mathbb{R}[X]$. Note that each $f \in \mathbb{R}[X]$ induces an equivariant function from $(X, -)$ to $(\mathbb{C}, -)$, so there is a natural $\mathbb{R}$-algebra homomorphism $\varepsilon\colon \mathbb{R}[X] \to A_X$.

For *any* space with involution $(Y, -)$, we claim that the following three statements are equivalent:

(1) *There exists a morphism varphi: $Y \multimap X$.*

(2) *There exists an $\mathbb{R}$-algebra homomorphism $h\colon A_X \to A_Y$.*

(3) *There exists an $\mathbb{R}$-algebra homomorphism $g\colon \mathbb{R}[X] \to A_Y$.*

Once these equivalences are established, we can apply them to the spaces $Y = S^{m-1}$. Taking the supremum of the values of m for which these statements hold, we get immediately

$$s'(X) = s'(A_X) = s'(\mathbb{R}[X]), \tag{4.36}$$

which gives more than what we want!

Let us now complete the proof of 4.35 by checking the equivalence of (1), (2) and (3).

(1) $\Longrightarrow$ (2) is clear, as we can take $h = \varphi^*$.

(2) $\Longrightarrow$ (3) is also clear, as we can take $g = h \circ \varepsilon$.

(3) $\Longrightarrow$ (1). Given g as in (3), let $\xi_i \in \mathbb{R}[X]$ $(1 \leq i \leq n)$ be the coordinate functions on X. For $y \in Y$, we can define

$$\varphi(y) = (g(\xi_1)(y), \ldots, g(\xi_n)(y)) \in \mathbb{C}^n.$$

This point lies in the affine variety X, since, for any polynomial $f(x_1, \dots, x_n) \in \mathfrak{A}$,

$$f(\xi_1, \dots, \xi_n) = 0 \implies f(g(\xi_1), \dots, g(\xi_n)) = 0$$
$$\implies f(g(\xi_1)(y), \dots, g(\xi_n)(y)) = 0.$$

It is routine to check that $\varphi \colon Y \to X$ is continuous and equivariant, and that the induced map $\varphi^* \colon A_X \to A_Y$ "extends" the given homomorphism $g \colon \mathbb{R}[X] \to A_Y$. The first part of this statement already yields the desired condition (1). □

Let us record the following nice consequence of 4.35.

Corollary 4.37. *For the affine variety $X \subseteq \mathbb{C}^n$ above, the following statements are equivalent:*

(1) $s(\mathbb{R}[X]) < \infty$.

(2) $s(X) = s(A_X) < \infty$.

(3) $s'(X) = s'(A_X) < \infty$.

(4) *X has no real points.*

Proof. (1) ⇒ (2) follows in light of the $\mathbb{R}$-algebra homomorphism $\varphi \colon \mathbb{R}[X] \to A_X$. (2) ⇒ (3) is clear since $s'(X) \le s(X)$. (3) ⇒ (4) follows since $s'(X)$ would be ∞ if $(X, -)$ has any fixed point. Finally, (4) ⇒ (1) follows from the Weak Real Nullstellensatz 4.7. □

Note that the first part of the above proof yields also the following additional information:

$$s(X) = s(A_X) \le s(\mathbb{R}[X]). \tag{4.38}$$

However, unlike the situation in 4.36, the inequality here may no longer be an equality, as we shall see in the following result.

Proposition 4.39. *Let $\gamma(x) \in \mathbb{R}[x] \setminus \mathbb{R}$ be such that $\gamma(a) \ge 1$ for all $a \in \mathbb{R}$. let $\mathfrak{A}$ be the principal ideal generated by $\gamma(x)^2 + y^2$ in $\mathbb{R}[x, y]$, and let $X = V_{\mathbb{C}}(\mathfrak{A})$. Then $s(X) = s(A_X) = 2$, but $s(\mathbb{R}[X]) = 3$.*

Proof. First, we note that X is the union of the two curves $y = \pm i\gamma(x)$, which intersect at the points $\{(c, 0) \in \mathbb{C}^2 \colon \gamma(c) = 0\}$. Thus, X is connected, and $X \cap \mathbb{R}^2 = \varnothing$. From these, we have $s(X) = 2$. Next, let $\theta(x) = \gamma(x)^2 - 1$. Since $\theta(\mathbb{R}) \ge 0$, we can write $\theta(x) = \theta_1(x)^2 + \theta_2(x)^2$ for suitable $\theta_j \in \mathbb{R}[x]$. Then, in $\mathbb{R}[X]$, we have

$$-1 = \theta_1(\overline{x})^2 + \theta_2(\overline{x})^2 + \overline{y}^2,$$

so $s(\mathbb{R}[X]) \le 3$. If equality *doesn't* hold here, we would have an equation

$$-1 = f_1(x, y)^2 + f_2(x, y)^2 + p(x, y)(\gamma(x)^2 + y^2) \in \mathbb{R}[x, y]. \tag{4.40}$$

We may now "recycle" a large part of the argument used in the proof of 4.12. First, we may assume that $f_j = g_j + yh_j$ $(g_j, h_j \in \mathbb{R}[x])$, whereupon $\deg_y p = 0$ (that is, $p \in \mathbb{R}[x]$). Comparing y-power terms in 4.40, we get

$$p + h_1^2 + h_2^2 = 0, \quad g_1h_1 + g_2h_2 = 0, \quad \text{and} \quad -1 = g_1^2 + g_2^2 + p\gamma^2.$$

Writing $g = g_1^2 + g_2^2$, we have (by the 2-square identity)

$$g(1+g) = (g_1^2 + g_2^2)(h_1^2 + h_2^2)\gamma^2 = (g_1h_2 - g_2h_1)^2\gamma^2.$$

Since g and $1+g$ are relatively prime, it follows that $g = k^2$ and $1+g = \ell^2$ for some $k, \ell \in \mathbb{R}[x]$. But then

$$1 = \ell^2 - k^2 = (\ell + k)(\ell - k)$$

implies that $\ell \pm k \in \mathbb{R}$ and so $\ell, k \in \mathbb{R}$. This clearly contradicts $1 + g = -p\gamma^2$, since $\gamma \in \mathbb{R}[x]$ is a nonconstant polynomial. We have thus shown that $s(\mathbb{R}[X]) = 3$. □

Much more can be said about the computation of the level and colevel of spaces and of algebras. We refer the reader to [DL] and [Pf$_5$] (and the literature contained therein) for other aspects of level theory that are not touched upon in this introductory section.

As a concluding remark, let us explain the role played by the field of real numbers $\mathbb{R}$ in the second half of this section. It is true that the Borsuk-Ulam Theorem was formulated and proved as a result on the topology of the real spheres, so it was convenient for us to work over the reals and consider primarily $\mathbb{R}$-algebras. However, most (if not all) of the results on the level can be extended to algebras over an *arbitrary* real-closed field k. This can be done in essentially two different ways. First, by Tarski's Principle (a powerful theorem on the elimination of quantifiers in logic), any first-order statement true for one real-closed field is always true for another. Therefore, if we reformulate the version of the Borsuk-Ulam Theorem in 4.14 by using *polynomial mappings* from the $(n-1)$-sphere over k to the Euclidean space k^{n-1}, then the fact that 4.14 is true for $k = \mathbb{R}$ implies that it is true for all real-closed fields k. Of course, all we need out of 4.14 was only the case of polynomial mappings. This observation, suitably extended to other similar situations, will essentially enable us to "transfer" level theorems over $\mathbb{R}$ to the case of other real-closed ground fields. A second approach, not assuming Tarski's Principle, is to seek purely algebraic proofs of the algebraic versions of whatever topological theorems are needed for the desired applications. (It may even be legitimately argued that algebraic proofs are better, since, in many cases, the topological versions can be deduced from the algebraic ones by using Weierstrass's Approximation Theorem for continuous mappings.) In the case of the Borsuk-Ulam Theorem, this has been done successfully by Knebusch [Kn$_6$], with subsequent simplifications by Arason and Pfister

[AP_3]. In the latter paper, various significant generalizations of the fact that $s(A_n) = n$ were obtained over arbitrary ground fields k.

Harkening back to the theme of the first part of this section, we shall content ourselves with a quick proof of the following:

Theorem 4.41. *Let $A_n(R) = R[x_1, \ldots, x_n]/(1 + x_1^2 + \cdots + x_n^2)$, where R is a semireal commutative ring. Then $s(A_n(R)) = n$.*

Proof. By 4.10, there exists a homomorphism $\varphi\colon R \to k$ into a formally real field k. After replacing k by one of its real-closures, we may assume that k is real-closed. By Tarski's Principle[(10)], we have $s(A_n(k)) = n$. From this, it follows that $s(A_n(R)) = n$. □

5. Pythagoras Numbers of Commutative Rings

Since the study of sums of squares is interesting for rings as well as for fields, the computation of the Pythagoras numbers of rings has also been a favorite topic for research in quadratic form theory. In this section, we shall briefly present a few highlights from this area of study.

As in §4, we shall restrict our attention to commutative rings. Therefore, by a *ring* below, we shall mean again a *commutative ring.* The basic definitions needed to start off our study are the same as in the case of fields. For a ring A, we write $\sigma(A)$ for the set of sums of squares in A, and for $a \in A$, we define $\mathrm{len}(a)$ (or $\mathrm{len}_A(a)$, the *length of* a in A) to be ∞ if $a \notin \sigma(A)$, and to be n if a is a sum of n, but no fewer, squares in A. The *Pythagoras number* of A is defined, as in the field case, to be

$$P(A) = \sup\{\,\mathrm{len}(a)\colon\ a \in \sigma(A)\}. \tag{5.1}$$

For rings of finite level, the Pythagoras number is "almost" determined by the level, in the following sense.

Proposition 5.2 (Joly, Peters). *Let A be a ring with $s = s(A) < \infty$. Then $s \le P(A) \le s + 2$. If $2 \in \mathrm{U}(R)$ or s is even, then $P(A) \le s + 1$.*

Proof. $P(A) \ge s$ is clear, since $\mathrm{len}(-1) = s$. For any $a \in \sigma(A)$, write $a = \sum_{i=1}^n a_i^2$. For $c := 1 + a_1 + \cdots + a_n$, we have $c^2 = 1 + a + 2b$ for some $b \in A$, and so

$$a = c^2 - 1 - 2b = c^2 + b^2 - (1 + b)^2. \tag{5.3}$$

Since $\mathrm{len}(-1) \le s$, this yields $\mathrm{len}(a) \le 2 + s$.

If $s \in \mathrm{U}(A)$, we get $\mathrm{len}(a) \le s + 1$ (for any a) by the familiar argument in the proof of XI.5.6(2). Finally, suppose s is even, and consider again

(10) Alternatively, we can use here Knebusch's algebraic version of the Borsuk-Ulam Theorem, or Lang's Homomorphism Theorem 4.6.

$a \in \sigma(A)$. Since $-1 \in \sigma(A)$, we also have $-a \in \sigma(A)$, so by 5.3, we can express $-a$ in the form $x^2 + y^2 - (1+y)^2$ in A. Writing $-1 = z_1^2 + \cdots + z_s^2$, we have

$$a = (1+y)^2 + (x^2+y^2)(z_1^2 + \cdots + z_s^2).$$

Since s is *even*, the classical 2-square identity shows that the second term on the RHS above is a sum of s squares. Therefore, $\text{len}(a) \leq 1 + s$, as desired. □

Since the number $s(A)$ above can be any positive integer according to Theorem 4.15, 5.2 comes very close to showing that the Pythagoras number $P(A)$ can also be any prescribed positive integer.(11) However, 5.2 alone doesn't quite do the job. Instead, we can appeal directly to the following result from [CDLR].

Proposition 5.4. *Let $A = \mathbb{R}[x_1, \ldots, x_n]$, where the generators $\{x_i\}$ are subject to the relations $x_i x_j x_k = 0$ for all i, j, k. Then $P(A) = n$.*

Proof. Clearly, $x_1^2 + \cdots + x_n^2 \in A$ has length n. Thus, it suffices to show that *any* $f \in \sigma(A)$ has length $\leq n$. Let

$$f = a + \lambda(x) + \varphi(x) = \sum \left(a_i + \lambda_i(x) + \varphi_i(x)\right)^2, \tag{5.5}$$

where a, $a_i \in \mathbb{R}$, $\lambda(x)$, $\lambda_i(x)$ are linear forms, and $\varphi(x)$, $\varphi_i(x)$ are quadratic forms. We argue in two cases.

Case 1: $a = 0$. Since $\sum a_i^2 = 0$, we have $a_i = 0$ for all i. Thus, 5.5 simplifies to $f = \varphi(x) = \sum \lambda_i(x)^2$. This means that f is a positive semidefinite quadratic form, so by the diagonalization theorem, f is a sum of no more than n squares of linear forms over $\mathbb{R}$.

Case 2: $a \neq 0$. In this case, it turns out that f is a perfect square in A! To show this, let us try to find a quadratic form $\sigma(x)$ such that

$$f = \left(\sqrt{a} + \frac{\lambda(x)}{2\sqrt{a}} + \sigma(x)\right)^2 = a + \lambda(x) + \frac{\lambda(x)^2}{4a} + 2\sqrt{a}\,\sigma(x).$$

(Note that $a = \sum a_i^2 > 0$ here.) For this equation to hold, we need only choose

$$\sigma(x) = \left(2\sqrt{a}\right)^{-1}\left(\varphi(x) - \frac{\lambda(x)^2}{4a}\right).$$ □

Note that the ring A constructed above is a finite-dimensional algebra over $\mathbb{R}$. In fact,

$$\dim_{\mathbb{R}} A = 1 + n + \frac{n(n+1)}{2} = \frac{(n+1)(n+2)}{2}.$$

(11)Of course, in this section, we are not assuming Hoffmann's result that the Pythagoras number of a field can be any natural number.

However, A is a local ring with a nilpotent maximal ideal, so there are many 0-divisors in A. If we want to come up with an example (with a prescribed Pythagoras number) that is an *integral domain*, a different construction is needed. As it turned out, the following is true for the affine domain A_n defined in 4.11.

Theorem 5.6. *Let* $A_n = \mathbb{R}[x_1, \dots, x_n]/(1 + x_1^2 + \cdots + x_n^2)$. *Then, in the polynomial ring* $A_n[t]$, *we have* $\operatorname{len}(t) = n + 1$. *In particular,* 5.2 *implies that* $P(A_n[t]) = n + 1$.

The proof of this result in [DL] made use of the general theory of another invariant of a ring, called the *sublevel*. Here, we'll try to provide the shortest route to this result without developing any more machinery. Thus, we'll skip the notion of the sublevel altogether, and go right to the heart of the arguments for proving 5.6. The crucial step is to prove the following fact by using a little bit of topology.

Lemma 5.7. *Let* $f_1, \dots, f_n \in A_{S^{n-1}}$ *(the ring of equivariant* $\mathbb{C}$*-valued functions on* S^{n-1} *defined in §4). Then there exists* $z \in S^{n-1}$ *such that* $f_1(z), \dots, f_n(z) \in \mathbb{C}$ *are collinear on a line in the Gaussian plane through the origin.*

Proof. We can write (as in §4) $f_j(z) = p_j(z) + iq_j(z)$, where p_j, q_j are real functions on S^{n-1}, with p_j even and q_j odd. We may assume that $\{q_j\}$ (resp. $\{p_j\}$) have no common zero on S^{n-1} (for otherwise the desired conclusion would be obvious). After a "normalization", $\{q_j\}$ (resp. $\{p_j\}$) defines a continuous odd (resp. even) mapping q (resp. p) from S^{n-1} to S^{n-1}. Then q has an *odd* topological degree, and p has an *even* topological degree, so q and p *cannot be homotopic*. This implies that there exists $z \in S^{n-1}$ at which q and p are antipodal (i.e., $q(z) = -p(z)$). [For, if otherwise, we could have constructed a homotopy by

$$(5.8) \qquad H_t(z) = \frac{(1-t)\,q(z) + t\,p(z)}{\|(1-t)\,q(z) + t\,p(z)\|} \qquad (z \in S^{n-1},\ t \in [0,1]),$$

with $H_0 = q$ and $H_1 = p$.] This means that $q_j(z) = -\delta \cdot p_j(z)$ for some nonzero real number δ independent of j. Therefore, $\{f_j(z)\}$ all lie on the line $y + \delta x = 0$, as desired. □

For a quadratic form $q(x) = \sum_{i,j=1}^n a_{ij} x_i x_j$ over a ring A, let us say q is *isotropic* (over A) if $q(b_1, \dots, b_n) = 0$ for some *unimodular* n-tuple $(b_1, \dots, b_n) \in A^n$ (i.e., an n-tuple for which $\sum_j b_j A = A$). Otherwise, we say q is *anisotropic* over A. We record the following consequence of 5.7.

Corollary 5.9. *The diagonal form* $n\langle 1 \rangle$ *is anisotropic over* $A_{S^{n-1}}$, *and also over the ring* A_n *in* 5.6.

Proof. Since there is a homomorphism[12] from A_n to $A_{S^{n-1}}$, it is sufficient to handle $A_{S^{n-1}}$. Assume $n\langle 1\rangle$ is isotropic over $A_{S^{n-1}}$, so we have $f_1^2 + \cdots + f_n^2 = 0$ with $\sum_j f_j g_j = 1$, where $f_j, g_j \in A_{S^{n-1}}$. By 5.7, there exists $z \in S^{n-1}$ such that $f_j(z) = r_j e^{\theta i}$ for some $\theta \in \mathbb{R}$ and $r_j \in \mathbb{R}$. Now

$$0 = f_1(z)^2 + \cdots + f_n(z)^2 = (r_1^2 + \cdots + r_n^2)\, e^{2\theta i} \implies \text{all } r_j = 0,$$

and so $f_j(z) = 0$ for all j, which contradicts $\sum_j f_j g_j = 1$. □

Note that the anisotropicity of $n\langle 1\rangle$ over A_n is a *strengthening* of the fact that $s(A_n) = n$. For, if $s(A_n) < n$, then an equation $1 + f_2^2 + \cdots + f_n^2 = 0 \in A_n$ would have shown that $n\langle 1\rangle$ is isotropic over A_n.

Proof of 5.6. Assume, instead, $t = \sum_{j=1}^n \left(f_{j0} + f_{j1}t + \cdots + f_{jr}t^r\right)^2$, where $f_{jk} \in A_n$. Then

$$\sum_{j=1}^n f_{j0}^2 = 0 \quad \text{and} \quad 2\sum_{j=1}^n f_{j0}f_{j1} = 1,$$

which contradict the anisotropicity of $n\langle 1\rangle$ over A_n! □

Remark 5.10. Having proved that $P(A_n[t]) = n + 1$, we may legitimately wonder if also $P(A_n) = n + 1$. This is certainly not the case if $n = 1$, since $A_1 \cong \mathbb{C}$ has Pythagoras number 1 (not 2). It is *also* false for $n = 2$: in [CDLR], it is shown that $P(A_2)$ is 2 (not 3). It *may be* true that $P(A_n) = n + 1$ for $n \geq 3$, but so far this has been proven only in the case where $n = 2^k$ $(k \geq 2)$ in [CDLR].

In the second half of this section, we shall give some information on the Pythagoras numbers of affine algebras. It turns out that, in many cases, the Pythagoras numbers are infinite, but in the case of very small transcendence degrees, there are some interesting finiteness results. We shall start with the latter situation.

Theorem 5.11. *If A is an n-dimensional commutative algebra over a field k, then $P(A) \leq n \cdot P(k)$.*

Proof. If $\text{char}(k) = 2$, then any sum of squares in A is already a square in A, so $P(A) = 1$. In the following, we may thus assume $\text{char}(k) \neq 2$. We may also assume, of course, that $P(k) < \infty$. Fix a k-basis $w_1, \ldots, w_n$ for A and consider any $a \in \sigma(A)$, say $a = \sum_{i=1}^N a_i^2$. Writing $a_i = \sum_j \alpha_{ij} w_j$ $(\alpha_{ij} \in k)$, look at the quadratic form

$$q(x_1, \ldots, x_n) := \sum_{i=1}^N \left(\alpha_{i1}x_1 + \cdots + \alpha_{in}x_n\right)^2. \tag{5.12}$$

---

[12] See Step 1 in the proof of 4.30.

Diagonalizing this n-ary quadratic form over k, we can write

$$q(x_1, \ldots, x_n) = \sum_{j=1}^{n} \beta_j L_j(x)^2,$$

where $\beta_j \in k$ (possibly zero for some j) and $L_j(x)$ $(1 \leq j \leq n)$ are n linearly independent linear forms over k. Choosing $x = (x_1, \ldots, x_n) \in k^n$ such that $L_j(x) = 1$ and $L_i(x) = 0$ for $i \neq j$, we see that $\beta_j \in \sigma(k)$ for all j. We can thus write $\beta_j = \beta_{j1}^2 + \cdots + \beta_{jp}^2$, where $p = P(k)$ and $\beta_{ji} \in k$. Substituting w_j for x_j in 5.12, we get

$$a = q(w_1, \ldots, w_n) = \sum_{j=1}^{n} \left(\beta_{j1}^2 + \cdots + \beta_{jp}^2\right) L_j(y)^2,$$

which is a sum of np squares in A. This proves that $P(A) \leq np = n \cdot P(k)$, as desired. □

In the case where A is a finite field extension of the field k, the result 5.11 was first obtained by Pfister. The more general proof above is taken from [CDLR]. Note that the first part of this proof gave some information over rings too: if k is a commutative ring and A is a commutative k-algebra such that $A = \sum_{j=1}^{n} kw_j$ (for some $w_1, \ldots, w_n \in A$), then $P(A) \leq g_k(n)$, where $g_k(n)$ is the smallest integer such that any sum of squares of n-ary linear forms over k can be written as a sum of $g_k(n)$ squares of such linear forms. (Of course, if no such integer exists, $g_k(n)$ is taken to be ∞, in which case the inequality $P(A) \leq g_k(n)$ becomes Platonic.)

In the special case where k is a field of characteristic $\neq 2$, the quadratic form argument used in the proof of 5.11 gives a bound $g_k(n) \leq n \cdot P(k)$. This is, of course, only a very crude bound. For some specific fields k, better bounds may be available. For instance, over the rational field $k = \mathbb{Q}$, a classical result of L. Mordell showed that any n-ary positive semidefinite quadratic form can be expressed as a sum of $n+3$ squares of $\mathbb{Q}$-linear forms. Thus, $g_{\mathbb{Q}}(n) \leq n + 3$. Assuming this result, it follows that

Proposition 5.13. *For any commutative finite-dimensional* $\mathbb{Q}$*-algebra* A, $P(A) \leq \dim_{\mathbb{Q}} A + 3$.

The case of commutative finite-dimensional algebras over fields handled in 5.11 may be thought of as that of affine algebras of transcendence degree 0. The next case is that of affine algebras of transcendence degree 1. Here, assuming that the ground field k is real-closed, there is a nice finiteness result. However, the proof of this result requires a fairly deep theorem on positive semidefinite polynomials over a real-closed field. To make our exposition self-contained, we will need to first give an account for this theorem.

For some background information, recall that a polynomial $f(x) \in k[x_1, \ldots, x_n]$ (over a real-closed field k) is said to be *positive semidefinite* (or "psd" for short) if $f(k^n) \geq 0$ for the unique ordering on k. Clearly, any $f \in \sigma(k[x_1, \ldots, x_n])$ is psd, so it is natural to wonder about the converse. If either $n = 1$ or f is a quadratic form in $x_1, \ldots, x_n$, the converse is indeed true, as we have seen earlier, and we also know that two squares suffice in the first case, and n squares suffice in the second. Hilbert proved that the converse is also true in one more case; namely, when f is a *ternary quartic* form, and in this case he showed that three (but no fewer) squares suffice. Even more remarkably, Hilbert showed that, for n-ary forms of degree m, the converse of the statement above about psd forms holds *only* in the cases already mentioned; namely, only when $n = 2$, or $m = 2$, or $(n, m) = (3, 4)$. (Note that the case of binary forms boils down to that of polynomials in one variable upon dehomogenization.)

Nowadays, Hilbert's statement at the end of the last paragraph is very easy to verify. By obvious considerations, it suffices to handle only two crucial cases; namely, those of *ternary sextics* and *quaternary quartics*. For the first case, consider the following variation of the Motzkin polynomial[(13)]

$$S(x, y, z) = x^4y^2 + y^4z^2 + z^4x^2 - 3x^2y^2z^2. \tag{5.14}$$

This ternary sextic is psd in light of the Arithmetic-Geometric Inequality. To see that $S \notin \sigma(k[x, y, z])$, we can use the following term inspection method proposed by Choi-Lam in [CL$_1$] and [CL$_2$]. Assume, for the moment, that $S = \sum h_i^2$, where the $h_i \in k[x, y, z]$ are necessarily cubic forms. By term inspection, x^3, y^3 and z^3 cannot appear in h_i. From this, it follows further that xy^2, yz^2 and zx^2 also cannot appear. Thus, we must have

$$h_i = a_i x^2 y + b_i y^2 z + c_i z^2 x + d_i xyz.$$

But then a comparison of the coefficients of $x^2y^2z^2$ in the equation $S = \sum h_i^2$ yields $-3 = \sum d_i^2 \geq 0$, a blatant contradiction. Exactly the same method shows that the quaternary quartic

$$Q(w, x, y, z) = w^4 + x^2y^2 + y^2z^2 + z^2x^2 - 4wxyz \tag{5.15}$$

in psd, but is not in $\sigma\big(k[w, x, y, z]\big)$; see Exercise 11.

In retrospect, the only surprise seems to be the fact that such a mundane "high school algebra method" for checking that certain psd polynomials failed to be sums of polynomial squares was not used before the mid-1970s. Earlier methods used by Hilbert, Motzkin, Ellison, and Robinson were all significantly harder or more sophisticated in nature.

---

(13)The "original" Motzkin polynomial was a dehomogenization of the psd ternary sextic $S'(x, y, z) := z^6 + x^4y^2 + x^2y^4 - 3x^2y^2z^2$. The method used here for $S(x, y, z)$ applies equally well to show that $S'(x.y.z) \notin \sigma(k[x, y, z])$; see Exercise 11.

Based on his work on the comparison of psd forms and sums of squares of forms, Hilbert raised the question of whether any psd form (over $\mathbb{R}$) is a sum of squares in the rational function field $\mathbb{R}(x_1, \ldots, x_n)$.(14) This appeared as the "17th Problem" in Hilbert's famous list of open problems presented to the International Congress of Mathematicians at Paris in 1900. Hilbert's 17th Problem was solved (affirmatively) by Artin in 1927. Nowadays, this affirmative solution can be deduced without too much difficulty from Artin's Criterion for sums of squares in fields VIII.1.12 and Lang's Homomorphism Theorem 4.6. For a detailed proof, see, e.g., [L_5: §6] or [Pf_5: Ch. 6].

Coming back to sums of squares of *polynomials*, we shall now present the case of "biforms", where, again, in a special situation, psd (positive semidefinite) will imply sos (sums of squares). For two independent sets of indeterminates $y = \{y_1, \ldots, y_r\}$ and $x = \{x_1, \ldots, x_n\}$, a *biform* in y, x of bidegree (e, d) is a polynomial f that is a form in x of degree d when the y_i 's are viewed as constants, and a form in y of degree e when the x_j 's are viewed as constants. For such forms, we have the following wonderful result due to Jakubović [Jak] and Rosenblum and Rovnyak [RR], which seems to be some kind of "hybrid" of the binary case and the quadratic case in Hilbert's classical study of psd forms.

Theorem 5.16. *Let* $f(y, z; x_1, \ldots, x_n) = \sum a_{ij}(y, z)x_i x_j$ *be a biform of bidegree* $(m, 2)$ *over a real-closed field* k. *If* f *is psd, then* f *is a sum of* $2n$ *squares of biforms.*

This theorem was first proved in the context of matrix polynomials and operator-valued functions. Subsequently, it has found interesting applications in diverse areas such as the theory of nonlinear regularization, optimal control, and differential games. In this section, it will hardly be relevant for us to discuss any of these applications; instead, we shall focus on the application of 5.16 to the computation of the Pythagoras number of affine algebras.

Since there is no easily accessible proof of 5.16 in textbooks, we shall include a self-contained proof here. The proof presented below follows in broad outline that given in my paper [CLR] with Choi and Reznick, but incorporates some simplifications and streamlining that I worked out jointly with Z. D. Dai in a seminar at Berkeley. We begin with the following easy result on inner products of vectors.

Lemma 5.17. *Let* D_i, E_i $(1 \le i \le n)$ *be vectors in* k^{2d} *such that*

$$(5.18) \qquad D_i \cdot D_j = E_i \cdot E_j, \quad \text{and} \quad D_i \cdot E_j + D_j \cdot E_i = 0 \quad \text{for all } i, j.$$

(14) In case the psd form is defined over $\mathbb{Q}$, Hilbert also asked if it is a sum of squares in $\mathbb{Q}(x_1, \ldots, x_n)$.

Then, after an orthonormal change of basis, we can arrange that

$$D_i = (s_{i1}, t_{i1}, \ldots, s_{id}, t_{id}), \qquad E_i = (-t_{i1}, s_{i1}, \ldots, -t_{id}, s_{id}). \tag{5.19}$$

Proof. We induct on n. The case $n = 0$ is vacuous, so assume $n \geq 1$. By hypothesis, D_1 and E_1 are orthogonal, with the same length, say s. If $s = 0$, we are done by induction, so assume $s \neq 0$. Using $s^{-1}D_1$ and $s^{-1}E_1$ as part of an orthonormal basis, we can arrange that

$$D_1 = (s, 0, 0, \ldots, 0), \qquad D_i = (s_{i1},\, t_{i1},\, \overline{D}_i) \quad (i \geq 2),$$
$$E_1 = (0, s, 0, \ldots, 0), \qquad E_i = (s'_{i1},\, t'_{i1},\, \overline{E}_i) \quad (i \geq 2),$$

where $\overline{D}_i,\, \overline{E}_i \in k^{2(d-1)}$. Putting $j = 1$ in 5.18, we see that $t'_{i1} = s_{i1}$ and $s'_{i1} = -t_{i1}$ for $i \geq 2$. With this information, 5.18 yields the same inner product equations for $\overline{D}_i,\, \overline{E}_i$ $(2 \leq i \leq n)$, so the induction proceeds. □

Proof of 5.16. For convenience, we shall prove the equivalent version of 5.16 obtained by dehomogenizing the set of variables $\{y, z\}$. After setting $z = 1$ (and changing notations), we may then think of $f = \sum_{i,j=1}^{n} a_{ij}(y)x_i x_j$ as a quadratic form over $k[y]$ (such that $f(k^{n+1}) \geq 0$). We shall proceed by induction on n.

If $n = 1$, f is just $a_{11}(y)x_1^2$. Then $a_{11}(y)$ is psd, and hence a sum of two squares in $k[y]$, so we are done. For $n > 1$, we may assume $a_{ij}(y) = a_{ji}(y)$ and write

$$f = a_{11}(y)x_1^2 + 2x_1 \textstyle\sum_{j \geq 2} a_{1j}(y)x_j + \overline{f}(y;\, x_1, \ldots, x_n).$$

We may assume that $a_{11}(y) \neq 0$, for otherwise the fact that f is psd implies that *all* $a_{1j}(y) = 0$, so $f = \overline{f}$ and we are done by induction. Now $a_{11}(y)$ is psd, and so is the following discriminant:

$$D = a_{11}(y)\, \overline{f}(y;\, x_2, \ldots, x_n) - \left(\textstyle\sum_{j \geq 2} a_{1j}(y)x_j\right)^2.$$

By the inductive hypothesis, D is a sum of $2n - 2$ squares of linear forms in $\{x_2, \ldots, x_n\}$ over $k[y]$. Then

$$a_{11}(y) \cdot f = \left(a_{11}(y)x_1 + \textstyle\sum_{j \geq 2} a_{1j}(y)x_j\right)^2 + D(y;\, x_2, \ldots, x_n)$$

is a sum of $2n - 1$ (in particular, $2n$) squares; say

$$a_{11}(y) \cdot f = \textstyle\sum_{\ell=1}^{2n} \left(\sum_{i=1}^{n} a_i^{(\ell)}(y)x_i\right)^2. \tag{5.20}$$

Let $A_i(y)$ be the vector of polynomials $(a_i^{(1)}(y), \ldots, a_i^{(2n)}(y))$. Then a comparison of the two sides of 5.20 yields $a_{11}(y) \cdot a_{ij}(y) = A_i \cdot A_j$. If we can "get rid of" the extra factor $a_{11}(y)$ to get $a_{ij}(y) = B_i \cdot B_j$ for suitable vectors of polynomials $B_i(y) = (b_i^{(1)}(y), \ldots, b_i^{(2n)}(y))$, then we will have

$$f = \textstyle\sum_{\ell=1}^{2n} \left(\sum_{i=1}^{n} b_i^{(\ell)}(y)\, x_i\right)^2,$$

as desired. Thus, we are now reduced to proving the following.

Lemma 5.21. *Let $(a_{ij}(y))$ be a symmetric $n \times n$ matrix over $k[y]$, and let $a(y) \in k[y] \setminus \{0\}$ be psd. Suppose there exist vectors of polynomials*

$$A_i(y) = \left(a_i^{(1)}(y), \ldots, a_i^{(2d)}(y)\right) \qquad (1 \le i \le n)$$

such that $a(y) \cdot a_{ij}(y) = A_i(y) \cdot A_j(y)$ for all i, j. Then there exist vectors of polynomials $B_i(y) = (b_i^{(1)}(y), \ldots, b_i^{(2d)}(y))$ such that $a_{ij}(y) = B_i(y) \cdot B_j(y)$ for all i, j.

Proof. The strategy of the proof is to "peel off" the psd factors of $a(y)$ one at a tine. Thus, we need only treat the following two cases: (1) $a(y) = y^2$, and (2) $a(y)$ is an irreducible psd quadratic polynomial. The first case is immediate, because here each $a_i^{(\ell)}(y)$ must have a factor y. In the second case, we may assume, after a change of variable, that $a(y) = y^2 + 1$. Let A be the $n \times 2d$ matrix over $k[y]$ with $A_1(y), \ldots, A_n(y)$ as rows. Then $a(y) \cdot (a_{ij}(y)) = A \cdot A^t$. Dividing out the entries of A by $y^2 + 1$, we may write

$$A = (y^2 + 1)C + yD + E, \tag{5.22}$$

where C, D, E are also $n \times 2d$, *but D and E are scalar matrices.* Setting $y = i = \sqrt{-1}$ (over $k(i)$), we get $(iD + E) \cdot (iD + E)^t = 0$, so we have

$$DD^t = EE^t, \quad \text{and} \quad DE^t + ED^t = 0.$$

Thus, the rows of D and E satisfy the equations 5.18. By 5.17, there exists an *orthogonal* matrix $T \in \mathrm{O}_{2d}(k)$ such that $T^{-1}DT$ and $T^{-1}ET$ have rows as in 5.19. What this means is precisely that, if J is the $2d \times 2d$ matrix with d diagonal blocks of the form $\begin{pmatrix} 0 & 1 \\ -1 & 0 \end{pmatrix}$, then $(T^{-1}DT)J = T^{-1}ET$, or $(DT)J = ET$. From 5.22,

$$\begin{aligned} AT &= (y^2 + 1)CT + yDT + ET \\ &= (y^2 + 1)CT + DT\,(yI_{2d} + J). \end{aligned}$$

For $M := yI_{2d} + J$, we have clearly $M^tM = (y^2 + 1)\,I_{2d}$. Thus,

$$AT = (CT)M^tM + DTM = BM$$

for some $n \times 2d$ matrix B over $k[y]$. Now

$$\begin{aligned} (y^2 + 1) \cdot (a_{ij}(y)) &= AA^t = (AT)(AT)^t \\ &= (BM)(BM)^t = (y^2 + 1)BB^t, \end{aligned}$$

so cancellation yields $(a_{ij}(y)) = BB^t$, as desired! □

Having completed the proof of 5.16, we now come to deal with the Pythagoras number of affine curves.

Theorem 5.23. *Let A be a (commutative) affine algebra of transcendence degree 1 over a real-closed field k_0. Then $P(A) < \infty$.*

Proof. By Noether's Normalization Theorem, there exists an element $y \in A$ transcendental over k_0 such that A is an integral extension of the polynomial ring $k := k_0[y]$. Since A is finitely generated as a k_0-algebra, A is finitely generated as a k-module, say $A = \sum_{j=1}^n kw_j$, for some n. To estimate the Pythagoras number $P(A)$, we shall use the invariant $g_k(n)$ introduced after the proof of 5.11. For any f that is a sum of squares of linear forms in $x_1, \ldots, x_n$ over $k = k_0[y]$, f is certainly psd as a polynomial in $\{y, x_1, \ldots, x_n\}$, and f is homogeneous of degree 2 in $\{x_1, \ldots, x_n\}$. By 5.16,[(15)] f is then a sum of squares of $2n$ linear forms over k. Therefore, $g_k(n) \leq 2n$, and the remarks made after the proof of 5.11 show that $P(A) \leq g_k(n) \leq 2n < \infty$. □

Example 5.24. In a few cases, the method of proof used above can give a pretty good bound. For instance, if $A = k_0[x, y]$ with the relation $x^2 = p(y) \in k_0[y]$, then A is generated by 1 and x as a module over $k_0[y]$, so the work above yields a bound $P(A) \leq 4$. If $p(y)$ is of *odd* degree > 1, then $P(A)$ is in fact either 3 or 4. This can be shown by checking that $1 + x^2 + y^2$ is not a sum of two squares in A. Assume, instead,

$$1 + x^2 + y^2 = (f_1(y) + xg_1(y))^2 + (f_2(y) + xg_2(y))^2 \in A.$$

Using the relation $x^2 = p(y)$, we get

$$f_1(y)g_1(y) + f_2(y)g_2(y) = 0, \tag{5.25}$$

$$f_1(y)^2 + f_2(y)^2 + p(y)(g_1(y)^2 + g_2(y)^2) = 1 + y^2 + p(y). \tag{5.26}$$

By an obvious degree consideration, 5.26 implies that $g_1(y), g_2(y)$ must be scalars with $g_1^2 + g_2^2 = 1$, and so, from 5.25, $f_1(y), f_2(y)$ are linearly dependent. If, say $f_2 = \lambda f_1$ ($\lambda \in k_0$), then 5.26 leads to $1 + y^2 = (1 + \lambda^2)f_1(y)^2$, which is a square in $k_0[y]$, a contradiction. (This example is drawn from [CDLR].)

Example 5.27. Generally speaking, the Pythagoras number bound on $P(A)$ given in the proof of 5.23 is not very sharp, and it certainly depends on the choice of the algebra A. But in fact, the Pythagoras numbers of affine algebras of transcendence degree 1 (over a real-closed field k_0) cannot be bounded. Consider, for instance, a finite-dimensional k_0-algebra A_0 with $P(A_0) = n$ (which exists by 5.4). Then $A = A_0[y]$ has transcendence degree 1 over k_0, but the fact that we have a surjective ring homomorphism $A \to A_0$ (with $y \mapsto 0$, for instance) implies that $P(A) \geq P(A_0) = n$.

[(15)] We are using here the version of 5.16 with the variables $\{y, z\}$ dehomogenized. This was, of course, the version of 5.16 that we actually proved.

For transcendence degree 2 or higher, it turns out that the Pythagoras number of an affine algebra is often infinite. The first class of examples illustrating this phenomenon is the following.

Theorem 5.28 ([CDLR]). *Let k_0 be any commutative ring with $s(k_0) = \infty$ (that is, k_0 is semireal). Then $P(k_0[x_1, \ldots, x_n]) = \infty$ for any $n \geq 2$.*

Note that none of the hypotheses in this theorem can be relaxed. If $n = 1$, a counterexample is given by $\mathbb{R}[x_1]$, which has Pythagoras number 2. On the other hand, if k_0 has finite level s, then $P(k_0[x_1, \ldots, x_n]) \leq s+2 < \infty$ by 5.2.

To prove Theorem 5.28, we first give a sufficient condition on a real commutative domain B in order that $P(B[x]) = \infty$. (Recall that B is *real* if $B \neq 0$ and $b_1^2 + \cdots + b_n^2 = 0$ in B implies that all $b_i = 0$.)

Theorem 5.29. *Let B be a real commutative domain satisfying the following two conditions:*

(1) *For any n, the group of orthogonal $n \times n$ matrices over B acts transitively on vectors $(a_1, \ldots, a_n) \in B^n$ such that $a_1^2 + \cdots + a_n^2 = 1$.*

(2) *There exists an element $b \in B \setminus \{0\}$ such that, whenever $a_1^2 + \cdots + a_n^2 = 1$ (for any n), $b \nmid a_i$ for every $a_i \neq 0$.*

Then $P(B[x]) = \infty$.

Proof. For any $f(x) \in B[x]$ of length n, we shall construct another polynomial $F(x) \in B[x]$ of length $n+1$. This will show that $P(B[x]) = \infty$.

Fix the element b in (2), which is, of course, a nonunit. Replacing b by $2b$ if necessary, we may assume that $b \nmid 2$. Taking any integer r such that $2r > \deg f$, we define

$$(5.30)\quad g(x) = x(x - 2b)(x - 3b) \cdots (x - rb) \quad \text{and} \quad F(x) = 1 + f(x)g(x)^2.$$

Note that the roots $\{0, 2b, \ldots, rb\}$ of $g(x)$ are distinct, and $F(x)$ is a sum of $n+1$ squares in $B[x]$. *We finish by checking that* $\operatorname{len}(F(x)) = n+1$.

Assume, instead, that $F(x) = \psi_1(x)^2 + \cdots + \psi_n(x)^2$. Since $\deg F < 4r$ and A is real, $\deg \psi_j < 2r$ for all j. Setting $x = 0$, we have $1 = \sum \psi_j(0)^2$. Therefore, after an orthogonal transformation over B, we may assume that $\psi_1(0) = 1$ and $\psi_j(0) = 0$ for $j \geq 2$. Write $\psi_1(x) = 1 + x\varphi_1(x)$ and $\psi_j(x) = x\varphi_j(x)$ for $j \geq 2$. Evaluating $F(x)$ at jb $(j \geq 2)$, we get

$$1 = (1 + jb\,\varphi_1(jb))^2 + (jb\,\varphi_2(jb))^2 + \cdots + (jb\,\varphi_n(jb))^2.$$

By (2), $\varphi_2(jb) = \cdots = \varphi_n(jb) = 0$. Thus, each of $\psi_2, \ldots, \psi_n$ is divisible by g; say $\psi_j = g\sigma_j$ (for $j \geq 2$), so now

$$g(x)^2(f(x) - \sigma_2(x)^2 - \cdots - \sigma_n(x)^2) = \psi_1(x)^2 - 1$$
$$= (\psi_1(x) - 1)(2 + x\varphi_1(x)).$$

Recalling that $b \nmid 2$, we see that $2 + x\varphi_1(x)$ does not vanish on $0, 2b, \ldots, rb$. Thus, $g(x)^2 | (\psi_1(x) - 1)$. However, $\deg(\psi_1) < 2r = \deg(g^2)$, so $\psi_1(x) = 1$. This leads to $f(x) = \sigma_2(x)^2 + \cdots + \sigma_n(x)^2$, a contradiction. □

Let us now record some consequences of 5.29.

Corollary 5.31. *Let B be a real commutative ring that is not a field. Assume that, for any n, if $a_1^2 + \cdots + a_n^2 = 1$ in B, then all except one of the a_i's are zero. Then $P(B[x_1, \ldots, x_d]) = \infty$ for any $d \geq 1$. (In particular, $P(\mathbb{Z}[x_1, \ldots, x_d]) = \infty$ for any $d \geq 1$.)*

Proof. Since there is a surjective ring homomorphism from $B[x_1, \ldots, x_d]$ onto $B[x_1]$, it suffices to show that $P(B[x_1]) = \infty$. Clearly, (1) in 5.29 is satisfied, and (2) there is satisfied by taking b to be a nonunit in $B \setminus \{0\}$. Thus, 5.29 applies. □

The fact that $P(\mathbb{Z}[x]) = \infty$ contrasts with a theorem of H. Liese [Li] to the effect that $P(\mathbb{Z}[[x]]) = 5$. Liese's result, however, does not extend to more variables: in [CDLR], it is also shown that $P(\mathbb{Z}[[x_1, \ldots, x_d]]) = \infty$ for any $d \geq 2$.

As a second application of 5.29, we now give

Proof of 5.28. Let $A = k_0[x_1, \ldots, x_n]$, where $n \geq 2$ and k_0 is a semireal commutative ring. To show that $P(A) = \infty$, we may assume again that $n = 2$. First assume k_0 is a formally real field. Let $B = k_0[x_1]$ and $x = x_2$, so $A = B[x]$. We are done if we can check conditions (1) and (2) in 5.29 for B. Suppose $f_1(x_1)^2 + \cdots + f_m(x_1)^2 = 1$, where $f_j \in B$. By a degree consideration, all $f_i \in k_0$. By I.4.7, $(f_1, \ldots, f_m)$ can be brought to $(1, 0, \ldots, 0)$ by an orthogonal matrix over the field k_0. This checks (1) in 5.29. For (2) there, we can simply pick b to be x_1 (noting that any nonzero f_j above is a unit in k_0).

Now let k_0 be any semireal ring. By 4.10, there exists a ring homomorphism $\sigma \colon k_0 \to K$ where K is a formally real field. Starting with the polynomial $F_1(x_2) = 1$ and taking b to be x_1, we can construct inductively via 5.30 a sequence of polynomials $\{F_j(x_2) \colon j \geq 1\}$ in $B[x_2]$ (with $B = k_0[x_1]$), where each F_j is a sum of j squares in $B[x_2]$. These are universal polynomials in $\mathbb{Z}[x_1, x_2]$ (constructed independently of k_0). By the first paragraph above, we know that F_j has length j in $K[x_1, x_2]$. In view of the homomorphism $\sigma\colon k_0 \to K$, it follows that F_j also has length j in $k_0[x_1, x_2]$. Therefore, $P(k_0[x_1, x_2]) = \infty$, as desired. $\square$

Remark 5.32. Note that Theorem 5.28 settles a few other problems on the Pythagoras number of rings as well.

(A) If K is a field, it is unknown whether $P(K) < \infty$ would imply $P\big(K[y]\big) < \infty$. However, if K is a commutative ring, or even a PID, this implication is definitely false. In fact, $K = \mathbb{R}[x]$ is a PID with $P(K) = 2$, but $P\big(K[y]\big) = P\big(\mathbb{R}[x, y]\big) = \infty$ according to 5.28.

(B) A nonreal affine algebra need not have finite Pythagoras number. For instance, $B = \mathbb{R}[u, v, x, y]/(u^2+v^2)$ has infinite Pythagoras number, as it can be mapped by a ring homomorphism onto $A = \mathbb{R}[x, y]$ with $P(A) = \infty$.

(C) It is possible for all proper factor rings of a commutative ring A to have finite Pythagoras number, but $P(A) = \infty$. In fact, for $A = \mathbb{R}[x, y]$, we have $P(A) = \infty$ by 5.28, but any proper factor ring $\overline{A}$ of A has transcendence degree ≤ 1 over $\mathbb{R}$, so $P(\overline{A}) < \infty$ by 5.23 and 5.11.

(D) It is possible for a commutative ring A to have $P(A) = \infty$ and yet $P(A_{\mathfrak{p}}) \leq n$ for all prime ideals $\mathfrak{p} \subseteq A$, where n is a fixed number (independent of $\mathfrak{p}$). In fact, for the ring $A = \mathbb{R}[x, y]$ again, it can be checked that $P(A_{\mathfrak{p}}) \leq 4$ for all prime ideals $\mathfrak{p} \subseteq A$; see [CDLR: (4.6)].

In [CDLR], there are many more results on the Pythagoras number of other kinds of rings, for instance local rings of dimension ≥ 2, and affine algebras of dimension ≥ 3, etc. For instance, it is proved that *any real affine algebra (over a field) of dimension* ≥ 3 *has infinite Pythagoras number.* Due to limitation of space, however, these results will not be presented here.

Because of the fact that so many rings have infinite Pythagoras number, there have been several attempts at *redefining* the Pythagoras number for (at least some types of) rings, in order to increase the chance of getting finite invariants. For work in this direction, see, e.g., the papers of L. Mahé. For a quick summary of the ideas behind this line of work, and the principal results herein, see Chapter 7, §2 of Pfister's book [Pf_5].

6. Some Open Questions

In this closing section, we assemble a number of open questions in the algebraic theory of quadratic forms, in the hope that a list of such open questions will stimulate further research in this area. For obvious reasons, I shall focus only on problems that are directly relevant to at least some parts of this book. Just to go for a round number, I'll put down ten such problems. Hearty thanks are due to D. Hoffmann and D. Leep for their valuable input into the collection of the open problems below, many of which are indeed concerned with the main themes of this chapter (and Chapter XI), namely, the quadratic invariants of fields (and rings).

We start with a couple of open questions concerning the level. The first question in this direction is on the level of generic rings of the type studied in §4 of this chapter.

Question 6.1. *For any commutative ring k, let*

$$A_n(k) = k[x_1, \ldots, x_n]/\big(1 + x_1^2 + \cdots + x_n^2\big).$$

Is the level of $A_n(k)$ given by $\min\{n,\ s(k)\}$?

If the ring k is semireal (that is, $s(k) = \infty$), an affirmative answer to 6.1 is provided by the result 4.41 in this chapter. The question remains open for commutative rings k of finite level. In the case where k is a *field* with $s(k) \geq n$, the answer is also known to be "yes", by the work of Arason and Pfister [AP3]; but if $s(k) < n$, it seems to be unknown if $s\big(A_n(k)\big) = s(k)$. In all likelihood, the answer to 6.1 (for rings k of finite level) would seem to be "yes". However, a full proof of this will probably require the introduction of new techniques. This is why I have included 6.1 as an open question.

Our next question concerns the level of nonreal fields with a *finite* number of square classes.

Question 6.2. *Given any integer $k \geq 3$, does there exist a field F with $|\dot{F}/\dot{F}^2| < \infty$ and $s(F) = 2^k$?*

This question is of interest because of its simple but intrinsic nature as an open problem in field theory. Note that the existence of a field F as in 6.2 (for some $k \geq 3$) would give an example of a finitely generated Witt group $W(F)$ of exponent $2^{k+1} \geq 16$. In particular, this would defeat the "Elementary Type Conjecture" in XII.7.11, since this Conjecture would have implied that all Witt rings of finite type have exponent ≤ 8.

In the quantitative direction, K. Becher has recently shown (in [Bec]) that, if a field F exists as in Question 6.2, then $|\dot{F}/\dot{F}^2| \geq 2^9 = 512$. In other words, if a nonreal field F has less than 512 square classes, then $s(F) \leq 4$.

According to Exercise 6 in Ch. XI, if Question 6.2 has a "yes" answer for *some* $k > 3$, then it would also have a "yes" answer for $k = 3$. Thus, it is particularly crucial to determine if there exists a field F with $|\dot{F}/\dot{F}^2| < \infty$ and $s(F) = 8$. Unfortunately, even this ostensibly simple existence question has remained wide open.

Of course, similar problems for other invariants can also be posed for fields with a finite number of square classes. We shall content ourselves with just one more, which is motivated by Merkurjev's construction of fields with prescribed even u-invariants (see Remark 2.23(4)).

Question 6.3. *Given any integer $n \geq 5$ (or $n = 3$), does there exist a nonreal field F with $|\dot{F}/\dot{F}^2| < \infty$ and $u(F) = 2n$?*

The problem of the existence of fields with *odd* u-invariants has still largely remained a mystery. This leads us to pose the following

Question 6.4. *Given any odd integer $n \geq 11$, does there exist a nonreal field F with $u(F) = n$?*

The choice of the integer "11" is prompted by the fact that there does exist a nonreal field of u-invariant 9, by the work of Izhboldin: see Remark 2.23(7). Of course, smaller odd integers need not be considered, since there are no nonreal fields with u-invariant 3, 5, or 7, according to XI.6.8.

Another u-invariant open problem of long standing is the following:

Question 6.5. *If F is a nonreal field with transcendence degree n over the real field $\mathbb{R}$, is $u(F) \leq 2^n$?*

This question stems from the work of S. Lang, and is sometimes referred to as "Lang's Problem." The impetus for this problem seemed to have come from the following two sources:

(1) The answer would be "yes" if the field F in question has level 1. In this case, F would be a function field of transcendence degree n over the complex field $\mathbb{C}$, so an affirmative answer to 6.5 is provided by (the quadratic form special case of) the theorem of Tsen and Lang.

(2) Thanks to Pfister's result XI.4.10, we know already that n-fold Pfister forms over F are universal. Unfortunately, in general, this fact in itself does not guarantee that every 2^n-dimensional form over F is universal.

Of course, (2) above suffices to give an affirmative answer to 6.5 for $n = 1$. For $n \geq 2$, Lorenz [Lo] and Elman and Lam [EL3] have shown that $u(F) \leq (4^n - 2^n)/2$, which is "off" by approximately a factor of 2^{n-1}. For instance, in the special case $n = 2$, one gets only $u(F) \leq 6$, instead of the desired $u(F) \leq 4$. (For a sketch of a direct verification for $u(F) \leq 6$, see the hint on XI.Exercise 16.)

The next question has already been posed in earlier versions of this book, but a definitive answer has remained elusive.

Question 6.6. *If F is a nonreal field, does $u(F) < \infty \Longrightarrow u\big(F(x)\big) < \infty$? If so, can $u\big(F(x)\big)$ be estimated in terms of $u(F)$?*

The answer to this question would seem to depend largely on getting good estimates on $u(K)$ in terms of $u(F)$, for finite field extensions K/F. For instance, could it be true that $u(K) \leq 2u(F)$, independently of the value of $[K : F]$? Conjectural good bounds like this seem difficult to prove (or disprove!). Short of general results for arbitrary fields, one can try to investigate Question 6.6 for specific classes of nonreal fields F. The following special case of Question 6.6, for instance, has attracted considerable attention.

Question 6.7. *Let $\mathbb{Q}_p$ be the field of the p-adic numbers. What is the precise value of $u\big(\mathbb{Q}_p(x)\big)$?*

In the case where p is odd, the finiteness of $u\big(\mathbb{Q}_p(x)\big)$ was proved by Hoffmann–Van Geel (and independently by Merkurjev) using a result of Saltman. In [HVG], Hoffmann and Van Geel obtained the upper bound $u\big(\mathbb{Q}_p(x)\big) \leq 22$. More recently, Parimala and Suresh [PS] have shown that $u\big(\mathbb{Q}_p(x)\big) \leq 10$, for p odd. The latter upper bound applies, in fact, to function fields of transcendence degree 1 over $\mathbb{Q}_p$. For p odd, the conjectural value of $u\big(\mathbb{Q}_p(x)\big)$ is 8. For $p = 2$, however, the lack of knowledge is complete, as it seems to be still unknown whether the rational function field $\mathbb{Q}_2(x)$ has a finite u-invariant!

The case of function fields over number fields can be expected to be even more difficult. Just for the record, we state the following:

Question 6.8. *For a nonreal number field F, is $u(F(x_1, \ldots, x_n)) < \infty$?*

Similar questions concerning the Pythagoras number $P(F)$ are also largely unanswered. For instance, there seems to be no counterexample to the possible claim that $P(F) < \infty \Longrightarrow P\big(F(x)\big) < \infty$. To be more specific, let us focus on two concrete questions on purely transcendental extensions over the rational field and the field of the real numbers.

Question 6.9. *What are the precise values of the Pythagoras numbers*

$$P(\mathbb{R}(x_1, \ldots, x_n)) \quad \textit{and} \quad P(\mathbb{Q}(x_1, \ldots, x_n))?$$

Of course, $P(\mathbb{R}(x_1)) = 2$. For $n \geq 2$, $P\big(\mathbb{R}(x_1, \ldots, x_n)\big)$ is known to be in the interval $[n+2,\, 2^n]$, as we have explained in XI.5.9(4). For $n = 2$, this boils down to the computation $P\big(\mathbb{R}(x_1, x_2)\big) = 4$ due to Cassels-Ellison-Pfister [CEP], but nothing more seems to be known for $n \geq 3$.

As for the case of the rational ground field, Pourchet's result (mentioned in XI.5.9(3)) gave $P\big(\mathbb{Q}(x_1)\big) = 5$, and Colliot-Thélène and Jannsen showed that

$$P\big(\mathbb{Q}(x_1, x_2)\big) \leq 8, \quad \text{and} \quad P\big(\mathbb{Q}(x_1, x_2, x_3)\big) \leq 16.$$

In their paper [CTJ], it was further shown that, for $n \geq 2$, the upper bound

$$(*) \qquad P\big(\mathbb{Q}(x_1, \ldots, x_n)\big) \leq 2^{n+1}$$

would hold modulo Milnor's Conjecture and a conjecture of K. Kato on higher-dimensional cohomology groups. The upper bound $(*)$ is now one step closer to reality, since Milnor's Conjecture has been proved by Voevodsky et al. ([Vo], [OVV$_1$], [OVV$_2$]). Given Milnor's Conjecture (but not Kato's), Arason has pointed out that one can establish the weaker bound

$$(**) \qquad P(\mathbb{Q}(x_1, \ldots, x_n)) \leq 2^{n+2}$$

Thus, this upper bound for the Pythagors number may be viewed as having been proven, for all n. For more details on this, see Pfister's survey [Pf$_6$].

Finally, since the study of *function fields* constitutes an important direction of recent research in quadratic form theory, it seems fitting for us to close with at least one open problem concerning function fields of quadratic forms. We choose the following "Quadratic Zariski Problem" proposed by Jack Ohm [Oh], since it is well connected to some of the material we have developed in X.4 and XII.2.

Question 6.10. *Let σ and τ be quadratic forms over F of dimension $n \geq 3$. If $\sigma > \tau > \sigma$ (that is, σ and τ become isotropic over the function field of each other), does it follow that $F(\sigma) \cong F(\tau)$ over F?*

According to X.4.25, the hypothesis that $\sigma > \tau > \sigma$ is equivalent to the condition that $F(\sigma)$ and $F(\tau)$ are "stably isomorphic" over F; that is, there exists an F-isomorphism

$$F(\sigma)(x_1, \ldots, x_m) \cong F(\tau)(x_1, \ldots, x_m)$$

for some m. Thus, 6.10 is equivalent to asking *whether stable isomorphism between $F(\sigma)$ and $F(\tau)$ would imply isomorphism (over F).* In the language of algebraic geometry, this amounts to asking: *if two projective quadrics of the same dimension are stably birationally equivalent, are they already birationally equivalent*? For *general* varieties (rather than just quadrics), this is known as "Zariski's Problem", and it is well-known that the answer is "No" in general. However, the known counterexamples are not for quadric hypersurfaces; in other words, the answer to 6.10 is unknown for n-dimensional quadratic forms σ and τ. If one of these two forms is isotropic, it is easy to see that $\sigma > \tau > \sigma$ implies that both are isotropic, in which case X.4.1 yields

$$F(\sigma) \cong F(x_1, \ldots, x_{n-2}) \cong F(\tau).$$

Thus, the remaining case for 6.10 is that where the quadratic forms σ and τ are both *anisotropic* over F. In case $\dim \sigma = \dim \tau = 3$ or 4, there is no problem, since $\sigma > \tau > \sigma$ implies in fact the *similarity* of σ and τ (by X.4.31 and XII.2.2). The best (published) result to date is that 6.10 has an affirmative answer if $\dim \sigma = \dim \tau \leq 6$. For this result, and a detailed discussion of the Quadratic Zariski Problem 6.10 in general, see Hoffmann's paper [Ho$_7$], and Ohm's paper [Oh].

Exercises for Chapter XIII

1. Show that if a commutative ring A is real, then so is any localization $S^{-1}A$ of A at a multiplicative set $S \subseteq A$.
2. Prove the following claims (1) and (2) made in the text:

   (1) A preordering T in a commutative ring A is an ordering iff, for any $a, b \in A$,
   $$ab \in -T \Longrightarrow a \in T \text{ or } b \in T.$$
   (2) A maximal preordering T is always an ordering.
   (3) Give an example of an ordering P in a commutative ring that is properly contained in another ordering P'.
3. A commutative local ring $(R, \mathfrak{m})$ is called *residually real* if its residue class field $R/\mathfrak{m}$ is formally real. Under this assumption, show that

   (1) R is semireal but not necessarily real;
   (2) R is real if it is a valuation domain.
4. Let A be a valuation domain with quotient field F. Show that $s(A) = s(F)$.
5. A commutative domain is said to be a *Prüfer domain* if its localizations at maximal ideals are valuation domains. Show that a Prüfer domain A is semireal iff it is real, iff its quotient field F is formally real.
6. Let A be a PID with $2 \in \mathrm{U}(A)$. If the quotient field of A has level n, show that $s(A)$, $P(A) \in \{n, n+1\}$. Is $s(A)$ necessarily equal to n?
7. Let $n \geq m = 2^k$ be given integers. Give an example of an integral domain A with a quotient field F such that $s(F) = m$ and $s(A) = n$. (**Hint.** Let A_n be as defined in 4.11, and consider the ring $A = A_n[t_1, \ldots, t_m]/(t_1^2 + \cdots + t_m^2)$.)
8. Let $n \in \{1, 2, 4, 8\}$. If A is a commutative ring with level n, show that $n\langle 1\rangle$ is anisotropic over A (that is, it has no unimodular isotropic vector in A^n).
9. Give an example of an element a in an integrally closed integral domain A such that $a \notin \sigma(A)$ but a is a sum of two squares in the quotient field of A.
10. Show that the ring $\mathbb{Z}[i, x]$ (where $i^2 = -1$) has level 1 and Pythagoras number 3 (so, in general, Proposition 5.2 cannot be further improved).
11. Use the term inspection method to show that the psd (positive semidefinite) quaternary quartic $Q(w, x, y, z)$ in 5.15 is *not* a sum of squares

in $k[w,x,y,z]$ for any real-closed field k. Do the same for the psd ternary sextic

$$S'(x,y,z) = z^6 + x^4y^2 + x^2y^4 - 3x^2y^2z^2 \in k[x,y,z].$$

12. The "original" Motzkin polynomial $M(x,y)$ (first introduced in XI.5.10) is the following dehomogenization of the ternary sextic in Ex. 11:

$$M(x,y) := S'(1,x,y) = 1 + x^4y^2 + x^2y^4 - 3x^2y^2 \in k[x,y].$$

Show that $(1+x^2)M(x,y)$ is a sum of three squares in the polynomial ring $k[x,y]$. From this, deduce that $M(x,y)$ is a sum of four squares in $k(x)[y]$ (and hence in $k(x,y)$). [**Hint.** Check that $(1+x^2)M(x,y) = q_1^2 + q_2^2 + q_3^2$, where

$$q_1 = 1 - x^2y^2, \quad q_2 = x(1-y^2), \quad \text{and} \quad q_3 = xy(1-x^2).$$

Now multiply by $1 + x^2$ and use the 2-square identity.]

Bibliography

(As a disclaimer, we should make it absolutely clear that the following list is not at all a general bibliography on quadratic form theory. Rather, it contains only the papers that are cited in our text, plus a few others that are included for their general relevance. Many important papers in quadratic form theory published in the last 70 years are not included in this very incomplete bibliography.)

[AO] H. Ahmad and J. Ohm: *Function fields of Pfister neighbors,* J. Algebra **178**(1995), 653–664.

[Al] A. A. Albert: *Tensor products of quaternion algebras,* Proc. Amer. Math. Soc. **35**(1972), 65–66.

[Ara_1] J. K. Arason: *Cohomologische Invarianten quadratischer Formen,* J. Algebra **36**(1975), 448–491.

[Ara_2] J. K. Arason: *A proof of Merkurjev's theorem,* Canad. Math. Soc. Conf. Proc., vol. **4**(1984), 121–130.

[AE] J. K. Arason and R. Elman: *Powers of the fundamental ideal in the Witt ring,* J. Algebra **239**(2001), 150–160.

[AP_1] J. K. Arason and A. Pfister: *Beweis des Krullschen Durchschnittsatzes für den Wittring,* Invent. Math. **12**(1971), 173–176.

[AP_2] J. K. Arason and A. Pfister: *Zur Theorie der quadratischen Formen über formalreellen Körpern,* Math. Zeit. **153**(1977), 289-296.

[AP_3] J. K. Arason and A. Pfister: *Quadratische Formen über affinen Algebren und ein algebraischer Beweis des Satzes von Borsuk-Ulam,* J. Reine Angew. Math. **331**(1982), 181–184.

[AS] E. Artin and O. Schreier: *Algebraische Konstruktion reeller Körper,* Abh. Math. Sem. Univ. Hamburg **5**(1927), 85–99.

[Ar] E. Artin: *Über die Zerlegung definiter Funktionen in Quadrate,* Abh. Math. Sem. Univ. Hamburg **5**(1927), 100–115.

[Ba] R. Baer: *Die Automorphismusgruppe eines algebraisch abgeschlossen Körpers der Charakteristik* 0, Math. Zeit. **117**(1970), 7–17.

[Bae] R. Baeza: *Quadratic Forms over Semilocal Rings,* Lecture Notes in Math., Vol. **655**, Springer-Verlag, Berlin-Heidelberg-New York, 1978.

[Be] E. Becker: *Euclidische Körper und Euclidische Hüllen von Körpern,* J. Reine Angew. Math. **268/269**(1974), 41–52.

[BK] E. Becker and E. Köpping: *Reduzierte quadratische Formen und Semiordnungen reeller Körper,* Abh. Math. Sem. Univ. Hamburg **46**(1977), 143–177.

[Bec] K. J. Becher: *On the number of square classes of a field of finite level,* Proc. Conf. on Quadratic Forms and Related Topics (Baton Rouge, Louisiana), Documenta Math. (extra volume), 2001, 65–84.

[Bm] L. Berman: *The Kaplansky radical and values of binary quadratic forms over fields,* Doctoral Dissertation, University of Calif., Berkeley, 1978.

[BLM] B. J. Birch, D. J. Lewis, and T. G. Murphy: *Simultaneous quadratic forms,* Amer. J. Math. **84**(1962), 110–115.

[Br] L. Bröcker: *Über die Anzahl der Anordnungen eines kommutativen Körpers,* Arch. Math. **29**(1977), 458–464.

[BDS] L. Bröcker, A. Dress, and R. Scharlau: *An (almost trivial) local-global principle for the representation of* -1 *as a sum of squares in an arbitrary commutative ring,* in "Ordered Fields and Real Algebraic Geometry" (San Francisco, CA, 1981), pp. 99–106, Contemp. Math. **8**(1982), Amer. Math. Soc., Providence, RI.

[Ca_1] J.W.S. Cassels: *On the representation of rational functions as sums of squares,* Acta Arith. **9**(1964), 79–82.

[Ca_2] J.W.S. Cassels: *Local Fields,* Student Texts, Vol. **3**, London Math. Soc., Cambridge Univ. Press, 1986, Cambridge.

[CEP] J.W.S. Cassels, W.J. Ellison, and A. Pfister: *On sums of squares and on elliptic curves over function fields,* J. Number Theory **3**(1971), 125–149.

[Ch] C. Chevalley: *Démonstration d'une hypothèse de M. Artin,* Abh. Math. Sem. Univ. Hamburg **11**(1936), 73–75.

[CL_1] M. D. Choi and T. Y. Lam: *An old question of Hilbert,* Proc. Conference on Quadratic Forms (ed. G. Orzech), Queen's Papers in Pure and Applied Math. **46**(1976), 385–405.

[CL_2] M. D. Choi and T. Y. Lam: *Extremal positive semidefinite forms,* Math. Ann. **231**(1977), 1–18.

[CDLR] M. D. Choi, Z. D. Dai, T. Y. Lam, and B. Reznick: *The Pythagoras number of some affine algebras and local algebras,* J. Reine Angew. Math. **336**(1982), 45–82.

[CLRR] M. D. Choi, T. Y. Lam, B. Reznick, and A. Rosenberg: *Sums of squares in some integral domains,* J. Algebra **65**(1980), 234–256.

[CT] J.-L. Colliot-Thélène: *The Noether-Lefshetz theorem and sums of* 4 *squares in the rational function field* $R(x, y)$, Compos. Math. **86**(1993), 235-243.

[CTOP] J.-L. Colliot-Thélène, M. Ojanguren, and R. Parimala: *Quadratic forms over fraction fields of two-dimensional Henselian rings and Brauer groups of related schemes,* Algebra, Arithmetic and Geometry, Parts I, II (Mumbai, 2000), 185–217, Tata Inst. Fund. Res. Stud. Math. **16**, Bombay, 2002.

[CTJ] J.-L. Colliot-Thélène and U. Jannsen: *Sommes des carrés dans les corps de fonctions,* C.R. Acad. Sci. Paris, Sér. I **312**(1991), 759–762.

[CP] P. E. Conner and R. Perlis: *A Survey of Trace Forms of Algebraic Number Fields,* Series in Pure Math., Vol. 2, World Scientific, Singapore, 1984.

[Co_1] C. M. Cordes: *The Witt group and the equivalence of fields with respect to quadratic forms,* J. Algebra **26**(1973), 400–421.

[Co_2] C. M. Cordes: *Quadratic forms over nonformally real fields with a finite number of quaternion algebras,* Pacific J. Math. **63**(1976), 357–365.

[CR] M. Coste and M.-F. Roy: *La topologie du spectre réel,* in "Ordered Fields and Real Algebraic Geometry" (San Francisco, CA, 1981), pp. 27–59, Contemp. Math. **8**(1982), Amer. Math. Soc., Providence, RI.

[Cr] T. C. Craven: *The Boolean space of orderings of a field,* Trans. Amer. Math. Soc. **209**(1975), 225–235.

[Cu] J. Cunningham: *Quadratic Forms,* Seminar Notes, University of Kentucky, 1970.

[Cz] A. Czogala: *On reciprocity equivalence of quadratic number fields,* Acta Arith. **58**(1991), 27–46.

[DL] Z. D. Dai and T. Y. Lam: *Levels in algebra and topology,* Comment. Math. Helv. **59**(1984), 376–424.

[DLP] Z. D. Dai, T. Y. Lam, and C. K. Peng: *Levels in algebra and topology,* Bull. Amer. Math. Soc. (New Series) **3**(1980), 845–848.

[Dem] V. B. Dem'yanov: *Pairs of quadratic forms over a complete field with discrete norm with a finite field of residue classes,* (in Russian) Izv. Akad. Nauk SSSR. Ser. Mat. **20**(1956), 307–324.

[DM_1] M. Dickmann and F. Miraglia: *On quadratic forms whose total signature is zero mod 2^n,* Invent. Math. **133**(1998), 243–278.

[DM_2] M. Dickmann and F. Miraglia: *Lam's conjecture,* Algebra Colloq. **10**(2003), 149–176.

[DD] J. Diller and A. Dress: *Zur Galoistheorie pythagoreischer Körper,* Arch. Math. **16**(1965), 148–152.

[DEK] C. Drees, M. Epkenhans, and M. Krüskemper: *On the computation of the trace forms of some Galois extensions,* J. Algebra **192**(1997), 209–234.

[EL_0] R. Elman and T. Y. Lam: *Determination of k_n $(n \geq 3)$ for global fields,* Proc. Amer. Math. Soc. **31**(1972), 427–428.

[EL_1] R. Elman and T. Y. Lam: *Pfister forms and the K-theory of fields,* J. Algebra **23**(1972), 181–213.

[EL_2] R. Elman and T. Y. Lam: *Quadratic forms over formally real fields and pythagorean fields,* Amer. J. Math. **94**(1972), 1155–1194.

[EL_3] R. Elman and T. Y. Lam: *Quadratic forms and the u-invariant,* I, Math. Zeit. **131**(1973), 283–304.

[EL_4] R. Elman and T. Y. Lam: *Quadratic forms and the u-invariant,* II, Invent. Math. **21**(1973), 125–137.

[EL_5] R. Elman and T. Y. Lam: *Classification theorems for quadratic forms over fields,* Comment. Math. Helv. **49**(1974), 373–381.

[EL_6] R. Elman and T. Y. Lam: *Quadratic forms under algebraic extensions,* Math. Ann. **219**(1976), 21–42.

[ELW_1] R. Elman, T. Y. Lam, and A. Wadsworth: *Amenable fields and Pfister extensions,* Proc. Conf. on Quadratic Forms (ed. G. Orzech), Queen's Papers in Pure and Applied Math. **46**(1976), 445–492.

[ELW_2] R. Elman, T. Y. Lam, and A. Wadsworth: *Orderings under field extensions,* J. Reine Angew. Math. **306**(1979), 7–27.

[ELTW] R. Elman, T. Y. Lam, J.-P. Tignol, and A. Wadsworth: *Witt rings and Brauer groups under multiquadratic extensions,* Amer. J. Math. **105**(1983), 1119–1170.

[F] F. G. Frobenius: *Über lineare Substitutionen und lineare Formen,* J. Reine Angew. Math. (1877).

[Fr] A. Fröhlich: *Quadratic forms "à la" local theory,* Proc. Cambridge Philos. Soc. **63**(1967), 579–586.

[GF] H. Gross and H. R. Fischer: *Non real fields k and infinite dimensional k-vector spaces,* Math. Ann. **159**(1965), 285–308.

[Ha] D. K. Harrison: *Witt Rings,* University of Kentucky Lecture Notes, Lexington, Kentucky, 1970.

[Ho_1] D. W. Hoffmann: *Function Fields of Quadratic Forms,* Doctoral Dissertation, University of Calif., Berkeley, 1992.

[Ho_2] D. W. Hoffmann: *Isotropy of 5-dimensional quadratic forms over the function field of a quadric,* K-Theory and Algebraic Geometry: Connections with Quadratic Forms and Division Algebras (Santa Barbara, CA, 1992), 217–225, Proc. Sympos. Pure Math., **58**, Part 2, Amer. Math. Soc., Providence, RI, 1995.

[Ho_3] D. W. Hoffmann: *On 6-dimensional quadratic forms isotropic over the function field of a quadric,* Comm. Algebra **22**(1994), 1999–2014.

[Ho_4] D. W. Hoffmann: *Isotropy of quadratic forms over the function field of a quadric,* Math. Zeit. **220**(1995), 461–476.

[Ho_5] D. W. Hoffmann: *On quadratic forms of height two and a theorem of Wadsworth,* Trans. Amer. Math. Soc. **348**(1996), 3267–3281.

[Ho_6] D. W. Hoffmann: *On the dimensions of anisotropic quadratic forms in I^4,* Invent. Math. **131**(1998), 185–198.

[Ho_7] D. W. Hoffmann: *Similarity of quadratic forms and half-neighbors,* J. Algebra **204**(1998), 255–280.

[Ho_8] D. W. Hoffmann: *On Elman and Lam's filtration of the u-invariant,* J. Reine Angew. Math. **495**(1998), 175–186.

[Ho_9] D. W. Hoffmann: *Pythagoras numbers of fields,* J. Amer. Math. Soc. **12**(1999), 839–848.

[Ho_{10}] D. W. Hoffmann: *Isotropy of quadratic forms and field invariants,* Quadratic Forms and Their Applications (Dublin, 1999), 73–102, Contemp. Math. **272**, Amer. Math. Soc., Providence, RI, 2000.

[HVG] D. W. Hoffmann and J. Van Geel: *Zeros and norm groups of quadratic forms over function fields in one variable over a local non-dyadic field,* J. Ramanujan Math. Soc. **13**(1998), 85–110.

[Hor] E. Hornix: *Formally real fields with prescribed invariants in the theory of quadratic forms,* Indag. Math. (N.S.) **2**(1991), 65–78.

[Iz] O. T. Izhboldin: *Fields of u-invariant 9,* Ann. Math. **154**(2001), 529–587.

[IK_1] O. T. Izhboldin and N. Karpenko: *Isotropy of six-dimensional forms over function fields of quadrics,* J. Algebra **209**(1996), 65–93.

[IK_2] O. T. Izhboldin and N. Karpenko: *Isotropy of virtual Albert forms over function fields of quadrics,* Math. Nachr. **206**(1999), 111–122.

[Ja] N. Jacobson: *Some applications of Jordan norms to involutorial simple associative algebras,* Adv. Math. **48**(1983), 149–165.

[Jak] V. A. Jakubović: *Factorization of symmetric matrix polynomials,* Dokl. Akad. Nauk SSSR **194**(1970), 532–535.

[Jaw_1] P. Jaworski: *Witt rings of fields of formal power series in two variables,* Ann. Math. Sil. **2**(1986), 13–29.

[Jaw_2] P. Jaworski: *Witt rings of fields of quotients of two-dimensional regular local rings,* Math. Zeit. **211**(1992), 533–546.

[Kah] B. Kahn: *La conjecture de Milnor (d'après V. Voevodsky).* Séminaire Bourbaki, Vol. 1996/97. Astérisque **245**(1997), Exp. No. 834, 379–418.

[Ka_1] I. Kaplansky: *Quadratic forms,* J. Math. Soc. Japan **5**(1953), 200–207.

[Ka_2] I. Kaplansky: *Fröhlich's local quadratic forms,* J. Reine Angew. Math. **239/240**(1969), 74–77.

[Kar_1] N. Karpenko: *On the first Witt index of quadratic forms,* Invent. Math. **153**(2003), 455–462.

[Kar_2] N. Karpenko: *Third proof of second gap in dimensions of quadratic forms from* I^n, Contemp. Math., to appear.

[Kar_3] N. Karpenko: *Holes in* I^n, to appear in Ann. Sci. École Norm. Sup.; see Preprint 128 on the Linear Algebraic Groups Preprint Server at the University of Bielefeld (http://www.mathematik.uni-bielefeld.de/LAG/).

[KM] N. Karpenko and A. Merkurjev: *Essential dimensions of quadrics,* Invent. Math. **153**(2003), 361–372.

[Ker] I. Kersten: *Brauergruppen von Körpern,* Vieweg, Wisbaden, 1990

[Kn_1] M. Knebusch: *Ein Satz über die Werte von quadratischen Formen über Körpern,* Invent. Math. **12**(1971), 300–303.

[Kn_2] M. Knebusch: *Specialization of quadratic and symmetric bilinear forms, and a norm theorem,* Acta Arith. **24**(1973), 279–299.

[Kn_3] M. Knebusch: *Some open problems,* Proc. of Conf. on Quadratic Forms (ed. G. Orzech), Queen's Papers in Pure and Applied Math. **46**(1976), 361–370.

[Kn_4] M. Knebusch: *Generic splitting of quadratic forms. I,* Proc. London Math. Soc. **33**(1976), 65–93.

[Kn_5] M. Knebusch: *Generic splitting of quadratic forms. II,* Proc. London Math. Soc. **34**(1977), 1–31.

[Kn_6] M. Knebusch: *An algebraic proof of the Borsuk-Ulam theorem for polynomial mappings,* Proc. Amer. Math. Soc. **84**(1982), 29–32.

[KRW] M. Knebusch, A. Rosenberg, and R. Ware: *Structure of Witt rings and quotients of Abelian group rings,* Amer. J. Math. **94**(1972), 119–155.

[KS] M. Knebusch and W. Scharlau: *Über das Verhalten der Witt-Gruppe bei galoischen Körpererweiterungen,* Math. Ann. **193**(1971), 189–196.

[Knu] M. A. Knus: *Quadratic and Hermitian Forms over Rings,* Grundl. Math. Wiss., Vol. **294**, Springer-Verlag, Berlin-Heidelberg-New York, 1991.

[Ku_1] M. Kula: *Fields with prescribed quadratic form schemes,* Math. Zeit. **167**(1979), 201–212.

[Ku_2] M. Kula: *Fields with nontrivial Kaplansky's radical and finite square class number,* Acta Arith. **37**(1981), 411–418.

[Ku_3] M. Kula: *Fields and quadratic form schemes,* Ann. Math. Sil. **13**(1985), 7–22.

[Lag] A. Laghribi: *Formes quadratiques de dimension* 6, Math. Nachr. **204**(1999), 125–139.

[L_1] T. Y. Lam: *Ten Lectures on Quadratic Forms over Fields,* Proc. of Conf. on Quadratic Forms (ed. G. Orzech), Queen's Papers in Pure and Applied Math. **46**(1976), 1–102.

[L_2] T. Y. Lam: *The theory of ordered fields,* Proc. of the Third Algebra and Ring Theory Conf. (ed. B. McDonald), Lecture Notes in Pure and Applied Math. **55**(1980), 1–152, Marcel Dekker, N.Y.

[L_3] T. Y. Lam: *Orderings, Valuations and Quadratic Forms,* CBMS Regional Conference Series in Math., Vol. **52**, Amer. Math. Soc., Providence, RI, 1983.

[L_4] T. Y. Lam: *An introduction to real algebra,* Rocky Mountain J. Math. **14**(1984), 769–814.

[L_5] T. Y. Lam: *On the diagonalization of quadratic forms,* Math. Magazine **72**(1999), 231–235.

[L_6] T. Y. Lam: *Fields of u-invariant* 6 *after A. Merkuryev,* in "Ring Theory 1989" (in honor of S. A. Amitsur), ed. L. Rowen, Israel Math. Conf. Proc. **1**(1989), pp. 12–30, Weizmann Science Press, Israel.

[L_7] T. Y. Lam: *Some consequences of Merkurjev's work on function fields,* unpublished manuscript, May, 1989.

[LLT] T. Y. Lam, D. Leep, and J.-P. Tignol: *Biquaternion algebras and quartic extensions,* Publ. Math. IHES **77**(1993), 63–102.

[Lp_1] D. Leep: *Systems of quadratic forms,* J. Reine Angew. Math. **350**(1984), 109–116.

[Lp_2] D. Leep: *The Amer-Brumer theorem over arbitrary fields*, preprint, 2004.

[LeM] D. Leep and M. Marshall: *Isomorphisms and automorphisms of Witt rings,* Canad. Math. Bull. **31**(1988), 250–256.

[LM] D. Leep and A. Merkurjev: *Growth of the u-invariant under algebraic extensions,* in "Recent Advances in Real Algebraic Geometry and Quadratic Forms", Contemp. Math. **155**, 327–332, Amer. Math. Soc., Providence, RI, 1994.

[Le_1] D. W. Lewis: *Witt rings as integral rings,* Invent. Math. **90**(1987), 631–633.

[Le_2] D. W. Lewis: *New proofs for the structure theorems for Witt rings,* Exposition. Math. **7**(1989), 83–8

[Le_3] D. W. Lewis: *Units in Witt rings,* Comm. Algebra **18**(1990), 3295–3306.

[Le_4] D. W. Lewis: *Annihilating polynomials for quadratic forms,* Int. J. Math. Sci. **27**(2001), 449–455.

[Li] H. Liese: *Quadratsummen in* $\mathbb{Z}$ *und* $\mathbb{Z}[[x]]$, Staatsexamens-Arbeit, Universität Münster, 1975, unpublished.

[Lo] F. Lorenz: *Quadratische Formen über Körpern,* Lecture Notes in Math., Vol. **130**, Springer-Verlag, Berlin-Heidelberg-New York, 1970.

[LL] F. Lorenz and J. Leicht: *Die Primideale des Wittschen Ringes,* Invent. Math. **10**(1970), 82–88.

[MS] P. Mammone and D. Shapiro: *The Albert quadratic form for an algebra of degree four,* Proc. Amer. Math. Soc. **105**(1989), 525–530.

[Ma_1] M. Marshall: *Abstract Witt Rings,* Queen's Papers in Pure and Applied Math., No. **57**, Queen's University, Kingston, Ontario, Canada, 1980.

[Ma_2] M. Marshall: *Exponentials and logarithms on Witt rings*, Pacific J. Math. **127**(1987), 127–140.

[Me_1] A. S. Merkurjev: *On the norm residue symbol of degree* 2 (in Russian), Dokl. Akad. Nauk SSSR **261**(1981), 542–547. (English translation: Soviet Math. Dokl. **24**(1982), 546–551.)

[Me_2] A. S. Merkurjev: *Kaplansky's conjecture in the theory of quadratic forms* (in Russian), Zap. Nauchn. Sem. Leningrad. Otdel. Mat. Inst. Steklov. (LOMI) **175**(1989), Koltsa i Moduli **3**, 75–89, 163–164. (English translation: J. Soviet Math. **57**(1991), 3489–3497.)

[Me_3] A. S. Merkurjev: *Simple algebras and quadratic forms* (in Russian), Izv. Akad. Nauk SSSR Ser. Mat. **55**(1991), 218–224. (English translation: Math. USSR-Izv. **38**(1992), 215–221.)

[Mi] J. Milnor: *Algebraic K-theory and quadratic forms,* Invent. Math. **9**(1970), 318–344.

[MH] J. Milnor and D. Husemoller: *Symmetric Bilinear Forms,* Springer-Verlag, Berlin-Heidelberg-New York, 1973.

[MW] J. Mináč and A. Wadsworth: *The u-invariant for algebraic extensions,* in "*K*-Theory and Algebraic Geometry: Connections with Quadratic Forms and Division Algebras" (Santa Barbara, CA, 1992), Proc. Sympos. Pure Math., **58**, Part 2, 333–358, Amer. Math. Soc., Providence, RI, 1995.

[Mon] J.-P. Monnier: *On Lam's conjecture concerning signatures of quadratic forms,* Arch. Math. **75**(2000), 198–206.

[Mor] F. Morel: *Voevodsky's proof of Milnor's Conjecture,* Bull. Amer. Math. Soc. (N.S.) **35**(1998), 123-143.

[Oh] J. Ohm: *The Zariski problem for function fields of quadratic forms,* Proc. Amer. Math. Soc. **124**(1996), 1679–1685.

[O'M] O. T. O'Meara: *Introduction to Quadratic Forms,* Grundl. Math. Wiss., Vol. **117**, Springer-Verlag, Berlin-Heidelberg-New York, 1963. (Reprinted in "Classics in Mathematics" Series, 2000.)

[OVV_1] D. Orlov, V. Vishik, and V. Voevodsky: *Motivic cohomology of Pfister quadrics and Milnor's conjecture on quadratic forms,* preprint.

[OVV_2] D. Orlov, V. Vishik, and V. Voevodsky: *An exact sequence for Milnor's K-theory with applications to quadratic forms,* electronic preprint, http://arXiv.org/abs/math/0101023.

[PS] R. Parimala and V. Suresh: *Isotropy of quadratic forms over function fields of p-adic curves,* Inst. Hautes Études Sci. Publ. Math. **88**(1998), 129–150.

[Pf_1] A. Pfister: *Zur Darstellung von −1 als Summe von Quadraten in einem Körper,* J. London Math. Soc. **40**(1965), 159–165.

[Pf_2] A. Pfister: *Multiplikative quadratische Formen,* Arch. Math. **16**(1965), 363–370.

[Pf_3] A. Pfister: *Quadratische Formen in beliebigen Körpern,* Invent. Math. **1**(1966), 116–132.

[Pf_4] A. Pfister: *Zur Darstellung definiter Funktionen als Summe von Quadraten,* Invent. Math. **4**(1967), 229–237.

[Pf_5] A. Pfister: *Quadratic Forms with Applications to Algebraic Geometry and Topology,* London Math. Soc. Lecture Notes Series, Vol. **217**, Cambridge Univ. Press, Cambridge, 1995.

[Pf_6] A. Pfister: *On the Milnor conjectures: history, influence, applications,* Jahresber. Deutsch. Math.-Verein. **102**(2000), 15–41.

[Po] Y. Pourchet: *Sur la reprśentation en somme de carrés des polynômes à une indéterminée sur un corps de nombres algébriques,* Acta Arith. **19**(1971), 89–104.

[Pr] A. Prestel: *Lectures on Formally Real Fields,* Lecture Notes in Math., Vol. **1093**, Springer-Verlag, Berlin-Heidelberg-New York, 1984.

[RW] A. Rosenberg and R. Ware: *The zero-dimensional Galois cohomology of Witt rings,* Invent. Math. **1**(1970), 65–72.

[RR] M. Rosenblum and J. Rovnyak: *The factorization problem for nonnegative operator valued functions,* Bull. Amer. Math. Soc. **77**(1971), 287–318.

[Ro] M. Rost: *Quadratic forms isotropic over the function field of a conic,* Math. Ann. **288**(1990), 511–513.

[Sc_1] W. Scharlau: *Zur Pfisterschen Theorie der quadratischen Formen,* Invent. Math. **6**(1969), 327–328.

[Sc_2] W. Scharlau: *Induction theorems and the structure of the Witt group,* Invent. Math. **11**(1970), 37–44.

[Sc_3] W. Scharlau: *Quadratic reciprocity laws,* J. Number Theory **4**(1970), 78–97.

[Sc_4] W. Scharlau: *Quadratic and Hermitian Forms,* Grundl. Math. Wiss., Vol. **270**, Springer-Verlag, Berlin-Heidelberg-New York, 1985.

[Se_1] J.-P. Serre: *Extensions de corps ordonnés,* C. R. Acad. Sci. Paris **229**(1949), 576–577.

[Se_2] J.-P. Serre: *L'invariant de Witt de la forme* $\mathrm{Tr}(x^2)$, Comment. Math. Helv. **59**(1984), 651–676.

[Sh] D. B. Shapiro: *Compositions of Quadratic Forms,* de Gruyter Expositions in Math., Vol. **32**, W. de Gruyter & Co., Berlin, 2000.

[Sp_1] T. A. Springer: *Sur les formes quadratiques d'indice zéro,* C. R. Acad. Sci. **234**(1952), 1517–1519.

[Sp_2] T. A. Springer: *Quadratic forms over a field with a discrete valuation,* Indag. Math. **17**(1955), 352–362.

[Sz_1] K. Szymiczek: *Quadratic forms over fields of finite square number,* Acta Arith. **28**(1975/76), 195–221.

[Sz_2] K. Szymiczek: *Matching Witts locally and globally,* Math. Slovaca **41**(1991), 315–330.

[Sz_3] K. Szymiczek: *Bilinear Algebra: An Introduction to the Algebraic Theory of Quadratic Forms,* Gordon and Breach Science Publishers, Amsterdam, 1997.

[Ta] O. Taussky: *A determinantal identity for quaternions and a new eight square identity,* J. Math. Anal. Appl. **15**(1966), 162–164.

[Ti] J.-P. Tignol: *Réduction de l'indice d'une algèbre simple centrale sur le corps des fonctions d'une quadrique,* in "Algebra, Groups and Geometry", Bull. Soc. Math. Belg. (Sér. A) **42**(1990), 735–745.

[Vi_1] A. Vishik: *On the dimension of anisotropic forms in* I^n, Max-Planck-Institut für Mathematik in Bonn, preprint MPI 2000-2001, 1–41.

[Vi_2] A. Vishik: *Motives of quadrics with applications to the theory of quadratic forms,* Proc. Summer School "Geometric Methods in the Algebraic Theory of Quadratic Forms", Lens, June, 2000, Lecture Notes in Math. **1835**, Springer-Verlag,B Berlin-Heidelberg-New York.

[Vo] V. Voevodsky: *The Milnor Conjecture,* preprint, December, 1996.

[Wad$_1$] A. Wadsworth: *Similarity of quadratic forms and isomorphism of their function fields,* Trans. Amer. Math. Soc. **208**(1975), 352–358.

[Wad$_2$] A. Wadsworth: *Merkurjev's elementary proof of Merkurjev's theorem,* Applications of Algebraic K-Theory to Algebraic Geometry and Number Theory, Parts I, II (Boulder, CO, 1983), 741–776, Contemp. Math., **55**, Amer. Math. Soc., Providence, RI, 1986.

[Wa] C. T. C. Wall: *Graded Brauer groups,* J. Reine Angew. Math. **213**(1964), 187–199.

[War$_1$] R. Ware: *When are Witt rings group rings?* Pacific J. Math. **49**(1973), 279–284.

[War$_2$] R. Ware: *Some remarks on the map between Witt rings of an algebraic extension,* Proc. of Conf. on Quadratic Forms (ed. G. Orzech), Queen's Papers in Pure and Applied Math. **46**(1976), 634–649.

[War$_3$] R. Ware: *When are Witt rings group rings? II,* Pacific J. Math. **76**(1978), 541–564.

[Wh] G. Whaples: *Algebraic extensions of arbitrary fields,* Duke Math. J. **24**(1957), 201–204.

[Wi] E. Witt: *Theorie der quadratischen Formen in beliebigen Körpern,* J. Reine Angew. Math. **176**(1937), 31–44.

[ZE] H. Zassenhaus and W. Eichhorn: *Herleitung von Acht- und Sechzehn- Quadrate-Identitäten mit Hilfe von Eigenschaften der verallgemeinerten Quaternion und der Cayley-Dickson'schen Zahlen,* Arch. Math. **17**(1966), 492–496.

Index

Titles in This Series

67 **T.Y. Lam,** Introduction to quadratic forms over fields, 2004

66 **Yuli Eidelman, Vitali Milman, and Antonis Tsolomitis,** Functional analysis, An introduction, 2004

65 **S. Ramanan,** Global calculus, 2004

64 **A. A. Kirillov,** Lectures on the orbit method, 2004

63 **Steven Dale Cutkosky,** Resolution of singularities, 2004

62 **T. W. Körner,** A companion to analysis: A second first and first second course in analysis, 2004

61 **Thomas A. Ivey and J. M. Landsberg,** Cartan for beginners: Differential geometry via moving frames and exterior differential systems, 2003

60 **Alberto Candel and Lawrence Conlon,** Foliations II, 2003

59 **Steven H. Weintraub,** Representation theory of finite groups: algebra and arithmetic, 2003

58 **Cédric Villani,** Topics in optimal transportation, 2003

57 **Robert Plato,** Concise numerical mathematics, 2003

56 **E. B. Vinberg,** A course in algebra, 2003

55 **C. Herbert Clemens,** A scrapbook of complex curve theory, second edition, 2003

54 **Alexander Barvinok,** A course in convexity, 2002

53 **Henryk Iwaniec,** Spectral methods of automorphic forms, 2002

52 **Ilka Agricola and Thomas Friedrich,** Global analysis: Differential forms in analysis, geometry and physics, 2002

51 **Y. A. Abramovich and C. D. Aliprantis,** Problems in operator theory, 2002

50 **Y. A. Abramovich and C. D. Aliprantis,** An invitation to operator theory, 2002

49 **John R. Harper,** Secondary cohomology operations, 2002

48 **Y. Eliashberg and N. Mishachev,** Introduction to the *h*-principle, 2002

47 **A. Yu. Kitaev, A. H. Shen, and M. N. Vyalyi,** Classical and quantum computation, 2002

46 **Joseph L. Taylor,** Several complex variables with connections to algebraic geometry and Lie groups, 2002

45 **Inder K. Rana,** An introduction to measure and integration, second edition, 2002

44 **Jim Agler and John E. McCarthy,** Pick interpolation and Hilbert function spaces, 2002

43 **N. V. Krylov,** Introduction to the theory of random processes, 2002

42 **Jin Hong and Seok-Jin Kang,** Introduction to quantum groups and crystal bases, 2002

41 **Georgi V. Smirnov,** Introduction to the theory of differential inclusions, 2002

40 **Robert E. Greene and Steven G. Krantz,** Function theory of one complex variable, 2002

39 **Larry C. Grove,** Classical groups and geometric algebra, 2002

38 **Elton P. Hsu,** Stochastic analysis on manifolds, 2002

37 **Hershel M. Farkas and Irwin Kra,** Theta constants, Riemann surfaces and the modular group, 2001

36 **Martin Schechter,** Principles of functional analysis, second edition, 2002

35 **James F. Davis and Paul Kirk,** Lecture notes in algebraic topology, 2001

34 **Sigurdur Helgason,** Differential geometry, Lie groups, and symmetric spaces, 2001

33 **Dmitri Burago, Yuri Burago, and Sergei Ivanov,** A course in metric geometry, 2001

32 **Robert G. Bartle,** A modern theory of integration, 2001

31 **Ralf Korn and Elke Korn,** Option pricing and portfolio optimization: Modern methods of financial mathematics, 2001

30 **J. C. McConnell and J. C. Robson,** Noncommutative Noetherian rings, 2001

TITLES IN THIS SERIES

29 **Javier Duoandikoetxea,** Fourier analysis, 2001
28 **Liviu I. Nicolaescu,** Notes on Seiberg-Witten theory, 2000
27 **Thierry Aubin,** A course in differential geometry, 2001
26 **Rolf Berndt,** An introduction to symplectic geometry, 2001
25 **Thomas Friedrich,** Dirac operators in Riemannian geometry, 2000
24 **Helmut Koch,** Number theory: Algebraic numbers and functions, 2000
23 **Alberto Candel and Lawrence Conlon,** Foliations I, 2000
22 **Günter R. Krause and Thomas H. Lenagan,** Growth of algebras and Gelfand-Kirillov dimension, 2000
21 **John B. Conway,** A course in operator theory, 2000
20 **Robert E. Gompf and András I. Stipsicz,** 4-manifolds and Kirby calculus, 1999
19 **Lawrence C. Evans,** Partial differential equations, 1998
18 **Winfried Just and Martin Weese,** Discovering modern set theory. II: Set-theoretic tools for every mathematician, 1997
17 **Henryk Iwaniec,** Topics in classical automorphic forms, 1997
16 **Richard V. Kadison and John R. Ringrose,** Fundamentals of the theory of operator algebras. Volume II: Advanced theory, 1997
15 **Richard V. Kadison and John R. Ringrose,** Fundamentals of the theory of operator algebras. Volume I: Elementary theory, 1997
14 **Elliott H. Lieb and Michael Loss,** Analysis, 1997
13 **Paul C. Shields,** The ergodic theory of discrete sample paths, 1996
12 **N. V. Krylov,** Lectures on elliptic and parabolic equations in Hölder spaces, 1996
11 **Jacques Dixmier,** Enveloping algebras, 1996 Printing
10 **Barry Simon,** Representations of finite and compact groups, 1996
9 **Dino Lorenzini,** An invitation to arithmetic geometry, 1996
8 **Winfried Just and Martin Weese,** Discovering modern set theory. I: The basics, 1996
7 **Gerald J. Janusz,** Algebraic number fields, second edition, 1996
6 **Jens Carsten Jantzen,** Lectures on quantum groups, 1996
5 **Rick Miranda,** Algebraic curves and Riemann surfaces, 1995
4 **Russell A. Gordon,** The integrals of Lebesgue, Denjoy, Perron, and Henstock, 1994
3 **William W. Adams and Philippe Loustaunau,** An introduction to Gröbner bases, 1994
2 **Jack Graver, Brigitte Servatius, and Herman Servatius,** Combinatorial rigidity, 1993
1 **Ethan Akin,** The general topology of dynamical systems, 1993